Syntheses in Limnogeology

The aim of this book series is to focus on syntheses or summaries of modern and/or ancient lake systems worldwide. Individual books will present as much information as is available for a particular lake basin or system of basins to offer readers one distinct reference as a guide to conduct further work in these areas. The books will synthesize the tectonics, basin evolution, paleohydrology, and paleoclimate of these basins and provide unbiased new interpretations or provide information on both sides of controversial issues. In addition, some books in the series will synthesize special topics in limnogeology, such as historical records of pollution in lake sediments and global paleoclimate signatures from lake sediment records.

More information about this series at http://www.springer.com/series/10029

Michael R. Rosen • David B. Finkelstein
Lisa Park Boush • Sila Pla-Pueyo
Editors

Limnogeology: Progress, Challenges and Opportunities

A Tribute to Elizabeth Gierlowski-Kordesch

Springer

Editors
Michael R. Rosen
United States Geological Survey California
Water Science Center
Carson City, NV, USA

David B. Finkelstein
Department of Geosciences
Hobart and William Smith Colleges
Geneva, NY, USA

Lisa Park Boush
Department of Geosciences
University of Connecticut
Storrs, CT, USA

Sila Pla-Pueyo
Department of Didactics of Experimental
Sciences Faculty of Educational Sciences
University of Granada
Granada, Spain

ISSN 2211-2731 ISSN 2211-274X (electronic)
Syntheses in Limnogeology
ISBN 978-3-030-66575-3 ISBN 978-3-030-66576-0 (eBook)
https://doi.org/10.1007/978-3-030-66576-0

This Springer imprint is published by the registered company Springer Nature Switzerland AG
The registered company address is: Gewerbestrasse 11, 6330 Cham, Switzerland

Preface

This book has been a labour of love for the editors, authors, and reviewers to show recognition of the contributions provided to the limnogeology community by Beth Gierlwoski-Kordesch over her career. When it was decided to write this volume, four years ago, the willingness of Beth's colleagues to contribute to the book was outstanding.

As can be seen in the 17 chapters contributed from all over the world, Beth had a huge influence in the limnogeology community. The editors hope that the readers of the book will obtain new ideas and syntheses of different processes and techniques used in limnogeology and be able to transfer them to their own work. The different chapters provide insight on paleoclimate, paleohydrology, palaeontology, geochemistry, sedimentology and diagenesis, and geophysics that illustrate the multidisciplinary nature of limnogeology.

The editors extend a huge thanks to the reviewers who read individual chapters. More than 35 colleagues reviewed the chapters, some more than once, to enable a high level of scholarship in each chapter. We would also like to extend our thanks to Springer Nature for help on the logistics of assembling and producing the book.

We hope that this book will be a lasting and useful tribute to Beth that can be used by limnogeologists all over the world for years to come.

Carson City, NV, USA	Michael R. Rosen
Storrs, CT, USA	Lisa Park Boush
Geneva, NY, USA	David B. Finkelstein
Granada, Spain	Sila Pla-Pueyo

Prologue

Dra. Elizabeth Gierlowski-Kordesch's impact on Spanish and the Global Limnogeology Community

Beth was a wonderful person and a great mentor. If there is something to highlight about her, apart from her love for her daughter, carbonates and her work addiction, it is her generosity and ability to host students in Ohio or even outside of the borders of her country, travelling wherever it was necessary to offer her help.

Her vitality and scientific activism led her to become involved with local scientists (such as B. Valero-Garcés, L. Cabrera, C. Arenas-Abad, E. Sanz Montero, M. Fregenal, A.M. Alonso-Zarza, J.P. Calvo, S. Pla-Pueyo, C. Viseras, etc.), support PhD students, facilitate the publication of articles, organise workshops and scientific meetings all around the globe (Fig. 1). Very few people have had such an impact on the development of a discipline, limnogeology, and on the formation of a generation of geologists, not just in Spain but in other countries, although her influence on Spanish scientists was particularly remarkable.

Her special relationship with Spanish geologists dates back to the 1980s, during her postdoctoral stage at the Freie Universitat of Berlín (Germany). Her research project brought her to the palustrine and lacustrine successions of the Lower Cretaceous in the Cuenca Mountain Range, thus sparking a relationship with our country that would last forever.

The Spanish scientific community working on lakes have a lot to thank her for. Few people have been so generous with their time and their knowledge, especially with students. Beth was always willing to correct once and again a manuscript, until the language in which it was written could be called proper English; she was always ready to write a recommendation letter, or to hear first-hand the interpretation of an outcrop from a PhD student in front of their poster in a conference; happy to sit down and resolve any problem, big or small, until the students would learn to do it by themselves. She demanded much and gave much of herself in the process. She never said no. There were always words of support followed by decisive actions to facilitate the mobility of new scientists.

Fig. 1 Beth with some of her Spanish colleagues (from left to right) C. Arenas-Abad and L. Cabrera and her Mexican colleague I. Israde-Alcántara during a fieldtrip in Acambay, Mexico, in 2006

Lakes were her excuse, but her grand passion was scientific knowledge. Her scientific rigour, her clear and straightforward speech, her critical and entrepreneurial character, and her perseverance to always discover something else in the geological record were her identity card as a scientist.

Beth was one of the most outstanding scientists in the task of impulsing and evidencing the enormous, and up to that point, underestimated abundance of lacustrine basins in the geological record, and the importance of limnogeology as a scientific discipline during the 1980s and 1990s. She worked tirelessly towards this goal through her participation in two successive International Geological Correlation Program projects funded by the IUGS: the IGCP 219 Comparative Lacustrine Sedimentology in Space and Time (1984–1989) and the IGCP 324 Global Paleoenvironmental Archives in Lacustrine Systems (1990–1995) projects.

She was relentless in organising meetings and congresses, editing special volumes, supporting any kind of networking and meetings among scientists, and training students. She was part of the group of scientists that decided to found the International Association of Limnogeology (IAL) to provide continuity to the tasks performed in projects IGCP 219 and 324, host and support the scientists working in this discipline. She was the heart of IAL from 1995 to 2016. Her generosity led her to dedicate a huge amount of time and effort to create and sustain this professional society in which everyone working on lakes feels comfortable. She would literally chase us to attend IAL meetings, she encouraged us to prepare and propose sessions and she corrected and edited our manuscripts, doing whatever was necessary to keep the network alive.

Beth was a tireless fighter, a citizen and a scientist of a globalised world. There were no languages or bureaucratic formalisms for her to prevent the freedom of communication and ideas exchanged among scientists. Before the Big Data era, she was a pioneer in promoting the compilation and synthesis of our knowledge about lakes of any age and any location on the planet. She was the soul of the invaluable task of gathering and publishing an insurmountable database hosting all the known lacustrine basins on the planet, a task that she carried out together with Kerry Kelts. As a result of their titanic effort, we now have two volumes available, essential references that have no equal with any other compilation carried out for any other depositional system. It was an almost heroic task, ahead of its time, highlighting the necessity of making available to the world the existent data in a common format.

We will always remember her outdoors, doing fieldwork, telling us time and again to look at the rocks, that we could find answers (and questions!) in the facies, in the continuous back and forth between detailed description, integration of outcrops in three-dimensional models, basin-scale interpretations, and back to facies again.

Thank you, Beth, for helping so many scientists to become established in their research careers, for spreading the passion and love for lakes and continental carbonates.

Thank you, Beth, for teaching us how to do our work better, to understand that limnogeology must be a global task, shared among scientists all over the world, with a great deal of analytic techniques, but anchored in the outcrop, the logs, and the sedimentary facies.

Thank you, Beth, for having been part of our lives.

El impacto de la Dra. Elizabeth Gierlowski-Kordesch sobre la comunidad española e internacional de limnogeología.

Beth era una bella persona y una gran mentora. Si hay algo que destacar, aparte del amor por su hija, los carbonatos y la adicción al trabajo, es su generosidad y habilidad para acoger estudiantes en Ohio o allende las fronteras de su país, viajando donde fuera necesario para prestar su ayuda.

Su vitalidad y su activismo científico le hicieron involucrarse con los científicos locales (tales como B. Valero-Garcés, L. Cabrera, C. Arenas-Abad, E. Sanz Montero, M. Fregenal, A.M. Alonso-Zarza, J.P. Calvo, S. Pla-Pueyo, C. Viseras…), apoyar a los doctorandos, facilitar la publicación de artículos, organizar talleres y congresos por todo el planeta (Fig. 1). Pocas personas han tenido un impacto tan notable como Beth en el desarrollo de una disciplina, la limnogeología, y en la formación especializada de una generación de geólogos, no sólo en España, aunque su incidencia en nuestro país fue particularmente notable.

Su especial relación con los científicos españoles se remonta a su periodo postdoctoral en la Universidad Freie de Berlín (Alemania) en los años 80; su proyecto postdoctoral la llevó a recalar en España para trabajar en las sucesiones de carbonatos lacustres y palustres del Cretácico Inferior de la Serranía de Cuenca, comenzando así una relación con nuestro país que ya duraría para siempre.

La comunidad científica española que trabaja en lagos tenemos mucho que agradecerle. Pocas personas han sido tan generosas con su tiempo y con su conocimiento, en especial con los estudiantes, como ella. Beth siempre estaba dispuesta a corregir una vez más un manuscrito, una y mil veces hasta que realmente podía llamarse inglés al idioma en el que estaba escrito; siempre dispuesta a escribir una carta de recomendación, o a escuchar de primera mano la descripción de unos afloramientos y su interpretación por parte de un doctorando delante de un poster en un congreso; a sentarse a resolver cada pequeño y gran problema de un doctorando hasta que aprendiera a hacerlo por sí mismo. Exigía mucho y daba mucho de sí misma en el proceso. Nunca decía no; siempre había palabras de apoyo seguidas de acciones decisivas para facilitar la movilidad de los nuevos científicos.

Su excusa eran los lagos, pero su gran pasión era el conocimiento científico. Su rigor, su discurso claro y preciso, su carácter emprendedor y crítico y su empeño por descubrir siempre algo más en el registro rocoso, eran las señas de identidad de su perfil como científica.

Beth fue una de las científicas más destacadas en la tarea de impulsar y evidenciar la enorme, y hasta aquel momento subestimada, abundancia de cuencas lacustres en el registro geológico y la importancia de la limnogeología como disciplina científica durante la década de los 80 y 90, con su activa presencia y su incansable trabajo en los dos proyectos sucesivos del auspiciados por la IUGS que permitieron este impulso: el IGCP 219 Comparative Lacustrine Sedimentology in Space and Time (1984-1989) y el IGCP 324 Global Paleoenvironmental Archives in Lacustrine Systems (1990-1995).

En la organización de reuniones y congresos, edición de volúmenes especiales, apoyo de todo tipo de redes y encuentros entre científicos, formación de estudiantes, Beth era inagotable. Ella formaba parte del grupo de limnogeólogos que decidió fundar la International Association of Limnogeology (IAL) para dar continuidad a la tarea que se había llevado a cabo en los IGCP 219 y 324, acoger y apoyar a los científicos que trabajaban en esta disciplina. Ella fue el corazón de la IAL desde 1995 hasta 2016. Su generosidad la llevó a dedicarle mucho tiempo y esfuerzo a la creación y mantenimiento de esta sociedad profesional en la que todos los que trabajamos en lagos nos sentimos a gusto. Literalmente nos perseguía para que acudiéramos a los congresos y reuniones de la asociación, nos estimulaba a preparar y proponer sesiones, nos corregía y editaba los trabajos, lo que fuera necesario para mantener viva la red.

Beth fue una luchadora incansable, una ciudadana y científica de un mundo globalizado. Para ella no había lenguas ni formalismos burocráticos que debieran impedir la libre comunicación e intercambio de ideas entre los científicos. Antes de la era de Big Data ella fue pionera en promover la compilación y síntesis de nuestro conocimiento sobre los lagos de cualquier edad y cualquier lugar del planeta. Beth fue el alma de la impagable tarea de recopilar y publicar una inmensa base de datos que alberga todas las cuencas lacustres conocidas del planeta, una tarea que sacó adelante junto con Kerry Kelts. Resultado del titánico esfuerzo de ambos disponemos de dos volúmenes de obligada referencia, que no tienen parangón con ningún otro tipo de recopilación que haya podido ser realizada para ningún otro sistema sedimentario. Fue un trabajo casi heroico, que se adelantaba a su tiempo al incidir en la necesidad de poner a disposición de todo el mundo los datos existentes en formatos comunes.

La recordaremos siempre en el campo, repitiendo que había que mirar a las rocas, que en la descripción detallada de las facies podíamos encontrar las respuestas – ¡y las preguntas!. En ese continuo ir y venir entre la descripción detallada, la integración de los afloramientos en modelos tridimensionales, las interpretaciones de procesos a escala de cuenca y más allá… Y vuelta a las facies, sustento de cualquier interpretación.

Gracias, Beth, por haber ayudado a tantos a establecerse en su carrera investigadora, gracias por haber extendido la pasión y el amor por los lagos y los carbonatos continentales.

Gracias, Beth, por enseñarnos a realizar mejor nuestro trabajo, a entender que la limnogeología tiene que ser global, una tarea compartida entre científicos de todo el mundo, con numerosas técnicas analíticas pero anclada en el afloramiento, el sondeo y las facies sedimentarias.

Gracias, Beth, por haber formado parte de nuestras vidas.

Departamento de Didáctica de las Ciencias Experimentales, Facultad de Ciencias de la Educación, Universidad de Granada
Granada, Spain — S. Pla-Pueyo,

Departamento de Geodinámica, Estratigrafía y Paleontología, Facultad de Ciencias Geológicas, Universidad Complutense de Madrid
Madrid, Spain — M. Fregenal,

Departamento de Estratigrafía, Facultad de Ciencias, e Instituto Universitario de Investigación en Ciencias Ambientales (IUCA), Universidad de Zaragoza
Zaragoza, Spain — C. Arenas-Abad ,

Instituto Pirenaico de Ecología (IPE) – CSIC, Campus de Aula Dei
Zaragoza, Spain — B. Valero-Garcés,

Departamento de Mineralogía y Petrología, Facultad de Ciencias Geológicas, Universidad Complutense de Madrid, Instituto de Geociencias
Madrid, Spain — E. Sanz-Montero, A. M. Alonso-Zarza

Contents

Part I
Introduction

Introduction to Limnogeology: Progress, Challenges, and Opportunities: A Tribute to Elizabeth Gierlowski-Kordesch

Michael R. Rosen, Lisa Park Boush, David B. Finkelstein, and Sila Pla-Pueyo

Abstract Elizabeth Gierlowski-Kordesch (1956–2016) was a leader and innovator in the specialty field of limnogeology since its beginnings in the late 1980s. Her excitement for field work and examining sediments was contagious, and she was always testing new research ideas. Beth would have been thrilled with the diversity of papers presented in the volume and the wide array of techniques used to determine the history, geochemistry, paleontology, and paleoclimate preserved in the sediments in basins that are located on every continent except Australia and Antarctica. She would also have been delighted that half the chapters were first authored by highly cited women scientists. Beth spent her career teaching, mentoring, conducting research with students and colleagues, and planning limnogeology conferences, books, and field trips. Her contributions span deep-time lakes from North and South America, Africa, Asia, and Europe, starting with her work on the Lower Jurassic East Berlin Formation where she conducted her Ph.D. research. Her work with Kerry Kelts at the University of Minnesota produced two books summarizing global lake research. These volumes are still used by many researchers, particularly as a starting point in their limnogeological studies. Her collaboration with Springer Nature® resulted in the series entitled *Syntheses in Limnogeology*, a publication that likely would not exist without her enthusiasm and perseverance. The papers in this second volume in the series describe a variety of Jurassic to modern lakes that range from fresh to hypersaline, shallow to deep, vary in size from <1 km^2 to 100s of km^2, and are found in a number of tectonic settings. Various proxies,

M. R. Rosen (✉)
U.S. Geological Survey, California Water Science Center, Carson City, NV, USA
e-mail: mrosen@usgs.gov

L. P. Boush
Department of Geosciences, University of Connecticut, Storrs, CT, USA

D. B. Finkelstein
Geoscience Department, Hobart & William Smith Colleges, Geneva, NY, USA

S. Pla-Pueyo
Department of Didactics of Experimental Sciences, Faculty of Educational Sciences, University of Granada, Granada, Spain

M. R. Rosen et al. (eds.), *Limnogeology: Progress, Challenges and Opportunities*, Syntheses in Limnogeology, https://doi.org/10.1007/978-3-030-66576-0_1

including microfossils and trace fossils and analyses of lacustrine sedimentology, stratigraphy, and stable isotopes are used to evaluate the sediment cores and stratigraphic sections to evaluate human and climate influences on the environment, the effects of tectonic, seismic, and volcanic activity, and variations in hydrology. The contributions in this volume reflect the diverse research that Beth conducted herself and we hope is a fitting honor to one of the founding scientists of Limnogeology.

Keywords Beth Gierlowski-Kordesch · Limnogeology · Paleoclimate · Paleohydrology · Book summary

Introduction

Elizabeth (Beth to everyone who knew her) Gierlowski-Kordesch (1956–2016), born October 4, 1956, in Chicago, Illinois, USA, was an unstoppable dynamo, leader, mentor, and one of the founders of the field of Limnogeology. A graduate of Lourdes High School and the University of Chicago, she earned a Ph.D. from Case Western Reserve University and completed post-doctoral work at the Freie Universitaet in Berlin, Germany, before becoming a professor of Geological Sciences at Ohio University, Athens, Ohio, in 1989, a position she retained until her death on May 17, 2016. Beth was also a Fulbright Scholar in 2014, receiving a Specialist Program grant to teach a one-week class on limnogeology at the University of Buenos Aires and then study the lacustrine Santa Clara de Arriba Formation in the Triassic Cuyana rift basin of central Argentina with colleagues at CONICET in Mendoza. The paper on the Cuyana rift basin was published with her colleagues the following year (Benavente et al. 2015), with additional work on the basin in this volume honoring Beth's doctrine of utilizing multidisciplinary lines of evidence to interpret lacustrine sequences (Benavente et al. 2021).

Beth had numerous collaborators all over the world and worked on a variety of aspects of limnogeology, even contributing to a chapter on the evolution of Miocene Old-World Monkeys (Rasmussen et al. 2019). She is author on more than 30 peer-reviewed journal articles and six book chapters, as well as editor of several books. She also was a founding editor of the *Syntheses in Limnogeology* book series by Springer Nature® along with Michael Rosen. She conducted her Ph.D. on the Lower Jurassic East Berlin Formation, part of the Hartford Group (Newark Supergroup), located in the Hartford Basin in Connecticut and Massachusetts (Gierlowski-Kordesch 1985; Gierlowski-Kordesch and Rust 1994), and this work led her to recognize the importance of limnogeology in understanding the rock record. Through this research, she appreciated the need for an organized global effort to study lacustrine rocks across geologic time, which, in 1993, led to her cofounding the International Association of Limnogeology (IAL). She served as its chair until her death in 2016 (Fig. 1). She also was instrumental in founding the Limnogeology Division of the Geological Society of America and served as its first

Fig. 1 Beth with illustrious limnogeology scientists. From left to right: Tom Johnson (Bradley and Russell award winner), Kevin Bohacs (former IAL board member), Nanna Noe Nygaard (ILIC1 organizer), Daniel Ariztegui (former IAL board member), Paul Buchheim (former IAL board member, deceased), Antje Schwalb (ILIC5 organizer), Melanie Leng (former IAL board member), Robin Renaut (Russell Award winner), Beth is directly below Robin, Alan Carroll (former IAL board member), Jean-Jacques Tiercelin (Bradley Award winner, ILIC2 organizer, deceased). Kneeling is Luis Cabrera (ILIC4 organizer), and Andy Cohen (Bradley and Russell award winner, ILIC3 organizer). (Photo by Michael Rosen)

chair in 2002–2004. Beth was awarded the IAL Wilmot H. Bradley Medal in 2011 at the quadrennial ILIC5 meeting of the association held at Konstanz, Germany. She was also instrumental in negotiations with the International Paleolimnology Association to combine the two organizations conferences in 2018 to ensure a livelier meeting and greater participation by scientists. The meetings are now triennial and continue to be combined meetings.

Beth always put students and mentoring first. She advised eight masters students and several undergraduate honors students since 2000 based on available (incomplete) Ohio University records (Ohio University does not offer a Ph.D. degree) and postdocs that came to work specifically with her. She also was frequently contacted by students from around the world for help on lacustrine rocks, which led to much collaboration over time (Valero-Garcés et al. 1997; Pla-Pueyo et al. 2009; Vranjković et al. 2010; Benavente et al. 2015; Goswami et al. 2016, among others).

Beth also published and edited significant syntheses on lake deposits in book chapters and journals including a two-volume compendium of lake deposits throughout the world coedited with Kerry Kelts (Gierlowski-Kordesch and Kelts 1995, 2000), several articles on carbonate deposition in lake systems (e.g.,

Gierlowski-Kordesch 2010; Gierlowski-Kordesch et al. 2013), and a news brief that synthesized new ideas in limnogeology (Gierlowski-Kordesch, 2003; Rosen and Gierlowski-Kordesch 2014). Her publications are what make limnogeology a thriving field today. Her contributions, humor, and relentless nudging of her colleagues to get things done will surely be missed.

Research Presented in this Volume

This book is organized by continent. Only Australia and Antarctica are not represented in the volume (Fig. 2), although Beth had deep interest in Australian lakes and attended a 10-day field trip to the lakes of the Murray basin and Strzelecki Desert led by Jim Bowler, Andy Herczeg, and John Magee after the 1986 International Association of Sedimentology Congress in Canberra. This is where Michael Rosen met Beth for the first time; their friendship, collaboration, and discussions on various aspects of lakes that began in Canberra continued for the next 30 years (Fig. 3a, b). The book also illustrates Beth's passion for discussing and organizing limnogeology concepts and for mentoring and looking carefully at the rocks (Figs. 4a–d and 5a, b).

The single chapter on African lakes (Scott et al. 2021) effectively synthesizes the record of modern animal traces found on the margins of Lake Magadi and Nasikie Engida, two lakes located in the semiarid southern Kenya Rift, an

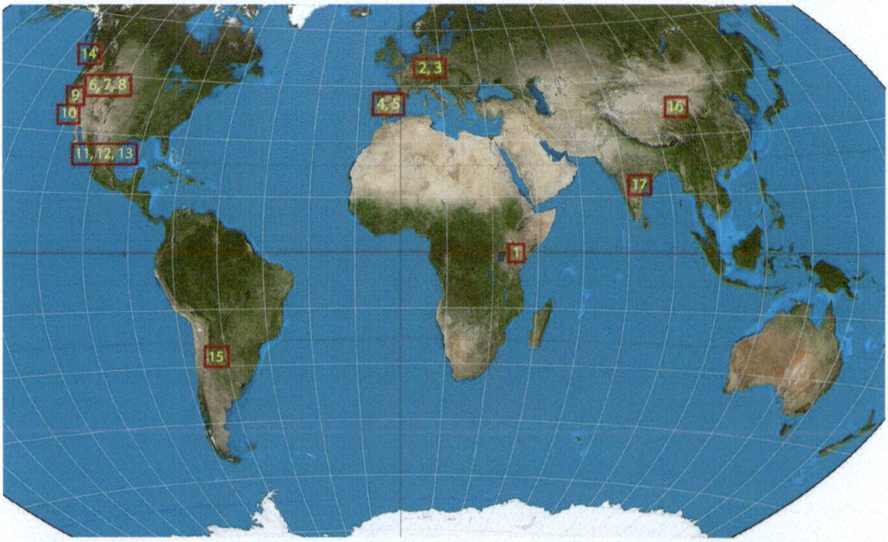

Fig. 2 Approximate locations of basins studied for this volume. Numbers correspond to chapter numbers for each basin. Base map is a Winkel triple projection from Wikimedia Commons, the free media repository. Imagery is a derivative of NASA's Blue Marble summer month composite

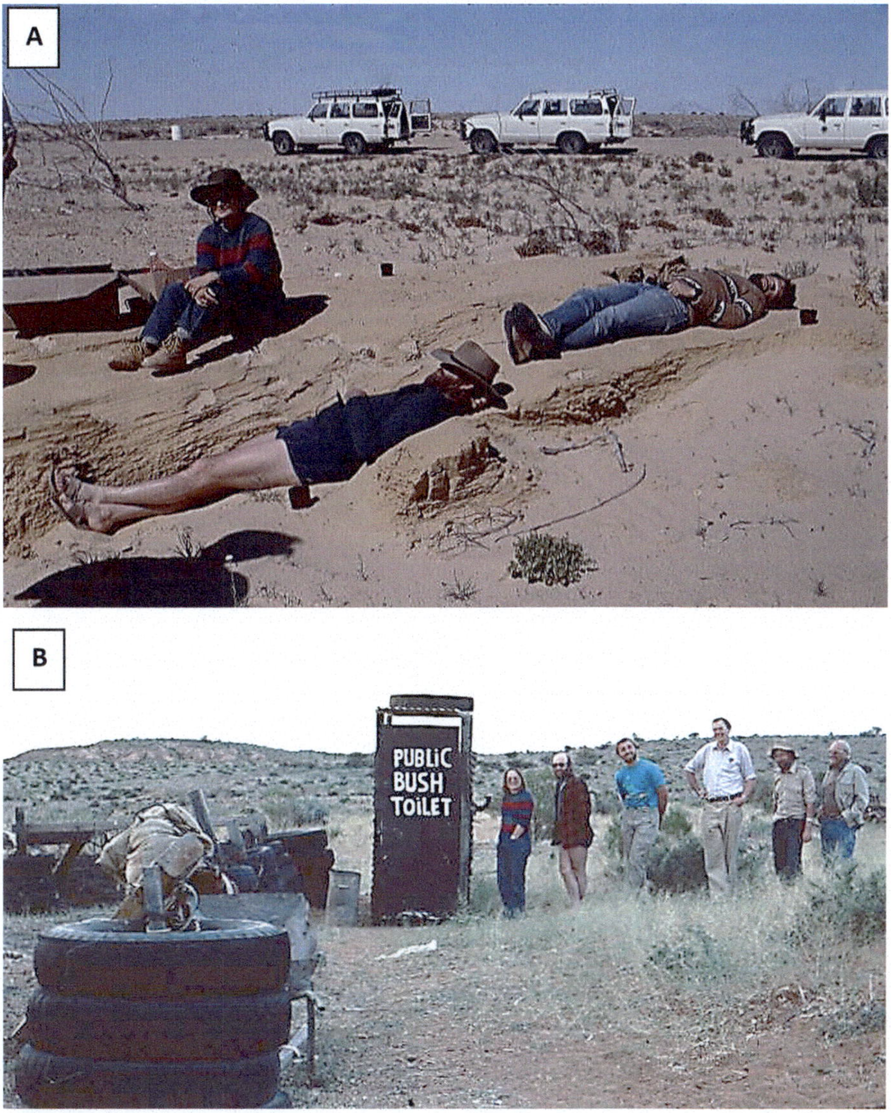

Fig. 3 (**a**) Beth sitting up somewhere in the Strzelecki Desert with more energy than trip leaders Andy Herczeg (upper slope) and John Magee (with hat on). (**b**) Always ready for some fun, Beth lines up with field trip participants (John Magee, trip leader, second in line, Will Schweller, next, and other participants) in the middle of nowhere to use a facility that has seen better days. (Photos by Michael Rosen)

example of the extreme conditions in continental rift settings. The high carbonate-bicarbonate and high sodium waters, with little calcium, are enhanced by abundant inflow from hot springs and magmatic-generated CO_2. Despite the extreme conditions in these lakes, macroorganisms exist along their margins.

Fig. 4 Photos of Beth involved in some of her favorite activities, (**a**) Organizing, meeting, and discussing the future of Limnogeology, pictured here at the International Association of Limnogeology board meeting held in Reno, Nevada, USA, in 2015. Board members at the time from left to right are Andy Cohen, Paul Buchheim (deceased), Jenni Scott, Kevin Bohacs, Beth, Michael Rosen (ILIC6 organizer), Adriana Mancuso, Tom Johnson, Alan Carroll. Photo by David Finkelstein. (**b**) Beth leading a field trip to the Mesozoic Newark Basin outcrops she studied for her dissertation. Pictured with Beth is Paul Olsen discussing the outcrop. Photo by Randy Steinen. (**c**) Beth taking a break with Mike Talbot (deceased) in Norway discussing carbonate rocks no doubt. Photo by Kevin Bohacs. (**d**) Beth on a field trip in Mexico looking at more recent lacustrine sedimentation. Beth is wearing a hat to the right. (Photo by Isabel Israde-Alcántara)

Fig. 5 Beth with colleagues and friends. (**a**) former postdoctoral colleague Sila Pla-Pueyo in the field in Ohio and (**b**) Kevin Bohacs enjoying some time at the 2012 GSA Meeting at Charlotte, North Carolina. (Photos provided by Sila Pla-Pueyo)

The authors use these modern biological traces preserved in the sediment to interpret Pleistocene and Holocene ichnites found in outcrops around the Lake Magadi basin. The subaerially produced Pleistocene and modern traces, particularly those preserved in chert, may be important to the interpretation of conditions involved in the precipitation of chemical sediments associated with these extreme environments.

The European lake systems included in the volume are from Germany (Moshayedi et al. 2021; Rösch et al. 2021) and Spain (Arenas-Abad et al. 2021; Calvo et al. 2021). These studies discuss rocks of Eocene to Pliocene age and a consideration of the climatic and human impact of lakes in southwestern Germany during the Holocene.

The chapter by Rösch et al. (2021) reviews and synthesizes studies by multiple authors on Holocene and modern lakes and examines the anthropogenic and natural environmental changes affecting the lakes and surrounding vegetation. The chapter uses historical and sedimentological records to distinguish between human causes (land clearance resulting in increases in sedimentation rates) and natural climate variability during the Holocene, indicating that human impacts were increasingly important from the Middle Holocene to present.

Moshayedi et al. (2021) examined the smallest extent of Eocene Lake "Prinz von Hessen," a pull-apart basin in the Sprendlinger Horst in southwest Germany. They examined lignite and mudstone beds that contain well-preserved pollen. Although there is some influence of climate change, most of the changes in flora were caused by tectonic activity, such as earthquake tremors resulting in the redeposition of the lignite and mudstone beds into the lake from earlier time periods.

In Spain, Arenas-Abad et al. (2021) synthesized the mechanisms of lacustrine and fluvial microbialite formation within the Neogene part of the Ebro basin. Through the evaluation of studies on microbiolite formation in a range of salinity regimes, they found that salinity likely controlled the types of microbes and their style of development, and that within laminated microbialites, the laminae are arranged into cycles that can record seasonal to multiannual changes in climate parameters. However, the gross morphology of microbialites was highly influenced by water depth and hydrodynamics.

Calvo et al. (2021) examined slump deposits in the Late Miocene lacustrine successions of the Cenajo and Las Minas basins in Spain. They concluded that each basin exhibits different causes for the slump features. In the Cenajo basin, slumping and folding occur on a large-scale, with an extensional tectonic feature that shows preferential directions of slumping within the basin. In contrast, in Las Minas basin, instability within the basin was likely caused by earthquakes that were probably coeval with the extrusion of volcanic rocks in the region. The authors conclude that careful examination of slump features can be used to determine specific tectonic features within a lacustrine basin.

As might be expected, eight of the nine chapters in the Americas section are written about lacustrine basins within North America, where Beth worked for most of her career. The remaining chapter in the section, with Beth as the lead author, focuses on the early diagenesis of carbonate marine rocks by fresh groundwater (meteoric water) (Gierlowski-Kordesch et al. 2021). Beth became interested in the role that groundwater played in lacustrine basins after discussions with Michael Rosen and reading his paper on the role of groundwater in playa basins (Rosen 1994). This greatly influenced her to study the role of groundwater in many systems, and she was convinced that it was an overlooked component of lacustrine systems and should be recognized for the role it plays in determining depositional environments and diagenesis in ancient lake basins.

Although Beth primarily worked in Paleozoic and Mesozoic lacustrine basins in the eastern U.S., all of the chapters in this volume are from the western U.S. and Mexico, showing the current trend of solving regional and global paleoclimate-driven hydrological issues in arid and semiarid environments.

The chapter by Park Boush et al. (2021) is the first detailed study of ostracodes from the Eocene Green River Formation in Fossil Basin, Wyoming. The study examines the distribution and preservation quality of two ostracode species. The authors used a combination of lithology and the presence of one species to identify shallower lake conditions and the second species to identify deepening conditions.

The next two chapters by Oviatt et al. (2021) and Nicoll (2021) summarize the history of Great Salt Lake, Utah, from the end of Lake Bonneville (~13,000 years ago) to the present and are syntheses of the new and exciting research being conducted on Great Salt Lake. The work by Oviatt et al. (2021) indicates that the depositional history and lake level change of Great Salt Lake are not continuous or slow or easily defined but punctuated by short-term intervals of high-frequency lake level changes. They suggest that Great Salt Lake paleo-lake level reconstructions should be more evidence-based, and not rely on preconceived notions of how changes should occur in the lake.

The synthesis of Nicoll (2021) shows that Great Salt Lake is not a dead lake and is a surprisingly diverse ecosystem that includes brine shrimp, microbialites and microbial mats, brine flies, and other invertebrates and microfauna linked to global migratory avian fauna along the Pacific Flyway. The lake also has complex geochemical signatures involving selenium and mercury as well as complex modes of carbonate and evaporite precipitation. This lake can act as a model for extraterrestrial systems and for ancient evaporite lakes because it is larger than most modern salt lakes and is comparable in size to many ancient lacustrine evaporitic lakes.

The chapter by Rosen et al. (2021) is the opposite extreme from Great Salt Lake with respect to size. Their chapter examines the role of river avulsion in determining the paleohydrology and paleoclimate record of Big Soda Lake, a small (0.6 km^2) but deep (currently 63 m deep) lake in Nevada. It is a saline maar lake that did not have a well-defined age. Rosen et al. (2021) determined the age of the lake to be almost 15,000 years old, much older than previous dating attempts. A 9-meter core taken in 2010 revealed a distinct change in diatom assemblage and a large shift in the oxygen and carbon stable isotope composition of authigenic calcite precipitated in the lake. The relatively rapid change from freshwater at the base to saline at 4.3 meters depth corresponded to the time when the Walker River was entering the Carson basin, and then avulsed back to Walker Lake around 5000 years ago. Although climate played a role in why avulsion occurred, the main reason for the depositional change was due to avulsion and not climate.

Starratt et al. (2021) present a diatom record from Lower Bear Lake in the San Bernardino Mountains of southern California that complements previous work done by multiple authors. They also extend the age model and provide a more accurate chronology for the lake. The authors compare the record to multiple local and regional records and determine that all the records are synoptic in nature and are not the result of microclimatic variability. They attribute pluvial periods in the lake to periods of increased atmospheric river activity, although the mechanism controlling the timing of the moisture has not yet been identified.

The next three chapters evaluate Pleistocene to modern paleoclimate variability and the impact of humans on lakes in central Mexico that have formed under various volcanic and extensional tectonic conditions (Israde-Alcántara et al. 2021; Soria-Caballero et al. 2021; Valero-Garcés et al. 2021). All the lakes have been altered by human disturbance over the past few thousand years via land drainage for agriculture and groundwater extraction for water supply and irrigation. These processes have lowered the ground surface around Mexico City by 8 m (Valero-Garcés et al. 2021).

The focus of the chapter by Israde-Alcántara et al. (2021) on Lake Zirahuén is a refinement of the paleoclimate record of this lava-damned basin, putting that record into a regional and global context. The waters of the lake are less-eutrophied than those of many lakes in the area and so provide a good record of Holocene climate change in central Mexico.

The study by Soria-Caballero et al. (2021) shows that Lake Zacapu records changes in sedimentary structures, which can be used to evaluate seismic and tectonic activity within the basin and across the region. Understanding the frequency and estimating the magnitude of these seismic events will allow managers and stakeholders to evaluate the risk of future events in this region of Mexico.

The chapter by Valero-Garcés et al. (2021) on Chalco Lake, a small remnant of a much larger Pleistocene lake, provides a detailed history of the Basin of Mexico, the location of Mexico City. The suite of relatively complete cores provides a record of millennial-scale sedimentological variability and climate change during past interglacials and rates of ecosystem responses that affect biodiversity. In addition, the record provides insights on the volcanic and tectonic evolution of the basin. The information contained in these cores is directly relevant to the people living in the Basin of Mexico, who face multiple challenges related to climate change, volcanic hazards, seismicity, and hydrology.

The last chapter from North America (Gierlowski-Kordesch et al. 2021) is not really about lacustrine sediments but instead looks at the role of meteoric groundwater as the main contributor to early diagenetic changes in carbonate sediments on the coast of Vancouver Island, British Columbia. Although this chapter may seem like it does not necessarily fit in the volume, it is an example of Beth's desire to use multidisciplinary hypotheses, in this case from hydrology, to solve lacustrine and geochemical problems. Beth, in collaboration and discussion with others, had already recognized that groundwater is an important factor that needs to be understood when studying modern and ancient lake sediments (López et al. 2014; Benavente et al. 2015; Benavente et al. 2021). Transferring this knowledge to the diagenesis of coastal rocks was the next step in this thinking.

The sole chapter from South America by Benavente et al. (2021) is a synthesis of the information needed for evaluating the paleohydrology of ancient lake basins and can be used as a teaching tool for scientists starting out in lacustrine research. The authors have integrated continental paleoenvironmental conditions (temperature, precipitation, hydrology, and hydrography) and used them to constrain the influences on terrestrial ecosystems. These lacustrine reconstructions are complicated by the variability of water sources, flow paths, and residence times of water, and paleogeographic effects that also complicate diagenesis. The complex response of lakes to changing inputs conditioned by upstream and downstream factors, including groundwater input, makes interpretations challenging. However, the authors show that such challenges can be addressed by thorough integration of paleogeography, stratigraphy, sedimentology, mineralogy, and stable isotope data within a framework that accounts for the many and convoluted controls on lake systems.

The chapters discussing lakes on the Asian continent include a detailed description of the sedimentology and depositional environments of modern Qinghai Lake (Zaixing and Chao 2021) that can be used as a framework for interpreting river inputs and delta deposits, as well as wind-driven and shoreline bar features of ancient lakes. The chapter uses these modern features as an analogy for the study in ancient continental rift lacustrine basins. The features described are useful for determining the depositional setting of ancient lake deposits and can be used in core studies to aid in the assessment of hydrocarbon potential and paleoenvironmental studies.

The last chapter, by Goswami and Ghosh (2021), demonstrates a further (and perhaps the penultimate) effort by Beth to help a Ph.D. student in India recognize the overbank sedimentary lacustrine features in her Jurassic lake fluvial-dominated system. The chapter by Goswami and Ghosh (2021) presents evidence that thin-carbonate deposits provide evidence of microbial communities forming in freshwater lakes on the floodplain of braided river system in the Pranhita-Godavari rift basin of India. These data provide evidence for microbial influence in these carbonate lenses, which are older than many other lacustrine microbialites that have been described. The data presented here will be important for further recognition of pre-Cenozoic freshwater lacustrine microbialites. As Beth's first lacustrine studies were on Mesozoic rift basins in the eastern U.S., this collaboration seems a perfect full circle in international collaboration, helping and mentoring of students, and essentially giving everything she had to advance the limnogeology community.

As this volume demonstrates, lacustrine basins have many challenges left to unravel. Beth Gierlowski-Kordesch was instrumental in coalescing and fostering a vibrant community around a variety of important process-related topics and mentoring many who continue her pioneering work in limnogeology.

Conclusion

The chapters in this volume honoring Beth address questions of paleohydrology, paleoclimate, and depositional environments in a variety of Mesozoic to modern lake settings. Together, they illustrate the passion that limnogeologists have for studying ancient and modern lake systems and focus on the challenges and opportunities in utilizing lacustrine systems to solve multidisciplinary research questions on problems in hydrology, paleontology, paleoclimate, and tectonic processes. The different basins represent various sedimentation types, geologic settings, lake types (shallow to deep, and small to large), as well as a range of sedimentary features from fresh to saline lakes. The chapters show that there are many challenges to be met in studying lacustrine environments, not the least of which is providing accurate chronology for basins studied. A large community of limnogeologists continues to meet these challenges. Without Beth's contribution to the limnogeology community, it would not be as vibrant and cohesive as it is today.

Acknowledgments The editors would like to thank all the authors and reviewers for their work to make this book a lasting testament to Beth's memory and leadership in limnogeology. Any use of trade, firm, or product names is for descriptive purposes only and does not imply endorsement by the U.S. Government.

References

Arenas-Abad, C., Martin-Bello, L., Pérez-Rivarés, F. J., Santos-Bueno, N., & Vázquez-Urbez, M. (2021). Lacustrine and fluvial microbialites in the Neogene of the Ebro Basin: A summary of up to date knowledge. In M. R. Rosen, D. Finkelstein, L. Park-Boush, & S. Pla-Pueyo (Eds.), *Limnogeology: Progress, challenges and opportunities: A tribute to Beth Gierlowski-Kordesch* (Syntheses in limnogeology no. 2). New York: Springer.

Benavente, C. A., Mancuso, A. C., Cabaleri, N., & Gierlowski-Kordesch, E. (2015). Comparison of lacustrine successions and their palaeohydrological implications in two sub-basins of the Triassic Cuyana rift, Argentina. *Sedimentology, 62*, 1771–1813.

Benavente, C. A., Mancuso, A. C., & Bohacs, K. (2021). Reconstructing paleoenvironmental conditions through integration of paleogeography, stratigraphy, sedimentology, mineralogy, and stable isotope data of lacustrine carbonates—An example from early middle Triassic strata of Southwest Gondwana, Cuyana rift, Argentina. In M. R. Rosen, D. Finkelstein, L. Park-Boush, & S. Pla-Pueyo (Eds.), *Limnogeology: Progress, challenges and opportunities: A tribute to Beth Gierlowski-Kordesch* (Syntheses in limnogeology no. 2). New York: Springer.

Calvo, J. P., Gómez-Gras, D., & Rodríguez-Pascua, M. A. (2021). Large-scale slumps and associated resedimented deposits in Miocene lake basins from SE Spain. In M. R. Rosen, D. Finkelstein, L. Park-Boush, & S. Pla-Pueyo (Eds.), *Limnogeology: Progress, challenges and opportunities: A tribute to Beth Gierlowski-Kordesch* (Syntheses in Limnogeology no. 2). New York: Springer.

Gierlowski-Kordesch, E. (1985). *Sedimentology and trace fossil paleoecology of the Lower Jurassic East Berlin Formation Hartford Basin Connecticut and Massachusetts Unpublished PhD dissertation*, Case Western Reserve University, Cleveland, 228 p.

Gierlowski-Kordesch, E. (2003). Limnogeology. Geotimes. http://www.geotimes.org/july03/high_limnogeo.html. Last accessed September 10. 2020.

Gierlowski-Kordesch, E. (2010). Chapter 1 lacustrine carbonates. In A. M. Alonso-Zarza & L. H. Tanner (Eds.), *Carbonates in continental settings: Facies, environments, and processes* (Developments in sedimentology 61) (pp. 1–101). Amsterdam: Elsevier. https://doi.org/10.1016/S0070-4571(09)06101-9.

Gierlowski-Kordesch, E., & Kelts, K. R. (Eds.). (1995). *Global geological record of Lake basins: Volume 1, world and regional geology*. Cambridge: Cambridge University Press. 461 p.

Gierlowski-Kordesch, E., & Kelts, K. R. (Eds.). (2000). *Lake Basins through space and time*. American Association of Petroleum Geologists, Studies in Geology (Book 46), Tulsa, 648 p.

Gierlowski-Kordesch, E., & Rust, B. R. (1994). The Jurassic East Berlin Formation, Hartford Basin, Newark Supergroup, (Connecticut and Massachusetts): a saline lake playa alluvial plain system. In R. Renaut & W. Last (Eds.), *Sedimentology and geochemistry of modern and Ancient Saline Lakes*., SEPM Special Publication #50, Society for Sedimentary Geology, Tulsa, 249–265.

Gierlowski-Kordesch, E., Finkelstein, D. B., Truchan Holland, J. J., & Kallini, K. D. (2013). Carbonate Lake deposits associated with distal siliciclastic perennial-river systems. *Journal of Sedimentary Research, 83*, 1114–1129. https://doi.org/10.2110/jsr.2013.81.

Gierlowski-Kordesch, E., Rothwell, G. W., Stockey, R. A., & Finkelstein, D. B. (2021). Submarine groundwater discharge as a catalyst for eodiagenetic carbonate cements within marine sedimentary basins. In M. R. Rosen, D. Finkelstein, L. Park-Boush, & S. Pla-Pueyo (Eds.), *Limnogeology: Progress, challenges and opportunities: A tribute to Beth Gierlowski-Kordesch* (Syntheses in Limnogeology No. 2). New York: Springer.

Goswami, S., & Ghosh, P. (2021). Freshwater microbialites in Early Jurassic fluvial strata of the Pranhita-Godavari Gondwana Basin, India. In M. R. Rosen, D. Finkelstein, L. Park-Boush, & S. Pla-Pueyo (Eds.), *Limnogeology: Progress, challenges and opportunities: A tribute to Beth Gierlowski-Kordesch* (Syntheses in Limnogeology No. 2). New York: Springer.

Goswami, S., Gierlowski-Kordesch, E., & Gosh, P. (2016). Sedimentology of the early Jurassic limestone beds of the Kota formation: Record of carbonate wetlands in a continental rift basin of India. *Journal of Paleolimnology, 59*, 21–38.

Israde-Alcántara, I., Vázquez-Castro, G., Aston, B., Davies, S., & Caballero Miranda, M. (2021). 12,000-year diatom-based palaeoenvironmental record from lago de Zirahuén, Mexico. In M. R. Rosen, D. Finkelstein, L. Park-Boush, & S. Pla-Pueyo (Eds.), *Limnogeology: Progress, challenges and opportunities: A tribute to Beth Gierlowski-Kordesch* (Syntheses in limnogeology no. 2). New York: Springer.

López, D. L., Gierlowski-Kordesch, E., & Hollenkamp, C. (2014). Geochemical mobility and bioavailability of heavy metals in a lake affected by acid mine drainage: Lake Hope, Vinton County, Ohio. *Water, Air, & Soil Pollution, 213*, 27–45.

Moshayedi, M., Lenz, O. K., Wilde, V., & Hinderer, M. (2021). Lake level fluctuations and allochthonous lignite deposition in the Eocene pull-apart basin "Prinz von Hessen" (Hesse, Germany) – A palynological study. In M. R. Rosen, D. Finkelstein, L. Park-Boush, & S. Pla-Pueyo (Eds.), *Limnogeology: Progress, challenges and opportunities: A tribute to Beth Gierlowski-Kordesch* (Syntheses in limnogeology no. 2). New York: Springer.

Nicoll, K. (2021). What's new about the Old Bonneville Basin? Fresh insights about the modern Limnogeology of Great Salt Lake. In M. R. Rosen, D. Finkelstein, L. Park-Boush, & S. Pla-Pueyo (Eds.), *Limnogeology: Progress, challenges and opportunities: A tribute to Beth Gierlowski-Kordesch* (Syntheses in limnogeology no. 2). New York: Springer.

Oviatt, C. G., Atwood, G., & Thompson, R. S. (2021). History of Great Salt Lake, Utah, USA, since the termination of Lake Bonneville. In M. R. Rosen, D. Finkelstein, L. Park-Boush, & S. Pla-Pueyo (Eds.), *Limnogeology: Progress, challenges and opportunities: A tribute to Beth Gierlowski-Kordesch* (Syntheses in limnogeology no. 2). New York: Springer.

Park Boush, L., Hall, C. M. S., Antonietto, L. S., & McFarland, A. J. (2021). Ecological response of ostracodes (Arthropoda, Crustacea) to lake level fluctuations in the Eocene Green River Formation, Fossil Basin, Wyoming, USA. In M. R. Rosen, D. Finkelstein, L. Park-Boush, &

S. Pla-Pueyo (Eds.), *Limnogeology: Progress, challenges and opportunities: A tribute to Beth Gierlowski-Kordesch* (Syntheses in limnogeology no. 2). New York: Springer.
Pla-Pueyo, S., Gierlowski-Kordesch, E. H., Viseras, C., & Soria, J. M. (2009). Major controls on sedimentation during the evolution of a continental basin: Pliocene–Pleistocene of the Guadix Basin (Betic Cordillera, southern Spain). *Sedimentary Geology, 219*, 97–114.
Rasmussen, D. T., Friscia, A. R., Gutierrez, M., Kappelman, J., Miller, E. R., Muteti, S., Reynoso, D., Rossie, J. B., Spell, T. L., Tabor, N. J., Gierlowski-Kordesch, E., Jacobs, B. F., Kyongo, B., Macharwas, M., & Muchemi, F. (2019). Primitive Old World monkey from the earliest Miocene of Kenya and the evolution of cercopithecoid bilophodonty. *Proceedings of the National Academy of Science, 116*, 6051–6056. www.pnas.org/cgi/doi/10.1073/pnas.1815423116.
Rösch, M., Feger, K.-H., Fischer, E., Hinderer, M., Kämpf, L., Kleinmann, A., Lechterbeck, J., Marinova, E., Schwalb, A., Tserendorj, G., & Wick, L. (2021). How changes of past vegetation and human impact are documented in lake sediments: Paleoenvironmental research in south-western Germany, a review. In M. R. Rosen, D. Finkelstein, L. Park-Boush, & S. Pla-Pueyo (Eds.), *Limnogeology: Progress, challenges and opportunities: A tribute to Beth Gierlowski-Kordesch* (Syntheses in limnogeology no. 2). New York: Springer.
Rosen, M. R. (1994). The importance of groundwater in playas: A review of playa classifications and the sedimentology and hydrology of playas. In M. R. Rosen (Ed.), *Paleoclimate and basin evolution of playa systems, Geological Society of America Special Paper No. 289,* Geological Society of America, Boulder, 1–18.
Rosen, M. R., & Gierlowski-Kordesch, E. (2014). Limnogeology, news in brief. *Environmental Earth Sciences, 73*, 913–917. https://doi.org/10.1007/s12665-014-3700-0.
Rosen, M. R., Reidy, L., Starratt, S., & Zimmerman, S. R. H. (2021). Middle Holocene hydrologic changes catalyzed by river avulsion in Big Soda Lake, Nevada, USA. In M. R. Rosen, D. Finkelstein, L. Park-Boush, & S. Pla-Pueyo (Eds.), *Limnogeology: Progress, challenges and opportunities: A tribute to Beth Gierlowski-Kordesch* (Syntheses in limnogeology no. 2). New York: Springer.
Scott, J. J., Buatois, L. A., Renaut, R. W., & Owen, R. B. (2021). Modern and ancient animal traces in hydrothermal and hypersaline extreme environments of the Kenya Rift Valley. In M. R. Rosen, D. Finkelstein, L. Park-Boush, & S. Pla-Pueyo (Eds.), *Limnogeology: Progress, challenges and opportunities: A tribute to Beth Gierlowski-Kordesch* (Syntheses in limnogeology no. 2). New York: Springer.
Soria-Caballero, D. C., Garduño-Monroy, V. H., Israde-Alcántara, I., Figueroa-Soto, A. G., Gómez-Vasconcelos, M. G., & Fagel, N. (2021). Sedimentary record of the Zacapu Basin, Michoacán, México and implications for P'urhépecha culture during the pre-classic and post-classic periods. In M. R. Rosen, D. Finkelstein, L. Park-Boush, & S. Pla-Plueyo (Eds.), *Limnogeology: Progress, challenges and opportunities: A tribute to Beth Gierlowski-Kordesch* (Syntheses in limnogeology no. 2). New York: Springer.
Starratt, S., Kirby, M. E., & Glover, K. (2021). Diatom record of Holocene moisture variability in the San Bernardino Mountains, California, USA. In M. R. Rosen, D. Finkelstein, L. Park-Boush, & S. Pla-Pueyo (Eds.), *Limnogeology: Progress, challenges and opportunities: A tribute to Beth Gierlowski-Kordesch* (Syntheses in limnogeology no. 2). New York: Springer.
Valero-Garcés, B. L., Gierlowski-Kordesch, E., & Bragonier, W. A. (1997). Pennsylvanian continental cyclothem development: No evidence of direct climatic control in the Upper Freeport formation (Allegheny Group) of Pennsylvania (northern Appalachian Basin). *Sedimentary Geology, 109*, 305–319.
Valero-Garcés, B., Stockhecke, M., Lozano-García, S., Ortega, B., Caballero, M., Fawcett, P., Werne, J. P., Brown, E., Sosa Najera, S., Pearthree, K., McGee, D., Hodgetts, A. G. E., & Martínez, R. (2021). Stratigraphy and sedimentology of the Upper Pleistocene to Holocene Lake Chalco drill cores (Mexico Basin). In M. R. Rosen, D. Finkelstein, L. Park-Boush, & S. Pla-Pueyo (Eds.), *Limnogeology: Progress, challenges and opportunities: A tribute to Beth Gierlowski-Kordesch* (Syntheses in limnogeology no. 2). New York: Springer.

Vranjković, A., Gierlowski-Kordesch, E., de Leuw, A., Pavelić, D., Mandic, O., Dragičević, I., Aljinović, D., & Harzhauser, M. (2010). Depositional model of freshwater lacustrine/palustrine carbonates (Early-Middle Miocene, Sinj Basin, Croatia). *Hrvatski geološki kongres Zagreb*, pp. 41–42.

Zaixing, J. and Chao, L. (2021). Sedimentary system of the modern Qinghai Lake. In M. R. Rosen, D. Finkelstein, L. Park-Boush, & S. Pla-Pueyo (Eds.), *Limnogeology: Progress, challenges and opportunities: A tribute to Beth Gierlowski-Kordesch* (Syntheses in limnogeology no. 2). New York: Springer.

Part II
African Lake

Modern and Ancient Animal Traces in the Extreme Environments of Lake Magadi and Nasikie Engida, Kenya Rift Valley

Jennifer J. Scott, Robin W. Renaut, Luis A. Buatois, R. Bernhart Owen, Emma P. McNulty, Mona Stockhecke, Kennie Leet, Tim K. Lowenstein, and M. Gabriela Mángano

Abstract Modern Lake Magadi and Nasikie Engida in the inner southern Kenya Rift exemplify the extreme conditions in some continental rift settings, with abundant inflow from hot springs and magmatic CO_2, producing high-carbonate-bicarbonate, high-sodium waters with little calcium in a semi-arid closed lake basin. Hypersalinity, hyperalkalinity, and hydrothermal conditions characterize the modern lake and lake margins, and also contributed to the environmental conditions represented by the Pleistocene to Holocene sedimentary record of the basin. Micro- and macro-organisms are restricted in diversity and distribution in the modern lake basin. This study documents modern animal traces present in lake-margin sites around Lake Magadi and Nasikie Engida, and those preserved in Pleistocene and Holocene sediment outcrops in the Magadi Basin. These findings are then applied to the interpretation of examples preserved in drill-cores (HSPDP-MAG14) that span the Pleistocene to recent sedimentary record of Lake Magadi. Observed associations among animal traces, substrates, and environmental conditions in the mod-

J. J. Scott (✉)
Department of Earth and Environmental Sciences, Mount Royal University, Calgary, AB, Canada
e-mail: jescott@mtroyal.ca

R. W. Renaut · L. A. Buatois · M. G. Mángano
Department of Geological Sciences, University of Saskatchewan, Saskatoon, SK, Canada

R. B. Owen
Department of Geography, Hong Kong Baptist University, Kowloon Tong, Hong Kong

E. P. McNulty · K. Leet · T. K. Lowenstein
Department of Geological Sciences, Binghamton University, Binghamton, NY, USA

M. Stockhecke
Large Lakes Observatory (LLO), University of Minnesota, Duluth, MN, USA

EAWAG Swiss Federal Institute of Aquatic Science and Technology, Dübendorf, Switzerland

M. R. Rosen et al. (eds.), *Limnogeology: Progress, Challenges and Opportunities*, Syntheses in Limnogeology, https://doi.org/10.1007/978-3-030-66576-0_2

ern sedimentary lake-margin environments help to interpret ancient lacustrine sediments in the geologic record.

Keywords Lake Magadi · Nasikie Engida · Vertebrate · Invertebrate · Ichnology · Chert · Trona · Hot springs · Hypersaline · Hyperalkaline · Alkaline · Hydrothermal · Lacustrine

Introduction

Extreme environmental conditions, such as hypersalinity, hyperalkalinity, and high temperatures, restrict the diversity of micro- and macro-organisms, and their biogenic structures that provide evidence for animal–sediment interactions in extreme continental settings preserved in the rock record. Lake Magadi and adjacent Nasikie Engida ("Little Magadi") in the semi-arid axial rift depression of the southern Kenya Rift Valley are among the most saline (up to ~320 g/L TDS: Total dissolved solids) and alkaline (pH: ~10.5) (Jones et al. 1977) lakes on Earth. Water flowing into in these lakes is supplied mainly by about 200 warm (~40 °C) to hot (~86 °C) springs around their margins (Baker 1958, 1963; Eugster and Jones 1968; Jones et al. 1977; Eugster 1980; Behr and Röhricht 2000; Darling 2001) in this semi-arid region (~450 mm rainfall annually), heated at relatively shallow depths by the asthenosphere (Crane 1981; Allen et al. 1989; Owen et al. 2018a). The high evaporation potential (~3500 mm per year), closed hydrology, bicarbonate-rich water in the drainage basin, and magmatic CO_2-charged springs lead to precipitation of sodium-carbonate-bicarbonate evaporite deposits (trona, nahcolite) (Eugster 1980, 1986; Lee et al. 2016, 2017; McNulty 2017; De Cort et al. 2019; Renaut et al. 2021). Several types of chert and silica gels, which formed under different conditions during the Neogene history of Lake Magadi (Hay 1968; Eugster and Jones 1968; Eugster 1969; Röhricht 1998; Behr and Röhricht 2000; Behr 2002; Brenna 2016; Leet et al. 2016), can also be attributed to the extreme conditions of the inner rift, with abundant aqueous silica in the hot spring waters and hyperalkaline lake (Jones et al. 1967, 1977).

Despite the extreme conditions, micro- and macro-organisms persist, and are locally abundant in areas with relatively less extreme conditions including areas fed by perennial and lower-salinity (~30 g/L TDS), lower-alkalinity (pH: ~9) spring waters. Benthic microbial mats associated with springs provide food for insects, especially in cooled hot-spring effluent such as at northwestern Nasikie Engida (e.g., Kambura et al. 2016). Alkaliphilic fish endemic to Lake Magadi (*Alcolapia grahami*) and lesser flamingos (*Phoeniconaias minor*) feed on planktonic cyanobacteria (e.g., *Arthrospira*) (Jenkin 1957; Bergman et al. 2003) that bloom in the fresher areas around springs and during seasonal rains. These extremophile macro-organisms interact with sedimentary substrates, and, together with some of the mammals and other birds also present at Magadi, leave traces of their behavior.

Preservation of their biogenic structures also depends on local sedimentary conditions, and the potential for substrate stability in environments where efflorescent salts are forming (Scott et al. 2010). Siliceous crusts and gels (Eugster and Jones 1968; Renaut et al. 1998, 2021; Behr and Röhricht 2000; Behr 2002), along with microbial mats, fine-grained clastics, and evaporite minerals, are present around the shorelines and in hot-spring effluent channels, and form the substrates for invertebrate and vertebrate traces in the heterogeneous shallow lacustrine to lake-margin environments (Scott 2010; Buatois et al. 2020).

This study investigates the distribution and composition of animal traces in modern and Quaternary shallow lacustrine and lake-margin sedimentary environments of several sub-basins at Lake Magadi and Nasikie Engida. The conditions that control the availability of resources (e.g., food) and the characteristics of substrates (e.g., matgrounds, silica-rich sediments), as well as the processes involved in trace preservation, are considered from modern environments in order to interpret the fossil examples. The aim of this study is to present a sample of the types of traces and substrates that can be expected from other, similar, extreme environments in the rock record, such as fossil sub-lacustrine burrow networks in chert associated with hydrothermal springs at Lake Baringo (Buatois et al. 2017). Sedimentary deposits that provide evidence for thermal springs, such as carbonate travertine, are uncommon in this semi-arid region, and are known only from a few Pleistocene examples in the Magadi basin (Owen et al. 2019). When reconstructing similar examples from the rock record, the distribution and composition of trace fossils may be the among the best indicators of favorable environments for macro-organisms and ecosystems based in otherwise stressful environments. Burrows in Pleistocene magadiite at Lake Magadi, for example, have shown that the siliceous precursor host to this enigmatic mineral might have precipitated in shallow, evaporating, saline lake water (Buatois et al. 2020). Both vertebrate and invertebrate traces are important for reconstructing extreme environments, and consideration of their distribution is essential (Gierlowski-Kordesch 1991). The subaerially produced modern and ancient traces presented here, particularly those preserved in chert, may also be especially significant for helping to interpret the conditions and processes involved in the deposition of chemical sediments often associated with extreme environments.

Geological Setting

Lake Magadi lies in a tectonically active region of the southern Kenya Rift Valley at ~605 m above sea level (a.s.l.) and ~1.25–2.0° S, in a topographically low point within the axial inner graben (Baker 1958, 1986). The Magadi Basin and its Pleistocene to Holocene lake deposits are floored by the Magadi Trachyte (~1 Ma), a widespread extrusive volcanic unit that covered the rift floor before later down-faulting along multiple normal faults in the inner north-south rift (Baker 1958, 1963, 1986; Baker and Wohlenberg 1971). A series of N-S trending normal faults dissect the inner rift floor into the multiple "lagoons" of Lake Magadi, together with some normal faults oriented at ~330°, due to the influence of Precambrian basement shear zones on the trajectory of normal fault propagation (Le Turdu et al. 1999; Morley 1999). Sites with warm- and hot-spring outflow tend to be present along the basement lineaments, where they intersect with normal faults in the shallower crust (Allen et al. 1989; Owen et al. 2019). Asthenosphere upwelling beneath the rift contributes to high temperature gradients within the relatively thin crust of the inner rift floor as well as the release of magmatic CO_2 (Lee et al. 2016, 2017), with the more-or-less modern configuration of the rift valley in place following deposition of the extrusive Magadi Trachyte (Baker 1986; Owen et al. 2019). Nasikie Engida, or "Little Magadi," a small shallow (1.6 m depth) lake fed mainly by hot springs in a small half-graben to the northwest of northern Lake Magadi, is separated from the main basin by a horst (Fig. 1). The hot springs at northwestern Nasikie Engida issue along normal faults several hundred meters upstream from the lake, at the intersection between the basin-bounding fault on the west, smaller faults to the east, and an inferred basement lineament that is oriented NW-SE (Owen et al. 2019).

The Pleistocene to Holocene stratigraphy of the Magadi Basin has been recently summarized by Owen et al. (2019), including evidence from outcrops (e.g., Baker 1958; Eugster 1969, 1980; Surdam and Eugster 1976; Behr and Röhricht 2000; Behr 2002) and from cores from the central Magadi Basin (MAG14-1A; MAG14-1C) and the northern sub-basin (MAG14-2A: 197 m deep) drilled during the Hominin Sites and Paleolakes Drilling Project (HSPDP) (Fig. 1). The basal parts of the cores represent freshwater lake deposits that lie unconformably on the Magadi Trachyte, and are overlain by mixed freshwater and saline lake deposits (mainly in MAG14-2A) that interfingered with alluvial fan, distal fan, and mudflat deposits (mainly in MAG14-1A) (McNulty 2017; Owen et al. 2018b, 2019). Bedded chert and siliceous mudstone along with some minor clastic intervals follow, and are overlain by fine-grained deeper-water lake deposits with higher total organic carbon (TOC), then evaporites interlayered with mud in the upper 60 m of core (McNulty 2017; Owen et al. 2018b, 2019).

Although direct correlations are not yet possible, the core succession corresponds well to the outcrop stratigraphy, with:

1. the Pleistocene Oloronga Beds representing a largely freshwater lake system as the first unit above the Magadi Trachyte, which in outcrop are unconformably overlain by pisolitic and laminar calcrete up to 40 cm thick across much of the axial rift depression;
2. the Pleistocene Green Beds (formerly called the Chert Series), which include lacustrine sediments, including bedded chert and intrusive cherts that form dikes and mounds commonly along N-S faults and fractures;
3. the lacustrine Late Pleistocene to early Holocene High Magadi Beds, composed of fluvial and lacustrine sands, silts and muds, some of which contain magadiite and Magadi-type chert; high lake levels are locally recorded by geomorphological features (paleoshoreline terraces, overflow channels) and thin stromatolitic limestone crusts upon bedrock (Baker 1958; Owen et al. 2019); and,

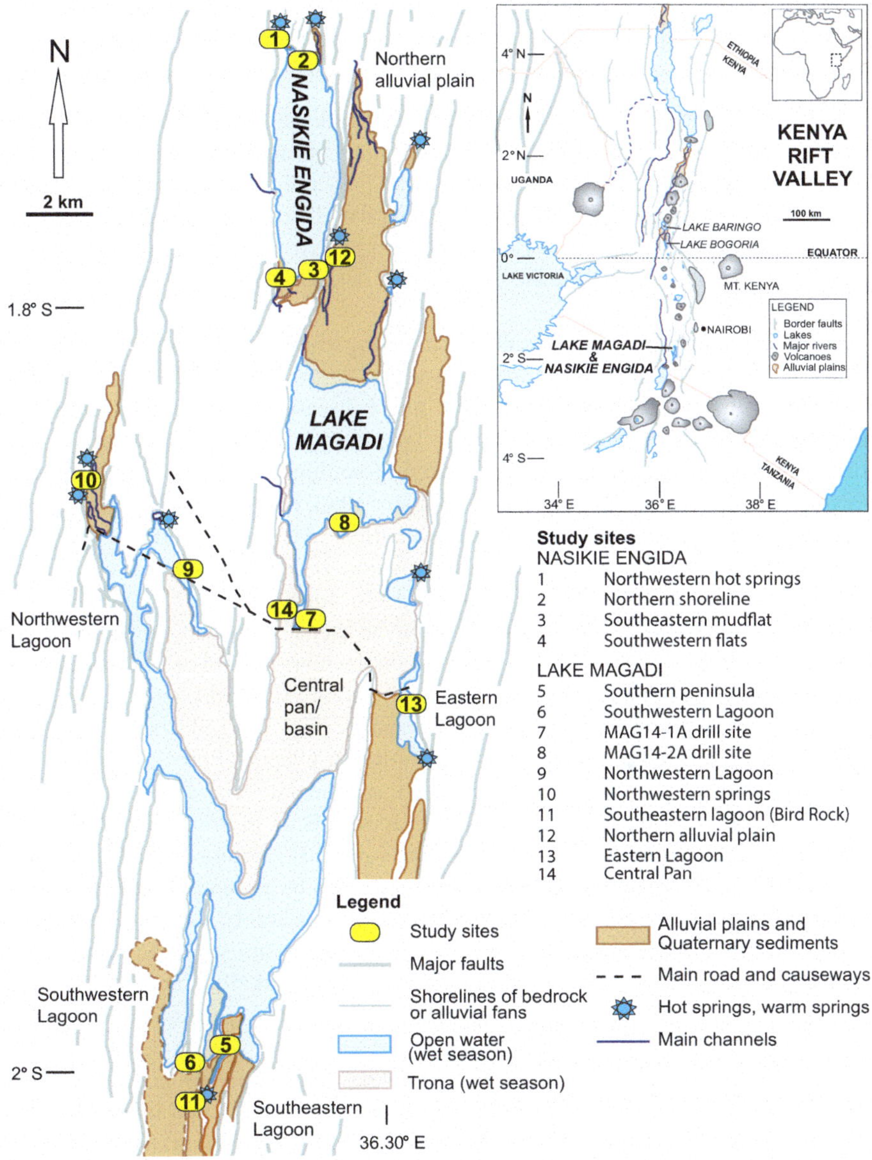

Fig. 1 Map of the Magadi area showing localities discussed in the text. In the dry season most of Lake Magadi is covered with trona except in marginal spring-fed lagoons

4. the Holocene to recent evaporites (trona, nahcolite) and muddy sediments in the axial lake and in spring-fed lake-margin lagoons (Baker 1958; Behr 2002; Owen et al. 2019).

Owen et al. (2019) used trace-element and bulk geochemistry together with U-series and $^{40}Ar/^{39}Ar$ geochronology and magnetostratigraphy to help correlate between the cores and outcrops. Field sites that preserve the trace fossils presented below include examples from the Green Beds, dated from ~190 to 160 ka (Owen et al. 2019), and the High Magadi Beds. The examples from the HSPDP-MAG14-1A core are from the interval below the Green Beds, dated from ~280 to ~400 ka (Owen et al. 2019).

Modern sedimentary environments at Lake Magadi are locally complex (Eugster 1980, 1986). Peripheral mudflats and open-water "lagoons" are fed both by thermal springs that supply little clastic sediment and by seasonal runoff that introduces fine-grained siliciclastic sediments. These lake-margin areas surround the central saline pan, which is often dry and composed of sodium carbonate salts broken by polygonal crack networks (Figs. 1 and 2; Eugster 1980; McNulty 2017). Lake brines are highly saline (up to ~320 g/L TDS) and alkaline (pH: ~10.5), with Na–CO_3–Cl–HCO_3 (and minor K, F, SO_4, and SiO_2) composition, reflecting chemical weathering of the volcanic bedrock (Surdam and Eugster 1976; Jones et al. 1977; Eugster 1980; Deocampo and Renaut 2016). During the rainy seasons, rainwater flows lakewards in ephemeral channels (Fig. 2g), falls directly on the margins and central parts of the evaporite pan, and recharges shallow cool springs (e.g., Eastern Lagoon).

Lake level can rise by ~50 cm every few years (Fig. 2e) reflecting changes in recharge and evaporation, resulting in dissolution or evaporite precipitation cycles in the axial central pan. Such changes affect sedimentation and produce organic muds interbedded with sodium carbonate evaporite beds in areas marginal to the deeper central pan (Fig. 2f; Eugster 1986; McNulty 2017), some linked to periodically stratified lakes. The lake brine is also recharged by groundwater-fed springs, which can be relatively fresh (1000–2000 ppm TDS) or highly saline (<35000 ppm TDS), depending on whether they are sourced by deep, hot water aquifers, or by dilute, cooler, shallow groundwater, or by concentrated, alkaline, and cold recirculated lake brine, or a mix of several fluids (Jones et al. 1977; Eugster 1980, 1986; Darling et al. 1990; Becht et al. 2006).

Nasikie Engida, a small shallow perennial lake fed mainly by hot springs, has saline to hypersaline water. The southern half of the lake is floored by sodium carbonate evaporite. In relatively dry years evaporite crusts (e.g., trona) form at the air/water interface in the southern quarter of the lake (Fig. 2a, b; McNulty 2017; Renaut et al. 2021). The salinity of the lake waters increases southward over 7 km from 30 g/L TDS along the northern shoreline to >260 g/L TDS along the southern shoreline where evaporites commonly precipitate. Steep shorelines and alluvial fan-deltas with distal low-gradient mudflats or ephemeral channel belts form along the margins of both lakes, with microbe-rich muds and trace-producing organisms present especially near cool, warm, or hot springs. More detailed descriptions of the local environments associated with animal traces are given below.

Fig. 2 Field photos of Lake Magadi and Nasikie Engida in 2006, 2007, and 2008. (**a**) Nasikie Engida with open water (from Site 3). Red color is from Bacteria and/or Archaea living in the hypersaline lake water. (**b**) Nasikie Engida with sodium-carbonate evaporites across the southern part of the lake (from Site 3). (**c**) Central basin area of Lake Magadi (near Site 7) with excavated channel and Tata Chemicals Magadi soda plant on the background. (**d**) The Northwestern Lagoon near the central pan showing polygons in sodium-carbonate crust (Site 9). The pink/red colour is from halophilic and alkaliphilic bacteria still alive in the wet salt crust. (**e**) The Northwestern Lagoon (Site 9) during the rainy season of 2006 with filamentous cyanobacterial bloom. Water depth is ~20 cm. (**f**) Cross-section of interbedded microbial mud and bedded trona evaporites in the Northwestern Lagoon (Site 9) area shown in (**d**). (**g**) View southwards of the northern ephemeral braidplain and spring-fed channel from the horst separating Lake Magadi from Nasikie Engida (Site 12). (**h**) Hot-spring outflow channel at northwestern Nasikie Engida (Site 1), approximately 200–300 m north of the lake shoreline

Methods

Modern sedimentary environments, substrates, and associated animal traces and their potential producers were investigated as part of a field survey of the Lake Magadi, Nasikie Engida, and Lake Bogoria basins mainly during July and August of 2007 and 2008 (Scott 2010). This "snapshot" of the environments and associated biology is supplemented by field observations during previous and subsequent seasons, mainly in 2006 and 2009, but also in 2011, 2013, 2014, and 2015. Localities investigated in the Magadi area are shown in Fig. 1. Field observations of sedimentary features (e.g., sedimentary structures, salt efflorescence) that contribute to the interpretation of sedimentary processes at each of the sites were recorded and photographed. Areas around the lake margins were visually surveyed for animal traces and their producers and recorded by photographs and field notes. Water samples were collected from most of the modern sites in 2007 and analyzed by Saskatchewan Research Council. Published water analyses (Eugster and Jones 1968; Jones et al. 1967, 1977; McNulty 2017; Renaut et al. 2021) are incorporated into the descriptions and discussion of environmental conditions at the modern sites. Outcrops of the Pleistocene Green Beds and High Magadi Beds were investigated at multiple sites, focusing mainly on lake-margin depositional settings. Sites preserving trace fossils were photographed, rock samples were collected, and stratigraphic sections were excavated and measured (Brenna 2016; Owen et al. 2019).

Study of cores obtained by the HSPDP (Cohen et al. 2016; Campisano et al. 2017) involved the detailed investigation of selected core sections preserving trace fossils from one of the cores retrieved from Lake Magadi in 2014 (MAG14-1A). Cores were prepared by cleaning the efflorescence that formed on the core surface during storage, logged at cm-scale detail, and imaged in high resolution using a GeoTek core imager. McNulty (2017) presented an analysis of sedimentary facies in the cores, and interpreted the depositional environments based on facies associations. Owen et al. (2018b, 2019) summarized the geochronology, bulk geochemistry, mineralogy, diatom, and pollen analyses undertaken on the MAG14 cores, based on research by McNulty (2017), Rabideaux (2018), and Muiruri (2018). The "archive" halves of the split cores were continuously scanned at 1 cm resolution for elemental composition using an ITRAX X-ray fluorescence core scanner at the Large Lakes Observatory (LLO), University of Minnesota–Duluth. Total percent by element were calibrated by a regression of weekly run NIST SRM standards and ratios of certain elements were calculated. Magnetic susceptibility (SI (× 10^−5)) was recorded from the split core at LacCore, University of Minnesota in Minneapolis.

Animal Traces and Lake-Margin Sedimentary Environments

Invertebrate Traces in Microbial Mats Associated with Hydrothermal Springs

Modern: Nasikie Engida Hot-Spring Effluent and Spring-Fed Lake Margin

Setting—The hottest springs in the modern Magadi region are present at northwestern Nasikie Engida, where water at up to 86 °C is expelled along a sub-basin-bounding normal fault with susequent channel flow toward the lake (Site 1, Fig. 1; Jones et al. 1977; Renaut et al. 2021). Cyanobacterial and bacterial mats, common at the hot-spring sites (Kambura et al. 2016), develop on exposed areas within and between the shallow effluent channels (Figs. 2h and 3a, b). Mud is present in shallow channel-margins, with coarse-grained sand accumulating in slightly higher areas (bars) within the channels. Cobbles and boulders, deeply corroded by alkaline water, line the border-fault margin of the spring outflow channel system. Although flow rates and volumes fluctuate, the spring system is perennial and supplies most of the water that collects in the lake (Jones et al. 1977; Renaut et al. 2021). The pH of the hot-spring outflow is relatively low at ~pH: 8.75–9.5, high enough to keep Si in solution, but it is relatively low compared to the lake waters of Nasikie Engida at ~pH: 9.5–10.5 (Eugster and Jones 1968; Jones et al. 1977; McNulty 2017). Similarly, salinity of the hot-spring water is relatively low (~30 g/L TDS) when compared to the closed-basin lake water (up to ~260 g/L TDS). The hot-spring water is also charged with mantle-derived CO_2 gas (Darling et al. 1995; Lee et al. 2016, 2017), providing an additional carbon source for the development of widespread mat-ground substrates that are utilized by macro-organisms capable of survival in high-temperature, high-alkalinity conditions. Insects typical of hydrothermal environments (e.g., ephydrid and chironomid dipteran flies, mites, spiders) (e.g., Brock et al. 1969; Wiegert and Mitchell 1973) were observed to inhabit the cyanobacterial mats, presumably where temperatures were cooler than ~50 °C, which is the upper temperature tolerance of insects known to persist in other hydrothermal ecosystems, such as at Yellowstone National Park, Wyoming (Brock 1970; Wiegert and Mitchell 1973).

Descriptions of traces—**(Type 1)** "Pock-mark structures": Small, circular to oval-shaped, shallow, open depressions in cyanobacterial mat; width ~2–3 mm; depth ~1–2 mm; margins of structure not sharp, and may be slightly raised surrounding depressions (Fig. 3c–f). **(Type 2)** Small, open vertical holes commonly in sets of two or as individual burrow openings; diameter ~3–4 mm; unlined/unwalled; roughly circular to oval-shaped holes (Fig. 3e).

Fig. 3 Microbial mats and traces in hot-spring effluent at northwest Nasikie Engida (Site 1). (**a**) Hydrothermal outflow channel with microbial mats in areas near flowing hot water. (**b**) Benthic microbial mat with gas-bubble structures (orange) and thousands of chironomid fly larvae (pink). (**c**) Thousands of regularly spaced and sized pock-mark structures or feeding pits produced by ephydrid fly adults in cyanobacterial mat with minor salt efflorescence. (**d**) Close-up of pock-mark structures showing ephydrid fly adults (arrow) feeding on the mat. (**e**) Small open burrows in pock-marked cyanobacterial mat. (**f**) Pock-marked surface with spiders hiding in depressions beneath salt efflorescence (arrows)

Remarks—The "pock-mark structures" are interpreted as feeding pits attributed to grazing of microbial mats by adult ephydrid flies, which were observed in direct association with the structures (Fig. 3d). The closely associated small burrows within the matground substrate (Fig. 3e) are attributed to ephydrid fly larvae. The mucilaginous cyanobacterial mats are used for feeding and also egg-laying when temperatures are ~35–40 °C; larvae feed throughout the mat once they have hatched (Brock et al. 1969). Chironomid larvae, also observed with microbial mats in spring-effluent channels (Fig. 3b), are sub-mm-sized and are unlikely to be the producers of the 3 mm diameter burrows (Fig. 3e). Ephydrids are well known from modern hydrothermal (e.g., Brock et al. 1969; Wiegert and Mitchell 1973) and hypersaline lake margins (Sanz-Montera et al. 2013), where they ingest cyanobacteria (Brock et al. 1969; Krivosheina 2008). However, to our knowledge, the pock-mark structures reported here have not yet been described elsewhere in the literature. Matground grazing by ephydrids in hypersaline lake margins of Spain leads to the destruction of the mats themselves, but meniscate-backfilled burrow networks are preserved in associated gypsum-rich substrates (Sanz-Montera et al. 2013). Cyanobacterial mats with ephydrid flies are also present in multiple other sites around the hypersaline lake margins and associated with warm springs elsewhere in the Magadi area, but the best example of the pock-mark feeding pits was observed in direct association with the hot-spring effluent at Nasikie Engida in 2007. The persistence of the grazing structures observed in the mats at northwestern Nasikie Engida may be related to high productivity in the mat relative to the degree of grazing (Wiegert and Fraleigh 1972; Sanz-Montera et al. 2013), which may be due to the perennial supply of water and carbon from CO_2-charged hydrothermal effluent (cf. Scott et al. 2020).

Ancient: Green Beds, Southeastern Peninsula, Lake Magadi

Setting—The Pleistocene Green Beds crop out along the Southeastern peninsula of Lake Magadi (Site 5, Fig. 1; section GB1 of Owen et al. 2019) preserves ~2.5 m of nodular chert, bedded chert with color-variable laminae, and tuffaceous siltstone (Behr and Röhricht 2000; Behr 2002; Brenna 2016; Owen et al. 2019). Thick-bedded cherts preserve heaved, domed, and cracked stromatolitic surface textures (Fig. 4a, b; Eugster 1986). Some contain calcified coccoid bacterial and cyanobacterial cells that formed within mats (Behr and Röhricht 2000). The presence of desiccation cracks (and possibly also subaqueous shrinkage cracks), large evaporite pseudomorphs (trona? gaylussite?), "pillow mounds" of chert and injected chert dikes (Behr and Röhricht 2000), amalgamated pillow mounds that may represent bioherms or discharge sites (Brenna 2016), as well as vertebrate and invertebrate trace fossils (Scott 2010), together provide evidence that the Green Beds were deposited, at least in part, at the sediment/water/air interface, and the silica was likely fed in part by hydrothermal springs. The southeastern peninsula locality (Site 5) sits at the basinward position of a horst with multiple chert-filled dykes and is segmented by a normal fault that runs along its length (Owen et al. 2019)). The chert dykes, attributed to tectonic activity (Behr and Röhricht 2000; Behr 2002), were recently dated at ~160 to 190 ka, and are broadly similar in age to the bedded cherts with trace fossils (Owen et al. 2019)). Trace fossils are observed in multiple localities of the Green Beds, in as much as ~50% of the outcrops (*Skolithos* of Röhricht 1998), but it is unknown whether the pock-mark structures are preserved in other localities.

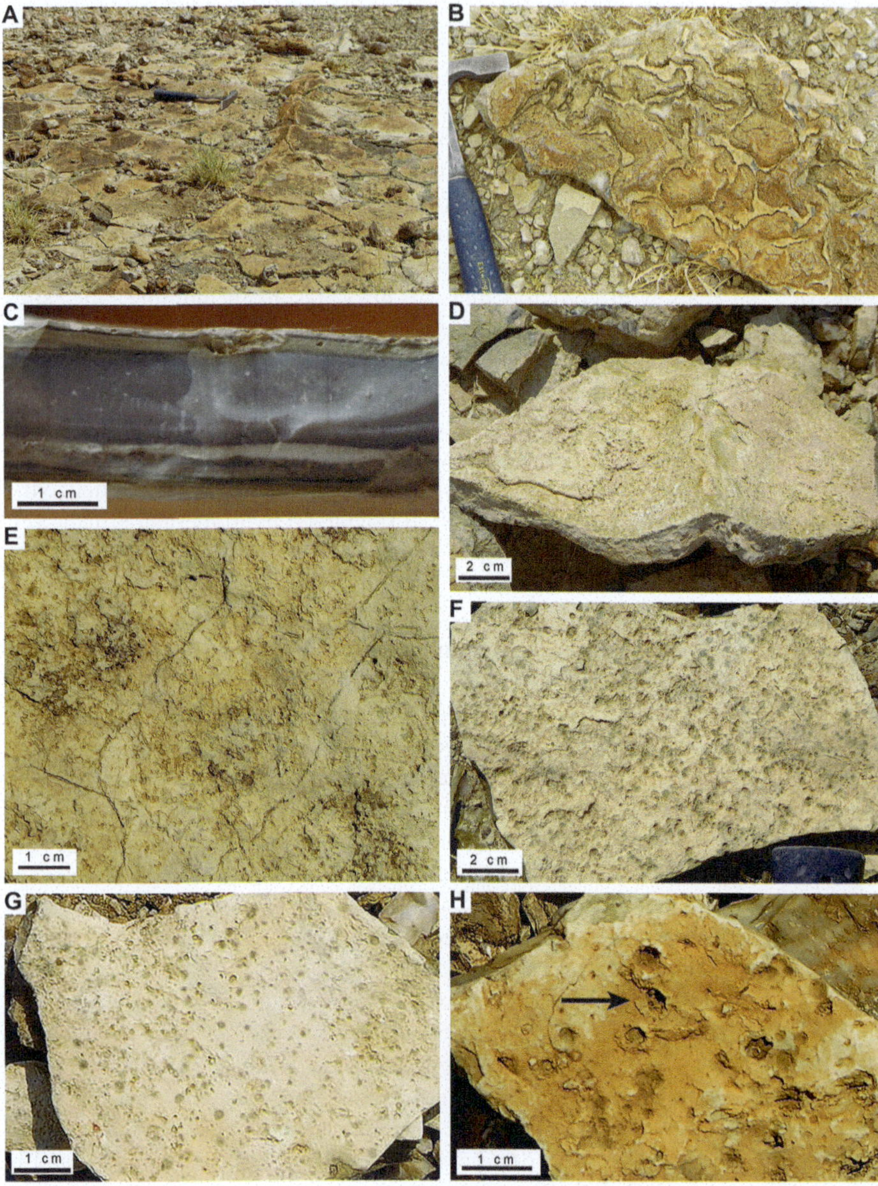

Fig. 4 Trace fossils and associated facies in the Green Beds, southeastern peninsula, Lake Magadi (Site 5). (**a**) Upper bedded chert surface at GB1 site. Hammer for scale. (**b**) Stromatolitic domes in Green Beds bedded chert. (**c**) Cross-section view of interlaminated Green Beds, with sub-mm holes in the upper layer. (**d**) Bedded chert surface showing laminated pink- and orange-coloured carbonate laminae above thick chert bed with peetee structure. Upper laminae may represent original cyanobacterial mats. (**e**) Surface with small-sized pock-mark structures preserved in negative epirelief on carbonate laminae on upper surface, along with surface trails (*Helminthoidichnites* isp.) similar to those produced by fly larvae at the northern Nasikie Engida shoreline and elsewhere

Descriptions of trace fossils—**(Type 1)** "Pock mark structures," a: Small, shallow, roughly circular depressions with or without slightly raised rims that may be only on one side; preserved in epirelief; may be irregular in shape; range from <2 mm diameter to ~4 mm; depth <1 mm; boundaries of depressions slightly sloped toward the center in apparently soft original substrate; other examples with sharp, more vertical boundaries may have been formed in more cohesive substrate; appear to be several "generations" of depressions, in places with several pock marks superimposed and slightly offset (Fig. 4e–g). **(Type 2)** "Pock mark structures," b: Medium-sized, shallow, roughly circular, and irregular depressions preserved in epirelief; may have raised ridges apparently "pushed" to one side of depression; examples with sharp margins ~5 mm in diameter; more irregular examples are up to 1 cm in diameter (Fig. 4h).

Remarks—Although the pock-mark structures in the Green Beds have been regarded as rain-drop impressions by other authors (e.g., Eugster 1969), we interpret these structures as feeding pits produced by adult ephydrid flies, comparable to the high-density depressions produced by modern ephydrid fly adults feeding on cyanobacterial mats in hot-spring effluent at northwestern Nasikie Engida (Fig. 3). Some of the Green Beds examples are nevertheless similar in general appearance to gas bubbles on the underside of cyanobacterial mats as illustrated by Sanz-Montera et al. (2013, their fig. 4), which may indeed be the origin of the larger (~5 mm diameter) very shallow (sub-mm) structures shown in Fig. 4g. Alternatively, these bubble-like diffuse structures could be relatively older pock-mark structures over which the microbial mat continued to grow, similar to the multi-generational feeding pits observed in modern examples at Nasikie Engida. The association of the pock-mark structures with other trace fossil types at the southeastern peninsula outcrop (Site 5, Fig. 1), such as larger vertical burrows (Scott 2010), corresponds well to the community (ichnocoenosis) observed near the hot-spring inflow to modern Nasikie Engida. Some of the closely associated small-sized vertical and oblique burrows (~2–3 mm deep; Fig. 4e, g) may have been produced by ephydrid larvae, which are also known to burrow in microbial mats sat hot-spring sites (Brock et al. 1969). The sub-mm-sized open pores observed in cross-section of the upper laminae of bedded chert (Fig. 4c) may be attributable to chironomid larvae. The traces illustrated by Röhricht (1998) and Brenna (2016) may have been produced by fly larvae, and compare well to the burrows from Nasikie Engida (Fig. 3e). Ephydrid flies are typical of high-temperature hydrothermal spring ecosystems once temperatures of the effluent cool to ~50 °C or less (Brock 1970) and hypersaline lake margins (Sanz-Montera et al. 2013). At the Nasikie Engida hot springs, ephydrids are

Fig. 4 (continued) (see Fig. 7c, d). These trails are interpreted to show overlapping among specimens (cf. Fig. 7d), not desiccation polygon junctions. (**f**) Close-up of pock-mark structures in upper laminae above chert. (**g**) Close-up of larger depressions, either older overgrown pock-mark structures or gas bubbles in cyanobacterial mat. Also present on this surface is the small open burrows comparable to those shown in Fig. 3e. (**h**) Close-up of medium-sized pock-mark structures with asymmetric push-up rims (arrow), possibly attributable to spider producers as in Fig. 3f closely associated with spiders that also inhabit the microbial mats while they hide and prey on ephydrids (Fig. 3f), which is typical in examples known from Yellowstone hydrothermal sites (e.g., Wiegert and Mitchell 1973). Spiders may be the producers of the slightly larger shallow depressions (pock-mark structures, b), closely associated with the high-density smaller pock-marks (Fig. 4h), as in the modern example from Nasikie Engida.

Vertebrate Footprints in Siliceous, Evaporite, and Microbial Mud Substrates

Modern: Nasikie Engida, Hot-Spring Effluent, and Southern Margin

Setting—As the hot-spring outflow from northwestern Nasikie Engida approaches the shoreline of the lake, it interacts with alkali trachyte (Fig. 2h) and fine-grained volcanic debris, the water temperature cools, flow rates decrease in shallow films that spread out over the marginal areas, and evaporation concentrates Si and Na, leading to the development of thin (~5 mm) sodium-aluminum-silica gels that desiccate into crusts (Fig. 5a; Eugster and Jones 1968) and finely crystalline crusts of trona (Renaut et al. 2021). Mammal footprints were observed in this type of substrate in different states of dryness, with corresponding differences in the depths of the impressions and the sharpness of their margins, with other mammal and bird footprints observed in nearby muddy substrates (Scott 2010).

Descriptions of traces—**(Type 1)** Mammal footprints (viverrid—mongoose): Small, four-toed footprints with clear claw impressions; measurements and description from hind footprint: total length ~3.5 cm; length without claws ~2.8 cm; total width ~2 cm; four digits impressed, all directed forward and well separated; interdigital space large; metatarsal pad chevron-shaped; all digits anterior to metatarsal pad; width of metatarsal pad almost as wide as all digits together (Fig. 5b, c; "m" in Fig. 5c). **(Type 2)** Mammal footprints (canid or felid): Medium-sized, canid footprints with four oval-shaped, forward-directed digit impressions; digits II and V set back from digits III and IV; digits III and IV even at anterior edge; large space between metatarsal pad and digits III and IV in hind footprint; metatarsal pad with two lobes at posterior and one lobe anterior; claw impressions not observed; length of hind footprint ~7 cm, width ~5.5 cm (Fig. 5d). **(Type 3)** Mammal footprints (small bovids—common duiker or domestic goat): Small-sized bovid footprints in moist, finely crystalline salt crust; anterior margin rounded but narrower than posterior margin; medial gap very narrow toward the anterior, wider toward the posterior but still narrow; length ~5–7 cm, width ~4–5 cm (Fig. 5c). **(Type 4)** Mammal footprints (large bovids—eland? buffalo?): Large bovid footprints in wet mud; impressions likely look larger due to soft, wet, muddy substrate; hooves rounded at anterior margin; large, sinuous medial gap between hooves; hind foot ~15 cm length and ~10 cm wide; front foot ~13 cm length and <10 cm wide (Fig. 5e). **(Type 5)** Mammal footprints (medium-sized bovids—gazelle): Medium-sized bovid footprints in moist mud with salt crust; footprint very heart-shaped with wide, rounded posterior

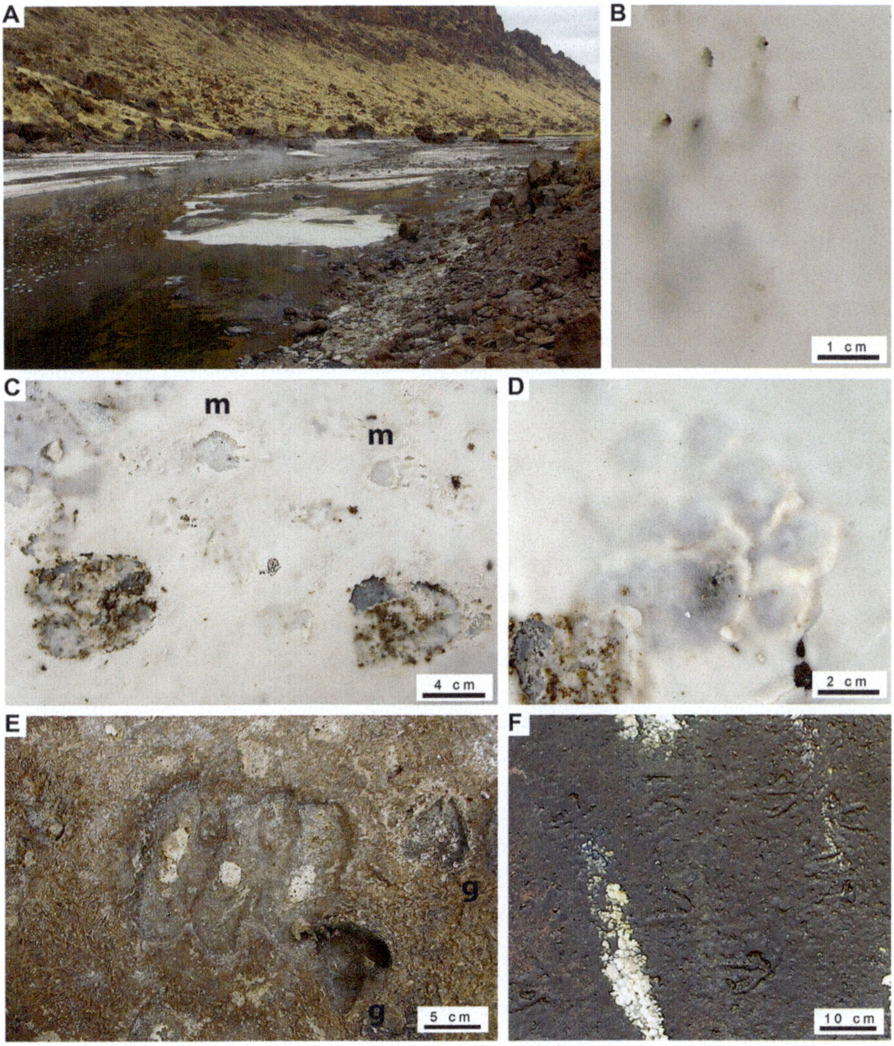

Fig. 5 Modern siliceous crusts at northwest Nasikie Engida (Site 1). (**a**) Thin, flat siliceous efflorescent crust forming within hot-spring effluent channel. (**b**) Close-up of silica crust (or mixed silica and sodium carbonate) with shallowly impressed mongoose footprints (double register) with distinct claw impressions and good pad impressions. (**c**) Close-up of crust with domestic goat footprints and mongoose footprints formed in sticky slightly desiccated crust. (**d**) Close-up of carnivore footprints (superimposed; double register) with push-up rims between digits and pads in wet crust. (**e**) Close-up of large bovid (?eland) footprints (double register) and sharply impressed gazelle footprints in microbial mud adjacent to flowing spring effluent. (**f**) Egyptian Goose footprints in cyanobacteria-rich mud with ephydrid fly adults

margins to hooves, and very pointy anterior margins; medial gap narrow, and complete from anterior to posterior, but wider toward anterior and posterior; front and hind not distinguished; length ~7 cm, width ~5.5 cm ("g" in Fig. 5e). **(Type 6)** Bird footprints (geese): Large, three-toed bird footprints with slightly triangular posterior margin and wide digits; webbing and claw impressions not observed; digit lengths similar; digits II and IV directed forward (Fig. 5f).

Remarks—In examples where the siliceous and/or sodium carbonate efflorescent substrate (Eugster and Jones 1968; Renaut et al. 2021) was impressed while still wet and soft, the footprints show push-up rims and good preservation of digit and pad impressions (Fig. 5d). As the substrate desiccates and becomes increasingly firm, claw impressions may sharply cut the crust while pads are shallowly impressed (Fig. 5b). When the crust becomes nearly completely desiccated, claw impressions may still cut the crust, and the impression margins are sharp where the uppermost crust is broken, yet the substrate is still slightly wet and cohesive below (Fig. 5c). If salt efflorescence continued to develop through time on this surface, it is expected that the surface would become irregular and preserve only isolated and disturbed depressions that may have originally shown detailed morphological features, as has been observed at Lake Bogoria (Scott et al. 2010). The final product, if preserved, may be similar to the fossil examples shown in Fig. 6b from the Pleistocene Green Beds at the southeastern peninsula of Lake Magadi (Site 6; see below). The mammal footprints impressed in the efflorescent crust were identified as: (1) mongoose (Fig. 5b, c); (2) canid or hyaenid (Fig. 5d); and (3) domestic goat (Fig. 5c). Larger bovid footprints (buffalo? eland?) were also associated with this site, impressed into microbe-rich mud substrate (Fig. 5e), along with Egyptian goose trackways (Fig. 5f). Two dead mongooses were found in the hot-spring effluent channels at Nasikie Engida; it is unknown if they were specifically using the relatively freshwater in some way (e.g., drinking, bathing). At Lake Bogoria in the central Kenya Rift, mammal tracks, including mongoose, are most abundant near the hot-spring site on the southern Loburu Delta, where relatively fresh and less alkaline boiling water issues from springs, providing an oasis-like habitat in an otherwise saline to hypersaline lake basin (Scott et al. 2009, 2020). During fieldwork at the hot-spring site at Nasikie Engida, several types of birds typically found in freshwater environments (e.g., Egyptian goose, hammerkop) were observed (Scott 2010).

Ancient: Green Beds, Southeastern Peninsula, Lake Magadi

Setting—Recently obtained U-series dates from the Pleistocene Green Beds indicate roughly time-equivalent deposition and injection of siliceous material to the trace-preserving sites between ~158 ka and ~191 ka (Owen et al. 2019)). The outcrops in the southern Magadi basin (Baker 1958; Surdam and Eugster 1976) comprise tuffaceous zeolitic sediments, mainly silts, as well as two main types of chert deposits that preserve trace fossils: (1) the bedded chert described above; and (2) nodular chert, with irregular chert nodules dispersed within silt-dominated clastic sediment (Behr and Röhricht 2000; Behr 2002; Scott 2010; Brenna 2016; Owen et al. 2019)). Outcrops of the Green Beds at the southeastern peninsula of Lake Magadi (Site 6, Fig. 1) preserve open invertebrate burrows in a structureless nodular chert substrate as well as shallowly impressed mammal footprints in bedded chert that caps the ~1 m thick deposits (Fig. 6a; Scott 2010). Here, we present the initial findings from this site, but more fieldwork is required to properly clean and document the footprint site. One of the surfaces that preserves possible small mammal footprints (Fig. 6d, e) and desiccation polygons is irregular in microtopography, and may drape a previously trampled bed or was affected by salt efflorescence, similar to the southern margin of modern Nasikie Engida. It is possible that some of the structures identified as "fluid escape structures" by Eugster (1969) are poorly preserved vertebrate footprints (Fig. 6e).

Descriptions of trace fossils—**(Type 1)** Vertebrate footprints, canid: Medium-sized shallow depressions attributable to two or three footprints (superimposed; double register) of a canid or hyaena; total length of "hind" impression ~6.5 cm including claw marks; total width ~4.5 cm; claw marks show tips of both front and hind impressions; preserved in concave epirelief (Fig. 6b). **(Type 2)** Vertebrate footprint, perissodactyl: Large shallow depression that may be a partially exposed rhino footprint; total length ~15 cm; total (extrapolated) width ~18 cm; if footprint, digit III and II or IV visible; digits broad and very rounded with digit III much wider; impression margins rounded; preserved in concave epirelief (Fig. 6c). **(Type 3)** ?3D burrow system: Small- and medium-sized, sharp-walled, open burrows; close association of smaller oblique and horizontal burrow (~4 mm diameter), which may branch from strongly oblique larger burrow (~1 cm diameter); also shows possible horizontally oriented chamber (~3 cm wide, ~1.5 cm high) (Fig. 6g). **(Type 4)** *Skolithos* A: Small, sharp-walled, unlined, circular, open, vertical burrows; diameter ~3–4 mm; burrows appear straight and not oblique within top 1 cm; depths >1 cm (only observed from top); no raised ridges surrounding burrows preserved (Fig. 6f). **(Type 5)** *Skolithos* B: Medium-sized, sharp-walled, unlined, circular to oval, open vertical and oblique burrows; diameter <1 cm; depths >1 cm, but only observed from top; no raised ridges surrounding burrows preserved (Fig. 6h).

Remarks—The bedded chert upper surface preserves multiple irregular depressions that are likely vertebrate footprints, but only relatively well-preserved examples can be attributed to their probable producers: (1) a medium-sized canid (Fig. 6b); and (2) a rhino (Fig. 6c). The footprinted upper surface also preserves evaporite (trona? gaylussite?) pseudomorphs; however, the best-preserved footprints, with claw marks (Fig. 6b), appear not to have been affected by disruptive salt efflorescence on the surface. The chert substrate was likely stabilized by microbes and/or simply the cohesive nature of a desiccating siliceous ooze or crust similar to that observed at northern Nasikie Engida (cf. Eugster and Jones 1968). Well-preserved microbial cells within bedded cherts elsewhere indicate that the substrate was soft and probably gelatinous (Behr 2002; Brenna 2016) prior to exposure and the impression of the vertebrate traces into a firm substrate. In places, the irregular surface topography, poor preservation of the footprints, and desiccation polygons signify that the surface may have developed irregular salt efflorescent crusts, followed by desiccation. The sharp-walled open burrows and possible 3D burrow networks preserved in the nodular chert facies (Fig. 6g) may have been produced by spiders, earwigs, and/or beetles, based on comparison with the modern traces known from the lake margins of southern and northern Nasikie Engida, and also with modern examples from Lake Bogoria in the central Kenya Rift (Scott et al. 2009; Scott 2010). All of the trace fossils preserved at this site indicate syn-sedimentary or early

Fig. 6 Vertebrate footprints and invertebrate trace fossils preserved in the Green Beds, southeastern peninsula, Lake Magadi (Site 6). (**a**) General view of the ichnofossiliferous Green Beds outcrop at the southeastern peninsula (Site 6). The nodular chert facies below and bedded chert and carbonate capping the outcrop. Field of view in foreground approximately 5 m across. (**b**) Close-up of upper surface with shallowly but distinctly impressed carnivore footprints with claw marks (arrows), probably produced by a canid. Note the finely detailed surface texture preserved and the desiccation cracks originating from the depression, following more predictable polygon patterns away from the tracks. (**c**) Close-up of large depression into bedded chert with crystal pseudo

post-depositional subaerial exposure of the bedded and nodular cherts, which may have become somewhat rigid substrates with continued exposure, as evidenced by the shallowly impressed but good quality canid tracks (Fig. 6b). The most comparable modern settings are those at the northern shoreline of Nasikie Engida, where siliceous gels coalesce in the eulittoral zone, the mouth of the hot-spring outflow channel at northwestern Nasikie Engida, and at the southern margin of Nasikie Engida, where clumpy masses of silica were observed within organic-rich mud beneath salt efflorescent crusts. Hydrothermal, hypersaline, and hyperalkaline conditions interact at these sites to produce similar substrates to those observed at the Green Beds localities at the southeastern peninsula.

Invertebrate Traces Associated with Lake-Margin Siliceous Sediments

Modern: Northern Shoreline of Nasikie Engida

Setting—The northern shoreline of Nasikie Engida is bordered by a high-gradient supralittoral zone with coarse-grained alluvium on the Magadi Trachyte, which sharply transitions to a low-gradient eulittoral–littoral zone where fine-grained sand, microbe-rich mud, and siliceous gels up to 15 cm thick are found in the shallow lake water (Site 2, Figs. 1 and 7a, b; Renaut et al. 2021). The siliceous ooze forming in the evaporating lake water in the littoral to eulittoral zone was also observed in several other sites around the shoreline of Nasikie Engida in 2006. Eugster and Jones (1968) also recognized siliceous Na-Al-silicate gels forming in association with hot springs and lake margins of Nasikie Engida, as well as at the hot springs at the Northeastern Lagoon of Lake Magadi (not visited in this study). Sodium-carbonate efflorescence develops on the surface of exposed areas. The northern shoreline at Nasikie Engida, in particular, is near the outflow of hot-spring effluent, with relatively low salinity and abundant Si in the alkaline, bicarbonate-rich water (Renaut et al. 2021; McNulty 2017). Evaporation, a decrease in

Fig. 6 (continued) morphs. The footprint is likely a perissodactyl mammal track, possibly produced by a rhino. (**d**) Close-up of the footprinted surface, with an irregular micro-topography, desiccation polygons, and a possible mammal track with sharp walls. Note the origin of a desiccation crack from the prominent digit (arrow), and well impressed other digits and track margin. The producer may have been a small perissodactyl, felid, or hyaenid. (**e**) Irregular surface of bedded chert with pseudomorphs and carbonate top laminae. The surface is broken in places, but also may preserve the fluid-escape structures of Eugster (1969) and possibly poor-quality footprints (arrow). (**f**) Small vertical burrows and horizontal burrow in the upper surface. (**g**) Large-sized open horizontal burrow (arrow) in nodular chert, possibly produced by earwigs in a relatively firm substrate (cf. Scott et al. 2009; Scott 2010). Note the irregular and angular nature of the chert nodules, potentially comparable to those in the HSPDP-MAG14-1A core in Fig. 9a, b. (**h**) Bioturbated green-coloured siltstone with chert nodules with hundreds of ~2 mm-scale open burrows and one larger open burrow (arrow), probably produced by ephydrid fly larvae or earwig nymphs, and earwigs or spiders, respectively

Fig. 7 The modern northern shoreline of Nasikie Engida (Site 2). (**a**) Modern shoreline of northern Nasikie Engida with siliceous gel forming in the littoral to eulittoral zone. Salt efflorescence, microbial sediments, and bioturbated sediments are present in the coarser-grained eulittoral zone. (**b**) Open films of water (centimeters-deep) in the eulittoral zone. (**c**) Overlapping simple trails and their dipteran producers in the cyanobacterial films in the shallow open water of the eulittoral zone. Note black-colored and salt-encrusted irregular microbial surface surrounding the pools. (**d**) Close-up of simple traces being produced by fly larvae (possibly Stratiomyidae) in the microbial-detritus covered surface. The traces show secondary successive branching produced from the flies re-using older traces and travelling along the same structure (arrows). Chironomid tubes lie on the surface

temperature, a decrease in pH (possibly triggered by the production of CO_2 during cyanobacterial respiration), or a mix of these processes, can lead to the supersaturation of Si and the formation of hydrous silica precipitates (Renaut et al. 1998, 2021; Behr 2002; Buatois et al. 2020). Benthic cyanobacteria are present in the shallow lake water, and provide food for some insects (flies, beetles), which are then preyed upon by other insects (e.g., earwigs, spiders) (Figs. 7 and 8). The microbial mats may also protect the siliceous gel substrates (Eugster and Jones 1968). The association of the setting and biology at Nasikie Engida thus favors the preservation of traces in siliceous and microbe-rich substrates.

Descriptions of traces—**(Type 1)** Surface trails (incipient *Helminthoidichnites* isp.): Small (~2–3 mm diameter), straight to slightly curving, simple horizontal trails; may cross-over other trails, but do not branch (Fig. 7c, d). **(Type 2)** Chironomid tube structures: Very small (~2 mm external diameter) tube structures constructed of organic detritus; < ~1 cm in length; concentrations of reworked tube structures oriented horizontally on organic detritus substrate; in situ tubes vertically oriented with open central holes of sub-millimeter diameter (Fig. 7d). **(Type 3)** 3D burrow

Fig. 8 Traces in exposed eulittoral to supralittoral zones at the northern shoreline of Nasikie Engida (Site 2). (**a**) Earwig burrows (nest) in medium-grained sand substrate in eulittoral zone. The larger open burrow was produced by the adult, shown in the photo, and the smaller burrows are produced by the earwig nymphs, also shown in the photo (arrow). Obliquely oriented cross-section view, with white evaporite crust marking the sediment surface. (**b**) Branching burrow network of earwig nymphs. Cross-section view with sediment surface at top of image. (**c**) Multiple spider burrows with excavated sediment above salt efflorescence in the drier supralittoral zone. (**d**) Slit-like burrow with spider and small amount of webbing at entrance. This burrow is possibly a re-used earwig burrow. (**e**) Circular and large open spider burrow with webbing and sediment aggregates in the excavation pile surrounding the opening. (**f**) Large-sized funnel-shaped open burrow, likely produced by a spider. (**g**) Excavation piles in the supralittoral zone, likely produced by ants as they excavate a larger structure beneath the surface. (**h**) Macrotermitinae mound (~1 m across) built around an isolated tree in the higher topography terrestrial zone adjacent to Nasikie Engida

networks (earwig): 3D burrow system of earwig adults and nymphs; burrows open, unlined/unwalled; orientations vertical, horizontal, and oblique; in this example, burrows dominantly horizontal; variable diameters, but constant diameter along length of individual branches; larger branches (adult) ~8 mm diameter; smaller branches (nymphs) ~3–4 mm wide (Fig. 8a, b). **(Type 4)** Open burrows (earwig?, spider): Large, slit-like or elliptical-shaped, oblique, open burrow; length unknown; width ~4 cm, height ~1.5 cm; tumulus not present in example, excavated material may have been incorporated into salt efflorescence; web present around burrow opening (Fig. 8d). **(Type 5)** Vertical burrows (incipient? *Skolithos* isp.; observed from surface only; spider): Large, circular, oblique, open burrow; length unknown, width ~3 cm; burrow opening surrounded by even, circular ring of pellet-like excavated material approximately 0.5 cm above the surface; spider web remnants around edges of the burrow (Fig. 8e). **(Type 6)** Funnel-shaped vertical burrows (spider): Large, shallow, funnel-shaped vertical burrows; width at surface <5 cm; width at base ~1.5–2 cm; depth ~3 cm; opening may be covered with web (Fig. 8f). **(Type 7)** 3D burrow network (Indeterminate, observed from surface only), ants: Large mounds of excavated, dry, pellet-like material; two adjacent cone-shaped mounds with one open, burrow (1 cm in diameter) in the center of depression in the center of mound; width of smaller mound ~10 cm, larger mound ~30 cm wide (Fig. 8g).

Remarks—The simple trails observed in very shallow water were likely produced by fly and beetle larvae, but the producers were not directly observed at this site except for the pupal cases of ephydrids, fly larvae (possibly stratiomyid), and chironomid tube cases (Fig. 7c, d). The traces produced in exposed sediments of fine- to medium-grained sand were produced by earwigs (Fig. 8a, b) and spiders (Fig. 8c–f), which were directly observed at the site and in some cases within their burrows. The traces of beetles, such as tiger beetles (Cicindelidae) or rove beetles (Staphylinidae), common in saline mudflats around Lake Magadi and Lake Bogoria (Scott et al. 2009; Scott 2010), were notably lacking during our fieldwork at northern Nasikie Engida. However, they were found at two sites at Lake Magadi (Northwestern Lagoon, Eastern Lagoon). The ichnocoenosis observed at the northern shoreline of Nasikie Engida in 2007 represents conditions deriving from the interaction between: (1) hydrothermal spring input the lake; (2) relatively fresh but high alkalinity water; (3) evaporation at the lake margin; and (4) a high-gradient supralittoral zone above the water table, where air-breathing predaceous insect burrowers dwell. The close association between siliceous gels forming in the evaporating lake water and air-breathing burrowers at this locality provides an analog that can be applied to interpreting examples in the rock record with chert or silica-rich substrates, such as the bedded cherts of the Green Beds (above) in the southern Lake Magadi basin or the bedded chert facies of the HSPDP-MAG14-1A drill-core (see below) and other examples from outcrop (Buatois et al. 2020).

Although Nasikie Engida contains water supplied mainly by perennial hot springs, lake level fluctuates due to seasonal and longer-term changes in rainfall and evaporation rates (McNulty 2017). During periods of even slight lake-level fall, the shallow lacustrine substrates are expected to show evidence of exposure, including bioturbation by the closely associated air-breathing burrowers (e.g., Buatois et al. 2020). Cross-cutting relationships such as this are common in lake-margin areas of closed-basin lakes with fluctuating shorelines, where the best evidence of lake-level fall in areas with limited clastic input is provided by traces and early diagenesis rather than by sedimentation (e.g., Scott et al. 2009; Scott and Smith 2015). Termites (Macrotermitinae) produce mounds and burrow systems in terrestrial and exposed lake-margin sediments when the water table falls during low lake levels (Scott et al. 2009; Owen et al. 2009). Termite mounds were observed in higher areas at Nasikie Engida (Fig. 8h), and it is expected that if lake level fell in this setting, termite tunnel systems would also cross-cut the silica-rich lake-margin sediments.

Ancient: MAG14-1A Core, Central Basin, Lake Magadi

Setting—Trace fossils are preserved in several intervals of the HSPDP drill-cores MAG14-1A and MAG14-2A, where they cross-cut both laminated and structureless mudstones deposited in association with sandstone and clast-supported conglomerates, which are interpreted as distal alluvial fan deposits (McNulty 2017). Other signs of exposure in the cores, locally associated with invertebrate burrows, include desiccation cracks and brecciated laminated mudstones representing lake-margin mudflats (McNulty 2017). In general, MAG14-1A preserves coarser-grained material associated with a more proximal position to alluvial fan environments that prograded from the basin margin, and can sometimes be correlated with perennial saline lake mudstones further basinward in core MAG14-2A (Fig. 1; McNulty 2017). Examples of trace fossils in the MAG-1A core are presented here in detail, but many more occurrences are known from the cores (McNulty 2017). The examples here include two core sections from the interval MAG14-1A 88.83–93.11 mbs (57Q-1; 58Q-1), deposited ~280 ka (Owen et al. 2019)). This interval is part of a thicker package that represents transitional environments between a distal alluvial fan setting with sandstone and conglomerate below, and interbedded laminated mudstones and bedded chert above (McNulty 2017). The lithofacies (F) fit best with: F5: bedded silty clay; F12: silicified mudstone; and F13: bedded (and nodular) chert, as illustrated by Owen et al. (2019)). The second interval presented here, from MAG14-1A 114.44–117.44 mbs (66Y-1; 67Y-1), deposited at ~400 ka (Owen et al. 2019)), represents a mudstone, siltstone, and sandstone unit within a thick package of desiccation-cracked mudstones representing a lake-margin mudflat, and also include F14: massive mud with roots and F15: sand and gravel (McNulty 2017; Owen et al. 2019)). These sediments and interpreted depositional settings are comparable to the modern marginal environments, particularly the shorelines of Nasikie Engida (see above), the Northwestern Lagoon of Lake Magadi (see below), and the distal mudflats of the northern alluvial plain. The MAG14 sediments with traces shown here share some similarities with the bedded chert facies and laminated and structureless mudstone of the Green Beds outcrop (GB1) in the southern Magadi basin (Owen et al. 2019)). The Green Beds in outcrop are stratigraphically higher, and younger by ~100–200 ka, than the examples presented here from the MAG14 cores (Owen et al. 2019)).

Descriptions of trace fossils—**(Type 1)** Horizontal, oblique, and vertical burrows, backfilled: Small-sized (4 mm diameter) unlined horizontal burrows backfilled from host with or without meniscae (Figs. 9a-1, d-1, and 10a-2). **(Type 2)**

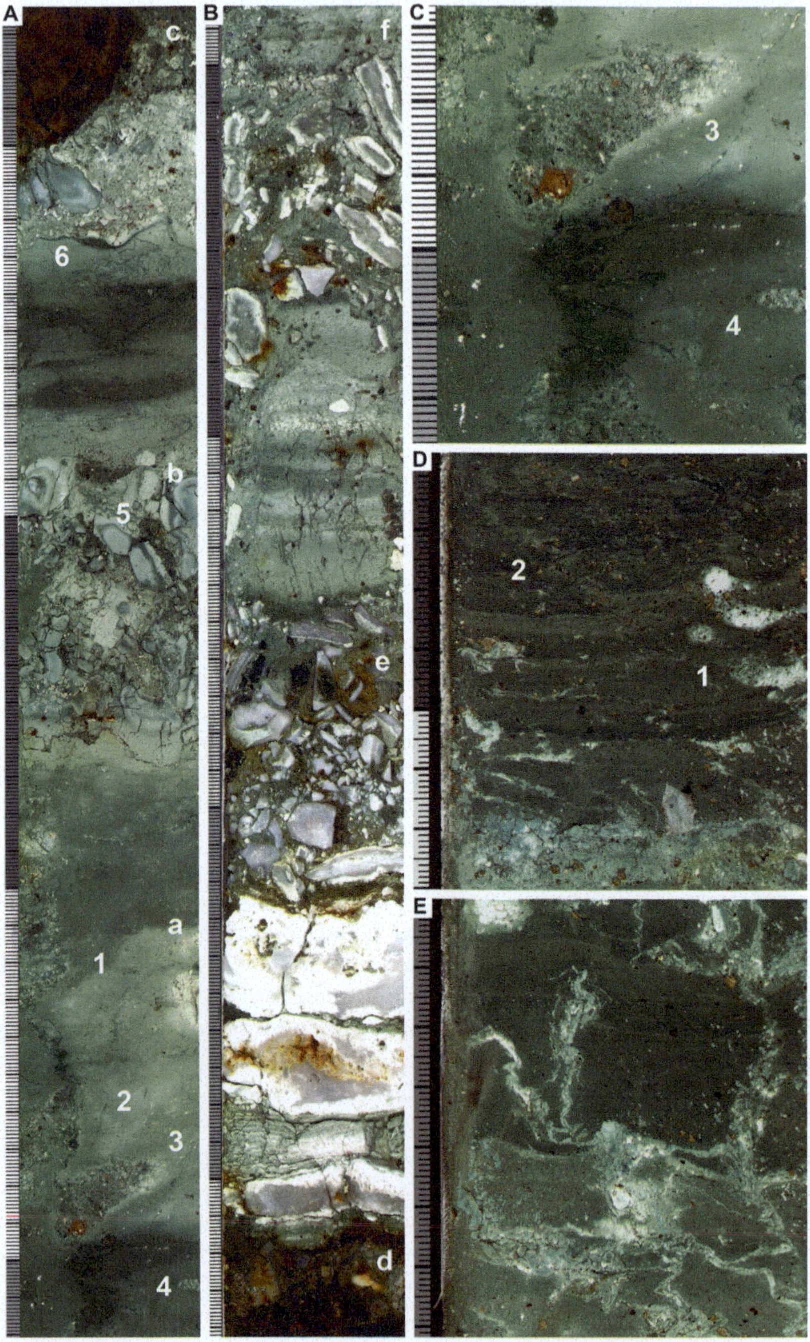

Fig. 9 Lake-margin to terrestrial traces associated with silica-rich mudstone from the MAG14-1A drill-core, central pan area, Lake Magadi (Site 7). Scales are in mm. (**a**) MAG14-1A-57Q-1-A: Evidence of exposure in successive silica-rich horizons. The lower bed (a) preserves small

Vertical burrows or roots (?*Skolithos*), small: Small-sized (2 mm diameter) vertical to oblique burrows or roots produced in soft- to firm-substrate, fill similar to host (Fig. 9a-2). **(Type 3)** Horizontal and oblique to vertical burrows, possible 3D network: Medium-sized (7 mm diameter) unlined burrows with a fill that is different from the host (poorly sorted sandstone into brecciated siliceous mudstone) (Fig. 9a-5). **(Type 4)** Vertical and horizontal burrows, pellet-filled: Medium-sized (4–5 mm diameter), pellet-backfilled horizontal and vertical burrows with fill from nearby host material, burrow margins distinct but not sharp; substrate appears to have been a mix of cohesive and soft sediment with the same burrow showing different substrate consistencies (Fig. 9a-4, c-4). **(Type 5)** Vertical to oblique burrows and cavities: Large burrows (~1.5 cm diameter) and associated cavities (~2–3 cm diameter) partially backfilled and partially passively filled with material different from the host or material same as the host; passive fill may be same as the host; sharp walls with oxidation rims in places (Figs. 9a, c and 10b-3, d-1).

Remarks—The burrows described here can be grouped into suites that represent their relative timing. The smaller burrows represent initial exposure (Type 1: small backfilled; Type 2: small vertical), and may be attributable to ephydrid fly larval burrows (cf. Sanz-Montera et al. 2013) and possibly first- or second-instar tiger beetle burrows (cf. Scott et al. 2009). Medium-sized pellet-filled burrows (Type 3) and sand-filled burrows (Type 4), as well as bioturbated horizons (Fig. 10a-a), represent slightly longer-term exposure and soft to firm substrates. Sharp-walled vertical burrows and cavities (Type 5) are deep-tier traces formed from longer-term exposure surfaces associated with lower lake levels, and were likely produced by termites and possibly also by spiders where fill is passive. The sharp-walled larger vertical burrows (e.g., Fig. 10b, d) demonstrate that the burrows sharply cross-cut oxidized and early cemented beds (Fig. 10d-b), and also pre-date compaction of the laminated mudstones that they sharply cross-cut (Fig. 10d-c). The timing,

Fig. 9 (continued) meniscate-backfilled burrows (1) and small vertical burrows and/or roots (2), as well as a horizontal (?pellet) backfilled burrow (3). The bed is cross-cut by terrestrial burrows (4) originating from an overlying exposure surface, possibly (c) or (d). Sand-filled burrows (5) are also present in (b), which contains silica-rich sediment probably brecciated and disturbed during drilling, but which shows earlier oxidation around the vertical sand-filled structure (5). A subsequent exposure surface (c) shows small desiccation cracks (6) below a disturbed bed rich in Si that contains a cobble-sized trachyte clast and may have been bioturbated and/or disrupted by salt-efflorescence, although the disturbance may also be due to drilling. The lighter green color is interpreted to represent oxidation in the depositional or early post-depositional environment. (**b**) MAG14-1A-57Q-1-A: Bedded chert above (d) is interpreted to represent deposition in a shallow lake to lake margin. The lower bed (e) preserves still-intact bedded chert as well as chert breccia clasts related to disruption during drilling. The second brecciated chert bed (f) preserves possible sand-filled burrows (7), but the bed was disturbed during drilling. (**c**) MAG14-57Q-1-A: Close-up showing the terrestrial vertical burrow (3) and horizontal backfilled burrow (4) cross-cutting bed (a). Note the sharp margins of the vertical burrow and lighter color in the oxidized margin of the burrow. The backfilled burrow (4) contains material brought in from above. (**d**) MAG14-1A-67Y-1-A: Meniscate-backfilled burrows (1) produced in soft substrate with small desiccation cracks (2) in mudstone overlying more siliceous and desiccation-cracked bed in (**e**). (**e**) MAG14-1A-67Y-1-A: Compacted and irregular desiccation cracks present in siliceous mudstone bed underlying the irregularly laminated more clastic unit with burrows in bed above, shown in (**d**)

Fig. 10 Lake-margin to terrestrial traces associated with silica-rich mudstones from the MAG14-1A drill-core, central pan area, Lake Magadi (Site 7). Scales are in mm. (**a**) MAG14-1A-58Q-1-A: Bioturbated and possibly efflorescence-disturbed sandy mudstone bed (a)

significance, and suggested producers of the traces are based in part on comparison with modern examples from Magadi and Nasikie Engida, as well as comparison to other trace fossils preserved in other cores from the Kenya Rift Valley (Potts et al. 2020).

The intervals with the traces illustrated here preserve diatom assemblages (stages D5 and D6) that include relatively high abundances of both freshwater benthic and saline benthic forms, along with *Aulacoseira* spp. and minor *Thalassiosira* spp. and *Cyclotella meneghiniana*, which also represent mixed freshwater and saline environments (Owen et al. 2004, 2019). The benthic diatoms, counted only for the MAG14-2A core, indicate both saline and freshwater substrates basinward from the exposed and bioturbated substrates illustrated here from MAG14-1A. This is especially notable when considered together with the peaks in percentage Si, which are precisely positioned with the exposure surfaces from which the burrows originate in sediments interpreted to have been deposited in lake-margin settings (Fig. 11a). The burrows repeatedly appear to originate from exposure surfaces in silica-rich mudstone in thin stacked bedsets representing minor lake flooding and exposure in this transitional environment (Figs. 9a–c and 11). This indicates that silica-enrichment in this setting may be associated with evaporation, similar to the northern shoreline of Nasikie Engida (Fig. 7a, b; Renaut et al. 2021). The siliceous muds in the core were then cross-cut by terrestrial burrows during the subsequent longer periods of exposure (Fig. 9a, c). In cases where only deep-tier burrows are present (e.g., 66Y-1), the biogenic structures cross-cut silica-rich beds present in the zone of oxidation, interpreted as representing the depth to the water table during low lake level (Fig. 10c-a).

As a whole, the burrows studied here comprise passively filled sharp-walled large and vertically oriented forms, such as those produced by spiders, earwigs and/

Fig. 10 (continued) with vague backfilled burrow (1). The bioturbated bed (sandy mudstone) overlies finer-grained material (mudstone) with vertical and horizontal burrows (2) produced in a soft substrate with variable states of cohesiveness. The burrow margins are sharp where they cross-cut the lighter, and probably more siliceous, material, and relatively indistinct where they cross-cut the darker, probably more clastic, material. (**b**) MAG14-1A-58Q-1-W: Large terrestrial burrow network with cavities and sharp burrow margins, probably produced by termites, and excavated into firm siliceous mudstone (3). (**c**) MAG14-1A-66Y-1-A: Exposure surface (e) and oxidized sediments with siliceous beds in flooding to exposure cycle, overlain in turn by deeper lacustrine facies. The lower black parallel-laminated mudstone represents oxygen-poor deeper lacustrine sediments, which are overlain by oxidized parallel-laminated mudstone with siliceous interbeds (a–d), and finally exposed in a rooted sandy mudstone associated with cobble-sized trachyte clasts. This succession represents the upper part of a lacustrine sequence. (**d**) MAG14-1A-66Y-1-A: The oxidized laminated mudstone is interbedded with siliceous horizons and cross-cut by a large vertical burrow (1), probably produced by termites but possibly by a spider. Note that the burrow disrupts the lower siliceous and iron-stained bed (b) and continues below it, but does not completely cross-cut the bed as would be expected for a desiccation crack. The overlying siliceous bed (c) shows evidence for differential compaction and oxidation associated with the breccia burrow fill, indicating that the timing of cementation occurred following bioturbation by prior to compaction by overlying sediments. (**e**) MAG14-1A-66Y-1-A: The exposure surface (e) comprises a sandy mudstone with well preserved roots (2) that overlies a disrupted siliceous bed (d) possibly deposited and disrupted in the eulittoral zone. The initiation of flooding in the next cycle is marked by a bedded chert layer (f)

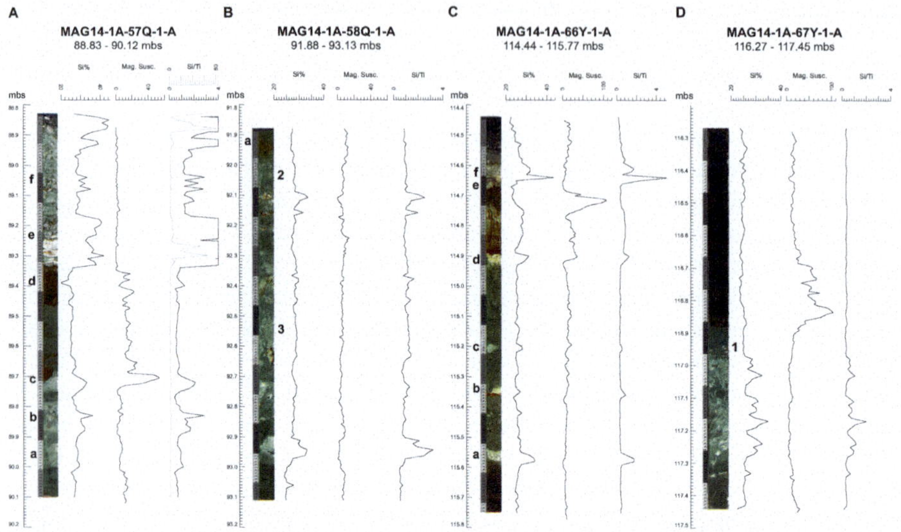

Fig. 11 Scans of selected core sections from MAG14-1A, presented with the Si%, magnetic susceptibility, and Si/Ti ratios for each of the core lengths. Note that the scale for magnetic susceptibility varies. (**a**) MAG14-1A-57Q-1-A: The exposure beds as referred to in Fig. 9 are labelled accordingly. Peaks in the Si concentration and Si/Ti ratio are directly associated with the bedded cherts, which are associated with signs of exposure in beds a, b, and c, and possibly in f. The lowermost bed (a), preserves small burrows produced in soft substrates that were likely not desiccated. This corresponds to a relatively indistinct Si high except in the position with evidence of early oxidation at the margin of the terrestrial burrow. Magnetic susceptibility indicates the presence of relatively coarse-grained clastic material associated with the siliceous exposure surfaces (c, d). (**b**) MAG14-1A-58Q-1-A: Peaks in Si concentration and the Si/Ti ratio are associated with exposure related to bioturbation, but there is a lack of distinct peaks directly associated with the bioturbated horizons (a), softground burrows, (2) and terrestrial burrows (3), except in the lower bed, where oxidation related to terrestrial bioturbation at depth may have led to the cementation of the bed. (**c**) MAG14-1A-66Y-1-A: The peaks in Si concentration and Si/Ti ratio in this deep lake to exposure cycle illustrate that the silica-rich beds are associated with the shallower environments in the shallowing upwards cycles as well as early oxidation. The beds are labelled as they are in Fig. 10. High magnetic susceptibility is associated with the more clastic-rich upper part of the cycle. (**d**) MAG14-1A-67Y-1-A: This core length also shows the association between the siliceous mudstone and evidence of exposure, in this case primarily evidenced by desiccation cracks. The meniscate-backfilled burrows shown in Fig. 9 (1) are present above the main siliceous unit and are present in the more clastic-rich bed overlying the desiccation-cracked mudstone. Here, the overlying lacustrine flooding event is associated with high values in the magnetic susceptibility curve, which is not necessarily due to clastic input but seems to be related to highs in Ca, Sr, and Mn (not shown)

or termites, and also flies and probably beetle or earwig burrows similar to those shown from the modern examples. In the modern environments studied in 2007 and 2008, only Nasikie Engida was observed to contain siliceous gels in the littoral to eulittoral zone. As noted by Owen et al. (2018b), silica could be supplied to lacustrine waters by high-temperature hydrothermal input, as well as dissolution of siliceous bedrock (e.g., trachyte), ash, and/or diatoms in high pH settings. The low Ca concentrations but abundant Na, the lack of calcite and a mix of freshwater and benthic diatoms from the intervals shown here (Owen et al. 2019)), along with siliceous substrates associated with exposure and trace fossils, together indicate that the assemblage from 57Q-1 may be most comparable to the northern shoreline of Nasikie Engida, where perennial hot springs feed the lake and trace diversity is relatively high. The deep-tier burrows preserved in 66Y-1 (Fig. 10c–e), almost certainly produced by termites, are associated with roots and a cobble-sized trachyte clast along with a muddy granular sandstone substrate that overlies laminated lacustrine deposits. In this example from the MAG14-1A core, the traces help to recognize a relatively significant drop in lake level, comparable in stratigraphic and paleoenvironmental significance to the deep-tier termite burrows preserved in the ODP core from the Olorgesailie Formation (Owen et al. 2009; Potts et al. 2020). In the example from MAG14-1A-66Y-1, integration of sedimentologic and ichnologic evidence indicates relatively wide fluctuations in water depths of the lake. The close investigation of the trace fossils preserved in the MAG14-1A core has the potential to provide valuable insights into the processes, timing, and depositional settings for the substrates, as well as the stratigraphic packaging of lake flooding to exposure cycles within the trace fossil-preserving intervals identified by McNulty (2017).

Traces Associated with Hyperalkaline Warm Springs

Modern: Northwestern Lagoon, Southeastern Lagoon, and Northern Alluvial Plain of Lake Magadi

Setting—Warm springs issue along several normal faults that bound the sub-basins of Lake Magadi, particularly where the normal faults intersect with underlying basement structures (Morley et al. 1999; Owen et al. 2019)). Simple invertebrate trails and vertebrate traces (footprints and nest-mounds) were observed near spring inflow at the Northwestern Lagoon (Figs. 12 and 13), Southeastern Lagoon (Fig. 14), and the northern alluvial plain (Fig. 15). The water that feeds the springs may be recycled within the closed basin from local groundwater, or it may derive from distant meteoric water travelling underground through bedrock, or from local meteoric water heated at depth, or a combination of these sources (Jones et al. 1977; Darling et al. 1990; Becht et al. 2006; Owen et al. 2018a). The springs are a perennial source of relatively fresh water (~30–35 g/L TDS) when compared to the hypersaline lake, and because of their moderate temperatures (~38–45 °C), benthic cyanobacterial mats can develop in the shallow water (Figs. 13b, c and 14c; Jones et al. 1977) and sedges may even grow near the spring vents (Fig. 12a, c). Substrates associated with the warm springs in the eulittoral to littoral zone are mainly microbial-mat-bound mud and fine-grained sand, with coarser-grained clastics and boulders rimming the high-gradient lake margins along normal faults. Braidplains with networks of ephemeral streams bring meteoric water to the lake basin from the northwest, north, and south, contributing fine-grained sand and silt to the sites, which also may be fed

Fig. 12 Modern traces associated with alkaline warm springs at the Northwestern Lagoon, Lake Magadi (Site 10). (**a**) Warm springs along the steep-gradient lake-bounding trachyte with sedges at spring output and microbe-rich salt efflorescence at margin. (**b**) Close-up of high-density chironomid tubes mainly in vertical life-position orientation. (**c**) Sedges at spring output with mammal tracks (domestic goat) sharply cross-cutting efflorescent crust on microbial mud below. (**d**) Opening to spider burrow with webbing in salt efflorescent crust near spring input

by warm alkaline springs. Salt efflorescence develops during evaporation of exposed substrates.

Descriptions of traces—**(Type 1)** Chironomid tubes: Small, open, vertically oriented tube structures of organic material; ~3 mm external diameter, ~1 mm internal diameter; approximately 2–3 cm long; tubes were present in dense concentrations and supported one another to remain vertical (Fig. 12b). **(Type 2)** Horizontal trails (incipient *Helminthoidichnites* isp. and incipient *Gordia* isp.): Small surface trails (2–2.5 mm diameter); straight to gently curving or looping; some examples of sharp turns (~120°); examples of cross-cutting same trail and other trails; margins of trails may be irregular in cyanobacterial mat and loose organic substrates (Figs. 13b and 15b, c). **(Type 3)** Fish nests: Circular excavated depressions in soft sediment to substrates of muddy sand, with internal rings that progressively decrease in diameter from ~15 cm and increase in depth (~3–8 cm) toward the center of the structure; excavated material rims the depression and separates the nests, which are typically found in closely spaced groups (Fig. 14d, e). **(Type 4)** Footprints (shorebirds): Medium-sized, three-toed footprints without webbing; digits straight and directed outward from metatarsal; posterior margin of prints triangular; total length ~3.5–4 cm; total width ~6 cm (Fig. 13d, e). **(Type 5)** Feeding structures (shorebirds), “bill-drag”: Surface “trails” apparently produced by shorebirds swinging narrow bill along substrate; “trails” are gently curved into semi-circles; ~12–16 cm in length and ~ 5 mm diameter (Fig. 13e). **(Type 6)** Feeding structures (shorebirds),

Fig. 13 Modern traces associated with alkaline warm springs at the Northwestern Lagoon, Lake Magadi (Site 10). (**a**) Spring-fed lake-margin with flamingo trampled area in microbe-bound mud with salt efflorescence developing on slightly exposed areas. (**b**) Simple fly larval trails in shallow water. (**c**) Close-up of flamingo footprints with clear webbing and digit impressions in microbe-bound mud with minor efflorescence where mud is exposed. Footprints randomly oriented and likely produced during feeding or drinking. (**d**) Close-up of microbial detritus on mud substrate with shorebird footprints and bill-probe marks. (**e**) Shorebird trackway and bill-drag feeding marks in sandy substrate coated with microbial detritus. (**f**) Degraded trample-ground associated with abandoned flamingo nesting or practice nest-building site. (**g**) Close-up of recent flamingo re-build of nest on degraded older nest-mound surface; salt efflorescence and tiger beetle bioturbation now reworking the firm and relatively dry substrate. (**h**) Close-up of tiger beetle larval burrows (3rd instar) in firm and moist older degraded nest-mound surface substrate

Fig. 14 Modern traces and trace producers at the Southeastern Lagoon near "Bird Rock," Lake Magadi (Site 11). (**a**) Yellow-billed storks, little egret, and black-winged stilt in foreground; lesser and greater flamingos in background. (**b**) Bird footprints (shorebird and stork) and bill-probe marks. (**c**) Warm-spring vent in shallow water, surrounded by benthic microbial mats and algae. (**d**) Fish-nest depressions and their producers (*Alcolapia grahami*). (**e**) High-density fish-nest depressions separated by excavation rims

probe-marks: Small (2–3 mm) holes with distinct margins; irregular and roughly circular to oval-shaped; some clear examples of crescent-shaped holes, usually in pairs with concave margins of crescents facing toward one another; often in groups of multiple holes (Figs. 13d and 14b). **(Type 7)** Footprints (flamingo): Large, three-toed, webbed footprints with rounded posterior margins; webbing to tips of digits; claws extend beyond webbing; tips of digits face only slightly outward; webbing not always impressed depending on substrate; total length ~7–8 cm; total width

Fig. 15 Spring-fed channels and pools along the western horst of the northern alluvial plain at Lake Magadi (Site 12). (**a**) View looking south from horst that separates Lake Magadi from Nasikie Engida. Wildebeest trampled ground and trackways crossing flats (lower left). (**b**) Larval fly trails in shallow spring water, along with ephydrid fly pupal cases. (**c**) Close-up of cross-cutting and re-used fly larval trails with ephydrid pupal cases in foreground. (**d**) Trampled microbe-rich mud in shallow spring-fed pool. (**e**) Wildebeest footprints in wet mud substrate between spring-fed channels. (**f**) Wildebeest trackways crossing the northern alluvial plain

~11 cm; examples in subaqueous substrates may be circular depressions (Fig. 13a, c). **(Type 8)** Flamingo nest-mounds and associated surfaces: Large mounds of organic-rich mud; wider at base than top with either flat top surface or with central depression (<4 cm deep); external portion of mounds porous and made of irregular lumps of mud; tops of mounds may have a smooth surface crust, likely of sodium evaporites and flamingo waste; may be associated with linear to irregular raised surfaces of muddy sediment with rounded margins, < 40 cm diameter (Fig. 13f, g). **(Type 9)** Footprints (long-legged waders), little egret: Medium-sized, unwebbed four-toed footprints; nodes along digits clearly impressed; claw impressions prominent; digits straight and ~6 mm in diameter; total length ~6.5–7 cm excluding hallux; total width ~7 cm; digits directed forward or outward (Fig. 14b). **(Type 10)** Footprints (bovids), wildebeest: Medium-sized hoof prints; hind foot total length ~8.5 cm; hind foot total width ~6.5 cm; tracks narrower at anterior than posterior; tips of hooves slightly pointed; outside margins slightly concave toward anterior; posterior margins of hooves rounded; medial gap straight and continues through track, widening slightly (<1 cm) at back (Fig. 15d, e; Scott 2010). **(Type 11)** Vertical burrows (incipient *Skolithos* isp.), tiger beetles: Small- to medium-sized (~2–4 mm diameter), straight, circular, open vertical burrows with sharp burrow boundaries; unlined/unwalled; slight, circular depression (~1 mm depth) around burrow opening (Fig. 13h).

Remarks—Traces produced by microbe-feeding insects (e.g., fly larvae) and their predators (e.g., tiger beetles, spiders), flamingos and other birds (e.g., storks, plovers, stilts), and alkaliphilic fish dominate the assemblage associated with alkaline warm springs at Lake Magadi. The flamingo nest mounds at the Northwestern Lagoon are degraded and were originally produced when water levels were higher than in 2007 (Fig. 13f, g); "practice mounds" (from 2007) are uncommon but present on older nest-mound surfaces (Fig. 13g). Mammal footprints (wildebeest, domestic goat) were also observed at these sites (Figs. 12c and 15). In the case of migratory wildebeest, their trails cut across the northern plain where spring waters flow along the western horst (Site 12, Fig. 1), but they may also be visiting sites with spring waters (Fig. 15a, d–f). Substrates in the spring-fed areas are typically wet and soupy, and remain under thin films of water to several centimeter depths. Microbial mats and clay provide some cohesion to the substrates, and contribute to the retention of good detail in vertebrate impressions (e.g., Fig. 13c). Salt efflorescence is less of an important destructive taphonomic factor in these sites when compared to the lake-margin flats away from spring input. Perennial input of the spring waters to these sites supports localized less extreme oasis-like environments for micro- and macro-organisms living within the shallow water. During rainy periods, planktonic filamentous cyanobacteria bloom across the flooded lagoons (Fig. 2e) and provide the primary food resource for vertebrate primary consumers, namely, lesser flamingos (e.g., Jenkin 1957) and alkaliphilic fish (*Alcolapia*; Bergman et al. 2003). During the dry periods, however, the braidplains near spring sites desiccate and standing water in the shrinking lake becomes increasingly hypersaline as evaporation continues, killing the cyanobacteria (Grant and Tindall 1986; Jones et al. 1998) and leading to the deposition of organic-rich mud interbedded with evaporite crusts (Fig. 2d, f) in areas near the central pan. Flamingos are migratory, and may leave the Magadi area until good conditions return, or remain restricted to the spring sites for food and water for drinking and bathing (Allen 1956; Brown 1973; Krienitz 2018). The fish are restricted to flooded areas near spring sites because of the relatively low pH (~9.5; Jones et al. 1977; Bergman et al. 2003), as are the invertebrates that require subaqueous conditions or moist substrates (e.g., fly and beetle larvae).

Traces Associated with Hypersaline Conditions

Modern: Eastern Lagoon and Central Pan, Lake Magadi; Southern Nasikie Engida Shoreline

Setting—Salinity becomes increasingly extreme distal to the sources of relatively freshwater, whether it is spring-fed or supplied from the ephemeral braidplain channels during rainy periods. The Eastern Lagoon of Lake Magadi near the causeway (Site 13; Figs. 1, 16, and 17) receives some meteoric water input during rainy periods, is at some distance from the hot springs (Fig. 1), and has some input of isolated cool-water meteoric springs along the lake-margin mudflats in places (Fig. 16a). The southern shoreline of Nasikie Engida (Fig. 18a–d) receives only direct meteoric precipitation, but can remain wet beneath salt crusts formed during the dry season because the inflow from northwestern Nasikie Engida hot springs is perennial with high flow rates (Jones et al. 1977; Renaut et al. 2021). The southwestern flats of Nasikie Engida border an alluvial plain, comprise coarse-grained clastic (sand- to

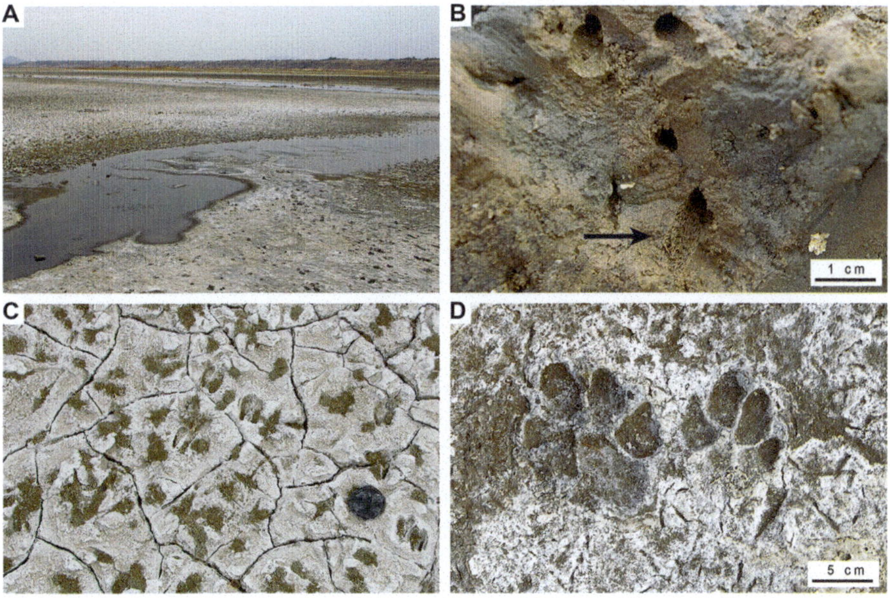

Fig. 16 Saline mudflats at the Eastern Lagoon, Lake Magadi (Site 13). (**a**) Mudflat eastern margin of the Eastern Lagoon with cool-spring outflow crossing the flat. (**b**) Close-up of open and partly backfilled vertical burrows leading to oxidation of oxygen-poor mud substrate. Elongate pellets (arrow) along the margin of the burrow in the lower center indicate that these burrows are probably produced by staphylinid beetles, which were also directly observed at the site. (**c**) Mammal tracks (domestic goat) and maribou stork (?) footprints in salt-encrusted desiccation-cracked firm mud substrate with reworked excavated tumuli from staphylinid burrows filling depressions. (**d**) Hyaena footprints in sandy mud substrate with salt efflorescence. Note triangular-shaped medial digits and lack of claw impressions (cf. Fig. 6d)

Fig. 17 Saline mudflats at the Eastern Lagoon near the causeway, Lake Magadi (Site 13). (**a**) Recent but abandoned flamingo nest-mounds covered in salt efflorescence. (**b**) Flamingo mud excavation around nest-mound, possibly representing a feeding structure of a greater flamingo. Now covered in salt efflorescence. (**c**) Overprinted flamingo footprint surface with footprints all directed in same direction, probably produced during display at roughly the same time as nest-building. (**d**) Small invertebrate open burrows produced in microbe-rich sandy sediment below salt efflorescence. Probably produced by ephydrid larvae, staphylinid larvae, or earwig nymphs. (**e**) Bioturbated mudflat surface with tumuli of excavated material, probably produced by staphylinid beetles. (**f**) Spider burrow with excavated material in well formed aggregates; small amount of webbing at burrow opening

pebble-sized) and muddy substrates, and are more laterally extensive than the salt-encrusted wet microbial mud of the southeastern flats (Fig. 18a, b). The central basin pan of Lake Magadi receives mostly direct meteoric precipitation, and bedded evaporites there contain minor evidence of fluctuating wetter/drier periods in evaporite crystal morphology and thin irregular interbeds of clastic silt (Fig. 2c; Scott 2010; McNulty 2017). The central pan evaporites may be cut by shoreline-normal

Fig. 18 Saline and hypersaline flats at Nasikie Engida (Sites 3 and 4) and central basin, Lake Magadi (Site 14). (**a**) Saline flat at distal toes of alluvium, southwest Nasikie Engida shoreline. (**b**) Grant's gazelle footprints in weaksly salt-encrusted muddy coarse-grained sand in flat adjacent to main zone of salt efflorescence. (**c**) Microbe-rich salt efflorescence at lake margin, southern shoreline of Nasikie Engida. (**d**) Microbe-rich ooze beneath salt crust at southern Nasikie Engida. Yellowish white clumps may be siliceous and/or bacterial colonies. (**e**) Channel (~1.5 across) cutting through western margin of the central pan of Lake Magadi, north of the causeway. Layers of trona crystals show transport of crystallized evaporites before induration of brown mud. Not clear if the channel was excavated as part of the Magadi Soda operation or an ephemeral channel. (**f**) Close up of large bovid (probably domestic cattle) footprint in firm mud substrate in the channel shown in (**e**)

meandering ephemeral mud-filled channels that show evidence of flow and transport of trona crystals (Fig. 18e). Open linear features cutting through the pan evaporites are also found parallel to shoreline with imbricated trona rafts (McNulty 2017). Salinities in these sites can reach up to ~320 g/L TDS in dry periods (Jones et al. 1977; McNulty 2017).

Descriptions of traces—**(Type 1)** Vertical burrows (cf. incipient *Polykladichnus* isp.): Small open burrows; diameter variable, ~2–3 mm; burrow openings irregular and roughly circular or oval-shaped; examined only in plan view (Fig. 17d). **(Type 2)** Vertical burrows (incipient *Skolithos* isp.): Small- to medium-sized (~2–4 mm diameter), straight, circular, open vertical burrows with sharp burrow boundaries; unlined/unwalled (not shown; Scott 2010). **(Type 3)** Vertical burrows (elongate pellet-filled burrows): Medium-sized (4–6 mm diameter), open or pellet-filled, oblique and vertical burrows; pellets are elongate and tiny (<1 mm diameter, < 3 mm long); burrow boundaries sharp; unlined, unwalled; associated with excavated pellet-like aggregates in tumuli surrounding burrow opening (Fig. 16b). **(Type 4)** Vertical burrows (?incipient *Skolithos* isp.), spider: Large (1.5–2 cm diameter), circular, open burrows with large circular tumuli (<2 cm high) of excavated material surrounding burrow opening; burrow boundaries not sharp; may have web covering burrow opening (Fig. 17f). **(Type 5)** Flamingo nest-mounds and associated surfaces: Large mounds of organic-rich mud; wider at base than top with either flat top surface or with central depression (<4 cm deep); external portion of mounds porous and made of irregular lumps of mud; tops of mounds may have a smooth surface crust, likely of sodium-carbonate evaporites and flamingo waste; may be associated with linear to irregular raised surfaces of muddy sediment with rounded margins, up to 40 cm diameter (Fig. 17a). **(Type 6)** Feeding and/or nest building structures (flamingos): Very large (<75 cm diameter), shallow, circular depressions usually with central, circular, rounded raised area (<30 cm diameter); depth of depressions ~5 cm; margins of depressions have slightly raised rims (~3 cm) (Fig. 17b). **(Type 7)** Footprints (flamingo): Large, three-toed, webbed footprints with rounded posterior margins; webbing to tips of digits; claws extend beyond webbing; tips of digits face only slightly outward; total length ~7–8 cm; total width ~11 cm; all impressions directed same direction and produced at same time (Fig. 17c). **(Type 8)** Footprints (hyaenid): Medium-sized four-toed canid-like footprints; small interdigital space; digit pads large and crowded; metatarsal pad large and circular, with one anterior lobe and broad, rounded single-lobed posterior margin; claws not impressed; total length ~7–7.5 cm; total width ~5.5 cm (Fig. 16d). **(Type 9)** Footprints (bovids, domestic goat): Medium-sized hoof prints; total length ~4–5 cm; total width ~3–4 cm; front print narrower and more pointed than hind print; tips of front prints pointed; tips of hind prints blunt; splayed in wet clay; overall shape of hind print square; overall shape of front print triangular to heart-shaped (Fig. 16c). **(Type 10)** Bird nests (scrape): Large (<30 cm outside diameter; ~20 cm diameter at top), circular, bird nest built from dried salt crust chips, gravel, and twigs; <5 cm high, with shallow depression over most of top surface (~15 cm diameter) (not shown; Scott 2010). **(Type 11)** Mammal footprints (bovids, gazelle): Medium-sized bovid footprints in moist mud with salt crust; footprint very heart-shaped with wide, rounded posterior margins to hooves, and very pointy anterior margins; medial gap narrow, and complete from anterior to posterior, but wider toward anterior and posterior; front and hind not distinguished; length ~7 cm, width ~5.5 cm (Fig. 18b). **(Type 12)** Footprints (bovids, domestic cow): Medium-sized, poorly impressed hoof prints in slightly indurated substrate; only anterior margins and outside margin impressed;

overall shape circular; total width ~5 cm; total length ~6–7 cm; rounded anterior margin and tips of hooves; outside edge convex; medial gap straight (Fig. 18f).

Remarks—Hypersalinity may be the most important factor in terms of restricting the diversity of cyanobacterial microbial food sources and their macro-organism consumers, when compared to hydrothermal and hyperalkaline conditions (Brock 1970; Grant and Ross 1986; Grant and Tindall 1986; Grant and Jones 2016; Mengistou 2016). The traces produced by organisms that can tolerate hypersaline conditions are subsequently affected by salt efflorescence, which may destroy surface features. Early cementation by zeolites (e.g., erionite, analcime), however, may help to preserve substrates from hypersaline settings, such as the flamingo nest-mound surfaces of the High Magadi Beds (see below) and Lake Bogoria (Scott et al. 2012). Some trace-producing insects, such as the staphylinids observed at the Eastern Lagoon, Lake Magadi, are well adapted to saline and hypersaline environments, and are able to feed on microbes that live in mats (e.g., Gerdes et al. 2008; Scott et al. 2020) and also interstitially within the upper few centimeters of sediment (Griffiths and Griffiths 1983; Garcia and Niell 1991). The saline mudflats at the Eastern Lagoon, Lake Magadi, are seasonally flooded and supplied with isolated sites of cool meteoric water that drain onto the flats. As in the other spring sites in the other lagoons of Lake Magadi, it is near these springs that insect burrows and mammal footprints are most abundant once the shallow lake waters recede. Flamingo traces produced when the flats were flooded, such as "practice" nest-mounds (Fig. 17a) and trampled-grounds likely produced during pre-mating display marches (Fig. 17c), develop salt efflorescent crusts once the site is abandoned. In the more extreme hypersaline settings of the central pan of Lake Magadi and the southern Nasikie Engida shoreline, isolated and rare mammal footprints (Fig. 18c, f) and possible predaceous insect burrows (earwig, spider) may be encountered in close association with salt efflorescence and bedded evaporites. With continued evaporite precipitation in these sites distal to spring input, the preservation potential of the traces is near nil. Footprints impressed in muddy coarser-grained clastics in hypersaline lake margins at distal alluvial fans, such as at southwestern Nasikie Engida (Fig. 18b), may have a higher chance of survival into the rock record because they are likely to be higher gradient and landward of the zone where capillary evaporation of saline groundwater (cf. Scott et al. 2010) or frequent replenishment of hypersaline lake water to the sediment-air interface.

Ancient: Fossil Flamingo Nests and Surfaces, Eastern Lagoon and Southern Lagoons, Lake Magadi

Setting—Degraded flamingo nest-mounds and associated irregular surfaces are preserved as trace fossils in the cemented High Magadi Beds (Late Pleistocene) around the Eastern Lagoon and southern lagoons of Lake Magadi (Fig. 19; Scott et al. 2012), forming "ichno-landscapes" (Buatois et al. 2016). The interpreted depositional setting for these traces is similar to the modern hypersaline mudflats at the Eastern Lagoon, where flamingos continue to build nest-mounds on the mud that

Fig. 19 Fossil flamingo nest-mounds and associated sediments in the High Magadi Beds, Lake Magadi. (**a**) Southern Lake Magadi (near site 6), with remnant degraded nest-mounds coated in fine-grained salt efflorescence. (**b**) Close-up of a salt-covered degraded remnants of the nest-mound surface at the site shown in (**a**). (**c**) Uncovered High Magadi Beds nest mound and associated surface buried under shallow sand cover at the Eastern Lagoon (Site 13). (**d**) Multiple nest-mound surfaces stacked in thick remnant of the High Magadi Beds in the Eastern Lagoon. Salt-efflorescence coats the surface of the outcrop, and highlights the oncoid-like pebbles in the lower part of the outcrop

drapes the fossil surfaces during periods when standing water in the lagoon is receding (Fig. 17a). The Northwestern Lagoon at modern Lake Magadi is a comparable setting, where sub-fossil degraded nest-mound surfaces are observed (Fig. 13f). At two of the High Magadi Beds fossil sites (southern Eastern Lagoon and Eastern Lagoon), several m-thick remnants of the thicker lake-margin sedimentary package indicate that those sites may have been re-used repeatedly as breeding sites (Fig. 19d; Scott et al. 2012), similar to margins of the Etosha Pan in Namibia (Berry 1972). In comparison to modern Lake Natron, the most stable flamingo breeding site today (Brown 1973; Krienitz 2018), the interpreted setting for the High Magadi Beds nest-mounds was likely a persistent set of extensive saline mudflats around a saline lake. Nest mounds at Lake Bogoria and at the Etosha Pan are most abundant near sites of relatively freshwater (fluvial input and springs) that the birds use for bathing and drinking (Allen 1956; Brown 1973; Berry 1972; Krienitz 2018; Scott et al. 2012).

Description of trace fossils—(**Type 1**) Flamingo nest-mounds and flamingo-modified surfaces: Circular mounds, irregularly shaped mounds, and linear to irregular, flamingo-modified surfaces in indurated reddish-brown silts; mounds very degraded and rounded at top; vary from ~10 to 20 cm diameter at top, from ~15 to

30 cm at base, and from 5 to 15 cm high; linear surfaces variable, but commonly ~15 cm in diameter (Fig. 19a–d).

Remarks—The fossil examples of nest mounds are generally small, salt-weathered remnants of the original nest mounds (cf. Eugster 1980; Scott et al. 2012). Finely crystalline sodium carbonate efflorescence develops on the High Magadi Beds nest-mounds and associated surfaces (Fig. 19; Buatois et al. 2016, their fig. 11.12), helping to distinguish them from younger sub-fossil degraded mounds that are also present in the modern mudflats present around Lake Magadi. The fossil beds are silts cemented by zeolites (e.g., erionite and analcime; Surdam and Eugster 1976), which forms either as a primary precipitate or as an alteration product from tuffaceous or clayey sediments when the Na concentration is higher than other cations, and the silica activity is higher than aluminum (Hay 1968; Eugster and Jones 1968; Surdam and Eugster 1976). Pleistocene analcime-cemented flamingo nest-mound surfaces present around Lake Bogoria (Scott et al. 2012) likely formed by the evaporative concentration of Na groundwater adjacent to the lake during low lake levels (Renaut 1993). This process is also probable for the High Magadi Beds flamingo nest-mound sites, and certainly the cementation had to have occurred following the production of the mounds from soft mud substrates, from which the flamingos build the mounds (Allen 1956; Brown 1973). Notably, the fossil surfaces at Magadi are not associated with fossil footprints of flamingos, just as at Lake Bogoria (Scott et al. 2012). The excessive trampling by flamingos of the surface during its formation, followed by later salt-weathering of the still somewhat porous sediment, precludes the likelihood of footprint preservation. Invertebrate trace fossils were not observed with the fossil nest-mounds at Lake Magadi, which may indicate that the preserved breeding sites were not close to relatively freshwater input, as in the modern examples where insect traces are encountered (e.g., Fig. 16).

Discussion: Significance of Animal Traces for Interpreting Conditions in Extreme Environments

A consideration of the distribution of ichnocoenoses—the assemblages of trace fossils produced by a community of organisms—when interpreting paleoenvironments is perhaps especially important for the extreme settings of saline lakes, like Lake Magadi and Nasikie Engida (Gierlowski-Kordesch 1991; Buatois et al. 2020). Local environmental conditions and sedimentary processes are widely variable at Lake Magadi and Nasikie Engida around the heterogeneous lake-margin areas, which ultimately are controlled by the geological setting of the basin: an inner continental rift with groundwater heated by the asthenosphere at shallow depths (Morley et al. 1999), supply of magmatic CO_2 (Lee et al. 2016, 2017), closed hydrology, and an arid to semi-arid equatorial climate. As shown by the modern examples here, the localized oasis-like conditions provided by spring waters contribute to the persistence of cyanobacterial food sources for primary trace-producing consumers (e.g., ephydrids, fish, flamingos), followed by their predators (e.g., spiders, earwigs, shorebirds, hyaenas), as is comparable with other hot-spring ecosystems (e.g., Brock 1970; Wiegert and Mitchell 1973). In the arid modern climate of the Magadi region, the springs provide a source of relatively freshwater for migratory mammals and birds, which may visit the relatively less-extreme areas of the lake periodically. The sites of oasis-like conditions are also sites where traces have a better likelihood of preservation in cohesive microbially bound, muddy, siliceous substrates in areas with relatively less destruction by salt efflorescence. The localized distribution of ichnocoenoses around springs at Lake Magadi and Nasikie Engida is especially highlighted by the extremely restrictive hypersaline conditions elsewhere.

In all of the modern and fossil examples shown here, the traces represent either syn-depositional or early post-depositional subaerial exposure of the substrates. The extreme conditions from these environments also correspond to the characteristics and composition of the substrates, whether they are bedded cherts, nodular chert, silica-rich mudstone, mudstone broken up by salt efflorescence, or siltstone cemented by zeolites like analcime. The cherts and siliceous mudstone, in particular, form in a variety of settings in the Magadi region and the conditions involved in their formation are not well understood. The associations shown here in both the modern and ancient examples, along with other clues such as the preservation of microbial cells in the Green Beds (Behr and Röhricht 2000) or desiccation cracks (McNulty 2017) and other features (e.g., "labyrinth patterns"; Leet et al. 2016) in the MAG14 cores, contribute to understanding the local conditions related to the formation of at least some of the bedded and nodular cherts. Modern siliceous sediments are best known from Nasikie Engida, where the hottest, deepest sourced springs in the area provide most of the water held in that basin, and evaporative concentration of the lake waters can lead to the formation of siliceous gels in the littoral to eulittoral zone (Fig. 7; Jones et al. 1967; Eugster and Jones 1968; Renaut et al. 2021). Similarly, there is a close association between the well-developed benthic microbial mats and insects, fish, and birds in the lake-margin environments associated with perennial conditions suitable for the consistent presence of this food source—CO_2-charged perennial spring waters with relatively low pH and salinity (Lee et al. 2016, 2017; Jones et al. 1977; McNulty 2017). As at Lake Bogoria in the central Kenya Rift, the spring-fed and CO_2-charged mudflats lead to productive conditions for microbial food sources for trace-producing insects in an oasis-like area (Loburu) in a location with otherwise saline lake-margin conditions where lower diversity trace assemblages are produced (Scott et al. 2020). At Lake Baringo, a chert substrate developed at the Soro hot-spring site on the volcano Ol Kokwe in the center of the lake preserves insect larval tunnels produced when the microbe-rich siliceous sediment was soft (Buatois et al. 2017).

Chemical sedimentation is typical of continental extreme sedimentary environments, but depending on the key conditions in these environments (i.e., pH, temperature, salinity) and solute concentrations and compositions, different mineral suites will be preserved in the rock record (e.g., Benison 2019; Renaut et al. 2021). Abundant microbial life is typical in extreme settings, and has likely formed the basis for ecosystems in extreme settings throughout the Phanerozoic and possibly even earlier (Jones et al. 1998; Grant and Jones 2016). Many of these examples, in particular, are associated with hydrothermal input, such as the Early Devonian Rhynie Chert (Trewin et al. 2003; Dunlop and Garwood 2017; Channing 2017). Conditions of extreme hypersalinity can develop in closed basins with high subsidence rates and/or extremely high evaporation potential, or a combination with other factors which may or may not include hydrothermal input, such as in the Eocene Green River Formation (e.g., Smith et al. 2015; Lowenstein et al. 2017; Demicco and Lowenstein 2019). The interacting controls of hydrothermal input, hypersalinity, and hyperalkalinity at Lake Magadi and Nasikie Engida are a product of the tectonic setting, accentuated by climate (i.e., aridity). The clues provided by animal traces, such as those described in the examples here, can help to determine the sedimentary processes and predominant conditions, and their heterogeneity, in extreme environments in the rock record.

Conclusions

Modern and ancient animal traces are abundant in localized areas in the extreme environments of Lake Magadi and Nasikie Engida in the southern Kenya Rift Valley, with assemblages of simple insect trails and burrows associated with vertebrate traces (footprints, nest-mounds) produced mainly in microbial (especially cyanobacterial), siliceous, evaporitic, and zeolitic clastic substrates. The producers of the modern traces include flies (e.g., ephydrids), beetles (e.g., staphylinids, tiger beetles), earwigs, spiders, and termites (e.g., Macrotermitinae), as well as fish, birds (e.g., flamingos, plovers), and mammals. All either require microbial or primary consumer food sources adapted to the extreme conditions, or are able to move between sites where conditions are relatively less extreme. Similar trace fossils and sedimentary substrates and environments are represented by the fossil examples, where the lake-margin traces can provide clues about substrate conditions (e.g., soft siliceous gels and early subaerial exposure) or fluctuations in lake level leading to the overprinting of terrestrial trace fossils on lacustrine sediments (e.g., MAG14-1A core). The Pleistocene, Holocene, and modern examples presented here reflect the interacting controls of hydrothermal water, high salinity, and high alkalinity on sedimentary environments and their conditions, as well as the high-Si low-Ca input to the lake and high evaporation potential in the semi-arid climate of the southern rift.

Acknowledgments Beth Gierlowski-Kordesch inspired us in our approach to investigating modern and ancient traces in heterogeneous environments in the Kenya Rift, and her scientific contributions, encouragement, and friendship through the years will always be cherished. We are very grateful for scientific contributions and support provided by colleagues from the Hominin Sites and Paleolakes Drilling Project, the Smithsonian Institution Olorgesailie Drilling Project, and LacCore Laboratories at the University of Minnesota. Shangde Luo (National Cheng-Kung University) and Chuan-Chou Shen (National Taiwan University) are gratefully acknowledged for significant contributions by dating the cherts. J.J.S., R.W.R., and L.A.B. were supported by the

Natural Sciences and Engineering Research Council of Canada (NSERC to LAB: grant 311726-13; RWR: grant GP629) and a PhD scholarship to J.J.S. R.B.O. acknowledges the Hong Kong Research Grants Council for continued support (HKBU12300815 and 12304018). M.S. gratefully acknowledges support from the Swiss National Science Foundation grant P300P2 158501. Drilling of the MAG14 cores was funded by ICDP and NSF grants (EAR-1123942, BCS-1241859, EAR-1115118, and EAR-1338553). Analyses were supported by the Hong Kong Research Grants Council (HKBU-201912 and 12304018). We thank the National Museums of Kenya, the Kenyan National Council for Science and Technology, the Kenyan Ministry of Mines, and the National Environmental Management Authority of Kenya for providing permits. Tata Chemicals Magadi Limited and the Magadi Administrative District of Kajiado County provided local support during drilling and field research. This is publication #30 of the Hominin Sites and Paleolakes Drilling Project (HSPDP). Thank you to Editor Michael Rosen for his persistence and dedication to this volume in Beth's honor, and to reviewer Daniel Hembree for his constructive comments and helpful edits.

References

Allen, R. P. (1956). *The Flamingos: Their Life History and Survival*. New York: National Audubon Society, Research Report no. 5. 285 pp.

Allen, D. J., Darling, W. G., & Burgess, W. G. (1989). Geothermics and hydrogeology of the southern part of the Kenya Rift Valley with emphasis on the Magadi-Nakuru area. *British Geological Survey Research Report*, SD/89/1, 68 pp.

Baker, B. H. (1958). Geology of the Magadi area. *Geological Survey of Kenya Report, 42*, 81 pp.

Baker, B. H. (1963). Geology of the area south of Magadi. *Geological Survey of Kenya Report, 61*, 42 pp.

Baker, B. H. (1986). Tectonics and volcanism of the southern Kenya Rift Valley and its influence on rift sedimentation. In L. E. Frostick, R. W. Renaut, I. Reid, & J.-J. Tiercelin (Eds.), *Sedimentation in the African Rifts. Geological Society Special Publication, 25* (pp. 45–57). London: Geological Society.

Baker, B. H., & Wohlenberg, J. (1971). Structure and evolution of the Kenya Rift Valley. *Nature, 229*, 538–542.

Becht, R., Mwango, F., & Muno, F. A. (2006). Groundwater links between Kenyan Rift Valley lakes. In *Proceedings of the 11th World Lakes Conference* (vol. 2, pp. 7–13) Nairobi, 2005.

Behr, H.-J. (2002). Magadiite and magadi chert: A critical analysis of the silica sediments in the Lake Magadi Basin, Kenya. In R. W. Renaut & G. M. Ashley (Eds.), *Sedimentation in Continental Rifts. SEPM Special Publication, 73* (pp. 257–273). Tulsa: SEPM.

Behr, H. J., & Röhricht, C. (2000). Record of seismotectonic events in siliceous cyanobacterial sediments (Magadi cherts), Lake Magadi, Kenya. *International Journal of Earth Sciences, 89*, 268–283.

Benison, K. (2019). The physical and chemical sedimentology of two high-altitude acid salars in Chile: Sedimentary processes in an extreme environment. *Journal of Sedimentary Research, 89*, 147–167.

Bergman, A. N., Laurent, P., Otiang'a-Owiti, G., Bergman, H. L., Walsh, P. J., Wilson, P., & Wood, C. M. (2003). Physiological adaptations of the gut in the Lake Magadi tilapia, *Alcolapia grahami*, an alkaline- and saline-adapted teleost fish. *Comparative Biochemistry and Physiology Part A, 136*, 701–715.

Berry, H. H. (1972). Flamingo breeding on the Etosha Pan, South West Africa, during 1971. *Madoqua, Series I 5*, 5–31.

Brenna, B. L. (2016). *The Chemical, Physical, and Microbial Origins of Pleistocene Cherts at Lake Magadi, Kenya Rift Valley* (MSc thesis). University of Saskatchewan, 170 pp.

Brock, T. D. (1970). High temperature systems. *Annual Review of Ecology and Systematics, 1*, 191–220.

Brock, M. L., Wiegert, R. G., & Brock, T. D. (1969). Feeding by *Paracoenia* and *Ephydra* (Diptera: Ephydridae) on the microorganisms of hot springs. *Ecology, 50*, 192–200.

Brown, L. H. (1973). *The Mystery of the Flamingos* (121 pp). Nairobi: East African Publishing House.

Buatois, L. A., Labandeira, C. C., Mángano, G. M., Cohen, A., & Voigt, S. (2016). Chapter 11: The Mesozoic lacustrine revolution. In G. M. Mángano & L. A. Buatois (Eds.), *The Trace-Fossil Record of Major Evolutionary Events: Volume 2: Mesozoic and Cenozoic. Topics in Geobiology, 40* (pp. 179–263). Dordrecht, Springer.

Buatois, L. A., Renaut, R. W., Scott, J. J., & Owen, R. B. (2017). An unusual occurrence of the trace fossil *Vagorichnus* preserved in hydrothermal silica at Lake Baringo, Kenya Rift Valley: Taphonomic and paleoenvironmental significance. *Palaeogeography, Palaeoclimatology, Palaeoecology, 485*, 843–853.

Buatois, L. A., Renaut, R. W., Owen, R. B., Behrensmeyer, A. K., & Scott, J. J. (2020). Animal bioturbation preserved in Pleistocene magadiite at Lake Magadi, Kenya Rift Valley, and its implications for the depositional environment of bedded magadiite. *Scientific Reports, 10*, 6794, 13 pp. https://doi.org/10.1038/s41598-020-63505-7

Campisano, C., Cohen, A., Arrowsmith, J., Asrat, A., Behrensmeyer, A. K., Brown, E. T., Deino, A. L., Deocampo, D. M., Feibel, C. S., Kingston, J. D., Lamb, H. F., Lowenstein, T. K., Noren, A., Olago, D. O., Owen, R. B., Pelletier, J. D., & Potts, R. (2017). The Hominin Sites and Paleolakes Drilling Project: High-resolution paleoclimate records from the East African rift system and their implications for understanding the environmental context of hominin evolution. *PaleoAnthropology*, 1–43. https://doi.org/10.4207/PA.2017.ART104

Channing, A. (2017). A review of active hot-spring analogues of Rhynie; environments habitats and ecosystems. *Philosophical Transactions of the Royal Society, B, 373: 2016-490.* 12 pp. https://doi.org/10.1098/rstb.2016.0490

Cohen, A., Campisano, C., Arrowsmith, R., Asrat, A., Behrensmeyer, A. K., Deino, A., Feibel, C., Hill, A., Johnson, R., Kingston, J., Lamb, H., Lowenstein, T., Noren, A., Olago, D., Owen, R. B., Potts, R., Reed, K., Renaut, R., Schabitz, F., Tiercelin, J.-J., Trauth, M. H., Wynn, J., Ivory, S., Brady, K., O'Grady, R., Rodysill, J., Githiri, J., Russell, J., Foerster, V., Dommain, R., Rucina, S., Deocampo, D., Russell, J., Billingsley, A., Beck, C., Dorenbeck, G., Dullo, L., Feary, D., Garello, D., Gromig, R., Johnson, T., Junginger, A., Karanja, M., Kimburi, E., Mbuthia, A., McNulty, E., Muiruri, V., Nambiro, E., Negash, E. W., Njagi, D., Wilson, J. N., Rabideaux, N., Raub, T., Seir, M. J., Smith, P., Urban, J., Warren, M., Yadeta, M., Yost, C., & Zinaye, B. (2016). The Hominin Sites and Paleolakes Drilling Project: Inferring the environmental context of human evolution from eastern African rift lake deposits. *Scientific Drilling, 21*, 1–16. https://doi.org/10.5194/sd-21-1-2016

Crane, K. (1981). Thermal variations in the Gregory Rift of southern Kenya (?). *Tectonophysics, 74*, 239–262.

Darling, W. G. (2001). Magadi and Suguta: The contrasting hydrochemistry of two soda lake areas in the Kenya Rift Valley. In R. Cidu & F. Frau (Eds.), *Proceedings of the 10th International Symposium on Water-Rock Interaction (WRI-10)* (pp. 95–98). Lisse: Swets & Zeitlinger (Balkema).

Darling, W. G., Allen, D. J., & Armannsson, H. (1990). Indirect detection of subsurface outflow from a Rift Valley lake. *Journal of Hydrology, 113*, 297–305.

Darling, W. G., Griesshaber, E., Andrews, J. N., Armannsson, H., & O'Nions, R. K. (1995). The origin of hydrothermal and other gases in the Kenya Rift Valley. *Geochimica et Cosmochimica Acta, 59*, 2501–2512.

De Cort, G., Mees, F., Renaut, R. W., Sinnesael, M., Van der Meeren, T., Goderis, S., Keppens, E., Mbuthia, A., & Verschuren, D. (2019). Late-Holocene sedimentation and sodium carbonate deposition in hypersaline, alkaline Nasikie Engida, southern Kenya Rift Valley. *Journal of Paleolimnology, 62*, 279–300.

Demicco, R. V., & Lowenstein, T. K. (2019). When "evaporites" are not formed by evaporation: The role of temperature and pCO2 on saline deposits of the Eocene Green River Formation, Colorado, USA. *Geological Society of America (GSA) Bulletin*, 16 pp. https://doi.org/10.1130/B35303.1

Deocampo, D. M., & Renaut, R. W. (2016). Geochemistry of African soda lakes. In M. Schagerl (Ed.), *Soda Lakes of East Africa* (pp. 77–96). Cham: Springer.

Dunlop, J. A., & Garwood, R. J. (2017). Terrestrial invertebrates in the Rhynie chert ecosystem. *Philosophical Trsansactions of the Royal Society, B, 373*. https://doi.org/10.1098/rstb.2016.0489

Eugster, H. P. (1969). Inorganic bedded cherts from the Magadi area, Kenya. *Contributions to Mineralogy and Petrology, 22*, 1–31.

Eugster, H. P. (1980). Lake Magadi, Kenya, and its precursors. In A. Nissenbaum (Ed.), *Hypersaline Brines and Evaporitic Environments. Developments in Sedimentology, 28* (pp. 195–232). Burlington: Elsevier.

Eugster, H. P. (1986). Lake Magadi, Kenya: A model for rift valley hydrochemistry and sedimentation? In L. E. Frostick, R. W. Renaut, I. Reid, & J.-J. Tiercelin (Eds.), *Sedimentation in the African Rifts. Geological Society Special Publication, 25* (pp. 177–189). London: Geological Society.

Eugster, H. P., & Jones, B. F. (1968). Gels composed of sodium-aluminium-silicate, Lake Magadi, Kenya. *Science, 161*, 160–164.

Garcia, C. M., & Niell, F. X. (1991). Burrowing beetles of the genus *Bledius* (Staphylinidae) as agents of bioturbation in the emergent areas and shores of an athalassic inland lake (Fuente de Piedra, southern of Spain). *Hydrobiologia, 215*, 163–173.

Gerdes, G., Porada, H., & Bouougri, E. H. (2008). Bio-sedimentary structures evolving from the interaction of microbial mats, burrowing beetles and the physical environment of Tunisian coastal sabkhas. *Senckenbergiana Maritima, 38*, 45–58.

Gierlowski-Kordesch, E. (1991). Ichnology of an ephemeral lacustrine/alluvial plain system: Jurassic East Berlin Formation, Hartford Basin, USA. *Ichnos, 1*, 221–232.

Grant, W. D., & Jones, B. E. (2016). Bacteria, Archaea and viruses of soda lakes. In M. Schagerl (Ed.), *Soda Lakes of East Africa* (pp. 97–148). Cham: Springer.

Grant, W. D., & Ross, H. N. M. (1986). The ecology and taxonomy of halobacteria. *FEMS Microbiology Reviews, 39*, 9–15.

Grant, W. D., & Tindall, B. J. (1986). The alkaline saline environment. In R. A. Herbert & G. A. Codd (Eds.), *Microbes in Extreme Environments* (pp. 25–54). London: Academic Press.

Griffiths, C. L., & Griffiths, R. J. (1983). Biology and distribution of the littoral rove beetle *Psamathobledius punctatissimus* (Le Conte) (Coleoptera: Staphylinidae). *Hydrobiologia, 101*, 203–214.

Hay, R. (1968). Chert and its sodium-silicate precursors in sodium-carbonate lakes of East Africa. *Contributions to Mineralogy and Petrology, 17*, 255–274.

Jenkin, P. M. (1957). The filter-feeding and food of flamingoes (Phoenicopteri). *Philosophical Transactions of the Royal Society of London, B, 240*, 401–493.

Jones, B. J., Rettig, S. L., & Eugster, H. P. (1967). Silica in alkaline brines. *Science, 158*, 1310–1314.

Jones, B. F., Eugster, H. P., & Rettig, S. L. (1977). Hydrochemistry of the Lake Magadi basin, Kenya. *Geochimica et Cosmochimica Acta, 41*, 53–72.

Jones, B. E., Grant, W. D., Duckworth, A. W., & Owenson, G. G. (1998). Microbial diversity of soda lakes. *Extremophiles, 2*, 191–200.

Kambura, A. K., Mwirichia, R. K., Kasili, R. W., Karanja, E. N., Makonde, H. M., & Bogas, H. I. (2016). Bacteria and Archaea diversity within the hot springs of Lake Magadi and Little Magadi in Kenya. *Biomed Central (BMC) Microbiology, 16*, 136, 12 pp. https://doi.org/10.1186/s12866-016-0748-x

Krienitz, L. (2018). *Lesser Flamingos: Descendants of Phoenix* (249 pp). Berlin, Heidelberg: Springer-Verlag.

Krivosheina, M. G. (2008). On insect feeding on cyanobacteria. *Journal of Paleontology, 42*, 596–599.

Lee, H., Muirhead, J. D., Fischer, T. P., Ebinger, C. J., Kattenhorn, S. A., Sharp, Z. D., & Kianji, G. (2016). Massive and prolonged deep carbon emissions associated with continental rifting. *Nature Geoscience, 9*, 145–149.

Lee, H., Fischer, T. P., Muirhead, J. D., Ebinger, C. J., Kattenhorn, S. A., Sharp, Z. D., Kianji, G., Takahata, N., & Sano, Y. (2017). Incipient rifting accompanied by the release of subcontinental lithospheric mantle volatiles in the Magadi and Natron basin, East Africa. *Journal of Volcanology and Geothermal Research, 346*, 118–133.

Leet, K., Lowenstein, T. K., Owen, R. B., Renaut, R. W., Deocampo, D. M., Cohen, A. S., McNulty, E. P., Muiruri, V. M., Rabideaux, N. M., Billingsley, A. L., & Mbuthia, A. (2016). Origins of Magadi-type chert: New clues from the HSPDP Lake Magadi drill cores. *Geological Society of America, Abstracts with Programs, 48*(7), 42–49.

Le Turdu, C., Tiercelin, J. J., Richert, J. P., Rolet, J., Xavier, J. P., Renaut, R. W., Lezzar, K. E., & Coussement, C. (1999). Influence of pre-existing oblique discontinuities on the geometry and evolution of extensional fault patterns: Evidence from the Kenya rift using SPOT imagery. In C. K. Morley (Ed.), *Geoscience of Rift Systems – Evolution of East Africa. AAPG Studies in Geology, 44* (pp. 173–191). Tulsa: The American Association of Petroleum Geologists.

Lowenstein, T. K., Jagniecki, E. A., Carroll, A. R., Smith, M. E., Renaut, R. W., & Owen, R. B. (2017). The Green River salt mystery: What was the source of the hyperalkaline lake waters? *Earth-Science Reviews, 173*, 295–306.

McNulty, E. (2017). *Lake Magadi and the Soda Lake Cycle: A Study of the Modern Sodium Carbonates and of Late Pleistocene and Holocene Lacustrine Core Sediments* (MSc thesis). Binghamton University, Graduate Dissertations and Theses, 25, 125 pp.

Mengistou, S. (2016). Invertebrates of East African soda lakes. In M. Schagerl (Ed.), *Soda Lakes of East Africa* (pp. 205–226). Cham: Springer.

Morley, C. K. (1999). Influence of preexisting fabrics on rift structure. In C. K. Morley (Ed.), *Geoscience of Rift Systems–Evolution of East Africa. AAPG Studies in Geology, 44* (pp. 151–160). Tulsa: The American Association of Petroleum Geologists.

Morley, C. K., Ngenoh, D. K., & Ego, J. K. (1999). Introduction to the East African rift system. In C. K. Morley (Ed.), *Geoscience of Rift Systems–Evolution of East Africa. AAPG Studies in Geology, 44* (pp. 1–18). Tulsa: The American Association of Petroleum Geologists.

Muiruri, V. (2018). *Late Quaternary Diatom and Palynomorph Stratigraphies and Palaeoenvironments of the Koora Graben and Lake Magadi Basin, Kenya Rift Valley* (PhD thesis). Hong Kong Baptist University, 303 pp.

Owen, R. B., Renaut, R. W., Hover, V. C., Ashley, G. M., & Muasya, A. M. (2004). Swamps, springs, and diatoms: Wetlands of the semi-arid Bogoria-Baringo Rift, Kenya. *Hydrobiologia, 518*, 59–78.

Owen, R. B., Renaut, R. W., Scott, J. J., Potts, R., & Behrensmeyer, A. K. (2009). Wetland sedimentation and associated diatoms in the Pleistocene Olorgesailie Basin, southern Kenya Rift Valley. *Sedimentary Geology, 222*, 124–137.

Owen, R. B., Renaut, R. W., & Lowenstein, T. K. (2018a). Spatial and temporal geochemical variability in lacustrine sedimentation in the East African Rift System: Evidence from the Kenya Rift and regional analyses. *Sedimentology, 65*, 1697–1730.

Owen, R. B., Muiruri, V. M., Lowenstein, T. K., Renaut, R. W., Rabideaux, N., Luo, S., Deino, A. L., Sier, M. J., Dupont-Nivet, G., McNulty, E. P., Leet, K., Cohen, A. S., Campisano, C., Deocampo, D., Shen, C.-C., Billingsley, A., & Mbuthia, A. (2018b). Progressive aridification in East Africa over the last half million years and implications for human evolution. *Proceedings of the National Academy of Sciences, 115*(44), 11174–11179.

Owen, R.B., Renaut, R.W., Muiruri, V.M., Rabideaux, N.M., Lowenstein, T.K., McNulty, E.P., Leet, K., Deocampo, D., Luo, S., Deino, A.L., Cohen, A., Sier, M.J., Campisano, C., Shen, C.-C., Billingsley, A., Mbuthia, A., Stockhecke, M. 2019. Quaternary history of the Lake Magadi Basin, southern Kenya Rift: Tectonic and climatic controls. *Palaeogeography, Palaeoclimatology, Palaeoecology, 518*, 97–118.

Potts, R., Dommain, R., Moerman, J. W., Behrensmeyer, A. K., Deino, A. L., Beverly, E. J., Brown, E. T., Deocampo, D., Kinyanjui, R., Lupien, R., Owen, R. B., Rabideaux, N., Russell, J. M., Stockhecke, M., Riedl, S., deMenocal, P., Faith, J. T., Garcin, Y., Noren, A., Scott, J. J.,

Western, D., Bright, J., Clark, J. B., Cohen, A. S., Heil, C. W., Keller, C. B., King, J., Levin, N. E., Brady, K., Muiruri, V., Renaut, R. W., Rucina, S. M., & Uno, K. (2020). Increased ecological resource variability during a critical transition in hominin evolution. *Science Advances, 6*(43), eabc8975, 14 pp. https://doi.org/10.1126/sciadv.abc8975

Rabideaux, N. (2018). *Late Quaternary East African Environmental Change based on Mineralogical and Geochemical Analysis of Outcrop and Core Material from the Southern Kenya Rift* (PhD thesis). Georgia State University, 884 pp.

Renaut, R. W. (1993). Zeolitic diagenesis of late quaternary fluviolacustrine sediments and associated calcrete formation in the Lake Bogoria basin, Kenya Rift Valley. *Sedimentology, 40*, 271–301.

Renaut, R. W., Jones, B., & Tiercelin, J.-J. (1998). Rapid *in situ* silicification of microbes at Loburu hot springs, Lake Bogoria, Kenya Rift Valley. *Sedimentology, 45*, 1083–1103.

Renaut, R. W., Owen, R. B., Lowenstein, T. K., De Cort, G., McNulty, E., Scott, J. J., & Mbuthia, A. (2021). The role of hydrothermal fluids in sedimentation in saline alkaline lakes: Evidence from Nasikie Engida, Kenya Rift Valley. *Sedimentology, 68*, 108–134.

Röhricht, C. (1998). *Lithologie und der Chertserien des Magadi Beckens, Lake Magadi, Kenia* (PhD thesis). University of Gottingen.

Sanz-Montera, E., Calvo, J.-P., Garcia del Cura, M. A., Ornosa, C., Outerelo, R., & Rodriguez-Aranda, J. P. (2013). The rise of the diptera-microbial mat interactions during the Cenozoic: Consequences for the sedimentary record of saline lakes. *Terra Nova, 25*, 465–471.

Scott, J. J. (2010). *Saline Lake Ichnology: Kenya Rift Valley and Eocene Green River Formation, Wyoming* (PhD thesis). University of Saskatchewan, 547 pp.

Scott, J. J., & Smith, M. E. (2015). Trace fossils of the Eocene Green River lake basins, Wyoming, Utah, and Colorado. In M. E. Smith & A. R. Carroll (Eds.), *Stratigraphy and Paleolimnology of the Green River Formation, Western USA* (pp. 317–354). Dordrecht: Springer-Verlag.

Scott, J. J., Renaut, R. W., Buatois, L. A., & Owen, R. B. (2009). Biogenic structures in exhumed surfaces around saline lakes: An example from Lake Bogoria, Kenya Rift Valley. *Palaeogeography, Palaeoclimatology, Palaeoecology, 272*, 176–198.

Scott, J. J., Renaut, R. W., & Owen, R. B. (2010). Taphonomic controls on animal tracks at saline, alkaline Lake Bogoria, Kenya Rift Valley: Impact of salt efflorescence and clay mineralogy. *Journal of Sedimentary Research, 80*, 639–665.

Scott, J. J., Renaut, R. W., & Owen, R. B. (2012). Impacts of flamingos on saline lake margin and shallow lacustrine sediments in the Kenya Rift Valley. *Sedimentary Geology, 277–278*, 32–51.

Scott, J. J., Buatois, L. A., Mángano, M. G., Renaut, R. W., & Owen, R. B. (2020). Bioturbation in matgrounds at Lake Bogoria in the Kenya Rift Valley: Implications for interpreting the heterogeneous early Cambrian sea floor. *Lethaia, 53*, 62–71.

Smith, M. E., Carroll, A. R., & Scott, J. J. (2015). Stratigraphic expression of climate, tectonism, and geomorphic forcing in an underfilled lake basin: Wilkins Peak Member of the Green River Formation. In M. E. Smith & A. R. Carroll (Eds.), *Stratigraphy and Paleolimnology of the Green River Formation, Western USA* (pp. 61–102). Dordrecht: Springer-Verlag.

Surdam, R. C., & Eugster, H. P. (1976). Mineral reactions in the sedimentary deposits of the Lake Magadi region, Kenya. *Geological Society of America Bulletin, 87*, 1739–1752.

Trewin, N. H., Fayers, S. R., & Kelman, R. (2003). Subaqueous silicification of the contents of small ponds in an Early Devonian hot-springs complex, Rhynie, Scotland. *Canadian Journal of Earth Sciences, 40*, 1697–1712.

Wiegert, R. G., & Fraleigh, P. C. (1972). Ecology of Yellowstone thermal effluent systems: Net primary production and species diversity of a successional blue-green algal mat. *Limnology and Oceanography, 17*, 215–228.

Wiegert, R. G., & Mitchell, R. (1973). Ecology of Yellowstone thermal effluent systems: Intersects of blue-green algae, grazing flies (*Paracoenia*, Ephydridae) and water mites (*Partnuniella*, Hydrachnellae). *Hydrobiologia, 41*, 251–271.

Part III
European Lakes

Lake-Level Fluctuations and Allochthonous Lignite Deposition in the Eocene Pull-Apart Basin "Prinz von Hessen" (Hesse, Germany) – A Palynological Study

Maryam Moshayedi, Olaf K. Lenz, Volker Wilde, and Matthias Hinderer

Abstract High-resolution palynological analysis of 52 core samples from a distinct part of the lacustrine filling of the Eocene pull-apart basin "Grube Prinz von Hessen" in Southwest Germany has been applied to recognize driving factors responsible for changes in vegetation and environment. The 15 m studied interval is characterized by a specific alternation of lignite beds and mudstones which were deposited when the lake had reached its smallest extent. The diverse and well-preserved palynoflora shows that the shoreline and the marginal swamp around the lake were dominated by herbaceous and woody swamp communities, whereas the forest association in the hinterland was mainly dominated by Juglandaceae and Fagaceae. Non-metric multidimensional scaling and Q- and R-mode cluster analyses of the microflora reveal strong differences between the palynomorph assemblages of lignite and mudstone beds. The situation in the vicinity of the lake was influenced by tectonic activity, such as earthquake tremor, resulting in the redeposition of lignite and mudstone. Nevertheless, the same subordinate trend in the vegetation is obvious in the microflora of both lithologies. Slight qualitative changes

Supplementary Information The online version of this chapter (https://doi.org/10.1007/978-3-030-66576-0_3) contains supplementary material, which is available to authorized users.

M. Moshayedi (✉) · M. Hinderer
Technische Universität Darmstadt, Institute of Applied Geosciences, Applied Sedimentology, Darmstadt, Germany
e-mail: moshayedi@geo.tu-darmstadt.de

O. K. Lenz
Technische Universität Darmstadt, Institute of Applied Geosciences, Applied Sedimentology, Darmstadt, Germany

Senckenberg Gesellschaft für Naturforschung, General Directorate, Frankfurt am Main, Germany

V. Wilde
Senckenberg Forschungsinstitut und Naturmuseum, Sektion Paläobotanik, Frankfurt am Main, Germany

M. R. Rosen et al. (eds.), *Limnogeology: Progress, Challenges and Opportunities*, Syntheses in Limnogeology, https://doi.org/10.1007/978-3-030-66576-0_3

may be explained by slightly decreasing temperature and precipitation during the deposition of the studied section. Therefore, changes in the paleoenvironment and the ecosystems were mainly controlled by tectonic activity, but some influence of climate change may also be noted.

Keywords Central Germany · Paleogene · Palynology · Lacustrine sediments · Climate change · Tectonic activity

Introduction

The Sprendlinger Horst in Southwest Germany consists of a Paleozoic metamorphic and magmatic basement with a Rotliegend (Lower Permian) sedimentary cover (Marell 1989), and represents the northern extension of the Odenwald basement which is flanking the Upper Rhine Graben to the northeast (Fig. 1). Several small, isolated basins filled by lacustrine sediments of Paleogene age are known from the area, some of which have been studied from drill cores. Most of them represent the filling of maar-type volcanic structures (e.g. Jacoby 1997; Harms et al. 1999; Jacoby et al. 2000; Felder and Harms 2004). This includes the Eocene maar lake of Messel, which is well known for the perfect preservation of fossils and was in the focus of numerous paleoenvironmental studies (e.g., Schaal and Ziegler 1988; Gruber and Micklich 2007; Lenz et al. 2007, 2011; Collinson et al. 2012; Smith et al. 2018).

Another Eocene record on the Sprendlinger Horst which has been cored is the lacustrine filling of Lake "Prinz von Hessen" (PvH), 5 km northeast of Darmstadt (Hesse, Germany) and 2 km southwest of the maar structure of Messel (Fig. 1). Nevertheless, in contrast to other Paleogene records on the Sprendlinger Horst which are of phreatomagmatic origin and contain sediments of deep meromictic lakes, the structure at Lake PvH represents a more shallow pull-apart basin (Felder et al. 2001; Hofmann et al. 2005; Felder and Gaupp 2006; Moshayedi et al. 2018).

Because qualitative and quantitative analyses of pollen and spores are especially suited for the reconstruction of plant communities and their evolution (Frederiksen 1996), our palynological study of the lacustrine succession at PvH provides a rare insight into the evolution of vegetation around a basin that was mainly influenced by tectonic activity and, therefore, experienced a complex pattern of deposition and subsidence. We have shown not only the long-term evolution of the lake, but also that the ecosystem around it had been controlled by tectonic activity resulting in phases of relatively rapid subsidence and deposition of organic-rich mudstone and bituminous shale (Moshayedi et al. 2018). Shallow water phases with the deposition of lignite can be related to periods of quiescent tectonic activity and limited subsidence. However, long-term changes in the composition of the forest vegetation at Lake PvH can be attributed to a change in humidity (Moshayedi et al. 2018).

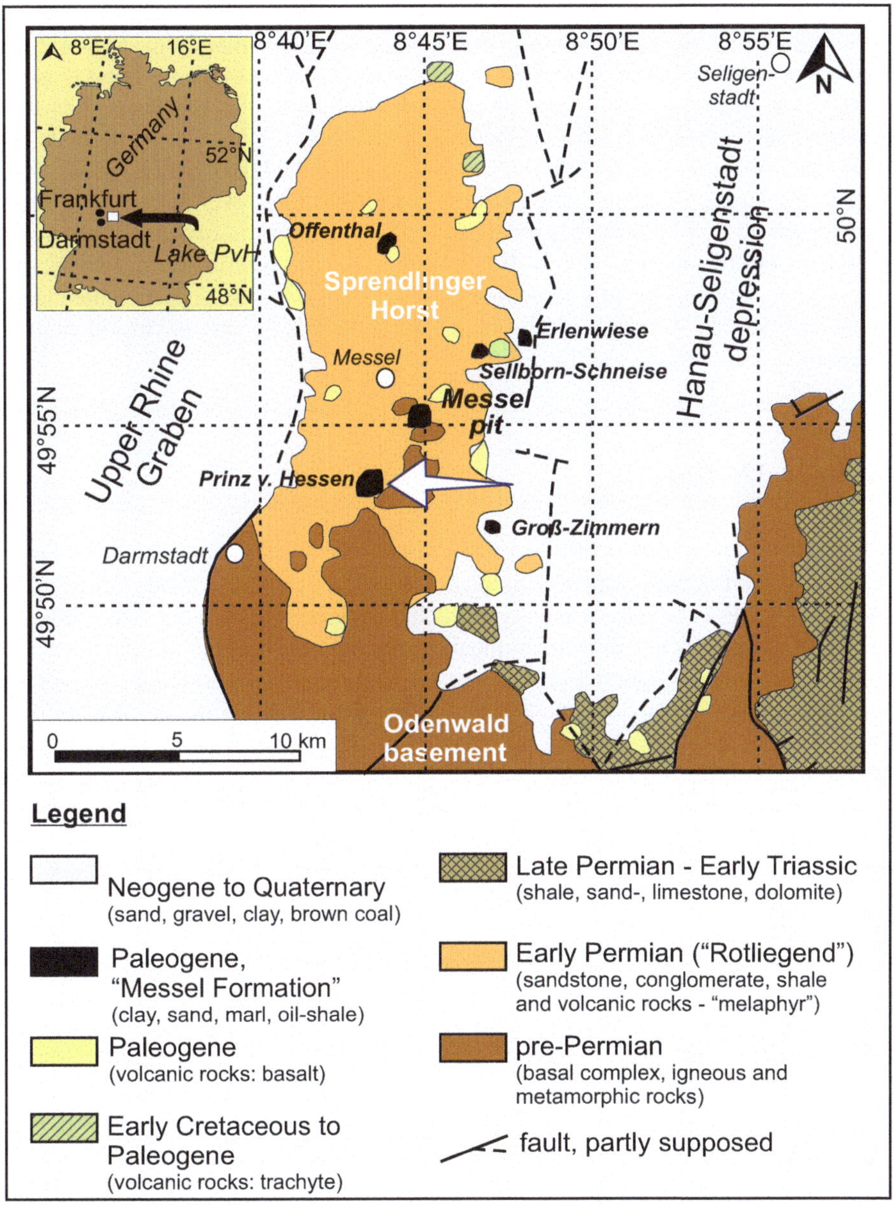

Fig. 1 Geological map of the Sprendlinger Horst (Southwest Germany) showing the location of Eocene Lake Prinz von Hessen (PvH) and other Paleogene sites in the area. (Modified after Harms et al. 1999 and Lenz et al. 2007)

Therefore, a combination of both, regional tectonic activity and climate change (humidity), had a significant impact on the paleoenvironment and the evolution of this ecosystem (Moshayedi et al. 2018).

Here, we present a high-resolution palynological study of a specific part of the lacustrine succession of Lake PvH, which is characterized by alternating beds of lignite and mudstone, representing a shallow phase of the lake's evolution. Orbital control of changes in annual precipitation and related lake-level fluctuations have been suggested as a possible trigger for the regular alternation of lignite and mudstone beds (Hofmann et al. 2005). We have tested if this alternation in lithofacies may serve as a record for orbitally controlled vegetational change in the Paleogene greenhouse system, or if other factors influenced the sedimentary environment and the composition of the vegetation.

Geological Setting

In contrast to the lacustrine fillings of maar-type volcanic structures, such as those at Messel or Offenthal (e.g., Harms et al. 1999; Jacoby et al. 2000; Felder and Harms 2004), PvH is special among the Paleogene deposits on the Sprendlinger Horst because it does not show the typical succession of lithozones as described for maar lakes by Pirrung (1998). Furthermore, the isotopic values of siderite (δ^{13}Carbon and δ^{18}Oxygen) in the sediments at PvH are significantly lower than typical values for long-lived deep and meromictic maar lakes (Felder and Gaupp 2006). These values fit those typical for open holomictic systems on river deltas or alluvial plains in which the position of the redox boundary is near to the sediment surface (Felder and Gaupp 2006). Therefore, the sedimentary succession discussed here for Lake PvH was deposited in a significantly shallower setting than a typical deep maar lake, and probably filled a small pull-apart basin (Felder et al. 2001; Felder and Gaupp 2006; Moshayedi et al. 2018).

A continuous core was drilled in 1997 in the centre of the lake basin (49°53′56.64″N, 8°43′48.31″E) which has a diameter of 600 to 800 m and was surrounded by sediments of Permian age (Felder et al. 2001). The 150 m long core revealed in the lower 55 m a clast supported, poorly sorted breccia with a mixed allochthonous succession, composed of Permian (Rotliegend) sandstones, granodiorite, and other fragments of the crystalline basement (Felder et al. 2001). In the upper 20 m of the breccia frequent lapilli appear. This part is followed by a 95 m succession of lacustrine sediments that were separated in a previous study in five lithozones (Moshayedi et al. 2018).

Previous Work

Sedimentology of the Lacustrine Succession

Based on lithological changes and the description of the fresh core by Felder et al. (2001) the 95 m thick lacustrine succession at PvH has been divided into five lithozones from bottom to top (LZ1–5; Fig. 2) by Moshayedi et al. (2018).

In LZ1, between 95 m and 64.80 m (Fig. 2), the sediments are characterized by layers of well bedded to laminated calcareous sand, silt, and clay suggesting proximity to a source of clastic material. The sediment was transported to the lake probably by river(s) and different grain size depended on changes in discharge rates. Occasional remobilization of the sediment produced graded bedding. LZ1 is comparable with a fluvial–lacustrine facies association (Carroll and Bohacs 1999), typical for an overfilled lake, in which the influx of water and sediment generally exceeds potential accommodation (Moshayedi et al. 2018).

LZ2, between 64.80 and 55.40 m (Fig. 2), is characterized by frequent changes in lithology and sedimentological processes including the deposition of bituminous shale with interbedded turbidites, grainflows, and mudflows. The first beds of laminated bituminous shale occur at a depth of 60.55 m. A high total sulphur content with an average of 2.5% and the lamination indicate that during this stage the lake was deep enough for a permanent chemocline and sub-oxic to euxinic conditions in its deep hypolimnion (Hofmann et al. 2005). In LZ2 and also in the succeeding LZ 3 the deposition of laminated organic-rich bituminous shale and mudstone (Fig. 2) with TOC values up to 45% (Hofmann et al. 2005) characterizes a fluctuating profundal (deepwater) facies association with changing lake level (Moshayedi et al. 2018). In such balanced-filled lakes tectonic accommodation equals water and sediment fill (Carroll and Bohacs 1999; Bohacs et al. 2000).

Laminated bituminous shale without bioturbation is typical for the succeeding LZ3, between 55.40 and 44.70 m, and indicates an open lake and increased water depth (Fig. 2; Moshayedi et al. 2018). Anaerobic conditions in the sediment at the lake bottom and a poorly oxygenated lower part of the water column are indicated by a total sulphur content in the bituminous shale of <1% (Hofmann et al. 2005). In samples above 47.5 m an increasing amount of detrital quartz (Hofmann et al. 2005) may indicate shallowing of the lake.

LZ4, between 44.70 and 30.82 m (Fig. 2), which is in the focus of the present study, shows an alternation of at least 20 lignite layers, with thicknesses ranging between 5 and 100 cm, and interbedded grey-green mudstone (Fig. 3). The TOC content of the lignite is between 50% and 60% (Hofmann et al. 2005). The lignite consists primarily of woody organic matter of the huminite group, which indicates a source either from terrestrial higher plants or from aquatic macrophytes which lived in the shallow littoral zone (Hofmann et al. 2005). Compared to the lignite, the interbedded massive mudstone is poor in organic matter (<10% TOC; Hofmann et al. 2005). LZ4 indicates a period with a very low lake level and a change to a more fluvial–lacustrine facies (Moshayedi et al. 2018).

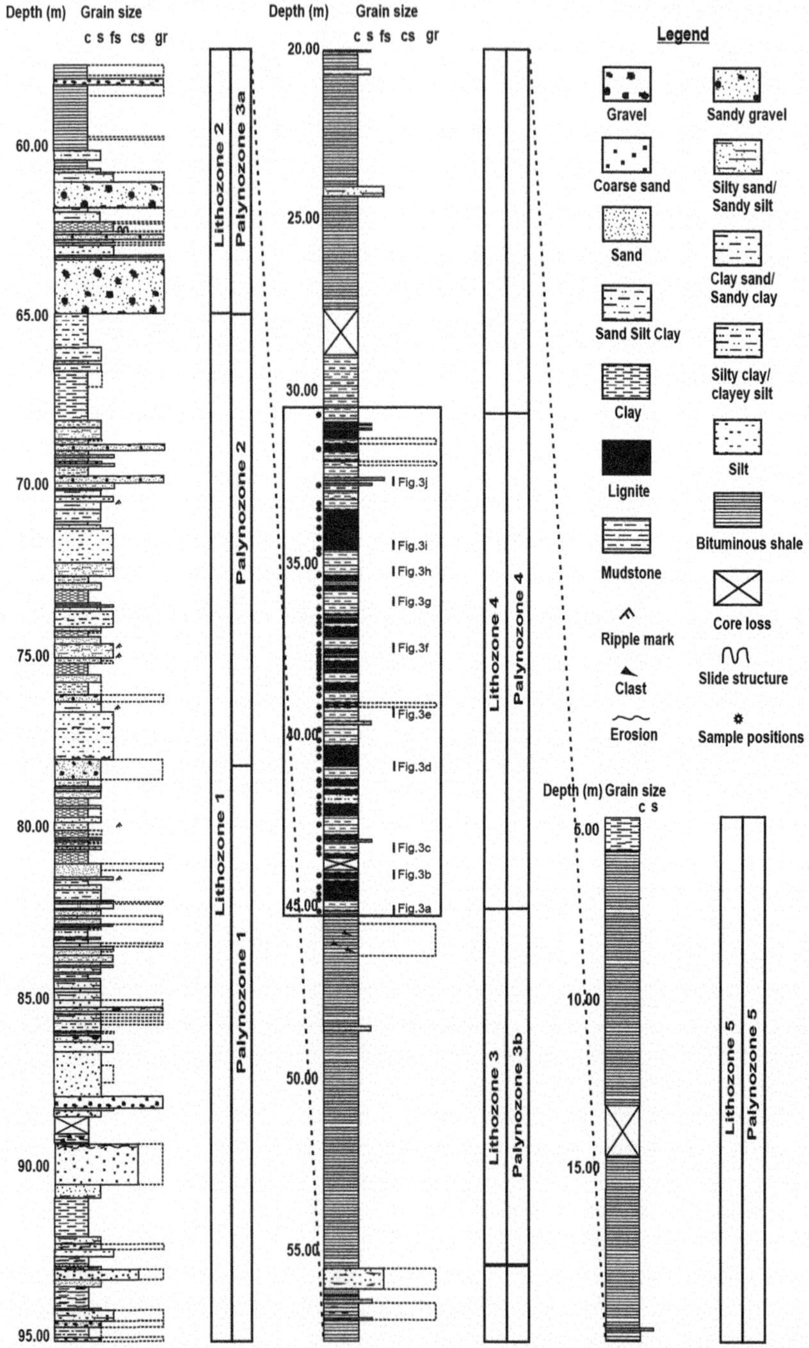

Fig. 2 Generalized section of the "Prinz von Hessen" core showing lithozones and palynozones of Moshayedi et al. (2018). The part of the core that has been studied for the present paper is framed. Within the frame, the position of samples is indicated by black dots, and the position of core images selected for Fig. 3 is shown by grey bars

Fig. 3 Selected images of the core illustrating the succession of LZ 4 in the core from PvH between 44.70 and 30.82 m depth (**a**) 45.16–44.18 m. Laminated, partly massive mudstone with a mudstone clast between 45.00 and 44.90 m, (**b**) 44.16–43.79 m. Alternation of mudstone and reworked lignite that contains a high proportion of clastic material (e.g., 44.05–44.00 m), (**c**) 43.35–43.00 m. Massive mudstone including a lignite clast (43.15 m), (**d**) 41.04–40.67 m. Crinkly laminated to massive mudstone with some lignite clasts and streaks of <1 cm thickness (41.02–40.90 m), (**e**) 39.46–39.08 m. Horizontally bedded mudstone, undisturbed, mostly laminated, including rare streaks of organic material, (**f**) 37.56–37.19 m. Reworked mudstone and

In LZ5, between 30.82 and 4.67 m (Fig. 2), massive mudstone and sometimes laminated bituminous shale with TOC values <10% (Hofmann et al. 2005) indicate a similar depositional environment as for LZ3. The lignite layers that are typical of LZ4 disappear completely (Fig. 2), but lignite clasts can be found in some of the mudstones and shales (Hofmann et al. 2005). LZ5 represents a further profundal facies association comparable with LZ2 and 3 (Moshayedi et al. 2018).

The abrupt changes of lithofacies separating the different lithozones suggest that tectonic activity has influenced the evolution of the basin by rapid uplift and subsidence (Moshayedi et al. 2018). This is especially true for the change between LZ's 1 and 2, which records considerable deepening of the lake within a few metres of sediment. The frequent occurrence of mass flows and turbidites, especially at the transitions between the individual facies associations and LZ's, may relate to strong tectonic events such as earthquakes during the history of the lake (Moshayedi et al. 2018).

Palynology

In a palynological study that was designed to unveil both long- and short-term changes in the vegetation over the entire time of deposition of the lacustrine sediments at PvH, five vegetational stages during the deposition time, resulting in five palynozones (PZ1 to 5; Fig. 2), were recognized (Moshayedi et al. 2018). They show an almost perfect correlation to the lithozones, indicating that sedimentary processes, tectonic activity and the deposition of palynomorphs are depending on each other (Moshayedi et al. 2018).

In PZ's 1 and 2, which encompass the fluvial-lacustrine facies of LZ 1, widespread fern meadows flourishing under wet conditions at the shoreline or in marginal swamps are represented by high numbers of fern spores. They strongly increase between PZ1 and PZ2 without a significant change in lithology and depositional environment. Therefore, this increase is probably not related to taphonomic processes and may indicate a change to more humid conditions from PZ1 to PZ2 (Moshayedi et al. 2018). This is supported by a decrease of pinaceous pollen, because Pinaceae are regarded as more tolerant of less humid conditions and well-drained soils than most of the plants that form components of the regular PvH forest (Moshayedi et al. 2018). The forest was characterized by elements of a climax stage of the Paleogene European vegetation, typical for inland sites of Western and

Fig. 3 (continued) lignite clasts, (**g**) 36.35–35.97 m. Mudstone with convolute bedding and incorporated lignite clasts, (**h**) 35.15–34.78 m. Reworked massive mudstone including frequent laminated mudstone clasts up to gravel size, (**i**) 34.62–34.24 m. Crumbly lignite including a 5 cm thick irregularly shaped lens of yellow sand, (**j**) 32.52–32.14 m. Alternation of laminated and massive mudstone and yellow, sometimes graded beds of sand and silt with varying dip directions between 32.52 and 32.33 m and massive clasts of light sandy material between 32.32 and 32.15 m. The pictures are photographs of FIS/HLUG (S. Borges, W. Schiller and M. Stryj) from core B/97- BK9

Central Europe during the greenhouse period of the Eocene (Mai 1981, 1995; Schaarschmidt 1988; Wilde 1989; Collinson et al. 2012), such as Fagaceae or Juglandaceae. Different species of Cupressaceae indicate that some swamp forest elements existed at the edge of the lake (Moshayedi et al. 2018).

Compared with the palynomorph assemblages of the fluvial–lacustrine facies association of LZ1 the strong decline or even disappearance of most fern spores is noteworthy in PZ3 (fluctuating profundal facies association; LZ 2). This abrupt decrease is correlated with the change in lithology and therefore coupled to a rapid deepening of the lake (Moshayedi et al. 2018). Therefore, tectonic activity and related changes in the paleoenvironment at Lake PvH were probably responsible for changes in the vegetation (Moshayedi et al. 2018). A strong peak of Restionaceae as a typical herbaceous swamp element in PZ3 is further evidence for changes in the vegetation at the lakeside. They point to a higher and oscillating lake level (Moshayedi et al. 2018). A higher lake level and extended shallow water areas at the lakeside are also indicated by the distribution of submersed and floating plants such as Hydrocharitaceae and Nymphaeaceae (Moshayedi et al. 2018).

A specific vegetational composition is represented in PZ4, encompassing the alternation of lignite and mudstone beds of LZ4, which are in the focus of our high-resolution study presented here. The high abundance of herbaceous elements in the palynoflora and the onset of lignite deposition indicate that wet areas were widely distributed around the lake in areas newly formed due to a drop in lake level (Moshayedi et al. 2018).

In PZ5 lake level increased and a profundal facies developed again (LZ5). Cupressaceae from the edge of the lake and Fagaceae from the climax vegetation in the vicinity of the lake still are dominating elements during PZ5. While they remained, the decrease of Sapotaceae pollen and relicts of a warm Cretaceous vegetation (Normapolles group) from the lower to the upper zones and the simultaneous increase of some juglandaceous taxa point to slight changes in the composition of the vegetation, which probably were related to climate change (Moshayedi et al. 2018).

Stratigraphy

In the 1920s, lignite was mined from the uppermost part of the succession at PvH in an area ca. 250 m east of the drilling site. The presence of *Eurohippus parvulus parvulus*, an early equid, in the mined lignite succession, points to a late middle Eocene to middle late Eocene age (European mammal chronology zones MP13-MP16; Franzen 2006). This comparatively thick lignite was not recovered in the core (Moshayedi et al. 2018) and probably represented the final silting-up phase of Lake PvH. Nevertheless, the palynological study of the lacustrine succession revealed a typical Lower Eocene palynomorph assemblage and proved that the lower part of the succession at PvH is considerably older (Moshayedi et al. 2018).

Therefore, the lake basin has probably existed for a significantly longer time (~6–8 Ma) than the nearby Eocene maar lakes including Messel (~1 Ma, Lenz et al. 2015).

Methods

Sampling and Sample Processing

For high-resolution palynological analysis, 52 samples were selected from the part of the core between 45 and 30 m depth (LZ 4, Fig. 2), which is mainly characterized by an alternation of c. 20 lignite beds and interbedded mudstone. Compared to the overview of Moshayedi et al. (2018), the number of analyzed samples for this core segment has been increased by 43 samples. The sample interval is approximately 10 cm and at least one sample in each bed has been analyzed (Fig. 2).

Palynological preparation followed the standard procedures as described by Kaiser and Ashraf (1974) including treatment with hydrochloric acid (HCl), hydrofluoric acid (HF), and potassium hydroxide (KOH). To remove flocculating organic matter and improve the transparency of the palynomorphs, the residue was briefly oxidized with nitric acid (HNO_3) or hydrogen peroxide (H_2O_2) after sieving with a mesh size of 10 μm. Remaining sample material and slides are stored at the Senckenberg Forschungsinstitut und Naturmuseum, Sektion Paläobotanik, Frankfurt am Main, Germany.

Quantitative Palynological Analysis

Numerical analyses of palynological data are based on quantitative palynomorph counts. At least 300 individual palynomorphs per sample were identified and counted at 400 times magnification to obtain a representative dataset for statistical analysis. A complete list of all palynomorphs encountered during the present study together with the raw data table is included in the supplementary material. Identification of palynomorphs is based on the systematic-taxonomic studies of Thomson and Pflug (1953), Thiele-Pfeiffer (1988), Nickel (1996), and Lenz (2005). Nevertheless, a relatively high proportion of 10–20% of the total assemblages is poorly preserved and has been counted as 'Varia'. Morphologically similar sporomorph taxa, which have been assigned to the same parent plant family, were lumped to minimize potential errors in the identification and counting of individual species.

The pollen diagram shows the abundance of the most important palynomorphs in percentages. They are arranged according to their weighted average value (WA regression, Ter Braak and Looman 1996) in relation to depth using the software C2 1.7.6 (Juggins 2007). This arrangement led to a structured pollen diagram showing the major patterns of compositional variation in relation to depth and revealed the different steps in the progression of the plant community (Janssen and Birks 1994). Pollen and spores were calculated to 100%, whereas algae, such as *Botryococcus* and *Ovoidites*, were added as additional percentages (in % of the total sum of pollen and spores).

Statistical Analysis

For statistical analyses we used Wisconsin double standardized raw data values (Bray and Curtis 1957; Cottam et al. 1978; Gauch and Scruggs 1979; Oksanen 2007). Hence, the species counts are standardized by dividing the empirical value by the maximum counting value across all samples for each species; then, sample counts are standardized by finding the proportion of each species versus the total sum for each sample. This equalizes the effects of rare and abundant taxa, and removes the influence of sample size on the analysis (Bray and Curtis 1957; Cottam et al. 1978). Nevertheless, rare species with a maximum value <1.5%, which do not show any significant pattern throughout the pollen diagram, were excluded from statistical analyses. Remains of algae were included in the numerical analysis because their appearance in parts of the section is important for the definition of individual palyno-phases.

To illustrate compositional differences and ecological trends, and to visualize the level of similarity between samples, non-metric multidimensional scaling (NMDS) with the standardized raw data values and the Bray-Curtis dissimilarity has been implemented (Bray and Curtis 1957; Hair et al. 2010) using the software package PAST 3.06 (Hammer et al. 2001). This ordination method was chosen as the appropriate multivariate model, because NMDS is the most robust unconstrained ordination method in ecology (Minchin 1987) and has been successfully applied to palynological data in previous studies (e.g., Jardine and Harrington 2008; Mander et al. 2010; Broothaerts et al. 2014). NMDS avoids the assumption of a linear or unimodal response model between the palynomorph taxa and the underlying environmental gradients and also avoids the requirement that data must be normally distributed.

NMDS of the complete data set revealed significant differences between the palynomorph assemblages from the mudstone beds and the lignite. Therefore, further analyses included separate cluster and NMDS analyses of both mudstone and lignite samples. For robust zonation of the pollen assemblages in the lignite and mudstone samples, two-way cluster analyses (Q- and R-mode) were applied resulting in a cluster matrix to identify samples with similar palynomorph contents (Q-mode) and to reveal which pollen-spore taxa group together (R-mode). Bootstrapped Q-mode cluster analysis was established by constrained cluster analysis using the unweighted pair-group average (UPGMA) method and Eucledian distance, whereas R-mode cluster analysis applied the unconstrained UPGMA method with the Euclidean distance (software PAST 3.06, Hammer et al. 2001). We have chosen bootstrapped constrained analysis for Q-mode defined sample clusters to group only stratigraphically adjacent samples during the clustering procedure to identify a zonation in the lithological succession and to test if the multiple phases (clusters) are robust or just produced by change (detailed results of the bootstrapping analysis are presented in the supplementary material).

Results

Sedimentology

In the sedimentary record of core PvH between 45 and 30.50 m, lignite beds with a thickness of 1 cm to 1 m alternate with clay-rich, slightly bituminous mudstones (Felder et al. 2001). The following description of this sequence LZ4 of Moshayedi et al. (2018) is based on the photo documentation and the lithological description of the fresh core by Felder et al. (2001). Figure 3 shows only representative parts of the succession; however, the complete photo documentation of LZ4 is included in the online supplementary material.

Generally, the succession of LZ4 is characterized by the lack of any bioturbation. There are no traces of rooting beneath any of the lignite layers (Hofmann et al. 2005; Moshayedi et al. 2018). Up to a depth of 44.70 m the lower part of LZ4 consists of a laminated, partly massive mudstone, sometimes containing clasts of mudstone (Fig. 3a) and bituminous shale, which indicate reworking and redeposition of sediment from the older LZ3 (Moshayedi et al. 2018). The first cm-thick lignite beds, which alternate with laminated to massive, horizontally bedded mudstone of several centimetres in thickness, occur between 44.70 and 44.15 m.

Between 44.15 and 42.92 m laminated mudstone dominates, in which 1–5 cm thick lignite beds are irregularly intercalated. The mudstone is predominantly horizontally laminated, but the individual laminae may be crinkly. Sometimes a dip of up to 10° can be observed (e.g., 43.80 m, Fig. 3b). In particular, the lignite shows traces of reworking, because lignite beds either contain a high portion of clastic material (44.03 m, Fig. 3b) or show erosional bases (43.15 m; Fig. 3c). Above a lignite bed between 42.92 and 42.78 m depth, laminated to massive mudstones dominate again with some interbedded lignite streaks of <1 cm thickness up to a depth of 42.20 m. The stratification is primarily horizontal, but loop structures and convolute bedding are common.

Up to a depth of 37.40 m an alternation of 9 up to 30 cm thick lignite beds and horizontally laminated mudstone follows. Here, the mudstone is in part recognizable as primary and undisturbed (Fig. 3e). Nevertheless, the mudstone often incorporates lignite clasts (40.90–41.05 m, Fig. 3d), which indicate reworking and transport. In the thick lignite beds, transport and reworking of the organic material is not clearly recognizable, but some of them are also characterized by an erosive base (e.g., 41.60 m and 41.35 m, see supplementary material). Between 37.40 and 37.02 m, a laminated mudstone follows, which shows reworking including the incorporation of massive mudstone as well as lignite clasts (Fig. 3f). In the succeeding part of the section up to 36.40 m an alternation of 5 up to 15 cm thick lignite beds and horizontally laminated mudstone occurs again. In contrast, the massive mudstone between 36.40 m and 35.40 m is characterized by in-situ displacement and mixing of the sediment, indicated by convolute bedding and the incorporation of lignite clasts (Fig. 3g). Following a lignite bed between 35.40 m and 35.24 m with an erosional base a massive mudstone including frequent laminated mudstone clasts up to gravel size occurs between 35.24 and 34.50 m (Fig. 3h). In the lower part also a 10 cm thick lignite clast is incorporated in the mudstone (Fig. 3h). Between 34.50 m and 33.35 m a more than 1 m thick crumbly lignite occurs, which includes up to 5 cm thick irregularly shaped lenses of yellow sandy material (Fig. 3i). In the succeeding 40 cm a laminated to massive mudstone appears, which includes also cm-thick yellow sandy lenses.

A change in the depositional system occurs between 32.95 m and 32.32 m with an alternation of laminated and massive mudstone and yellow, sometimes graded beds of sand and silt. Parts of this succession are characterized by varying dip directions (32.45; Fig. 3j). In the following section between 32.32 and 31.70 m thick clasts and lenses of light sandy sediments (32.25; Fig. 3j), mudstone and lignite appear in a mudstone matrix. The succeeding crumbly lignite between 31.70 m 30.85 m, representing the top of LZ4 (Moshayedi et al. 2018), is also characterized by up to 10 cm thick clasts of laminated to massive mudstone. LZ3 above 30.85 m is represented by massive to laminated dark bituminous shale.

Palynology

Differences Between Palynomorph Assemblages from Lignites and Mudstones

The pollen diagram (Fig. 4) shows that most of the palynomorphs occur over the entire interval of LZ4, but some taxa vary significantly in their proportional representation, because there is a strong correlation between lithology and palynomorph content. A Mann-Whitney-U-test reveals that among the 49 taxa or groupings of taxa, which are presented in the pollen diagram, 8 are characterized by significant abundance differences in lignite and mudstone palynomorph assemblages (Table 1). Some palynomorphs are more dominant in the mudstone beds, e.g., the swamp forest elements *Inaperturopollenites* spp. (Cupressaceae) and *Nyssapollenites* spp. (Nyssaceae), the spores of *Tegumentisporis* spp. (Selaginellaceae) and forest elements, such as *Tricolpo(ro)pollenites parmularius* (Eucommiaceae) and *Porocolpopollenites* spp. (Symplocaceae), while others, such as *Ericipites spp.* (Ericaceae) or fern spores (*Punctatosporites paleogenicus*; Polypodiaceae), are mostly associated with lignite samples (Table 1). Furthermore, other fern spores such as those of *Polypodiaceoisporites* spp. (Polypodicaeae) and *Baculatisporites* spp. (Osmundaceae) are confined or nearly restricted to the mudstone samples and

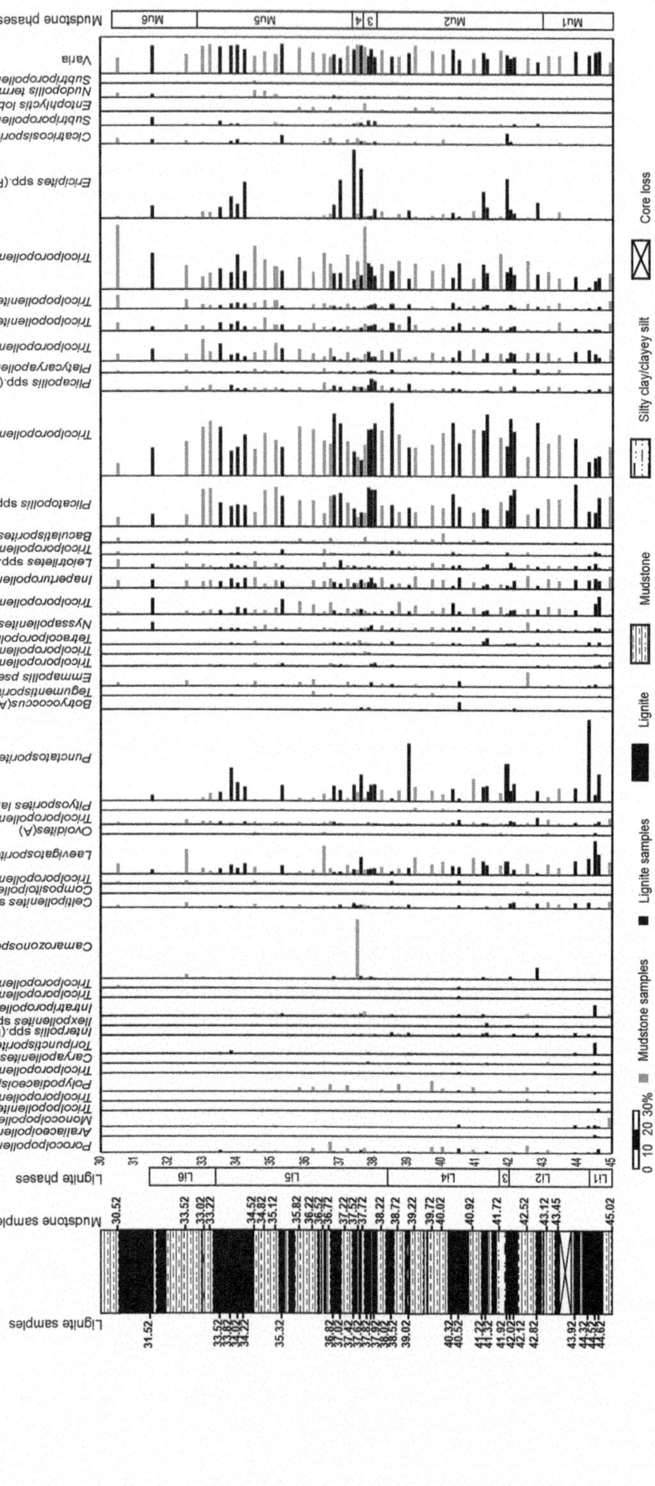

Fig. 4 Pollen diagram of 52 samples from core PvH between 30.52 and 45.02 m depth (LZ 4 of Moshayedi et al. 2018). (A) algae, (P) pollen, (S) spores, (F) fungi. Black bars: lignite samples. Grey bars: mudstone sample

Table 1 Comparison of species abundances of selected taxa in lignite and mudstone samples. To assess statistical differences Mann-Whitney U-test is used. *P*-values ≤0.05 (upper 8 species) indicate a significant difference between the abundance of a species in the lignite and mudstone assemblage, whereas *p*-values >0.05 point to an equal distribution

	Lignite				Mudstone					
Species	Max value [%]	Mean value [%]	Median value [%]	Standard deviation [%]	Max value [%]	Mean value [%]	Median value [%]	Standard deviation [%]	Mann-Whitney test, *u*-value	Mann-Whitney test, *p*-value
Tricolporopollenites parmularius	17.65	8.62	7.985	3.64	31.52	13.54	12.365	6.61	165	0.001
Porocolpopollenites sp.	2.12	0.34	0.155	0.54	4.8	1.12	0.645	1.25	185	0.004
Punctatosporites paleogenicus	40.37	9.3	7.43	8.41	11.07	2.79	1.975	2.76	101.5	<0.001
Tegumentisporis spp.	0.54	0.06	0	0.15	1.99	0.39	0.365	0.52	201	0.002
Emmapollis pseudoemmaensis	2.70	0.3	0	0.64	7.03	1.16	0.655	1.49	154	0.0004
Nyssapollenites kruschii spp.	3.92	0.87	0.58	0.89	4.3	1.34	1.28	0.9	215.5	0.02
Inaperturopollenites spp.	5.66	3.09	2.8	1.41	6.65	3.88	3.565	1.29	236	0.05
Ericipites spp.	33.86	7.31	4.385	8.67	3.56	1.35	1.1	1.16	185	0.005
Tricolporopollenites cingulum	36.12	19.59	21.15	7.87	26.95	17.78	17.325	5.41	304	0.54
Celtipollenites spp.	2.9	0.95	0.795	0.95	3.43	0.88	0.58	0.94	331.5	0.91
Plicatopollis spp.	21.22	12.21	10.425	4.39	19.37	10.88	11.045	4.67	272.5	0.23
Plicapollis spp.	5.98	1.68	1.285	1.42	4.23	1.64	1.475	1.02	315.5	0.69
Platycaryapollenites spp.	2.42	0.53	0.4	0.64	2.09	0.56	0.4	0.63	335.5	0.97
Tricolpopollenites liblarensis	7.14	2.26	1.975	1.48	6.36	2.41	2.185	1.4	305.5	0.56
Tetracolporopollenites spp.	3.45	0.77	0.79	0.73	2.02	0.64	0.48	0.43	307	0.58

therefore excluded from the Mann-Whitney-U-test. Therefore, more than 20% of the taxa show significant differences in abundance between lignite and mudstone. However, most palynomorphs from the forest vegetation in the vicinity of the lake, such as the dominating taxa *Tricolporopollenites cingulum* (Fagaceae) or *Plicatopollis* spp. (Juglandaceae), are independent in their abundance from the lithological changes (Table 1).

The difference between the composition of the pollen assemblages also is seen in a NMDS plot (Fig. 5). Samples from lignite and mudstone-dominated beds are characterized by substantially different palynomorph assemblages, because mudstone samples are generally concentrated on the left side while lignite samples are commonly found on the right side of the ordination space (Fig. 5). However, it is obvious that the lignite beds are fundamentally different in origin from the mudstones. Therefore, besides a different source, sorting of pollen and spores during transport and deposition must be considered (Galloway et al. 2015), which prevents a direct comparison between the palynomorph assemblages of lignite and mudstone samples. For this reason, we performed separate analyses for both lithologies.

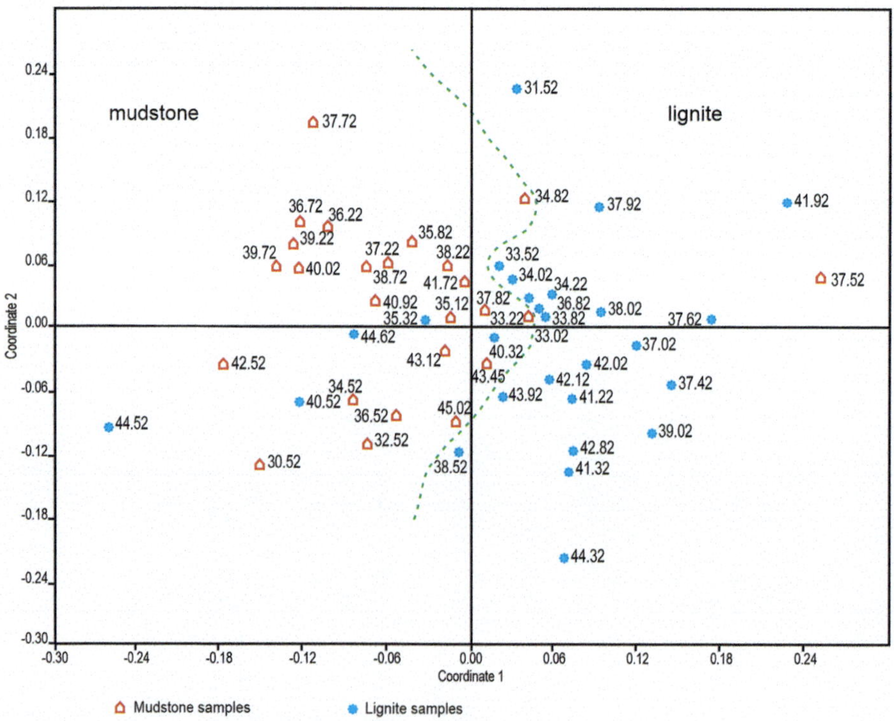

Fig. 5 Non-metric multidimensional scaling (NMDS) plot of palynological data of 52 samples from the succession of lignite and mudstone beds in LZ4 of Moshayedi et al. (2018) using the Bray-Curtis dissimilarity and the Wisconsin double standardized raw-data values. The scatter plot shows the arrangement of samples (numbers correspond to depth in m). Stars: Lignite samples. Pentagons: Mudstone samples

The Palynological Composition of the Lignite Samples

There are at least 20 lignite beds with thicknesses ranging from 5 to 100 cm in the core (Figs. 2 and 3). The bootstrapped Q-mode cluster analysis reveals that in the 27 lignite samples, seven robust palyno-phases can be distinguished (lignite/Li phases, Figs. 4 and 6a).

Phase Li 1

The 2 samples from phase Li 1 plot to the right side of the NMDS ordination space and appear clearly separate from the other lignite samples (Fig. 6b) indicating significant compositional differences. Both samples are from the lower part of the succession between 44.62 and 44.52 m. Most remarkable are the high abundances of *Intratriporopollenites* spp. (Tiliaceae) and *Toripunctisporites* sp. (fern spore of unknown botanical affinity) with up to 6% (Figs. 4 and 6a). Other fern spores such as *Laevigatosporites* spp. (Polypodiaceae, up to 17%), *Punctatosporites paleogenicus* (up to 14%), and *Leiotriletes* spp. (Schizaeaceae, up to 3%) are also abundant in this phase. Fern spores compose a total of up to 40% of these assemblages (see supplementary material raw data table).

Phase Li 2

Five samples, between 44.32 and 42.02 m, are combined in phase Li 2 and plot in the upper left part of the NMDS ordination space on the negative side of NMDS axis 1 (Fig. 6b). The most characteristic taxa are *Interpollis* spp. (unknown botanical affinity, up to 2%) and *Celtipollenites* spp. (Ulmaceae, up to 3%). *Punctatosporites paleogenicus* (up to 41%) reaches its maximum for the lignite layers, but other fern spores such as *Leiotrilites* spp. (up to 4%), *Laevigatosporites* spp. (up to 9%) and the lycopod spores of *Camarozonosporites* spp. (6%) are frequent at least in some samples (Figs. 4 and 6a). Fern spores account for a total of up to 60% of these assemblages.

Phase Li 3

One sample from a depth of 41.92 m is characterized by a unique combination of very high values for the spores of *Cicatricosisporites* spp. (Schizaeaceae, 6%) and pollen of *Ericipites* spp. (20%). It plots on the negative lower side of the NDMS ordination space away from the other lignite samples (Figs. 4 and 6b).

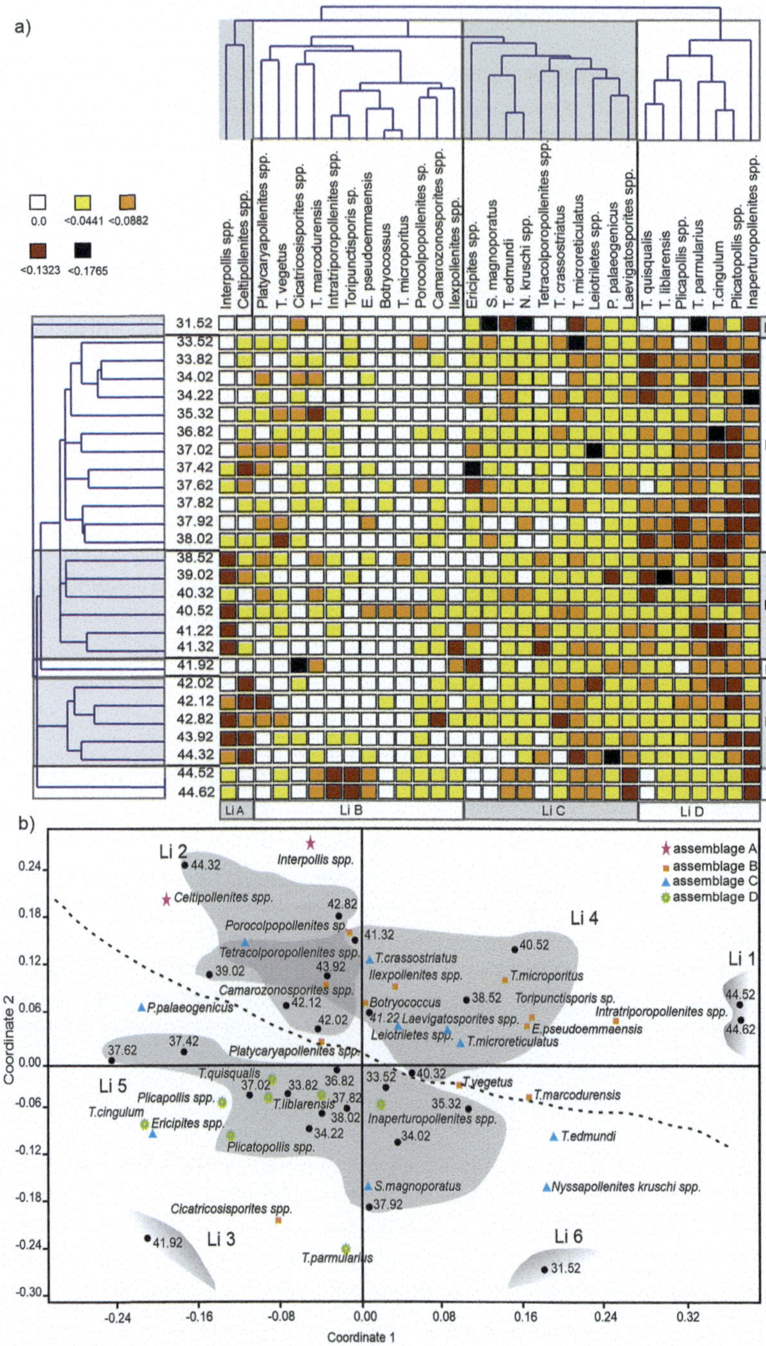

Fig. 6 Numerical analyses of palynological data from lignite samples of LZ4. (**a**) A combined constrained Q- and unconstrained R-mode cluster analysis of Wisconsin double standardized raw-

Phase Li 4

Phase Li 4 includes 6 samples between 41.32 and 38.52 m, which plot in the upper part of the NMDS ordination space (Fig. 6b). Compared to Li 2, *Interpollis* spp. (up to 2%) is relatively frequent, but *Celtipollenites* spp. is of a lower value (Fig. 6a). Other elements show peak abundances including *Tetracolporopollenites* spp. (Sapotaceae, 4% in 41.32 m), *Tricolporopollenites liblarensis* (Fagaceae, 8% in 39.02 m) or the coccal green algae *Botryococcus* (1%).

Phase Li 5

This phase consists of 12 samples between 38.02 and 33.52 m. In the NMDS plot they occur in the lower part of the ordination space mainly at the negative side of NDMS axis 2 (Fig. 6b). This position shows differences in the composition of the palynoflora compared to the older phases Li 2 and Li 4. Notable is the disappearance of *Interpollis* spp., along with an increase of *Plicatopollis* spp. (up to 18%) and *Inaperturopollenites* spp. (up to 5%). Characteristic and frequent palynomorph taxa are *Ericipites* spp. (up to 35%) and *Tricolporopollenites cingulum* (up to 31%), which reach their highest values for the studied core interval.

Phase Li 6

The uppermost sample (31.52 m) is similar to the samples from Li 1 and Li 3, and characterized by a specific palynomorph assemblage. The sample plots in the ordination space in the lower right, clearly separate from all other lignite samples (Fig. 6b). The sample is characterized by *Subtriporopollenites magnoporatus* (unknown botanical affinity, 4%), *Nyssapollenites kruschii* spp. (4%), *Tricolporopollenites edmundii* (Mastixiaceae 8%) and *Tricolpo(ro)pollenites*

Fig. 6 (continued) data values of 27 lignite samples (numbers correspond to depth in m) and 31 palynomorph taxa that occur with at least 1.5% in one sample. Different abundance classes based on the Wisconsin double standardized raw data values are shown in different colours. The darker the colour the higher is the proportion of specific palynomorphs within the palynomorph assemblage of a sample; white boxes indicate the absence of a specific taxon. For cluster analysis the unweighted pair-group average (UPGMA) method has been applied together with a Eucledian distance. As a result, 6 palynological phases (Li 1 to Li 6) can be distinguished in the lignite succession, which are related to abundance variations in 4 palynomorph assemblages (Li A to Li D). (**b**) Non-metric multidimensional scaling (NMDS) of palynological data from 27 lignite samples from LZ 4 of the Eocene Lake PvH using the Bray-Curtis dissimilarity and the Wisconsin double standardized raw-data values. The different palynomorph assemblages (Li A to Li D) are indicated by different symbols, whereas the different palynophases (Li 1 to Li 6) appear in grey. The dotted line indicates the general separation of the succession as based on differences in the composition of the microflora with an older part including phases Li 2 and 4 and a younger part with phase Li 5. Phases Li 1, Li 3 and Li 6 represent individual samples which are characterized by specific palynomorph assemblages clearly different from assemblages of the other lignite samples *parmularius*. (18%). All of these taxa reach their maximum for the succession of lignite beds here (Figs. 4 and 6a).

General Changes in the Composition of the Vegetation Recorded in Lignite Samples

The NMDS plot reveals that a younger and an older part can generally be separated in the succession of lignitic layers based on differences in palynomorph assemblages. Samples from Li 1, 2 and 4, between 44.62 and 38.52 m, which plot in the upper part of the ordination space of the NMDS (Fig. 6b), characterize the older series. They are composed mainly of palynomorph taxa that are clustered in the assemblages Li A, B, and C of the R-mode cluster analysis (Fig. 6a). In contrast, samples from Li 5 between 38.02 and 31.52 m plot in the lower part of the NMDS ordination space and are mainly composed of palynomorphs that are clustered in assemblage Li D along with some elements of assemblage Li C (e.g., *Ericipites* spp. or *S. magnoporatus*; Fig. 6a, b). Therefore, a change in the general composition of the vegetation during the deposition of the lignite is obvious. Furthermore, a single sample from the lower part (Li 3) and another one from the top of the record (Li 6) are characterized by a specific microfloral composition clearly different from the other lignite samples.

One of the main differences between the lower phases (Li 1, 2 and 4) and the upper phase Li 5 is the disappearance of *Interpollis* spp. and *Celtipollenites* spp., which appear regularly in low numbers in the older part and cluster in assemblage Li A (Figs. 4 and 6a). Pollen and spores that are clustered in assemblage Li B occur generally in low numbers in the lignite succession, except for peak abundances in the two lowermost samples. Assemblage Li C is composed of palynomorphs that occur regularly with low-to-high numbers and include *Ericipites* spp. or *Punctatosporites paleogenicus*. These taxa vary in frequency throughout the succession of lignites (Fig. 4) and are independent from the general change in the overall vegetation. The dominant palynomorph taxa are clustered in assemblage Li D. Elements of a forest vegetation, such as juglandaceous pollen (*Plicapollis* spp., *Plicatopollis* spp.) and pollen with fagalean affinity, e.g., *Tricolporopollenites cingulum* and *Tricolpopollenites liblarensis,* as well as pollen from a swamp forest (*Inaperturopollenites* spp.), are generally found in high numbers in the lignite. But NMDS and cluster analysis show that these elements slightly increase in the younger phase Li 5 (Fig. 6).

The Palynological Composition of the Mudstone Samples

Bootstrapped Q-mode cluster analysis of 25 samples from the mudstones results in 6 robust palynomorph phases (mudstone/Mu phases, Fig. 7a).

Phase Mu 1

Mu1 includes 3 samples between 45.02 and 43.12 m, which plot in the upper left part of the NMDS ordination space on the negative side of axis 1 (Fig. 7b). Characteristic taxa are *Monocolpopollenites* spp. (Palmae, up to 5%), *Celtipollenites* spp. (up to 4%), *Tricolporopollenites crassostriatus* (Solanaceae or Gentianaceae, up to 4%) and *Tricolporopollenites vegetus* (Hamamelidaceae, up to 3%), all of which reach their maximum for the mudstone samples (Figs. 4 and 7a).

Phase Mu 2

Mu 2 is composed of 8 samples collected between 42.52 and 38.22 m. The samples generally plot on the right side of the NMDS ordination space on the positive side of axis 1 (Fig. 7b). Only the lowermost sample (42.52 m) plots in the lower left of the NMDS ordination space, clearly separate from the other mudstone samples. Compared to the older phase Mu 1, a distinct change in the composition of the microflora is obvious. Most remarkable is the rare, but regular, appearance of palynomorphs that cluster in assemblage Mu B of the R-mode cluster analysis (Fig. 7a). These include *Polypodiaceoisporites* spp. (up to 6%), *Tegumentisporis* spp. (Selaginellaceae, up to 2%) or the pinaceous pollen *Pityosporites labdacus* (up to 2%). Various species that are combined in assemblage Mu C decrease in proportion compared to phase Mu 1, such as *Celtipollenites* spp., or disappear completely, e.g., *Monocolpopollenites* sp. (Figs. 4 and 7a).

The separation of the sample in 42.52 m in the NMDS plot from the other samples of phase Mu2 is based on the occurrence of *Emmapollis pseudoemmaensis* (Chloranthaceae, 7%) and *Nyssapollenites kruschii* (5%), which attain their maximum for the succession of mudstone beds here.

Phase Mu 3

A single sample (37.72 m) plots in the upper right corner of the NMDS ordination space far away from all other mudstone samples, also indicating a specific palynomorph composition (Fig. 7b). The maximum of *Tricolpo(ro)pollenites parmularius* (31%) together with the maximum of *Intratriporopollenites* spp. (2%) in the succession is noteworthy. Furthermore, the freshwater fungus *Entophlyctis lobata* (4%), a saprophytic cytrid that is often associated with algal blooms (Geitler 1962; Bradley 1967), is abundant.

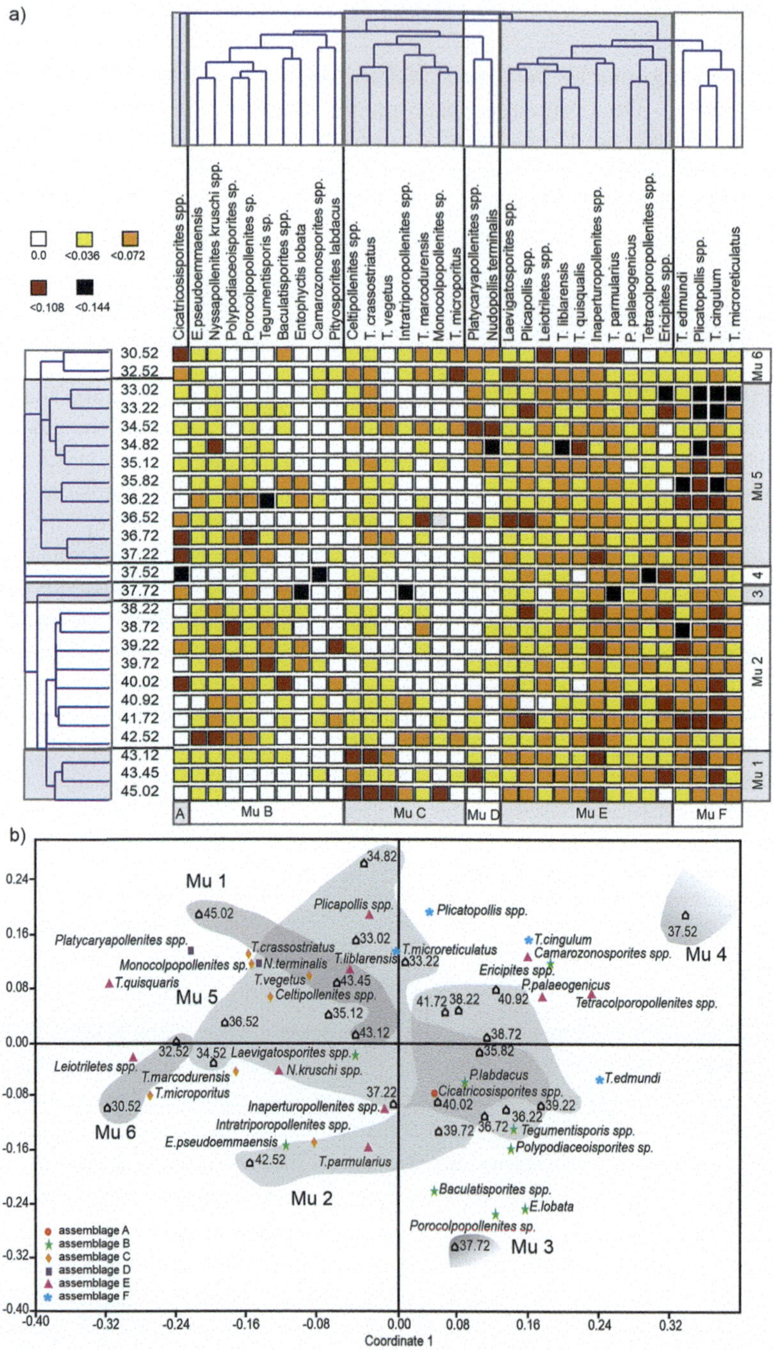

Fig. 7 Numerical analyses of palynological data from mudstone samples of LZ4. (**a**) A combined constrained Q- and unconstrained R-mode cluster analysis of Wisconsin double standardized raw-

Phase Mu 4

One sample (37.52 m) plots in the NMDS ordination space on the negative end of axis 2 (Fig. 7b). Significant differences are obvious when compared to the sample 20 cm below especially with regard to a maximum of *Camarozonosporites* spp. (29%), which is almost absent from all other mudstone samples (Fig. 4). Other palynomorphs characteristic for this sample are *Cicatricosisporites* spp. (3%) and *Tetracolporopollenites* spp. (Sapotaceae, 2%) (Fig. 7a).

Phase Mu 5

The 10 samples clustered in assemblage Mu 5, between 37.22 and 33.02 m, plot in the centre of the NMDS ordination space on both, the negative and positive sides of axis 1 (Fig. 7b). There is a wide overlap in the ordination space with samples from Mu 1 and 2 indicating a similar palynomorph composition. Only a few samples are slightly different.

The samples from 36.52 m and 34.52 m depth, which are plotted on the negative end of NMDS axis 1 (Fig. 7b), are characterized by peak abundances of rare elements such as *Platycaryapollenites* spp. (Juglandaceae, 2%), *Tricolporopollenites marcodurensis* (Vitaceae, 3%), whereas the sample in 34.82 m, plotted on the positive end of NMDS axis 2 (Fig. 7b), is characterized by *Nudopollis terminalis* (4%) and *Tricolporopollenites liblarensis* (7%), both showing their maximum for the mudstones (Figs. 4 and 7a).

In the two uppermost samples of Mu 5 in 33.22 and 33.02 m depth, the dominant pollen of the forest vegetation *Tricolporopollenites cingulum* (up to 27%), *Plicatopollis* spp. (up to 19%) and *Tricolporopollenites microreticulatus* (up to 11%) reach their maximum for the mudstone assemblages.

Fig. 7 (continued) data values of 25 mudstone samples (numbers correspond to the depth in m) and 33 palynomorph taxa with a maximum of at least 1.5% in one sample. Different abundance classes based on the Wisconsin double standardized raw data values are shown in different colours. The darker the colour the higher is the proportion of specific palynomorphs within the palynomorph assemblage of a sample; white boxes indicate the absence of a specific taxon. For cluster analysis, the unweighted pair-group average (UPGMA) method has been applied together with a Eucledian distance. As a result, 6 palynological phases (Mu 1 to Mu 6) can be distinguished in the mudstone succession, which are related to abundance variations in 6 palynomorph assemblages (Mu A to Mu F). (**b**) Non-metric multidimensional scaling (NMDS) plot of palynological data of 25 mudstone samples from LZ 4 of the Eocene Lake PvH using the Bray-Curtis dissimilarity and the Wisconsin double standardized raw-data values. The different palynomorph assemblages (Mu A to Mu F) are indicated by different symbols, whereas the different palynophases (Mu 1 to Mu 6) are shaded in grey. Phases Mu1 and Mu 6 plot on the left side of the ordination space, indicating a similar composition of the palynomorph assemblages, whereas phases Mu 2, Mu 3, and Mu 4 are generally on the right side of the ordination space. Phase Mu 5 as a transitional phase between the older phases Mu 2 zo Mu 4 and the youngest phase Mu 6 plot in the centre of the ordination space

Phase Mu 6

Two samples from the top of the succession (32.52 and 30.52 m) plot in the NMDS ordination space on the left side on the negative end of axis 1 (Fig. 7b). There is a notable decrease of elements that dominate the palynomorph assemblage in the upper samples of phase Mu 5, such as *Tricolporopollenites cingulum* or *Plicatopollis* spp. (Figs. 4 and 7a). In contrast, there is an increase in the fern spore *Cicatricosisporites* spp. and elements associated with cluster Mu E. *Tricolpo(ro) pollenites parmularius* is especially common (32%) in the uppermost sample.

General Changes in the Composition of the Vegetation Recorded in the Mudstones

The NMDS plot reveals that, similar to the succession of lignite beds, the succession of mudstones in LZ4 can be arranged in a way that indicates a step-wise change in the development of the vegetation. However, a clear separation in a younger and older part as for the lignite assemblages, each with a specific palynological composition, is not recognizable in the mudstone samples, because phases Mu 2 and Mu 5, mainly representing the older respectively the younger part of the mudstone succession, overlap in the NMDS plot, which points to a similar palynological composition (Fig. 7). However, phase Mu 1, including the oldest mudstones in the succession between 45.02 and 43.12 m, plots on the left side of the NMDS ordination space and is mainly composed of palynomorph taxa that group in assemblages Mu C, E and F of the R-mode cluster analysis (Fig. 7a). In contrast, samples from Mu 2 to Mu 4 between 42.52 and 35.12 m, plot, clearly separated from phase Mu 1, generally on the right side of the NMDS ordination space and are mainly characterized by palynomorphs that cluster in assemblages Mu A and Mu B (Fig. 7a). This indicates a change in the composition of the vegetation between phases Mu 1 and Mu 2 to Mu 4. Furthermore, the upper samples of phase Mu 5 and both samples from Mu 6 plot on the negative side of NMDS axis 1 in an area together with the oldest phase Mu 1 (Fig. 7b). This shows that the vegetation in the youngest samples had a composition similar to the vegetation at the beginning of the deposition of the mudstones (= base of LZ4).

Cicatricosisporites spp., the only taxon of Mu A, has a unique distribution in the mudstone succession with peak abundances in phases Mu 4 and Mu 6. Therefore, it is clearly separated from all other taxa in the R-mode cluster analysis (Fig. 7a). Pollen and spores that cluster in assemblage Mu B occur regularly in low-to-medium numbers in phases Mu 2 to Mu 5 and are almost restricted to them (Fig. 7a). These plot in the NMDS ordination space generally on lower right side (Fig. 7b). The occurrence of polypodiaceous spores (*Polpodiaceoisporites* ssp.) is remarkable in Mu B because they are completely missing in samples from the lignite beds. Their occurrence indicates a habitat outside of the peat-forming environments for these ferns. In contrast to Mu B, assemblage Mu C is more prevalent in the oldest and youngest phases Mu 1 and Mu 6. Assemblage Mu D is characterized by only two taxa *Platycaryapollenites* spp. and *Nudopollis terminalis*. Both are typical elements of the youngest phases (Mu 5 and Mu 6) with *N. terminalis* being almost restricted to these phases. Assemblages Mu E and Mu F are composed of the main elements of the forest vegetation and include pollen of the Fagaceae (*Tricolporopollenites cingulum, Tricolpopollenites liblarensis*) and of the juglandaceous alliance (*Plicatopollis* spp. *Plicapollis* spp.). These elements occur regularly in medium-to-high values in the mudstone. Nevertheless, the values vary independent from the general trends in the composition of the vegetation.

Discussion

Sedimentology

According to Hofmann et al. (2005), the lignite beds from the core of Lake PvH consist primarily of woody organic matter of the huminite group, indicating that terrestrial higher plants or aquatic macrophytes, which lived in the shallow littoral zone of the lake, contributed organic material. Neither the lignites nor the underlying mudstones show traces of rooting and most of the lignite beds are characterized by erosional contacts at the base (Fig. 3b). Furthermore, lignite clasts are often incorporated in the mudstones (Fig. 3d, f), which indicates an allochthonous origin of the lignitic material by transport from peat-forming environments into the lake. This is also confirmed by the relatively low TOC content of 50–60% and a high content of clastic material in the lignite (Hofmann et al. 2005). Lignite accumulation may therefore have included repeated erosional and redepositional events (Hofmann et al. 2005; Moshayedi et al. 2018).

Compared to the lignite beds, the gray-green mudstones are low in organic matter (<10% TOC) and consist mostly of laminated mudstones, which indicate deposition in a low energy environment. The setting has been interpreted as a shallow water and well-oxygenated site (Hofmann et al. 2005). The same is true for the upper part of the studied succession with finely layered and laminated silt and sand.

Similar to the lignite beds, the clastic material has, at least in part, been redeposited (Fig. 3). For example, oppositely dipping finely layered to laminated packages occur between 32 and 33 m depth (Fig. 3j). This may be a result of disturbance and redeposition due to tectonic activity. Convolute bedding as a result of the in-place deformation of unconsolidated mudstones may also indicate tectonic activity (Fig. 3g). Furthermore, individual clasts of remobilized mudstone within the mudstone succession (Fig. 3a, h) as well as within lignite beds (Fig. 3i) show that the mudstone has mainly been reworked.

However, parts of the succession are characterized by horizontally laminated mudstones which show no traces of deformation or redeposition (Fig. 3e). Therefore, during periods of reduced tectonic activity, mudstones were primarily derived from surface transport of clastic material into the lake. In the upper part of the sequence between 34.50 and 33.35 m, in which a massive lignite occurs (Fig. 3i), lignite-forming swamp areas may have reached their maximum extension in the vicinity of the lake. Accordingly, the deposition of lignitic material predominated over clastic material derived from surface transport.

Vegetational Trends in the Lignite Succession

The pollen diagram and the two-way cluster analyses do not show any taxonomic turnover across the lignite samples throughout the succession (Figs. 4 and 6a). Changes are mainly restricted to the quantitative composition of the pollen assemblages. However, the NMDS plot indicates that there are slight qualitative changes in the lignite record, e.g., the disappearance of the Normapolles element *Interpollis* spp. in the younger samples (Fig. 6b). Therefore, the lignite succession can be subdivided in a lower part (e.g., Li 2, Li 4) and an upper part (e.g., Li 5), each represented by a specific palynomorph composition (Fig. 6b).

Fern spores, such as Polypodiaceae and Schizaeaceae as well as spores of Lycopodiaceae, are especially abundant in the lignite beds (Figs. 4 and 6a). This points to a wide distribution of herbaceous pteridophytes growing under wet conditions at the shoreline of the lake and in marginal swamps (e.g., Lenz and Riegel 2001; Lenz et al. 2011). Spores of ferns and fern allies occur together with pollen of a forest association mainly dominated by Juglandaceae and Fagaceae representing the typical Paleogene paratropical inland flora of Europe and North America (Mai 1981, 1995; Manchester 1989). Furthermore, a peat-forming swamp with trees such as Cupressaceae und Nyssaceae existed at least in some areas at or near to the lake.

Slight shifts in abundance in the plant assemblages between the lower and upper phases are documented by (1) a general decrease of *Laevigatosporites* spp. and *Punctatosporites paleogenicus* from the lower to the upper part and (2) the regular appearance of *Cicatricosisporites* spp. only in the upper beds. This suggests that compositional changes occurred in the vegetation at the lake side during the time of deposition which may have been related to a general climate change between the lower and upper part of the succession. A slight decrease in temperature from the older to the younger part is interpreted from the disappearance of some more (sub) tropical elements (Sanderson and Donoghue 1994; Plunkett et al. 1996; Couvreur et al. 2011) including palms and Sapotaceae which lived in the forest in areas around the lake (Fig. 4). On the other hand, the Ericaceae (*Ericipites* spp.) as swamp elements occur with notable changes in frequency throughout the lignites, but appear more abundantly in the younger phase Li 5 (Figs. 4 and 6a). Ericaceae today mainly grow in more temperate regions of the world and in cooler mountainous areas of tropical latitudes (Kron 1996; Kron et al. 2002; Luteyn 2002), frequently preferring wet ground. Their common appearance in the younger part of the lignite succession may point to a slight decrease in temperature. The same is true for the disappearance of the extinct Normapolles element *Interpollis* spp. in the younger phases, which is thought to have preferred warm conditions (Daly and Jolley 2015).

The disappearance of pollen of *Carya* and *Celtis,* which both today prefer more moist habitats and, for example, are important elements of forested wetlands in the Southern United States (Tiner 2017), is noteworthy. This may indicate that less humid climate conditions prevailed in the younger part in phase Li 5. As a consequence, the general vegetation change in the lignite succession of Lake PvH can possibly be related to a gradual change toward slightly less humid and slightly cooler conditions.

The lower phases (Li 1 to Li 4) also include remains of freshwater algae (Fig. 4). Colonies of the coccal green alga *Botryococcus* are represented by low numbers in four samples, but disappear completely in the upper part (Fig. 4). *Ovoidites*, a cyst of Zygnemataceae, is found only in one sample of phase Li 1 (Fig. 4). Even if the overall plankton production was probably very low and/or the well-mixed oxic water column prevented the preservation of plankton (Hofmann et al. 2005), the occurrence of algae in the lower phases and their complete disappearance in the upper phases may have been related to changes in water chemistry and nutrient supply in response to decreasing precipitation. Decreasing nutrient availability also may be indicated by an increasing abundance of Ericaceae in the younger part, as these plants frequently prefer nutrient-poor soils (Riegel and Wilde 2016).

Some samples in the sequence of lignite beds are characterized by specific palynomorph assemblages and clearly distinguished as 'outliers' from the other assemblages in phases Li 2, 4 and 5 in the Q-mode cluster analysis (Li 1, 3, 6) (Fig. 6a). The same is seen in the NMDS plot (Fig. 6b). Nevertheless, the two outliers from Li 1 and the single sample from Li 7, which come from the base and the top of the lignite succession, respectively, fit the general trend of vegetational change. Samples of Li 1 plot along with the older samples from phases Li 2 and Li 4 in the upper half of the ordination space of the NMDS, while the sample of Li 7 is found in the lower part of the NMDS plot together with the samples from the younger phases Li 5 and Li 6 (Fig. 6b). However, the fact that these samples have a palynomorph composition distinctly different from the other samples may be related to the fact that the lignite beds are of allochthonous origin. Tectonic activity in the small pull-apart basin probably led to repeated earthquake tremor (Moshayedi et al. 2018), resulting in the mobilization of organic detritus or peat and its transport into the centre of the lake. This material may have had different sources around the lakeside in which the composition of the local vegetation differed. While the contribution of the zonal forest vegetation to the pollen assemblage did not change, the contribution of local communities increased in these outlier samples. This could explain the abnormal high proportion of fern spores of *Toripunctisporis* sp. in phase Li1 and of *Cicatricosiporites* sp. in the sample from phase Li 3, or the increase in *Nyssa* pollen in phase Li6.

Vegetational Trends in the Mudstone Succession

Six different palynomorph phases in the mudstone succession can be distinguished based on constrained Q-mode cluster analysis (Fig. 7). The NMDS plot shows that the oldest part of the succession (phase Mu 1) and the youngest phases (upper part of Mu 5 and Mu 6) are characterized by a similar palynomorph assemblage. These phases plot together on the left side of the ordination space. In contrast, phases Mu 2 to Mu 4 plot generally on the right side of the NMDS ordination space, indicating slight changes in the composition of the vegetation. Nevertheless, these changes are a consequence of the occurrence of rare elements such as *Porocolpopollenites* sp. or the pinaceous pollen *Pityosporites labdacus*.

Generally the vegetation, which is represented by the palynomorph assemblages in the mudstones, is similar to the associations found in the lignite beds. A diverse herbaceous vegetation dominated by ferns, such as Polypodiaceae, Osmundaceae and Schizaeaeceae and to lesser extent by Selaginellaceae, was associated with a forest vegetation dominated by Juglandaceae and Fagaceae as well as a *Nyssa/Taxodium* swamp community. However, the general trend toward slightly less humid and slightly cooler conditions during deposition, recognized in the succession of the lignites, is also obvious in the microflora of the mudstone beds. Warm elements, such as Symplocaceae (*Porocolpopollenites* sp.), Sapotaceae or palms, which occur regularly at least in parts of the older phases (Mu 1 to Mu 5) disappear in the younger part of the succession (Mu 6, Fig. 7a). Palms, which are mainly distributed in tropical to subtropical climates, have been widely used as indicators for a warm and humid climate (e.g., Greenwood and Wing 1995; Morley 2000, 2003; Walther et al. 2007). In contrast, the Normapolles taxon *Nudopollis terminalis* appears regularly in the upper phases (Mu 5 to Mu 6). The *Nudopollis* parent plants, possibly of juglandaceous affinity, preferred warmer and dryer environments compared to other Normapolles elements (Daly and Jolley 2015). A trend to less humid conditions is supported by the disappearance of the freshwater fungus *Entophlyctis lobata* in the upper beds. Generally, fungi and their spores are common in Paleogene strata during warmer periods at higher latitudes (Elsik 1996), but they are most abundant under relatively humid conditions (Kuhry 1985).

Compared to the lignite samples, which generally can be subdivided in a lower and upper series, the oldest (Mu 1) and the youngest samples (upper part of Mu 5, Mu 6) of the mudstone succession are characterized by relatively similar palynomorph assemblages. Therefore, if changes in pollen spectra were the result of climate the vegetation at the end of the deposition time was distributed under climatic conditions similar to those at the beginning of mudstone deposition. Hence, this could indicate an overall climate cycle.

The NMDS analysis of mudstone beds shows that some samples (Mu 3, 4) are characterized by specific palynomorph assemblages and plot separate from all others in the ordination space (Fig. 7b). Similar to the "outlier samples" of the lignite succession, these mudstone assemblages follow the general trend and plot in the ordination space next to the successive phases, either on the negative or positive side of the first NMDS axis. Thus, the contribution of abundant pollen from the zonal forest vegetation in these samples is the same compared to the other samples. Differences in the microflora are restricted to changes in the local contribution of the parent vegetation. For example, phase Mu 4 (37.52 m depth) is characterized by a maximum occurrence of *Camarozonosporites* spp. (Lycopodiaceae) which is almost missing in all the other samples. This may indicate that these spores were transported into the lake due to tectonic activity and related redeposition of the sediment from a local source around with abundant growth of lycopods. However, because redeposition cannot be unequivocally proven, other possible factors affecting transport, such as wind direction or intensity, are not excluded (Friis-Christensen and Svensmark 1997; Kern et al. 2013).

Comparison of Lignite and Mudstone Palynomorph Assemblages

Comparing the palynomorph assemblages of lignite and mudstone samples significant differences may be noted (Table 1), because more than 20% of the palynomorph taxa, such as pollen of Cupressaceae and Ericaceae, are significantly more abundant either in mudstone or in lignite samples.

However, in particular the arboreal pollen, which originates from the regional forest, is independent of the lithology in their frequency distribution. This concerns, e.g., fagaceous pollen, such as *Tricolpopollenites liblarensis* or *Tricolporopollenites cingulum*, as well as juglandaceous pollen, such as *Plicatopollis* spp., *Plicapollis* spp. or *Platycaryapollenites* spp. In particular *Plicatopollis* spp. and *T. cingulum* are dominating the palynomorph assemblages at Lake PvH with values up to 21% and 36% (Table 1). Although both elements are more prevalent in the lignites, these differences are not statistically significant (Table 1). Both the Juglandaceae and the Fagaceae were major elements of the vegetation of the Central European paratropical rainforest during the Lower and Middle Eocene in the region (Nickel 1996; Lenz et al. 2011; Riegel et al. 2015; Riegel and Wilde 2016). At Lake PvH the Fagaceae dominated during the deposition of the studied sediments of LZ 4, but both the juglandaceous and fagaceous pollen show strong frequency fluctuations similar as in Messel. However, an orbital influence as found at Messel (Lenz et al. 2011, 2015, 2017) cannot be proven at Lake PvH because a suitable age model is lacking. Since there is no difference in the general abundance of pollen from main elements of the forest vegetation between lignite and mudstone samples independent from facies changes, a uniform pollen rain from the forest around the lake is documented in the respective sediments at Lake PvH.

Significant differences in palynomorph abundance between lignite and mudstone samples, however, are evident for elements of the local vegetation from the edge of the lake. This may be related to taphonomical processes bringing pollen into the depositional area and to a lesser extent to the changing contribution of different vegetation communities during the deposition time of the lignite and mudstone beds. For example, a *Nyssa/Taxodium* swamp forest community has generally been widespread on the edge of the lake, because the respective pollen can be found in both lignite and mudstone samples. However, since pollen of Cupressaceae and Nyssaceae was not only transported by wind but probably also by runoff from a larger catchment area together with clastic material *Nyssapollenites* spp. and *Inaperturopollenites* spp. occur in the mudstone with significantly higher values. The same applies possibly to pollen of Eucommiacae (*Tricolpo(ro)pollenites parmularius*) and Symplocaceae (*Porocolpopollenites* sp.). A strong influence of surface transport is also supported by the fact that the fern spores *Polypodiaceoisporites* spp. (Polypodicaeae) and *Baculatisporites* spp. (Osmundaceae) are restricted to mudstone samples, because fern spores are generally transported in higher numbers by fluvial and surface transport (DeBusk Jr 1997).

In contrast, pollen of Ericaceae as one of the main elements of the peat-forming vegetation is significantly more abundant in the lignites. Lignitic material is a regular component of the mudstones. This explains that the strong frequency fluctuations of the ericaceous pollen between lignite and mudstone samples are not necessarily related to changes in the abundance of these plants around Lake PvH, but purely on the degree of redeposition of lignitic material. The same applies to the fern spore *Punctatosporites paleogenicus*, whose parent plants must be regarded as important elements of the lignite-forming community.

Controls of Vegetation Changes

The frequencies of palynomorph taxa from the mudstone beds and the lignite layers are not directly comparable because sources were different and processes acting during transport and deposition varied. For example, the values of *Tricolporopollenites cingulum* are generally higher in the lignite, while the values for *Tricolpo(ro)pollenites parmularius* are significantly higher in the mudstone (Table 1). However, the frequency trends for six of the dominant elements in the microflora (Fig. 8) almost coincide with a minimum of *Tricolporopollenites cingulum* between 37 and 38 m or maxima of *Tricolpo(ro)pollenites parmularius* between 37 and 38 m and at *c*. 42 m. This illustrates general trends in the development of the vegetation during the deposition of the studied succession.

An overall cyclic trend is possibly suggested by the frequency distribution of the background taxa including a similar microfloral composition at the base and at the top of the studied succession (Fig. 7). Hofmann et al. (2005) assumed that the deposition of the lignites may have been the result of lake level influencing auto-cyclic controls, such as processes that were strictly related to the progressive delivery of sediment to a water-filled depression or related to a combination of auto-cyclic, climate-related and tectonic control. However, systematic variation in abundance of the palynomorphs is not correlated to lithology. The complete absence of rooting (Hofmann et al. 2005) and erosional structures at the base of some lignites support

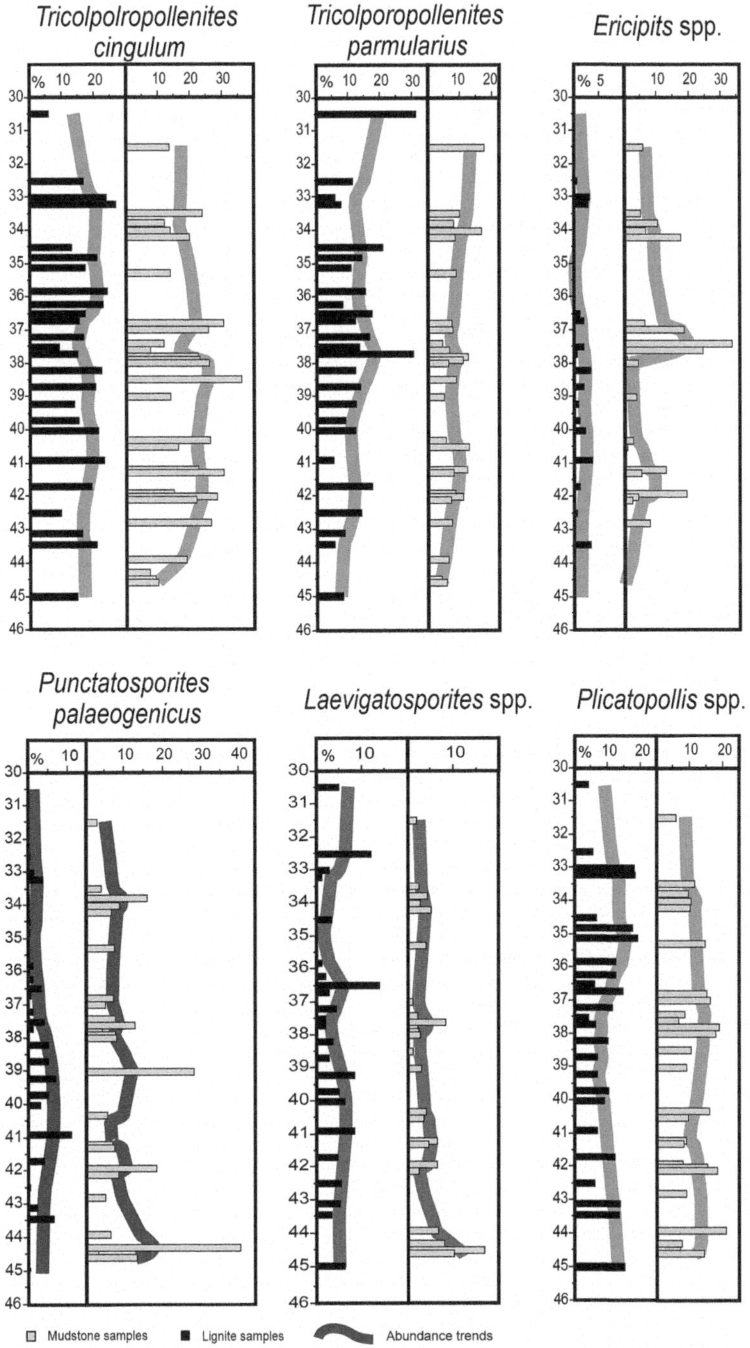

Fig. 8 Abundance variation of six palynomorph taxa common to lignite (black bars) and mudstone (grey bars) samples between 30.52 and 45.02 m depth (LZ4). The grey lines represent the smoothed abundance trends using a 12-point-Gaussian kernel according to Hammer et al. (2001)

an allochthonous origin for the organic material. This is a strong argument for discontinuities in the sequence in combination with repeated erosion and redeposition of organic material. An orbital control of lake-level fluctuations and related deposition of plant matter (biomass) as suggested by Hofmann et al. (2005) cannot be proven by our study. The palynological investigation of the lignite and mudstone samples may therefore suggest that climate change influenced the composition of the vegetation over time.

Nevertheless, compared to the influence of tectonic activity as recorded by sedimentological evidence (Moshayedi et al. 2018) climate change is of minor importance for the composition of the palynomorph assemblages.

The structure at PvH is aligned in SW-NE direction with three other Eocene basins (Messel, Erlenwiese, Sellborn-Schneise; Fig. 1) along a major concealed fault, the Messel Fault Zone (MZN; Mezger et al. 2013). Therefore, tectonic events, such as small earthquakes, may have occurred frequently during the history of the lake and led to remobilization and transport of marginal deposits into more central parts of the basin at Lake PvH. Figure 9 is based mainly on tectonic activity as the main factor influencing facies and paleoenvironment changes at PvH.

The model distinguishes between 4 scenarios:

(a) *Quiescent tectonic activity*: Predominant surface transport of fine-grained clastic material leads to the deposition of laminated mudstones without lignite clasts (Fig. 3e).
(b) *Tectonic activity with redeposition of lignite and mudstone*: In case of increased tectonic activity, eg, as a result of earthquakes, lignitic material is transported from areas of peat accumulation at the edge of the lake into the basin. At the same time clastic sediments will also be redeposited and in situ displacement occurs (Fig. 3b, d, f).
(c) *Tectonic activity and redeposition of mudstones*: During phases of reduced peat formation, tectonic activity primarily leads to redeposition and mixing of mudstones with few lignitic clasts (Fig. 3a, g, h).
(d) *Tectonic activity and redeposition of lignites*: During the phases of enhanced peat formation at the edge of the lake, tectonic activity primarily leads to redeposition of lignite (Fig. 3i) including some clastic material.

The model (Fig. 9) also includes minor changes in the composition of the vegetation, especially affecting the local habitats in swampy areas at the edge of the lake which are influenced by lake level. At the same time the regional forest vegetation

Fig. 9 (continued) (**c**) tectonic activity and preferential redeposition of mudstone, (**d**) tectonic activity and preferential redeposition of lignitic material (see text for details). For each of the scenarios examples are given by references to core images presented in Fig. 3. Changes in the vegetation during the (re)deposition of lignite and mudstone are restricted to the local flora at the edge of the lake (shaded). Different Li (lignite) and Mu (mudstone) phases are specified, but it should be noted that the respective assemblages are not directly associated with one of the tectonic scenarios (see text)

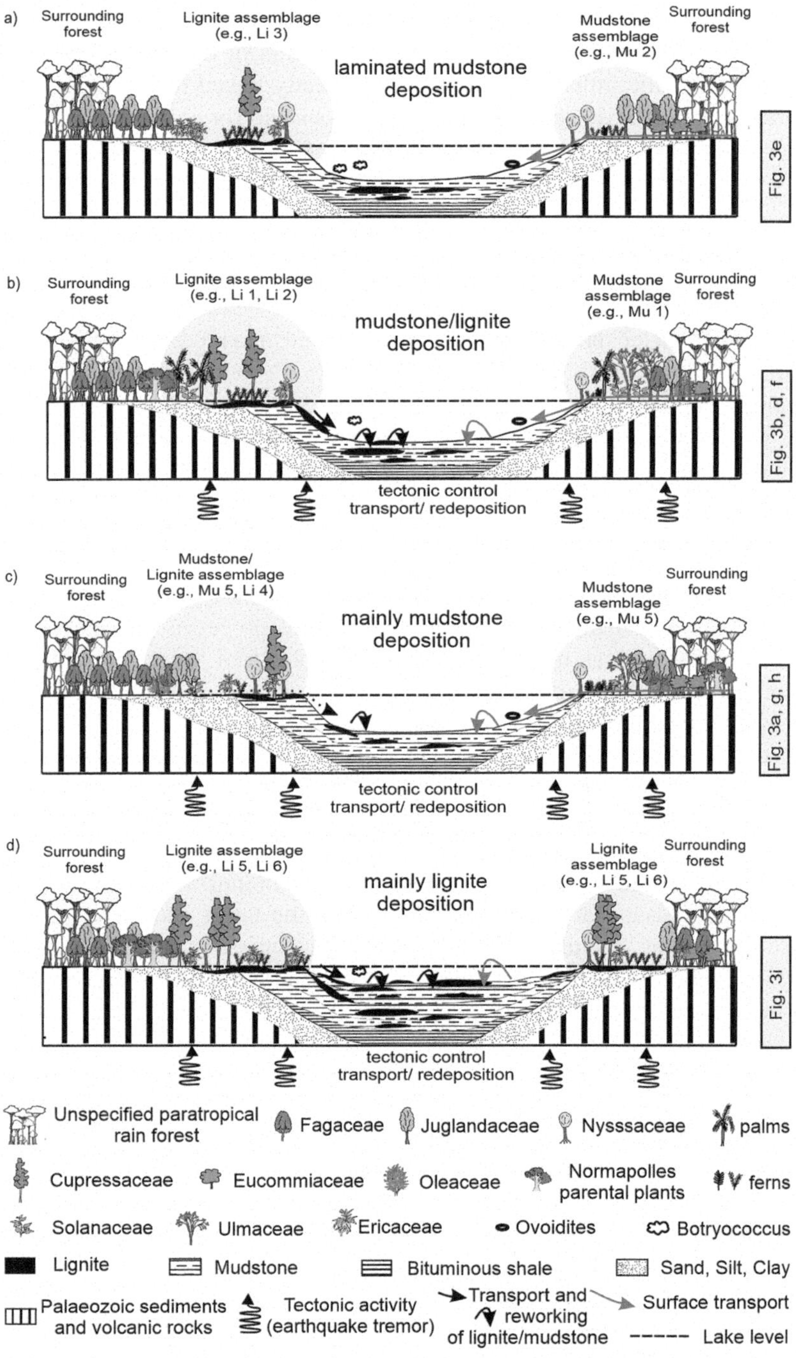

Fig. 9 Four different scenarios for the basin at PvH with regard to tectonic activity and vegetation: (**a**) Quiescent tectonic activity, (**b**) tectonic activity with redeposition of both lignite and mudstone,

persists. The different vegetational phases (Li and Mu phases) which are summarized in Fig. 9 show different phases in the evolution of the vegetation, probably reflecting a climatic influence. They are not directly linked to one of the tectonic scenarios and here is no specific vegetation associated, for example, with times of quiescent tectonic activity.

Conclusions

The studied 15 meter succession of alternating lignite and mudstone beds comprises only a small part of the lacustrine sequence at Lake 'Prinz von Hessen', but includes the time during which the lake probably had its smallest extent and lowest water level, which led to the establishment of extensive swamp environments and peat accumulation. High-resolution palynology, in combination with statistical analyses, shows that trends in the composition of the vegetation can clearly be identified during this lake phase, although they are masked by distinct facies changes (lignite versus clastic beds). The difference between lignites and clastic beds is reflected by significant differences in abundance for more than 20% of the palynomorphs. However, this is primarily related to taphonomical processes. While pollen and spores from the lignites mainly represent the more local aspect of peat forming environments around the lake, the mudstone record is derived from surface transport covering a larger catchment area.

Nevertheless, our study shows that the general vegetational trends were consistently expressed throughout both lithologies. Therefore, the study provides insight into the evolution of the vegetation during a short interval of the Paleogene greenhouse phase. A general climate change to slightly cooler and less humid conditions resulted only in minor changes in the qualitative composition of the vegetation. Only few taxa disappear between samples from the lower and upper part of the studied succession. It is therefore obvious that the vegetation, in general, had been robust in composition during the time recorded in the studied lacustrine succession. Only the communities at the lake shore were seriously affected by climate change.

However, our study shows quantitative changes in the vegetation, especially with regard to dominant elements as recorded in the palynoflora and probably recording changes in the dominance of different plant families. This applies especially to the Fagaceae and the Juglandaceae, which were among the most widespread elements of the Central European forests during the Paleogene (Mai 1981). These families alternated in their predominance in the forest vegetation at PvH in a similar way as described for the vegetation at nearby Lake Messel (Lenz et al. 2015). Such changes in the dominance of individual plant families may also have been related to orbitally controlled climate change at PvH; however, this cannot be unequivocally proven without an exact age model for the succession.

Frequent disturbance of the mudstones and an allochthonous origin of the lignitic material suggest that tectonic activity may have had some impact on the paleoenvironment of the lake basin and its surroundings. This is not only evident for the studied short interval, but also for the whole lacustrine sequence of PvH as described by Moshayedi et al. (2018). Tectonic activity is expressed by frequent earthquake tremor as identified by disturbed bedding, frequent evidence for redeposition and several discontinuities in the sedimentary record (Moshayedi et al. 2018). In summary, a combination of both, tectonic activity, and, to a minor extent, climate change, were the driving factors for changes in paleoenvironment and the paleoecosystems which may have been related to changes in lake level.

Acknowledgements and Funding Our research has been carried out as a project funded by the Deutscher Akademischer Austauschdienst (DAAD) under the grant 110207-54900347 including the PhD appointment of the first author. This generous support is gratefully acknowledged. The help of Karin Schmidt, Paleobotanical Section, Senckenberg Research Institute and Natural History Museum Frankfurt, in sampling and sample preparation is also acknowledged. Core photographs of FIS/HLUG from the core B/97-BK9 were kindly supplied by the Department for Messel Research and Mammalogy, Senckenberg Research Institute and Natural History Museum Frankfurt. Finally, we thank Robert A. Gastaldo and Carlos Jaramillo for their constructive comments and suggestions, which greatly helped to improve the paper.

References

Bohacs, K. M., Carroll, A. R., Neal, J. E., & Mankiewicz, P. J. (2000). Lake-basin type, source potential, and hydrocarbon character: An integrated sequence-stratigraphic-geochemical framework. In E. H. Gierlowski-Kordesch & K. R. Kelts (Eds.), *Lake basins through space and time* (AAPG studies in geology) (Vol. 46, pp. 3–34).

Bradley, W. H. (1967). Two aquatic fungi (Chytridiales) of Eocene Age from the Green River Formation of Wyoming. *American Journal of Botany, 54*, 577–582.

Bray, J. R., & Curtis, J. T. (1957). An ordination of the upland forest communities of southern Wisconsin. *Ecological Monographs, 27*, 325–349.

Broothaerts, N., Verstraeten, G., Kasse, C., Bohncke, S., Notebaert, B., & Vandenberghe, J. (2014). Reconstruction and semi-quantification of human impact in the Dijle catchment, Central Belgium: A palynological and statistical approach. *Quaternary Science Reviews, 102*, 96–110.

Carroll, A. R., & Bohacs, K. M. (1999). Stratigraphic classification of ancient lakes: Balancing tectonic and climatic controls. *Geology, 27*, 99–102.

Collinson, M. E., Smith, S. Y., Manchester, S. R., Wilde, V., Howard, L. E., Robson, B. E., Ford, D. S., Marone, F., Fife, J. L., & Stampanoni, M. (2012). The value of X-ray approaches in the study of the Messel fruit and seed flora. *Palaeobiodiversity and Palaeoenvironments, 92*(4), 403–416.

Cottam, G., Goff, F. G., & Whittaker, R. H. (1978). Wisconsin comparative ordination. In R. H. Whittaker (Ed.), *Ordination of plant communities*. Handbook of Vegetation Science (Vol. 5-2, pp. 185–213). Dordrecht: Springer.

Couvreur, T. L. P., Forest, F., & Baker, W. J. (2011). Origin and global diversification patterns of tropical rain forest: Inferences from a complete genus-level phylogeny of palms. *BMC Biology, 9*, 1–12.

Daly, R. J., & Jolley, D. W. (2015). What was the nature and role of Normapolles angiosperms? A case study from the earliest Cenozoic of Eastern Europe. *Palaeogeography, Palaeoclimatology, Palaeoecology, 418*, 141–149.

DeBusk, G. H., Jr. (1997). The distribution of pollen in the surface sediments of Lake Malawi, Africa, and the transport of pollen in large lakes. *Review of Palaeobotany and Palynology, 97*, 123–153.

Elsik, W. C. (1996). Chapter 10, Fungi. In J. Jansonius & D. C. McGregor (Eds.), *Palynology: Principles and applications* (Vol. 1, pp. 296–305). Houston, TX: American Association of Stratigraphic Palynologists Foundation.

Felder, M., & Gaupp, R. (2006). The δ13C and δ18O signatures of siderite – A tool to discriminate mixis patterns in ancient lakes. *Zeitschrift der Deutschen Gesellschaft für Geowissenschaften, 157*, 397–410.

Felder, M., & Harms, F. J. (2004). Lithologie und genetische Interpretation der vulkano-sedimentären Ablagerungen aus der Grube Messel anhand der Forschungsbohrung Messel 2001 und weiterer Bohrungen (Eozän, Messel-Formation, Sprendlinger Horst, Südhessen). *Courier Forschungsinstitut Senckenberg, 252*, 151–206.

Felder, M., Harms, F.J. & Liebig, V. (2001). Lithologische Beschreibung der Forschungsbohrungen Groβ-Zimmern, Prinz von Hessen und Offenthal sowie zweier Lagerstättenbohrungen bei Eppertshausen (Sprendlinger Horst, Eozän, Messel-Formation, Süd-Hessen). *Geologisches Jahrbuch Hessen, 128*, 29–82.

Franzen, J. L. (2006). Eurohippus parvulus parvulus (Mammalia, Equidae) aus der Grube Prinz von Hessen bei Darmstadt (Süd-Hessen, Deutschland). *Senckenbergiana Lethaea, 86*, 265–269.

Frederiksen, N. O. (1996). Chapter 29, Vegetational history. Introduction. In J. Jansonius & D. C. McGregor (Eds.), *Palynology: Principles and applications* (Vol. 3, pp. 1129–1131). Houston, TX: American Association of Stratigraphic Palynologists Foundation.

Friis-Christensen, E., & Svensmark, H. (1997). What do we really know about the sun-climate connection? *Advances in Space Research, 20*, 913–921.

Galloway, J. M., Tullius, D. N., Evenchick, C. A., Swindles, G. T., Hadlari, T., & Embry, A. (2015). Early Cretaceous vegetation and climate change at high latitude: Palynological evidence from Isachsen Formation, Arctic Canada. *Cretaceous Research, 56*, 399–420.

Gauch, H. G., & Scruggs, W. M. (1979). Variants of polar ordination. *Vegetatio, 40*, 147–153.

Geitler, L. (1962). Entwicklungsgeschichte der Chytridiale Entophlyctis apiculata auf der Protococcale Hypnomonas lobata. *Österreichische botanische Zeitschrift, 109*, 138–149.

Greenwood, D. R., & Wing, S. L. (1995). Eocene continental climates and latitudinal temperature gradients. *Geology, 23*, 1044–1048.

Gruber, G., & Micklich, N. (2007). Messel—Treasures of the Eocene. *Darmstadt: Hessisches Landesmuseum p, 158*.

Hair, J. R. J. F., Black, W. C., Babin, B. J., & Anderson, R. E. (2010). *Multivariate data analysis* (7th ed., p. 761). London: Pearson.

Hammer, Ø., Harper, D. A. T., & Ryan, P. D. (2001). PAST: Paleontological statistics software package for education and data analysis. *Palaeontologia Electronica, 4*(1).: https://palaeo-electronica.org/2001_1/past/issue1_01.htm.

Harms, F. J., Aderhold, G., Hoffmann, I., Nix, T., & Rosenberg, F. (1999). Erläuterungen zur Grube Messel bei Darmstadt, Südhessen. *Schriftenreihe der Deutschen Geologischen Gesellschaft, 8*, 181–222.

Hofmann, P., Duckensell, M., Chpitsglous, A., & Schwark, L. (2005). Geochemical and organic petrological characterization of the organic matter of lacustrine Eocene oil shales (Prinz von Hessen, Germany): Reconstruction of the depositional environment. *Paléo, 33*, 155–168.

Jacoby, W. (1997). Tektonik und Eoza¨ ner Vulkanismus des Sprendlinger Horstes, NE-Flanke des Oberrheingrabens. *Schriftenreihe dt. geol. Ges, 2*, 66–67.

Jacoby, W., Wallner, H., & Smilde, P. (2000). Tektonik und Vulkanismus entlang der Messeler-Sto¨ rungszone auf dem Sprendlinger Horst: geophysikalische Ergebnisse. *Zeitschrift der Deutschen Gesellschaft für Geowissenschaften, 151–154*, 493–510.

Jardine, P. E., & Harrington, G. J. (2008). The Red Hills Mine palynoflora: A diverse swamp assemblage from the Late Paleocene of Mississippi, U.S.A. *Palynology, 32*, 183–204.

Janssen, C. R., & Birks, H. J. B. (1994). Recurrent groups of pollen types in time. *Review of Palaeobotany and Palynology, 82*, 165–173.

Juggins, S. (2007). C2 Software for ecological and palaeoecological data analysis and visualization. *User guide Version, 1*(5), 73.

Kaiser, M. L., & Ashraf, R. (1974). Gewinnung und Präparation fossiler Pollen und Sporen sowie anderer Palynomorphae unter besonderer Berücksichtigung der Siebmethode. *Geologisches Jahrbuch, 25*, 85–114.
Kern, A. K., Harzhauser, M., Soliman, A., Piller, W. E., & Mandic, O. (2013). High-resolution analysis of upper Miocene lake deposits: Evidence for the influence of Gleissberg-band solar forcing. *Palaeogeography, Palaeoclimatology, Palaeoecology, 370*, 167–183.
Kron, K. A. (1996). Phylogenetic relationships of Empetraceae, Epacridaceae, Ericaceae. Monotropaceae, and Pyrolaceae: Evidence from nuclear ribosomal 18s sequence. *Annals of Botany, 77*, 293–303.
Kron, K. A., Powell, E. A., & Luteyn, J. L. (2002). Phylogenetic relationships within the blueberry tribe (Vaccinieae, Ericaceae) based on sequence data from matK and nuclear ribosomal ITS regions, with comments on the placement of Satyria. *American Journal of Botany, 89*, 327–336.
Kuhry, P. (1985). Transgression of a raised bog across a coversand ridge originally covered with an oak-lime forest. Palynological study of a middle Holocene local vegetational succession in the Amtsven (Northwest Germany). *Review of Palaeobotany and Palynology, 44*, 303–353.
Lenz, O. K. (2005). Palynologie und Paläoökologie eines Küstenmoores aus dem Mittleren Eozän Mitteleuropas Die Wulfersdorfer Flözgruppe aus dem Tagebau Helmstedt, Niedersachsen. *Palaeontographica B, 271*, 1–157.
Lenz, O. K., & Riegel, W. (2001). Isopollen maps as a tool for the Reconstruction of a coastal swamp from the Middle Eocene at Helmstedt (Northern Germany). *Facies, 45*, 177–194.
Lenz, O. K., Wilde, V., & Riegel, W. (2007). Recolonization of a Middle Eocene volcanic site: Quantitative palynology of the initial phase of the maar lake of Messel (Germany). *Review of Palaeobotany and Palynology, 145*, 217–242.
Lenz, O. K., Wilde, V., & Riegel, W. (2011). Short-term fluctuation in vegetation and phytoplankton during the middle Eocene greenhouse climate: A 640-kyr record from the Messel oil shale (Germany). *International Journal of Earth Sciences, 100*, 1851–1874.
Lenz, O. K., Wilde, V., Mertz, D. F., & Riegel, W. (2015). New palynology-based astronomical and revised 40Ar/39Ar ages for the Eocene maar lake of Messel (Germany). *International Journal of Earth Science, 104*, 873–889.
Lenz, O. K., Wilde, V., & Riegel, W. (2017). ENSO- and solar-driven sub-Milankovitch cyclicity in the Palaeogene greenhouse world; high-resolution pollen records from Eocene Lake Messel, Germany. *Journal of the Geological Society, 174*, 110–128.
Luteyn, J. L. (2002). Diversity, adaptation, and endemism in neotropical Ericaceae: Biogeographical patterns in the Vaccinieae. *The Botanical Review, 68*, 55–87.
Mai, D. H. (1981). Entwicklung und klimatische Differenzierung der Laubwaldflora Mitteleuropas im Tertiär. *Flora, 171*, 525–582.
Mai, D. H. (1995). *Tertiäre Vegetationsgeschichte Europas—Methoden und Ergebnisse* (p. 691). Jena: Gustav Fischer Verlag.
Manchester, S. R. (1989). Attached reproductive and vegetative remains of the extinct American–European genus Cedrelospermum (Ulmaceae) from the early Tertiary of Utah and Colorado. *American Journal of Botany, 76*, 256–276.
Mander, L., Kürschner, W. M., & McElwain, J. C. (2010). An explanation for conflicting records of Triassic–Jurassic plant diversity. *Proceedings of the National Academy of Sciences of the United States of America, 107*, 15351–15356.
Marell, D. (1989). Das Rotliegende zwischen Odenwald und Taunus.- Geologische Abh. *Hessen, 89*, 1–128.
Mezger, J. E., Felder, M., & Harms, F. J. (2013). Crystalline rocks in the maar deposits of Messel: Key to understand the geometries of the Messel Fault Zone and diatreme and the post-eruptional development of the basin fill. *Zeitschrift der Deutschen Gesellschaft für Geowissenschaften, 164*, 639–662.
Minchin, P. R. (1987). An evaluation of the relative robustness of techniques for ecological ordination. *Vegetatio, 69*, 89–107.

Morley, R. J. (2000). *Origin and evolution of tropical rain forests* (p. 362). Chichester: Wiley.

Morley, R. J. (2003). Interplate dispersal paths for megathermal angiosperms. *Perspectives in Plant Ecology. Evolution and Systematics, 6*, 5–20.

Moshayedi, M., Lenz, O. K., Wilde, V., & Hinderer, M. (2018). Controls on sedimentation and vegetation in an Eocene pull-apart basin (Prinz von Hessen, Germany): Evidence from palynology. *Journal of the Geological Society, 175*, 757–773.

Nickel, B. (1996). Die mitteleozäne Mikroflora von Eckfeld bei Manderscheid/Eifel. *Mainzer Naturwissenschaftliches Archiv. Beiheft, 18*, 1–121.

Oksanen, J. (2007). Standardization methods for community ecology. Documentation and user guide for package. *Vegan, 1*, 8–6.

Pirrung, M. (1998). Zur Entstehung isolierter alttertiärer Seesedimente in zentraleuropäischen Vulkanfeldern. *Mainzer Naturwissenschaftliches Archiv Beiheft, 20*, 1–117.

Plunkett, G. M., Soltis, D. E., & Soltis, P. S. (1996). Higher level relationships of Apiales (Apiaceae and Araliaceae) based on phylogenetic analysis of rbcL sequences. *American Journal of Botany, 83*, 499–515.

Riegel, W., Lenz, O. K., & Wilde, V. (2015). From open estuary to meandering river in a greenhouse world: An ecological case study from Middle Eocene of Helmstedt, Northen Germany. *PALAIOS, 30*, 304–326.

Riegel, W., & Wilde, V. (2016). An early Eocene Sphagnum bog at Schöningen, northern Germany. *International Journal of Coal Geology, 159*, 57–70.

Sanderson, M. J., & Donoghue, M. J. (1994). Shifts in diversification rate with the origin of angiosperms. *Science, 264*, 1590–1595.

Schaarschmidt, F. (1988). Der Wald, fossile Pflanzen als Zeugen eines warmen Klimas. In S. Schaal & W. Ziegler (Eds.), *Messel – ein Schaufenster in die Geschichte der Erde und des Lebens. Waldemar Kramer* (pp. 27–52). Frankfurt am Main.

Schaal, S., & Ziegler, W. (1988). *Messel-Ein Schaufensten in die Geschichte der Erde und des Lebens, Kramer* (p. 315). New York.

Smith, K. T., Schaal, S. F. K., & Habersetzer, J. (2018). *Messel – An ancient greenhouse ecosystem*. Senckenberg-Buch 80, Schweizerbart, Stuttgart, 315p.

Ter Braak, C. J. F., & Looman, C. W. N. (1996). Regression. In R. H. G. Jongman, C. J. F. Ter Braak, & O. F. R. Tongeren (Eds.), *Data analysis in community and landscape ecology* (pp. 29–77). Cambridge: Cambridge University Press.

Thiele-Pfeiffer, H. (1988). Die Mikroflora aus dem mitteleozänen Ölschiefer von Messel bei Darmstadt. *Palaeontographica B, 211*, 1–86.

Thomson, P. W., & Pflug, H. (1953). Pollen und Sporen des mitteleuropäischen Tertiärs. Gesamtübersicht über die stratigraphisch und paläontologisch wichtigen Formen. *Palaeontographica B, 94*, 1–138.

Tiner, R. W. (2017). Practical considerations for wetland identification and boundary delineation. In *Wetlands* (pp. 129–154). CRC Press.

Walther, G. R., Gritti, E. S., Berger, S., Hickler, T., Tang, Z. Y., & Sykes, M. T. (2007). Palms tracking climate change. *Global Ecology and Biogeography, 16*, 801–809.

Wilde, V. (1989). Untersuchungen zur Systematik der Blattreste aus dem Mitteleozän der Grube Messel bei Darmstadt (Hessen, Bundesrepublik Deutschland). *Courier Forschungsinstitut Senckenberg, 115*, 1–123.

How Changes of Past Vegetation and Human Impact Are Documented in Lake Sediments: Paleoenvironmental Research in Southwestern Germany, a Review

Manfred Rösch, Karl-Heinz Feger, Elske Fischer, Matthias Hinderer, Lucas Kämpf, Angelika Kleinmann, Jutta Lechterbeck, Elena Marinova, Antje Schwalb, Gegeensuvd Tserendorj, and Lucia Wick

Abstract During the last four decades, sediment cores were collected from the deepest zones of >30 small lakes in south-west Germany to investigate archaeological questions about settlement and land use history in Central Europe. The primary

---

M. Rösch (✉)
Institut für Ur- und Frühgeschichte und Vorderasiatische Archäologie, Universität Heidelberg, Heidelberg, Germany

K.-H. Feger
Institut für Bodenkunde und Standortslehre, TU Dresden, Tharandt, Germany

E. Fischer · E. Marinova · G. Tserendorj
Landesamt für Denkmalpflege im RP Stuttgart, Gaienhofen-Hemmenhofen, Germany

M. Hinderer
Institute for Applied Geosciences, TU Darmstadt, Darmstadt, Germany

L. Kämpf
Institut für Bodenkunde und Standortslehre, TU Dresden, Tharandt, Germany

CNRS – Laboratoire de Chrono-Environment, Université de Franche-Comté Besancon, Besancon, France

A. Kleinmann
Ritter-Eccart-Str. 5, Herbertingen, Germany

J. Lechterbeck
Arkeologisk museum, Universitetet i Stavanger, Stavanger, Norway

A. Schwalb
Institute of Geosystems and Bioindication, TU Braunschweig, Braunschweig, Germany

L. Wick
Integrative Prehistory and Archaeological Science (IPAS), University of Basel, Basel, Switzerland

M. R. Rosen et al. (eds.), *Limnogeology: Progress, Challenges and Opportunities*, Syntheses in Limnogeology, https://doi.org/10.1007/978-3-030-66576-0_4

aim was to establish well-dated pollen profiles with very high temporal resolution. A major focus was to reconstruct the extent and intensity of land clearing during the Neolithic, the Iron Age, the Roman Period as well as through the Medieval timeframe in context with intermittent recovery periods during political and economic crises. In southwestern Germany, lake sites are not evenly distributed, but rather restricted to the formerly glaciated landscapes of the Northern Alpine foreland, including Lake Constance and the Black Forest. The most important results are as follows:

1. Central basinal sediments of small lakes are the best archives to study environmental change, past human impact, and vegetation.
2. The lakes reflect the environment and its change, triggered by natural and anthropogenic factors.
3. Lakes in areas with silicic bedrock have much lower sedimentation rates than lakes in the carbonate-rich North Alpine foreland in Germany.
4. Human impact causes sedimentation rates to increase by different rates, depending on the magnitude of impact and system sensitivity; in most cases, lamination disappears from the sedimentary record.
5. The small lakes integrate local landscape history, whereas the larger lakes integrate over larger catchments and record regional signals.
6. During the Late Weichselian and early Holocene, the landscape history of Southwestern Germany was rather uniform on a regional scale.
7. Landscape history from the middle Holocene onwards was more and more impacted by man, resulting in a variety of local and regional effects that are recorded in these lakes.
8. Overall, climate and natural factors determined the environmental settings, and people preferred regions with fertile soils and warm climate for settling and farming.

Keywords Sediment cores from small lakes · Southwest Germany · Late Weichselian and Holocene landscape history · High-resolution core analysis

Lakes in Southwestern Germany

In Southwestern Germany, there are two regions where glacial lakes have formed: the Alpine foreland and the Black Forest (Schwarzwald) (Fig. 1). Of the 59 natural lakes in Southwestern Germany, 45 are situated in the Alpine foreland and 13 in the Black Forest. The largest pre-alpine lake is Lake Constance (Bodensee), which was formed by the former Rhine Glacier. It is presently surrounded by small glacial and kettle lakes. The fringes of the maximum Weichselian ice extend reached 50 km farther to the north and east beyond the modern lake shore. The formerly glaciated areas are nowadays hilly landscapes (Allgäu and Oberschwaben) with numerous small lakes and mires as well as wetland archaeological sites. The Alpine Rhine flows through Lake Constance and causes a high rate of siltation (Hinderer 2001;

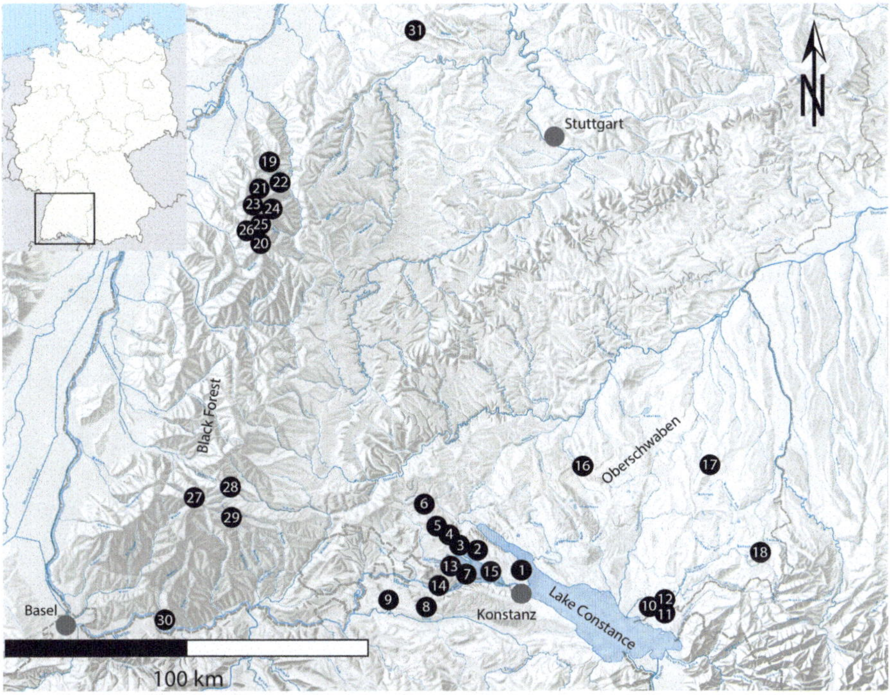

Fig. 1 Map of Southwestern Germany with the lakes discussed in this paper. Numbers according to Table 1

Fig. 2a). The surrounding lakes are disconnected from the major river network of the Rhine and experienced low sedimentation rates dominated by autochthonous processes, resulting in sediments consisting mainly of organic and calcareous matter (Fig. 2b). These small lakes provide excellent sediment archives to reconstruct the local paleoenvironmental history and the role of human disturbance. This is also partly true for Lake Constance, where sediments from the deep basin reflect both anthropogenic impact and northern hemisphere climate change (Hanisch et al. 2009; Schwalb et al. 2013). In addition to glacial lakes, a few karstic lakes were studied, with Aalkistensee as the most prominent example.

Lake Formation

The pre-Alpine lakes formed during melting of the Rhine Glacier lobe between ca. 19 and 17 ka BP (Preusser et al. 2011). Prior to the Younger Dryas, sedimentation was dominated by allochthonous fluvial and aeolian sediment input (Wessels 1998). Autochthonous sedimentation of lake marls started during the Bølling and Allerød interstadials and reached a maximum in the Early Holocene (ca. 9.5 ka BP). This

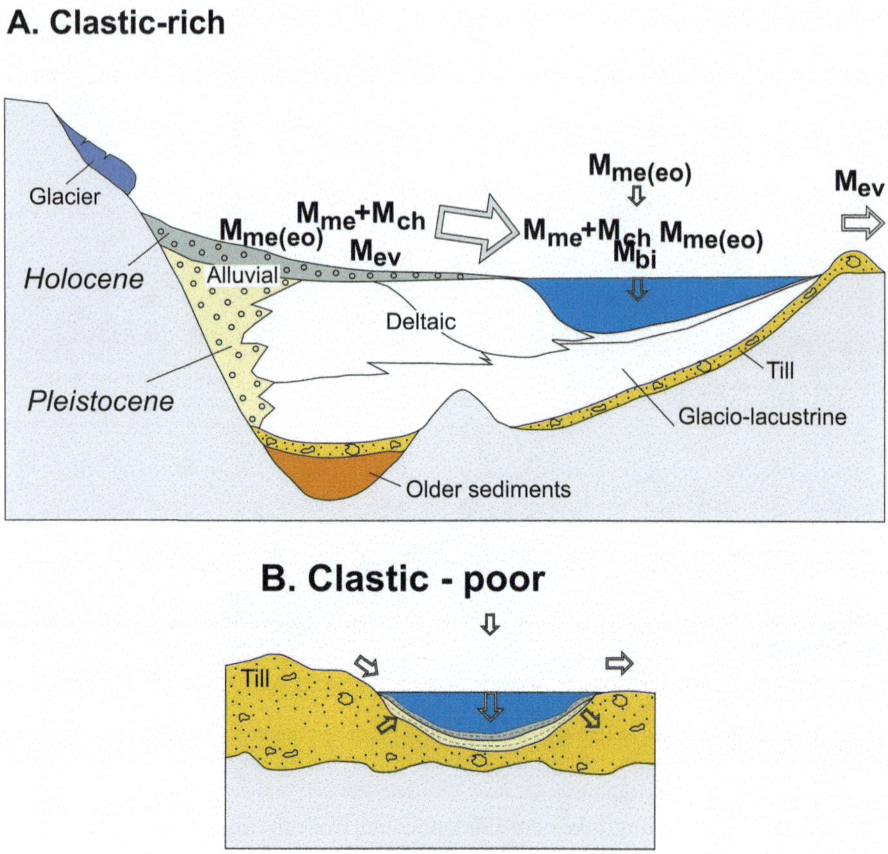

Fig. 2 Modes of sedimentation in perialpine lakes and mass balance → Hinderer 2001; M = mass, me = mechanical, ev = evaporitic, ch = chemical, bio = biochemical, (eo) = eolian, gr = groundwater

period, characterized by authochthonous sedimentation, is dominated by lacustrine marl indicating an enhanced input of dissolved carbonate from chemical weathering of minerals provided by glaciogenic sediments (Müller 1966). In Lake Constance, marl formation decreased after ca. 8 ka BP (Wessels 1998). In small lakes the littoral sediments have higher carbonate contents than those from the center of a lake, because macrophytes trigger formation of carbonate as, for example, in the Steißlinger See (Lechterbeck 2001; Lechterbeck et al. 2014).

The small cirque lakes in the Black Forest region represent the remnants of a local glaciation during the last Ice Age (Lang 2005). Because the bedrock consists of granites, gneisses, and sandstones, lake marl could not develop during that time. During the Late Weichselian, thin clastic layers were deposited, followed by fine-detritus gyttja and later dy, a sediment of dystrophic lakes containing abundant humic acids. This change was triggered by the paludification and acidification of the

Table 1 The studied lakes, geographical position, and physical properties

No.	Site	N	E	m a.s.l.	Size (ha)	Max. depth (m)	Core length (m)	Pollen levels	Radiocarbon dates	annually laminated	Covered time (millennia)	References
Bodensee												
1	Mainau-Obere Güll	47°42′20”	9°11′01”	394	47,300	150/2	14	985	19		15–0	Rösch and Wick (2018)
2	Mindelsee	47°45′20”	9°01′22”	406	100	14	6	402	15		15–0	Rösch (2013), Rösch et al. (2014a, 2014b)
3	Buchensee-Südost	47°46′01”	8°59′05”	431	1.6	2	9	797	8		15–0	Rösch and Wick (2019)
4	Böhringer See	47°45′48”	8°56′18”	409	5.1	9	7	525	15		15–0	Lechterbeck and Rösch (2020)
5	Litzelsee	47°46′08”	8°55′50”	413	1.3	8	7	449	28		15–0	Rösch and Lechterbeck (2016)
6	Steisslinger See	47°47′57”	8°55′01”	450	11.3	16	6	464		x	15–0	Lechterbeck (2001)
7	Hornstaad-Bodensee	47°41′45”	9°00′31”	394	6300	45/2	14	862	25		15–0	Rösch (1992, 1993, 1997)
8	Nussbaumer See	47°37′01”	8°49′05”	434	25	7	10	182	22		15–0	Haas and Hadorn (1998), Rösch (1983)
9	Husemer See	47°37′22”	8°42′21”	409	15.4	16	8.4	59		x	15–0	Rösch (unpubl.)
10	Schleinsee	47°36′48”	9°38′07”	474	14.9	11.6	10.3	452	29	x	15–0	Müller (1962), Geyh et al. (1971), Kleinmann und Merkt (in prep.)

(continued)

Table 1 (continued)

No.	Site	N	E	m a.s.l.	Size (ha)	Max. depth (m)	Core length (m)	Pollen levels	Radiocarbon dates	annually laminated	Covered time (millennia)	References
11	Degersee	47°36′45”	9°39′02”	478	31.9	11.5	10.4	519	24	x	15–0	Kleinmann et al. (2015), Merkt and Müller (1978)
12	Muttelsee	47°37′07”	9°40′15”	492	8.1	6.5	9.5	189		x	15–0	Merkt und Müller (unpubl)
13	Zeller See/ Bodensee	47°43′14”	8°58′25”	395	6300	25	6.5	68			15–0	Rösch (1991), Giovanoli et al. (1990)
14	Wangen/ Bodensee	47°39′39”	8°56′22”	395	6300	45/2	12	177			15–5	Rösch (2002)
15	Gnadensee	47°42′35”	9°03′41”	395	6300	20	6	350			15–0	Ryabogina and Rösch (unpubl.)
Oberschwaben/Allgäu												
16	Königseggsee	47°55′57”	9°26′58”	627	15.6	9.6	7.1	408	12		15–0	Fischer and Rösch (2017)
17	Stadtsee/Bad Waldsee	47°55′24”	9°45′30”	583	14	12	9.0	223	36	x	5–0	Merkt und Müller (unpubl.), Fischer et al. (2010)
18	Großer Ursee	47°45′11”	10°01′32”	695	19.4	9.8	12.0	1200	21		15–0	Rösch et al. (2020)
Schwarzwald												
19	Herrenwieser See	48°40′09”	8°17′47	830	1.2	9.5	6.8	369	44		15–0	Rösch (2012)
20	Glaswaldsee	48°25′37”	8°15′46”	839	2.9	11	1.8	153	16		15–0	Rösch and Heumüller (2008)
21	Mummelsee	48°35′56”	8°12′06”	1028	3.3	17	2.6	193	12		15–0	Tserendorj (unpubl.)

No.	Site	N	E	m a.s.l.	Size (ha)	Max. depth (m)	Core length (m)	Pollen levels	Radiocarbon dates	annually laminated	Covered time (millennia)	References
22	Schurmsee	48°36′51”	8°19′13”	795	1.6	13	5.0	244	19		15–0	Tserendorj (unpubl.)
23	Wilder See am Ruhestein	48°34′14”	8°14′25	910	2.1	11.5	3.9	259	16		15–0	Rösch (2009a, 2009b)
24	Huzenbacher See	48°34′32”	8°20′58”	747	2.5	7.5	3.1	364	25		15–0	Rösch and Tserendorj (2011a, 2011b)
25	Buhlbachsee	48°30′06”	8°14′45”	790	2.2	4.5	9.0	318	19		15–0	Tserendorj (unpubl.)
26	Ellbachsee	48°29′03”	8°18′21”	770	2.9	2	10.2	85			15–0	Rösch (unpubl.)
27	Feldsee	47°52′17”	8°01′59”	1109	9	34	4.7	39	13		15–0	Knopf et al. (2019)
28	Titisee	47°53′39”	8°08′43”	846	107	39	7.8	301	13		15–0	Knopf et al. (2019)
29	Schluchsee	47°49′08”	8°09′09”	930	538	61	12.1	850	18		15–0	Knopf et al. (2019)
30	Bergsee	47°34′20”	7°56′09”	382	6.7	14	33.0	633	10		15–0	Knopf et al. (2016), Kämpf et al. (2017)
Gäu												
31	Aalkistensee/ Maulbronn	48°59′41”	8°45′41”	227	12	2.0	3.9	57.0	8		4–0	Rösch et al. (2017)

surroundings and took place in the fifth millennium BP, coincident with the expansion of *Abies* across the region.

Karstic lakes usually form in sinkholes and depressions. Although these lakes are small and record only sections of the Holocene, they provide paleoeocological research targets of lake sediment archives outside the Black Forest in the northwest and the perialpine lakes in the southeast in a landscape often chosen by prehistoric humans as preferred areas for settlements and agricultural land use. This makes these lakes valuable as archives for paleoenvironmental studies. One of these exceptional lakes is Aalkistensee close to the UNESCO World Heritage site monastery Maulbronn, which formed as a sinkhole above collapsed gypsum beds (Rösch et al. 2017). It is located in the Early Neolithic settled lowlands characterized by fertile soils and a warm and dry climate. There, prehistoric sediments were found accidentally while coring the sediments of what was thought to have been a former monastery's pond constructed sometime between the twelfth and fifteenth centuries BCE (Rösch et al. 2017).

A Brief History of Paleo-Lake Research in Southwest Germany

During the first six decades of research in southwest Germany from about 1920 to 1980, palynologists preferred to work in mires or at lake shores, because of the lack of suitable coring equipment and the availability of abundant organic carbon in peat suitable for radiocarbon dating (Gaillard et al. 2018).

One of the pioneers of lake sediment research in Germany was Josef Merkt, a geologist and paleolimnologist (Müller 1973). He was one of the first to analyze thinly laminated lake sediments using large thin sections (Merkt 1971). Since the late 1970's, when Gerhard Lang became Professor for Geobotany at the University of Bern, new coring methods were introduced to vegetation history and applied in projects in Switzerland and nearby regions; this included a modified Livingstone sampler (Merkt and Streif 1970), and operating from a raft to collect lake cores (Lang 1985; Ammann 1989; Rösch 1983, 1995). When modern paleoecological work in southern Germany started in the early 1980s in collaboration with archaeologists, primary efforts focused still on mires (Rösch 1985, 1990; Hölzer 1987; Küster 1988; Smettan 1988; Liese-Kleiber 1993, Rösch 2000). Many radiocarbon dates were needed in order to establish robust chronologies and age control, and prior to the emergence of accelerator mass spectrometry (AMS), large amounts of carbon were needed. Therefore, peat was preferred for analyses as it allowed the direct dating of bulk sediment. Lake sediments did not contain sufficient carbon material and often suffered from a reservoir effect, especially when carbonate-rich sediments were present. To obtain better time resolution, the cores were sampled continuously.

An exceptional opportunity to obtain cores from Lake Constance opened with the strong winters in 1985/86 and 1986/87, when the Bodensee-Untersee was frozen. Cores were retrieved from the broad shallow water zone in front of the archaeological excavation site of Hornstaad-Hörnle dating to the early fortieth century BCE (Rösch 1992, 1993, 1997). At the same time, a paleolimnological working group, consisting of Federico Giovanoli and coworkers, University of Konstanz, took several cores from Bodensee-Untersee (Giovanoli et al. 1990).

Modern paleoecological investigations in Oberschwaben started with work in mires (Liese-Kleiber 1993). In the eastern Bodensee region, a team of the Lower Saxony State Geological Survey and the Federal Institute for Geosciences and Natural Resources, including Josef Merkt and Helmut Müller, investigated annually laminated sediments from Schleinsee, Degersee, and Muttelsee (Kleinmann et al. 2015; Kleinmann and Rammelmair 2003; Clark et al. 1989; Merkt and Müller 1978, 1995; Geyh et al. 1971, 1974; Merkt 1971; Müller 1962). Of the many lakes in Oberschwaben, several were cored in search of laminated sediments, including Illmensee, Ruschweiler See, Steeger See, Obersee, Bibersee, Olzreuter See, and Niklassee (Merkt, Müller, Streif unpubl.). But only few were studied in detail: Königseggsee, Stadtsee in Bad Waldsee, and Großer Ursee, the latter situated further to the east in the western Allgäu (Homann et al. 1990; Fischer and Rösch 2017; Rösch et al. 2008; Fischer et al. 2010; Rösch and Hahn 2016; Rösch et al. 2020).

Compared with the Alpine foreland, the Black Forest was thought to have been settled quite late, probably not before the Middle Ages (Wilmanns 2001). Due to its anticipated lack of agricultural aptitude, the Black Forest was not in the focus of archaeological research until Iron Age hillforts and remains of Celtic iron mining were found (Damminger and Wieland 2003; Gassmann and Wieland 2005; Gassmann et al. 2006; Rösch et al. 2009).

Eventually, the question whether these settlements and activities made a lasting impact on landscape and vegetation triggered a number of paleoenvironmental studies. After initial work in various mires (Rösch et al. 2009), all seven extant cirque lakes of the northern Black Forest were sampled; in the first campaign this was accomplished with logistical support by the University of Bern. Pollen analysis of sediment cores from the deepest parts of the lakes clearly shows substantial deforestation and an intensification of land use during the Iron Age, long before the Medieval Age (Rösch 2009a, 2009b; Rösch 2009/2010). In the same region, paleoecological work on soils and sediments from the Medieval and early industrial times, with the goal to study the environmental impact by local glass production and ore processing industries, as well as deposition of air pollutants originating from long-distance transport such as inorganic acids, organic pollutants, and heavy metals (Feger 1986, 1993). Hinderer et al. (1998) modeled long-term acidification of the cirque lake Herrenwieser See and calibrated the hydrochemical model with a diatom pH reconstruction. Moreover, the impact of past land-use (deforestation, forest devastation, modern afforestation with monocultural spruce) on biogeochemical fluxes was quantified in selected other cirque lakes of the northern and southern Black Forest (Feger 1986, 1993; Zeitvogel and Feger 1990). Collaborations using multi-proxy records have continued through today.

Recent work in the southern Black Forest locations are at higher elevations, up to 1500 m, an area that hosts fewer, but larger and deeper lakes: Titisee, Feldsee, Schluchsee and Bergsee (Fig. 1). After 2010, sediment cores were collected from these lake centers in water depths of up to 50 m and deeper, using a UWITEC corer (Fischer and Rösch 2017). The Bergsee is different from the other lakes in the Southern Black forest, as it is situated at low elevation at the southernmost fringes of the Black Forest, close to the Hochrhein Valley, and was not glaciated during the Weichselian, but during the Saalian (Becker et al. 2006); it should, therefore, contain deposits of Eemian age from before the last glaciation. Because these sediments are buried under >30 m of Weichselian and Holocene material, their recovery turned out to be difficult and has not yet been successful (Kämpf et al. 2017).

More recent work in the Bodensee region started in the 1990s, when Jürgen Schneider and his team from Göttingen University began their investigations at Steißlinger See with its annually laminated sediments (Kerig and Lechterbeck 2004; Lechterbeck 2001; Wolf 1994; Eusterhues 2000). In 2003, a 14 m littoral core of Bodensee-Obersee near the Isle of Mainau was collected from a boat with a Merkt/Streif piston corer. Research continued, with the help of Josef Merkt and Willi Tanner of the University of Bern, and cores were collected from the centers of small glacial lakes in the western Lake Constance region including Mindelsee, Buchenseen, Böhringer See, and Litzelsee (Wick and Rösch 2006; Rösch 2013; Rösch et al. 2014a; Rösch and Lechterbeck 2016; Rösch and Wick 2018, 2019; Lechterbeck and Rösch 2020).

Material and Methods

From the beginning, it was clear that work with pollen records should be linked to sediment facies as well. Therefore, in addition to necessary radiometric dating, sedimentological and geochemical questions and methods were always considered and included by building an extensive network of researchers from different disciplines and institutions. Most lake sediments discussed here are continuous postglacial records of the last 16 millennia, covering the Late Weichselian and Holocene (Table 1 and Fig. 1). The sediment types differ between regions and lakes.

Each core was obtained in sections of either 1 or 2 m length. Usually an overlapping parallel core was taken about 1 m away to bridge the eventually disturbed or missing section ends, and to provide a continuous undisturbed sediment sequence.

Using sedimentological methods such as x-ray diffraction and loss-on-ignition (LOI), a master (composite) core was constructed providing a complete sequence (Berglund 1986). Sampling for pollen and other proxies was performed with L-shaped metal profiles that cover the entire core length. These were subsequently continuously subsampled in intervals of 1 cm or less. For each pollen sample, a minimum sum of 1000 arboreal pollen grains were counted. A high pollen count is necessary to ensure statistical significance even for less common pollen types (Birks

2005). Dating was done by ^{14}C AMS on terrestrial macrofossils that were picked after the sediment was wet-sieved. Age models were constructed using Oxcal (Bronk Ramsey 2009), which interpolates between the samples. The larger the distance between radiocarbon dates, the broader the confidence intervals of age are in the space between the dates.

Results: Past Environments and Human Impact in Southwest Germany

Five out of the 31 lakes (Table 1) contained annually laminated sediments deposited over several millennia (Table 2). These annually laminated sediments cover different sections of the Holocene, beginning in the Boreal or Atlantic and ending in the Subatlantic at the latest. An exception is Stadtsee of Bad Waldsee, where lamination continues in the medieval period and modern times.

With respect to vegetation and landscape history as recorded in lake sediments, two stages are differentiated: (1) the Late Weichselian together with the Early Holocene, and (2) the Middle to Late Holocene. The first corresponds to the first three phases of a glacial-interglacial cycle (Iversen 1958; Andersen 1966) and is very uniform and synchronous across the entire area, as far as the chronology based on many radiocarbon dates tells us. The reforestation starts with a dwarf shrub stage characterized by *Salix*, *Betula nana* and *Ephedra*, followed by a shrub stage with *Juniperus* and *Hippophaë* prior to the development of a *Betula* forest. Later *Betula* was partly replaced by *Pinus*. In contrast to northern central Europe, evidence for deforestation during the Younger Dryas is weak. The Holocene starts with a second expansion of *Betula*, followed by *Corylus*, *Ulmus*, *Quercus*, *Tilia*, *Acer* and, somewhat later, *Fraxinus*. Here, key differences between the considered landscapes concerning the importance of *Ulmus*, *Quercus*, *Tilia*, and *Fraxinus* (Table 3) become apparent: *Quercus* is most abundant in the northern Black Forest, *Ulmus* and *Fraxinus* are abundant in Oberschwaben/Allgäu but less abundant in the Black Forest. *Tilia* is abundant at Lake Constance, but also in the southern Black Forest.

Table 2 Lakes with annually laminated sediments

No.	Lake	Annually laminated from (B.P., ca.)	Annually laminated to (B.P., ca.)	Reference
1	Steisslinger See	10,000	1000	Lechterbeck (2001)
2	Husemer See	7500	2200	Rösch (unpubl.)
3	Schleinsee	9500	5500	Müller (1962), Geyh et al. (1971)
4	Degersee	5500	2600	Kleinmann et al. (2015), Merkt and Müller (1978)
5	Stadtsee/Bad Waldsee	4400	100	Fischer et al. (2010)

Table 3 Maximum values of important arboreals (*Corylus, Quercus, Ulmus, Tilia, Fraxinus*) in pollen profiles

Site	Corylus	Quercus	Ulmus	Tilia	Fraxinus
Hornstaad	60	20	20	10	10
Mainau-Obere Güll	50	20	25	12	8
Mindelsee	50	25	22	10	18
Buchensee-Südost	50	30	18	8	12
Böhringer See	30	32	18	5	5
Litzelsee	57	35	12	6	6
Steisslinger See	50	30	18	6	15
Nussbaumer See	80	20	20	8	13
Husemer See	55	20	35	12	6
Degersee	65	12	30	10	15
Zeller See/Bodensee	25	30	10	15	12
Wangen/Bodensee	65	20	20	15	8
Gnadensee	60	25	15	10	12
Königseggsee	55	25	20	5	10
Großer Ursee	60	<20	30	10	20
Herrenwieser See	60	35	10	5	10
Glaswaldsee	85	50	13	7	16
Mummelsee	47	24	25	13	7
Schurmsee	75	43	11	5	10
Wilder See am Ruhestein	65	36	12	5	10
Huzenbacher See	72	42	10	6	8
Buhlbachsee	60	29	8	6	8
Ellbachsee	57	31	10	5	10
Feldsee	55	20	15	7	12
Titisee	55	25	10	9	10
Schluchsee	60	25	10	10	12
Bergsee	60	30	12	10	10
Aalkistensee/Maulbronn	30	50	7	<3	12
Average Bodensee	**54**	**25**	**20**	**10**	**11**
Average Oberschwaben/Allgäu	**58**	**22**	**25**	**8**	**15**
Average Nordschwarzwald	**58**	**32**	**11**	**6**	**9**
Average Südschwarzwald	**58**	**25**	**12**	**9**	**11**

These differences are a function of the soil quality and the elevation (Ellenberg 1978; Lang 1990, 1994).

With the expansion of the shade-tolerant trees *Fagus sylvatica*, *Abies alba,* and *Picea abies,* the regional differences among the study areas became more pronounced, not only in terms of abundance of tree species, but also in terms of timing of their expansion (Figs. 3 and 4). At Lake Constance, in the western as well as in the eastern parts including the southwestern Allgäu, *Fagus* spread first, around 7000 B.P. A clearly pronounced expansion of *Abies* and *Picea* is missing. In Oberschwaben, the *Fagus* expansion occurred 1000 years later, at about 6000 BP,

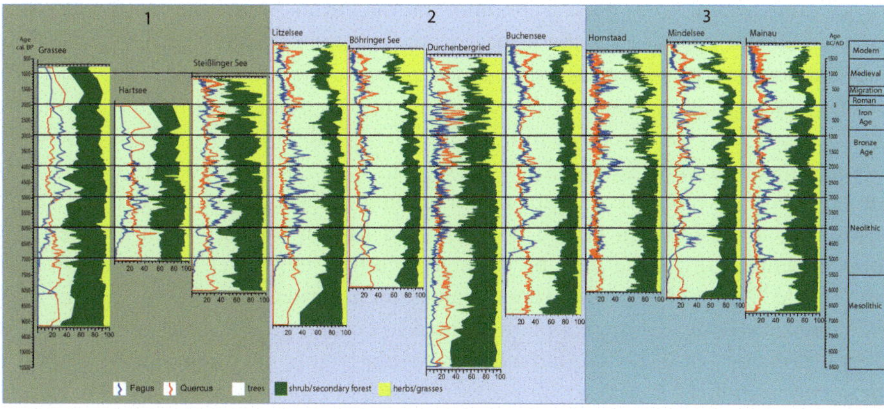

Fig. 3 Main pollen diagrams of the western Bodensee region. Included are two diagrams from two mires in the nearby Hegau in the west of Bodensee, Grassee, and Hartsee

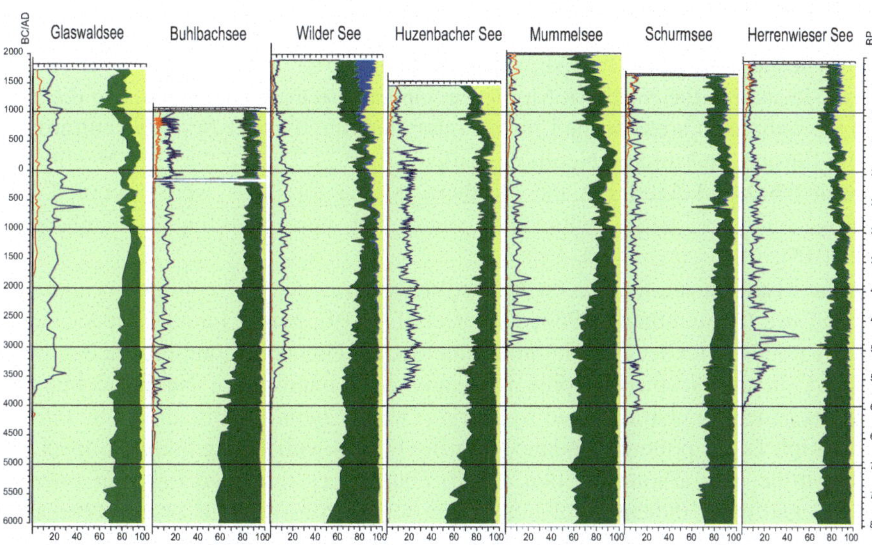

Fig. 4 Main pollen diagrams of northern Black Forest (Schwarzwald).
Areas: light green: trees, dark green: shrubs, blue: dwarf shrubs, yellow: herbs and grasses; lines: blue: *Abies alba*, red: *Picea abies*

coinciding approximately with the beginning of the final *Ulmus* decline. In the Allgäu region *Fagus* increased gradually after 7000 BP, more or less together with *Abies*, which was less abundant. *Picea* increased as well, but weaker and slower. In the Black Forest, the expansion of *Abies* was stronger and faster than the *Fagus* expansion and started around 6000 BP, coinciding with the final *Ulmus* decline. *Fagus* was more abundant at Bodensee and in Oberschwaben/Allgäu than in the

Black Forest, where *Abies* played an important role, even dominating until 4500 BP, when significant traces of human impact were documented (Rösch 2015).

In the northern Black Forest *Picea* did not exceed 5% prior to the late Medieval period. A natural occurrence of *Picea* in this region is doubtful. In the southern Black Forest *Picea* was already a minor component of the forest in higher elevations during prehistoric times (Table 3). In all regions *Carpinus* arrived late (during the Iron Age) and played only a minor role, reaching highest percentages (up to10%) during the early Medieval period with reduced human impact, when people had stopped maintaining the forest by regularly cutting the lower tree layer of the "Mittelwald." The "Mittelwald" forest management strategy was known and in use since the early Bronze Age; coppicing aimed to sustain growth of a forest with an upper tree canopy of old oaks, and a lower, regularly coppiced small tree and shrub canopy consisting of *Carpinus*, *Fagus*, *Corylus* and others (Bärnthol 2003; Hasel and Schwartz 2002).

In the lowlands of this region, Neolithic groups started practicing agriculture since the second half of the eighth millennium BP. This caused deforestation and the development of a cultural landscape. Between 7500 and 6300 BP, Neolithic cultures were restricted to landscapes with a warm and dry climate and fertile soils (Preuss 2008). The lakes discussed here show little human impact during this period because they are located in the Northern Alpine foreland and in the Black Forest, far distant from the settlement areas of the Linear Pottery culture, the first Neolithic culture in middle Europe, and middle Neolithic settlement areas. During the Late Neolithic, between 6300 and 4300 BP, the Northern Alpine forelands were occupied by Neolithic cultures, partly with lake shore and mire dwellings (Mainberger et al. 2015).

In the Black Forest, there is no archaeological evidence of Neolithic occupation and land use (Schlichtherle 1988; Knopf et al. 2019). Minor human impact indicated by pollen evidence is ambiguous and often explained as long-distance pollen transport. In the Northern Alpine forelands, high intensity of Neolithic human impact is generally characterized by *Corylus* and/or *Betula* dominance, *low intensity* by high *Fagus* pollen abundance. During phases with strong human impact in the Neolithic, non-arboreal pollen (NAP) abundances remained low, but microcharcoal and other indicators of human impact, such as ruderals or even cereals, may occur. During the Neolithic slash-and-burn agriculture (Rösch et al. 2017; Rösch et al. 2014a), new plots opened up, were cultivated for a short time and then left again for reforestation, and cultivation plots were opened elsewhere. This would explain the similarity of the overall pattern as well as the lack of synchrony between the single *Fagus* and *Corylus*/*Betula* maximum abundances in different pollen records. Furthermore, human activity and agriculture is indicated by the presence of cereal pollen grains and some apophytes such as *Artemisia* and *Plantago lanceolata*. Alternative hypotheses to explain this kind of human impact consider small-scale fields and moderate modification of the forest structure (Jacomet et al. 2017). Lake sediments also provide an opportunity to evaluate the role of paleofire by closer examination and statistical evaluation of the macrocharcoal records (Whitlock and Anderson 2003).

Since the Bronze Age extensive plough (ard) cultivation was practiced (Tegtmeier 1993), resulting in permanently open arable fields without trees and shrubs or their stubs and roots. People managed the land with the application of a short fallow systems such as the so-called field-grass cultivation. These permanently open areas, together with further reduction of the forest canopy by animal browsing, resulted in decreasing arboreal and increasing non-arboreal pollen sums. Because some principles of land use such as extensive ard/mouldboard plough cultivation, forest grazing, short fallow systems, remained unchanged from the Bronze Age until the nineteenth century A.D., the arboreal/non-arboreal pollen relation can be considered as indicator for the magnitude of human impact. In all profiles and regions, the human impact during Medieval and early Modern Times seems to have been stronger than during Bronze and Iron Age (Table 4). There are clear differences in the magnitude of human impact between the different regions. Human impact was minor in the Black Forest, higher in Northern Alpine foreland, and highest in the western Bodensee region. The most prominent human impact was observed in Aalkistensee, with more than 50% NAP during the Early Bronze Age, indicating a landscape devoid of forests.

During the Medieval and Modern Time period, the differences between the regions remained, and in some cases they were even more pronounced. In the areas of the western Bodensee and the Kraichgau, a maximum NAP average of 45% and 65% (Aalkistensee) indicates open landscapes without forest, whereas in Oberschwaben/Allgäu (maximum of 30% NAP) and northern Black Forest (maximum of 22% NAP), obviously some woodland survived. These differences were caused by differences in landscape resources such as soil conditions, climate, and infrastructure. The medieval pattern has been maintained until today when modern population density was reached.

The degree of deforestation indicated by the non-arboreal/arboreal pollen ratio (NAP/AP) is not constant for centuries or even millennia, but fluctuates considerably (Figs. 3 and 4). Without human impact Central Europe would be totally covered by woodland but agriculture and human settlements caused deforestation. A spontaneous reforestation, together with a decrease of anthropogenic indicators, is a result of a decrease in land use pressure (Behre 1986). In its early stages, reforestation is clearly indicated by an increase of *Betula* pollen.

Field abandonment phases in southern Germany are chronologically not evenly distributed (Rieckhoff and Rösch 2019), and there are phases with high and phases with low land use pressure (Fig. 4). From the Late Bronze Age to the middle Latène period, land use pressure was variable. During the late Latène period, land use pressure was strong in the beginning, then declined to reach a prehistoric minimum, and increased again during the Roman period. During the first centuries BCE and CE, respectively, in archaeological terms the last century of the late pre-Roman Iron Age, land use activities decreased to nearly zero for several decades. After a phase with strong human impact during the Roman period, human impact decreased again from the third century CE onwards and remained very low during the migration and Merovingian period, but increased strongly during the early Medieval period, from the seventh/eighth century CE onwards. During the late Medieval and early Modern

Table 4 Intensity of human impact in the pollen profiles shown by the Cereal sum and *Plantago lanceolata*

	Prehistoric				Medieval			
Site	NAP	Age	Cereal sum	*Plantago lanceolata*	NAP	Age	Cereal sum	*Plantago lanceolata*
Hornstaad	40	−2	5	10	50	9	8	10
Mainau-Obere Güll	40	−3	<2	<3	40	11	<2	<3
Mindelsee	20	−3	<2	6/−15	60	19	<2	5
Buchensee-Südost	15	−3	<2	5/−18	40	19	<4	5
Böhringer See	25	−2	1	3/−15	50	19	5	6/13
Litzelsee	17	−6	<1	5/−5	35	19	<2	5/8
Steisslinger See	35	−4	<1	6/−17	50	12	<4	5/15
Nussbaumer See	25	−3	1	3	60	18	6	10
Husemer See	40	−17	<4	<3	45	18	<3	<3
Degersee	15		<1	<3	40		<2	<3
Wangen/Bodensee	12		<2	<3	25		5	8
Gnadensee	15		<1	<2	35		<2	<3
Königseggsee	20	−8	<1	<3	20	12	<2	<3
Stadtsee/Bad Waldsee	25	−3	<3	<5				
Großer Ursee	20	−2	<1	<3	35	19	<3	<4
Herrenwieser See	18	−2	<1	3	22	10	<1	<3
Glaswaldsee	18	−5	<1	2	25	12	<1	3
Mummelsee	18	−4	<1	2	30	18	<1	<3
Schurmsee	20	−6	<1	1	20	15	<1	<3
Wilder See am Ruhestein	15	−1	<1		20	11	1	2
Huzenbacher See	12	−1	<1	1	25	12	1	<3
Buhlbachsee	8	−2	<1	1	13	6	<1	2
Schluchsee	20	−2	<1	<2	30	8	<1	<2
Bergsee	25	2	<1	<2	25	8	1	<2
Aalkistensee/ Maulbronn	55	−20	<2	<5	65	15	7	<5
Average Bodensee	**25**		**1.5**	**4**	**45**		**2.5**	**5**
Average Oberschwaben/ Allgäu	**22**		**1**	**3**	**30**		**2**	**3**
Average Nordschwarzwald	**16**		**<1**	**<2**	**22**		**<1**	**2**
Average Südschwarzwald	**23**		**<1**	**<2**	**28**		**<1**	**<2**

Times, a last and rather weak initial reforestation phase is indicated by shortly increasing *Betula* pollen percentages, which could be explained by climatic, demographic and economic crises during the 14th and 17th centuries A.D.

Lakes, Sediments, Climate, and Environment

Holocene sediment accumulation rates differed considerably in southwestern Germany. Sediment cores collected from the littoral areas of Lake Constance were up to 14 m long. The total length of sediment cores taken from the center of other lakes in the Northern Alpine foreland ranged between less than 4 (Buchensee) and 12 m (Großer Ursee). In the Black Forest, the range was <2 m (Glaswaldsee) to 10 m (Ellbachsee). The exceptionally long 33 m sequence from Bergsee dates back to about 44 ka BP (Becker et al. 2006). Sedimentation rates in the Black Forest lakes were lower than in the lakes of Northern Alpine foreland because of smaller catchment areas, poorer soils, and lack of biogenic chalk precipitation (Table 5).

Today, most of the studied lakes are not in a natural state, because the water level has been modified by humans and the trophic state was changed by nutrient input from the catchment. In the northern Black Forest, modern dams were constructed in order to provide enough water to transport timber downstream. The lake level of Schluchsee was raised by 40 m in 1930 to produce electricity, thereby enlarging not only the depth, but also the lake surface, from about 100 to 514 ha. Aalkistensee was thought to be an artificial lake constructed by the medieval monastery of Maulbronn, but as the studies show, the monks enlarged an already existing natural lake that was present during the Neolithic/Bronze Age transition (Rösch et al. 2017).

Sediment types also show a regional differentiation; sediment cores from Lake Constance consist of lake marl, mineral components, and rather low organic matter contents (Wessels 1995, 1998; Schwalb et al. 2013). The advantage of Lake Constance, being a large lake, is that its sediments record larger scale environmental information rather than small lakes that may be affected by catchment-specific characteristics (Schwalb et al. 2013). Therefore, Sediments from large lakes such as Lake Constance may reflect a clearer climate signature than smaller lakes, where global signatures are often overprinted by local processes, especially those triggered by anthropogenic activities. Oxygen and carbon isotopes from benthic ostracode species archived in a sediment core from Lake Constance, covering the last approximately 16 cal ka, for example, track the general North Atlantic and European temperature history since deglaciation. Schwalb et al. (2013) suggested that around 9 cal ka, the northward retreat of the Northern Hemisphere Ice Sheets, and consequently the polar front, left the Alpine region affected by a more maritime climate. By about 7 cal ka, the North Atlantic region had probably warmed sufficiently to

Table 5 Archaeological and historical periods in southwestern Germany

Cultures/dynasties	Cultural chronology	Millennia B.C./A.D.	B.C./A.D.	B.P.
			2000	0
			1900	−100
			1800	−200
	Early Modern Times		1700	−300
			1600	−400
		2nd	1500	−500
	Late Medieval		1400	−600
			1300	−700
Staufer	High Medieval		1200	−800
Salier			1100	−900
			1000	−1000
Ottonen	Early Medieval		900	−1100
Karolinger			800	−1200
	Merovingian		700	−1300
Merowinger			600	−1400
	Migration period	1st	500	−1500
Alamannen			400	−1600
			300	−1700
Römer	Roman period		200	−1800
			100	−1900
		A.D.	0	−2000
Latene D		B.C.	−100	−2100
Latene C	Late Pre-Roman Iron Age		−200	−2200
Latene B			−300	−2300
Latene A			−400	−2400
Hallstatt D	Early Pre-Roman Iron Age	1st	−500	−2500
			−600	−2600
Hallstatt C			−700	−2700
			−800	−2800
Urnfield culture (Late Bronze Age)	B		−900	−2900
	Hallstatt		−1000	−3000
	A		−1100	−3100
	Bronze		−1200	−3200
Burial Mount Culture	D		−1300	−3300
	C		−1400	−3400
	B	2nd	−1500	−3500
			−1600	−3600
Arbon	Bronze A2		−1700	−3700
			−1800	−3800

(continued)

Table 5 (continued)

Cultures/dynasties	Cultural chronology	Millennia B.C./A.D.	B.C./A.D.	B.P.
Singen	Earliest Bronze Age		–1900	–3900
			–2000	–4000
			–2100	–4100
Bell Beaker Culture			–2200	–4200
			–2300	–4300
	Final Neolithic		–2400	–4400
Corded Ware Culture		3rd	–2500	–4500
			–2600	–4600
			–2700	–4700
			–2800	–4800
Goldberg III Culture			–2900	–4900
	Late Neolithic		–3000	–5000
			–3100	–5100
Horgen Culture			–3200	–5200
			–3300	–5300
		4th	–3400	–5400
Michelsberg Pfyn/ Altheim			–3500	–5500
			–3600	–5600
	Younger		–3700	–5700
			–3800	–5800
Schussenried/ Hornstaad			–3900	–5900
			–4000	–6000
	Neolithic		–4100	–6100
Bischheim/ Schwieber-dingen/ Aichbühl			–4200	–6200
			–4300	–6300
Rössen			–4400	–6400
Stichbandkeramik	Middle Neolithic	5th	–4500	–6500
			–4600	–6600
Großgartach			–4700	–6700
			–4800	–6800
Hinkelstein			–4900	–6900
			–5000	–7000
Linear-bandkeramik			–5100	–7100
	Old Neolithic		–5200	–7200
		6th	–5300	–7300
			–5400	–7400
			–5500	–7500
			–5600	–7600

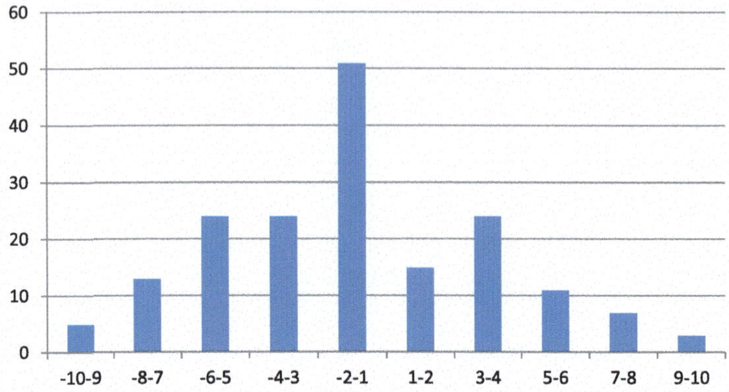

Fig. 5 Frequency distribution of reforestation phases in southern Central Europe from 1000 BC to 1000 AD in intervals of two centuries

increase precipitation in Central Europe and consequently increased detrital-clastic runoff to Lake Constance (Schwalb et al. 2013) (Fig. 5).

Sediment cores from small lakes in the Northern Alpine foreland consist of organic-rich mud with a high proportion of carbonate. The content in minerogenic components is generally low but variable and sometimes increases in correlation with human impact. Sedimentation in the cirque lakes of the northern Black Forest started with organic-rich mud on top of glacial clays, indicating oligotrophic conditions. The decrease and disappearance of *Isoetes lacustris* and *Isoetes echinospora,* plants related to the taxonomic group of lycopods that grow in the shallow water of the littoral zone of oligotrophic lakes suggest the onset of bog growth and soil acidification associated with raw humus formation in the catchment after 5 cal ka that caused a shift from an oligotrophic to a dystrophic state. The dark waters of these dystrophic lakes were likely too opaque to support the growth of submerged macrophytes. Because of the crystalline bedrock, lakes in the southern Black Forest also produced sediments that lack carbonate or have very low carbonate content. However, the bedrock here is less acidic than in the northern Black Forest. Therefore, the water remained clear and free of humic acids; there was no sedimentation change from mud to dy, and *Isoetes* species did not disappear (Jacob et al. 2009).

The interactions between sedimentation and anthropogenic events within the catchment causing vegetation changes and the climatic, biotic, or anthropogenic causes of environmental change are topics we have not yet fully understood. For example, the main increase of the mercury values in lake sediments from the northern Black Forest coincides with the main vegetation change from mixed oak forest to *Abies* forest, but it is not clear if this is based on coincidence or causality (Rydberg et al. 2015; Schütze et al. 2018). The pollen data are furthermore useful for paleodemographic, climatic, and environmental modelling.

Conclusions

As a result of nearly 40 years of paleo-environmental work on southwestern German lakes, our knowledge of vegetation and landscape changes under human influence has considerably improved, providing insights into changes of woodland development, events and human activities in the catchment, and their effects on lake sedimentation processes, as well as qualitative and quantitative changes of land use. Differences in bedrock geology, soils, and climate are important, but increasing human impacts, detectable since the Neolithic, were strongly responsible for shaping vegetation and landscape. Whereas the precise onset of human impact during the Early Neolithic is hardly detectable in the lakes discussed, because it took place far away in landscapes lacking lakes, the anthropogenic transformation of the landscape during the late Neolithic in the Northern Alpine foreland is clearly visible as a replacement of primary forest by coppiced forest. The extended ard plough and field-grass economy since the Bronze Age is even better recognized by an increase of non-arboreal pollen.

Southwestern Germany is regarded as one of the best-investigated areas of Europe regarding vegetation and landscape history during the Holocene. Most of the data sets from this region are available in the European Pollen Database (http://www.europeanpollendatabase.net/index.php). These data are often used to test new ecological evaluation methods, or to reconstruct land cover and climate of the past. Archaeologists also use the proxy data to constrain variables in paleo-demographic models (Shennan 2018), which has inspired new ideas about Neolithic subsistence strategies (Schier 2009, 2017). As these methods of synthesis were applied to pollen data and radiocarbon dates from the Lake Constance region, it was for the first time possible to directly link the vegetation record with the archaeological record (Lechterbeck et al. 2014). Research shows that even rare taxa in the pollen records, mostly overlooked or even neglected in the past, can be of great paleo-ecological importance and thus critical for refining the vegetation history in detail (Rösch and Lechterbeck 2016; Rösch 2016, 2017).

Future paleoecological research on lakes in southwestern Germany and surrounding regions is planned to include analysis of aquatic bioindicators and persistent organic pollutants such as polycyclic aromatic hydrocarbons from combustion processes, and will calibrate the proxy records with historic documents back to Medieval Time (Hollert et al. 2018). Such a study will start soon at Stadtsee Bad Waldsee, where annually laminated lake sediments exist until Modern Times. The aim is to identify the effects of population growth or decrease, farming intensity, economic production, trade activity, climate change and catastrophic events such as fire and floods on past surface water quality. These integrated investigations could then provide detailed information about the long-term response of surface waters to changing emissions and socio-economic development, as well as contribute to understanding the quasi-natural reference conditions of waters.

Acknowledgments The research was funded by DFG grants PL 95/39-2, PL95/48-1, RO 2282/4, RO 2282/6, RO 2282/8 and RO2282/14. The team of the laboratory of Archaeobotany Hemmenhofen was involved with different kinds of technical work such as coring, sampling and sample preparation, evaluation of data. Brigitta Ammann and Willi Tanner, University of Bern, provided technical support during field work. Critical and very constructive comments to this paper were made by Josef Merkt, who always supported his colleagues and early career scientists by collecting cores as well as by providing methodological and scientific advice. We especially appreciated the careful revisions by an anonymous editor and Michael Rosen that significantly improved the manuscript.

References

Ammann, B. (1989). Late-quaternary palynology at Lobsigensee-regional vegetation history and local lake development. *Diss Bot, 137*(Vaduz).

Andersen, S. T. (1966). Interglacial vegetational succession and lake development in Denmark. *Palaeobotanist, Ranchi, 15*, 117–127.

Bärnthol R (2003) Nieder- und Mittelwald in Franken: Waldwirtschaftsformen aus dem Mittelalter (= Schriften und Kataloge des Fränkischen Freilandmuseums). Fränkisches Freilandmuseum, Bad Windsheim.

Becker, A., Ammann, B., Anselmetti, F., Hirt, A. M., Magny, M., Millet, L., Rachoud, A. M., Sampietro, G., & Wüthrich, C. (2006). Palaeoenvironmental studies on Lake Bergsee, Black Forest, Germany. *Neues Jahrbuch für Geologie und Paläontologie, 240*, 405–445.

Behre, K. E. (1986). *Anthropogenic indicators in pollen diagrams*. Rotterdam/Boston: Balkema.

Berglund, B. E. (1986). *Handbook of holocene palaeoecology and palaeohydrology*. Chichester: Wiley.

Birks, H. J. B. (2005). Fifty years of quaternary pollen analysis in Fennoscandia 1954–2004. *Grana, 44*, 1–22.

Bronk Ramsey, C. (2009). Bayesian analysis of radiocarbon dates. *Radiocarbon, 51*(1), 337–360.

Clark, J. S., Merkt, J., & Müller, H. (1989). Post-glacial fire, vegetation, and human history on the northern alpine forelands, South-Western Germany. *Journal of Ecology, 77*, 897–925.

Damminger, F., & Wieland, G. (2003). Ausgrabungen auf dem Rudersberg, Stadt Calw. *Archäologische Ausgrabungen in Baden-Württemberg, 2002*(Stuttgart 2003), 92–95.

Ellenberg, H. (1978). *Vegetation Mitteleuropas mit den Alpen in ökologischer Sicht*. Stuttgart: Ulmer.

Eusterhues, K. (2000). Die Sedimente des Steißlinger Sees (Hegau, Süddeutschland) – Ein Archiv für zeitlich hoch aufgelöste geochemische Untersuchungen zu Umweltveränderungen im Holozän. https://ediss.uni-goettingen.de/bitstream/handle/11858/00-1735-0000-0006-B353-3/eusterhues.pdf?sequence=1 vom 22.01.2018.

Feger, K. H. (1986). Biogeochemische Untersuchungen an Gewässern im Schwarzwald unter besonderer Berücksichtigung atmogener Stoffeinträge (Dissertation). – Freiburger Bodenkundl. Abh. 17.

Feger, K. H. (1993). Influence of soil development and management practices on freshwater acidification in Central European forest ecosystems. In S. CEW & R. W. Wright (Eds.), *Acidification of freshwater ecosystems: Implications for the future* (pp. 67–82). Wiley.

Fischer, E., Rösch, M., Sillmann, M., Ehrmann, O., Liese-Kleiber, H., Voigt, R., Stobbe, A., Kalis, A. E. J., Stephan, E., Schatz, K., & Posluschny, A. (2010). Landnutzung im Umkreis der Zentralorte Hohenasperg, Heuneburg und Ipf. Archäobotanische und archäozoologische Untersuchungen und Modellberechnungen zum Ertragspotential von Ackerbau und Viehhaltung. In D. Krausse (Ed.), *"Fürstensitze" und Zentralorte der frühen Kelten, Forschungen und Berichte zur Vor- und Frühgeschichte in Baden-Württemberg 120, Teil 2* (pp. 195–265).

Fischer, E., & Rösch, M. (2017). Schwarzwald: Urwald oder Siedlungsraum? *Archäologie in Deutschland, 1*(/2017), 75.

Gaillard, M. J., Berglund, B. E., Birks, H. J. B., Edwards, K. J., & Bittmann, F. (2018). "Think horizontally, act vertically": The centenary (1916–2016) of pollen analysis and the legacy of Lennart von Post. *Vegetation History and Archaeobotany, 27*, 267–269. https://doi.org/10.1007/s00334-017-0656-5 .

Gassmann, G., & Wieland, G. (2005). Frühkeltische Eisenproduktion im Nordschwarzwald – Rennöfen des 5. Jahrhunderts v. Chr. bei Neuenbürg-Waldrennach, Enzkreis. *Archäologische Ausgrabungen in Baden-Württemberg, 2004*(Stuttgart 2005), 102–107.

Gassmann, G., Wieland, G., & Rösch, M. (2006). Das Neuenbürger Erzrevier im Nordschwarzwald als Wirtschaftsraum während der Späthallstatt- und Frühlatènezeit. *Germania, 84*(2), 73–306.

Geyh, M. A., Merkt, H., & Müller, H. (1971). Sediment-, Pollen- und Isotopenanalysen an jahreszeitlich geschichteten Ablagerungen im zentralen Teil des Schleinsees. *Archiv für Hydrobiologie, 69*(3), 366–399.

Geyh, M. A., Merkt, H., Müller, H., & Streif, H. (1974). Reconstructions paleoclimatiques et paléoecologiques à partir de l'étude des sediments lacustres de l'Allemagne méridionale. *Société hydrotechnique de France, compte rendu des treizièmes Journées de l'Hydraulique (Paris 1974)*, 1–7.

Giovanoli, F., Hofmann, W., Neukirch, S., Niessen, F., Ostendorp, W., Schmitz, W., Schröder, G., & Sturm, M. (1990). Beiträge zur Landschafts- und Siedlungsgeschichte am Bodensee-Untersee: Paläolimnologische Untersuchungen. *Siedlungsarchäologie im Alpenvorland, Berichte der römisch-Germanischen Kommission, 71*, 245–308.

Haas, J. N. & Hadorn, P. (1998). Die Vegetations- und ulturlandschaftsgeschichte des Seebachtals von der Mittelsteinzeit bis zum Frühmittelalter anhand von Pollenanalysen. In: A. Hasenfratz & M. Schnyder: Das Seebachtal – Eine Archäologische und paläoökologische Bestandesaufnahme. Forschungen im Seebachtal 1. Archäologie im Thurgau 4: 221–255.

Hanisch, S., Wessels, M., Niessen, F., & Schwalb, A. (2009). Late quaternary lake response toclimate change and anthropogenic impact: Biomarker evidence from Lake Constance sediments. *Journal of Paleolimnology, 41*, 393–406.

Hasel, K., & Schwartz, E. (2002). *Forstgeschichte. Ein Grundriss für Studium und Praxis.* Remagen: Kessel.

Hinderer, M. (2001). Late Quaternary denudation of the Alps, valley and lake fillings and modern river loads. *Geodinamica Acta, 14*, 231–263. Elsevier.

Hinderer, M., Jüttner, I., Winkler, R., Steinberg, C. E. W., & Kettrup, A. (1998). Comparing trends in lake acidification using hydrochemical modelling and paleolimnology: The case of the Herrenwieser See, Black Forest, Germany, 5 Abb. *The Science of the Total Environment, 218*, 113–121. Elsevier.

Hollert, H., Crawford, S. E., Brack, W., Brinkmann, M., Fischer, E., Hartmann, K., Keiter, S., Ottermanns, R., Ouellet, J. D., Rinke, K., Rösch, M., Roß-Nickoll, M., Schäffer, A., Schüth, C., Schulze, T., Schwarz, A., Seiler, T. B., Wessels, M., Hinderer, M., & Schwalb, A. (2018). Looking back – Looking forward: A novel multi-time slice weight-of-evidence approach for defining reference conditions to assess the impact of human activities on lake systems. *Science of the Total Environment, 626*, 1036–1046.

Hölzer, A. & Hölzer, A. (1987). Paläoökologische Mooruntersuchungen an der Hornisgrinde im Nordschwarzwald. *Carolinea 45*, 43–50

Homann, M., Merkt, J., & Müller, H. (1990). Über Alter und Entwicklung des Königseggsees bei Hoßkirch (Oberschwaben). – Jh. Geol. *Landesamt Baden-Württemberg, 32*, 247–254. 2 Abb.; Freiburg.

Iversen, J. (1958). The bearing of glacial and interglacial epochs on the formation and extinction of plant taxa. *Uppsala Univ. Årsskr.*, 210–215.

Jacob, F., Feger, K. H., Klinger, T., & Rösch, M. (2009). Akkumulation von amorphem Silizium in holozänen Seesedimenten des Herrenwieser Sees (Nordschwarzwald). Berichte der Deutschen Bodenkundlichen Gesellschaft, Jahrestagung der DBG, Bonn, 5. - 13. Sept. 2009.

Jacomet, S., Ebersbach, R., Akeret, Ö., Antolín, F., Baum, T., Bogaard, A., Brombacher, C., Bleicher, N. K., Heitz-Weniger, A., Hüster-Plogmann, H., Gross, E., Kühn, M., Rentzel, P., Steiner, B. L., Wick, L., & Schibler, J. M. (2017). On-site data cast doubts on the hypothesis of shifting cultivation in the late Neolithic (c. 4300-2400 cal. BC): Landscape management as an alternative paradigm. *The Holocene, 26*(11), 1858–1874.

Kämpf, L., Wick, L., Rius, D., Duprat-Qualid, F., Millet, L., & Feger, K.-H. (2017). Spuren menschlicher Landnutzung in Sedimenten des Bergsees (Südschwarzwald). In J. Lechterbeck, E. Fischer, & Kontrapunkte (Eds.), *Festschrift für Manfred Rösch, Universitätsforschungen zur prähistorischen Archäologie* (Vol. 300, pp. 41–50).

Kerig, T., & Lechterbeck, J. (2004). Laminated sediments, human impact, and a multivariate approach: A case study in linking palynology and archaeology (Steisslingen, Southwest Germany). *Quaternary International, 113*, 19–39. https://doi.org/10.1016/S1040-6182(03)00078-8 .

Kleinmann, A., & Rammelmair, D. (2003). Hochauflösende EDXRF-Scans von mittelholozänen gewarvten Sedimenten aus dem Schleinsee und Belauer See. *Terra nostra, 2003/6*, 245–249. Berlin.

Kleinmann, A., Merkt, J., & Müller, H. (2015). Sedimente des Degersees: Ein Umweltarchiv – Sedimentologie und Palynologie. In M. Mainberger, J. Merkt, & A. Kleinmann (Eds.), *Pfahlbausiedlungen am Degersee – Archäologische und naturwissenschaftliche Untersuchungen. Berichte zu Ufer- und Moorsiedlungen Südwestdeutschlands VI, Materialhefte zur Archäologie in Baden-Württemberg* (Vol. 102 (Darmstadt), pp. 409–471).

Knopf, T., Bosch, S., Kämpf, L., Wagner, H., Fischer, E., Wick, L., Millet, L., Rius, D., Duprat-Qualid, F., Rösch, M., Feger, K. H., & Bräuning, A. (2016). Archäologische und naturwissenschaftlichen Untersuchungen zur Landnutzungsgeschichte des Südschwarzwaldes. *Archäologische Ausgrabungen in Baden-Württemberg, 2015*(Stuttgart), 50–55.

Knopf, T., Fischer, E., Kämpf, L., Wagner, H., Wick, L., Duprat-Oualid, F., Floss, H., Frey, T., Loy, A. K., Millet, L., Rius, D., Bräuning, A., Feger, K. H., Rösch, M. (2019). Zur Landnutzungsgeschichte des Südschwarzwalds – Archäologische und naturwissenschaftliche Untersuchungen. Fundberichte aus Baden-Württemberg 39, Landesamt für Denkmalpflege im Regierungspräsidium Stuttgart, Stuttgart. pp. 19–201.

Küster, H. (1988). *Vom Werden einer Kulturlandschaft. Quellen u. Forsch. z. prähistorischen und provinzialrömischen Archäologie* (p. 3). Weinheim: Acta humaniora. VCH.

Lang, G. (1985). *Swiss Lake and Mire Environments during the last 15 000 Years*. Dissertationes Botanicae 87, Vaduz.

Lang, G. (1990). *Die Vegetation des westlichen Bodenseegebietes*. Pflanzensoziologie 17, [2]Jena/Stuttgart/New York.

Lang, G. (1994). *Quartäre Vegetationsgeschichte Europas*. Jena/Stuttgart/New York: Fischer.

Lang, G. (2005) *Seen und Moore des Schwarzwaldes als Zeugen spätglazialen und holozänen Vegetationswandels*. Andrias 16, Karlsruhe.

Lechterbeck, J. (2001). "Human Impact" oder "Climatic Change"? Zur Vegetationsgeschichte des Spätglazialsund Holozäns in hochauflösenden Pollenanalysen laminierter Sedimente des Steißlinger Sees (Südwestdeutschland). Tübingen.

Lechterbeck, J., Edinborough, K., Kerig, T., Fyfe, R., Roberts, N., & Shennan, S. (2014). Is Neolithic land use correlated with demography? An evaluation of pollen-derived land cover and radiocarbon-inferred demographic change from Central Europe. *The Holocene, 24(10)*, 1297–1307. https://doi.org/10.1177/0959683614540952.

Lechterbeck J, Rösch M (2020) Böhringer See, western Lake Constance (Germany): an 8500 year record of vegetation change. Grana (online first) https://doi.org/10.1080/00173134.2020.1784265.

Liese-Kleiber, H. (1993). Pollenanalysen zur Geschichte der Siedlungslandschaft des Federsees vom Neolithikum bis ins ausgehende Mittelalter. *Dissertationes Botanicae, 196*(Berlin/ Stuttgart), 347–368.

Mainberger M, Merkt J, Kleinmann A (2015) Pfahlbausiedlungen am Degersee – Archäologische und naturwissenschaftliche Untersuchungen. Berichte zu Ufer- und Moorsiedlungen Südwestdeutschlands VI, Materialhefte zur Archäologie in Baden-Württemberg 102: 535 S.; Darmstadt.

Merkt, J. (1971). Zuverlässige Auszählungen von Jahresschichten in Seesedimenten mit Hilfe von Gross-Dünnschliffen. *Archiv für Hydrobiologie, 69*, 145–154.

Merkt, J., & Streif, H. (1970). Stechrohr-Bohrgeräte für limnische und marine Lockersedimente. *Geologisches Jahrbuch, 88*, 137–148.

Merkt, J., & Müller, H. (1978). Paläolimnologie des Schleinsees. In Geologisches Landesamt Baden-Württemberg (Ed.), *Erläuterungen zur Geologischen Karte von Baden-Württemberg (Stuttgart)* (pp. 29–31).

Merkt J, Müller H (1995): Laminated sediments from Neolithic to the Hallstatt period in South Germany. – PACT II, 1: 101–116, 8 figs.; Brussels.

Müller, G. (1966). Die Sedimentbildung im Bodensee. *Naturwiss, 53*, 237–247.

Müller, H. (1962). Pollenanalytische Untersuchung eines Quartärprofils durch die spät- und nacheiszeitlichen Ablagerungen des Schleinsees (Südwestdeutschland). *Geologisches Jahrbuch, 79*(Hannover), 493–526.

Müller H (1973): Anregungen zu paläolimnologischen Untersuchungen im Rahmen der Urgeschichtsforschung. – Informationsblätter zu Naturwissenschaften der Ur- und Frühgeschichte 4, Geologie 5: 1–15; Göttingen.

Preuss J (2008) Das Neolithikum in Mitteleuropa. Kulturen – Wirtschaft – Umwelt vom 6. bis 3. Jahrtausend v.u.Z. – Übersichten zum Stand der Forschung. [2]Beier und Beran, Weißbach.

Preusser, F., Graf, H. R., Keller, O., Krayss, E., & Schlüchter, C. (2011). Quaternary glaciation of northern Switzerland. *E&G Quaternary Science Journal, 60*, 282–305.

Rieckhoff, S., & Rösch, M. (2019). Ein keltischer exodus? Archäologisch-botanische Überlegungen zum Übergang Eisenzeit – Römische Kaiserzeit in Südwestdeutschland. In R. Karl & J. Leskovar (Eds.), *Interpretierte Eisenzeiten. Fallstudien, Methoden, Theorie. Tagungsbeiträge der 8. Linzer Gespräche zur interpretativen Eisenzeitarchäologie. Studien zur Kulturgeschichte von Oberösterreich 49, Linz* (pp. 57–87).

Rösch, M. (1983). *Geschichte der Nussbaumer Seen (Kt.Thurgau) und ihrer Umgebung seit dem Ausgang der letzten Eiszeit aufgrund quartärbotanischer, stratigraphischer und sedimentologischer Untersuchungen. Mitteilungen der Thurgauischen Naturforschenden Gesellschaft* (p. 45). Frauenfeld: Huber.

Rösch, M. (1985). Ein Pollenprofil aus dem Feuenried bei Überlingen am Ried: Stratigraphische und landschaftsgeschichtliche Bedeutung für das Holozän im Bodenseegebiet. In *Berichte zu Ufer- und Moorsiedlungen Südwestdeutschlands 2, Materialhefte zur Vor- und Frühgeschichte Baden-Württembergs 7 (Stuttgart)* (pp. 43–79).

Rösch, M. (1990). Vegetationsgeschichtliche Untersuchungen im Durchenbergried. In *Siedlungsarchäologie im Alpenvorland 2, Forschungen und Berichte zur Vor- und Frühgeschichte in Baden-Württemberg* (Vol. 37 (Stuttgart), pp. 9–56).

Rösch, M. (1991). Ein Pollenprofil aus dem Profundal der Radolfzeller Bucht (Bodensee-Untersee). Fundberichte aus Baden-Württemberg 16 (Stuttgart) 57–62.

Rösch, M. (1992). Human impact as registered in the pollen record: Some results from the western Lake Constance region, Southern Germany. *Vegetation History and Archaeobotany, 1*, 101–109.

Rösch, M. (1993). Prehistoric land use as recorded in a lake-shore core at Lake Constance. *Vegetation History and Archaeobotany, 2*, 213–232.

Rösch, M. (1995). Geschichte des Nussbaumersees aus botanisch-ökologischer Sicht. In A. Schläfli (Ed.), *Naturmonographie Die Nussbaumer Seen, Schriftenreihe der Kartause Ittingen 5/Mitteilungen der Thurgauischen Naturforschenden Gesellschaft 53* (pp. 43–59). Frauenfeld: Huber.

Rösch, M. (1997). Holocene sediment accumulation in the shallow water zone of Lake Constance. Archiv für Hydrobiologie. *Supplementum 107 (Monographic Studies), 4*(Stuttgart), 541–562.
Rösch, M. (2000). Long-term human impact as registered in an upland pollen profile from the southern Black Forest, South-Western Germany. *Vegetation History and Archaeobotany, 9*, 205–218.
Rösch, M. (2002). Ein Pollenprofil von Wangen, Hinterhorn, Gemeinde Öhningen, Kreis Konstanz. *Fundberichte aus Baden-Württemberg, 26*, 7–19.
Rösch, M. (2009a). Zur vorgeschichtlichen Besiedlung und Landnutzung im nördlichen Schwarzwald aufgrund vegetationsgeschichtlicher Untersuchungen in zwei Karseen. *Mitteilungen des Vereins für Forstliche Standortskunde und Forstpflanzenzüchtung, 46*, 69–82.
Rösch, M. (2009b). Botanical evidence for prehistoric and medieval land use in the Black Forest. In *Medieval Rural Settlement in Marginal Landscapes* (pp. 335–343). Ruralia VII (Turnhout, Belgium).
Rösch, M. (2009/2010). Der Nordschwarzwald – das Ruhrgebiet der Kelten? Neue Ergebnisse zur Landnutzung seit über 3000 Jahren. *Alemannisches Jahrbuch (Freiburg)*, 155–169.
Rösch, M. (2012). Vegetation und Waldnutzung im Nordschwarzwald während sechs Jahrtausenden anhand von Profundalkernen aus dem Herrenwieser See. *Standort.Wald Mitteilungen des Vereins für forstliche Standortskunde und Forstpflanzenzüchtung, 47*, 43–64.
Rösch M (2013) Change of land use during the last two millennia as indicated in the Pollen record of a Profundal Core from Mindelsee, Lake Constance Region, Southwest Germany. In: Von Sylt bis Castanas, Festschrift für Helmut Johannes Kroll, Offa 69/70 (Neumünster) 355–370.
Rösch, M. (2015). Nationalpark – Natur – Weißtanne – Fichte: Sechs Jahrtausende Wald und Mensch im Nordschwarzwald. *Denkmalpflege in Baden-Württemberg, 44*(3), 154–159.
Rösch, M. (2016). Weinbau am Bodensee im Spiegel der Rebpollen. In T. Knubben & A. Schmauder (Eds.), *Seewein – Weinkultur am Bodensee* (pp. 51–59). Thorbecke: Ostfildern.
Rösch M (2017) Evidence for rare crop weeds of the Caucalidion group in Southwestern Germany since the Bronze Age - Paleo-ecological implications. Vegetation History and Archaeobotany 26 (online_first).
Rösch M, Heumüller M (2008) Vom Korn der frühen Jahre – Sieben Jahrtausende Ackerbau und Kulturlandschaft. Archäologische Informationen Baden-Württemberg 55, Esslingen.
Rösch, M., Fischer, E., Müller, H., Sillmann, M., & Stika, H. P. (2008). Botanische Untersuchungen zur eisenzeitlichen Landnutzung im südlichen Mitteleuropa. *Forschungen und Berichte zur Vor- und Frühgeschichte in Baden-Württemberg, 101*(Festschrift Jörg Biel), 319–347.
Rösch M, Gassmann G, Wieland G (2009) Keltische Montanindustrie im Schwarzwald – eine Spurensuche. In: Kelten am Rhein, Proceedings of the Thirteenth International Congress of Celtic Studies, erster Teil, Archäologie, Ethnizität und Romanisierung, Beihefte Bonner Jahrbücher 58,1, 263–278.
Rösch, M., & Tserendorj, G. (2011a). Florengeschichtliche Beobachtungen im Nordschwarzwald (Südwestdeutschland). *Hercynia N.F., 44*, 53–71.
Rösch, M., & Tserendorj, G. (2011b). Der Nordschwarzwald – früher besiedelt als gedacht? Pollenprofile belegen ausgedehnte vorgeschichtliche Besiedlung und Landnutzung. *Denkmalpflege in Baden-Württemberg, 40*(2), 66–73.
Rösch, M., Kleinmann, A., Lechterbeck, J., & Wick, L. (2014a). Botanical off-site and on-site data as indicators of different land use systems: A discussion with examples from Southwest Germany. In F. Bittmann, R. Gerlach, M. Rösch, & W. Schier (Eds.), *Farming in the forest – Ecology and economy of fire in prehistoric agriculture, Vegetation History and Archaeobotany* (Vol. 23 (Suppl. 1), pp. 121–133).
Rösch, M., Kleinmann, A., Lechterbeck, J., & Wick, L. (2014b). Erratum to: Botanical off-site and on-site data as indicators of different land use systems: A discussion with examples from Southwest Germany. *Vegetation History and Archaeobotany, 23*, 647–648.
Rösch, M., & Hahn, S. (2016). Besiedlung und Landnutzung im Allgäu von der Jungsteinzeit bis zur Neuzeit – ein interdisziplinäres Forschungsprojekt. *Archäologische Ausgrabungen in Baden-Württemberg, 2015*(Stuttgart), 45–50.

Rösch, M., & Lechterbeck, J. (2016). Seven Millennia of human impact as reflected in a high resolution pollen profile from the profundal sediments of Litzelsee,Lake Constance region, Germany. *Vegetation History and Archaeobotany, 25*, 339–358.

Rösch, M., Fischer, E., & Kury, B. (2017). Die Maulbronner Klosterweiher - Spiegel von vier Jahrtausenden Kulturlandschaftsgeschichte. *Denkmalpflege in Baden-Württemberg, 46*(4), 282–287.

Rösch, M., & Wick, L. (2018). 41. Western Lake Constance (Germany): Überlinger See, Mainau. *Grana, 58*, 78. https://doi.org/10.1080/00173134.2018.1509123.

Rösch, M., & Wick, L. (2019). 43. Buchensee (Lake Constance region, Germany). *Grana, 58*(4), 308–310. https://doi.org/10.1080/00173134.2019.1569127.

Rösch M, Stojakowits P, Friedmann A (2020) Does site elevation determine the onset and intensity of human impact? Pollen evidence from southern Germany. Vegetation History and Archaeobotany (online first https://doi.org/10.1007/s00334-020-00780-4).

Rydberg, J., Rösch, M., Heinz, E., & Biester, H. (2015). Influence of catchment vegetation on mercury accumulation in lake sediments from a long-term perspective. *Science of the Total Environment, 538*, 896–904.

Schier, W. (2009). Extensiver Brandfeldbau und die Ausbreitung der neolithischen Wirtschaftsweise in Mitteleuropa und Südskandinavien am Ende des 5. Jahrtausends v. Chr. *Prähistorische Zeitschrift, 84*, 15–43.

Schier, W. (2017). Die tertiäre Neolithisierung – Fakt oder Fiktion? In J. Lechterbeck & E. Fischer (Eds.), *Kontrapunkte, Festschrift für Manfred Rösch, Universitätsforschungen zur prähistorischen Archäologie 300* (pp. 129–145). Bonn: Habelt.

Schlichtherle H (1988) Das Jung- und Endneolithikum in Baden-Württemberg. In: Planck D (ed) Archäologie in Württemberg. Ergebnisse und Perpsektiven archäologischer Forschung von der Altsteinzeit bis zur Neuzeit, Theiss, Stuttgart pp. 91–110.

Schütze, M., Tserendorj, G., Pérez-Rodríguez, M., Rösch, M., & Biester, H. (2018). Prediction of holocene mercury accumulation trends by combining palynological and geochemical records of Lake Sediments (Black Forest, Germany). *Geosciences, 8*, 358. https://doi.org/10.3390/geosciences8100358, 1–21 (online first).

Schwalb, A., Dean, W. E., Güde, H., Hanisch, S., Sobek, S., & Wessels, M. (2013). Benthic ostracode $\delta^{13}C$ as sensor for early Holocene establishment of modern circulation patterns in Central Europe. *Quaternary Science Reviews, 66*, 112–122. https://doi.org/10.1016/j.quascirev.2012.10.032.

Shennan, S. (2018). *The first farmers of Europe An evolutionary perspective*. Cambridge: Cambridge University Press.

Smettan H (1988) Naturwissenschaftliche Untersuchungen im Kupfermoor – ein Beitrag zur Moorentwicklung sowie zur Vegetations- und Siedlungsgeschichte der Haller Ebene. In: Küster HJ (Hg) Der prähistorische Mensch und seine Umwelt, Festschrift für Udelgard Körber-Grohne, Forschungen und Berichte zur Vor- und Frühgeschichte in Baden-Württemberg 31, 81–101.

Tegtmeier, U. (1993). *Neolithische und bronzezeitliche Pflugspuren in Norddeutschland und den Niederlanden*. Bonn.

Wessels, M. (1995). Bodensee-Sedimente als Abbild von Umweltveränderungen im Spät- und Postglazial, Göttinger Arb. *Geol. Paläont., 66*, 87.

Wessels, M. (1998). Late-Glacial and postglacial sediments in Lake Constance (Germany) and their palaeolimnological implications. *Arch. Hydrobiol. Spec. Issues Advanc. Limnol., 53*, 411–449.

Whitlock, C., & Anderson, R. S. (2003). Fire history reconstructions based on sediment records from lakes and wetlands. In T. T. Veblen, W. L. Baker, G. Montenegro, & T. Swetman (Eds.), *Fire and climatic change in temperate ecosystems of the western Americas* (p. 3/31). New York: Springer.

Wick, L., & Rösch, M. (2006). Von der Natur- zur Kulturlandschaft – Ein Forschungsprojekt zur jungsteinzeitlichen und bronzezeitlichen Landnutzung am Bodensee. *Denkmalpflege in Baden-Württemberg, 35*(4), 225–233.

Wilmanns, O. (2001). *Exkursionsführer Schwarzwald - Eine Einführung in Landschaft und Vegetation*. Stuttgart.

Wolf, U. (1994). *Nähr- und Schadstoffbelastung kleiner Seen in Baden Württemberg unter Berücksichtigung der Sedimentationsgeschichte*. Göttingen: Die Rolle der kleinen Stehgewässer im regionalen Verbund.

Zeitvogel, W., & Feger, K. H. (1990). Pollenanalytische und nutzungsgeschichtliche Untersuchungen zur Rekonstruktion des historischen Verlaufs der Boden- und Gewässerversauerung im Nordschwarzwald. - Allg. Forst- u. *Jagdz., 161*, 136–144.

Large-Scale Slumps and Associated Resedimented Deposits in Miocene Lake Basins from SE Spain

José P. Calvo, David Gómez-Gras, and Miguel A. Rodríguez-Pascua

Abstract Slump, breccias, and turbidite deposits interbedded with diatomaceous marlstone and limestone occur in the upper part of the Late Miocene lacustrine succession of the Cenajo and Las Minas Basins. The basins display half-graben geometries where the northern margins consist of a network of large and deep, listric normal faults. The largest single deformational unit, reaching up 40 m in thickness, crops out in the Cenajo Basin and consists of soft-sediment deformed slump deposits displaying a variety of fold styles and geometries. This represents possibly the largest slump body hitherto described in fossil lake systems. The amplitude of the folds varies from meters in major folds to centimeters in the intrafolial minor ones. Brittle deformation of the slump strata is also common and cross cuts the ductile fabrics. Measured major fold axial traces indicate preferential N-S to NW-SE slide movement. The predominant strain in the deformed deposits is simple shear. Outcrops in the Cenajo Basin allow observation of extensional structures at the head slide, large sheath, and recumbent folds in the intermediate zone and tight compression structures at the toe, where folds and thrusts overrun the slightly deformed down slope strata. In contrast, the slump deposits in the Las Minas Basin are seen to pass down-dip into breccia and turbidite beds, which is indicative of flow transformation and internal disaggregation of the folded beds with distance. A number of laterally linked turbidite facies are recognized in the uppermost resedimented unit of the Las Minas Basin. The occurrence of these deposits points to marked instability in the lake basins. This period was preceded and followed by pelagic sedimentation of varve-like laminated, diatom-rich deposits. Most probably, basin instability

J. P. Calvo (✉)
Department of Petrology & Geochemistry, Universidad Complutense, Madrid, Spain
e-mail: jpcalvo@ucm.es

D. Gómez-Gras
Department of Geology, Universidad Autónoma de Barcelona, Barcelona, Spain
e-mail: David.Gomez@uab.cat

M. A. Rodríguez-Pascua
IGME – Geological Survey of Spain, Madrid, Spain
e-mail: ma.rodriguez@igme.es

M. R. Rosen et al. (eds.), *Limnogeology: Progress, Challenges and Opportunities*, Syntheses in Limnogeology, https://doi.org/10.1007/978-3-030-66576-0_5

was related to earthquake shocks, perhaps coeval to the extrusion of volcanic lamproite rocks in the region during the late Tortonian.

Keywords Slump deposits · Breccias · Debris flows · Turbidites · Lake basins · Soft-sediment deformation structures · Earthquakes · Miocene

Introduction

Resedimentation processes in lakes are related either to autogenic trigger mechanisms, e.g., lake basin morphology, differential slope gradients, sediment dispersal and high accumulation rates, or with external factors, often catastrophic events, such as earthquakes, volcanic eruptions, and powerful storms (Talbot and Allen 1996; Osleger et al. 2009; Owen and Moretti 2011; García-Tortosa et al. 2011; Hilbert-Wolf and Roberts 2015; Weinberger et al. 2016; Alsop et al. 2017). Common occurrence of resedimented deposits in modern lakes is clearly shown by high-resolution seismic systems used in hydrocarbon prospecting (Katz 1990; Morley 1999; Scholz 2002), investigation of Quaternary lake-level fluctuations related to climate change (Seltzer et al. 1998; Van Daele et al. 2011), and research and reconstruction of historical and recent earthquakes (e.g., Niemi and Ben-Avraham 1994; Schnellmann et al. 2002).

Although the quality of seismic imaging has improved considerably, yielding a detailed picture of the sedimentary bodies derived from mass movements, image resolution is however often insufficient to analyze and interpret small-scale sedimentary features of the resulting deformed deposits. This is usually supplemented by boreholes obtained by piston coring and/or drilling, yet the observation field is relatively small (Monecke et al. 2004; Schnellmann et al. 2005; Osleger et al. 2009).

The study of large-scale, well-exposed outcrop examples provides a good opportunity to study sedimentary features and facies relationships in such deformational units. Good exposures of large synsedimentary slide units usually correspond to marine formations, e.g., the giant slide Mesozoic deposits from Alexander Island, Antarctica (Macdonald et al. 1993), the Permian Vischkuil Formation from the Karoo Basin in South Africa (Van der Merwe et al. 2011; Brooks et al. 2018), and some examples from the northern Apennines (Italy) and south-central Pyrenees (Spain) (Ogata et al. 2012). These are examples of mass transport deposits exposed at outcrop at a scale comparable to subsurface examples. However, exposures of such large-scale deformational units in fossil lake successions are scarce. Up to 30 m-thick slump units involving massive to laminated mudstone and deltaic sandstone were described by Link and Osborne (1978) in the Pliocene Ridge Basin, California. Decameter-thick slump structures were also recognized in the Tierras Blancas Basin of the Mexican Volcanic Belt by Rodríguez-Pascua et al. (2010). Many other studies pointed out the occurrence of slide and slump units in both recent and ancient lake successions (e.g., Buatois and Mángano 1995; Ilgar and

Nemec 2005; Gibert et al. 2005; Moretti and Sabato 2007; García-Tortosa et al. 2011; Gibert et al. 2011), yet the deformational structures therein described usually range from decimeters to a few meters in thickness, most of them dealing with terrigenous lake systems. Slump deposits and/or other resedimented facies have been barely described in fossil lake carbonate systems except for minor gravity-driven sediments in bench (steep gradient) lake margins (Platt and Wright 1991). In this context, the size of the slumps is usually small (centimeters to a few meters in thickness) and the slumped sediment is composed typically of laminated carbonate. Some examples come from the early Cretaceous Peterson Limestone in northern USA (Glass and Wilkinson 1980), where thickness of the slumpized laminated beds rarely exceeds a few tens of centimeters, and from the Pleistocene diatomaceous deposits of the Tierras Blancas Basin in central Mexico where a slump structure reaches up to several tens of meters in thickness (Rodríguez-Pascua et al. 2010).

This paper presents detailed sedimentological and structural analyses of a large-scale, up to 40 m-thick slump deposit developed in the Upper Miocene Cenajo and Las Minas lake basins in SE Spain (Calvo and Elizaga 1994). The deformed deposits in the slump consist mainly of laminated carbonate and diatomaceous marlstone and are well-exposed along several hundreds of meters (Fig. 1), thus allowing observation of the transit to debris flows and/or turbidite beds. In view of its size, the described slump deposit is probably the largest one hitherto described in fossil lake systems. Breccia debris flow and thick turbidite beds occur beneath and above the

Fig. 1 General view of the strongly folded slump beds and associated resedimented deposits in cliffs at the left side of the Segura River in the Cenajo Basin; both the lower and upper boundaries of the mega-slump are marked by dashed lines (thickness of the slump unit between the two lines is about 40 m). The section is oriented NW (left side of the picture) – SE (right side of the picture)

slump unit, respectively. The resedimentation processes represented by the slump, breccia debris flow, and turbidite deposits are interpreted to be most probably related to earthquake shaking as a response to fault activity in the northern margin of the basins.

Geologic Setting

During the late Miocene, several continental extensional basins were formed in the external side of the Betic Chain, the so-called Prebetic Zone, in SE Spain (Sanz de Galdeano and Vera 1992; Rodríguez-Pascua et al. 2000) (Fig. 2a). The basins are located throughout a paleotectonic dextral transfer zone related to the emplacement of a major tectonic thrust, the Cazorla-Alcaraz Arc. This tectonic structure is characterized by a system of NW verging fault propagation folds related to imbricated thrusts. Three main faults with NW-SE directions (Pozohondo, Liétor, and Socovos-Calasparra faults) limit the Neogene basins, which are called Las Minas, Cenajo, Elche de la Sierra, Gallego, Híjar, Hellín, and Pozohondo basins (Fig. 2b). The resedimented deposits studied in this paper occur nearby the northernmost faults limiting the Cenajo and Las Minas basins, which cover an area of ~40 km^2 and ~250 km^2, respectively (Fig. 3).

The basins are elongate and formed as rapidly subsiding troughs during the late Vallesian and late Turolian (Tortonian-Messinian of the marine chronostratigraphic scale) (Calvo and Elizaga 1994). Half-graben geometry can be deduced for both the Cenajo and Las Minas basins (Rodríguez-Pascua et al. 2000). The northern basin margins consist of a network of large and deep, listric normal faults. The activity of these bounding faults increased during the latter stages of basin evolution creating

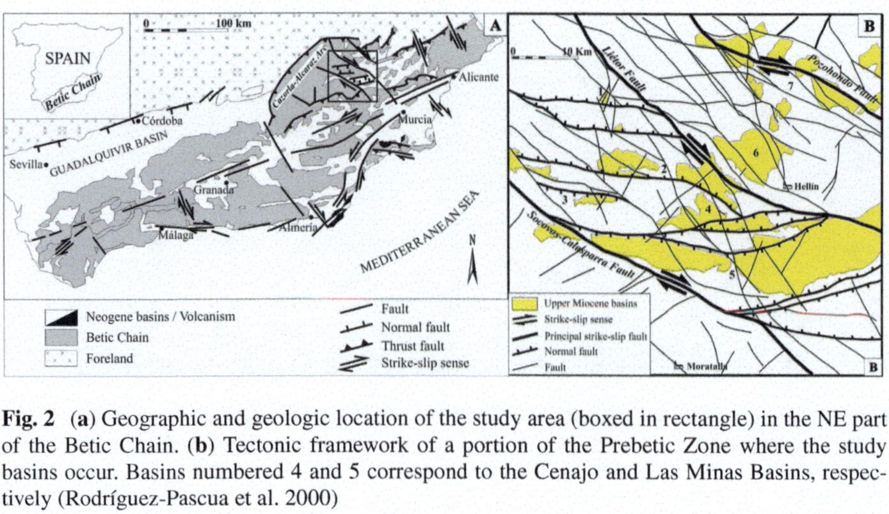

Fig. 2 (**a**) Geographic and geologic location of the study area (boxed in rectangle) in the NE part of the Betic Chain. (**b**) Tectonic framework of a portion of the Prebetic Zone where the study basins occur. Basins numbered 4 and 5 correspond to the Cenajo and Las Minas Basins, respectively (Rodríguez-Pascua et al. 2000)

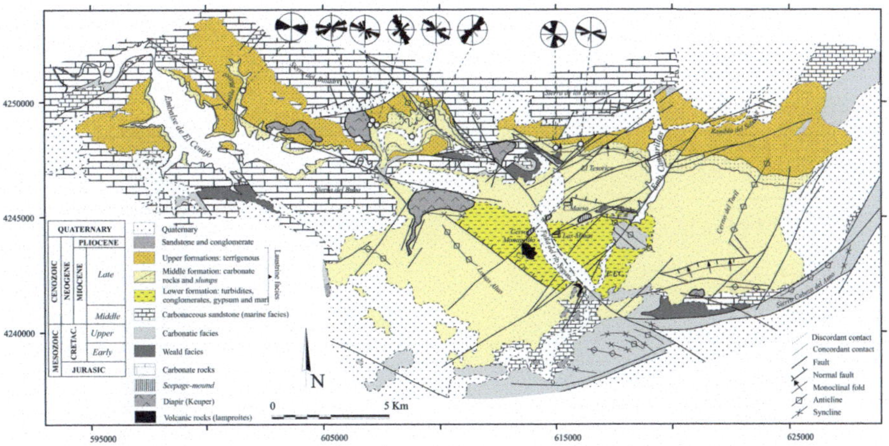

Fig. 3 Geologic map of the Las Minas and Cenajo Basins. Structural data gathered from the slump bodies and associated resedimented deposits are represented in the stereonets shown at the uppermost part of the map

accommodation space available for the deposition of deep lake facies to the north (Ortí et al. 2014).

The sedimentary successions of the Cenajo and Las Minas basins are very similar. They are composed of up to 500 m-thick terrestrial, mainly lacustrine deposits (Fig. 4). Servant-Vildary et al. (1990) found diatoms, e.g., several species of *Nitzschia*, *Diploides*, *Cocconeis*, *Thalassionema*, etc., indicative of episodic marine flooding at early stages of the sedimentary infilling of the Las Minas Basin while the remaining succession was deposited in shallow to moderately deep lake environments (Bellanca et al. 1989; Servant-Vildary et al. 1990; Calvo and Elizaga 1994; Calvo et al. 2000; Ortí et al. 2014; Pozo et al. 2016). Thick diatomaceous marlstone beds at the upper part of the stratigraphic succession of the Cenajo Basin are largely formed of frustules of centric forms of freshwater *Cyclotella* species (Foucault et al. 1987; Calvo et al. 1998).

Late Miocene clastic, evaporite, carbonate (both dolomite and limestone), and diatomaceous marlstone deposits overlie unconformably Mesozoic formations as well as middle Miocene marine carbonate strata. The lowermost deposits consist of reworked turbidite beds sourced from shallow lake carbonate platforms (Elizaga 1994) and marlstone–limestone cycles indicative of shoreface to foreshore lake subenvironments (Calvo et al. 2000). The carbonates are overlain by a monotonous succession of mudstone, laminated marlstone (locally paper-shale), and both detrital and chemically precipitated gypsum with intervening dolomite beds (Rosell et al. 2011; Ortí et al. 2014). Authigenic native sulphur and dolomite occur within the marlstone and gypsum beds (Fig. 4) (Bellanca et al. 1989; Lindtke et al. 2011). The evaporite sequence is overlain by a thick package of alternating laminated limestone and diatomaceous marlstone, which displays varve-like lamination and contains abundant centric diatoms and siliceous sponge spicules (Calvo et al. 1998).

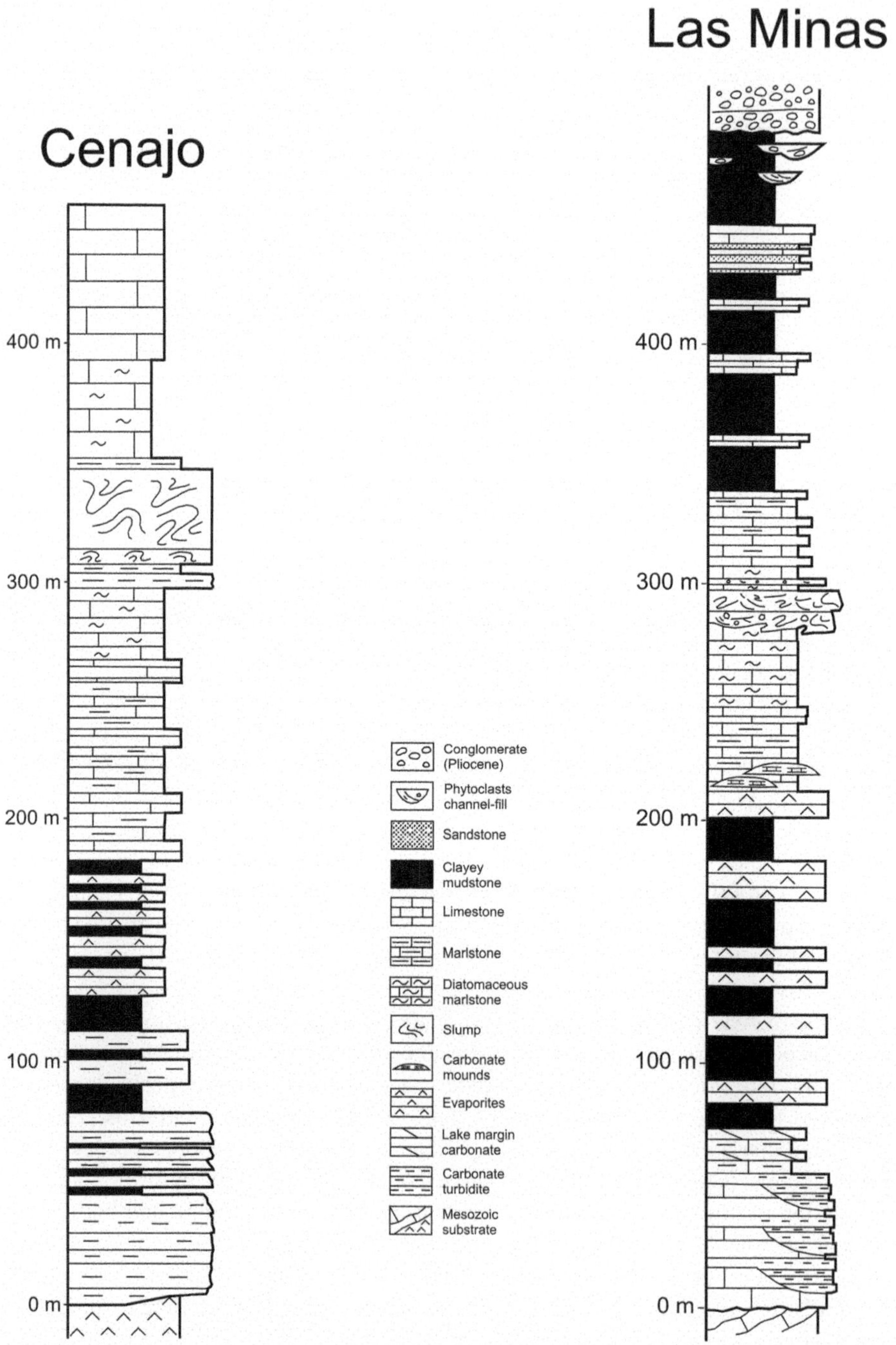

Fig. 4 Lithostratigraphic logs of the sedimentary infilling of the Cenajo and Las Minas Basins; note location of the slump units at a similar stratigraphic position in both basins

Intercalations of carbonate turbidites are frequent. The latter facies characterize deepening of the lake basins, the higher depth resulting in stratification of the water column and further good preservation of the laminated diatomaceous deposits. The slump and associated resedimented deposits occur intercalated in the upper part of the Miocene succession (Fig. 4).

Soft-sediment deformation structures occur throughout most of the sedimentary succession infilling the basins (Rodríguez-Pascua et al. 2000). The structures were in most cases interpreted as seismites, thus providing evidence of the tectonic instability of the basins during the late Miocene. In this context, the slump deposits were interpreted as triggered by earthquakes (Elizaga and Calvo 1988). Elizaga (1994) suggested that the seismicity could be related to extrusive lamproitic volcanism in the region.

Dating of the Late Vallesian to Turolian stratigraphic succession of the studied basins was based on occurrence of faunas of vertebrates (Calvo et al. 1978; Calvo and Elizaga 1994) and radiometric analysis of volcanic rocks. Except for some micromammal remains, the diatomaceous marlstone deposits located at the upper part of the succession did not provide palaeontological evidence for precise dating despite their high content in diatoms (which did not offer accurate biostratigraphic information; Foucault et al. 1987) and abundant remains of fishes, amphibians, and phytolites. However, the uppermost lake deposits include some beds that contain clasts and grains of the lamproitic volcanic rocks that form the Cerro del Monagrillo volcano (see Fig. 3 for location). These rocks were dated by analyzing phlogopite crystals using the single-crystal laser fusion Ar/Ar method (Bellon et al. 1981; Rosell et al. 2011). These authors determined an age of 7.65 ± 0.09 Ma for the youngest lamproitic lavas, thus in a similar range to that obtained by Nobel et al. (1981) in the same locality (7.2 ± 0.3 Ma).

Materials and Methods

The structural analysis of the slump deposits was carried out by exhaustive measurement of fold axe planes, folds vergence, dip and continuity of fold hinges, in order to determine the structural styles of the slump folds. Dip and strike of faults and/or fissure planes indicative of brittle behavior of the deformed beds were also measured, with estimation of the compressional or extensional nature of the faults (see Martinsen and Bakken 1990; Van der Merwe et al. 2011; Alsop et al. 2017). A total of 298 values related to ductile and brittle deformation features (141 in the Cenajo Basin; 157 in the Las Minas Basin) were obtained from nine measurement points in each basin. The data were treated statistically and represented in stereonets and rose diagrams, which helped in paleoslope determinations and paleoflow reconstructions. The rationale for the structural analysis of the slump units followed that used by Bradley and Hanson (1998), including the reference frames, i.e., the present-day and the stratigraphic reference frames, that must be applied to restore the slump deposits to the initial position.

The description of the resedimented clastic deposits associated with the slump unit is according to the turbidite types defined by Mutti (1992) and Mutti et al. (1999). Sedimentological analysis of turbidite beds in the Cenajo and the Las Minas Basins allows distinction of F1, F2, F3, F5, F6, F8, and F9 turbidite facies types of those authors. They occur mainly beneath the basal décollement zone but also overlying the slump units. Petrographic observation of 27 samples collected from both resedimented and intercalated deposits was carried out in order to discriminate the autochthonous or allochthonous nature of clasts and grains and to characterize the fabric of the deposits.

Resedimented Beds in the Cenajo Basin

In the Cenajo Basin, the resedimented units occur as superposed sheets separated by some thin undeformed beds (Figs. 1 and 5). Total measured thickness of the set of resedimented deposits reaches up to 55 m (Fig. 5); single beds can be followed laterally for a few hundreds of meters. The lowermost unit consists of 2.5 m of matrix-supported conglomerate (F2 turbidite facies of Mutti 1992) where maximum clast size reaches up to 12 cm. Quartzite and carbonate pebbles of extrabasinal origin are recognized in the conglomerate. This bed is covered by about 13 m of alternating laminated marlstone and thin turbidite beds, sharply overlain by a 7.60 m-thick coarse-grained, matrix-supported breccia bed containing clasts of varied size and shape. Interestingly, many fragments, in particular the coarsest ones, are composed of locally silicified, folded limestone and diatomaceous marlstone that accordingly are considered autochthonous components. The matrix content ranges from 35% to 50% (Fig. 6).

The top of the breccia bed is overlain by 0.50 m-thick laminated marlstone, which is partly eroded by the basal conglomerate of the slump deposit. The slump sheet is almost 40 m thick and consists of a continuum of strongly folded limestone and diatomaceous marlstone beds showing local faulting (Figs. 5 and 7). Fine-grained sandstone beds representative of more marginal lake environments are also present. The top of the slump deposit is capped by laminated fine-grained sandstone to siltstone characteristic of the F9 turbidite facies defined by Mutti (1992).

The section containing resedimented deposits in the Cenajo Basin ends with an up to 3.15 m-thick sequence of turbidite beds (F3, F6, and F9 turbidite facies sensu Mutti 1992) (Fig. 5). The uppermost part of the sedimentary infilling of the basin is formed of 110 m-thick alternating diatomaceous marlstone and limestone beds where no apparent deformational structures are found (Bellanca et al. 1989).

From northwest to southeast, several parts of the slump can be distinguished. The slump head shows extensional features, mainly a scar zone where spaced-out tilted blocks occur bounded by small-to-medium-scale listric faults. These faults are in turn locally associated with antithetic fractures. A few tens of meters further to the east, the slump deposit displays a variety of fold styles, e.g., sheath and recumbent folds showing interlimb angles $<15°$ and round hinges, normal and reverse faults

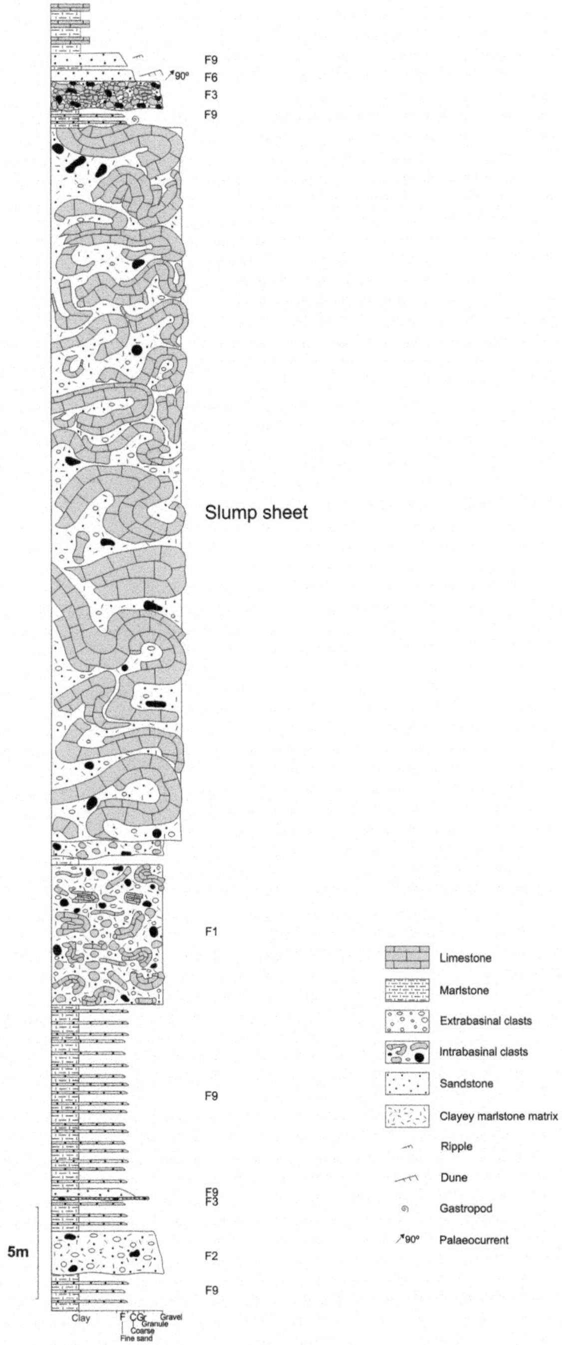

Fig. 5 Log showing the resedimented deposits of the Cenajo Basin

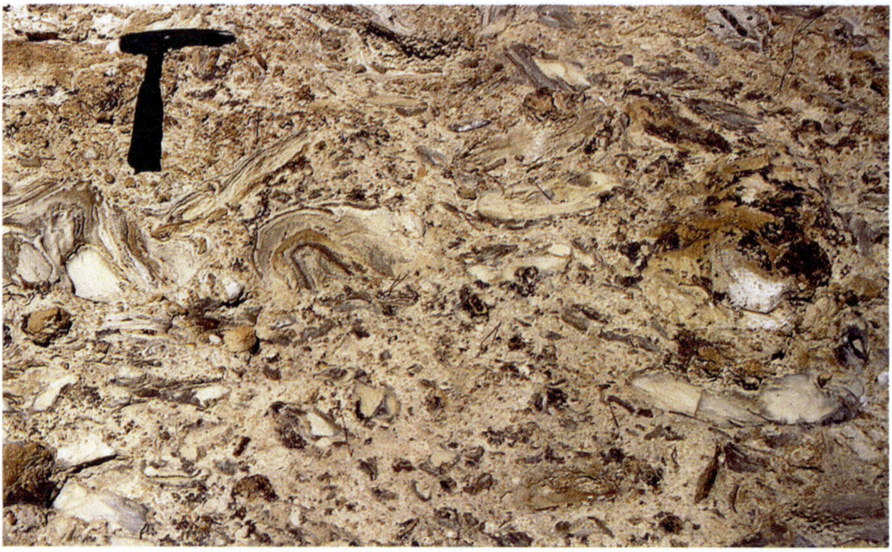

Fig. 6 Close-up view of coarse-grained, matrix-supported breccia occurring below the slumped unit (Facies F1 of Mutti 1992). Note that clasts are mostly composed of autochthonous folded limestone and diatomaceous marlstone with chert fragments; length of hammer for scale 28 cm

and thrusts, and sheets of non-deformed beds and coarse-grained breccias. A large overturned fold (Figs. 1 and 7) highlights the strong deformation undergone by the lacustrine beds and indicates that sliding developed apparently from north to south (left to right in the figures). Detailed structural analysis of the fold axes and vergences and associated brittle to ductile deformational structures ($n = 141$) indicate a main E-W to NE-SW orientation (see stereonet and rose diagram in Fig. 8). Core of the slumped beds is that pictured in Fig. 7. The basal décollement zone is erosive onto the underlying laminated diatomaceous marlstone but overlaps smoothly the thick conglomerate bed below. The toe of the slump is formed of compressional structures characterized by tight synforms and antiforms with vertical fold planes. Interestingly, the folded deposits of the toe (compression) zone overrun slightly deformed down slope strata indicating that the mass movement was able to ramp out from its original shear surface and deposit in an unconfined manner over the lake floor. This geometry characterizes frontally emerged slides, where larger down dip translation occurs (Frey-Martinez et al. 2006). In the Cenajo slump, the folded deposits of the toe zone pass laterally into highly dipping strata resembling foreset beds, though they consist actually of overturned folds truncated by slump slabs at their tops (see details of boxed area in Fig. 7).

The upper part of the slump sheet shows large-scale hummock geometry where turbidite deposits and pelagic diatomaceous and limestone beds were accommodated (Fig. 1). However, sharp contact between turbidites and deformed beds is observed in outcrops (Fig. 9).

Fig. 7 Toe (compression) zone of the slump body in the Cenajo Basin (the picture shows detailed view of the right side of Fig. 1). Note overrunning of the folded beds on slightly deformed down slope strata and the foreset appearance of beds in the right side of the picture. The measured thickness of the slump unit is about 40 m. The section is oriented NW (left side of the picture) – SE (right side of the picture). The boxed area shows closer view of the dipping strata with truncation of overturned folds by slump slab

Resedimented Beds in the Las Minas Basin

A significant occurrence of slump bodies and associated resedimented deposits is observed at about 280 m up in the stratigraphic succession of the Las Minas Basin (Fig. 4). Nine stratigraphic sections showing good exposures of the resedimented units were studied in the northern margin of the Las Minas Basin (Fig. 10); the deformational units, showing intercalations of non-deformed deposits, reach up to a maximum thickness of 35 m in that part of the basin.

Largely developed slump bodies can be observed in sections A to D. In the westernmost location, the resedimented deposits start with some coarse-clastic decimeter-thick beds intercalated in laminated carbonate that are sharply overlain

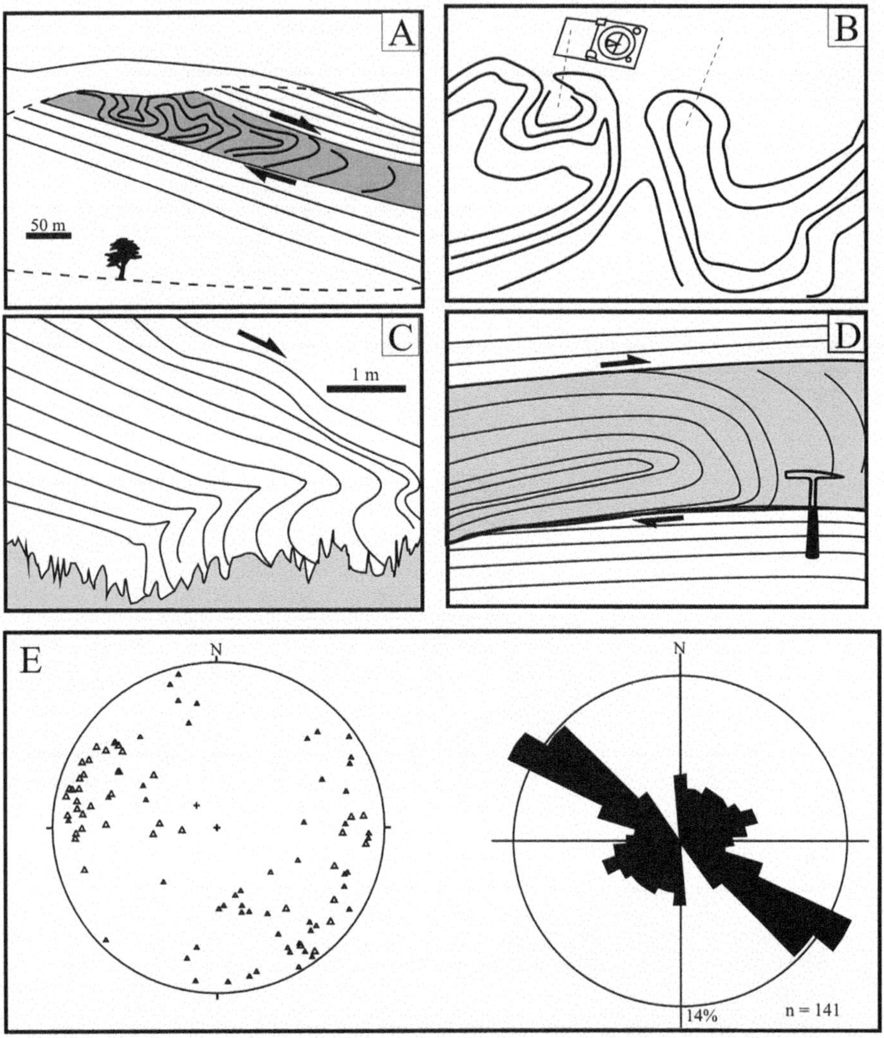

Fig. 8 Close-up view of different fold styles in the slump unit occurring in the Cenajo Basin: (**a**) basal detachment; (**b**) superposed folding; (**c**) blind thrust structure; (**d**) recumbent fold in the shear zone of the basal detachment. The total set of data obtained from fold axes in the Cenajo Basin is represented in the stereonet and rose diagram

by a 6 m-thick slump sheet characterized by strongly folded carbonate beds (Fig. 11). This lowermost slump unit changes into thinner and finer-grained conglomerate with distance (Fig. 10). Thus, about 1 km far from the outcrop represented as section A, the resedimented deposits consist of a lower, lobe-shaped coarse-grained conglomerate bed overlain by a thick set of slumped beds (Fig. 12; section B in Fig. 10). Matrix-supported breccia beds including large slabs and/or smaller pieces of alluvial conglomerate and sandstone mixed with lacustrine

Fig. 9 Outcrop view showing the sharp contact between thinly bedded turbidites (Facies F9 of Mutti 1992) and strongly folded beds at top of the slump. Thickness of the slumpized beds reaches up to 2 m in the picture

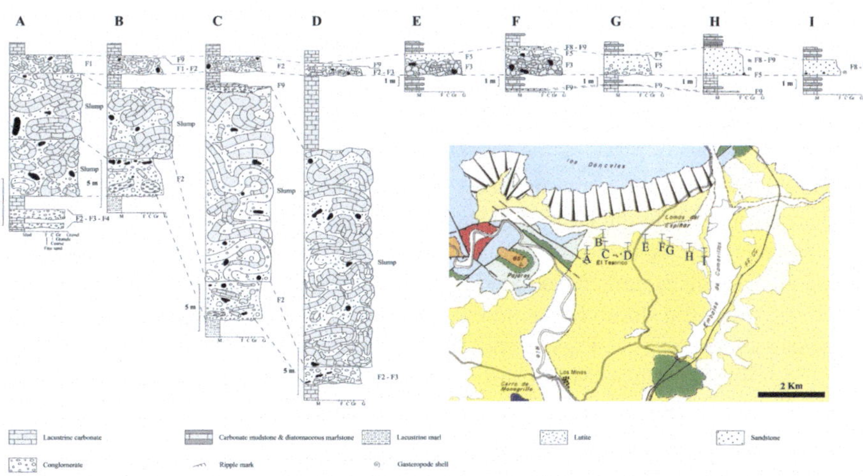

Fig. 10 Logs of the resedimented units along the northernmost part of the Las Minas Basin (see location of the spaced-out exposures in the inset map)

calcareous marlstone and limestone clasts are apparent in the lower parts of sections C and D (Fig. 10). Similar to that observed in the Cenajo breccia, the marlstone and limestone fragments occur locally as contorted and/or folded fragments.

The lower resedimented unit is overlain by another slump of approximately the same thickness in section A (Fig. 10). However, a marked change in the relative

Fig. 11 View of slump bodies in the north-westernmost part of the Las Minas Basin. The high thickness of the slump deposits in this location is due to superposition of two slump sheets. The picture corresponds to section A in Fig. 10. The measured thickness of the resedimented units reach up 18 m. The section is oriented WNW (left side of the picture) – ESE (right side of the picture)

thickness of one and other slump sheets can be observed towards the east, the lower one decreasing in both thickness and clast size while the second slump unit increases in thickness reaching up to 22 m in section D. The structural analysis of the fold axes and vergences and associated brittle to ductile deformational structures (n = 157) within the slump body indicates a main E-W to NW-SE orientation of the fold axes (see graphics in Figs. 3 and 13), thus suggesting that the mass movement was predominantly toward south and southwest. This is in agreement with the location of the basin depocenter during later evolutionary stages of the Las Minas Basin (Elizaga 1994; Rodríguez-Pascua 2001).

The two resedimented units do not crop out farther so its sedimentary and structural evolution with distance cannot be described and the interpretation would be speculative. In contrast, the uppermost part of section A consists of a 1.80 m-thick coarse-grained conglomerate bed mainly composed of fragments of Miocene lacustrine limestone and minor quartzite and carbonate clasts recycled from Mesozoic formations of the basin margin. The sandstone beds also contain abundant grains reworked from the lake formations. The sedimentary features of the bed can be followed in detail from W to E along 3.5 km in the northern part of the Las Minas Basin (Fig. 10). An outcrop view of the resedimented deposit is shown in Fig. 14

Fig. 12 Outcrop view of lobe-shaped basal conglomerate overlain by a slump unit formed of limestone slabs and folds. The picture corresponds to section B in Fig. 10. The dashed line outlines the geometry of the conglomerate body. Person for scale is 1.74 m tall. The section is oriented WNW (left side of the picture) – ESE (right side of the picture)

(section E in the figure). The primary changes recognized in this deposit are a marked decrease of grain size and change in the turbidite facies association while the thickness of the deposit does not vary significantly.

The three sedimentary units distinguished in section A can be followed eastward although the lower slump sheets do not crop out in sections E to I (Fig. 10). Sedimentological characterization of the turbidite facies, i.e., F1 to F9 of Mutti (1992) and Mutti et al. (1999), is written in the figure. Turbidite facies F1, F2, and F3 were described in previous chapters. Turbidite facies F5 consists mainly of massive, poorly sorted, coarse- to medium-grained sandstone. Beds showing these features occur at the upper part of the resedimented deposits in the Las Minas Basin (Fig. 10) usually overlying clast-supported F3 facies. Turbidite facies F6 consists of medium-grained sandstone showing cross-bedding and other current structures. Turbidite facies F8 and F9 include fine- to very fine-grained sandstone and siltstone showing graded bedding and relatively good sorting. These facies lack tractional features, which is consistent with their occurrence as the most distal deposits of the uppermost resedimented unit in the Las Minas Basin (see sections H and I in Fig. 10).

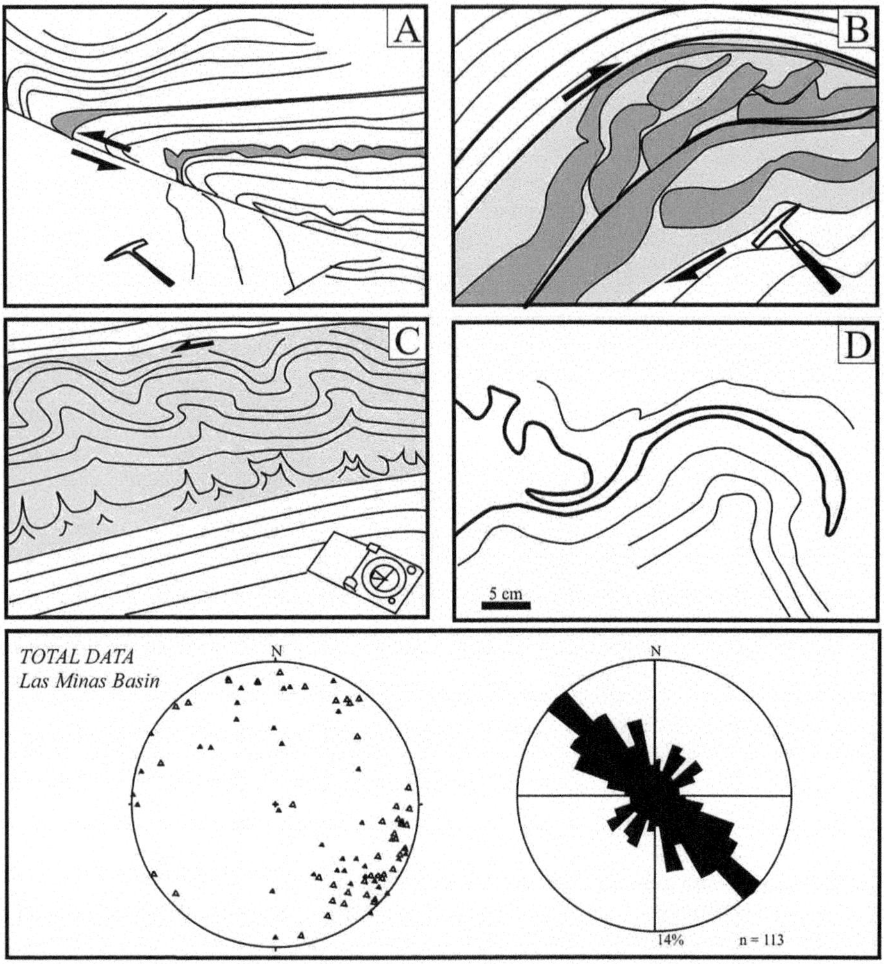

Fig. 13 Close-up view of different fold styles of the slump in the Las Minas Basin: (**a**) basal detachment; (**b**) thrust duplex structure; (**c**) convolute bedding showing shear stress; (**d**) superposed folding. The total set of data obtained from fold axes in the Las Minas Basin are represented in the stereonet and rose diagram

Discussion

Slump Deposits

The studied deposits, especially those described in the Cenajo Basin and at the lower resedimented units of the Las Minas Basin match well the definition of slump deposits as "laterally displaced sediment masses bounded by a basal shear plane and with evident contortion and rotation of contained strata" (Stow 1986).

Fig. 14 Outcrop view of a 2.60 m-thick turbidite bed in the northern part of the Las Minas Basin (section E in Fig. 10). The turbidite deposit is mainly formed of Facies F3 of Mutti (1992); note the sharp contact with underlying laminated marlstone (white) and porcelanite (dark green) beds

The low degree of internal disaggregation and the basal sheared unit (Fig. 1) lead to these units being interpreted as a slide deposit that moved over a basal sheet or décollement surface (e.g., Vanneste et al. 2006; Alves 2010; Masson et al. 2010; Baeten et al. 2014), marking a clear boundary that separates the deformed contorted (hanging wall) strata of the deformed mass from underlying undeformed (footwall strata) (Hampton et al. 1996; Frey-Martinez et al. 2006). The distribution of fold data (Figs. 8 and 13) allows to interpreting if the folds remained attached and verging in a consistent direction or they became detached (Martinsen 1994); in the latter case, they would have been subject to rotation and a wider distribution of data would be expected (Van der Merwe et al. 2011). In the sections where the slumps were observed (Figs. 1, 7, and 11), both descriptions and measurements dealt mainly with 2D features. The orientations of fault planes, fold axial planes, hinge lines, and fold axes show some dispersion (see rose diagrams in Fig. 3), but they are consistent enough to interpret the dominant movement direction of slump movement. Data spread is comparable to previous compiled data on slumps by Martinsen et al. (2003), Strachan and Alsop (2006), and Strachan (2008). The structural data as a whole point out to down dip movement from the northern basin margins to basin depocenters located south and south-west, which is consistent with the asymmetry of the half-graben geometry of both the Cenajo and Las Minas Basins (Rodríguez-Pascua et al. 2000).

Given the small areal extent of the Cenajo Basin (Fig. 3), the slump deposits in that location are considered to have moved through a short transport distance (no longer than a few hundreds of meters). This resulted in a low degree of internal disaggregation and a relatively small size (from meters to a few tens of meters) of the fold geometries. The series of normal faults and/or tilted blocks with scar planes in the slump occurring at the Cenajo Basin is evidence of deformation under extensional conditions (Farrell 1984; Martinsen and Bakken 1990; Wignall and Best 2004; Posamentier and Walker 2006; Alonso et al. 2008). Laterally, the slump consists of opposite verging folds cut by a series of thrusts that indicate compressional deformation at the distal (toe) end of the slide (Gawthorpe and Clemmey 1985; Moore and Shannon 1991) (Fig. 7). Regarding fold and fault relationships, the fact that the fault planes dissect folds in areas of high ductile strain suggests that the ductile phase corresponding to folding largely predates the brittle phase (Martinsen and Bakken 1990; Van der Merwe et al. 2011; Alsop et al. 2017). Observation of the distal evolution of the slump deposits in the studied basins is not easy because of lack of appropriate outcrops. However, the lobe-shaped basal conglomerate described in section B of the Las Minas Basin (Fig. 12) probably represents the transformation of the distal end of a slide into a disaggregated, clastic deposit. This is supported by the fact that the lower slump unit in section A (Fig. 10) thins toward section B, but in contrast the upper slump unit clearly thickens in between the two sections. Published models based on modern, subsurface, and outcrop investigations show that folds in a slide/slump become more disaggregated toward the distal end of the slide (Farrell 1984; Bryn et al. 2005; Strachan and Alsop 2006; Dasgupta 2008; Piper and Normark 2009; Ogata et al. 2012; Shanmugam 2012). Taking in mind this evidence, breccia deposits discussed below are described and interpreted as genetically linked to the distal part of the slumps (Fig. 15), and could be represented similar to the intermediate stages of a slide from slope failure to turbidites (see Fig. 11 of Bryn et al. 2005).

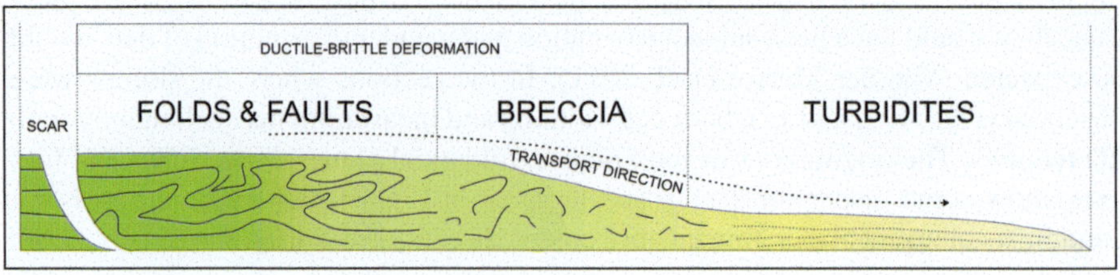

Fig. 15 Idealized sketch showing the lateral evolution from slump deposits into breccias and turbidite beds; the design of the picture was supported by field observations in the northern part of the Las Minas Basin (see Fig. 10 of this paper)

Breccia

In the Cenajo Basin, a breccia deposit underlies and is clearly differentiated of the slump deposit occurring a few meters above (Fig. 5). The breccia comprises poorly sorted to randomly distributed contorted limestone and diatomaceous marlstone clasts supported by fine-grained (marly) matrix (Fig. 6). This suggests rapid or "en masse" deposition from a cohesive flow and is interpreted as a debris-flow deposit (e.g., Mutti 1992; Mutti et al. 1999; Haughton et al. 2003; Talling et al. 2004; Haughton et al. 2009; Sumner et al. 2009; Talling et al. 2012), similar to that concluded by Strachan (2008) from the Little Manly Slump in New Zealand. In the latter study case, fragmentation of the slump unit into a debris flow deposit could have resulted from local tensile stresses and/or flow transformation of the slumping into a more fluidal flow type (Strachan 2008).

Moreover, abundance of clasts consisting of pieces of folded limestone and diatomaceous marlstone (Fig. 6) clearly provides evidence of their close relationship to slump bodies (e.g., Ogata et al. 2014). Thus, angular to subrounded clasts in the breccias are thought to have resulted from the fragmentation of original bedding within the slump and the formation of a debrite texture (sensu Mulder and Alexander 2001). The petrology of the breccia is broadly similar to the surrounding deposits, which supports a mainly intrabasinal source for the clastic fraction being consistent with disaggregation of slumpized beds. Accordingly, our interpretation of the Cenajo section points out to two superimposed resedimented units, the lowermost being represented by the breccia bed at the cliffs of the Segura River and the second and upper one corresponding to the slump body largely exposed in the section. In both cases, transport distance was short (i.e., tens to very few hundred meters) as it can be deduced from the small size of the Cenajo Basin. Moreover, our interpretation of the breccias as genetically linked to a lower slump unit would indicate that sediment disaggregation can be strong despite not so long transport (Strachan 2008; Alsop et al. 2017).

Breccia deposits in the Las Minas Basin occur mainly in the lower resedimented rock unit laterally linked to slump deposits (see sections A, B, C, and D in Fig. 10). The occurrence of local contorted and/or marlstone and limestone fragments within the breccias clearly indicates that the deposit formed by disaggregation of the slump unit. Thinning of the breccias with distance also provides evidence of the genetic relationship between the two resedimented bodies. The breccias can be interpreted as a result of rapid or "en masse" deposition from a cohesive debris flow or transport and deposition by hyperconcentrated flows (Facies F1 and/or F2 of Mutti 1992 and Mutti et al. 1999). Difference between cohesive and non-cohesive debris-flow deposits is not clear in the studied deposits.

Turbidites

Clast-supported F3 facies (Fig. 14) is laterally related to the F1 and F2 facies described above and is mostly related to more dilute flow moving relatively coarse-grained sediment, thus indicating loss of cohesiveness of transport medium from W to E in the northern part of the Las Minas Basin (Figs. 10 and 15).

Deposition of the F5 facies took place by granular flow with high excess pore pressure segregated from the coarse-grained F3 facies (Mutti et al. 1999). The F5 turbidites represent distal facies of the resedimented deposits with progressive thinning of the facies with distance. Presence of current structures in F6, a turbidite type only occurring in the uppermost resedimented deposits of the Cenajo Basin (Fig. 5), is indicative of bypass and reworking from long-lived and large-volume turbulent flows (Mutti et al. 1999). Lack of tractional features in turbidite facies F8 can be interpreted as resulting from high rates of sediment fallout preventing traction (Mutti et al. 1999). Accordingly, they represent the waning stages of a turbidite current.

Stages of Resedimentation Processes/Events

Three main stages of resedimentation processes/events can be defined after the lithostratigraphic and sedimentological analyses of the Late Miocene sedimentary record in the Cenajo and Las Minas Basins. Following accumulation on the lake floor of sediments formed mainly of laminated diatomite and marlstone under low-energy conditions (marked as "lake sediments" in Fig. 16), a first event of instability resulted in the formation of slump bodies and associated breccias. These deposits were mostly accumulated in depocenters aligned parallel (W–E to NW-SE) to the basin margins; in the Cenajo Basin, the first resedimentation event is represented by superimposed breccias (Stage A, Fig. 16) while transition from slump bodies to breccia deposits is observable in the Las Minas Basin.

The second major event is clearly recognizable in both basins because of the large development of slump bodies. The orientations of the slump folds and associated soft-sediment deformation structures are variable but a main NW-SE direction can be determined (Stage B, Fig. 16). The accumulation of the resedimented deposits was likely coincident with maximum depth of the lake depocenters. The slump sheets resulted from slope failures that initiated at single points and then translated downslope until cessation and relaxation (Alsop and Marco 2013).

The third resedimentation event (Stage C, Fig. 16) includes deposition of a rather complete spectrum of turbidite facies, from cohesive coarse- to dilute fine-grained flow deposits from W to E in the Las Minas Basin.

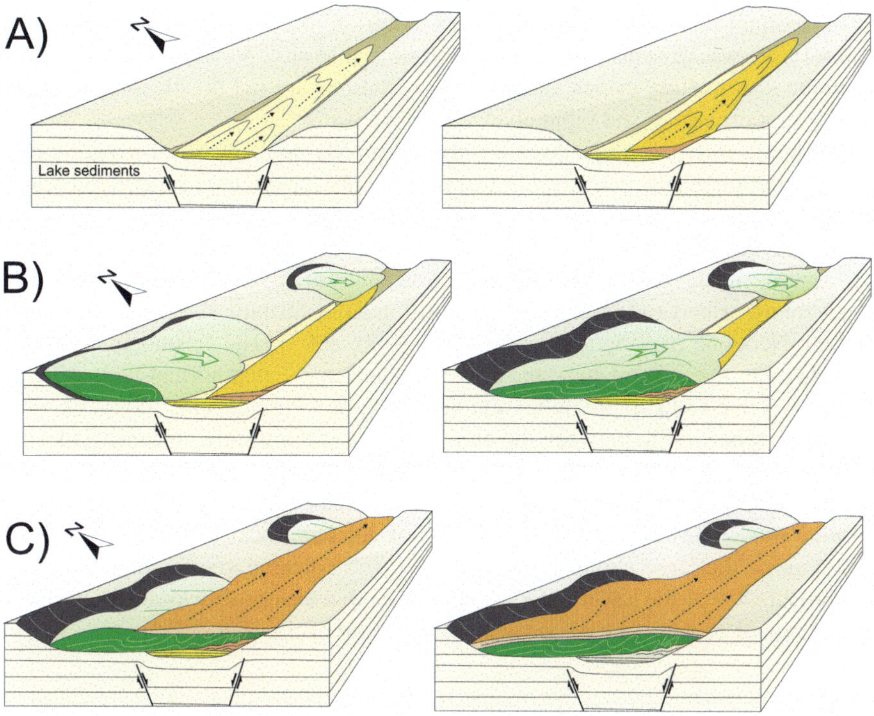

Fig. 16 Stages of the sedimentological evolution of the slump deposits and associated resedimented beds (see text for explanation)

Triggering of Resedimentation Events

Usually, resedimentation processes in lakes represent catastrophic events where variable amount of sediment and/or rock masses move as a result of a variety of triggering mechanisms, whether autogenic or external to the lake basins. Thus, overloading of sediment in lake platforms as well as wave-induced (i.e., seiches) cyclical and/or impulsive stresses, changes in lake level, etc., constitute reliable autogenic mechanisms that could result in mass flow/sliding processes (Talbot and Allen 1996; Masson et al. 2010; Gibert et al. 2011). On the other hand, earthquakes have been invoked as a common natural process leading to soft-sediment deformation, including slumps/slides at various scales (Schnellmann et al. 2002; Rodríguez-Pascua et al. 2003; Moretti and Sabato 2007).

Effect of water waves can be discarded as a trigger mechanism of resedimentation of the studied deposits because of the small areal extent, especially of the Cenajo Basin. Sedimentation rates could be relatively high during the accumulation of diatomaceous marlstone but less in which concerns lacustrine limestone; annual sedimentation rates of about 0.5–0.75 mm in diatomaceous marlstone were recorded by Calvo et al. (1998) and Rodríguez-Pascua (2001), thus its effect in eventual overloading cannot totally be ruled out. Recognition of lake level changes by Bellanca et al. (1989) was based on the fluctuation of stable isotope values in limestone–diatomite sedimentary cycles although the variations are not drastic and cannot be used to support strong and rapid changes in lake water.

We interpret that earthquakes were the most reliable mechanism triggering the resedimentation events described in this paper. This statement is supported by long seismogenic history of the basins in which the resedimented units occur. In the Cenajo and Las Minas Basins, a range of soft-sediment deformation structures, i.e., sand dykes, mushroom-like forms, pillow structures, mixed layers, loop bedding, etc., were described and interpreted as resulting from earthquake events throughout the Late Neogene (Rodríguez-Pascua et al. 2000, 2003). The seismogenic activity was related to the movement of the strike-slip and associated normal faults (Fig. 2) that were active and still are in the region (Rodríguez-Pascua et al. 2009).

A close relationship between the emplacement of lamproitic volcanic rocks and the large slump body occurring in the Cenajo Basin was suggested by Elizaga and Calvo (1988). The presence of clasts of volcanic rocks in the slump deposits and the turbidite sandstones indicates that extrusion of volcanic rocks took place before the formation of the large slump body. However, the volcanic vents probably caused resedimentation of the lower breccia deposits so that tremor shocks associated with the volcanism could account for the destabilization of the lake sediments.

Concluding Remarks

The slump units and associated resedimented deposits in the Cenajo and Las Minas lake basins show unique exposures of the head, body, and toe of a large-scale slide formed by tectonic instability of the basin margins. Well-preserved and largely developed folds in the intermediate zone of the slump body of the Cenajo Basin provide evidence that the mass movement was able to ramp out from its original shear surface and deposit in an unconfined manner over the lake floor. Exposures of the resedimented deposits in the Las Minas Basin allow recognition of sedimentary structures related to flow transformation during down-dip transport and provide a good picture of disaggregation phenomena leading to formation of debris flow and/or turbidite deposits.

Acknowledgments This paper is primarily dedicated to our colleague Elisabeth Gierlowski-Kordesch acknowledging her big efforts in spreading out the discipline of Paleolimnology. Likewise, the authors want to render tribute to our colleague Emilio Elizaga, who unfortunately passed away in 1992 after building a geological, stratigraphic, and sedimentological framework of the basins that were crucial for further studies. We acknowledge our colleagues from the ETH Zurich Flavio Anselmetti, Katryn Monecke, Michi Schnelmann, Amfried Becker as well as our friend Giovanna Scopelitti for their collaboration in fieldwork and exchanging ideas on resedimentation processes in lakes. Luis Gibert (University of Barcelona) is thanked for providing details on slump deposits in some western USA Neogene lake systems. Dave A. Osleger (University of

California, Davis) contributed to improve the scientific content of the paper. Assistance of Miquel Poyatos (University of Oslo, Norway) in the interpretation of some facies and correction of the text is greatly appreciated. We thank Javier Elorza (University of the Basque Country, Spain) for their help in discussing field observations and Marta Roigé (Universitat Autonoma de Barcelona) for technical support in the preparation of the manuscript. We are deeply indebted to Lisa Park Boush and David Finkelstein for their excellent reviews that have contributed to improve a first version of the paper. Last but no least we thank Michael Rosen for his great job as an Editor of the volume in which this paper is included.

References

Alonso, J. L., Gallastegui, J., García-Sansegundo, J., Farias, P., Fernádez, L. R., & Ramos, V. A. (2008). Extensional tectonics and gravitational collapse in an Ordovician passive margin: The Western Argentine Precordillera. *Gondwana Research, 13*, 204–215.

Alsop, G. I., & Marco, S. (2013). Seismogenic slump folds formed by gravity-driven tectonics down a negligible subaqueous slope. *Tectonophysics, 605*, 48–69.

Alsop, G. I., Weinberger, R., Marco, S., & Levi, T. (2017). Identifying soft-sediment deformation in rocks. *Journal of Structural Geology, 94*, 98–115.

Alves, T. M. (2010). 3D seismic examples of differential compaction in mass-transport deposits and their effect on post-failure strata. *Marine Geology, 271*, 212–224.

Baeten, N. J., Laberg, J. S., Vanneste, M., Forsberg, C. F., Kvalstad, T. J., Forwick, M., & Haflidason, H. (2014). Origin of shallow submarine mass movements and their glide planes—Sedimentological and geotechnical analyses from the continental slope off northern Norway. *Journal of Geophysical Research: Earth Surface, 119*, 2335–2360.

Bellanca, A., Calvo, J. P., Censi, P., Elízaga, E., & Neri, R. (1989). Evolution of diatomite – carbonate lacustrine cycles of Miocene age, Southeastern Spain: Petrology and isotope geochemistry. *Journal of Sedimentary Petrology, 59*, 45–52.

Bellon, H., Bizon, G., Calvo, J. P., Elizaga, E., Gaudant, J., & López, N. (1981). Le volcán du Cerro del Mongraillo (province de Murcie): âge radiométrique et corrélation avec les sédiments néogènes du Bassin de Hellín. *Comptes Rendus Acad. Sciences Paris, 2ème Série, 2*, 1035–1038.

Bradley, D., & Hanson, L. (1998). Paleoslope analysis of slump folds in the Devonian flysch of Maine. *Journal of the Geological, 106*, 305–318.

Brooks, H. L., Hodgson, D. M., Brunt, R. L., Peakall, J., & Flint, S. S. (2018). Exhumed lateral margins and increasing flow confinement of a submarine landslide complex. *Sedimentology, 65*, 1067–1096.

Bryn, P., Berg, K., Forsberg, C. F., Solheim, A., & Kvalstad, T. J. (2005). Explaining the Storegga slide. *Marine and Petroleum Geology, 22*, 11–19.

Buatois, L. A., & Mángano, M. G. (1995). Sedimentary dynamics and evolutionary history of a Late Carboniferous Gondwanic lake in North-Western Argentina. *Sedimentology, 42*, 415–436.

Calvo, J. P., & Elizaga, E. (1994). The Cenajo and Las Minas – Camarillas basins (Miocene), southeastern Spain. In E. Gierlowski-Kordesch & K. Kelts (Eds.), *Global geological record of Lake Basins, (Vol. 1*, pp. 319–324). Cambridge University Press, Cambridge.

Calvo, J. P., Elizaga, E., López, N., Robles, F., & Usera, J. (1978). El Mioceno Superior continental del Prebético Externo. Evolución del Estrecho Nord-Bético. *Boletín Geológico y Minero, 89*, 9–28.

Calvo, J. P., Rodríguez-Pascua, M. A., Martín-Velázquez, S., Jiménez, S., & De Vicente, G. (1998). Microdeformation of lacustrine laminite sequences from Late Miocene formations of SE Spain: An interpretation of loop bedding. *Sedimentology, 45*, 279–292.

Calvo, J. P., Gómez-Gras, D., Alonso-Zarza, A. M., & Jiménez, S. (2000). Architecture of a bench-type carbonate lake margin and its relation to fluvially dominated deltas, Las Minas Basin, Upper Miocene, Spain. *Journal of Sedimentary Research, 70*, 240–254.

Dasgupta, P. (2008). Experimental decipherment of the soft sediment deformation observed in the upper part of the Talchir Formation (Lower Permian), Jharia Basin, India. *Sedimentary Geology, 205*, 100–110.

Elizaga, E. (1994). *Análisis de facies sedimentarias y petrología de los depósitos lacustres de edad Neógeno superior de la Zona Prebética, Albacete, España* (p. 216). Albacete, Spain: Instituto de Estudios Albacetenses.

Elizaga, E., & Calvo, J. P. (1988). Evolución sedimentaria de las cuencas lacustres neógenas de la zona Prebética (Albacete, España). Relación, posición y efectos del vulcanismo durante la evolución. Interés minero. *Boletín Geológico y Minero, 99*, 3–12.

Farrell, S. G. (1984). A dislocation model applied to slump structures, Ainsa Basin, South Central Pyrenees. *Journal of Structural Geology, 6*, 727–736.

Foucault, A., Calvo, J. P., Elizaga, E., Rouchy, J. M., & Servant-Vildary, S. (1987). Place des dépôts lacustres d'âge miocène supérieur de la region d'Hellín (Province d'Albacete, Espagne). *Comptes Rendus de l'Académie des Sciences Paris, 2ème Série Paris, 305*, 1163–1166.

Frey-Martinez, J., Cartwright, J., & James, D. (2006). Frontally confined versus frontally emergent submarine landslides: A 3D seismic characterization. *Marine and Petroleum Geology, 23*, 585–604.

García-Tortosa, F. J., Alfaro, P., Gibert, L., & Scott, G. (2011). Seismically induced slump on an extremely gentle slope (<1°) of the Pleistocene Tecopa paleolake (California). *Geology, 39*, 1055–1058.

Gawthorpe, R. L., & Clemmey, H. (1985). Geometry of submarine slides in the Bowland Basin (Dinantian) and their relation to debris flows. *Journal of Geological Society, 142*, 555–565.

Gibert, L., Sanz de Galdeano, C., Alfaro, P., Scott, G., & Lopez-Garrido, A. C. (2005). Seismic-induced slump in early Pleistocene deltaic sediments of the Baza Basin (SE Spain). *Sedimentary Geology, 179*, 279–294.

Gibert, L., Alfaro, P., García-Tortosa, F. J., & Scott, G. (2011). Superposed deformed beds produced by single earthquakes (Tecopa Basin, California): Insights into paleoseismology. *Sedimentary Geology, 235*, 148–159.

Glass, S. W., & Wilkinson, B. H. (1980). The Peterson limestone – Early cretaceous lacustrine carbonate deposition in western Wyoming and southeasten Idaho. *Sedimentary Geology, 27*, 143–160.

Hampton, M. A., Lee, H. J., & Locat, J. (1996). Submarine landslides. *Reviews of Geophysics, 34*, 33–59. https://doi.org/10.1029/95RG03287.

Haughton, P. D., Barker, S. P., & McCaffrey, W. D. (2003). 'Linked' debrites in sand-rich turbidite systems – Origin and significance. *Sedimentology, 50*, 459–482.

Haughton, P. D., Davis, C., McCaffrey, W., & Baker, S. (2009). Hybrid sediment gravity flow deposits – Classification, origin and significance. *Marine and Petroleum Geology, 26*, 1900–1918.

Hilbert-Wolf, H. L., & Roberts, E. M. (2015). Giant seismites and megablock uplift in the East African rift: Evidence for late Pleistocene large magnitude earthquakes. *PLoS One, 10*(6), e0129051. https://doi.org/10.1371/journal.pone.0129051.

Ilgar, A., & Nemec, W. (2005). Early Miocene lacustrine deposits and sequence stratigraphy of the Ermenek Basin, Central Taurides, Turkey. *Sedimentary Geology, 173*, 233–275.

Katz, B. J. (Ed.). (1990). Lacustrine basin exploration. American Association of Petroleum Geologists 50, 340pp.

Lindtke, J., Ziegenbalg, S. B., Bunner, B., Rouchy, J. M., Pierre, C., & Peckmann, J. (2011). Authigenesis of native Sulphur and dolomite in a lacustrine evaporitic setting (Hellín basin, Late Miocene, SE Spain). *Geological Magazine, 148*, 655–669.

Link, M. H., & Osborne, R. H. (1978). Lacustrine facies in the Pliocene Ridge Basin Group: Ridge Basin, California. In A. Matter & M. E. Tucker (Eds.), *Modern and ancient Lake sediments*

(International Association of Sedimentologists Special Publs), (vol 2, pp. 169–187). Oxford, Great Britain. Blackwell Publ. Co.

Macdonald, D. I. M., Moncrieff, A. C. M., & Butterworth, P. J. (1993). Giant slide deposits from a Mesozoic fore-arc basin, Alexander Island, Antarctica. *Geology, 21*, 1047–1050.

Martinsen, O. J. (1994). Evolution of an incised valley-fill, the Pine Ridge Sandstone of southeastern Wyoming, USA: Systematic sedimentary response to relative sea-level change. In R. W. Dalrymple, B. A. Zaitlin, & R. Boyd (Eds.), *Incised Valley systems: Origin and sedimentary sequences* (Spec Publ Soc Econ Palaentol Mineral), (Vol. 51, pp. 109–128). Tulsa, Oklahoma: SEPM.

Martinsen, O. J., & Bakken, B. (1990). Extensional and compressional zones in slumps and slides in the Namurian of County Clare, Ireland. *Journal of the Geological London, 147*, 153–164.

Martinsen, O. J., Lien, T., Walker, R. G., & Collinson, J. D. (2003). Facies and sequential organization of a mudstone dominant slope and basin floor succession: The Gull Island formation, Shannon Basin, Western Ireland. *Marine and Petroleum Geology, 20*, 789–807.

Masson, D. G., Wynn, R. B., & Talling, P. J. (2010). Large landslides on passive continental margins: Processes, hypotheses and outstanding questions, in submarine mass movements and their consequences. In D. C. Mosher (Ed.), *Advances in natural and technological hazards research* (Vol. 28, pp. 153–165). Dordrecht, Netherlands: Springer.

Monecke, K., Anselmetti, F., Becker, A., Sturm, M., & Giardini, D. (2004). The record of historic earthquakes in lake sediments of Central Switzerland. *Tectonophysics, 394*, 21–40.

Moore, J. G., & Shannon, P. M. (1991). Slump structures in the late tertiary of the Porcupine Basin, offshore Ireland. *Marine and Petroleum Geology, 8*, 184–197.

Moretti, M., & Sabato, L. (2007). Recognition of trigger mechanisms for soft-sediment deformation in the Pleistocene lacustrine deposits of the Sant'Arcangelo Basin (Southern Italy): Seismic shock vs. overloading. *Sedimentary Geology, 196*, 31–45.

Morley, C. K. (Ed.). (1999). *Geoscience of rift systems – evolution of East Africa*. American Association of Petroleum Geologists, Studies in Geology 44.

Mulder, T., & Alexander, J. (2001). The physical character of subaqueous sedimentary density flows and their deposits. *Sedimentology, 48*, 269–299.

Mutti, E. (1992). *Turbidite Sandstones* (275 p). San Donato Milanese: AGIP – Istituto di Geologia Università di Parma.

Mutti, E., Tinterri, R., Remacha, E., Mavilla, N., Angela, S., & Fava, L. (1999). *An introduction to the analysis of ancient turbidite basins from an outcrop perspective*. Amer. Assoc. Petrol. Geol., Cont. Educ. Course Note Series 39.

Niemi, T. M., & Ben-Avraham, Z. (1994). Evidence for Jericho earthquakes from slumped sediments of the Jordan River delta in the Dead Sea. *Geology, 22*, 395–398.

Nobel, F. A., Andriessen, P. A. M., Hebeda, P. H., Priem, H. N. A., & Rondeel, E. H. (1981). Isotopic dating of the postalpine neogene volcanism in the Betic cordilleras, Southern Spain. *Geologie en Mijnbow, 60*, 209–214.

Ogata, K., Pini, G. A., Carè, D., Zélic, M., & Dellisanti, F. (2012). Progressive development of block-in-matrix in a shale-dominated shear zone: Insights from the Bobbio Tectonic Window (Northern Apennines, Italy). *Tectonics, 31*. https://doi.org/10.1029/2011TC002924.

Ogata, K., Pogacnik, Z., Pini, G. A., Tunis, G., Festa, A., Camerlenghi, A., & Rebesco, M. (2014). The carbonate mass transport deposits of the Paleogene Friuli Basin (Italy/Slovenia): Internal anatomy and inferred genetic processes. *Marine Geology, 356*, 88–110.

Ortí, F., Rosell, L., Gibert, L., Moragas, M., Playà, E., Inglés, M., Rouchy, J. M., Calvo, J. P., & Gimeno, D. (2014). Evaporite sedimentation in tectonically active basins: The lacustrine Las Minas gypsum unit (late Tortonian, SE Spain). *Sedimentary Geology, 311*, 17–42.

Osleger, D. A., Heyvaert, A. C., Stoner, J. S., & Verosub, K. L. (2009). Lacustrine turbidites as indicators of Holocene storminess and climate: Lake Tahoe, California and Nevada. *Journal of Paleolimnology, 42*, 103–122.

Owen, G., & Moretti, M. (2011). Identifying triggers for liquefaction-induced soft-sediment deformation in sands. *Sedimentary Geology, 235*, 141–147.

Piper, D. J., & Normark, W. R. (2009). Processes that initiate turbidity currents and their influence on turbidites: A marine geology prespective. *Journal of Sedimentary Research, 79*, 347–362.

Platt, N. H., & Wright, V. P. (1991). Lacustrine carbonates: Facies models, facies distribution and hydrocarbon aspects. In P. Anadón, L. Cabrera, & K. Kelts (Eds.), *Lacustrine facies analysis* (International association Sedimentologists special Publs), (Vol. 13, pp. 57–74). Oxford, Great Britain. Blackwell Sci. Co.

Posamentier, H. W., & Walker, R. G. (2006). Deep-water turbidites and submarine fans. In H. W. Posamentier & R. G. Walker (Eds.), *Facies models revisited* (SEPM Special Publ), (Vol. 84, pp. 399–520). Tulsa, Oklahoma: SEPM.

Pozo, M., Calvo, J. P., Scopelitti, G., & González-Acebrón, L. (2016). Seepage carbonate mounds in Caenozoic sedimentary sequences from the Las Minas Basin, SE Spain. *Sedimentary Geology, 334*, 1–20.

Rodríguez-Pascua, M. A. (2001). *Paleosismicidad y sismotectónica de las cuencas lacustres neógenas del Prebético de Albacete* (p. 285). Albacete, Spain: Instituto Estudios Albacetenses.

Rodríguez-Pascua, M. A., Calvo, J. P., & Gómez-Gras, D. (2000). Soft-sediment deformation structures interpreted as seismites in lacustrine sediments of the Prebetic Zone, SE Spain, and their potential use as indicators of earthquake magnitudes during the Late Miocene. *Sedimentary Geology, 135*, 117–135.

Rodríguez-Pascua, M. A., Calvo, J. P., Davenport, C., & Gómez-Gras, D. (2003). Sedimentary record of seismic events, with examples from recent and fossil lakes. In B. Valero-Garcés (Ed.), *Limnogeology in Spain: a Tribute to Kerry Kelts* (pp. 253–282). Biblioteca de Ciencias CSIC. Madrid, Spain.

Rodríguez-Pascua, M. A., Bischoff, J., Garduño-Monroy, V. H., Pérez-López, R., Giner-Robles, J. L., Israde-Alcántara, I., Calvo, J. P., & Williams, R. V. (2009). Estimation of the tectonic slip-rate from quaternary lacustrine facies within the intraplate Albacete province (SE Spain). *Sedimentary Geology, 222*, 89–97.

Rodríguez-Pascua, M. A., Garduño-Monroy, V. H., Israde-Alcántara, I., & Pérez-López, R. (2010). Estimation of the paleoepicentral area from the spatial gradient of deformation in lacustrine seismite (Tierras Blancas Basin, Mexico). *Quaternary International, 219*, 66–78.

Rosell, L., Ortí, F., Gibert, L., Deino, A., & Gimeno, D. (2011). *Las Minas de Hellín gypsum: Cyclicity and age*. Paper presented at the IAS-2011 Meeting of Sedimentology, Zaragoza, Spain, abstracts volume.

Sanz de Galdeano, C., & Vera, J. A. (1992). Stratigraphic record and palaeogeographical context of the Neogene basins in the Betic Cordillera, Spain. *Basin Research, 4*, 21–36.

Schnellmann, M., Anselmetti, F., Giardini, D., McKenzie, J. A., & Ward, S. N. (2002). Prehistoric earthquake history revealed by lacustrine slump deposits. *Geology, 30*, 1131–1134.

Schnellmann, M., Anselmetti, F., Giardini, D., & McKenzie, J. A. (2005). Mass movement-induced fold-and-thrust belt structures in unconsolidated sediments in Lake Lucerne (Switzerland). *Sedimentology, 52*, 271–289.

Scholz, C. A. (2002). Applications of seismic sequence stratigraphy in lacustrine basins. In W. M. Last & J. P. Smoll (Eds.), *Basin analysis, coring and chronological techniques, tracking environmental changes using lake sediments* (Vol. 1, pp. 7–22). Dordrecht, The Netherlands: Kluwer Academic Publishers.

Seltzer, G. O., Baker, P., Cross, C., Dunbar, R., & Fritz, S. (1998). High-resolution seismic reflection profiles from Lake Titicaca, Peru – Bolivia: Evidence for Holocene aridity in the tropical Andes. *Geology, 26*, 167–170.

Servant-Vildary, S., Rouchy, J. M., Pierre, C., & Foucault, A. (1990). Marine and continental water contributions to a hypersaline basin using diatom ecology, sedimentology and stable isotopes: An example in the late Miocene of the Mediterranean (Hellin Basin, southern Spain). *Palaeogeography, Palaeoclimatology, Palaeoecology, 79*, 189–204.

Shanmugam, G. (2012). *New perspectives on deep-water sandstones: Origin, recognition, initiation, and reservoir quality*. Elsevier, 9, Amsterdam.

Stow, D. A. V. (1986). Deep clastic seas. In H. G. Reading (Ed.), *Sedimentary environments and facies* (Vol. 2, pp. 399–444). Oxford: Blackwell.

Strachan, L. J. (2008). Flow transformation in slumps: A case study from the Waitemata Basin, New Zealand. *Sedimentology, 55*, 1311–1332.

Strachan, L. J., & Alsop, G. I. (2006). Slump folds as indicators of palaeoslope: A case study from the Fisherstreet Slump of County Clare, Ireland. *Basin Research, 18*, 451–470.

Sumner, E. J., Talling, P. J., & Amy, L. A. (2009). Deposits of flows transitional between turbidity current and debris flow. *Geology, 37*, 991–994.

Talbot, M. R., & Allen, P. A. (1996). Lakes. In H. G. Reading (Ed.), *Sedimentary environments: Processes, facies and stratigraphy* (pp. 83–124). Oxford: Blackwell Science.

Talling, P. J., Amy, L. A., Wynn, R. B., Pekall, J., & Robinson, M. (2004). Beds comprising debrite sandwiched within co-genetic turbidite: Origin and widespread occurrence in distal depositional environments. *Sedimentology, 51*, 163–194.

Talling, P. J., Masson, D. G., Sumner, E. J., & Malgenisi, G. (2012). Subaqueous sediment density flows: Depositional processes and deposit types. *Sedimentology, 59*, 1937–2003.

Van Daele, M., van Melden, A., Moernaut, J., Beck, C., Audemard, F., Sanchez, J., Jouanne, F., Carrillo, E., Malavé, G., Lemus, A., & De Batist, M. (2011). Reconstruction of late Quaternary Sea- and lake-level changes in a tectonically active marginal basin using seismic stratigraphy: The Gulf of Cariaco, NW Venezuela. *Marine Geology, 279*, 37–51.

Van der Merwe, W. C., Hodgson, D. M., & Flint, S. S. (2011). Origin and terminal architecture of a submarine slide: A case study from the Permian Vischkuil formation, Karoo Basin, South Africa. *Sedimentology, 58*, 2012–2038.

Vanneste, M., Mienert, J., & Bünz, S. (2006). The Hinlopen slide: A giant, submarine slope failure on the northern Svalbard margin, Arctic Ocean. *Earth and Planetary Science Letters, 245*, 373–388.

Weinberger, R., Alsop, G. I., Levi, T., & Eyal, I. (2016). Coseismic horizontal slip revealed by sheared clastic dikes in the Dead Sea basin. *Geological Society of America Bulletin, 128*, 1193–1206.

Wignall, P. B., & Best, J. L. (2004). Sedimentology and kinematics of a large retrogressive growth-fault system in Upper Carboniferous deltaic sediments, western Ireland. *Sedimentology, 51*, 1343–1358.

Lacustrine and Fluvial Carbonate Microbialites in the Neogene of the Ebro Basin, Spain: A Summary of Up-to-Date Knowledge

Concha Arenas-Abad, Leticia Martin-Bello, F. Javier Pérez-Rivarés, Nerea Santos-Bueno, and Marta Vázquez-Urbez

Abstract This contribution provides a comparison between laminated microbial structures formed in saline lacustrine environments (lower and middle Miocene in the Sierra de Alcubierre and Montes de Castejón) and fluvial to fluvio-lacustrine environments (middle to upper Miocene in the Muela de Borja and Montolar hill), in the Ebro Basin. Saline and freshwater conditions were deduced from sedimentological features and stable isotope composition ($\delta^{13}C$ and $\delta^{18}O$) of the carbonate facies. Primarily the focus is on laminae and lamination: shape, thickness, textural features, and lamina arrangement.

In the saline carbonate lacustrine record, stromatolites and rare oncolites consist of calcite and dolomite, and occur as thin planar, domed, and stratiform stromatolites. These forms are always associated with laminated limestones and dolostones that include ripple and horizontal lamination, and hummocky cross-stratification. In the microbialites, micrite, microsparite, and sparite define three types of simple laminae that generally are grouped into composite dark and light laminae. The laminae are thin, flat to slightly undulate, with very smooth cross-sectional shapes. Preserved microbial components are rare and, when present, consist of loose filamentous bodies or rare tubes sub-perpendicular to lamination.

This chapter is dedicated to our colleague Elisabeth Gierlowski-Korsdesch, who would have enjoyed visiting the Neogene record of the Ebro Basin

C. Arenas-Abad (✉) · L. Martin-Bello · N. Santos-Bueno
Division of Stratigraphy, Department of Earth Sciences, University of Zaragoza, Zaragoza, Spain

Geotransfer Group and Institute for Research on Environmental Sciences of Aragón (IUCA), University of Zaragoza, Zaragoza, Spain
e-mail: carenas@unizar.es

F. J. Pérez-Rivarés · M. Vázquez-Urbez
Division of Stratigraphy, Department of Earth Sciences, University of Zaragoza, Zaragoza, Spain

M. R. Rosen et al. (eds.), *Limnogeology: Progress, Challenges and Opportunities*, Syntheses in Limnogeology, https://doi.org/10.1007/978-3-030-66576-0_6

In the freshwater environment (open-lake and fluvial record) abundant oncolites, calcite-coated stems and minor stromatolites consist of low-Mg calcite. These components formed in gently channeled streams and open lakes, with tufaceous palustrine fringes. The shape of oncolites reflects the morphology of the nuclei. The laminae consist of micrite, microsparite, and sparite, forming simple and composite laminae. The laminae are generally thick, gently undulate to wavy in cross section. There are multiple ranges of lamination. Microbial bush-shaped and fan-shaped bodies, set perpendicular to sub-perpendicular to lamination, are abundant, which makes the laminae wavy to crinkly and porous.

It is suggested that salinity exerted some control on the types of microbes and/or their style of development, which impacted the geometry of the laminae. In all cases, the lamina arrangement into cycles can record seasonal to multi-annual changes in climate parameters. In contrast, gross morphology of microbialites was highly influenced by depth and hydrodynamics, and largely by the nucleus morphology in the case of oncolites and stem coatings.

Keywords Stromatolites · Oncolites · Calcite-coated stems · Microbial lamination · Lacustrine and Fluvial carbonates · Saline and Fresh water · Neogene · Ebro Basin

Introduction

Laminated microbial structures (i.e., stromatolites and oncolites, sensu Riding 2011) are common in a variety of marine and continental settings, and can be found throughout the geological record, from the Arcaic to present day (Awramik 1992; Bosak et al. 2013; Riding 2000; Walter 1972, 1976). Many scientists have focused on the environmental significance of the variable external morphology and internal growth forms in the corresponding deposits (Anadón and Zamarreño 1981; Andres and Reid 2006; Casanova 1986, 1994; Hofmann 1973; Tosti and Riding 2017; Zamarreño et al. 1997). Others have discussed the meaning of lamination in terms of changes of depositional and climatic parameters, but published works are much fewer (Bertrand-Sarfati et al. 1994; Bosak et al. 2013; Casanova 1994; Frantz et al. 2014; Hofmann 1973). Compared to the marine record, the number of studies dealing with continental laminated microbialites is much smaller, despite the non-marine realm offering a wide array of depositional situations to be studied. Actually, understanding microbialite formation in the continental record is still open to debate, mainly because of the great variety of both intrinsic and extrinsic parameters involved and the diverse scales of observation (Frantz et al. 2014; Storrie-Lombardi and Awramik 2006). For example, the continental record allows comparison between freshwater and saline microbial laminated structures through space and/or time, an issue that merits more detailed studies.

This chapter focuses on stromatolites, oncolites, and the associated calcite-coated stems that formed in two distinct depositional settings through the Neogene in the Ebro Basin (northeast of Spain). It is a summary of up-to-date knowledge on the Neogene Ebro Basin microbialites that includes a comparison between the laminated microbial structures formed in saline lacustrine environments (lower and middle Miocene in the Sierra de Alcubierre and Montes de Castejón) and fluvial to fluvio-lacustrine environments (middle and upper Miocene in Muela de Borja and Montolar hill), and also a comparison with other lacustrine and fluvial records. Textural and geometric features of the laminae, along with microbial components and lamina arrangement, are used to compare the impact of environmental and climatic parameters in each depositional setting.

In order to make the comparison through the Ebro Basin, for the Sierra de Alcubierre and Montes de Castejón regions, previous work by Arenas and Pardo (1999) and recent work by Martín-Bello et al. (2019) are cited and adapted herein to describe the main features of stromatolites and their lamination. In the case of the Muela de Borja and Montolar regions, work by Arenas et al. (2000), Vázquez-Urbez (2008), and Vázquez-Urbez et al. (2013) constitute the essential information for further detailed description, interpretation, and comparison of oncolites and their lamination.

Geological Setting

The Ebro Basin, a Cenozoic Alpine, intermontane basin in the northeastern part of the Iberian Peninsula, formed as a result of the collision of the Iberian and Eurasian plates. It is bounded by the Pyrenean, Iberian, and Catalan Coastal Ranges (Fig. 1). The basin fill has a wedge-shaped geometry that widens to the north, except in the western sector where the fill forms a symmetric trough (Muñoz et al. 2002; Muñoz-Jiménez and Casas-Sainz 1997; Pardo et al. 2004). It is considered the latest foreland basin of the Pyrenean Ranges. The sedimentary record comprises continental and marine deposits until the Eocene, i.e., during the time that the basin was connected to the Atlantic Ocean. The marine deposits include platform carbonates and marls, coastal and deltaic detrital facies, and evaporites. From the late Eocene, the basin became fully continental (Riba et al. 1983) and gathered approximately 5.5 km in thickness of alluvial, fluvial, and lacustrine deposits. In the middle to late Miocene the basin drainage opened to the Mediterranean Sea, then starting a period of emptying that continues nowadays (Arche et al. 2010; García Castellanos et al. 2003; Urgeles et al. 2011; Vázquez-Urbez et al. 2005, 2013).

The sedimentary record has been divided into eight genetic stratigraphic units, named 1–8 (Fig. 1; Muñoz et al. 2002; Pardo et al. 2004). Except for unit 8, ages of these units have been determined by paleontological sites and magnetostratigraphic studies (Fig. 1). Units 1–3 are Paleogene, unit 4 is Oligocene-Miocene, and units 5–8 are Miocene in age. Figure 2 shows the paleogeographic evolution of the basin during the Neogene.

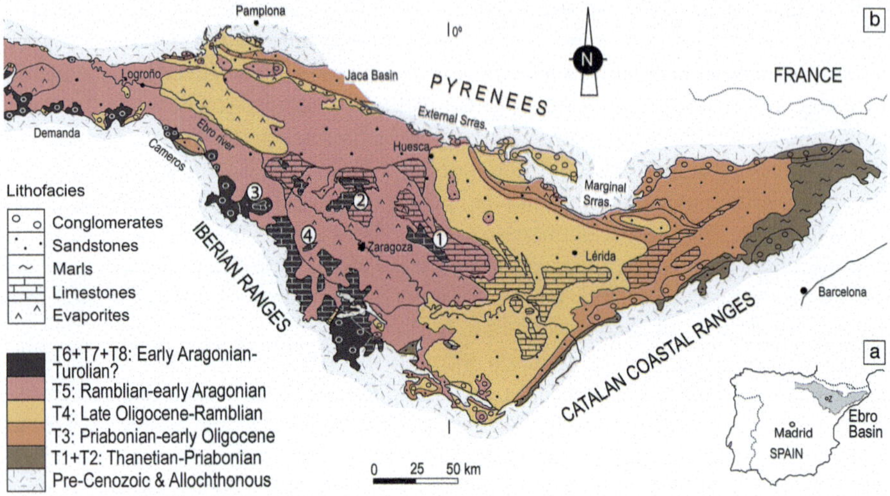

Fig. 1 (**a**) Location of the Ebro Basin in the Iberian Peninsula. (**b**) Geological map of the Ebro Basin, with genetic stratigraphic units and lithostratigraphy (adapted from Muñoz et al. 2002 and Pardo et al. 2004). 1, Sierra de Alcubierre; 2, Montes de Castejón; 3, Muela de Borja; 4, Montolar hill

This work focuses on the microbialites of the Miocene lacustrine and fluvio-lacustrine successions of a part of the central region of the Ebro Basin: units 5, 6, and 7 in the Sierra de Alcubierre (SA) and Montes de Castejón (MC), and units 7 and 8 in the Muela de Borja (MB) (Figs. 1 and 3a, b). Other outcrops in the Ebro Basin comparable to those of the Muela de Borja that include laminated microbialites overlying unit 7 (likely of unit 8) are located approximately 30 km west of the city of Zaragoza, at the hill named Montolar (MO) (Fig. 2). The sediment source areas of the associated alluvial and fluvial systems were the Pyrenees for the first two areas, and mainly the Iberian Ranges, for the latter two (Fig. 2). In the four uplands the strata are nearly horizontal or have small dip. Unit 8 deposition occurred when the Ebro Basin was opened to the Mediterranean sea (Vázquez-Urbez et al. 2013).

Stratigraphy and Sedimentology of the Neogene: Units 5, 6, 7, and 8

Stratigraphy

Distributions of units and lithofacies are shown in Fig. 3a, b. Dating of units 5–7 in the studied area was based on paleontological data and magnetostratigraphic studies (compiled in Pérez Rivarés 2016; Pérez-Rivarés et al. 2018). Unit 8 lacks

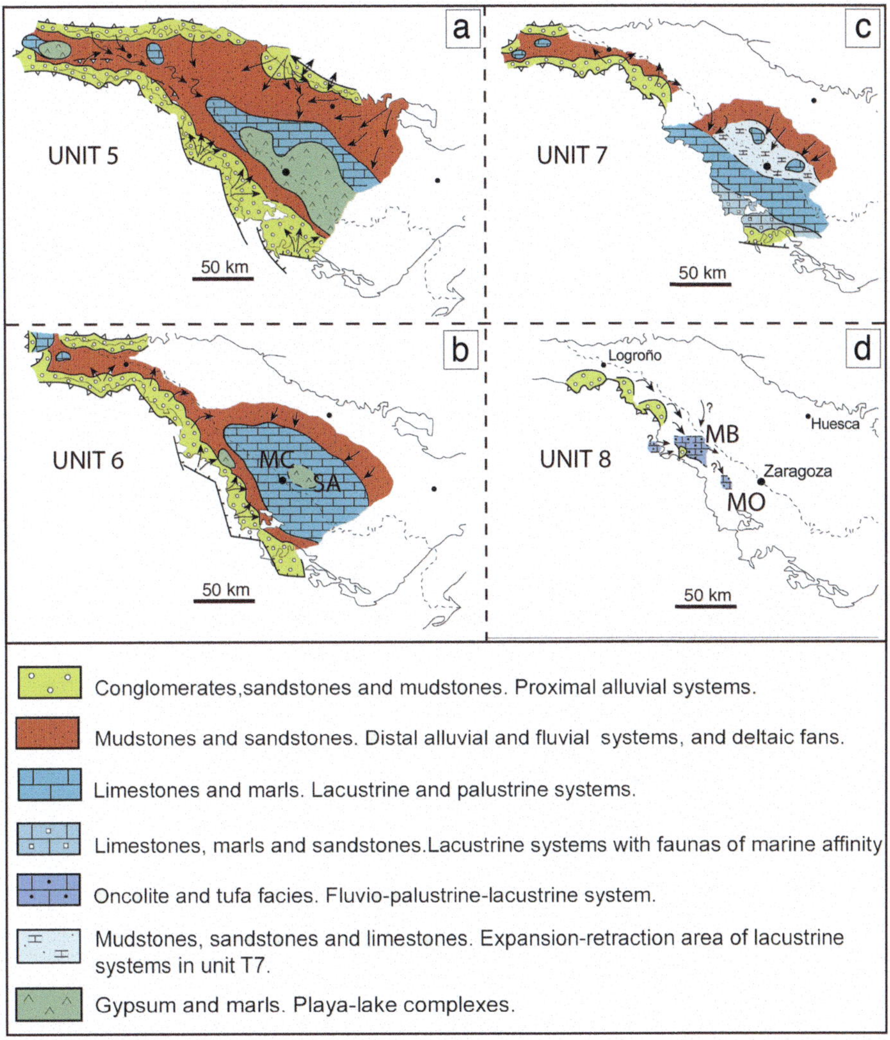

Fig. 2 Paleogeographic maps at the time of deposition of the four Neogene units in the Ebro Basin. (Modified from Pardo et al. (2004)). Uplands: SA Sierra de Alcubierre, MC Montes de Castejón, MB Muela de Borja, MO Montolar area

chronological data, but its age is assumed to be middle-late Miocene given its location on unit 7 (Arenas et al. 2000; Muñoz et al. 2002). The main stratigraphic features of units 5–8 in the studied areas of the basin in this contribution are shown in Fig. 4 and are briefly described below. In the central part of the basin, the boundaries between these units are sharp lithological changes and/or inflections in the sequential evolution, and locally involve erosional surfaces (Pardo et al. 2004; Pérez-Rivarés et al. 2018).

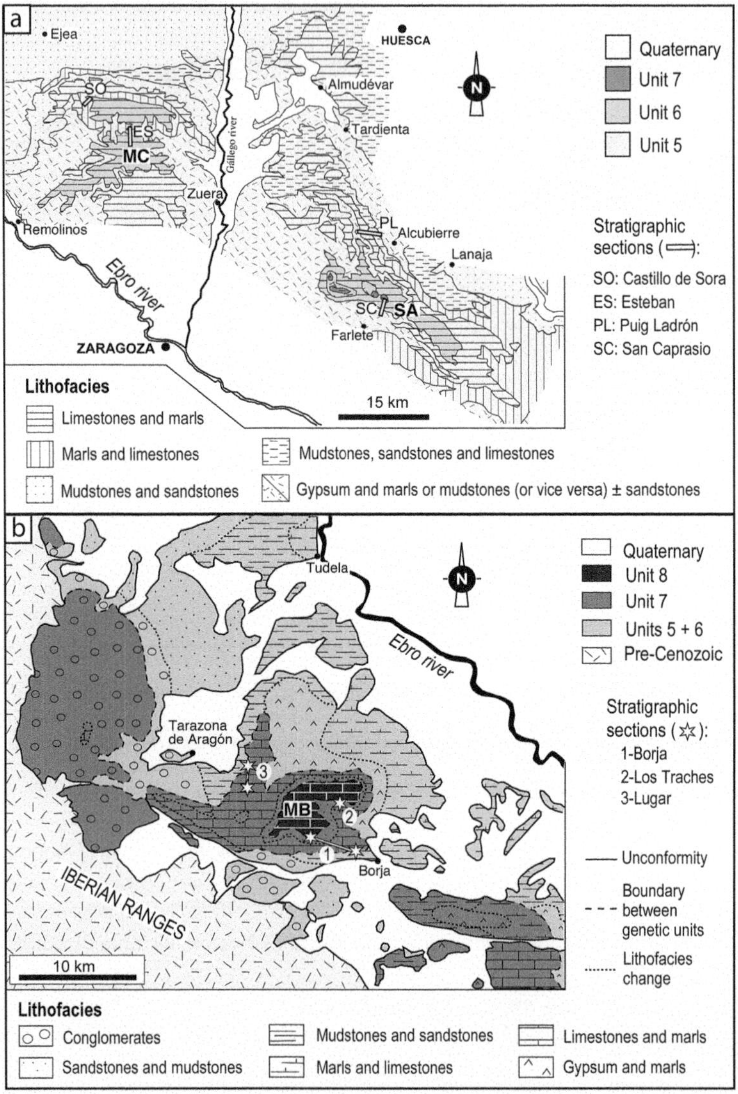

Fig. 3 Geological maps, with stratigraphic genetic units and lithofacies in: (**a**) the Sierra de Alcubierre (SA) and Montes de Castejón (MC), and (**b**) the Muela de Borja (MB). (**a**) is based on Arenas and Pardo (1999), and (**b**) is based on Pérez et al. (1988) and Vázquez-Urbez et al. (2013)

Unit 5 (C6AAr–C5Cn), ca. 330 m thick, consists mainly of gypsum and marls with limestones intercalations (Zaragoza Formation). These lithologies grade laterally to the east and north to dominant limestones and marls (Alcubierre Formation), which further north pass into mudstones and sandstones (Uncastillo and Sariñena Formations, to the north of the River Ebro).

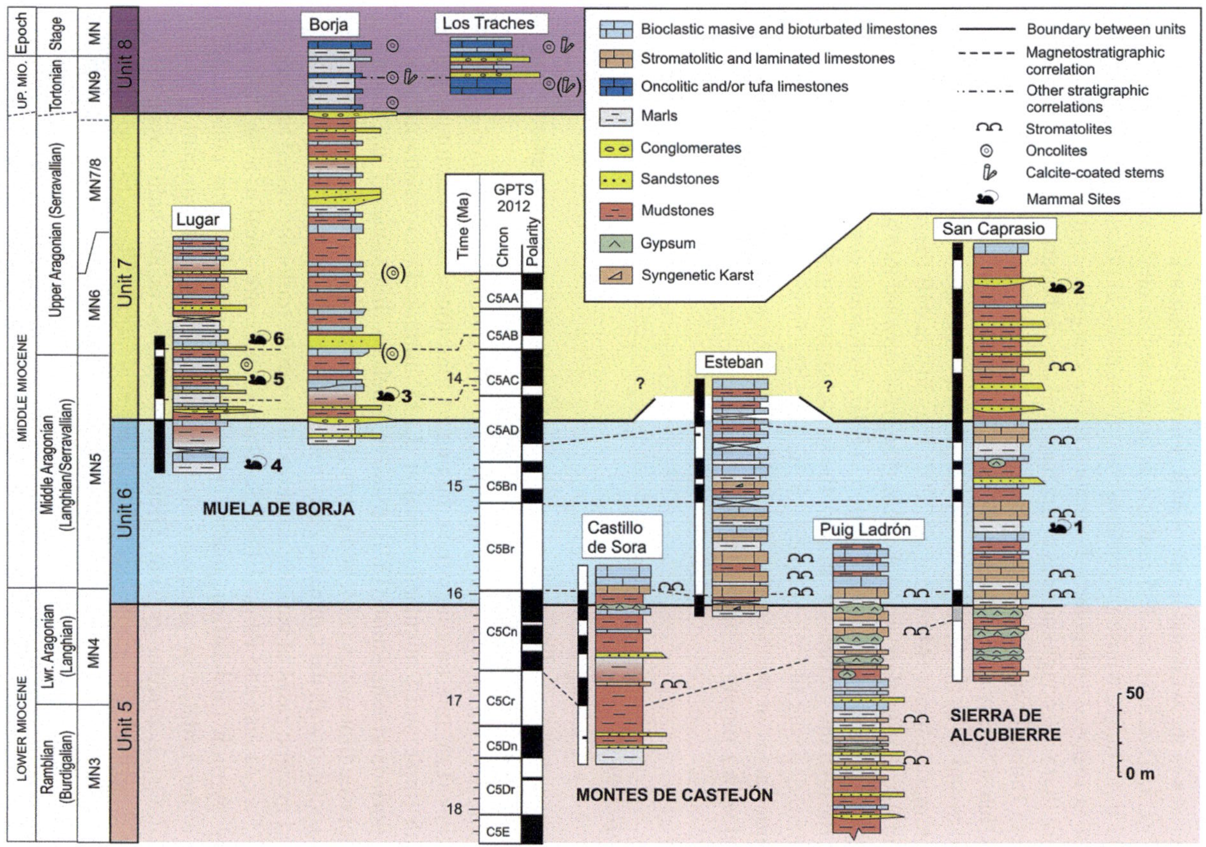

Fig. 4 Representative stratigraphic sections in the Sierra de Alcubierre, Montes de Castejón and Muela de Borja uplands, Neogene of the Ebro Basin. Note the abundance of stromatolites and oncolites, and calcite-coated stems. Stratigraphy compiled from Arenas and Pardo (2000) and Vázquez-Urbez et al. (2013). Magnetostratigraphy based on Pérez-Rivarés et al. (2018). A maximum difference of approximately 300 ka is detected between the boundaries of units in the north and south domains. Fossil mammal Sites: 1, SC-109 and 2, SC-212 (Agustí et al. 2011); 3, Borja (Cuenca et al. 1992); 4, Tarazona de Aragón (Astibia 1986); 5, ME-20, and 6, ME-30 (Murelaga et al. 2008)

Unit 6 (C5Cn–C5AD), ca. 140 m thick, is mostly formed of limestones and marls with minor fine sandstone intercalations (in SA, MC and MO, Alcubierre Formation, and in MB, Tudela Formation). The most important lateral changes correspond to the presence of gypsum deposits in the southwestern part of SA (Perdiguera Member) and to the north and east of the MB region (Monteagudo Gypsum), and to conglomerates to the west and south of MB (Fitero Formation).

Unit 7 (C5AD–C5r3), ca. 185 m thick, is composed of mudstones, sandstones, marls, and limestones, whose proportions vary from place to place. In SA only the lower half is exposed (Alcubierre Formation). The complete unit is only seen in MB, where there are conglomerates at the base and the carbonates include oncolite limestones (Tudela Formation).

Unit 8 (latest Serravalian to unknown end), only crops out in MO and MB, where it is ca. 50 m and 55 m thick, respectively. In westward regions it is formed of proximal conglomerate facies (Muñoz et al. 2002; Fig. 2d). In MB the unit is formed of coarse and fine detrital deposits at the base that grade upward into a dominant limestone sequence (Muela de Borja Formation in MB), with some conglomerates intercalations through the carbonates. In MO, the sequence is formed only by limestone and rare marl deposits and minor breccia at the base. The limestones of both uplands have unique facies in the Neogene of the Ebro Basin: oncolite and tufa deposits associated with a variety of bioclastic limestones.

Sedimentology

A wide array of sedimentary facies have been described, including coarse to fine clastics, carbonate, and sulphate deposits (Table 1).

In the *Sierra de Alcubierre and Montes de Castejón*, the carbonate facies include: stromatolites (Ls), rare oncolites (Lo), laminated limestones and dolostones (Ll), either with ripple and lenticular lamination (Ll.1), hummocky cross-stratification (Ll.2) or horizontal lamination (Ll.3), massive bioclastic limestones (Lm), bioturbated bioclastic limestones (Lb), and marls (M). Stromatolites and oncolites occur always associated with the laminated carbonate facies; both Ll and Ls are formed of calcite and in some cases of variable amounts of dolomite (dolomite varied between 0% and 75%; Arenas and Pardo 1999). Sulphate facies correspond to a variety of secondary gypsum (alabastrine, macrocrystalline, and granular) that constitutes nodules, either isolated or grouped into layers (Gn), horizontal lamination and ripples (Gl, Gr). Lenticular gypsum (Glen) is rare. Glauberite and halite layers (H) have been described in borehole logs (cf., Salvany et al. 2007). Clastic facies include sandstones (Sm, St, Sr, Sh) and mudstones (Fg, Fo). The carbonate facies are associated vertically into simple sequences that represent (1) distal alluvial to lacustrine expansion, (2) deepening-shallowing processes, or (3) shallowing processes (Fig. 5b, I, II, and III in this work. Facies associations A, B, and C in Arenas and Pardo 1999).

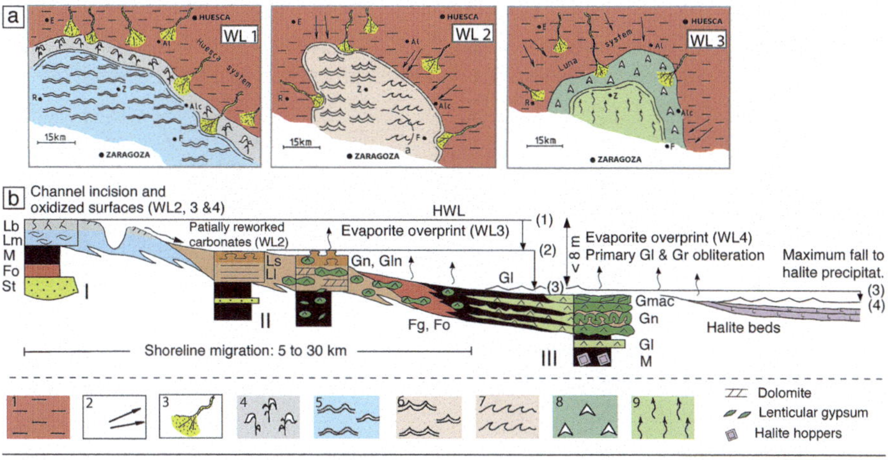

Fig. 5 (**a**) Paleogeographic maps and (**b**) simplified sedimentary facies model for lacustrine systems of units 5, 6, and lower half of 7 in the central part of the Ebro Basin (Sierra de Alcubierre and Montes de Castejón; modified from Arenas et al. 2007a, b). The scheme shows the different sequences of facies and processes resulting from a continuous fall of lake water level from freshwater (HWL = WL 1) to saline carbonate (WL 2), sulphate (WL 3) and halite (WL 4) depositional environments. Sequence I: Distal alluvial to carbonate lacustrine expansion; sequence II: shallowing upward in saline carbonate environment; sequence III: shallowing upward in sulphate lacustrine environment and evaporite overprint. The maps represent the distribution of depositional subenvironments during certain time intervals correspondent to the upper part of unit 5 (a: WL 1; c: WL 3) and lower part of unit 6 (b: WL 2; Line **a** indicates the approximate northern boundary of sulphate areas during WL 3). Legend for maps: 1: alluvial plain (distal sector of alluvial systems), with 2: sheet flows, and 3: channelled flows; 4: palustrine margin; 5: freshwater carbonate depositional environments; 6: saline carbonate depositional environments; 7: surge-dominated area; 8: saline mud flat; 9: sulphate depositional environments. Localities: Alc Alcubierre, Al Almudévar, E Ejea de los Caballeros, F Farlete, R Remolinos, Z Zuera

In the lacustrine facies model proposed for units 5 and 6 in SA and MC, and lower half of unit 7 in SA, stromatolites and rare oncolites formed during oscillations between high water level (freshwater carbonate-depositing conditions, Lm, Lb) and low water level (sulphate-depositing conditions, Gh, Gr); sulphate deposition has not been reported in unit 7 (Fig. 5a). These lake-level variations are consistent with lithofacies distribution through space and with the differences in the $\delta^{13}C$ and $\delta^{18}O$ between facies, with higher values in Ll and Ls than in M, Lm, and Lb (Fig. 6). Mean $\delta^{13}C$ and $\delta^{18}O$ values (‰ PDB) were, respectively, −1.7 ± 1.1 and −3.3 ± 2.6 for Ll and Ls (N = 112), and −3.0 ± 1.1 and −6.1 ± 1 for Lm, Lb and M (N=41) (values reported in Arenas and Pardo 1999). Therefore, facies Ll and Ls were interpreted to have formed in saline carbonate conditions, at places with dolomicrite precipitation (Arenas et al. 1997; Arenas and Pardo 1999). These stromatolites are abundant throughout the succession, show a variety of external and internal morphologies, and are associated with storm and fair-weather surge activity and colonization of erosional, commonly exposed, surfaces (Arenas and Pardo 1999; Martin-Bello et al. 2019).

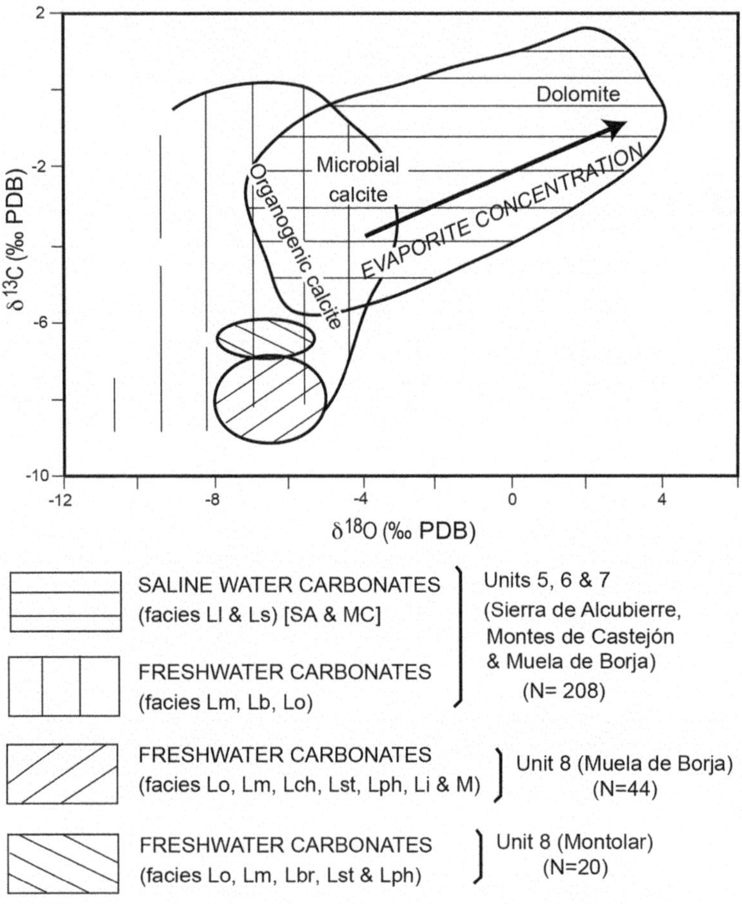

Fig. 6 Stable isotope composition ($\delta^{13}C$ and $\delta^{18}O$) of carbonate facies of the Neogene in the Ebro Basin. Diverse facies of all the studied areas. Note tight groups of unit 8 in Muela de Borja and Montolar hill. (Compiled from Arenas and Pardo (1999), Arenas et al. (2000), Vázquez-Urbez (2008) and Vázquez-Urbez et al. (2013))

In the *Muela de Borja* region, carbonate facies are formed of calcite and consist of tufa, oncolites (Lo), rare stromatolites (Ls), intraclastic limestones (Li), bioclastic limestones (Lm) with variable degree of bioturbation (Lb), and marls (M), most of which are coincident with those in SA and MC. Tufa facies are only present in unit 8 and include: up-growing stem boundstones (Lst) and calcite-coated phytoclast rudstones (Lph). Oncolites and rare stromatolites are present in units 7 and 8. Oncolites in unit 7 are within marl and marly limestone layers. In unit 8, oncolites, and calcite-coated phytoclasts and stems are ubiquitous. Clastic facies are formed of conglomerates (Co, Ct), sandstones (Sm, St, Sf) and mudstones (Fm), as summarized in Table 1. In the *Montolar hill*, the carbonate facies of unit 8 coincide with those in MB, except for the presence of moss boundstones (Lbr). In MO, moss phytoherms constitute mounds that are linked to stem boundstones and rudstones.

Table 1 Main textural and structural characteristics of lithofacies in the studied area. Modified from Arenas et al. (2007b)

Unit	Facies	Subfacies	Texture	Physical sediment. structures & geometry	Biological & diagenetic features & components	Interpretation
Units 5, 6 ,7 (Sierra de Alcubierre and Montes de Castejón)	Marls: M	Laminated marls Massive marls		Parallel lamination or lenticular stratification None		Settle-out, mostly in offshore lake areas linked to runoff periods
	Laminated Limestones: Ll	Ll 1 Ll 2 Ll 3	Micrite and dolomicrite with mm to cm detrital laminae and lenses	Lenticular or wavy stratification Hummocky cross-stratification Graded parallel lamination	Associated Associated Associated	Wave influence above or below storm surge level Shore sheet flows or inner turbidite-like currents (Ll2) Moderate salinity waters
	Stromatolitic Limestones & Oncolites: Ls, Lo	Thin planar stromatolites, Domed stromatolites, Stratiform stromatolites and Oncolites	Boundstones	Microscopic alternations of light and dark micrite and microspar laminae	Rod and coccoid calcite bodies. Rare filament cyanbacteria, fungi and EPS	Shallow / marginal lacustrine areas. Moderate salinity waters in S. Alcubierre and M. Castejón
	Bioclastic Limestones	Massive: Lm	Mudstones & wackestones	Uncommon and poor, lamination	If present, weak bioturbation	Fresh water, shallow lacustrine areas. Permanent water supply
	Bioturbated Limestones: Lb	Depending on dia-genetic processes	Mudstones & wackestones	Desiccation cracks	Nodules, breccias, etc.	Palustrine conditions. Shallowing of previous fresh water ponded areas
	Nodular Gypsum: Gn		Alabastrine	Commonly, massive	Nodules, forming beds or isolated	Evaporite processes in saline mud flats
	Lenticular Gypsum: Glen		Lenses	Massive, rarely forming laminae		Gypsum precipitation in water lake and interstitially within the sediment
	Rippled & Laminated Gypsum: Gr, Gl		Alabastrine	Parallel, lenticular & rippled lamination		Gypsum precipitation in hypersaline lake water
	Glauberite & Halite: H		Cubic & chevron			Precipitation of Na-rich SO_4 and Cl in very shallow salinas and interstitially
	Mudstones: F	Green, grey & ochre Fg, Fo		None, massive Parallel lamination	Mottling	Nearshore lake areas or alluvial plains surrounding lacustrine areas
	Sandstones: Sm, Sh, Sr & St			Massive, parallel, rippled, & trough-cross stratification		Sheet flows and channels of alluvial plain near or within the lake areas.

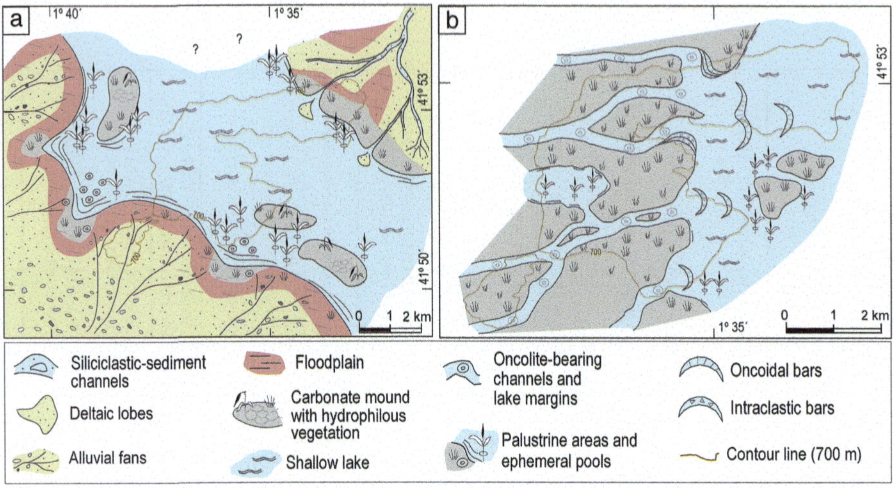

Fig. 7 Paleogeographic maps during deposition of lower half of unit 7 (**a**) and for the most common context during deposition unit 8 (**b**) in the Muela de Borja. (Modified from Vázquez-Urbez et al. (2013))

In unit 7 oncolites are restricted to distal alluvial to lacustrine-palustrine sequences (facies association 4 of Vázquez-Urbez et al. 2013; Figs. 2c and 7a). In unit 8, oncolites and the associated calcite-coated phytoclasts are ubiquitous and occur at the base of lacustrine-palustrine and fluvial carbonate sequences. Stem phytoherm tufa occurs associated with calcite-coated phytoclast tufa. Stromatolites are limited to the top of some oncolite layers.

Two sedimentary facies models were proposed for the Muela de Borja region. One explains the uppermost part of unit 6 and the entire unit 7, which corresponds to a shallow lacustrine system with palustrine fringes, with fluvial and alluvial inputs from southern and westward provenance (Fig. 7a). Oncolites formed in very shallow channels that overspilt producing overbank oncolite layers on the flood plain (e.g., as described in Oligocene deposits of Mallorca by Arenas et al. 2007a). The other model corresponds to unit 8; it represents a carbonate fluvial-lacustrine-palustrine system with shallow and uniform slope toward a lake body located to the east (Fig. 7b; Vázquez-Urbez et al. 2013). Overall, oncolites and calcite-coated phytoclasts formed in shallow, low sinuosity channels and lake margins, both fringed by extensive palustrine areas in which hydrophilous plants were coated by calcite, producing phytoherms. Breakage of these calcite-coated stems yielded phytoclasts. Stromatolites represent shallowing and calm conditions after fluvial oncolite and phytoclast accumulation (i.e., as in facies association 10a of Vázquez-Urbez et al. 2013).

In the *Montolar hill*, a fluvio-lacustrine facies model was proposed to explain the sedimentation of deposits that overlie unit 7 (Fig. 12 of Arenas et al. 2000). The model has some similarities with that of unit 8 in MB, e.g., in relation to oncolite-forming low-sinuosity channels, extensive palustrine fringes, and shallow lacustrine areas. The most important sedimentary differences are the absence of intercalated detrital deposits and the presence of moss cascades beside the channels. The MO fluvio-lacustrine system (at 503 m above sea level) probably received surface water inputs from outlets from MB (approximately at 740–800 m above sea level; Fig. 2d). This is consistent with a fluvio-lacustrine system running southeastward on its way to enter the Mediterranean Sea.

These models are consistent with the low values of the stable isotopic composition and poor correlation between C and O (Fig. 6). In MB, with mean values of $\delta^{13}C = -7.9 \pm 0.4$, and $\delta^{18}O = -6.8 \pm 0.7‰$ PDB ($N = 44$) (Vázquez-Urbez et al. 2013). In MO, the mean values are: $\delta^{13}C = -6.4 \pm 0.2$, and $\delta^{18}O = -7.2 \pm 0.4‰$ PDB ($N = 20$) (Arenas et al. 2000).

Saline Lacustrine Microbialites

Stromatolites and rare oncolites in units 5, 6, and 7 in SA and MC occur: (1) at the base of deepening-shallowing sequences, and (2) through and at the top of shallowing upward sequences (Martín-Bello et al. 2019). Several individual stromatolites are commonly found adjacent to each other in the same layer.

Gross Morphology of Stromatolites and Oncolites

According to the length to height ratio (L:H) and the shape, three main external geometry-based types are distinguished in stromatolites (Martín-Bello et al. 2019; Fig. 8): thin planar, domed, and stratiform.

- Thin planar stromatolites are 1 mm to 10 cm thick and dm to 2 m, exceptionally 6 m, long. The internal structure consists of smooth, flat to undulatory, laterally continuous laminae, which can form small domes, cones, and non-branching columns (Fig. 8a–d).
- Domed and stratiform stromatolites are commonly laterally related to each other, and only differ in their lateral extent. The domed stromatolites are 10–30 cm high, with a L:H of <1.5 (Fig. 8e). The stratiform stromatolites are 10–30 cm high and 4–30 m long, with a $L:H \gg 100$ (Fig. 8f). Internal growth forms include domes and columns, with increasing width upward and with highly enveloping laminae.

Oncolites are 3–6 cm in diameter and primarily occur at the base of some domed and stratiform stromatolites.

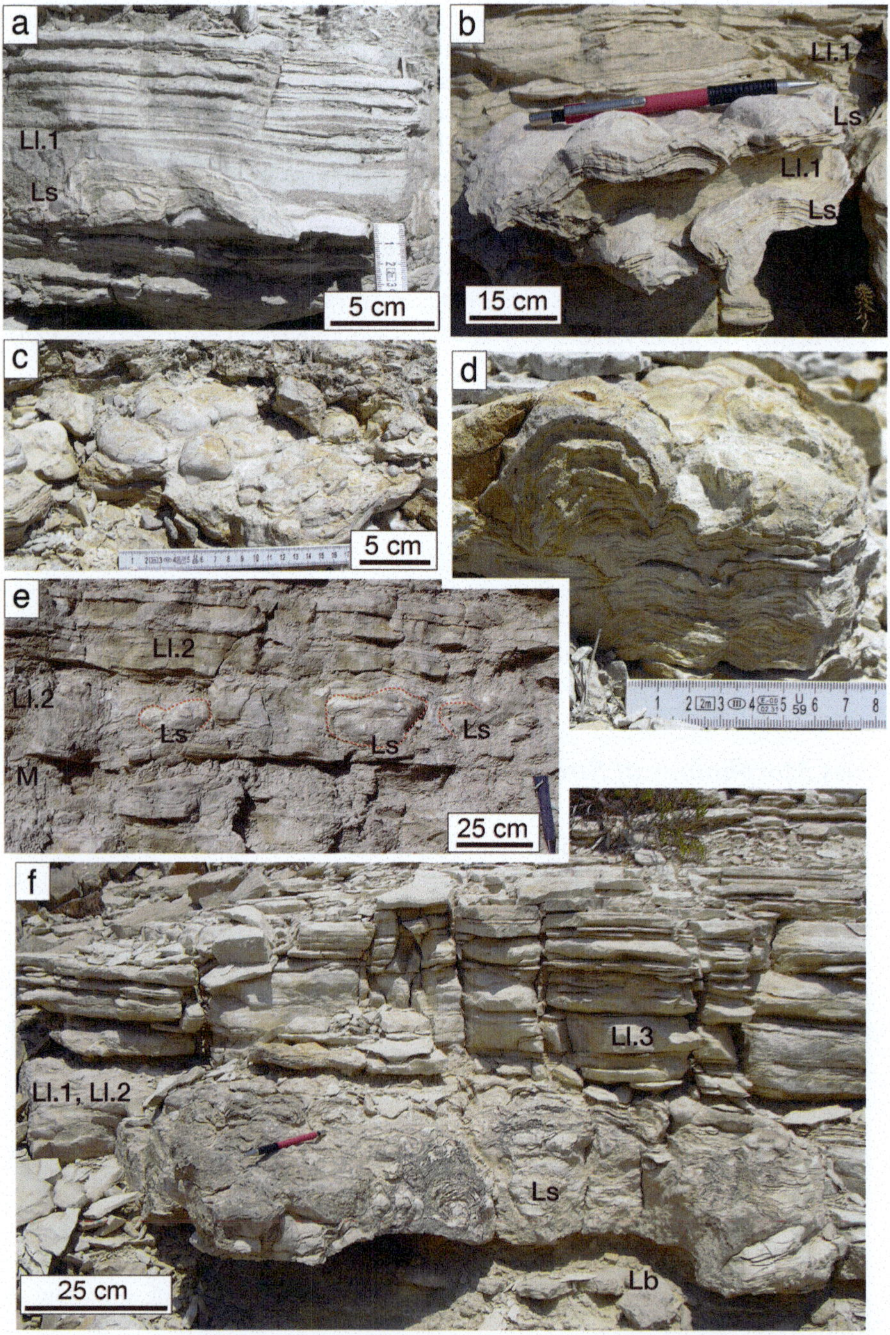

Fig. 8 Stromatolites in the Sierra de Alcubierre, associated with laminated limestones (units 5 and 6). (**a**, **b**, **c**, **d**) Thin planar stromatolites. In d), with flat laminae passing to domed laminae upward. (**e**) Domed stromatolites (enclosed by red lines). (**f**) Stratiform stromatolite. (**g**, **h**, **i**, **j**) Polished

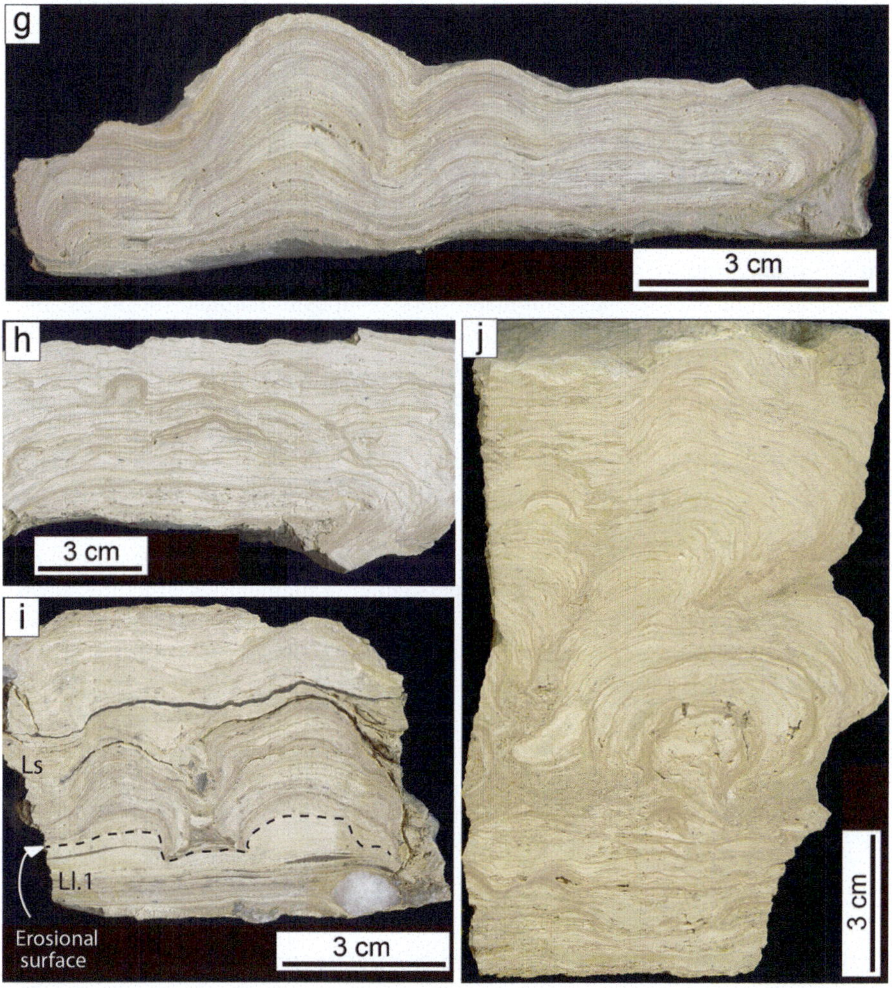

Fig. 8 (continued) sections of stromatolites. Note smoothness of the laminae in all of them. (**g, h**) Thin planar stromatolites (**i**) Portion of domed stromatolite. Note the erosional surface between laminated limestones and stromatolite (dashed line). (**j**) Portion of a stratiform stromatolite. Note the discontinuous laminae forming domes and columns

Mesoscopic and Microscopic Features of Lamination in Stromatolites and Oncolites

Lamination is distinguished by changes in color, primarily light versus dark, cream to beige, pale brown, and yellowish. The laminae, with variable lateral continuity, have gentle lateral changes in thickness and are always smooth, with neither crinkled nor festooned shapes. In cross section, the lamina shapes vary from flat to undulatory to gently to steeply convex (Fig. 8g–j).

Martin-Bello et al. (2019) described three types of simple laminae: (1) light dense micrite laminae, (2) light porous micrite to microsparite laminae, and (3) dark dense micrite laminae. These laminae commonly occur in that order from base to top, and form alternations of light and dark laminae. The passage from light to dark is mostly gradual, whereas the opposite is always sharp. These laminae are 0.1–1.9 mm thick (Figs. 9 and 10a, c). A fourth type is the fibrous laminae, up to 0.3 mm thick, which commonly alternate with either of the micrite types, and pass laterally to the dark micrite laminae. The simple micrite and microsparite laminae can be grouped into two different types of composite laminae in which either the dark or the light laminae are dominant: dark composite laminae (up to 2.8 mm thick) and light composite laminae (up to 6.4 mm thick). Most laminae that are visible to the naked eye correspond to composite laminae (e.g., in Fig. 8g–i). Three patterns of laminations are evident in the stromatolites: simple alternating lamination (Figs. 9b–d and 10a–b), cyclothemic lamination (Fig. 9b), and composite alternating lamination (Fig. 9a, b, e, f). The stromatolites in SA and MC are composed of one or more types of these lamination patterns (Figs. 9 and 10).

Microbial components in the stromatolites and oncolites of SA and MC are rare, as noted by Arenas (1993) and Martin-Bello et al. (2019). The dark dense and the light porous laminae contain rare and disperse micrite filamentous structures attributable to filamentous bacteria, likely cyanobacteria (Fig. 10). These filaments are set perpendicular to sub-perpendicular to lamination and pass across several simple laminae (Fig. 10c). In a very few cases, radially arranged micrite filaments are grouped in hemi-domic forms up to 1 mm high, which are referable to cyanobacterial mats (Fig. 10a, b). Isolated calcite tubes and scant elongated structures are present (Fig. 10d, e). Calcified thin films attributable to extracellular polymeric substances are among rhombohedral to sub-rhombohedral calcite micrite and dolomicrite crystals in the dark micrite laminae (Fig. 10f).

Freshwater Microbial Laminated Structures

In the fluvial and fluvio-lacustrine records in MB and MO, microbial laminated structures are present in the form of oncolites, calcite-coatings around phytoclasts, calcite-coatings around in situ stems, and stromatolites (Fig. 11). These components constitute distinct deposits that are briefly described below.

Oncolite limestones are rudstones and floatstones with matrix that form tabular and lenticular deposits, 0.15–2.1 m thick and meters to tens of meters in lateral extent, with flat to highly undulate, at places channel-shaped, bases (Fig. 11a, c, e). Locally, they also constitute sigmoidal strata. Commonly these limestones show fining-upward evolution relative to oncolite size. Although sedimentary structures

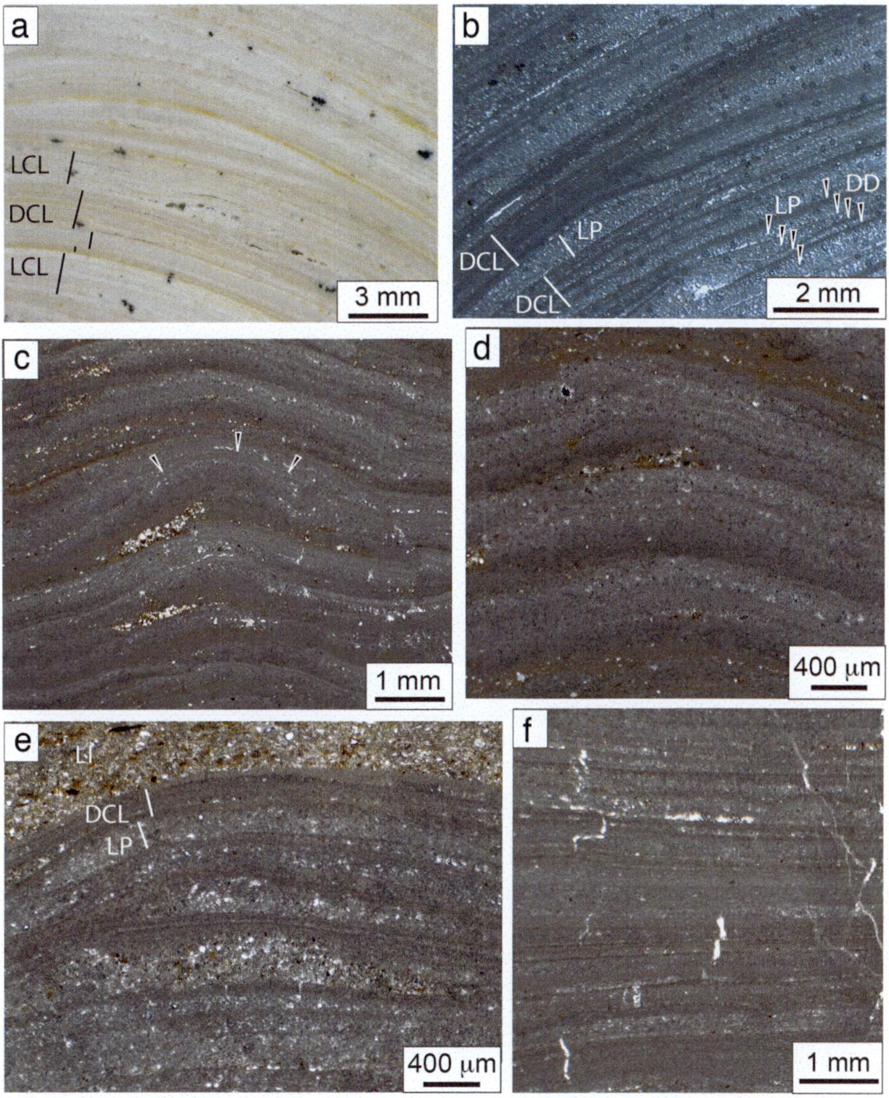

Fig. 9 Photomicrographs of stromatolites in the Sierra de Alcubierre (lens and optical microscope). (**a**, **b**) Smooth laminae arranged in alternating light and dark laminae. Note the simple and composite laminae. (**c**, **d**) Cyclothemic lamination consisting of cycles of light dense micrite, light porous microsparite and dark dense micrite. Arrows in (**c**) point to a convex-up form. (**e**) Alternating dark composite laminae and light simple laminae. Note the presence of siliciclastic intervals within the stromatolite and on top of it. (**f**) Alternating composite lamination. DD Dark Dense lamina, LP Light Porous lamina, DCL Dark Composite lamina, LCL Light Composite lamina

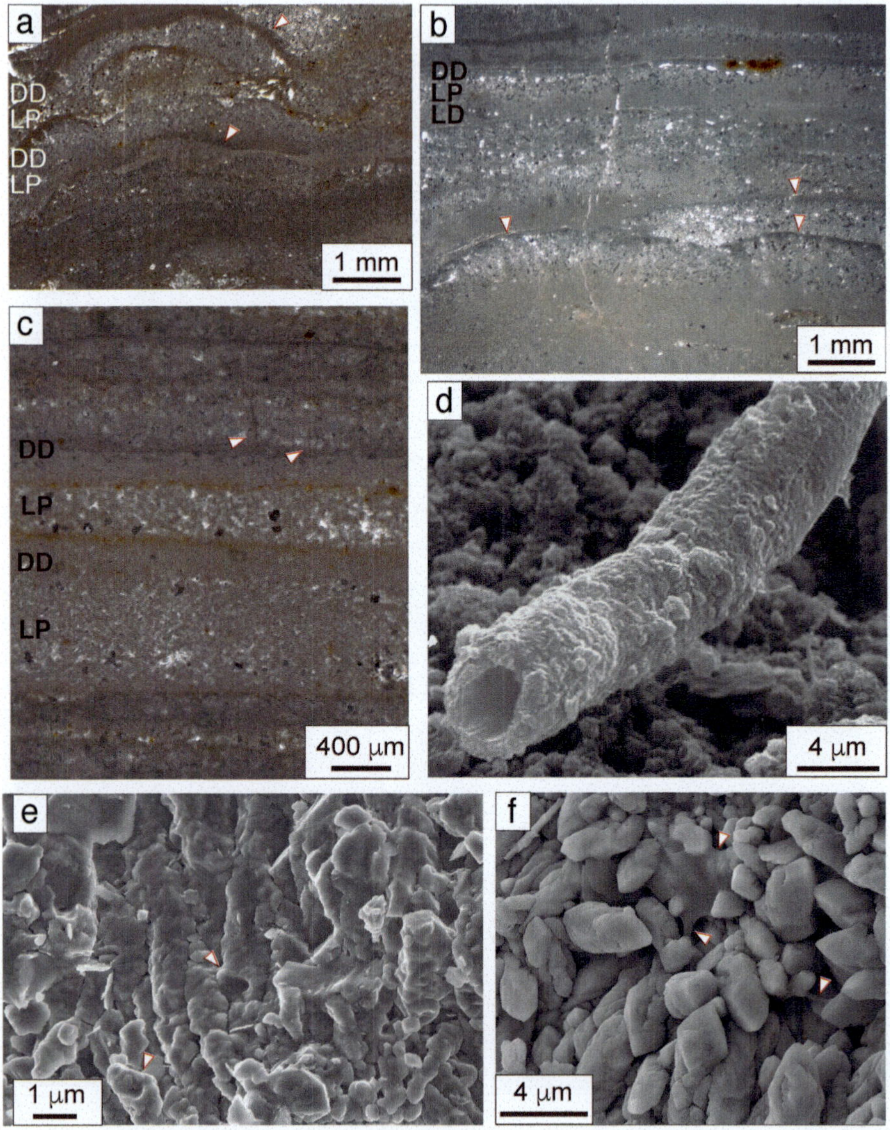

Fig. 10 Lamination and microbial evidence of stromatolites in the Sierra de Alcubierre (from optical microscope and scanning electron microscope). (**a**) Successive simple laminae arranged as small domes (arrowed), reflecting the shape of cyanobacterial bodies. (**b**) Detail of successive simple laminae forming convex-up bodies (arrowed) of cyanobacterial origin. (**c**) Smooth laminae containing micritic filamentous bodies perpendicular to lamination (arrowed). (**d**) Isolated calcite tube from cyanobacteria. Inner diameter is approximately 3 um long. In (**a**, **b**, **c**), LD Light Dense lamina, DD Dark Dense lamina, LP Light Porous lamina. (**e**, **f**) Texture of dense laminae in SEM. Note in (e), elongated bodies resembling cyanobacterial filaments or tubes (arrowed), and in (f), EPS among crystals (arrowed)

Fig. 11 Field and hand sample images of oncolites and calcite-coated stems in the Muela de Borja and Montolar area (units 7 and 8). (**a**) Oncolites in marlstones of unit 7. Note the dominant spherical shapes. (**b**) Deposit consisting of calcite-coated stems (phytoclasts) and oncolites, in unit 8. (**c**) Oncolite rudstone of unit 8 formed of varied shapes of oncolites, with fining-upward size-evolution.

are scant, at places, they have horizontal, undulate, and planar and trough cross-stratification, the later in sets up to 1.2 m thick (Fig. 11c).

Phytoclastic tufa are rudstones and floatstones with matrix consisting of fragments of calcite-coated non-oriented stems and often including oncolites, intraclasts, and fragments of stromatolites. These limestones form tabular and lenticular bodies, 0.05–1.5 m thick, and meters to tens of meters in lateral extent (Fig. 11b). Commonly, these limestones are structureless or have minor horizontal stratification.

Stem phytoherm tufa are boundstones consisting of in situ up-growing calcite-coated stems (Fig. 11g). They occur within strata of phytoclastic tufa in the form of palisades and bunches up to 25 cm thick and decimeter to 2 m in lateral extent, although large bodies exist locally.

Oncolites occur: (1) in unit 7, within marl layers, as overbank deposits, and (2) in unit 8, in a wide variety of environmental settings, mostly at the base of and through deepening-shallowing sequences, representing deposition in fluvial channels and shallow lacustrine areas with gentle agitation (Vázquez-Urbez et al. 2013). Stem phytoherm tufa are associated with calcite-coated phytoclast tufa in unit 8. Stromatolites are rare and occur only on top of oncoid and phytoclast limestones in unit 8.

Gross Morphology of Oncolites, Calcite-Coated Stems, and Stromatolites

According to the length to width ratio (L:W), the following shape-based types of oncolites are found (Fig. 11a–f):

1. Spheres, up to 6 cm in diameter, with $L:W1:W2 \approx 1$.
2. Ellipsoidal and conical bodies, up to 7.5 cm wide and 10.5 cm long, with $L:W1 \geq 2$, and $W1 \geq W2$.
3. Cylinders, up to 10 cm long, with $L:W > 5$, with $W1 \approx W2$.
4. Discoid bodies, up to 5 cm long, with $L:W1 \geq 2$, with $W1 >>> W2$.

Fig. 11 (continued) Note trough cross-stratification. (**d**) Detail of oncolites of unit 8. (**e**) Deposit consisting of oncolites and calcite-coated stems (phytoclasts) of unit 8. Note that long shapes are dominant. (**f**) Oncolite hand samples of units 7 and 8, exhibiting three main shapes that evoke the nucleus shape, i.e., sub-equidimensional particle, stem, and gastropod. (**g**) Palisade of calcite-coated upgrowing stems forming a small phytoherm within phytoclastic rudstones (unit 8). (**h**) Thin undulate stromatolites within phytoclastic rudstone of unit 8. Note that the stromatolite layer developed on calcite-coated, low-laying to horizontally laying parallel stems. Image (h) reprinted with permission from SEPM (J. Sed. Research, 83, p. 574)

Spherical oncolites are dominant in unit 7, whereas cylindrical, conical, and ellipsoidal oncolites are more abundant in unit 8. Discoid bodies are rare. The calcite coatings are up to 5.5 cm thick. The nuclei of these oncoids are particles of different nature and shape: intraclast, extraclast, gastropod or any other shelly organisms, stem, leaf, and even composite grains. The nucleus shape largely influences the external shape of the oncolites, or at least that of the younger laminae (Fig. 12a–c). The nuclei can be empty or filled with micrite or sparite.

The features of the calcite-coated stems, i.e. stem sizes and laminated coatings, are identical in the phytoclastic tufa and the phytoherm tufa. The inner diameter of the cavities left after plant decayed ranges from millimeters up to 2 cm wide. The length of these stems varies from a few centimeters to 25 cm, and the laminated coatings are up to 4 cm thick (Fig. 14a, b).

The stromatolites form thin planar bodies up to 8 cm thick and decimeters in lateral extent, with undulate bases and tops, in most cases due to the irregular shape of the substrate (Figs. 11h and 13a).

Mesoscopic and Microscopic Features of Lamination in Oncolites, Calcite-Coated Stems, and Stromatolites

Oncolites

In the oncolites, the laminae are primarily continuous and of equal thickness around the nucleus, although some specimens have highly asymmetrical cortices, to the point that some laminae wedge up to disappear (Fig. 12a, b). In cross sections, the lamina shapes are smooth, undulate, crinkled, and festooned, or even delineate discontinuous domes linked by subcontinuous laminae (Fig. 12a, c). In a number of specimens, different stages of growth are marked by changes in color and/or irregular surfaces that represent interruptions in the microbial growth and in many cases involve erosion of some laminae (Fig. 12c).

In thin sections, the laminae show a variety of cross-sectional shapes varying from slightly undulate to domed to columnar. The oncolites are made of alternating simple laminae consisting of light microsparite and minor sparite, and dark micrite. The dark laminae are thinner than the light laminae, with minimum values of 50μm–100μm and maximum values of 100μm–1.8 mm, respectively (Fig. 12d–f). However, in some specimens of unit 7, the laminae are thinner, between 12μm–40μm and 460μm–1.4 mm. In all cases, the simple laminae are grouped into composite dark and light laminae that alternate through time and correspond to the laminae visible to the naked eye (Fig. 12e).

Microbial evidence is ubiquitous and includes filamentous, tube-shaped, and spherical micrite bodies of microbial origin (Fig. 12f–h). The filaments are loose micritic bodies set sub-perpendicular to lamination. The calcite tubes are grouped into fan-shaped bodies and palisades of parallel tubes, up 30–260μm wide and

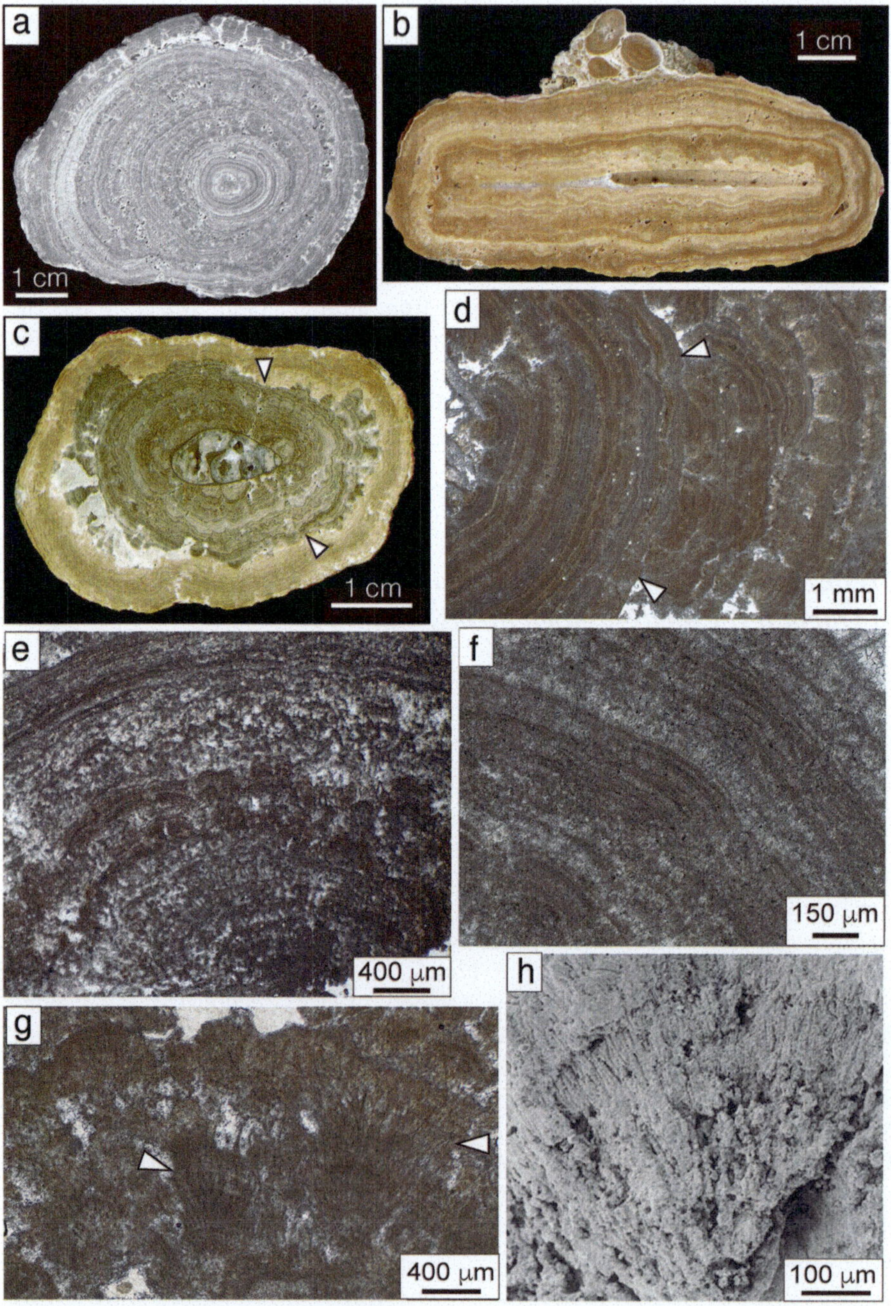

Fig. 12 Oncolite images from hand-sample cross sections (**a**, **b**, **c**) and photomicrographs (optical microscope: **d**, **e**, **f**, **g**; scanning electron microscope: **h**). (**a**, **b**, **c**) Sub-spherical, elongate, and ovoid sections evoking the nuclei shapes. Note in (**a**) the lateral change in the thickness coating; in (**b**) the alternating change in color of groups of laminae; in (**c**) the gastropod in the nucleus and the irregular erosional surface (arrowed) that coincides with a change in color of the coating. (**d**)

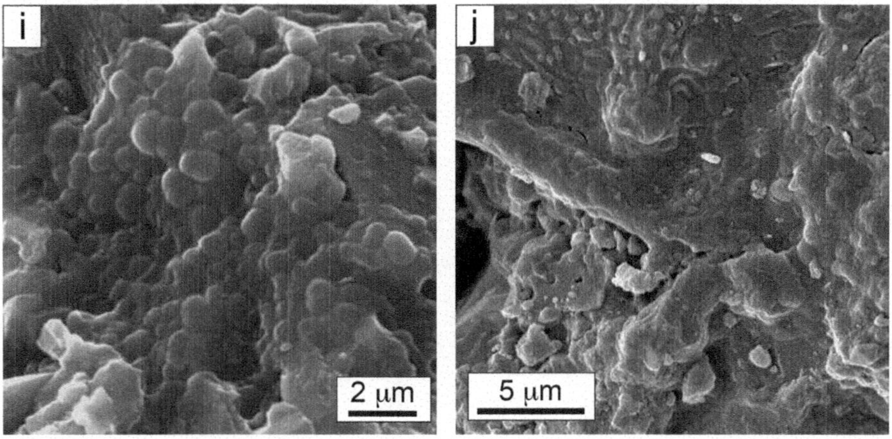

Fig. 12 (continued) Alternating composite lamination. Note the sharp change from smooth to domed-shaped laminae through time (arrowed). (**e**, **f**) Detail of alternating composite lamination: groups of dominant dark laminae alternate with groups of dominant light laminae. Note the gradual passage upward from the light to the dark composite laminae upward. (**g**) Undulate lamination consisting of fan-shaped micritic bodies formed of radially arranged micritic filaments from cyanobacteria. (**h**) Fan-like body with radially arranged calcite tubes in an oncolite-lamina. It resembles the characteristic pattern of *Rivularia haematites*-colony (cf., Hägele et al. 2006). (**i**, **j**) Spherical and elongate bodies within extracellular polymeric substances from bacterial activity. Images (h, i) reprinted with permission from SEPM (J. Sed. Research, 83, p. 577)

2 mm high (Fig. 12g). They correspond to calcified filaments themselves, and the tubes to calcification around microbial filaments, likely from cyanobacteria. In general, these large fan-shaped and palisadic structures give high porosity to the oncolites, as the spaces between the microbial elements remain mostly empty (Fig. 12h). The spherical bodies are $\approx 1\,\mu m$ wide and occur disperse and grouped within a film of EPS (Extracellular Polymeric Substances) on some filaments; they are interpreted as calcified coccoid bacteria (Fig. 12i, j).

Calcite-Coated Stems

The characteristics of the calcite-coated stems that concern texture and thickness of the laminae, lamina arrangement, and microbial components are identical in the phytoclastic tufa and in the phytoherm tufa. The laminae are concentric and continuous around the stems. They consist of alternating light and dark calcite laminae; each simple lamina can be up to 3 mm thick. The lamina shapes are varied: domed, undulate, and smooth.

- Laminae with domed cross-sectional shapes are formed of dense groups of micrite filaments that are set sub-perpendicular to the lamination, and their height is up to 3 mm. These domes occur isolated or laterally linked through thin laminae (Fig. 13e).

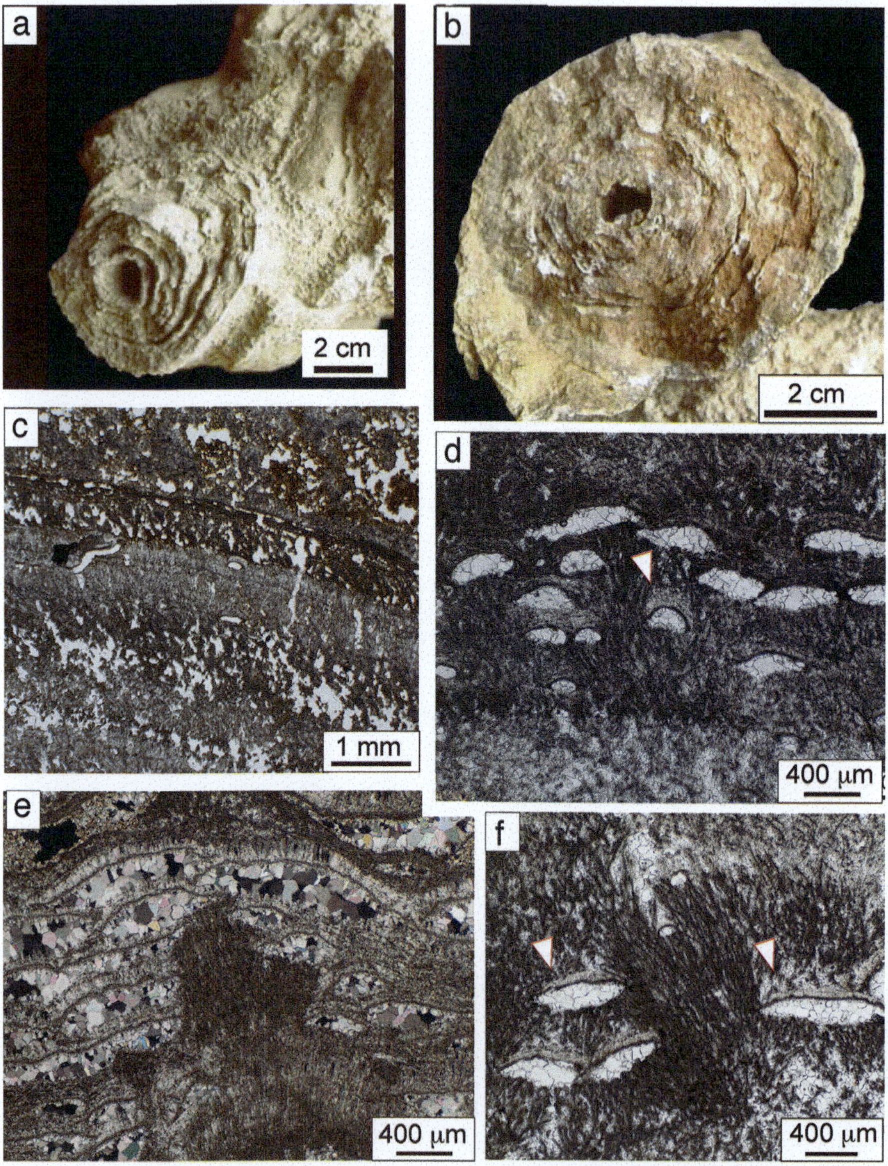

Fig. 13 Images of calcite-coated stems in cross sections (**a**, **b**) and microphotographs of lamination (c–f, optical microscope). (**a**, **b**) Concentric lamination around cavities from decayed stems. (**c**) Detail of alternating dark and light, highly porous laminae. (**d**) Light and dark laminae. The dark lamina consists of palisades of cyanobacterial filaments that contain flat-convex-shaped bodies (from insect larvae) filled by coarse sparite cement (arrowed) that are set parallel to lamina boundaries. (**e**, **f**) Detail of the flat-convex bodies filled by sparite cement within a micrite lamina. Note that thin micrite and microsparite laminae coat the upper part of the flat-convex bodies (arrowed), and that they crosscut the cyanobacterial filamentous bodies

- Undulate laminae result from mimicking the substrate irregularities. Their thickness ranges between 0.5 mm and 0.8 mm. They consist of micrite and sparite, with long crystals perpendicular to lamination. In some cases, the wavy structure is related to the presence of lenses (flat-convex) filled by sparite and outlined by micrite walls all around (Fig. 13d, e, f).
- Smooth laminae are made of micrite and occasional microsparite, minimum 0.2 mm to 0.5 mm thick and up to 2–3 mm thick. These laminae contain disperse micrite filamentous bodies, as well as their molds, with 8–10μm wide and 50–200μm long (Fig. 13c).

Stromatolites

The internal structure of stromatolites consists of alternating light sparite and dark micrite laminae (Fig. 14b, c). The laminae are continuous laterally and have smooth, undulate, and very irregular cross-sectional shapes. The smooth laminae are flat and parallel, 0.3–1 mm thick. The undulate and irregular laminae can form small domes,

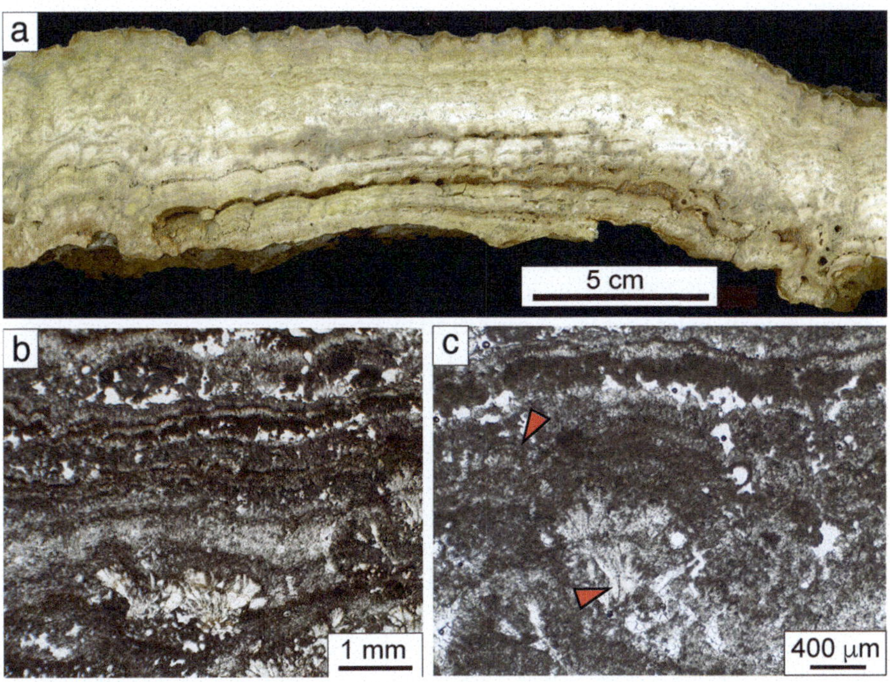

Fig. 14 Images of stromatolites in unit 8. (**a**) Cross section of a thin planar stromatolite. Note the undulate lamination. (**b**, **c**) Photomicrographs (optical microscope) showing alternating light and dark, festoon laminae. Fan-shaped bodies consisting of radially arranged filaments are present (arrowed). Image (**b**) reprinted with permission from SEPM (J. Sed. Research, 83, p. 577)

with lateral variations in thickness from 0.5 mm to 1 mm thick. In some cases, there are flat-convex bodies parallel to lamination that can be empty or filled by sparite.

Evidence of microbes includes the presence of micritic filamentous bodies sub-perpendicular to the laminae. These filaments can be dispersed or grouped forming bushes or domes (Fig. 14b, c). In scanning electron microscopy, calcified filaments are dispersed, 10–20μm wide and approximately 300μm long.

Discussion

Impact of the Depositional Environmental Factors on Microbialite Morphology

Stromatolites and Rare Oncolites in SA and MC

These microbialites have been interpreted to have formed in a saline carbonate lake context with common surge activity. Evidence of paleosalinity changes includes (1) the spatial distribution of these carbonate facies relative to the other carbonate and sulphate lithofacies, (2) the stable isotope composition and the content of dolomite of the various carbonate facies (Figs. 6 and 15), and (3), in some cases, the interstitial growth of gypsum nodules and lenses within the microbial structures and the associated laminated limestones (Arenas et al. 1997; Arenas and Pardo 1999). Other evidence of the shallow saline context is the trace fossils, which are limited to simple forms of invertebrate traces and very small vertical burrows, probably produced by insect larvae in the laminated limestones facies (J. J. Scott 2007, pers. com.). These features have been described as typical of very shallow saline lakes, as in Lake Gosiute, in the Wilkins Peak Member of the Eocene Green River Formation (Pietras and Carroll 2006; Scott and Smith 2015; Smoot 1983).

Surge activity in SA and MC is evidenced by ripples and ripple lamination, hummock cross-stratification, erosional surfaces, and intraclast rudstones, all of them formative features of the laminated limestone facies. The stromatolites associated with them also include erosional features that were caused by waves and return bottom-currents. These are indicated by the in situ erosional surfaces and the stromatolite fragments within the laminated limestones (Martin-Bello et al. 2019). Lacustrine stromatolites associated with surge activity have not been extensively reported in other basins, with the exception of a few works in the Green River Formation (Graf et al. 2015) and the late Archean in Western Australia (Awramik and Buchheim 2009).

Changes in the external geometry of lacustrine stromatolites mostly reflect variations in water level (Casanova 1994; Chidsey et al. 2015; Muniz and Bosence 2015; Vennin et al. 2018). In the Ebro Basin, lake-level variations are reflected through the vertical succession of sedimentary facies (i.e., facies associations), with specific sedimentary structures of the laminated limestones, and primarily through the dis-

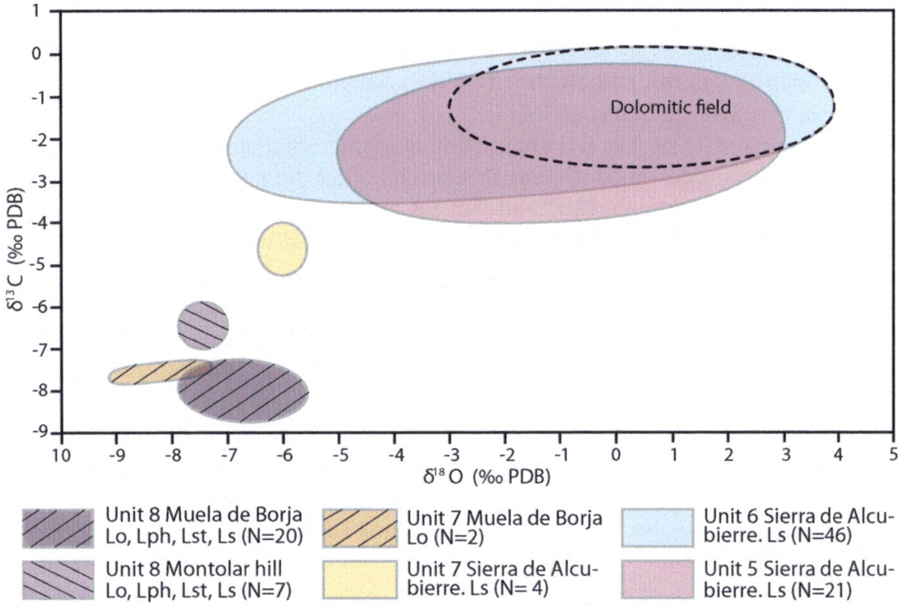

Fig. 15 Stable isotope composition ($\delta^{13}C$ and $\delta^{18}O$) of laminated microbialites distributed by units in the Sierra de Alcubierre, Muela de Borja and Montolar hill. Facies codes explained in Table 1. (Compiled from Arenas and Pardo (1999), Arenas et al. (2000), Vázquez-Urbez (2008) and Vázquez-Urbez et al. (2013))

tinct external geometry of stromatolites in these sequences (cf., Martin-Bello et al. 2019).

Thin planar stromatolites are associated with low lake level, while domed, columnar, and stratiform stromatolites are developed in deeper conditions (Della Porta 2015; Roche et al. 2018; Vennin et al. 2018). Low synoptic relief, highly enveloping and overturned laminae in stromatolites are indicative of low lake level or even desiccation during the development of many planar stromatolites of the Ebro Basin. In contrast, internal growth forms with higher synoptic relief within domed and stratiform stromatolites represent higher lake levels (Casanova 1994; Martin-Bello et al. 2019). Similar interpretations have been proposed in the Eocene Green River Formation lakes (Awramik and Buchheim 2015) and the Quaternary Lake Bonneville (Bouton et al. 2016).

As for hydrodynamics, the stromatolites of the Ebro Basin are associated with different wave-related sedimentary structures and erosional features affecting the stromatolites at different scales. Arenas and Pardo (1998) and Martin-Bello et al. (2019) suggested that these features were produced by fair weather waves that broke the partially lithified microbial mats (e.g., at the time of ripple formation in the laminated facies, Ll.1), or by storm waves that caused erosion and thus breakage of the microbial mats and of previous stromatolite and laminated deposits (e.g., at the time of forming hummocky cross-stratification, Ll.2). There are few examples that have reported this context in association with stromatolites, for example, in the Precambrian (Awramik and Buchheim 2009; Sakurai et al. 2015) and Cenozoic (Roche et al. 2018; Vennin et al. 2018).

Much remains to be learned about laminated microbial structures in lacustrine systems. For example, it is intriguing that stromatolites did not develop in the freshwater environments of units 5, 6, and 7 of SA and MC, i e., during periods of high lake levels (Fig. 5a). Stromatolites are also lacking in freshwater facies of the Green River Formation (e.g., Graf et al. 2015). Moreover, although unit 7 of this study deposited when overall salinity diminished relative to underlying units (Arenas and Pardo 1999), stromatolites are scant and are located on laminated limestones, at the top of shallowing sequences that represent saline conditions. The reasons for such a decrease in stromatolite abundance through time are open to debate. It is plausible that the proliferation of freshwater fauna and flora and frequent entrance of fine-sediment from fluvial supply made the lake not a likely place for the microbial mats to develop. In addition, saturation levels with respect to calcite have a significant influence on the cyanobacterial sheath calcification (Berrendero et al. 2016; Dupraz et al. 2009; Riding and Liang 2005).

Oncolites, Calcite-Coated Stems, and Stromatolites in MB and MO

These microbial laminated structures formed in freshwater fluvial, fluvio-lacustrine, and lacustrine environments with high content in calcium and bicarbonate (Arenas et al. 2000; Vázquez-Urbez 2008; Vázquez-Urbez et al. 2013). Freshwater conditions are identified by the sedimentological features that suggest the open character of the sedimentary systems, by the stable isotope composition of the carbonate facies (Figs. 6 and 15), by the low concentration of minor and trace elements (Vázquez-Urbez et al. 2013) and by the flora and fauna (charophytes and mollusks, above all). The values of the stable isotopic composition of the microbialite facies of unit 8 are mostly grouped and have poor correlation between C and O (Fig. 15). These values are close to those found in Oligocene fluvial stromatolites from the Marseille Graben (Casanova and Nury 1989) and very similar to some Eocene fluvial and fluvio-lacustrine stromatolites in the Eastern Ebro Basin (Zamarreño et al. 1997). The values are also consistent with those of Quaternary temperate European fluvial systems (as compiled by Andrews 2006).

Oncolites and calcite-coated phytoclasts formed in shallow low-sinuosity channels and slow flowing pools, where aquatic fauna and macrophytes thrived. The later were coated by calcite and originated up-growing stem phytoherms. Diverse shelly fauna and macrophytes, either coated or uncoated by calcite, provided the grains for the oncolites to form and for phytoclast accumulations. Tufa and oncolites are typical deposits in many modern streams and some lake shores in which water is rich in Ca-HCO_3^-, and where hydrophilous plants and other aquatic organisms are coated by calcite (Arenas et al. 2014; Arp et al. 2001; Ford and Pedley 1996; Golubić et al. 2008; Hägele et al. 2006; Pentecost 2005). Ancient lacustrine and fluvio-lacustrine systems that include similar structures have been studied in many places around the world (e.g., documented by Ford and Pedley 1996 and Capezzuoli et al. 2014).

In the case of oncolites, the external shape can give some clues on the hydrodynamic behavior (Casanova 1994; Hägele et al. 2006; Verrecchia et al. 1997). However, in many cases, it is the nucleus shape that guides the morphology of the younger laminae, and in some cases the full external geometry of the oncolites (Arenas et al. 2007a, 2015; Astibia et al. 2012). For example, oncolites developed around plant stems are cylindrical, while those around helicoid spiral gastropods are conical, as in MB and MO. In addition, the nature of the nuclei gives some clues on the features of the depositional environment. For instance, spherical to subspherical oncolites, dominant in unit 7, formed around rounded nuclei consisting of micrite, which suggests that these micrite grains derived from erosion of hardened lime mud deposits, and indeed that the vegetation cover in that area was not abundant, compared to unit 8 where many oncolite nuclei were stems (Vázquez-Urbez 2008).

The lateral variations of the laminated cortices within oncolites, i.e., the symmetrical or asymmetrical character of the coatings, provide information on the depositional environmental conditions. Considering that the laminae are continuous around the nuclei, it has been suggested that symmetrical coatings reflect relatively constant or periodic movement of the oncolites (Vázquez-Urbez 2008; Zamarreño et al. 1997). In contrast, the different thickness, for example, higher on the upper sides, can be due to the different growth rate of cyanobacteria on each side because of light and current action differences (Zamarreño et al. 1997). In the examples studied here, some oncolites of unit 7 are slightly asymmetrical, which suggests gentle differences in such parameters. In contrast, many oncolite cortices of unit 8 are quasi-symmetrical in thickness, which can be interpreted as indicative of relatively frequent movement of the particles due to relatively constant hydraulic conditions (e.g., water velocity, by either uni- or bi-directional flows, and depth) and lightness.

In the case of in situ calcite-coated stem bioherms, there are a few asymmetrical cortices relative to thickness. These changes correspond to the different water flow velocity (e.g., upstream side versus downstream side of the stem) and lightness (cf., Pedley 2009).

The presence of sharp changes in color and/or erosional surfaces denotes interruptions in the microbial growth and, therefore, changes in the "normal" conditions of oncolite formation, e.g., due to sharp increases in flow velocity. Erosional features in oncolites of the studied areas are patent in several specimens of unit 7 in MB, and the laminae that follow the irregular erosional surfaces represent a drastic change in the lamination color or style (e.g. Fig. 12c, d). Moreover, stromatolite fragments are common within the phytoclast limestones and the intraclast limestones. Together, these facts are indicative of sudden or sharp changes in hydraulics, e.g., flow regime, which is also consistent with overbank accumulation of oncolites in unit 7 of MB (cf., Arenas et al. 2007a).

In contrast, the laminated coatings around stems and stromatolites in the studied cases here do not contain common or large erosional discontinuities in their growth, which underlines the importance of stable hydraulic conditions for continuous lamination to be formed (e.g., as noted in modern fluvial environments by Pedley et al. 2009; Arenas et al. 2014). The only noticeable disturbance in microbial growth is represented by the great abundance of cavities attributable to the activity of larvae of aquatic insects (Fig. 13d, e, f), which crosscut the filamentous mats, as found in other microbialites (Brasier et al. 2010; Jones and Renaut 1996; Ramos et al. 2001; Zamarreño et al. 1997). Thriving aquatic insects indeed may suggest stable hydraulic conditions.

Laminae and Lamination: Environmental and Temporal Significance

Despite lamination being quite a simple structure from a textural point of view, the environmental interpretation and temporal significance of the laminae are still open to debate. This is mainly due to the large number of parameters involved and the several ranges of lamina arrangement or cyclicity that may be present (Arenas and Jones 2017).

Lamina Arrangement and Texture

The most common and simplest pattern of lamina arrangement in many laminated micritic microbialites (*fine-grained stromatolites*, sensu Riding 2000) is the alternation of dark and/or dense laminae and light and/or porous laminae, which has been recognized in many depositional settings (Casanova, 1986, 1994; Hofmann 1973; Monty 1976; Riding 2011; Suárez-González et al. 2014, among many others). However, the arrangement of laminae can be observed at different scales, and thus different orders of cyclicity are present; in summary, simple alternation, composite alternation, or cyclothemic lamination, as described by Monty (1976). The environmental causes of such cyclicity involve changes in temperature, precipitation, and evaporation (Casanova 1994; Golubić et al. 2000). These changes can promote variations in the water chemical composition, in particular the saturation levels respect to minerals, but also induce variations in the microbial development, in the abundance of certain microbes or even in the types of microbes of the microbial mat through time (Dupraz et al. 2009). Likewise, the duration of the laminae and the lamination patterns have been a matter of debate, since the lamination is marked by the discontinuity in the rate of the accreting processes, and these processes can be of different nature and duration (Golubić et al. 2000; Hofmann 1973; Pope et al. 2000). The most commonly accepted idea is that each couplet represents seasonal or multi-annual cyclicity (e.g., Casanova 1994; Petryshyn et al. 2012; Tang et al. 2014).

In the saline lacustrine stromatolites studied here, lamina arrangement includes simple and composite alternating laminations and cyclothemic lamination (Martin-Bello et al. 2019). Based on the cyclic textural changes coupled with cyclic stable isotopic variations through successive composite laminae, it has been suggested that the light laminae formed during periods of higher precipitation/evaporation ratio (P/E) and the dark laminae during periods of lower P/E. These cyclic changes can represent seasonal to multi-anual variations (Martin-Bello et al. 2019).

In the fluvial and fluvio-lacustrine environments, the textural features of the laminae in stromatolites and oncolites also show different ranges of cyclicity, comparable or similar to those described in the lacustrine stromatolites (e.g., Arenas et al. 2015; Lindqvist 1994). However, in some studied specimens here and elsewhere, the most common cycle consists of a light and/or porous composite lamina passing gradually upward to a dark and/or dense composite lamina (Fig. 12e, f). This passage also involves a morphological change from festooned or wavy simple laminae (bush or fan-shaped microbial bodies) to slightly wavy and even flat simple laminae upward (with loose microbial filaments, or filaments grouped alongside forming palisades) (cf., Casanova and Nury 1989; Zamarreño et al. 1997). These cycles can be up to a few millimeters thick. The composite character of these laminae are evident even to the naked eye in some cases. The duration of such cycles in the studied specimens in MB and MO is unknown, but each may represent 1 year. This temporal assumption is based on the fact that calcite precipitation is faster in fluvial environments than in lacustrine environments, as shown by studies of modern fluvial stromatolites. These studies show that composite laminae consisting of 4–6 simple laminae simple can form in a few months (Arenas and Jones 2017; Gradzinski 2010; Vázquez-Urbez et al. 2010). Moreover, the oxygen isotope composition in these modern cycles is related to seasonal changes in water temperature (e.g., Chafetz et al. 1991; Osácar et al. 2016). Likewise, ancient oncolites in mid latitude regions show similar changes in the isotopic composition through successive laminae (e.g., Anzalone et al. 2007; Arenas et al. 2015). Based on these results, in the studied area, each cycle consisting of two composite laminae should reflect seasonal variations in temperature and/or precipitation during a year.

In some calcite coatings on stems and stromatolites in MB and MO, there are cycles of upward-increasing number of insect-made structures that coincide with the denser or darker part of the laminae (Fig. 13d). The cavities are generated when insects, once having got a micrite coating around, decay or emerge as nymphs, and then the voids remain empty or become filled by early calcite cement (e.g., Fig. 13d). In many modern environments, the dark and/or dense micrite laminae in microbialites are reflecting calcite precipitation in warm conditions, when calcite saturation levels are higher (e.g., Arenas and Jones 2017; Arp et al. 2001). In a 100-ka old laminated tufa of central Greece, Brasier et al. (2010) found that the seasonality is reflected by alternating porous and dense laminae. The insect cavities (from chironomids) concentrated at the top of the dense laminae, which are interpreted to have formed in spring to early summer, as indicated by the lower $\delta^{18}O$ values respect to the porous laminae. In these microbialites, most part of the summer deposit is missing due to the lack of water. Regardless of this discontinuity, these chironomid cavities are within the warm-season deposits. Similar features are found in modern fluvial stromatolites (Gradzinski 2010). Therefore, the presence or varying abundance of such insect structures in laminated microbialites may be used as tentative indicators of temperature and hydrological conditions on a short-term scale. In the Eocene Lake Gosiute (Green River Formation, Wyoming), Leggit and Cushman (2001) found caddisfly-dominated (Trichoptera) carbonate mounds. They proposed that the larval cases grew on hardgrounds that, once became calcified, provided sites for the benthic microbial mat to grow. The authors suggested that microbial-caddisfly couplets may represent one-year cycle and that caddisfly pupation and aggregation behavior regularly interrupted the carbonate mound growth.

Shape and Thickness of the Laminae

Lamina shape and thickness reflect the response of the microbial growth and concurrent mineral precipitation to repetitive changes in the magnitude of a wide array of physical and chemical parameters over different time spans (Arenas et al. 2015).

The examples in this study show that in saline lacustrine carbonate environments the microbialitic laminae tend to be thin and smooth (cf., Pope et al. 2000), while in freshwater fluvial and fluvio-lacustrine environments the laminae show greater diversity of morphological types and their thickness is generally greater. A similar, but less marked, morphological difference occurs between the lacustrine and the fluvial stromatolites and oncolites, both in the Oligocene of Marsella (Casanova and Nury 1989) and in the Oligocene of the Ebro Basin (Zamarreño et al. 1997). In these two cases, the lacustrine microbialites that formed in closed environments showed smoother laminae, although the lake water was not considered as saline as in the Miocene.

In the Miocene record studied here, the smoothness of the laminae and the small lateral variations in laminae thickness concur with the poor presence of microbial filamentous bodies, which when present are mostly dispersed and not really abundant (compare Figs. 9 and 10 with Figs. 12, 13, and 14). The lacustrine stromatolites of the Eocene Green River Formation have many similarities with stromatolites of units 5 and 6 in the Ebro Basin. In both basins, stromatolites developed in saline shallow carbonate lakes and have planar and domed morphologies consisting of smooth, slightly convex laminae, for example, as compared with stromatolites in the Tipton Member (Surdam and Wray 1976) and in the upper portion of the Douglas Creek Member and the Parachute Creek Member (Chidsey et al. 2015).

In some marine laminated microbialites, the laminae are also smooth, even in association with wave structures (e.g. ripples), and the mineralogy is usually more complex than in freshwater environments. In the case of Porto Pi (Mallorca, Spain), the late Miocene microbialites that developed in intertidal to subtidal conditions consist of dolomicrite and occur as moderate- to giant-sized domes up to 5 m thick, in which the uppermost laminated portion is up to 2 m high (Suárez-González et al. 2017, 2019). Both agglutinated and micritic stromatolites are present and, in all studied specimens, the laminae are very smooth and continuous at different scales

(Suárez-González et al. 2019). Similarly, in other Messinian deposits along the western Mediterranean coast, the dolomitic stromatolites formed large domes characterized by smooth lamination and poor microbial evidence (Braga et al. 1995; Riding et al. 1991).

Holocene stromatolites from Walker Lake in western Nevada (USA)—a remnant of the Pleistocene Lake Lahontan complex—formed on the shallow gentle slope of a closed-basin alkaline lake. These stromatolites form domes consisting of fine-grained texture and submillimeter lamination, with lamina cross sections essentially smooth and with little evidence of microbial components (Petryshyn et al. 2012). These authors concluded that local climate, hydrology, and geochemistry were the primary drivers of lamination formation versus the diurnal or seasonal cycles of microbial mat metabolism.

In contrast, in freshwater environments the mineralogy of the laminated microbialites is usually simple. In some of them, the variety of lamina shapes has allowed to distinguish morphological types of lamination (e.g., Arenas et al. 2000, 2007a, 2015; Casanova and Nury 1989; Lindqvist 1994). Moreover, these morphological types can occur within the same microbialite body (e.g., in an oncolite). In fluvial and fluvio-lacustrine environments, the morphology of the laminae is largely related to distinct morphological types of microbes and microbial arrangements (e.g., flat, wavy, and columnar lamination; Arenas et al. 2007a). These morphological types are interpreted to represent either different microbial communities or variable development of the same microbial communities, in both cases as a response to changing environmental conditions. Among others, Casanova (1986, 1994), Casanova and Nury (1989), Freytet (1992), Zamarreño et al. (1997), and Arenas et al. (2000) showed that lamina shape is linked, at least partially, to the morphology of the microbial components, both in stromatolites and in oncolites.

Some modern oncolites that form in freshwater open lakes and streams with gentle water movement show that wavy laminae are associated with *Rivularia haematites* growth, for example, in Kelly Lake, Canada (Ferris et al. 1997), in a marshland in France (Caudwell et al. 2001), and in the river Alz, southern Germany (Hägele et al. 2006). This is due to cyanobacteria acting as templates for mineral precipitation. Nonetheless, regardless of this template condition, crystal shape is another important factor controlling the shape of microbialitic laminae (cf. Frantz et al. 2014). Modern stromatolites formed in fast flowing water in several karstic-fed streams show extensive deposits consisting of continuous, thick, and variable lamina shapes (Arenas et al. 2014; Berrendero et al. 2016; Gradzinski 2010). For example, in the tufa-depositing River Piedra (northeastern Spain), the distribution of the different cyanobacteria and the related sedimentary structures can vary as a function of the water flow conditions. Thus, the different fluvial environments can show different structures, with calcite coatings around the cyanobacteria being dependent on changes in calcite saturation index (Berrendero et al. 2016). For example, in fast flow conditions, *Phomidium incrustatum* is the dominant cyanobacterium and produce extensive laminae consisting of tight calcite tubes. In contrast, in slow flowing water, lamination is uneven and the dominant cyanobacterium is

Phormidium aerugineo-caeruleum, which commonly produces loose, randomly oriented calcite tubes. The correlative changes in saturation index occur within a permanent saturation state with respect to calcite, between ca. 0.9 and 0.6, respectively (Arenas et al. 2014).

If the morphological variability of the laminae were related only to the varying shape of the microbial bodies or microbial arrangements, this feature could be used as an indicator of microbial diversity and morphology, and thus indirectly of environmental conditions, such as salinity variations and hydrodynamics. Therefore, it is suggested that smooth versus undulate to wavy microbialitic laminae can be indicative of salinity-dependent factors or parameters, such as the microbiota (primarily, cyanobacteria) and the mineral precipitation.

Microbial Components: The Dark Side of Ancient Saline Carbonate Lakes

Evidence of microbial components in laminated microbialites of saline environments, including continental and marine settings, is not abundant in the geological record. However, modern saline carbonate lakes show abundant microbes and microbial structures (Browne et al. 2000; Dupraz et al. 2004; Renaut and Long 1989). This is the dark side of ancient saline carbonate microbialites. Several authors have postulated that the rarity of microbial components found in the ancient laminated microbialites is due to poor preservation potential, perhaps in relation to the poor calcification of phototrophic organisms; microbialite mineralization may be associated more with heterotrophic communities in the microbial mats (Chagas et al. 2016). However, low diversity in harsh conditions is also seen in other biota groups in many environmental situations (Browne et al. 2000). In the ancient record this is an intriguing issue. In recent environments and from experimental studies, several types of microbially mediated mineralization have been discussed, for instance, concerning the chemical conditions that are needed for sheath impregnation or physicochemical precipitation around cyanobacteria in a microbial mat (Arp et al. 2001; Dupraz et al. 2004, 2009; Merz-Preiß and Riding 1999). In the Miocene Ebro Basin specimens, the scant filament bacterial bodies suggest that the microbial populations in the microbial mats of these saline environments had low taxonomical diversity and/or the microbe calcification was not favored by the chemical conditions or by the specific taxons. In contrast, higher microbial diversity and high saturation levels in calcite and rapid precipitation favor calcification of the microbial components in freshwater (open) lake and fluvial environments.

Conclusions

Microbial laminated structures, e.g. stromatolites and oncolites, have the potential to bring further insights into multi-scale reconstruction of paleoenvironments and paleoclimatic changes through time in lacustrine and fluvial basins. In the Ebro Basin, the evolution from saline lacustrine to freshwater open-lake and fluvial environments through the Neogene is shown by changes in gross morphology of the microbialites, and variations in lamina shape and thickness within the stromatolites, oncolites, and calcite-coated stems. The following conclusions are highlighted:

External morphology and internal growth forms are the main features that provide information on water lake-level variations and hydrodynamics. Lamina shape, continuity, and texture inform on relative salinity levels and water depth, and microbial (bacterial) components.

Smooth microbialitic laminae with gentle lateral variations of thickness seem to be more common in saline environments. In saline environments, gentle changes in lamina shape and thickness of stromatolites may not involve concomitant changes in morphology of the microbial components, as in the Neogene examples studied here. Stressful environments for the microbial mat organisms are reflected through the lamina smoothness and low diversity of microbial morphology, likely in correspondence with low taxonomic variation.

In contrast, wavy, crinkled, or festooned laminae are abundant in oncolites and stromatolites of freshwater environments, in particular in fluvial settings. In a number of fluvial settings and open lakes, the shape of the laminae is conditioned by the morphology of the microbial bodies and their associations. In these settings, high saturation levels in calcite and rapid precipitation favor calcification of the microbial components, e.g., cyanobacteria.

Acknowledgments The information provided in this study is partially based on research funded since 1989 by numerous projects and three contracts of the Spanish Government, and one predoctoral research from the Aragón Government, Spain. Recent finding is from CGL2013-42867-P and FPI contract BES-2014-069389 of the Spanish Government and European Regional Funds. The results form part of the activities of the Geotransfer scientific group (Aragón Government, Operating Program FEDER Aragón 2014–2020). The Scanning Electron Microscopy, Optical Microscopy, and Rock Preparation services of the University of Zaragoza, Spain (Servicios de Apoyo a la Investigación) provided technical facilities. We are grateful to Dr. A. R. Carroll and Dr. E. Capezzuoli for their revisions, which improved the manuscript.

References

Agustí, J., Pérez-Rivarés, F. J., Cabrera, L., Garcés, M., Pardo, G., & Arenas, C. (2011). The Ramblian-Aragonian boundary and its significance for the European Neogene continental chronology. Contributions from the Ebro Basin record (NE Spain). *Geobios, 44*, 121–134.

Anadón, P., & Zamarreño, I. (1981). Paleogene nonmarine algal deposits of the Ebro Basin, Northeastern Spain. In C. Monty (Ed.), *Phanerozoic stromatolites* (pp. 140–154). Berlin: Springer-Verlag.

Andres, M. S., & Reid, R. P. (2006). Growth morphologies of modern marine stromatolites: A case study from Highborne Cay, Bahamas. *Sedimentary Geology, 185*(3–4), 319–328.
Andrews, J. E. (2006). Palaeoclimatic records from stable isotopes in riverine tufas: Synthesis and review. *Earth-Science Reviews, 75*, 85–104.
Anzalone, E., Ferreri, V., Sprovieri, M., & D'Argenio, B. (2007). Travertines as hydrologic archives: The case of the Pontecagnano deposits (southern Italy). *Advances in Water Resources, 30*, 2159–2175.
Arche, A., Evans, G., & Clavell, E. (2010). Some considerations on the initiation of the present SE Ebro river drainage system: Post-or pre-Messinian? *Journal of Iberian Geology, 36*, 73–85.
Arenas, C. (1993). *Sedimentología y paleogeografía del Terciario del margen pirenaico y sector central de la Cuenca del Ebro (zona aragonesa occidental)* (PhD thesis). Universidad de Zaragoza, Zaragoza. https://zaguan.unizar.es/record/70725
Arenas, C., & Jones, B. (2017). Temporal and environmental significance of microbial lamination: Insights from Recent fluvial stromatolites in the River Piedra, Spain. *Sedimentology, 64*, 1597–1629.
Arenas, C., & Pardo, G. (1998). Storm carbonate deposits in Miocene Lacustrine Systems of the Central Ebro Basin (Norheastern Spain). Extended abstracts AAPG Annual Convention and Exhibition, Salt Lake City, Utah, A30, 1–4.
Arenas, C., & Pardo, G. (1999). Latest Oligocene-Late Miocene lacustrine systems of the north-central part of the Ebro Basin (Spain): Sedimentary facies model and palaeogeographic synthesis. *Palaeogeography Palaeoclimatology Palaeoecology, 151*, 127–148.
Arenas, C., & Pardo, G. (2000). Neogene lacustrine deposits of the North-Central Ebro Basin, Northeastern Spain. In E. H. Gierlowski-Kordesch & K. R. Kelts (Eds.), *Lake basins through space and time* (Vol. 46, pp. 395–406). Tulsa, Okla: American Association Petroleum Geologists.
Arenas, C., Casanova, J., & Pardo, G. (1997). Stable-isotope characterization of the Miocene lacustrine systems of Los Monegros (Ebro Basin, Spain): Palaeogeographic and palaeoclimatic implications. *Palaeogeography Palaeoclimatology Palaeoecology, 128*(1–4), 133–155.
Arenas, C., Gutiérrez, F., Osácar, C., & Sancho, C. (2000). Sedimentology and geochemistry of fluvio-lacustrine tufa deposits controlled by evaporite solution subsidence in the central Ebro Depression, NE Spain. *Sedimentology, 47*, 883–909.
Arenas, C., Cabrera, L., & Ramos, E. (2007a). Sedimentology of tufa facies and continental microbialites from the Palaeogene of Mallorca Island (Spain). *Sedimentary Geology, 197*, 1–27.
Arenas, C., Pardo, G., Pérez-Rivarés, F. J., & Vázquez-Urbez, M. (2007b). From saline to freshwater lacustrine and fluvio-lacustrine carbonate systems: Neogene evolution of the central Ebro Basin. In C. Arenas, A. M. Alonso-Zarza, & F. Colombo (Eds.), *Geological field trips to the lacustrine deposits of the northeast of Spain, 4th International Limnogeology Congress, Geo-Guías* (Vol. 3, pp. 51–110). Zaragoza: Soc Geol Esp.
Arenas, C., Vázquez-Urbez, M., Auqué, L., Sancho, C., Osácar, M. C., & Pardo, G. (2014). Intrinsic and extrinsic controls of spatial and temporal variations in modern fluvial tufa sedimentation: A thirteen-year record from a semi-arid environment. *Sedimentology, 61*, 90–132.
Arenas, C., Piñuela, L., & García-Ramos, J. C. (2015). Climatic and tectonic controls on carbonate deposition in syn-rift siliciclastic fluvial systems: A case of microbialites and associated facies in the Late Jurassic. *Sedimentology, 62*, 1149–1183.
Arp, G., Wedemeyer, N., & Reitner, J. (2001). Fluvial tufa formation in hard-water creek. Deinschwanger Bach, Franconian Al, Germany. *Facies, 44*, 1–22.
Astibia, H. (1986). *Los macromamiferos del Mioceno medio de Tarazona de Aragón (depresión del Ebro)* [Tesis Doctoral]. Universidad del País Vasco, Bilbao.
Astibia, H., López-Martínez, N., Elorza, J., & Vicens, E. (2012). Increasing size and abundance of microbialites (oncoids) in connection with the K/T boundary in non-marine environments in the South Central Pyrenees. *Geologica Acta, 10*(3), 209–226.

Awramik, S. M. (1992). The history and significance of stromatolites. In M. Schidlowski (Ed.), *Early organic evolution: Implications for mineral and energy resources* (pp. 435–449). New York: Springer.

Awramik, S. M., & Buchheim, H. P. (2009). A giant, Late Archean lake system: The Meentheena Member (Tumbiana Formation; Fortescue Group), Western Australia. *Precambrian Research, 174*(3–4), 215–240.

Awramik, S. M., & Buchheim, H. P. (2015). Giant stromatolites of the Eocene Green River Formation (Colorado, USA). *Geology, 43*(8), 691–694.

Berrendero, E., Arenas, C., Mateo, P., & Jones, B. (2016). Cyanobacterial diversity and related sedimentary facies as a function of water flow conditions: Example from the Monasterio de Piedra Natural Park (Spain). *Sedimentary Geology, 337*, 12–28.

Bertrand-Sarfati, J., Freytet, P., & Plaziat, J. C. (1994). Microstructures in Tertiary nonmarine stromatolites (France). Comparison with Proterozoic. In J. Bertrand-Safarti & C. Monty (Eds.), *Phanerozoic stromatolites II* (pp. 155–191). Dordrecht: Kluwer Academic Publishers.

Bosak, T., Knoll, A. H., & Petroff, A. P. (2013). The meaning of stromatolites. *Annual Review of Earth and Planetary Sciences, 41*, 21–44.

Bouton, A., Vennin, E., Boulle, J., Pace, A., Bourillot, R., Thomazo, C., Brayard, A., Désaubliaux, G., Goslar, T., Yokoyama, Y., Dupraz, C., & Visscher, P. (2016). Linking the distribution of microbial deposits from the Great Salt Lake (Utach, USA) to tectonic and climatic processes. *Biogeosciences, 13*, 5511–5526.

Braga, J. C., Martín, J. M., & Riding, R. (1995). Controls on microbial dome fabric development along a carbonate-siliciclastic shelfbasin transect, Miocene, SE Spain. *PALAIOS, 10*, 347–361.

Brasier, A. T., Andrews, J. E., Marca-Bell, A. D., & Dennis, P. F. (2010). Depositional continuity of seasonally laminated tufas: Implications for $\delta^{18}O$ based palaeotemperatures. *Global and Planetary Change, 71*, 160–197.

Browne, K. M., Golubić, S., & Seong-Joo, L. (2000). Shallow marine microbial carbonate deposits. In R. E. Riding & S. M. Awramik (Eds.), *Microbial sediments* (pp. 233–249). Berlin: Springer-Verlag.

Capezzuoli, E., Gandin, A., & Pedley, M. (2014). Decoding tufa and travertine (fresh water carbonates) in the sedimentary record: The state of the art. *Sedimentology, 61*(1), 1–21.

Casanova, J. (1986). *Les stromatolites continentaux: Paléoecologie, paléohydrologie, paléoclimatologie. Application au Rift Gregory* (PhD thesis). Université Marseille-Luminy, Marseille.

Casanova, J. (1994). Stromatolites from the east African rift: A synopsis. In J. Bertrand-Sarfati & C. Monty (Eds.), *Phanerozoic stromatolites II* (pp. 193–226). Dordrecht: Kluwer Academic Publishers.

Casanova, J., & Nury, D. (1989). Biosédimentologie des stromatolites fluvio-lacustres du fossé oligocene de Marseille. *Bulletin de la Société Géologique de France, 5*(6), 1173–1184.

Caudwell, C., Lang, J., & Pascal, A. (2001). Lamination of swampy-rivulets Rivularia haematites stromatolites in a temperate climate. *Sedimentary Geology, 143*, 125–147.

Chafetz, H. S., Utech, N. M., & Fitzmaurice, S. P. (1991). Differences in the $\delta\ ^{18}O$ and $\delta\ ^{13}C$ signatures of seasonal laminae comprising travertine stromatolites. *Journal of Sedimentary Petrology, 61*, 1015–1028.

Chagas, A., Webb, G. E., Burne, R. V., & Southama, G. (2016). Modern lacustrine microbialites: Towards a synthesis of aqueous and carbonate geochemistry and mineralogy. *Earth-Science Reviews, 162*, 338–363.

Chidsey, T., Vanden Berg, M. D., & Eby, D. (2015). Petrography and characterization of microbial carbonates and associated facies from modern Great Salt Lake and Uinta Basin's Eocene Green River Formation in Utah, USA. In D. W. J. Bosence, K. A. Gibbons, D. P. Le Heron, W. A. Morgan, T. Pritchard, & B. A. Vining (Eds.), *Microbial carbonates in space and time: Implications for global exploration and production. Special publications 418* (pp. 261–286). London: Geological Society.

Cuenca, G., Canudo, J. I., Laplana, C., & Andres, J. A. (1992). Bio y cronoestratigrafía con mamíferos en la Cuenca Terciaria de Ebro: ensayo de síntesis. *Acta Geologica Hispánica, 27*(1–2), 127–143.

Della Porta, G. (2015). Carbonate build-ups in lacustrine, hydrothermal and fluvial settings: Comparing depositional geometry, fabric types and geochemical signature. In D. W. J. Bosence, K. A. Gibbons, D. P. Le Heron, W. A. Morgan, T. Pritchard, & B. A. Vining (Eds.), *Microabial carbonates in space and time: Implications for global exploration and production. Special publications, 418* (pp. 17–68). London: Geological Society.

Dupraz, C., Visscher, P. T., Baumgartner, L. K., & Reid, R. P. (2004). Microbe–mineral interactions: Early carbonate precipitation in a hypersaline lake (Eleuthera Island, Bahamas). *Sedimentology, 51*, 745–765.

Dupraz, C., Reid, R. P., Braissant, O., Decho, A. W., Norman, R. S., & Visscher, P. T. (2009). Processes of carbonate precipitation in modern microbial mats. *Earth-Science Reviews, 96*(3), 141–162.

Ferris, F. G., Thompson, J. B., & Beveridge, T. J. (1997). Modern freshwater microbialites from Kelly Lake, British Columbia, Canada. *PALAIOS, 12*(3), 213–219.

Ford, T. D., & Pedley, H. M. (1996). A review of tufa and travertine deposits of the world. *Earth-Science Reviews, 41*(3–4), 17–175.

Frantz, C. M., Petryshyn, V. A., Marenco, P. J., Tripati, A., Berelson, W. M., & Corsetti, F. A. (2014). Dramatic local environmental change during the Early Eocene Climatic Optimum detected using high resolution chemical analyses of Green River Formation stromatolites. *Palaeogeography Palaeoclimatology Palaeoecology, 405*, 1–15.

Freytet, P. (1992). Exemples de fossilisation de restes végétaux (algues, feuilles) par la calcite, en milieu fluviatile et lacustre, dans l'actuel et dans l'ancien. *Bulletin de la Société Botanique de France, 139*, 69–74.

García Castellanos, D., Vergés, J., Gaspar Escribano, J., & Cloetingh, S. (2003). Interplay between tectonics, climate, and fluvial transport during the Cenozoic evolution of the Ebro Basin (NE Iberia). *Journal of Geophysical Research. Solid Earth, 108*(B7).

Golubić, S., Seong-Joo, L., & Browne, K. M. (2000). Cyanobacteria: Architects of sedimentary structures. In R. E. Riding & S. M. Awramik (Eds.), *Microbial sediments* (pp. 57–67). Berlin: Springer-Verlag.

Golubić, S., Violante, C., Plenković-Moraj, A., & Grgasović, T. (2008). Travertines and calcareous tufa deposits: An insight into diagenesis. *Croatian Geological Survey, 61*, 363–378.

Gradzinski, M. (2010). Factors controlling growth of modern tufa: Results of a field experiment. In M. Pedley & M. Rogerson (Eds.), *Tufas and speleothems: Unravelling the microbial and physical controls. Special Publications 336* (pp. 143–191). London: Geological Society.

Graf, J. W., Carroll, A. R., & Smith, M. E. (2015). Lacustrine sedimentology, stratigraphy and stable isotope geochemistry of the Tipton Member of the Green River Formation. In M. E. Smith & A. R. Carroll (Eds.), *Stratigraphy and paleolimnology of the green river formation, Western USA. Synthesis in limnogeology* (Vol. 1, pp. 31–60). Dordrecht: Springer.

Hägele, D., Leinfelder, R., Grau, J., Burmeister, E. G., & Struck, U. (2006). Oncoids from the river Alz (southern Germany): Tiny ecosystems in a phosphorus-limited environment. *Palaeogeography Palaeoclimatology Palaeoecology, 237*, 378–395.

Hofmann, H. J. (1973). Stromatolites: Characteristics and utility. *Earth-Science Reviews, 9*, 339–373.

Jones, B., & Renaut, R. W. (1996). Morphology and growth of aragonite crystals in hot-spring travertines at Lake Bogoria, Kenya Rift Valley. *Sedimentology, 43*, 323–340.

Leggit, V. L., & Cushman, R. A. (2001). Complex caddisfly-dominated bioherms from the Eocene Green River Formation. *Sedimentary Geology, 145*, 377–396.

Lindqvist, J. K. (1994). Lacustrine stromatolites and oncoids. Manuherikia Group (Miocene), New Zealand. In J. Bertrand-Sarfati & C. Monty (Eds.), *Phanerozoic stromatolites II* (pp. 227–254). Dordrecht: Kluwer Academic Publishers.

Martin-Bello, L., Arenas, C., & Jones, B. (2019). Lacustrine stromatolites: Useful structures for environmental interpretation. An example from the Miocene Ebro Basin. *Sedimentology, 66*. https://doi.org/10.1111/sed.12577.

Merz-Preiß, M., & Riding, R. (1999). Cyanobacterial tufa calcification in two freshwater streams: Ambient environment, chemical thresholds and biological processes. *Sedimentary Geology, 126*, 103–124.

Monty, C. L. V. (1976). The origin and development of cryptalgal fabrics. In *Developments in sedimentology* (Vol. 20, pp. 193–249). Amsterdam: Elsevier.

Muniz, M. C., & Bosence, D. W. J. (2015). Pre-salt microbialites from the Campos Basin (offshore Brazil): Image log facies, facies model and cyclicity in lacustrine carbonates. In D. W. J. Bosence, K. A. Gibbons, D. P. Le Heron, W. A. Morgan, T. Pritchard, & B. A. Vining (Eds.), *Microbial carbonates in space and time: Implications for global exploration and production. Special Publications 418* (pp. 221–242). London: Geological Society.

Muñoz, A., Arenas, C., González, A., Luzón, A., Pérez, A., Villena, J., & Pardo, G. (2002). Ebro Basin (northeastern Spain). In T. Moreno & W. Gibbons (Eds.), *Geology of Spain* (pp. 301–309). London: Geological Society.

Muñoz-Jiménez, A., & Casas-Sainz, A. M. (1997). The Rioja trough (N Spain): Tectosedimentary evolution of a symmetric foreland basin. *Basin Research, 9*, 65–85.

Murelaga, X., Pérez-Rivarés, F. J., Vázquez-Urbez, M., & Zuluaga, M. C. (2008). Nuevos datos bioestratigráficos y paleoecológicos del Mioceno medio (Aragoniense) del área de Tarazona de Aragón (Cuenca del Ebro, provincia de Zaragoza, España). *Ameghiniana, 45*(2), 393–406.

Osácar, M. C., Arenas, C., Auqué, L. F., Pardo, G., Sancho, C., & Vazquez-Urbez, M. (2016). Discerning the interactions between environmental parameters reflected in $\delta^{13}C$ and $\delta^{18}O$ of recent fluvial tufas: Lessons from a Mediterranean climate region. *Sedimentary Geology, 345*, 126–144.

Pardo, G., Arenas, C., González, A., Luzón, A., Muñoz, A., Pérez, A., Pérez-Rivarés, F. J., Vázquez-Urbez, M., & Vilena, J. (2004). La cuenca del Ebro. In J. A. Vera (Ed.), *Geología de España* (pp. 533–543). Madrid: Sociedad Geológica de España-IGME.

Pedley, M. (2009). Tufas and travertines of the Mediterranean region: A testing ground for freshwater carbonate concepts and developments. *Sedimentology, 56*(1), 221–246.

Pedley, M., Rogerson, M., & Middleton, R. (2009). Freshwater calcite precipitates from in vitro mesocosm flume experiments: A case for biomediation of tufas. *Sedimentology, 56*, 511–527.

Pentecost, A. (2005). *Travertine*. Dordrecht: Springer, Science & Business Media.

Pérez Rivarés, F. J. (2016). *Estudio magnetoestratigráfico del Mioceno del sector central de la Cuenca del Ebro: Cronología, correlación y análisis de la ciclicidad sedimentaria* (PhD thesis). Universidad de Zaragoza, Zaragoza.

Pérez, A., Muñoz, A., Pardo, G., Villena, J., & Arenas, C. (1988). Las unidades tectosedimentarias del Neógeno del borde ibérico de la Depresión del Ebro (sector central). In A. Pérez, A. Muñoz, & J. A. Sánchez (Eds.), *Sistemas lacustres neógenos del margen ibérico de la Cuenca del Ebro : guia de campo : III Reunión del Grupo Español de Trabajo, Zaragoza, Noviembre 1988* (pp. 7–20). Zaragoza: Secretariado de publicaciones. Universidad de Zaragoza.

Pérez-Rivarés, F. J., Arenas, C., Pardo, G., & Garcés, M. (2018). Temporal aspects of genetic stratigraphic units in continental sedimentary basins: Examples from the Ebro basin, Spain. *Earth-Science Reviews, 178*, 136–153.

Petryshyn, V. A., Corsetti, F. A., Berelson, W. M., Beaumont, W., & Lund, S. P. (2012). Stromatolite lamination frequency, Walker Lake, Nevada: Implications for stromatolites as biosignatures. *Geology, 40*(6), 499–502.

Pietras, J. T., & Carroll, A. R. (2006). High-resolution stratigraphy of an underfilled lake basin: Wilkins Peak member, Eocene Green River Formation, Wyoming, USA. *Journal of Sedimentary Research, 76*, 1197–1214.

Pope, M., Grotzinger, J., & Schreiber, B. C. (2000). Evaporitic subtidal stromatolites produced by in situ precipitation: Textures, facies associations, and temporal significance. *Journal of Sedimentary Research, 70*, 1139–1151.

Ramos, E., Cabrera, L., Hagemann, H., Pickel, W., & Zamarreño, I. (2001). Palaeogene lacustrine record in Mallorca (NW Mediterranean, Spain): Depositional, palaeographic and palaeoclimatic implications for the ancient southeastern Iberian margin. *Palaeogeography Palaeoclimatology Palaeoecology, 172*, 1–37.

Renaut, R. W., & Long, P. R. (1989). Sedimentology of the saline lakes of the Cariboo plateau, interior British Columbia, Canada. *Sedimentary Geology, 64*(4), 239–264.

Riba, O., Reguant, S., & Villena, J. (1983). Ensayo de síntesis estratigráfica y evolutiva de la cuenca terciaria del Ebro. In L. Homenaje & J. M. Ríos (Eds.), *Geología de España 2* (pp. 131–159). Madrid: IGME.

Riding, R. (2000). Microbial carbonates: The geological record of calcified bacterial–algal mats and biofilms. *Sedimentology, 47*, 179–214.

Riding, R. (2011). Microbialites, stromatolites, and thrombolites. In *Encyclopedia of geobiology* (pp. 635–665). Dordrecht: Springer.

Riding, R., & Liang, L. (2005). Geobiology of microbial carbonates: Metazoan and seawater saturation state influences on secular trends during the Phanerozoic. *Palaeogeography Palaeoclimatology Palaeoecology, 219*, 101–115.

Riding, R., Braga, J. C., & Martín, J. M. (1991). Oolite stromatolites and thrombolites, Miocene, Spain: Analogues of Recent giant Bahamian examples. *Sedimentary Geology, 71*, 121–127.

Roche, A., Vennin, E., Bouton, A., Oliver, N., Wattinne, A., Bundeleva, I., Deconinck, J. F., Virgone, A., Gaucher, E. C., & Visscher, P. (2018). Oligo-Miocene lacustrine microbial and metazoan buildups from the Limagne Basin (French Massif Central). *Palaeogeography Palaeoclimatology Palaeoecology, 504*, 34–59.

Sakurai, R., Ito, M., Ueno, M., Kitajima, K., & Maruyama, S. (2015). Facies architecture and sequence-stratigraphic features of the Tumbiana Formation in the Pilbara Craton, northwestern Australia: Implications for depositional environments of oxygenic stromatolites during the Late Archean. *Precambrian Research, 138*, 255–273.

Salvany, J. M., García-Veigas, J., & Orí, F. (2007). Glauberite-halite association of the Zaragoza Gypsum Formation (Lower Miocene, Ebro Basin, NE Spain). *Sedimentology, 54*, 443–467.

Scott, J. J., & Smith, M. E. (2015). Trace fossils of the Eocene green river lake basins, Wyoming, Utah, and Colorado. In M. E. Smith & A. R. Carroll (Eds.), *Stratigraphy and Paleolimnology of the Green River Formation, Western USA. Synthesis in Limnogeology, 1* (pp. 313–350). Dordrecht: Springer.

Smoot, J. P. (1983). Depositional subenvironments in an arid closed basin; the Wilkins Peak Member of the Green River Formation (Eocene), Wyoming, USA. *Sedimentology, 30*, 801–827.

Storrie-Lombardi, M. C., & Awramik, S. M. (2006). A sideways view of stromatolites: Complexity metrics for stromatolite laminae. In: *Instruments, methods, and missions for astrobiology IX. Procedings* (vol. 6309, p. 63090P). SPIE Optics + Photonics, San Diego. https://doi.org/10.1117/12.679869

Suárez-González, P., Quijada, I. E., Benito, M. I., Mas, R., Merinero, R., & Riding, R. (2014). Origin and significance of lamination in Lower Cretaceous stromatolites and proposal for a quantitative approach. *Sedimentary Geology, 300*, 11–27.

Suárez-González, P., Arenas-Abad, C., Pomar, L., & Benito, M. I. (2017). "Giant microbialites": From thrombolites to stromatolites and back again. Upper Miocene, Mallorca (Spain). In: *Lyell Meeting 2017; Sticking together: microbes and their role in forming sediments, London, March 2017* (pp. 104–105). Abstract book. The Geological Society, London

Suárez-González, P., Arenas, C., Benito, M. I., & Pomar, L. (2019). Interplay between biotic and environmental conditions in pre-salt Messinian microbialites of the western Mediterranean (Upper Miocene, Mallorca, Spain). *Palaeogeography, Palaeoclimatology, Palaeoecology, 533*.

Surdam, R. C., & Wray, J. L. (1976). Lacustrine stromatolites, Eocene Green River Formation, Wyoming. In M. R. Walter (Ed.), *Stromatolites, developments in sedimentology* (Vol. 20, pp. 535–541). Amsterdam: Elsevier.

Tang, D., Shi, X., & Jiang, G. (2014). Sunspot cycles recorded in Mesoproterozoic carbonate biolaminites. *Precambrian Research, 248*, 1–16.

Tosti, F., & Riding, R. (2017). Fine-grained agglutinated elongate columnar stromatolites: Tieling Formation, ca 1420 Ma, North China. *Sedimentology, 64*(4), 871–902.
Urgeles, R., Camerlenghi, A., García-Castellanos, D., De Mol, B., Garcés, M., Vergés, J., Haslamk, I., & Hardman, M. (2011). New constraints on the Messinian sealevel drawdown from 3D seismic data of the Ebro Margin, western Mediterranean. *Basin Research, 23*, 123–145.
Vázquez-Urbez, M. (2008). *Caracterización y significado ambiental de depósitos tobáceos neógenos en la Cuenca del Ebro. Comparación con ambientes cuaternarios.* (PhD thesis). Universidad de Zaragoza, Zaragoza
Vázquez-Urbez, M., Osácar-Soriano, M. C., Arenas-Abad, M. C., Sancho-Marcén, C., & Auqué-Sanz, L. F. (2005). Variabilidad de la señal isotópica (δC^{13} y δO^{18}) del sistema tobáceo actual del Parque del Monasterio de Piedra (provincia de Zaragoza). *Geotemas, Madrid, 8*, 119–123.
Vázquez-Urbez, M., Arenas, C., Sancho, C., Osácar, C., Auqué, L., & Pardo, G. (2010). Factors controlling present-day tufa dynamics in the Monasterio de Piedra Natural Park (Iberian Range, Spain): Depositional environmental settings, sedimentation rates and hydrochemistry. *International Journal of Earth Sciences, 99*(5), 1027–1049.
Vázquez-Urbez, M., Arenas, C., Pardo, G., & Pérez-Rivarés, J. (2013). The effect of drainage reorganization and climate on the sedimentologic evolution of intermontane lake systems: The final fill stage of the Tertiary Ebro Basin (Spain). *Journal of Sedimentary Research, 83*(8), 562–590.
Vennin, E., Bouton, A., Bourillot, R., Pace, A., Roche, A., Brayard, A., Thomazo, C., Virgone, A., Gaucher, E., Desaubliaux, G., & Visscher, P. (2018). The lacustrine microbial carbonate factory of the successive Lake Bonneville and Great Salt Lake, Utah, USA. *Sedimentology, 66*(1), 165–204. https://doi.org/10.1111/sed.12499.
Verrecchia, E. P., Freytet, P., Julien, J., & Baltzer, F. (1997). The unusual hydrodynamic behaviour of freshwater oncolites. *Sedimentary Geology, 113*, 225–243.
Walter, M. R. (1972). Stromatolites and the biostratigraphy of the Australian Precambrian and Cambrian. *Special Publications of Palaeontology, 11*, 256.
Walter, M. R. (Ed.). (1976). *Stromatolites. Developments in sedimentology 20.* Amsterdam: Elsevier Scientific Publishing Company.
Zamarreño, I., Anadón, P., & Utrilla, R. (1997). Sedimentology and isotopic composition of Upper Palaeocene to Eocene non-marine stromatolites, eastern Ebro Basin, NE Spain. *Sedimentology, 44*, 159–176.

Part IV
North America

Ecological Response of Ostracodes (Arthropoda, Crustacea) to Lake-Level Fluctuations in the Eocene Green River Formation, Fossil Basin, Wyoming, USA

Lisa E. Park Boush, Christine M. S. Hall, Lucas S. Antonietto, and Andrew J. McFarland

Abstract The Eocene Green River Formation is one of the best-known *Konservat lagerstätten* and comprises lacustrine strata that were deposited during the Early Eocene Climatic Optimum (EECO). Two species of ostracodes, *Pseudoeucypris pagei* (Swain, Journal of Paleontology, 23:172–181, 1949) and *Hemicyprinotus watsonensis* (Swain, Journal of Paleontology, 23:172–181, 1949), were recovered from 16 intervals at three sites from Fossil Basin, Wyoming, USA. Population density per sample was quantified as the number of ostracode valves per cm^2. Analysis of kerogen content shows a significant difference between preservation based upon lithology, with ostracodes being more commonly preserved in kerogen-poor micrites but more abundant when preserved in kerogen-rich micrites. The nature of preservation also was correlated to lithology with ostracodes being preserved as whole carapaces more frequently in kerogen-poor micrites and being more broken and disarticulated in kerogen-rich micrites. Species tracked lithology as well, with the epiphytic species *P. pagei* occurring exclusively in kerogen-poor micrites and dolomicrites, while the benthic species *H. watsonenis* occurred in kerogen-rich micrites as well as some kerogen-poor micrites. The present study demonstrates how ostracodes are preserved differently within the basin and the utility of species occurrences in tracking changing lake environments, as similarly reported in other Green River Formation basins. Here, we interpret that the presence of *P. pagei* and the rocks it is preserved in indicate shallower lake conditions, while shifts to *H. watsonensis* are indicative of deepening lake conditions.

Keywords *Konservat lagerstätten* · Ostracoda · Fossil Lake · Taphonomy

---

L. E. Park Boush (✉) · C. M. S. Hall · L. S. Antonietto
University of Connecticut, Department of Geosciences, Storrs, CT, USA
e-mail: lisa.park_boush@uconn.edu; christine.hall@uconn.edu

A. J. McFarland
University of Akron, Department of Geology and Geography, Akron, OH, USA

M. R. Rosen et al. (eds.), *Limnogeology: Progress, Challenges and Opportunities*, Syntheses in Limnogeology, https://doi.org/10.1007/978-3-030-66576-0_7

Introduction

The Green River Formation is a thick complex of lacustrine strata deposited in two major basins–the Greater Green River Basin and the Uinta-Piceance Creek Basin–which are separated by the Uinta Uplift. The Greater Green River Basin (GGRB) consists of four sub-basins, including the Fossil, Bridger, Washakie, and Sand Wash basins that are partitioned from one another by the Rock Springs uplift, which trends north–south, and a number of smaller structures trending east–west (Smith et al. 2008). The Green River Formation lakes represent some of many Eocene lake basins or lake systems spread across North America, Southeast Asia, Australia, and Western Europe formed within rift zones or due to thrusting that were likely influenced by climatological changes during this time period (Gierlowski-Kordesch et al. 2008 and Gierlowski-Kordesch and Park 2004). The Green River Formation, deposited from 53.5 to 48.5 Ma, represents a 5 million year period that coincides with the Early Eocene Climatic Optimum, one of the warmest periods in the Cenozoic (Smith et al. 2003), which, along with tectonic activity greatly influenced the basin's hydrodynamics, particularly with respect to lake levels and open and closed basin morphologies. Much of the previous work on the Green River Formation has focused on reconstructing the depositional history of the GGRB (Aswasereelert et al. 2013; Baddouh et al. 2016; Buchheim et al. 2011, 2015; Carroll and Bohacs 2001; Machlus et al. 2015; Pietras and Carroll 2006; Smith et al. 2003, 2008; Smith and Carroll 2015; and references therein). These studies characterized the dynamic history of the GGRB within the context of tectonic and climatic regimes, as well as of coeval units in the nearby Piceance Creek and Uinta basins in Colorado and Utah, respectively.

The GGRB began as a foreland basin during the Sevier Orogeny and was divided into its various sub-basins by uplifts of Precambrian basement during the Laramide Orogeny (Dickinson et al. 1988). This led to the staggered formation of Fossil Lake in the Fossil sub-basin, Lake Gosiute in the other sub-basins of the GGRB, and Lake Uinta in the Piceance Creek and Uinta basins (Beck et al. 1988; Buchheim et al. 2015; Pietras and Carroll 2006; Rhodes et al. 2002; Roehler 1992a). These lakes were connected to one another at various times throughout their history (Oriel and Tracey Jr. 1970; Smith et al. 2008). This study focuses on the smallest of these, Fossil Lake.

Fossil Basin.—Fossil Basin is a structural basin created by thrust-faulted ridges of north–south trending Paleozoic–Mesozoic rocks (Buchheim and Eugster 1998). It is bordered by the Tunp Range to the northwest, Oyster Ridge to the east, and unnamed ridges near the Crawford Mountains to the southwest (Buchheim and Eugster 1998). In this context, the micritic/dolomicritic Green River Formation was deposited in a north to south trending asymmetric syncline (Buchheim et al. 2011; Buchheim and Eugster 1998). The stratigraphy of Fossil Basin was most recently revised by Buchheim et al. (2011) into three units, formally named from oldest to youngest: the Road Hollow, Fossil Butte, and Angelo Members. Mammal fossils have placed the Green River Formation into the early Eocene, and the K-spar volcanic tuff that occurs near the top of the Fossil Butte Member (FBM) has been dated at ~51.97 ±0.09 Ma (Gazin 1952; Buchheim and Eugster 1998; Smith et al. 2003, 2008, 2010; Smith and Carroll 2015). Fossil Basin strata have been correlated to the Greater Green River Basin by Buchheim et al. (2015) who determined that the Road Hollow Member is equivalent to the main body of the Wasatch Formation and Upper Limestone Bed, the Fossil Butte Member is equivalent to the Tipton Member, and the Angelo Member is equivalent to the Wilkins Peak Member with evidence that supports hydrological connectivity between Fossil Basin and the other sub-basins of the GGRB as suggested by Oriel and Tracey Jr. (1970).

The Road Hollow Member is composed of bioturbated micrites, laminated calcimicrites, siltstones, and sandstones. The various siliciclastic layers may represent hydrologic connections between Fossil Lake and Lake Gosiute via streams but no paleocurrent data have been found to confirm this interpretation (Buchheim et al. 2015). The overlying Fossil Butte Member contains laminated calcimicrites, likely deposited under deep, still water at a time when the lake was at its largest extent (Buchheim 1994; Buchheim et al. 2011). Anoxia, as indicated by the dark color of the deposits, has been suggested to be the reason for the excellent fossil preservation in the Fossil Butte Member (Buchheim et al. 2011). The Angelo Member is composed primarily of dolomicrites, sodium bicarbonate salts, and chert (Buchheim 1994; Buchheim et al. 2011, 2002). It has been interpreted to have been deposited during the gradual shrinking of Fossil Lake, when it became more saline through evaporation, inferred from the deposition of Magadi-type chert (Buchheim et al. 2011).

The Green River Formation is renowned for its exceptionally well-preserved lacustrine and terrestrial fossils that include fish and plants, as well as rarer birds, bats, alligators, and insects (Ferber and Wells 1995; Grande 1984; Grande and Buchheim 1994; Ksepka and Clarke 2010; McGrew 1975; Roehler 1992a, 1992b). This exceptional level of preservation – sometimes displaying soft tissues – is due to various environmental factors related to anoxia and the subsequent limitation of biological degradation, which are due to water depth and oxygen stratification. The flora recovered from the Green River Formation resembles that of the present-day southeastern United States, suggesting a warm, humid, equitable climate (McGrew 1975).

Among the various faunal elements found within the Green River Formation are the Ostracoda. These microcrustaceans with bivalved carapaces made of low-Mg calcite have specific environmental tolerances, abundant distribution, and high preservation potential, providing an effective means by which to evaluate the biologic response to climatic and tectonic-driven hydrologic changes within aquatic environments (Boomer et al. 2003). Swain identified and described many of the ostracodes found in the Green River Formation during extensive studies in the late 1940s and 1950s, creating ostracode zones primarily focused on the Uinta and Piceance Creek Basins (Swain 1949, 1956, 1964 and Swain et al. 1971). Taylor (1972) initially described the paleoecological aspects of the ostracode fossils from the Green River Formation and found that Lake Gosiute ostracodes generally were more abundant in rocks with higher clay content, which he interpreted as being associated with quieter waters. Ingalls and Park (2010) identified six species of ostracodes in Lake Gosiute determining that ostracode distribution and species richness is correlated to lake basin type.

This study is the first detailed study of the ostracodes in Fossil Basin and examines the changes in ostracode occurrences within the Green River Formation in relation to lake environments of Fossil Lake in the Fossil Basin. Specifically, we address species distributions and preservation within the context of lake history.

Materials and Methods

Stratigraphic sections were measured and described at Road Hollow (41°39′43″ N, 110°48′08.8″ W), Bear River Gulch (41°43′31″ N, 110°50′24.0″ W), and Smith Hollow Quarry (41°48′04″ N, 110°43′28″ W) (Fig. 1), near the town of Kemmerer, Wyoming, USA. These sections span the entirety of the Green River Formation in Fossil Basin and include the interfingering Sandstone Tongue of the Wasatch Formation. Rock samples were collected throughout each section (>150 total samples), with a focus on laminated beds. Units containing ostracode fossils were sampled at one-centimeter intervals and split to expose bedding planes to obtain a high-resolution record of abundance and preservation style. Fossil ostracodes were examined under a binocular dissecting microscope. To quantify abundance, a 1 cm by 1 cm grid was randomly placed on the surface, and the number of individual valves within the grid was counted. This was repeated on each ostracode-bearing surface; the number of grid squares per sample ranged from 4 to 73, depending on the slab surface area. Ostracode density was calculated by averaging the grid square counts on each sampled surface (Appendices A, B, and C). Changes in lamination counts per centimeter of stratigraphic thickness were compared to changes in ostracode species throughout the measured sections to examine the correlation between ostracode diversity dynamics and changes in lake sedimentation. The taphonomy and preservational mode of ostracode valves–whether they were casts, molds, recrystallized, or unaltered–was examined for all specimens.

Results

Stratigraphy

The stratigraphy of Fossil Basin has recently been revised by Buchheim et al. (2011) into three units, formally named the Road Hollow, Fossil Butte, and Angelo Members. Our stratigraphic sections were measured in three locations: Road Hollow, Bear River Gulch, and Smith Hollow Quarry.

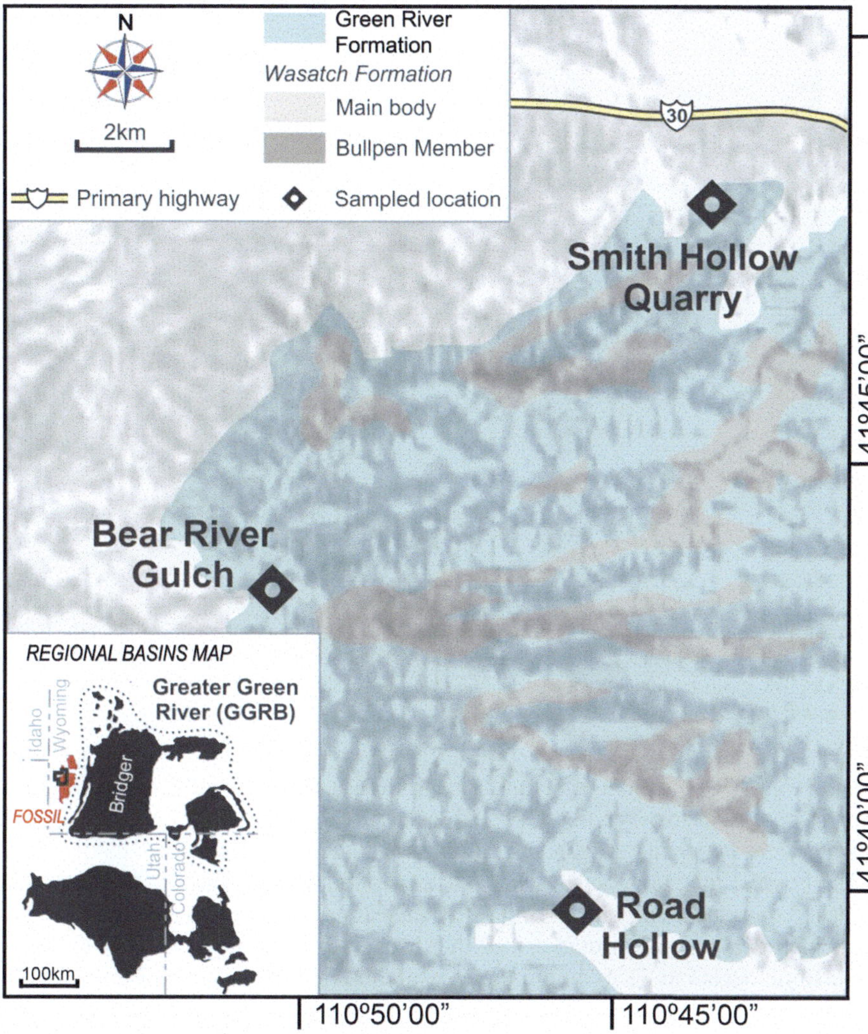

Fig. 1 Distribution of the Green River Formation within the Fossil Basin, indicating the locations of the sites sampled for this study. Map inset shows the Greater Green River Basin of Wyoming, Colorado, and Utah. Scale as shown. Map based on Buchheim et al. (2015). Dotted line represents the GGRB. Study area is denoted by black box in inset

Road Hollow Section.—The Road Hollow section measured approximately 115 m thick and includes the Road Hollow and Fossil Butte Members of the Green River Formation as well as the Bullpen Member of the Wasatch Formation (Fig. 2). The basal part of the measured section (0–44 m) represents the Upper Road Hollow Member and is dominated by bioturbated, structureless calcimicrites, with interbedded siltstones, sandstones, and kerogen-rich laminated micrites (KRLM). A kerogen-poor structureless micrite bed bearing the ostracode species *Pseudoeucypris*

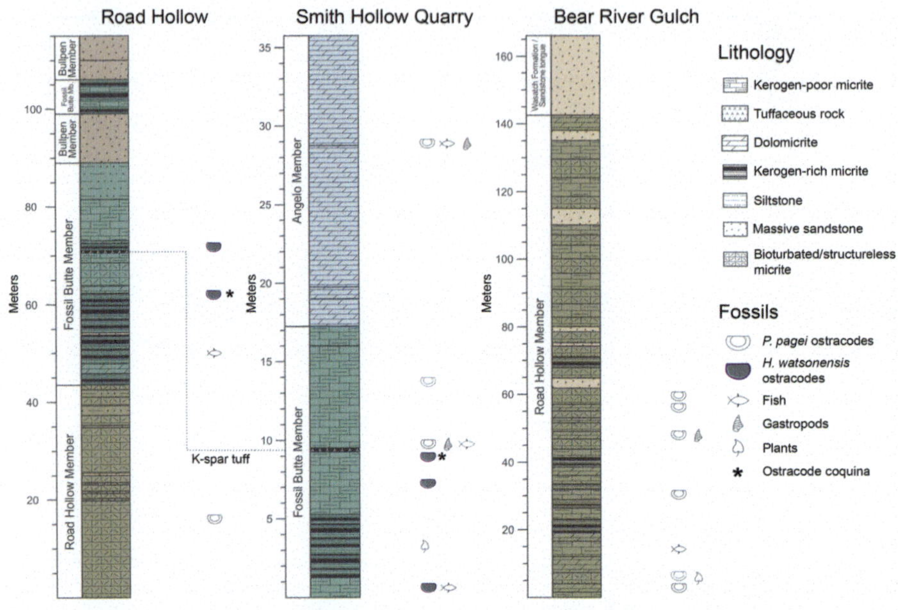

Fig. 2 Stratigraphic columns of the Green River and Wasatch formations as measured at the Road Hollow, Smith Hollow Quarry, and Bear River Gulch sites showing lithologies and levels with ostracode, fish, gastropod, and plant fossils. The positions of the K-spar tuff and ostracode marker bed are noted

pagei (Swain 1949) occurs at about 19 m, followed by interbedded siltstones and kerogen-poor micrites. The upper portion of the Road Hollow Member in this section alternates between kerogen-rich micrites and siliciclastic units.

The Fossil Butte Member overlies the Road Hollow and contains kerogen-rich laminated micrites, kerogen-poor micrites, sandstones, and dolostones. The K-spar tuff, which is a prominent marker bed that is used to correlate sections within the Fossil Butte occurs at 71 m within the measured section (Fig. 2). Fish fossils were rare but recovered at 50 m within kerogen-poor laminated micrites (KPLM) interbedded between kerogen-rich micrites. Some microlaminated micrite layers contain ostracode fossils (Fig. 2). An ostracode coquina, described in Buchheim et al. (2015) as the 'ostracodal limestone' (OSLS) is designated as a critical marker bed that occurs below the K-spar tuff within the Fossil Butte Member and can be traced from the Fossil Basin to the Gosiute Basin. At the Road Hollow section, an 8.5 m-thick ostracode-rich layer exists at 62 m and approximately 10 m below the K-spar tuff (51.97 Ma ± 0.09, Smith et al. 2010) which may correspond with the OSLS even though it is stratigraphically lower than it appears in other sections. Above the K-spar tuff, the Fossil Butte Member consists of mostly kerogen-poor micrites and siltstones, with approximately 8 m of kerogen-poor micrite overlain by approximately 7 m of siltstone. Two 9–10-meter-thick beds of sandstone from the Bullpen Member of the Wasatch Formation interbed with kerogen-rich and kerogen-poor micrites of the Fossil Butte member at the top of the measured section (Fig. 2).

Smith Hollow Quarry Section.—The measured section at Smith Hollow Quarry is approximately 36 m thick and contains micrites and dolostones that span the Fossil Butte and Angelo Members (Fig. 2). The lower part of the section is dominated by kerogen-rich microlaminated micrites of the Fossil Butte Member, but also includes thin beds of siliciclastics; these units are generally volcanic tuffs, and while relatively thin, are numerous throughout the section. Kerogen-poor micrites occur above and below the K-spar tuff, which occurs at ~9.5 m. The upper part of the section is dominated by dolomicrites of the Angelo Member.

Fish, gastropod, and plant fossils are present in both kerogen-poor and kerogen-rich microlaminated micrites (Fig. 2). Abundant fish fossils, for which the Green River Formation is known, are found within this section, mostly *Knightia* spp., but larger species such as *Diplomystus* spp., *Mioplosus* spp., *Notogoneus* spp., *Priscacara* spp., and *Phareodus* spp. also are found. Ostracode fossils from this section are found in kerogen-poor, microlaminated micrites, and dolomicrites—with *Hemicyprinotus watsonensis* (Swain 1964) in the lower section occurring in kerogen-poor laminated micrites and *P. pagei* occurring in sections above in both kerogen-poor laminated micrites and dolomicrites. At Smith Hollow Quarry, the ostracode coquina (OSLS) occurs in the lower 3 cm of a 0.8–1-meter-thick bed which lies directly below the K-spar tuff.

Bear River Gulch Section.—The Bear River Gulch section is 166 m thick and contains micrites, dolostones, siltstones, and sandstones of the Road Hollow Member of the Green River Formation and sandstones of the Sandstone Tongue of the Wasatch Formation (Fig. 2). The lower part of the section contains microlaminated and structureless micrites and dolomicrites. Toward the top of the section, alternating microlaminated micrites and bioturbated micrites are capped by coarse-grained, flaggy sandstones of the Wasatch Formation. The units toward the top of the Road Hollow Member at the Bear River Gulch section contain several interbedded, kerogen-rich, kerogen-poor, structureless, and laminated micrites, along with sandstones and other siliciclastics. Ostracodes were not recovered from these beds at this site. The contacts between structureless and laminated micrites are gradational, whereas contacts between micrites and siliciclastics are abrupt. Based on stratigraphic correlations of key beds as well as the lack of dolomicrites that are characteristic of the Angelo Member, the Bear River Gulch section is assigned to the Road Hollow Member. Ostracodes are found in kerogen-poor, microlaminated as well as structureless micrites and dolomicrites in the lower part of the section. Gastropod and plant fossils occur in dolomicrites and bioturbated micrites, respectively. Fish fossils occur in a kerogen-poor microlaminated micrite.

Ostracodes

Two ostracode species were recovered from 16 beds in the Road Hollow, Fossil Butte, and Angelo members in Fossil Basin—the eurytypic plant-dwelling species *Pseudoeucypris pagei* (Swain 1949)—which is associated with organic and carbonate-rich depositional settings, and *Hemicyprinotus watsonensis* (Swain

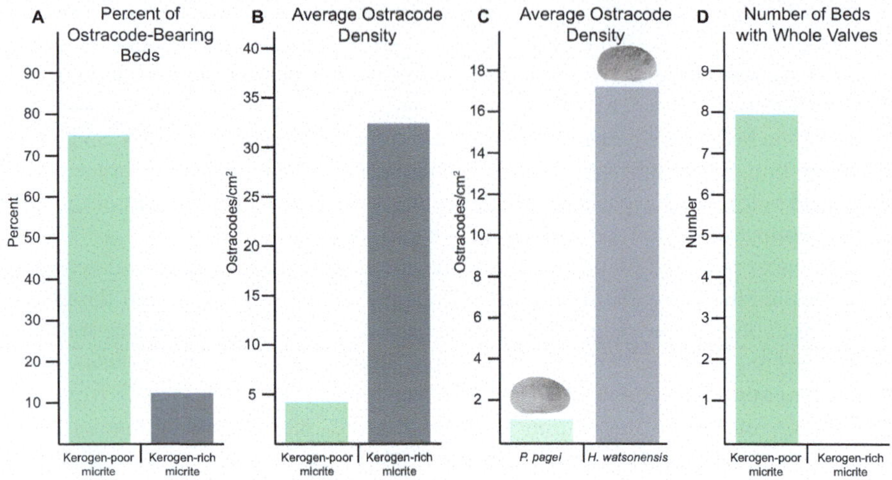

Fig. 3 (**a**) Ratio of ostracode-bearing beds from the three sites that are found in the kerogen-poor and kerogen-rich micrite lithofacies. (**b**) Average ostracode density of all ostracode-bearing beds in the kerogen-poor and kerogen-rich micrite lithofacies. (**c**) Average ostracode density of all ostracode-bearing beds by species. SEM image of each species as indicated. (**d**) Number of bedding surfaces with whole ostracode valves preserved

1964)—which is also a eurytypic plant dweller that occurs in a broad range of lacustrine environments (Kaesler and Taylor 1971; Swain 1999; Swain et al. 1971). An additional mud-dwelling, but rarer species, *Cypridea* (*Bisulcocypridea*) *bisulcata* (Swain 1949), was reported from carbonate- and organic-rich environments within Fossil Basin by Swain et al. (1971) but was not recovered in this study.

The ostracode fossils recovered vary with lithology (Fig. 3). Ostracodes are found in both kerogen-rich and kerogen-poor micrites. Of the ostracode-bearing beds, approximately 75% of the ostracode material is from kerogen-poor micrites (Fig. 3a). Average ostracode density (ostracodes/cm^2) is higher in kerogen-rich micrites (~32/cm^2) than in the kerogen-poor micrites (~4/cm^2) (Fig. 3b). In the kerogen-poor facies, valves are primarily preserved by calcite recrystallization and are unbroken; but in the kerogen-rich facies, there is only rare preservation of original shell material. These fossils are typically compressed and often broken due to compaction. Rare three-dimensional preservation occurs in kerogen-poor, microlaminated, and structureless micrites (Fig. 3d). The ostracode coquina layer that is found below the K-spar tuff in both the Road Hollow and Smith Hollow Quarry sections is dominated by three-dimensionally preserved *H. watsonensis* valves within a kerogen-rich, laminated micrite.

Ostracodes recovered from the lower units of kerogen-poor micrites at the Road Hollow Section are dominated by *P. pagei* while the upper units of kerogen-rich micrites contain *H. watsonensis*. At Smith Hollow Quarry, there is a shift from *H. watsonensis* which occur in the lower units that are dominated by kerogen-rich micrites, to *P. pagei* which are found in kerogen-poor, microlaminated micrites and thin beds of dolomicrites in units above the K-spar tuff. Ostracodes recovered from the Angelo Member are exclusively *P. pagei* and are found in kerogen-poor micro-laminated micrites (Fig. 2).

While there is a density difference between kerogen-poor and kerogen-rich micrite facies, there also are density differences with respect to species, sites, and stratigraphy. On average, ostracode-bearing kerogen-poor micrites have an ostracode density of 4.2/cm^2, while kerogen-rich micrites have almost an order of magnitude greater average density of 32.3/cm^2 (Fig. 3b). These facies patterns also are reflected in the density differences within Green River Formation members, with the Road Hollow Member having an average ostracode density of 1.3/cm^2, the Fossil Butte Member having 13.1/cm^2, and the Angelo Member having 2.1/cm^2. The Fossil Butte Member is the only member with kerogen-rich micrite beds that bear ostracodes. Ostracodes were not recovered from this lithology within the Road Hollow or Angelo members. The Road Hollow site is the only site with ostracode-bearing kerogen-rich micrite beds within the Fossil Butte Member. It has the highest average density of the three sites at 21.7/cm^2. At Smith Hollow Quarry, where ostracodes only occur in kerogen-poor micrites and dolomicrites, there is a lower average density of 6.0/cm^2. Similarly, at Bear River Gulch, average density is low at 1.5/cm^2 and ostracodes occur in kerogen-poor micrites. Finally, there is an average density difference between species, with *H. watsonensis* occurring at rate of 17.3/cm^2 and *P. pagei* at 1.2/cm^2 (Fig. 3c). In the kerogen-rich micrites containing *H. watsonensis*, average density is 32.3/cm^2 and is much lower in kerogen-poor micrites at 9.8/cm^2, again correlating to the differences in facies.

Discussion

Our measured sections indicate that Fossil Lake was a dynamic lacustrine system that responded to both climatological and tectonic changes within the basin (Buchheim et al. 2011, 2015). At its inception, Fossil Lake was likely an overfilled lake which became a fluctuating profundal (balanced-filled) lake, as represented by the finely laminated, kerogen-rich micrites that were deposited. Over the history of the lake, it progressed into an evaporative (underfilled) lake, as evidenced by the dolomicrites that dominate the upper units (Carroll and Bohacs 2001). Magadi-type cherts occur in this unit but are restricted to the north and central parts of Fossil Basin (Buchheim et al. 2011). It is possible for an underfilled lake to have fluvial inputs, which we find here expressed as wedges of coarser-grained sandstones of the Wasatch Member (i.e., the Sandstone Tongue) in the upper parts of the measured sections, especially within the sites located toward the margins of the lake basin. These changes in lake basin type, from overfilled to balanced-filled to underfilled, are similar to comparative, time-equivalent strata in other parts of the GGRB (Buchheim et al. 2011, 2015; Carrol et al. 2008; Rhodes et al. 2002; Smith et al. 2008).

The Road Hollow Member as observed at the Bear River Gulch and Road Hollow sites in Fossil Basin is representative of an overfilled lake and is the earliest stage of Fossil Lake (Fig. 2) (Buchheim et al. 2011). Micrites comprise the great majority of the rocks of this member, whereas dolomicrites are relatively rare. The Fossil Butte Member, observed here at the Smith Hollow Quarry and Road Hollow sections, is the best-known member from Fossil Lake (Buchheim et al. 2015) and the exceptionally well-preserved fossils and rocks appear as characteristic of a relatively deep, balance-filled, lake system (Fig. 2). This is consistent with the common kerogen-rich and kerogen-poor laminated micrites in the sections measured, and with dolomicrites appearing near the top of the FBM. Changes in organic content also occur from kerogen-rich to kerogen-poor micrites up-section, indicating that lake anoxia existed early in its history and perhaps became more ventilated through time. The transition from KRLM to KPLM also could be related to dilution by more abundant carbonate precipitation nearer the lake margins in addition to increased ventilation (Buchheim et al. 2015).

The Angelo Member represents the end of Fossil Lake's existence and is characteristic of an underfilled lake. This is consistent with the section measured here dominated by dolomicrites, with interfingering sandstone tongues appearing toward the top of the measured section of this member.

The sections measured in this study broadly correlate to the overall basin tectonics. As Sevier thrust-faulting was creating the basin during Road Hollow Member-time, the lake was expanding and depositing micrites. As basin subsidence slowed with the end of thrust-faulting, the lake became balance-filled at its largest extent, during Fossil Butte Member-time, developing an anoxic zone on the bottom that allowed for the deposition and preservation of kerogen-rich micrites and the exceptional preservation of fossils. As the lake began to shallow, it became more saline during Angelo Member-time. Fluvial deposition of sandstones toward the top of this member corresponds with the gradual disappearance of the lake at the beginning of the middle Eocene.

Ostracode Occurrences

The two ostracode species recovered in Fossil Basin were found to be associated with different lithologies and in different densities, suggesting that they might have had different environmental tolerances, even though they were both considered to be plant dwellers (Kaesler and Taylor 1971; Swain 1999; Swain et al. 1971). The absence of the rare species *C. bisulcata* in the measured sections may indicate that the depositional environments may have fluctuated with respect to oxygenation because *C. bisulcata* is known for being abundant in shallow, well-circulated waters (Ingalls and Park 2010; and references therein). While *P. pagei* and *H. watsonensis* have been reported to occur together in other parts of the GGRB, including in the Angelo Member (Oriel and Tracey Jr. 1970), they do not co-occur in Fossil Basin perhaps due to different depth preferences, as noted by their association with either kerogen-rich or kerogen-poor lithologies. *P. pagei is* reported elsewhere (e.g., Lake Uinta) to be most commonly found in limestones and also could tolerate high amounts of organic matter (Swain 1999; Swain et al. 1971). In Fossil Basin, it is found in kerogen-poor micrites as well as dolomicrites, but not in kerogen-rich micrites. While *P. pagei* can tolerate high organic matter, this may not be its preferred environment because in Lake Uinta, they are typically found in kerogen-poor lithologies, while *H. watsonensis* is found in more kerogen-rich environments (Ingalls and Park 2010), which is consistent with their distribution within Fossil Basin. *H. watsonensis* is generally found in rocks deposited in carbonate- and organic-rich environments and in great numbers, possibly as a result of mass mortalities and potentially indicative of deeper lake conditions, which is consistent with Swain et al. (1971) who purport that *H. watsonensis* was the only ostracode species of the larger Green River Formation ostracode fauna able to tolerate deeper, more stagnant environments.

Facies shifts and other proxies indicate that Fossil Lake fluctuated throughout its history which is reflected in the ostracode occurrences. At the Road Hollow locality, *P. pagei* dominates in the lower Road Hollow Member and later is replaced by *H. watsonensis* in the Fossil Butte Member. This transition to a species better able to tolerate deeper, more stagnant conditions may indicate that lake depth increased, and was expanding during this time, and/or that the lake was less ventilated. The opposite scenario can be seen in the upper Fossil Butte and Angelo members at the Smith Hollow Quarry site, which transition from *H. watsonenis* to *P. pagei* in correspondence with a facies shift to a shallower water environment (Fig. 2).

Although Fossil Basin is best known for its rich diversity and preservation of fish and other faunal and floral elements, ostracode diversity is comparatively low, especially with respect to other sub-basins of the Green River complex such as Lake Uinta which contains 18 species and Lake Gosiute which contains six species (Ingalls and Park 2010; Swain 1956; Swain et al. 1971). Reasons for this disparity may include basin depth and resulting oxygen levels as well as the ecological association of the two species found, *P. pagei* and *H. watsonensis*, and their plant-dwelling habitats. Other areas within Fossil Basin may have been shallower and may have had more species present, such as *C. bisulcata*, which also has been reported from this lake; however, the total number of species is still much lower than in other Green River sub-basins. Nektonic species, such as *Procyprois ravenridgensis* and *Potamocypris williamsi*, are not found at all, even though they are prevalent in other parts of the Green River (Ingalls and Park 2010; Swain 1956, 1964).

Preservation

Ostracode fossils are found to be mostly preserved by recrystallized calcite. This is similar to ostracode preservation in layers deposited in Lake Gosiute as documented by Ingalls and Park (2010) and could reflect similar conditions in both lakes, yielding similar taphonomic results. There are some specimens that appear to be original shell material, but a detailed analysis of the composition of the valves was not performed. Three-dimensional preservation is common in kerogen-poor units and some dolomicrite units.

The taphonomic conditions within Fossil Basin may reflect the oxygen levels in the lake. The kerogen-rich, microlaminated micrites are interpreted to have been deposited in deep, still, anoxic waters (Buchheim et al. 2015). The high abundance of *H. watsonensis* that occur within these beds is often represented by broken valves which could have been due to postmortem compaction in a deeper-water setting. In the sites dominated by kerogen-poor, bioturbated micrites, anoxia was not a factor. Ostracode fossils are found unbroken and without preferential orientation, and often with completely articulated carapaces, suggesting higher rates of deposition and rapid burial.

Conclusion

The two ostracode species—each with their ecological preferences— correlate to lake environmental indicators such as changes in lithology and bedding characteristics. Ostracodes within Fossil Basin are predominantly preserved by calcite recrystallization and are found with greater frequency in kerogen-poor laminated micrites but in greater abundance in kerogen-rich micrites. Valves that are found in kerogen-poor micrites tend to be better preserved, often three dimensional; valves that are found in kerogen-rich micrites tend to be compressed and broken, most likely due to postdepositional compression.

The ecological dynamics of Fossil Lake can be used to compare and contrast basin hydrodynamics with other basins of the Greater Green River Basin, which possibly could inform our understanding of the climatological and tectonic drivers on this system.

Acknowledgments The authors wish to thank the anonymous reviewers who greatly improved this manuscript and the following individuals: Francisco Moore (University of Akron; National Science Foundation), Arvid Aase (Fossil Butte National Monument), Jerome Montgomery (Green River Stone Company), and Blossom Frank (University of Akron). This project was part of AJM's Masters Thesis at the University of Akron and is dedicated to the memory of Elizabeth Gierlowski-Kordesch.

Appendix A.—Sampling, Lithologic Description, and Analysis of Fossil Content from the Bear River Gulch (GR09BRG) Outcrop

Sample				Fossil content		
Position (m)	Number	Lithology	Area (cm^2)	Ostracodes Number	Ostracodes Density	Additional
9.00		Quaternary alluvium (cover)				
11.80	S002A	Micrite (mud-supported), light cream, kerogen-poor, laminated				
12.10	S002B	Micrite, light cream to yellow orange, kerogen-poor, bioturbated				
12.60	S002C	Micrite (mud-supported), light tan to yellow orange, kerogen-poor, structureless	9	16	1.78	
13.85	S002D	Micrite, light grayish/yellowish tan, kerogen-poor, laminated				
14.25	S002E	Micrite, light brown to tan or gray, kerogen-rich, structureless				
15.95	S002F	Micrite, light tan/cream, kerogen-poor, structureless	5.5	17	3.09	
16.85	S002G	Dolomicrite (mud-supported), light cream fresh and weathered, structureless				
17.35	S002H	Micrite (mud-supported), light grayish/yellowish tan kerogen-poor, structureless				
17.65	S002I	Dolomicrite, light cream/brown				
18.25	S002J	Micrite (mud-supported), light cream to rusty tan, kerogen-poor, bioturbated	4	1	0.25	Plants
18.45	S002K	Dolomicrite (mud-supported), light cream/tan, structureless				
19.35	S002L	Micrite with small calcite veins(mud-supported), light tan to gray or, rarely, green, kerogen-poor, laminated	29	2	0.07	
19.95	S002M	Dolomicrite (mud-supported), light tan/cream, structureless				
21.75	S002N	Micrite (mud-supported), light cream/tan; kerogen-poor, laminated (>1 mm-thick laminations)				
22.45	S003A	Dolomicrite (mud-supported), light cream, structureless				

Sample				Fossil content		
Position			Area	Ostracodes		
(m)	Number	Lithology	(cm^2)	Number	Density	Additional
24.45	S003B	Micrite (mud-supported), light cream to light brown, kerogen-poor, laminated				
25.15	S003B FISH	Micrite, kerogen-poor, laminated				Fish
26.55	S004	Micrite, light cream, kerogen-poor, structureless				
26.85	S005	Micrite, light gray to light yellow/tan, kerogen-poor, laminated (>1 mm-thick laminations)				
30.85	S006	Dolomicrite (mud-supported) with small interbedded organic layers, light cream fresh to brown/gray, structureless				
31.35	S007	Siltstone with calcareous cement, light cream to cream/yellow, structureless				
32.60	S008	Micrite, creamy yellow to gray/brown, kerogen-poor, bioturbated				
33.50	S009	Dolomicrite, creamy yellow to gray/brown, bioturbated				
37.50	S010	Micrite, light cream/yellow to gray/brown with rusty orange spots, kerogen-poor, bioturbated				
38.00	S011	Micrite, light purple/gray to pale yellow, kerogen-poor, laminated, friable				
39.90	S012	Micrite (mud-supported), light cream, kerogen-poor, structureless				
40.50	S013	Micrite (mud-supported), chocolate brown to light tan, kerogen-rich laminated				
41.50	S014	Micrite, light cream/orange to gray to tan, kerogen-poor, laminated (<1 mm-thick laminations)				
42.25	S015	Micrite (mud-supported), chocolate brown to light tan, kerogen-rich, laminated				
45.65	S016	Micrite, light gray/creamy yellow/orange, kerogen-poor, laminated (1 mm-thick laminations)				
47.65	S017	Dolomicrite, light tan/cream with alternating rust color bands, laminated				

Sample				Fossil content		
Position (m)	Number	Lithology	Area (cm^2)	Ostracodes Number	Ostracodes Density	Additional
48.55	S018	Micrite, light gray/tan to rusty orange, kerogen-poor, laminated				
53.05	S019	Micrite, green to light tan, kerogen-rich, structureless				
54.40	S020	Dolomicrite (mud-supported), light gray to tan, bioturbated				Ichnofossils Gastropods
56.20	S021	Micrite, light tan/cream, kerogen-poor, structureless				
56.70	S022	Dolomicrite, light gray/tan				
57.60	S023	Micrite, light gray/tan, kerogen-poor, structureless				
58.10	S024	Dolomicrite, light gray/tan, bioturbated				
62.90	S025	Micrite, light gray/tan, kerogen-poor, mostly laminated but locally structureless				
63.20	S026	Micrite, chocolate/light brown, kerogen-rich, laminated				
65.40	S027	Micrite, light gray/tan, kerogen-poor, mostly structureless but locally laminated				
66.40	S028A, B	Dolomicrite, light gray/tan/yellow, weakly laminated	20	2	0.1	Gastropods (?)
67.90	S029	Micrite, light cream/tan, kerogen-poor, structureless	18.5	64	3.46	
71.50		Quaternary alluvium (cover)				
71.80		Sandstone				
73.90		Quaternary alluvium (cover)				
75.50		Sandstone				
83.50		Micrite, kerogen-rich, structureless, interbedded with micrite, kerogen-poor, laminated, and sandstone				
83.80		Sandstone				
87.80		Micrite, kerogen-rich, structureless, interbedded with micrite, kerogen-poor, laminated				
88.80		Sandstone				
120.60		Micrite, kerogen-rich, structureless, interbedded with micrite, kerogen-poor, laminated				
125.50		Sandstone				

Sample				Fossil content		
Position (m)	Number	Lithology	Area (cm^2)	Ostracodes		Additional
				Number	Density	
145.50		Micrite, kerogen-rich, structureless, interbedded with micrite, kerogen-poor, laminated, mud, kerogen-rich, and dolomicrite				
147.50		Sandstone				
150.50		Dolomicrite				

Appendix B.—Sampling, Lithologic Description, and Analysis of Fossil Content from the Road Hollow (GR09RH) Outcrop

Sample				Fossil content		
Position (m)	Number	Lithology	Area (cm^2)	Ostracodes		Additional
				Number	Density	
2.00	S012	Micrite, light brown/tan, structureless, well hardened				
4.00	S013	Mudstone with calcareous cement, light/dark gray to green				
4.80	S014	Micrite, light/dark gray, kerogen-poor, structureless				
6.30	S015	Micrite (mud-supported), light cream/gray, kerogen-poor, structureless				
9.05	S016	Micrite (mud-supported), light gray to tan, kerogen-poor, laminated (>1 mm-thick laminations)				
12.05	S017	Micrite, dark/light gray to green, structureless				
17.35	S018	Micrite (mud-supported), light gray/green/tan, kerogen-poor, laminated (<1 mm-thick laminations)				
20.85	S019	Micrite (mud-supported), light brown/tan, kerogen-poor, structureless	19.5	6	0.31	
21.45	S020	Sandstone, fine grained with calcareous cement, light brown to dark tan, laminated				
22.25		Micrite, light brown/gray, kerogen-poor, bioturbated to laminated				
22.35		Siltstone				

Sample				Fossil content		
Position (m)	Number	Lithology	Area (cm^2)	Ostracodes Number	Ostracodes Density	Additional
22.95		Micrite, laminated to structureless				
23.75		Micrite, kerogen-poor, structureless				
23.85		Siltstone				
24.45		Micrite, structureless				
25.25		Siltstone				
25.95		Micrite, structureless				
26.95		Micrite, kerogen-rich, structureless				
27.45		Sandstone, siliciclastic				
28.95		Micrite, tan to light cream, structureless				
30.45		Micrite, strongly laminated				
31.95		Micrite, structureless				
32.65		Micrite, structureless, with 10–20 cm siltstone level				
34.15		Micrite, structureless				
35.15		Micrite, structureless				
35.85	S023	Dolomicrite, light tan/gray, laminated or bioturbated				
36.60		Micrite, kerogen-poor, laminated				
37.10	S024A	Micrite, green to light gray, kerogen-rich, structureless				
37.15		Siltstone				
38.85	S024B	Micrite, kerogen-rich, structureless				
39.15		Siltstone				
43.35		Micrite, laminated or bioturbated				
45.95		Micrite, kerogen-rich, laminated (variable thick laminations)				
46.75	S025A	Micrite, chocolate/light brown, kerogen-rich, laminated, very friable				
48.55	S025B	Micrite, green to light tan, kerogen-rich, laminated				
50.95	S026	Dolomicrite (mud-supported), light tan/gray, kerogen-poor				
51.85	S027A	Micrite, brown to light tan, kerogen-rich, laminated or bioturbated				
52.35	S027B	Micrite, rusty/light tan, kerogen-poor, laminated				Fish
55.70	S027C	Micrite, light cream to tan, kerogen-rich, bioturbated				
55.73	S027D	Sandstone, fine grained, green/gray brown, flaggy				
64.23	S027E	Micrite, chocolate/light brown, kerogen-rich, laminated	10	506	50.6	

Sample				Fossil content		
Position (m)	Number	Lithology	Area (cm^2)	Ostracodes		Additional
				Number	Density	
66.23	S028	Dolostone (grain-supported), light/ dark tan, structureless				
67.98	S029A	Micrite, light tan/cream, kerogen-poor, structureless				
70.73	S029B	Micrite, green/gray/light tan, kerogen-rich, structureless				
72.63	S029C	Micrite, light/lighter gray, kerogen-poor, laminated				
73.53	S030	Tuff?				
75.33	S031A	Micrite, chocolate brown to light tan, kerogen-rich, laminated (<1 mm-thick laminations)	12	169	14.08	
76.03	S031B	Dolomicrite (mud-supported), light cream/tan, bioturbated, with ash/silt level 3 cm from bottom, bioturbated				
76.53	S032A	Micrite, light brown/tan kerogen-poor, laminated				
76.55	S032B	Siltstone				
83.15	S032C	Micrite, kerogen-rich, laminated interbedded with dolomicrite, kerogen-rich, and dolomicrite, tan to a light tan, kerogen-poor, laminated				
91.15	S033	Siltstone, light brown, heavily bioturbated, interbedded with silty/ fine-grained sandstone, coarsening upward				
102.65	S034	Sandstone with calcareous cement, moderately well sorted, subangular, light gray tan/tan, very weakly laminated				
104.65	S035A	Micrite, gray tan to light brown, kerogen-rich, structureless				
104.95	S035B	Micrite, light tan/yellow-tan, kerogen-poor, laminated				
107.25	S035C	Micrite, gray/light gray/tan, kerogen-rich, laminated				
107.35		Micrite, brown, kerogen-poor, laminated				
110.10	S035D	Micrite, blue, kerogen-rich, laminated				
110.35		Sandstone, moderately well sorted, subangular, light gray tan/tan, very weakly laminated				
113.35		Sandy micrite, kerogen-rich?				

Sample				Fossil content		
				Ostracodes		
Position (m)	Number	Lithology	Area (cm^2)	Number	Density	Additional
118.35	S037	Sandstone with calcareous cement, subangular, moderately well sorted, light tan/yellow-tan, very weakly laminated				

Appendix C.—Sampling, Lithologic Description, and Analysis of Fossil Content from the Smith Hollow Quarry (GR09FBFB) Outcrop

Sample				Fossil content		
				Ostracodes		
Position (m)	Number	Lithology	Area (cm^2)	Number	Density	Additional
1.000	1001	Micrite, brown to light tan, kerogen-poor, laminated (0.5–1 mm-thick laminations); evidence of phosphate deposits	10	154	15.4	Fish
1.030	1002	Tuff, dark brown to black, very fine grained, well hardened	10	2	0.2	Fish
1.670	001	Micrite, brown to light tan, kerogen-rich, laminated (0.5–1 mm-thick laminations), moderately to well hardened				Plants
1.680	002	Tuff, orange to light orange, fine grained, well hardened				
1.830	003	Micrite, brown to light tan, kerogen-rich, laminated (0.2–3 mm-thick laminations)				Plants
1.865	004	Tuff, light to dark orange, medium grained, well hardened				
1.905	005	Micrite, dark brown to light tan, kerogen-rich, laminated (0.2–1 mm-thick laminations), more organic at the first centimeter				Plants
1.970	006	Tuff, light/dark/burnt orange, reworked (evidence of void spaces or pockets)				
2.770	007	Micrite, dark brown/gray to light tan, kerogen-rich (oily), laminated (0.5–2 mm-thick laminations)				Ichnofossils?

Sample				Fossil content		
Position (m)	Number	Lithology	Area (cm²)	Ostracodes Number	Density	Additional
4.470	008	Micrite with siliciclastic layers, blue/gray to light tan, kerogen-rich (oil shale levels), laminated				Fish Plants Coprolites
4.610	009	Muddy siltstone, dark gray to yellow to light brown, with massive bedding, moderately hardened				Plants
5.040	010	Micrite, light tan and dark brown to buff and light gray, kerogen-rich, laminated				
5.175	011	Muddy siltstone, gray to light tan, with massive bedding, moderately hardened				Plants
5.425	012	Micrite, gray to light tan/orange, kerogen-rich, laminated				
5.470	013	Tuff (ashy), light yellow to buff, very fine grained				
6.470	014	Dolomicrite, yellow to buff, laminated or structureless				
8.070	015	Micrite, tan to yellow/orange, kerogen-poor, laminated				
8.345	016	Dolomicrite, creamy orange to white/orange, laminated to structureless, capped by 5-cm ash layer				
9.245	017	Micrite, light tan to slightly yellow tan, kerogen-poor, laminated (±1 mm-thick laminations), coquina at bottom	37.5 14.5	3 341	0.08 23.52	
9.385	018	K-spar tuff, light to rusty orange, fine grained, well hardened				
12.985	019	Micrite with some siliclastic layers, light tan to buff, kerogen-poor laminated (±1 mm-thick laminations)	73	12	0.16	Fish Gastropods
14.685	020	Dolomicrite, light brown to buff, laminated (2–3 mm-thick laminations) or structureless, well hardened	72.5	25	0.34	
14.865	021	Siltstone				
14.880	022	Tuff, light to dark brown to dark gray, very fine grained, laminated (±1 mm-thick laminations), well hardened				
15.305	023	Muddy dolomicrite, brown to dark gray, laminated (±1 mm-thick laminations), shimmery				

Sample				Fossil content		
Position (m)	Number	Lithology	Area (cm^2)	Ostracodes Number	Ostracodes Density	Additional
15.580	024	Tuff (ash between 1.5 and 5 cm from bottom part, reworked ash for next 20 cm, then ash again for 2.5 cm until top), dark to light brown, fine grained with eventual coarse grains of quartz, friable				
15.650	025	Micrite, dark brown to light tan and dark gray, kerogen-rich, laminated, with soft sediment deformation				
16.750	026	Siltstone, red and brown to light tan, coarse grained, kerogen-poor, laminated, with soft sediment deformation and calcareous cementing				
17.050	027	Sandstone, red and brown to light tan, fine grained, kerogen-poor, laminated, with soft sediment deformation and calcareous cementing				
17.210	028	Siltstone				
17.735	029	Dolomicrite, light gray to pale gray or green with some rust-colored clasts				
17.745	030	Tuff, orange to green or gray to light tan, fine grained but with coarser grains throughout, moderately hardened				
17.845	031	Dolomicrite with reworked ash layers, dark brown to tan, moderately hardened; large quartz crystals in the ash				
18.095	032	Volcaniclastic sandstone (first 22 cm) capped by 3 cm-thick laminations-ash layer, golden to light creamy orange, very fine grained, unbedded				
18.395	033	Dolowackestone, variably thick laminations, dark brown and dark gray to blue/gray and light reddish brown, laminated, well hardened				
18.495	034	Micrite, kerogen-rich, laminated				
18.670	035	Dolomicrite, gray with rust-colored stains to pale yellow, structureless, well hardened				

Sample				Fossil content		
Position (m)	Number	Lithology	Area (cm^2)	Ostracodes Number	Density	Additional
18.710	036	Tuff with chunks of calcareous mud, pale yellow to white, poorly hardened				
18.910	037	Siltstone, golden/pale yellow				
18.960	038	Tuff, light gray or pale/dark yellow, very fine grained, well hardened				
19.560	039	Dolomicrite, peach to dark reddish tan, laminated, well hardened				
20.460	040	Micrite with larger-than-average grains, light tan and brown or gray to white and pale yellow, laminated (undulating laminations), moderately well hardened				
21.060	041	Tuff (interbedding of ashes and siliciclastic sandstones), gray to golden yellow, very fine grained, well hardened				
21.860	042	Dolomicrite with coarser-grained calcite crystals, brown to light blue or gray, kerogen-rich (purple oil shale levels)				
25.260	043	Dolomicrite light gray/tan and pale green, with rust-colored stains				
25.290	044	Tuff, pale orange to orange, fine grained, laminated, moderately well hardened				
25.440	045	Dolomicrite, light gray/brown to pale yellow, laminated (>1 mm-thick laminations)				
26.440	046	Dolomicritic muddy oil shale, blue/gray, kerogen-rich, unbedded				
26.940	047	Dolomicrite, white or light cream to yellow or orange (with white marker bed), kerogen-poor, weakly laminated				
28.340	048	Dolomicrite (grain supported), light green to dark gray, with some rust coloration, kerogen-rich, structureless				
29.590	049	Micrite, light/yellower cream, kerogen-poor, laminated (<1 mm-thick laminations), then tuff between 100 and 103 cm of the layer, creamy to light orange to white, very fine grained, well hardened	15	32	2.13	Fish, gastropods

Sample				Fossil content		
Position (m)	Number	Lithology	Area (cm^2)	Ostracodes Number	Ostracodes Density	Additional
29.770	050	Micrite, light gray to orange tan, dark gray/rusty, kerogen-rich, laminated (>1 mm-thick laminations), well hardened				
30.070	051	Dolomicrite (mud-supported), light gray/tan, kerogen-poor, laminated (<1 mm-thick laminations), with salt casts				
30.130	052	Tuff, white or light tan to orange, very fine grained, well hardened				
30.880	053	Dolomicrite (mud-supported), light gray or very light green to brown and black, structureless, well hardened, then tuff between 30 and 35 cm of the layer, light gray or tan to yellow-orange or orange, very fine grained, weakly laminated, well hardened				
34.680	054	Dolomicrite (grain supported), yellow to paler yellow or light tan, structureless, well hardened				
34.920	055	Dolomicrite, gray to black or light tan, laminated (<1 mm-thick laminations)				

References

Aswasereelert, W., Meyers, S. R., Carroll, A. R., et al. (2013). Basin-scale cyclostratigraphy of the Green River Formation, Wyoming. *GSA Bulletin, 125*, 216–228.

Baddouh, M., Meyers, S. R., Carroll, A. R., et al. (2016). Lacustrine 87Sr/86Sr as a tracer to reconstruct Milankovitch forcing of the Eocene hydrologic cycle. *Earth and Planetary Science Letters, 448*, 62–68.

Beck, R. A., Vondra, C. F., Filkins, J. E., & Olander, J. D. (1988). Syntectonic sedimentation and Laramide basement thrusting, Cordilleran foreland; timing of deformation. In C. J. Schmidt & W. J. Perry (Eds.), *Interaction of the rocky mountain foreland and the cordilleran thrust belt* (Vol. 171, pp. 465–487). Boulder, Colorado: Geological Society of America. https://doi.org/10.1130/MEM171-p465.

Boomer, I., Horne, D. J., & Slipper, I. J. (2003). The use of ostracods in palaeoenvironmental studies, or what can you do with an ostracod shell? *The Paleontological Society Papers, 9*, 153–180.

Buchheim, H. P. (1994). Eocene Fossil Lake: A history of fluctuating salinity. In R. W. Renaut & W. M. Last (Eds.), *Sedimentology and geochemistry of modern and ancient Saline Lakes. SEPM Special Publication* (Vol. 50, pp. 239–247). Tulsa, OK: SEPM Society for Sedimentary Geology.

Buchheim, H. P., Biaggi, R. E., & Cushman, R. A., Jr. (2015). Stratigraphy and lnterbasinal correlations between Fossil and the Green River Basin, Wyoming. In M. E. Smith & A. R. Carroll (Eds.), *Stratigraphy and paleolimnology of the green river formation, Western USA. Syntheses in limnogeology* (Vol. 1, pp. 127–151). Dordrecht: Springer Netherlands.

Buchheim, H. P., Cushman, R. A., Jr., & Biaggi, R. E. (2011). Stratigraphic revision of the Green River Formation in Fossil Basin, Wyoming: Overfilled to underfilled lake evolution. *Rocky Mountain Geology, 46*, 165–181. https://doi.org/10.2113/gsrocky.46.2.165.

Buchheim, H. P., & Eugster, H. P. (1998). Eocene Fossil Lake: The Green River Formation of Fossil Basin, southwestern Wyoming. In J. K. Pittman & A. R. Carroll (Eds.), *Modern & ancient lake systems: New problems and perspectives* (Vol. 26, pp. 191–208). Salt Lake City, Utah: Utah Geological Association Guidebook.

Buchheim, H. P., Loewen, M. A., Cushman, R. A., Jr., & Biaggi, R. E. (2002). Stratigraphic revision of the Green River Formation in Fossil Basin, Wyoming: Three distinct phases of Fossil Lake. *GSA Abstracts with Programs, 34*, 479.

Carrol, A. R., Doebbert, A. C., Booth, A. L., et al. (2008). Capture of high-altitude precipitation by a low-altitude Eocene lake, western USA. *Geology, 36*, 791–794.

Carroll, A. R., & Bohacs, K. M. (2001). Lake-type controls on petroleum source rock potential in nonmarine basins. *AAPG Bulletin, 85*, 1033–1053. https://doi.org/10.1306/8626CA5F-173B-11D7-8645000102C1865D.

Dickinson, W. R., Klute, M. A., Hayes, M. J., et al. (1988). Paleogeographic and paleotectonic setting of Laramide sedimentary basins in the central Rocky Mountain region. *Geological Society of America Bulletin, 100*, 1023–1039.

Ferber, C. T., & Wells, N. A. (1995). Paleolimnology and taphonomy of some fish deposits in "Fossil" and "Uinta" lakes of the Eocene Green River Formation, Utah and Wyoming. *Palaeogeography, Palaeoclimatology, Palaeoecology, 117*, 185–210. https://doi.org/10.1016/0031-0182(94)00127-T.

Gazin, C. L. (1952). The lower Eocene Knight Formation of western Wyoming and its mammalian faunas. *Smithsonian Miscellaneous Collections, 117*, 1–82.

Gierlowski-Kordesch, E. H., Jacobson, A. D., Blum, J. D., & Valero Garcés, B. L. (2008). Watershed reconstruction of a Paleocene–Eocene lake basin using Sr isotopes in carbonate rocks. *GSA Bulletin, 120*, 85–95. https://doi.org/10.1130/B26070.1.

Gierlowski-Kordesch, E. H., & Park, L. E. (2004). Comparing species diversity in the modern and fossil record of lakes. *Journal of Geology, 112*, 703–717.

Grande, L. (1984). Paleontology of the Green River Formation, with a review of the fish fauna. Second edition. *The Geological Survey of Wyoming Bulletin, 63*, 1–334.

Grande, L., & Buchheim, H. P. (1994). Paleontological and sedimentological variation in early Eocene Fossil Lake. *Contributions to Geology, 30*, 33–56. https://doi.org/10.2113/gsrocky.30.1.33.

Ingalls, B. R., & Park, L. E. (2010). Biotic and taphonomic response to lake level fluctuations in the Greater Green River Basin (Eocene), Wyoming. *PALAIOS, 25*, 287–298. https://doi.org/10.2110/palo.2009.p09-048r.

Kaesler, R. L., & Taylor, R. S. (1971). Cluster analysis and ordination in paleoecology of Ostracoda from the Green River Formation (Eocene, USA). In H. J. Oertli (Ed.) *Colloque sur la Paléoécologie des Ostracodes. Pau – 20-27/VII/1970: Bulletin du Centre de Recherches Pau-SNPA* (Vol. 5, pp 153–165).

Ksepka, D. T., & Clarke, J. A. (2010). New fossil mousebird (Aves: Coliiformes) with feather preservation provides insight into the ecological diversity of an Eocene North American avifauna. *Zoological Journal of the Linnean Society, 160*, 685–706. https://doi.org/10.1111/j.1096-3642.2009.00626.x.

Machlus, M. L., Ramezani, J., Bowring, S. A., et al. (2015). A strategy for cross-calibrating U-Pb chronology and astrochronology of sedimentary sequences: An example from the Green River Formation, Wyoming, USA. *Earth and Planetary Science Letters, 413*, 70–78.

McGrew, P. O. (1975). Taphonomy of Eocene fish from Fossil Basin, Wyoming: Fieldiana. *Geology, 33*, 257–270.

Oriel, S. S., & Tracey, J. I., Jr. (1970). Uppermost Cretaceous and Tertiary stratigraphy of Fossil Basin, Southwestern Wyoming. *Geological Survey Professional Paper, 635*, 1–53.

Pietras, J. T., & Carroll, A. R. (2006). High-resolution stratigraphy of an underfilled lake basin: Wilkins Peak Member, Eocene Green River Formation, Wyoming, USA. *Journal of Sedimentary Research, 76*, 1197–1214. https://doi.org/10.2110/jsr.2006.096.

Rhodes, M. K., Carroll, A. R., Pietras, J. T., et al. (2002). Strontium isotope record of paleohydrology and continental weathering, Eocene Green River Formation, Wyoming. *Geology, 30*, 167–170. https://doi.org/10.1130/0091-7613(2002)030<0167:SIROPA>2.0.CO;2.

Roehler, H. W. (1992a). Correlation, composition, areal distribution, and thickness of Eocene stratigraphic units, greater Green River basin, Wyoming, Utah, and Colorado. *U.S. Geological Survey Professional Paper, 1506*, E1–E49.

Roehler, H. W. (1992b). Introduction to greater Green River Basin geology, physiography, and history of investigations. *U.S. Geological Survey Professional Paper, 1506*, A1–A14.

Smith, M. E., & Carroll, A. R. (Eds.). (2015). *Stratigraphy and paleolimnology of the green river formation, Western USA*. Dordrecht: Springer Netherlands.

Smith, M. E., Carroll, A. R., & Singer, B. S. (2008). Synoptic reconstruction of a major ancient lake system: Eocene Green River Formation, western United States. *GSA Bulletin, 120*, 54–84. https://doi.org/10.1130/B26073.1.

Smith, M. E., Singer, B., & Carroll, A. (2003). ^{40}Ar/^{39}Ar geochronology of the Eocene Green River Formation, Wyoming. *GSA Bulletin, 115*, 549–565. https://doi.org/10.1130/0016-7606(2003)115<0549:AGOTEG>2.0.CO;2.

Smith, M. E., Chamberlain, K. R., Singer, B. S., & Carroll, A. R. (2010). Eocene clocks agree: Coeval ^{40}Ar/^{39}Ar, U-Pb, and astronomical ages from the Green River Formation. *Geology, 38*, 527–530. https://doi.org/10.1130/G30630.1

Swain, F. M. (1949). Early Tertiary Ostracoda from the Western Interior United States. *Journal of Paleontology, 23*, 172–181.

Swain, F. M. (1956). Early Tertiary ostracode zones of Uinta Basin. *Geology and Economic Deposits of East Central Utah, Seventh Annual Field Conference, 1956*, 125–139.

Swain, F. M. (1964). Early Tertiary freshwater Ostracoda from Colorado, Nevada and Utah and their stratigraphic distribution. *Journal of Paleontology, 38*, 256–280.

Swain, F. M. (1999). Fossil nonmarine Ostracoda of the United States. *Developments in Palaeontology and Stratigraphy, 16*, 1–401.

Swain, F. M., Becker, J., & Dickinson, K. A. (1971). Paleoecology of Tertiary and fossil Quaternary non-marine Ostracoda from the Western Interior United States. In H. J. Oertli (Ed.) *Colloque sur la Paléoécologie des Ostracodes. Pau – 20–27/VII/1970: Bulletin du Centre de Recherches Pau-SNPA* (vol. 5, pp 461–477).

Taylor, R. S. (1972). *Paleoecology of ostracodes from the Luman Tongue and Tipton Member (early Eocene) of the Green River Formation, Wyoming* (Unpublished Ph.D. thesis). University of Kansas, Lawrence, 96 p.

History of Great Salt Lake, Utah, USA: since the Termination of Lake Bonneville

Charles G. Oviatt, Genevieve Atwood, and Robert S. Thompson

Abstract During the past half century or so diverse histories of Great Salt Lake have been written from differing perspectives and all of them have contributed ideas and essential data. The published literature, however, can be confusing and misleading. In this chapter, we review and provide context for a number of those publications. This chapter is intended as a summary of what is known, what is not known, and what cannot be known with precision about the history of the lake.

Great Salt Lake is the largest hydrographically closed lake in the Bonneville basin of northwestern Utah. It responds to both short-term weather and long-term climate. In the Lake Bonneville/Great Salt Lake lacustrine system, the end of Lake Bonneville at 13,000 yr BP marks the beginning of Great Salt Lake. The much larger and deeper lakes of the Bonneville lake cycle responded to the pluvial climate of oxygen isotope stage 2, but the warmer, drier climate of oxygen isotope stage 1 led to rapid fluctuations within a relatively narrow, well-documented elevation range, 5 m above and 9 m below the historical mean elevation of ~1280 m. Two exceptional but short-lived rises of Great Salt Lake to elevations higher than 5 m above ~1280 m have been documented —one during the Gilbert episode, which peaked about 11,600 yr BP near an elevation of 1295 m, and one to about 1289 m sometime after about 11,000 yr BP.

The historical Great Salt Lake hydrograph (the past 150 years) shows its labile behavior. Smooth-curve hydrographs based on estimates of lake level at time scales of decades, centuries, or millennia, such as those presented in previous publications, do not accurately portray the way lake level rises and falls, and a precise plot of post-Bonneville changes in level of Great Salt Lake would resemble the "jagged" historical record. The available sedimentary and geomorphic data are not conducive at this time to the production of a highly precise hydrograph, so we suggest that post-Bonneville lake-level history be portrayed, imprecisely but accurately, as con-

C. G. Oviatt (✉)
Department of Geology, Kansas State University, Manhattan, KS, USA
e-mail: joviatt@ksu.edu

G. Atwood
Earth Science Education, Salt Lake City, UT, USA

R. S. Thompson
U.S. Geological Survey, Denver Federal Center, Denver, CO, USA

M. R. Rosen et al. (eds.), *Limnogeology: Progress, Challenges and Opportunities*, Syntheses in Limnogeology, https://doi.org/10.1007/978-3-030-66576-0_8

fined generally between the elevation limits of 1285 and 1271 m, with an indication of the exceptional spikes in the lake level.

Keywords Great Salt Lake · Closed-basin lake · Shorezone processes · Holocene climate · Lake Bonneville

Introduction

The purpose of this chapter is to summarize what is known about the history of changing levels of Great Salt Lake (Fig. 1) in post-Bonneville time since about 13,000 years BP.[1,2] This period includes the Holocene.[3] Somewhat paradoxically, considering the differences in spatial and temporal scales, the history of lake-level fluctuations of older Lake Bonneville has been studied more thoroughly than that of relatively young Great Salt Lake. Two lines of evidence, from shorelines and offshore sediments, provide the basis for interpreting the post-Bonneville record of Great Salt Lake. Both have contributed to general understanding of lake history, both have inherent uncertainties, and it is difficult to align information from these two sources. We review previous work, present new data, and conclude that an accurate and precise hydrograph of post-Bonneville Great Salt Lake cannot be produced at this time. The upper and lower elevation limits of most post-Bonneville lake fluctuations, however, can be determined accurately and are presented in this chapter. We emphasize that, although the water budget for Great Salt Lake is dependent on climate, the short-term behavior of the lake is labile, and extreme rises and falls in level are caused by changes in weather.

[1] In this chapter, we give elevations of lake level in meters with conversions to feet in parentheses. The scientific community uses the metric system, whereas topographic maps and gaging results by the U.S. Geological Survey in the Great Salt Lake basin are given in feet, and most people who live in the basin use units of feet in their everyday lives. Elevation measurements are rounded to the nearest meter (and foot), except in a few cases where we report previously published elevations with higher precision.

[2] Unless otherwise noted in this chapter, all ages are reported in calibrated years before present (yr BP; "present" is AD 1950); calibrated years are approximately equal to calendar years. Calibrations of radiocarbon ages are based on CALIB7.0 and slight revision CALIB7.1 (Reimer et al. 2013; Stuiver and Reimer 1993).

[3] The Holocene Epoch began at the end of the Younger Dryas Stadial at 11,500 to 11,700 yr BP (Walker et al. 2009; Walker et al. 2012), and in the Bonneville basin, the Gilbert episode ended at about 11,500 yr BP (Oviatt 2014). The Holocene, which comprises most of post-Bonneville time, can be subdivided into three parts: early Holocene, ending about 8200 yr BP, middle Holocene ending about 4200 yr BP, and late Holocene ending at time zero (modern day) (Walker et al. 2012).

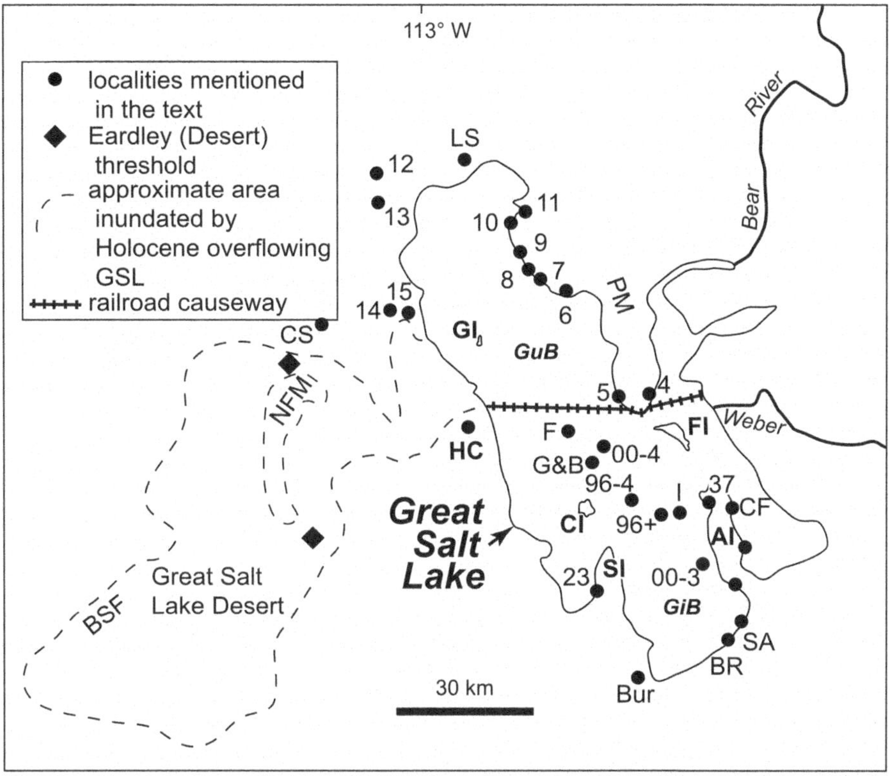

Fig. 1 Map of Great Salt Lake (GSL) and the surrounding region named locations: AIAntelope Island, BCBrigham City, BLBaileys Lake, BRBlack Rock, BSFBonneville Salt Flats, CFCamera Flats, CICarrington Island, CSCrescent Spring, FIFremont Island, GIGunnison Island, GiBGilbert Bay, GuBGunnison Bay, HCHomestead Cave, LSLocomotive Springs, NFMNewfoundland Mountains, PMPromontory Mountains, SASaltair, SIStansbury Island, SLCSalt Lake City, SPSeagull Point, WWendover. cores: 00-3 core GSL00-3 (Dinter and Pechmann 2004, 2014), 96-4 core GSL96-4 (Thompson et al. 2016), 96+ core GSL96+ (Thompson et al. 2016), core 00-4 core GSL00-4, F core F of Spencer et al. (1984), G&B core of Grey and Bennett (1972), I core I of Spencer et al. (1984), McKenzie and Eberli (1987). Numbered dots are approximate locations listed in Table 3 (from Murchison 1989)

Great Salt Lake in Chronological, Sedimentologic, and Geomorphic Contexts

Lake Bonneville and Great Salt Lake are two end members of one lake system in the same basin (Atwood et al. 2016). In our interpretation, Great Salt Lake began at 13,000 yr BP with the final regression of Lake Bonneville. Lake Bonneville was the Pleistocene lake resulting from the cooler and wetter climate of Oxygen Isotope Stage 2 (a glacial period), and Great Salt Lake resulted from the warmer and drier climate of Oxygen Isotope Stage 1 (an interglacial period). Lake Bonneville was

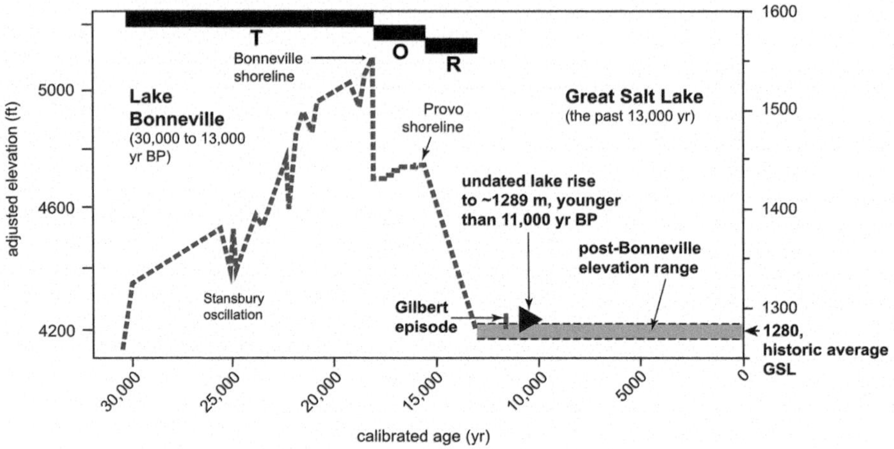

Fig. 2 Chronology of the Lake Bonneville and Great Salt Lake (GSL) system. The curve for Lake Bonneville is based on data from Oviatt et al. (2015); the information for the Great Salt Lake part of the diagram is from this chapter (Fig. 8 is an enlargement of Great Salt Lake history). Transgressive phase, overflowing phase, regressive phase

larger, deeper, and fresher than Great Salt Lake and is of interest for many reasons, including its connections to Pleistocene global-scale climate changes. But understanding the history, processes, and environments of Great Salt Lake is more societally relevant today as we attempt to adapt to ongoing climate change.

The history of fluctuations of Great Salt Lake (Figs. 1 and 2) interests many people, including archaeologists, environmental planners, hydrologists, meteorologists, and biologists. A partial list of scientific publications concerning Great Salt Lake includes Madsen and O'Connell 1982; Kay and Diaz 1985; Lall and Mann 1995; Simms and Stuart 2002; Baxter et al. 2005; Lall et al. 2006; Oliver et al. 2009; Mohammed and Tarboton 2012; Baskin 2014; DeRose et al. 2014; Lindsay et al. 2017; Newell et al. 2017; Bouton et al. 2016, 2019; Pace et al. 2016; Baxter 2018; Baxter et al. 2018; Homewood et al. 2018; Venin et al. 2018; Vanden Berg 2019; and Baxter and Butler 2020.

The Gilbert-episode Lake, which we consider to be a component of Great Salt Lake, reached an elevation of ~1295 m (~4250 ft), higher than that of any other post-Bonneville rise of Great Salt Lake. If the lake were to rise again as high as it did during the Gilbert episode, the Salt Lake City International Airport (~1289 m; ~4228 ft) would be submerged. The name "Gilbert-episode Lake" is used here rather than "Lake Gilbert" because the name "Lake Gilbert" has precedence, being applied to the Pleistocene lake in Grass Valley, NV (Hubbs and Miller 1948).

The Gilbert-episode Lake occupied the hydrographically closed basin, and apparently was the lake response to an increase in inputs from precipitation and river runoff and/or a decrease in temperature and evaporation. The possibility of a connection between the Gilbert episode and global-scale climate change (e.g., the Younger Dryas) is not well understood—the Gilbert episode occurred at the tail end of the Younger Dryas, but earlier in the Younger Dryas (that is, between the end of Lake Bonneville at about 13,000 yr B.P. and the Gilbert episode) that lake levels were likely low, similar to modern lake levels, although no shorelines of that age have been identified. Recent work suggests that isostatic rebound in response to the removal of the Lake Bonneville water load may have been completed by the time of the Gilbert episode (Oviatt 2014).

The Gilbert-episode Lake was brackish and supported ostracodes (Oviatt 2014) in the deepest part of the basin, that is, in the area of modern Great Salt Lake. However, in the Great Salt Lake Desert, where evaporation dominated over inflow, the Gilbert-episode Lake may have been more saline. Evaporation did not render Great Salt Lake saline enough for ostracodes to be replaced by brine shrimp until immediately after the Gilbert episode (Oviatt 2014; Thompson et al. 2016). We do not thoroughly discuss the Gilbert-episode lake in this chapter because it has been addressed elsewhere (Oviatt 2014; Oviatt et al. 2005).

To help put fluctuations of Great Salt Lake into context, we point out that the total elevation range of fluctuations of the still-water plane in late Holocene Great Salt Lake (~14 m) are only slightly greater than the measurement uncertainty of elevations of geomorphic features at the Bonneville and Provo shorelines (~2–3 m; Currey 1982; Chen and Maloof 2017). The total elevation range of shoreline evidence for Lake Bonneville spans almost 300 m, so measurement uncertainty might be ~1% of the total elevation range. In the case of shorelines of Great Salt Lake, measurement uncertainty could easily be 1/3–1/2 of the total elevation range.

Shorelines in the closed basin of Great Salt Lake so far have not provided precise evidence of ancient lake levels for reconstructions of lake history. The two Lake Bonneville shorelines that can be mapped consistently throughout the basin, the Bonneville and Provo shorelines, were controlled by overflow (Gilbert 1890, p. 171; Currey 1982; Miller 2016; Oviatt and Jewell 2016). Shorelines formed during the closed-basin phases of Lake Bonneville have not been mapped basinwide and have been difficult to precisely date –this generalization includes the Stansbury shoreline, one of the three named shorelines of Lake Bonneville (Gilbert 1890; Oviatt et al. 1990). Lake Bonneville occupied a hydrographically closed basin for about 14,000 of approximately 17,000 years of its existence (Oviatt 2015).

In regard to isostatic rebound in the Bonneville basin, the basin may have been near, or in isostatic equilibrium, during the transgressive phase of Lake Bonneville (Fig. 2; Bills et al. 1994; Bills, B., written communication, 2020) (in other words, isostatic depression of Earth's crust beneath the lake basin was essentially keeping up with rising lake level and increasing water load during the transgressive phase). This conclusion is consistent with the observations of Miller et al. (2013), who suggested the basin was close to isostatic equilibrium at the time the main Provo shoreline of Lake Bonneville had formed after more than 1000 yr of overflowing conditions and stable lake level following the essentially instantaneous removal of over 100 m of water during the Bonneville flood. The Gilbert-episode lake reached its highest level about 1500 yr after Lake Bonneville had finally regressed to elevations similar to those of modern Great Salt Lake (lake level had dropped during the noncatastrophic regressive phase of Lake Bonneville by ~200 m —from the Provo shoreline to Great Salt Lake elevations —by about 13,000 yr BP, and some rebound would have been occurring during the regressive phase). Therefore, isostatic rebound of the basin was probably complete prior to the Gilbert episode. Stratigraphic and geomorphic observations support this hypothesis (Oviatt 2014). If so, no isostatic rebound has occurred in the basin during the Holocene.

Previous workers have assumed (i.e., Murchison 1989) that the effects of isostatic rebound in response to the removal of the Lake Bonneville water load could be detected in the spatial distribution of elevations of Great Salt Lake shorelines. Crittenden (1963) attempted to address the question of whether isostatic rebound of the Bonneville basin is still occurring in modern times by presenting some data from releveling surveys along the old Central Pacific Railroad, which crossed the northern part of the Bonneville basin (Crittenden 1963, his Fig. 11). The survey data were inconclusive, and no independent observations have shown that Holocene or historical uplift, which could be attributed to isostatic rebound, has occurred. Considering this absence of evidence for late Holocene isostatic rebound in the basin and the preliminary observation that isostatic rebound in response to the removal of the Lake Bonneville water load was probably complete prior to the Gilbert episode, we are disinclined to think that rebound has influenced the elevations of shorelines of Great Salt Lake or the gradient of the mudflats in the Great Salt Lake Desert (Eardley 1962). Future work is needed to test this conclusion.

Twice during the AD 1980s (in 1986 and 1987, with a seasonal decline of roughly 0.3 m between the two rises; Fig. 3), Great Salt Lake reached its historical high elevation (Fig. 3), both times to about 1284 m (4212 ft). In a plot of lake level vs.

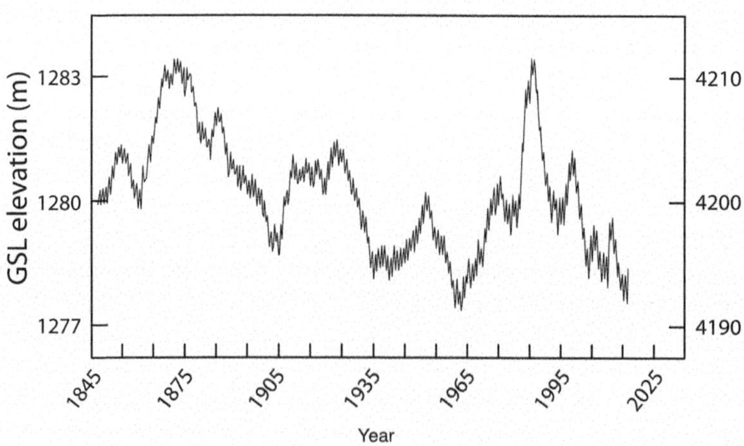

Fig. 3 Historical Great Salt Lake hydrograph (modified from a diagram drawn by R. Baskin, U.S. Geological Survey, written communication, 2017). The maximum elevation range in historical time has been about 6.4 m (21 ft), from a high of approximately 4212 ft in the 1870s and in 1986–87, to a low of about 4191 ft in 1963. The curve ends in 2017. More recent lake-level data can be obtained at this web site, https://ut.water.usgs.gov/greatsaltlake/. Note that lake elevations prior to AD 1875 are from G.K. Gilbert's reconstruction of lake levels based on anecdotal evidence (word-of-mouth information from local residents) (Gilbert 1890; Mabey 1986). Some of the long-term decline in lake level seen in this plot is likely related to human activities (Arnow and Stephens 1990)

time with a "zoomed-out" time scale, the pair of high-lake levels during the 1980s would be regarded as one episode of Great Salt Lake reaching its historical high. The previous high episode during the 1870s, which probably also consisted of more than one peak, rose to an elevation similar to the 1980s (Fig. 3) (Mabey 1986). The lake-level highs of the 1980s led to much discussion among the citizens of Utah about the lake's behavior and its impacts on urban developments and infrastructure along its shores and on its lakebed (Atwood 1986), and among scientists who were interested in the causes of the lake's behavior, especially the connections between weather and climate in the basin and changes in lake level (Kay and Diaz 1985). During the 1960s, Great Salt Lake shrank to one of its two historical low episodes declining to 1277.7 m (4191 ft) (the second historical low episode is ongoing in 2020 with a level of 1278.5 m [4193.ft] at the end of July; https://waterdata.usgs.gov/ut/nwis/uv?site_no=10010000). The low-lake levels during the 1960s led to speculation about the possibility of the lake drying up altogether (Stokes 1966).

Great Salt Lake is a terminal lake in a hydrographically closed or endorheic basin, meaning that it has no surface outlet and that it is the terminal sink for the rivers that enter it. Lake level is constantly changing as the balance between input and output changes (inputs are mainly from direct precipitation on the lake and runoff in rivers, and outputs are by evaporation; groundwater inputs and outputs are minor; Arnow and Stephens 1990). High-lake levels threaten highways, railroads, sewage-treatment plants, and septic systems, and lead to potential groundwater problems, including instability of airport runways (for example, in the 1980s water in the sediments beneath the runways at Salt Lake City International Airport was pumped to reduce the potential for liquefaction, even though the lake itself did not reach the airport). Low-lake levels, which mainly occur during times of regional drought, cause problems for boat marinas and industries that depend on the saline lake waters, and raise the possibility of exposing large areas of odoriferous muds (which could possibly contain trace amounts of toxic substances) that can be eroded by wind, contributing to air pollution. When Great Salt Lake hovers near its average level, near approximately 1280 m (4200 ft), little concern is voiced among the residents of the basin, but when the lake rises or falls far above or below its average, concern rises proportionately. Wurtsbaugh et al. (2017) point out that modern desiccation of many saline lakes around the world has been exacerbated by human consumption of water that would otherwise flow into terminal lakes.

History of Closed-Basin Lakes

Commonly in studies of lake history, the general goal is to determine still-water elevation at well-dated times in the past. In this regard, the word "shoreline," as used in a geomorphic and paleolake context, has a different meaning than "shoreline" in a modern-lake context. For an existing lake, the word refers to the interface of the surface of the water body with the land—the elevation of this interface shifts constantly with weather and wind changes (Street-Perrott and Harrison 1985). The vertical range of this shift may be many meters, especially in a closed-basin lake on time scales of more than a year.

In a geomorphic and paleolake sense, the word "shoreline" should be interpreted as "evidence of a shoreline." The geomorphic evidence almost never corresponds exactly with the elevation of the still-water plane, although their elevations are usually within a few meters of each other (Atwood 2006). Landforms produced by wave activity in the shore zone, some erosional and some depositional, appear as "lines" or "shorelines" on slopes above modern lake level. All shorelines, except for the one that exists at the particular point in time that marks "now," are "paleoshorelines," and in our opinions, it is unnecessary to insert the prefix "paleo-" every place the word "shoreline" is used.

If shoreline landforms can be mapped, dated, and placed in a stratigraphic context, they can be used to reconstruct lake history. For a closed-basin lake, in which the lake has undergone many fluctuations over tens to thousands of years, the geomorphic record will be complicated, with multiple shorelines at different elevations in different places controlled in part by local geomorphic and tectonic factors. Some shoreline depositional evidence produced by older lake cycles will be destroyed or covered by younger lake rises. Depositional or constructional shorelines (mostly barrier beaches) in the Great Salt Lake basin consist of gravel accumulations laid down during many times the wave zone passed over them as the lake surface repeatedly went up and down through that elevation range.

Shorelines are studied on the ground and on aerial imagery. Offshore sediments are studied in cores and outcrops. It takes considerable effort to accurately correlate between the two, and correlations are rarely unequivocal. Part of the difficulty in deciphering both the shoreline and sediment records of Great Salt Lake is a consequence of the substantial reworking and overprinting of sediment caused by high-frequency lake-level changes in the shallow lake. In the case of Great Salt Lake, shorelines are difficult to date reliably and to decipher stratigraphically. In contrast, if offshore sediments are stratigraphically intact and not reworked, they have the potential to be dated (although a carbon reservoir or other effects may cause difficulties with radiocarbon dating). Offshore sediments only provide evidence of relative lake-level changes (up or down); precise measures of lake level, water volume, and surface area cannot be derived from core investigations. Studies of offshore sediments in cores could yield information about high-magnitude changes in sedimentology and lake chemistry caused by yearly to millennial changes in water budget driven by changes in weather and climate.

Long-Term Persistence of Shallow Closed-Basin Lakes in the Bonneville Basin

Pleistocene Lake Bonneville is frequently described as the precursor of Great Salt Lake, and Great Salt Lake is referred to as a remnant of Lake Bonneville. These portrayals are correct if considered on short-time scales, but they are misleading on long-time scales because Lake Bonneville existed for only a small part of the long-term history of the basin. On a time scale of millions of years, Great Salt Lake is typical in the basin, and deep freshwater lakes are unusual; this is a valid generalization even though the basin hydrography has changed over time (for instance, the upper Bear River was diverted into the basin about 50,000 yr BP; Pederson et al. 2016). Analyses of the Burmester sediment core show that Lake Bonneville was one of only four deep lakes in the basin during the past ~800,000 yr. The Burmester core was collected in 1970 by A.J. Eardley (Eardley et al. 1973) a few kilometers south of Great Salt Lake and was reinterpreted by Oviatt et al. (1999). The preserved sediments in the Burmester core suggest that for more than 90% of the past 800,000 years, shallow lakes remained below the elevation of the core site (~1288 m; 4225 ft; Fig. 1), lower than the highest elevation of the Gilbert-episode Lake (1295 m; 4250 ft). PreBonneville shallow lakes that rose higher than the core site, such as the Cutler Dam lake cycle (Kaufman et al. 2001; Oviatt et al. 1987) and the Gilbert-episode lake, did not leave a stratigraphic record sampled by the core (the reason for this has not been determined, but it may be related to the rapidity of the lake cycles and the lack of a source of clastic sediment at the core site).

A core taken from Great Salt Lake between Carrington Island and Promontory Point (core GSL00-4, Fig. 1), has been interpreted (Balch et al. 2005) as indicating that at times during the past ~200,000 yr ostracodes and gastropods lived in wetland environments on what is now the floor of Great Salt Lake. For this to have happened, hypersaline lake water, which is toxic to ostracodes and gastropods, would have been replaced by brackish to freshwater. It could be argued that Great Salt Lake had "dried up" at these times because hypersaline water was not present at the core site, but the basin floor remained wet and "puddles" of hypersaline water may have existed somewhere. The fresh or brackish water in the wetlands was probably delivered by one or more of the rivers that enter the basin from the east (Bear, Weber, Jordan Rivers; Fig. 1), and/or by groundwater discharging from springs. At the site of cores GSL96-5&6 (referred to in this paper as GSL96+ after Thompson et al. 2016; Fig. 1), in one of the deepest parts of Great Salt Lake in the south arm between Stansbury Island and Antelope Island, wetland environments did not exist between about 40,000 yr BP and the initial rise of Lake Bonneville about 30,000 yr BP (Thompson et al. 2016).

Sediments of the Post-Bonneville Lake

No evidence has been found of wetland environments anywhere on the floor of Great Salt Lake during the past 13,000 yr (Grey and Bennett 1972, p. 4; Balch et al. 2005; Dinter and Pechmann 2014; examination of many sediment cores from Great Salt Lake by Oviatt and Thompson, 1995-present). Using the modern distribution of wetlands around the margins of Great Salt Lake (Sorenson et al. 2020) as an analog, post-Bonneville wetlands would have been distributed near the mouths of rivers and the margins of the hypersaline lake. During most of post-Bonneville time, brine-shrimp fecal-pellet muds have been deposited in Great Salt Lake, sediment indicative of hypersaline waters (Thompson et al. 2016). The fecal-pellet muds are remarkably uniform in composition and sedimentary structures, with wavy bedding and ripple laminations (Spencer et al. 1984; Thompson et al. 2016) that suggest the mud and sand were moved, at least periodically and locally, by lake-bottom currents. This suggests the lake may have been shallow enough during all of postGilbert-episode time for wind-driven waves at the lake surface to agitate the bottom. Other possible causes of bottom agitation are meteotsunamis, seiches, and seismic events, although they are episodic in occurrence.

Colman et al. (2002) interpreted seismic profiles of Great Salt Lake sediment as indicating truncation at "a very slight angle" stratigraphically beneath the Mazama ash and above the Lake Bonneville deposits. They thought the truncation might represent an "unconformity… the result of an early Holocene low lake stand, one significantly lower than present" (p. 76). However, in our examination of sediment cores from different locations in Gilbert Bay, we have seen no indication of such an unconformity. In core GSL96+ (Thompson et al. 2016, p. 238), a sediment hiatus at an early to middle Holocene stratigraphic level may be related to sediment loss during the coring operation rather than to an unconformity –a similar sediment hiatus is not seen in other cores taken in Gilbert Bay.

It is difficult to precisely and reproducibly date Great Salt Lake sediments. One of our many conclusions from years of working with the stratigraphy of Great Salt Lake sediment is that it is likely that on a 1000-yr or longer time scale, sediments are constantly reworked. The process of reworking may include partial oxidation of some organic matter, and suspension, and then redeposition of older organic and inorganic particles. In core GSL96+, Thompson et al. (2016) found a nearly uniform pattern with few outliers in a plot of the ~40 radiocarbon ages of total organic carbon in the sediments versus depth, but the uniform array was offset by approximately 1800 ^{14}C yr from independently determined ages at four places in the core, three of them in post-Bonneville time. The cause of the age offset is not fully understood (Thompson et al. 2016), but could be related to a radiocarbon reservoir in the water, or reworking of older organic matter, or some other undetermined process. If the sediments are contaminated with older materials from reworking, the results from geochemical, biological, or other analyses, might be smoothed or blended through time (both spatially and temporally), reducing the precision and resolution of the proxies. Bowen et al. (2019) documented variability in radiocarbon ages of different materials in sediments of Great Salt Lake that is consistent with the observations of Thompson et al. (2016).

As an example of dating problems in Great Salt Lake sediments, we provide here a list of radiocarbon ages of organic materials associated with the Mazama ash (Table 1). The age of the Mazama ash is well-known because numerous radiocarbon ages of associated materials have been obtained in western North America (Bacon 1983; Hallet et al. 1997), and its age has been determined by counting annual layers in the Greenland ice sheet (Zdanowicz et al. 1999). Its radiocarbon age is 6790 ± 100 ^{14}C yr BP, and its calibrated age is 7630 ± 150 cal yr BP. All of the radiocarbon ages from samples associated with the Mazama ash in cores from Great Salt Lake (Table 1) are older than the radiocarbon ages reported by Bacon (1983), Hallet et al. (1997), and Zdanowicz et al. (1999).

Table 1 Radiocarbon ages associated with the Mazama ash in Great Salt Lake sediments

Lab number	Core[a]	Depth (cm)	Material[b]	^{14}C Age[c]	References
WW1643	96–6	536[d]	pollen+	7440 ± 50	Thompson et al. (2016)
WW1726	96–6	538[d]	TOC	8530 ± 50	Thompson et al. (2016)
WW1644	96–6	539[d]	pollen+	7590 ± 60	Thompson et al. (2016)
WW2561	96–4	91.0[d]	TOC	8550 ± 50	Thompson et al. (2016)
WW2562	96–4	94.5[d]	TOC	8530 ± 40	Thompson et al. (2016)
WW1385	96–4	99.4[d]	TOC	8660 ± 50	Thompson et al. (2016)
W4667	Spencer F	(10-cm)[e]	organic matter	8800 ± 90	Spencer et al. (1984)
W4663	Spencer F	(10-cm)[e]	carbonate	9500 ± 90	Spencer et al. (1984)
W4673	Spencer I	(10-cm)[e]	organic matter	9940 ± 100	Spencer et al. (1984)
W4665	Spencer I	(10-cm)[e]	carbonate	10,820 ± 110	Spencer et al. (1984)
AA44475	GSL00-3B-4H-1	845[f]	charcoal	7186 ± 66	Dinter and Pechmann (2004); Dinter, written communication
regional Mazama ash average	–	–	–	6845 ± 50	Bacon (1983)
regional Mazama ash average	–	–	–	6730 ± 40	Hallet et al. (1997)
Mazama ash age from Greenland ice sheet layer counting	–	–	–	~6730 ^{14}C yr BP (7630 ± 150 cal yr BP)	Zdanowicz et al. (1999)

[a]Core locations are shown in Fig. 1
[b]Pollen+ = mostly pollen, but may include whatever materials survived the pollen-preparation techniques, such as possible older organic fragments; TOC = total inorganic carbon dispersed in lake sediments; charcoal = "nonwoodcharcoal presumed to derive primarily from grass and brush fires on slopes and in valleys surrounding the Great Salt Lake" (Dinter and Pechmann 2004, p. 2).
[c]Radiocarbon ages as reported by the dating laboratory.
[d]Sample thickness 1 cm; Mazama ash is present in core GSL96+ at a depth of 538 cm; in GSL96-4 at 92 cm; the ash is approximately 0.5 cm thick in both cores.
[e]10-cm sections of core "centered at the Mazama" –depth to Mazama ash not given (Spencer et al. 1984, p. 329).
[f]Core-sediment sample was 4 cm thick, 3 cm above ash to 1 cm below ash, including ash (Dave Dinter, written communication, 2018).

Previous Work

In the literature review below, we have selected some references that have been cited as reliable sources for information about post-Bonneville Great Salt Lake. Some previous references provide data that are relevant to an accurate understanding of lake history, but many of the previously published interpretations should probably be abandoned now. While we were engaged in this review, we realized that it is the natural tendency of researchers (or humans in general) to try to relate what they observe to their preconceptions. In the case of studies of the history of Great Salt Lake, which we are considering here, many preconceptions have been based on implied correlations with paleoclimatic proxy data in the region and not on geologic data directly from Great Salt Lake. Some hypotheses about Great Salt Lake that seemed relatively well supported to us and to others in the decades preceding 2020 have turned out, with closer scrutiny and new perspectives, to be less reliable than we thought.

During the twentieth and into the twenty-first century, many people contributed data and interpretations about Great Salt Lake shorelines (including Eardley et al. 1957; Ross 1973; Rudy 1973; Currey 1977; Currey and James 1982; Currey et al. 1984; Currey 1987; Currey et al. 1988; Murchison 1989; Atwood 1994, 2006; Atwood and Mabey 2000; Murchison and Mulvey 2000; Oliver et al. 2009; Oviatt 2014; Madsen et al. 2015; Oviatt et al. 2015). Published studies that utilized offshore sedimentary records to help understand the post-Bonneville behavior of the lake include Grey and Bennett (1972), Mehringer (1977, 1985), McKenzie and Eberli (1987), Balch et al. (2005), and Thompson et al. (2016). Some of these publications are discussed below.

Gilbert (1890)

Gilbert (1890) said little about the post-Bonneville history of Great Salt Lake. He suggested that post-Bonneville time might have been very short (Gilbert 1890, p. 261), but this assessment was based on his interpretation of the thicknesses of gravel wedges within the Bonneville deposits compared to the thicknesses of post-Bonneville alluvial deposits. It is now known that the inter-Bonneville gravel wedges, which he interpreted as alluvium that had accumulated slowly, are more likely to be lacustrine gravel units that accumulated rapidly (Oviatt 1997). Radiocarbon dating and stratigraphic information not available to Gilbert clearly show that Lake Bonneville ended about 13,000 yr BP (Fig. 2).

Gilbert (1890) focused primarily on Lake Bonneville, although he discussed historical variations of Great Salt Lake levels and contributed other observations about modern Great Salt Lake. Gilbert established the first gage to monitor lake level (at Black Rock in 1875; Mabey 1986). He extended the historical record from the 1870s back to the 1840s using "traditional" (anecdotal) information he had obtained from local residents and documented the 1870s highstand events (Mabey 1986).

Eardley et al. (1957)

A.J. Eardley (in Eardley et al. 1957) examined geomorphic features of Great Salt Lake and its environs, and developed an idea referred to in this chapter as the expansion/stabilization hypothesis to help explain lake behavior. Eardley described a topographic divide, which he called the Desert threshold (Fig. 1; it is now sometimes referred to as the "Eardley threshold"), between Great Salt Lake and the flat-floored desert to the west (the Great Salt Lake Desert). The hypothesis was that if Great Salt Lake were to rise and overflow into the Great Salt Lake Desert, which does not have significant input of surface water, the expansion of the lake's surface area would cause an increase in evaporative output and result in lake-level stabilization. As noted correctly by Currey et al. (1984), "... at the 4217-foot [1285 m] level ... the lake expands westward into the Great Salt Lake Desert (West Desert) and abruptly increases its surface area by nearly one-third," and this one-third increase would consist of very shallow waters that would easily heat up and evaporate.

Eardley's expansion/stabilization hypothesis is a reasonable theoretical idea, but the application of the principle in the Great Salt Lake basin has not been straightforward. Eardley attempted to relate the shoreline he had mapped at places in the Wasatch Front area, which he called the "Gilbert beach" or "Gilbert level," to lake stabilization caused by expansion into the Great Salt Lake Desert (Eardley et al. 1957), but later abandoned that idea (Eardley 1962). Currey (1982; Currey et al. 1984) recognized that the Desert threshold was double, one divide northeast of the Newfoundland Mountains, one southeast of the Newfoundland Mountains (Fig. 1), with the two divides having nearly the same elevation, 1285 m (4217 ft). Currey kept Eardley's hypothetical reasoning for lake stabilization but applied it to late Holocene Great Salt Lake rather than to the Gilbert-episode lake.

Grey and Bennett (1972)

Grey and Bennett (1972) analyzed carbonate sediments in a core from Great Salt Lake. The core was collected from a location about 7 km south of the railroad causeway and about 12 km west of Fremont Island (Grey and Bennett 1972; Fig. 1). Although Grey and Bennett (1972) did not indicate when the core was collected, the core was mentioned in Mehringer et al. (1971), so it was probably obtained prior to 1970. Grey and Bennett (1972) reviewed the geochemical controls on isotopes of carbon ($\delta^{13}C$), oxygen ($\delta^{18}O$), and sulfur ($\delta^{34}S$) in lake sediments, and built on those ideas to interpret their analytical results from the core. The oxygen isotope results of Grey and Bennett (1972) do not show significant trends younger than the Mazama ash (Fig. 4a). Lighter values of $\delta^{18}O$ suggest periods when the lake water was relatively dilute and when lake levels were generally high (Grey and Bennett 1972). Figure 4a shows possible relatively high-lake levels represented by the oxygen isotopes at depths of about 60 and 95 cm in the core (with approximate ages of 2500 and 4000 ^{14}C yr BP, respectively, estimated using the postMazama ash chronology of Grey and Bennett 1972).

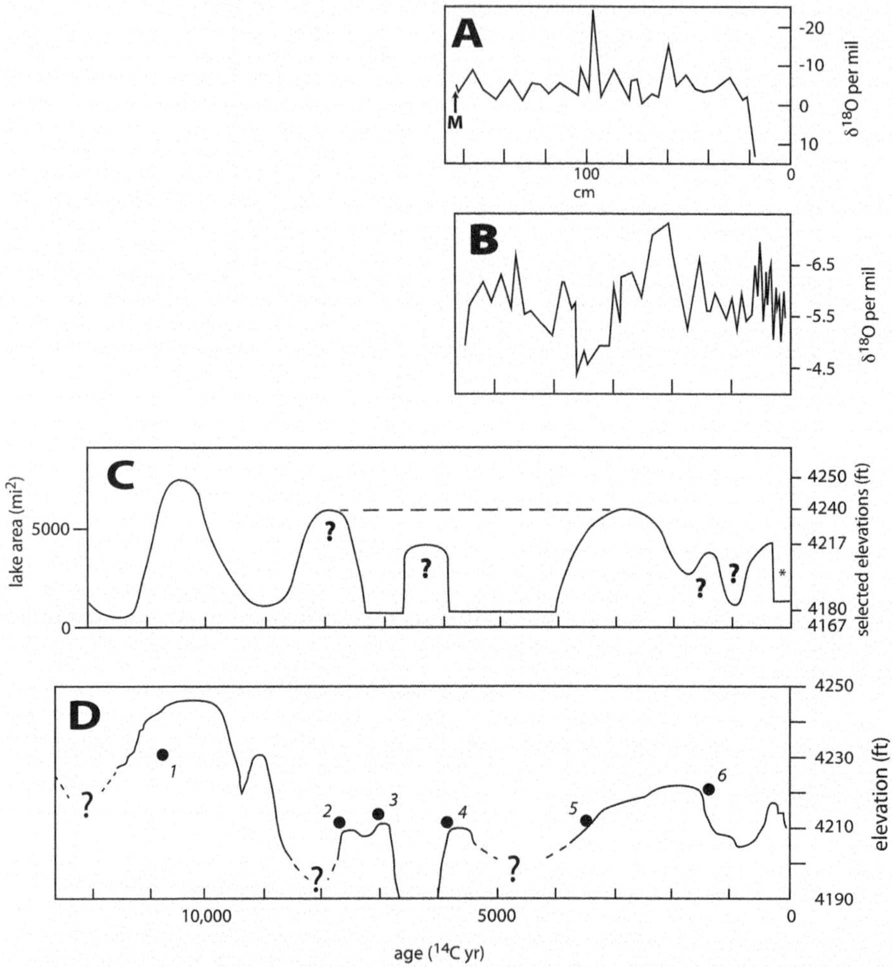

Fig. 4 Previously published interpretations of the post-Bonneville history of Great Salt Lake. For each record, the depth scale or radiocarbon time scales are from the original publication. These figures were drawn based on the published figures and should be considered approximate renditions of the original drawings. (**a**) Oxygen-isotope record modified from Grey and Bennett (1972); this is not a lake-level record; the vertical scale is reversed to emphasize those higher values of $\delta^{18}O$ correlate with lower lake levels; the horizontal scale is core depth. M = Mazama ash (7630 calyr BP). (**b**) Oxygen-isotope record modified from McKenzie and Eberli (1987); this is not a lake-level record; the vertical scale is reversed to emphasize those higher values of $\delta^{18}O$ correlate with lower lake levels; the horizontal scale is age in radiocarbon years (labeled in Fig. 4d). (**c**) Hydrograph modified from Currey et al. (1984); the asterisk marks historical Great Salt Lake, shown in Fig. 3; the horizontal scale is age in radiocarbon years; the vertical scale on the left is lake surface area (mi^2), and the vertical scale on the right is elevation (ft). (**d**) Hydrograph modified from Murchison (1989, his Fig. 40, p. 128); radiocarbon ages are numbered and summarized in Table 2; the horizontal scale is age in radiocarbon years

Table 2 Radiocarbon ages used by Murchison (1989) in construction of his lake-level chronology (Fig. 4d). Page numbers refer to Murchison (1989)

^{14}C age	Number in Fig. 4d	Lab number	Location	Material dated	Interpretation of Murchison (1989)	Alternative interpretation
10,990 ± 110	*1*	Beta-22431	Public Shooting Grounds	Gastropods	Deposition during Gilbert time	All ^{14}C ages of mollusk shells from the Public Shooting Grounds are likely to be incorrect because of a radiocarbon reservoir in the spring water feeding the wetlands (Oviatt 2014; Oviatt et al. 2005).
7650 ± 90	*2*	Beta-25290	Seagull Point, Antelope Island	"humate extract in wedge-shaped bodies of organic sediment" p. 78	"accumulated behind berms or some other barrier beaches" p. 78; "accumulation on a lagoon floor during a protracted lake rise and stabilization level" p. 79;	Organic sediment in a lagoon depression; the relationship between the organic sediment and lake level is unknown, and the lagoon depression could easily have formed long before the organic sediment was deposited
7070 ± 100	*3*	Beta-22432	West side Stansbury Island	"ooids cemented to oolite … from a plainar [sic] bed that overlies shore pebbles …" p. 74	"… probably a beach that was buried by ooid sands of a later Holocene transgression"	oolitic sand sampled in eolian sand dunes; the relationship between dated ooids and lake level is unknown; the radiocarbon age of ooids is complicated (Paradis et al. 2017)

(continued)

Table 2 (continued)

^{14}C age	Number in Fig. 4d	Lab number	Location	Material dated	Interpretation of Murchison (1989)	Alternative interpretation
5890 ± 120	*4*	Beta-26629	Seagull Point, Antelope Island	Humate extract in "organic marshy material" p. 79	Interpretation not given explicitly in the text, but in Fig. 40, p. 128 of Murchison (1989), the age is plotted adjacent to an upward-trending segment of the interpreted lake-level curve implying the age dates a lake-level rise	Organic sediment in a lagoon depression; the relationship between the organic sediment and lake level is unknown, and the lagoon depression could easily have formed long before the organic sediment was deposited
3450 ± 250	*5*	Beta-25289	Seagull Point, Antelope Island	Humate extract in "marsh sediment from a lagoon" p. 107	"the next identifiable lake rise at this site" p. 107; "The beach depositional evidence that suggests lake rise episode is a cemented gravel layer at 4221 ft … farther north and west in the exposure" p. 107	Organic sediment in a lagoon depression; the relationship between the organic sediment and lake level is unknown, and the lagoon depression could easily have formed long before the organic sediment was deposited
1400 ± 75	*6*	Beta-25580; ETH-3999	Camera Flats, Antelope Island	Humate extract from a 16-in. (40 cm) pit (p. 108)	In "a depression behind the 4221 ft (1286.5 m [should be 1286.6]) berm … thought to be a washover marsh from the late Holocene high" "a minimum limiting date for the late Holocene high" p. 108	Organic sediment in a lagoon depression; the relationship between the organic sediment and lake level is unknown, and the lagoon depression could easily have formed long before the organic sediment was deposited

Mehringer (1977, 1985)

Mehringer (1977) said that a core from Great Salt Lake (probably the same core analyzed by Grey and Bennett 1972) revealed "a history of lake oscillations with higher than present levels about 6000, 3600, and 2800 B.P." (Mehringer 1977, p. 121; presumably these are radiocarbon years interpolated in the core, based on the radiocarbon age of the Mazama ash, as it was known at that time). He also stated that there had been "probable flooding of the Great Salt Lake Desert, between 3500 and 2200 B.P." (based on analysis of a sediment core from Crescent Spring at the edge of the Great Salt Lake Desert near Hogup Cave; Fig. 1; Mehringer 1977, p. 121). Apparently, these interpretations were based on inferences about lake level from pollen data from the cores. The ground-surface elevation at Crescent Spring is 1297 m, and the core interval Mehringer referred to as between "3500 and 2200 B.P." is 1–2 m below the ground surface (Fig. 11 in Mehringer 1985). Therefore, he apparently suggested that Great Salt Lake had risen higher than 1295 m (higher than the Gilbert-episode lake) during the late Holocene. No known evidence exists from other localities of such a large late Holocene rise in Great Salt Lake. The core from Crescent Spring described by Mehringer (1977, 1985) has not been examined by other researchers.

Currey (1977)

In a field trip guidebook for a meeting of the Association of American Geographers, Currey (1977) summarized information he had gathered about shorelines at a state park on Antelope Island. He inferred the ages of the shorelines from general knowledge of Holocene climate in the western United States. He estimated the Gilbert shoreline age and elevation as 8000–9000 ^{14}C yr BP, 1303 m (4275 ft), the Fremont shoreline as 5200–6300 ^{14}C yr BP, 1290.8 m (4235 ft), and the Eardley shoreline as 2000–3800 ^{14}C yr BP, 1285.6 m (4218 ft). He also identified "bottom features" (that he interpreted as "desiccation" polygons) offshore from Antelope Island, which he thought formed about 6000 ^{14}C yr BP. Currey formalized much of this information in a later publication (Currey 1980).

Currey (1980)

In the 1980 Utah Geological Survey publication on Great Salt Lake, Currey (1980) mapped shorelines of Lake Bonneville and discussed post-Bonneville levels of Great Salt Lake. He reported slightly different ages for the Fremont and Eardley shorelines than he had given in 1977: Fremont 4000–5000 ^{14}C yr BP, and Eardley 2000–3000 ^{14}C yr BP, but did not cite the sources for this information.

Currey (1980) discussed the features he interpreted as desiccation polygons in shallow water on the floor of Great Salt Lake. Polygons are visible on aerial photographs and satellite images (Fig. 6 in Currey 1980), and many of them are now exposed because lake level is low (Baxter et al. 2018; Bouton et al. 2016; Vanden Berg 2019). Currey proposed that the polygons had formed during the middle Holocene, based on the widely accepted belief that middle Holocene climate in western North America was relatively warm and dry. Polygons have not been found on the floor of Great Salt Lake in deep water (Baxter et al. 2018; Clark et al. 2017; Colman et al. 2002; Dinter and Pechmann 2014; Thompson et al. 2016). Polygonal cracks in Great Salt Lake sediments have not been studied from a stratigraphic or sedimentological perspective, so their origin and age (or ages?) have not been determined, although there seems to be some connection with microbial processes (Vanden Berg 2019). Complete desiccation is not necessarily required to form polygonal cracks in muddy sediment (e.g., McMahon et al. 2016; Moscardelli et al. 2012), and until more work has been done on the Great Salt Lake polygons, it may be best not to refer to them as "desiccation" polygons, or to attribute their origin to inferred times when the lake bed dried up.

An untested idea is that the polygons formed not just once but at multiple times, reforming over and over again in the same locations while the bottom mud stayed moist but lake level fell low enough to expose the lake floor around its edges. This idea would be consistent with the observation that the polygons seem to cut the youngest sediment on the floor of the lake –if they had formed during a single event during the middle Holocene, they would not be visible through late Holocene sediments.

Currey and James (1982)

In a chapter, in a volume on archaeology in the Great Basin (Madsen and O'Connell 1982), Currey and James (1982) summarized what was known about late Pleistocene and Holocene paleoenvironments in the Great Salt Lake region. They provided a reconstruction of Great Salt Lake history based partly on inferences from information about regional paleoclimate and partly on information from the Great Salt Lake basin. The part of the summary of lake history that is based on examination of archaeological data is an important contribution to knowledge about late Holocene lake-level changes, but most of the conclusions about higher and older Holocene levels of Great Salt Lake were inferred from paleoclimate information described from other localities in the Rocky Mountain and Great Basin regions, and are probably not accurate.

Currey et al. (1984)

As Great Salt Lake rose during the early 1980s, the Utah Geological Survey asked Don Currey to publish his shoreline mapping (Currey 1982) in a format that could be easily read and interpreted by nonscientists. The map of five major shorelines of Lake Bonneville and Great Salt Lake by Currey et al. (1984) included a diagram showing fluctuations of Lake Bonneville and Great Salt Lake (the Great Salt Lake part of the diagram is redrawn in Fig. 4c in this chapter). Two vertical axes, one labeled lake-surface elevation and the other surface area, helped to show the role of surface area as a natural control on the rate of lake-level rise. Although no reference citations were provided in the discussion, the summary of the history of Great Salt Lake from Currey et al. (1984) has been relied upon by many people. The map and its accompanying text and figures have provided some scientific context for the events and governmental decisions of the 1980s. The map and hydrograph served their purpose to convey generalities of the Lake Bonneville and Great Salt Lake system as they were known at the time, but they were not part of a peer-reviewed scientific publication.

McKenzie and Eberli (1987)

McKenzie and Eberli (1985; 1987) analyzed carbon and oxygen isotopes and determined the percentages of carbonate minerals (aragonite, calcite, and dolomite) in sediments from core I (Fig. 1), taken in 1979 by Spencer et al. (1984). They interpreted oxygen-isotope results for the past 5500 radiocarbon years (Fig. 4b).

McKenzie and Eberli (1987) constructed a chronology for the core and isotope record based on some uncertain assumptions. As primary age control, they used an inference from Currey and James (1982) that a "playa stage," when they thought Great Salt Lake had almost dried up, had ended at 5500 ^{14}C yr BP. McKenzie and Eberli (1987) reported that the last appearance of "authigenic dolomite" in the core was at a depth of 1.825 m, and they assumed the "authigenic dolomite" marked the playa stage. Therefore, they placed an age of 5500 ^{14}C yr BP at that depth in the core (5500 ^{14}C yr BP is about 6300 cal yr BP). McKenzie and Eberli (1987) placed an additional time marker at a depth of 0.95 m, where "authigenic calcite" became the dominant carbonate mineral. They inferred that this change in mineralogy occurred because of a freshening of the lake and correlated it "… with the mid-neolacustral episode which occurred between 3500 and 2200 years B.P." (1987, p. 130).

McKenzie and Eberli (1987) did not explain what "authigenic dolomite" is and how it could be distinguished from other forms of dolomite in the sediment. The "playa stage" of Great Salt Lake during the middle Holocene and the thought that lake levels were generally higher between 3500 and 2200 ^{14}C yr BP (~3800–2200 yr BP) were based on the review by Currey and James (1982) (discussed above).

Considering these uncertainties, the chronology developed by McKenzie and Eberli (McKenzie and Eberli 1987; Fig. 4b) should be viewed with some skepticism.

In order to translate oxygen-isotope results to lake level at the time of sediment deposition, McKenzie and Eberli (1987) showed the record of historical lake-level changes compared with the oxygen-isotopic results they obtained from a short gravity core (the results are not shown in this chapter), from the same site as core I. The calculated sedimentation rate in the upper part of the section at the core site was faster than deeper in the section (~1.9 mm/yr near the top compared to ~0.26 mm/yr below about 0.5 m in the section), as would be expected. In the gravity core, their isotope samples were spaced about 2 cm apart. Highs and lows in the isotope results were interpreted to line up almost perfectly with highs and lows on the historical lake-level curve. However, the isotopic values that might have been obtained for samples between those spaced at about 2 cm are unknown, and if twice as many or three times as many data points were available, it seems unlikely that the isotope curve would be correlated nearly one-to-one with the lake-level curve.

Pedone (2002) has found a very rough correlation between the isotopic composition of Great Salt Lake waters and lake level, but a precise relationship has not been determined. Until further work establishes the relationship between lake level and oxygen isotopic composition of carbonates in sediments (taking into account the possibility of sediment reworking), the conclusions of McKenzie and Eberli (1987) should be viewed with caution. Their oxygen-isotope curve (Fig. 4b) should not be interpreted as a lake-level curve.

Merola et al. (1989)

Merola et al. (1989) described remote-sensing data related to the topography of the Great Salt Lake Desert (Fig. 1). In that paper, Currey interpreted rounded eolian dunes and abrupt changes in slope, or trimlines, at the margins of dunes on the mudflats, as evidence of what he called "hydroeolian planation" (Merola et al. 1989, p. 240). Hydroeolian planation is a term coined by Currey (1987) for a set of geomorphic processes that involved cyclically repeated erosion by shallow, wind-driven bodies of water on mudflat surfaces, followed by drying and deflation of those surfaces.

Undoubtedly the hydroeolian process is real, but Currey (in Merola et al. 1989) made an additional (and unsupported) interpretation where he envisioned the shallow water body that was involved in hydroeolian planation in the Great Salt Lake Desert as connected to Great Salt Lake. He thought a shallow extension of Great Salt Lake, controlled by overflow across the Desert threshold at 1285 m (4217 ft), was probably responsible for the formation of the rounded dunes.

Currey stated that "locally, the upper limit of shallow water under dynamic conditions was as much as … 6 m … higher [than 1286.6 m]" (Merola et al. 1989, p. 243). Although it is not clear what this statement was based on, if it is correct, the elevations of the bases of eolian dunes in the Great Salt Lake Desert would be found as high as ~1293 m (4242 ft). However, rounded dunes are present at numerous locations in the Great Salt Lake Desert at elevations as high as 1315 m. It is possible that expansions of Great Salt Lake across the Desert threshold produced these trimlines, but frequent rainstorms, and surface water produced by the rising water table at times of low evaporation (even seasonally), would also produce shallow puddles of water that would be blown around on the mudflat surface to erode the margins of dunes. Although Currey stated that the trimlines on dunes "… can be traced laterally into erosional and depositional evidence of the same highstand [1286.6 m] around Great Salt Lake" (Merola et al. 1989, p. 240), he did not describe specific locations where these shorelines could be observed, and we have not found them. Our conclusions are that ephemeral water bodies unrelated to Great Salt Lake could easily create the trimlines and that it is best not to regard the rounded dunes in the Great Salt Lake Desert as solid evidence of Great Salt Lake incursions.

Murchison (1989)

Murchison, a doctoral student of Currey's, collected data on eolian dust accumulation on gravel barriers around Great Salt Lake. Murchison also described post-Bonneville shorelines around Great Salt Lake and attempted to date the "Holocene high," as defined by Currey (1980, 1990). Murchison (1989); Fig. 4d built his post-Bonneville Great Salt Lake chronology around a series of radiocarbon ages (Table 2). However, he did not give detailed assessments of radiocarbon sample materials, and did not discuss what the radiocarbon ages meant in terms of the samples' geologic context and lake history. Apparently, he assumed that the elevation of the radiocarbon ages was the same as lake level at the time each sample formed.

Examples of Murchison's (1989) interpretations of radiocarbon ages come from two locations on Antelope Island (Fig. 1). At Seagull Point, three organic-rich beds in sediments in lagoon settings (depressions dammed by gravel barrier beaches) could be dated by radiocarbon (Fig. 24 in Murchison 1989; Murchison and Mulvey 2000). Figure 40 of Murchison (1989) shows two of the ages from Seagull Point plotted next to, and slightly preceding, lake-level highs (represented by rounded humps in the lake-level curve; ages numbered 2 and 4 in Fig. 4d; Table 2). Age number 5 (Fig. 4d) is plotted next to and in contact with a rising segment of the lake-level curve. The reasons for these placements are not explained. From an exposure at Camera Flats, also on the east side of Antelope Island (Fig. 1), Murchison (1989) obtained a radiocarbon age of 1400 yr (number 6, Fig. 4d; Table 2) for organic sediment collected in a lagoon depression. He plotted the age at about the same elevation as the rounded curve depicting a lake-level high during the late Holocene, but again the reason for this placement was not given.

Murchison (1989) thought that it was possible to trace post-Bonneville shorelines around Great Salt Lake and he suggested that the shorelines varied in elevation in a systematic way, roughly correlated with the isostatic deformation seen in Lake

Bonneville shorelines. Based on this interpretation, he thought the cause of some of the perceived variation in elevation was the lingering effect of isostatic rebound from the removal of the much larger and earlier water load of Lake Bonneville. However, if the isostatic rebound in response to the removal of the Lake Bonneville water load was completed prior to the Gilbert episode (see discussion above), no isostatic effects should be seen in shorelines formed after the Gilbert episode. In addition, we now accept that it has not been possible to trace post-Bonneville shorelines completely around Great Salt Lake (see discussion below).

Each of the radiocarbon ages reported by Murchison (1989) can be reasonably interpreted in ways that differ from Murchison's interpretations (Table 2). Murchison's radiocarbon ages say little about the post-Bonneville history of Great Salt Lake except that lake level was lower than the elevation of the radiocarbon samples at the times they were deposited; samples for Murchison's radiocarbon ages were all collected from elevations higher than the historical high elevation. Murchison's depiction of the post-Bonneville Great Salt Lake record as a smooth, rounded curve (Fig. 4d) is not consistent with the spikey pattern of historical lake-level changes (see discussion below). We caution that the Murchison (1989) reconstruction should no longer be thought of as an accurate representation of fluctuations of post-Bonneville Great Salt Lake.

Currey (1990)

In a review paper on the geomorphology and paleohydrology of the Bonneville basin, Currey (1990) published some ideas about post-Bonneville lake history with little information on the sources of those interpretations. In his Fig. 18, he showed the "Holocene high" at 1286.7 m (4221 ft) and the "late prehistoric high" at 1285.5 m (4218 ft). Currey (1990) probably based the interpretation that the "Holocene high" had an elevation of 1286.7 m on his interpretation of eolian landforms in the Great Salt Lake Desert (see our comments above on Merola et al. 1989). This paper has been more or less accepted by many people as an accurate reconstruction of the history of Great Salt Lake, but in the clearer view provided by hindsight (see discussion below), we recognize that Currey's (1990) interpretations oversimplified the behavior and history of Great Salt Lake.

Madsen (2000) and Broughton et al. (2000)

Important information about post-Bonneville paleoecology in the Bonneville basin came from excavations of cave-floor deposits at Homestead Cave in the Lakeside Mountains on the west side of Great Salt Lake (Fig. 1; Madsen 2000). Broughton (2000); Broughton et al. (2000); and Broughton and Smith (2016) noted that the abundance of bones of the fish Utah chub (*Gila atraria*) in the cave deposits increased in several strata dating to the late Holocene (approximately 3700 yr BP, and between approximately 1150 and 950 yr BP; Fig. 12.11 in Broughton and Smith 2016). The carcasses of *Gila atraria* were apparently deposited in Homestead Cave by raptors (probably owls), which forage within a fairly short distance of their roosting location (5.6 km; Broughton 2000; Broughton et al. 2000; Broughton and Smith 2016), Homestead Cave in this case. Presumably, the raptors had picked up the fish carcasses at the shore of the post-Bonneville lake near the cave.

It is unlikely Great Salt Lake became fresh enough at any time since the end of the Gilbert episode for *Gila atraria*, or any fish, to survive, but a possibility is that low-density fresh- or brackish-water lenses that floated on, and were blown across, the dense salt water of the lake allowed the fish to make it to the vicinity of Homestead Cave from freshwater rivers and wetlands along the eastern shore (Broughton and Smith 2016). Abundant dead fish were observed on the exposed floor of Great Salt Lake between Antelope Island and Fremont Island in 2010, suggesting that the appearance of fish, including the possible transport of carcasses into Great Salt Lake, is not unusual, at least along the east side of the lake; the possibility of transport of fish carcasses by raptors should not be discounted. Living fish were reported in Great Salt Lake during the high-lake levels of the 1980s. Freshwater lenses may have been common at times in the past when large quantities of freshwater were flushed onto the saline lake, probably when climate was generally more conducive to enhanced river flow prior to the construction of dams on inflowing rivers. It is important to recognize that the plot of abundance of fish bones in deposits at Homestead Cave during the Holocene (Fig. 12.11 in Broughton and Smith 2016) should not be interpreted as a lake-level curve.

Atwood (1994, 2006), Atwood et al. (2016), and Atwood and Mabey (2000)

As director of the Utah Geological Survey during the 1980s wet cycle, Genevieve Atwood responded to inquiries from decision makers about the magnitude and frequency of lake fluctuations, such as: how high will the lake flood? and when will it flood again? Inspired by those reasonable but unanswerable questions, Atwood, in collaboration with Don Currey and Don Mabey, began shoreline research on Antelope Island to better understand the Great Salt Lake flooding hazard.

Atwood (1994) and Atwood and Mabey (2000) reported on field work to identify and correlate shoreline evidence between the 1980s highs and the “Holocene High” of Currey and Murchison (roughly 1286 m; 4220 ft). Evidence of the 1986–87 high shoreline was easily identifiable and continuous around Antelope Island, as it included anthropogenic timber and trash as well as shoreline gravels. However, evidence of higher shorelines, including the “Holocene high” was discontinuous and ambiguous. Atwood and Mabey (2000) reported that at least four high-lake events left evidence above the 1980s high shoreline and below what appeared to be the “Holocene High” (Fig. 5).

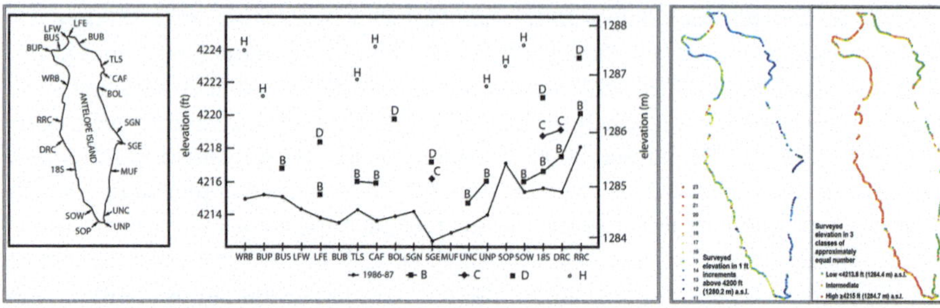

Fig. 5 The map and graph on the left show initial results from 1987 on the 1986–87 debris line on Antelope Island (modified from Atwood and Mabey 2000). B, C, D, and H indicate elevations of shoreline evidence in the Currey zone. The maps on the right show results averaged from 1228 survey locations around the perimeter of Antelope Island (Fig. 1.7 in Atwood et al. 2016). The colored patterns indicate contrasts of wave energy due to at least three coastal processes: longer fetch (distance across open water), strength of storm winds, and storm duration. Higher elevations on the west (windward) side of the island face into storm winds blowing across the main lake in contrast to lower elevations on the east (leeward) side of the island, with the mainland less than 15 km away. Practical insights from this work on Antelope Island and further work around Great Salt Lake (Atwood 2006) indicate only a few storms created the shoreline evidence of high lake levels in the 1980s, and probably none of the strongest storms happened on the very day of the highest still-water level as monitored at the Saltair gage. The 1980s debris line is a record of perhaps 10 days (at the most) of strong storm activity. Will the 1986–87 closed-basin lake shoreline, so important to the humans living in the basin then, be detectable from a geologic perspective even a century from now? Atwood (2006) identified only two places where the 1870s shoreline evidence could be distinguished from the 1986–87 shoreline. For shorelines in the Currey zone, evidence not only is discontinuous but also subtle compared to that of historical levels. For a threshold-controlled lake, the elevation of which could be steady for hundreds or thousands of storms, shorelines may be more distinct

Atwood's later research (2006; Atwood et al. 2016) questioned the assumption that evidence of specific Great Salt Lake shorelines defined horizontal datums. Evidence of the easily recognizable 1980s high shoreline was rarely at the monitored elevation of the 1980s high still-water lake level (1284 m, 4212 ft). Differences in elevation of the 1980s high shoreline surveyed at 1836 locations on Antelope Island and 608 locations at 10 places around Great Salt Lake document how coastal processes, particularly variability of storm-wave energy, cause significant variability in shoreline elevation. Shoreline evidence of the 1980s high shoreline on Antelope Island varied from 0.2 m below still-water lake level in vegetated areas to as much as 3.7 m above still-water lake level where headland topography focused wave energy. One-third of the surveyed elevations of shoreline evidence were over a meter above still-water elevation.

Atwood et al. (2016) shared the following insights of Great Salt Lake in the context of Lake Bonneville:

- Wind waves create the shorelines of Great Salt Lake: no wind, no waves, and no geomorphic work.

- For Great Salt Lake, greater shoreline superelevation (the difference between the elevation of the still-water lake surface and the elevation of shoreline evidence) correlates with longer fetch, windward versus leeward shores, shores that face into strong winds from the west and northwest, steeper shorezone slope, erosional coastal landforms, and shorezones devoid of vegetation.
- Patterns of shoreline superelevation along shores of Great Salt Lake can be used to infer storm-wind direction because Great Salt Lake is a fetch-limited lake, in contrast to fetch-dominated lakes, such as Lake Michigan or Lake Erie.

Atwood et al. (2016) cautioned that elevation variability for shorelines in shallow, closed-basin lakes is the rule, not the exception. One should be skeptical of consistent elevations reported for a time-defined post-Bonneville shoreline and be wary of attempts to correlate shoreline evidence based solely on elevation. Atwood (2006) reported on only one shoreline, the uniquely correlatable shoreline of the 1986–1987 highstand of Great Salt Lake, which had been monitored at its still-water elevation 1284 m (4212 ft), and which overprinted the shoreline evidence of the 1870s.

Miller et al. (2005)

Field work at Locomotive Springs (Fig. 1) has yielded information about the timing and elevation limits of late Holocene fluctuations of Great Salt Lake. At Locomotive Springs, organic-rich wetland deposits are interbedded with thin sand beds that contain ooids precipitated in the lake. By dating the interbedded organic beds, some idea can be obtained of the timing of Great Salt Lake incursions into the wetlands, but the elevations of the lake incursions cannot be determined precisely. Miller et al. (2005) reported some of this information at a meeting of the Geological Society of America, but the work has not been formally published.

At Locomotive Springs, a poorly defined barrier at about 1286 m (4220 ft) was occupied by Great Salt Lake as many as four times between 4700 and 2100 yr BP. Great Salt Lake inundated the wetland at elevations close to 1285.6 m (4218 ft) twice between 2100 and 1400 yr BP, and twice after 1400 yr BP. Based on this evidence, Great Salt Lake has not risen as high as 1286 m (4220 ft) during the past 500 years.

Hylland et al. (2012, 2014)

Hylland et al. (2012, 2014) studied the stratigraphy of Lake Bonneville and post-Lake Bonneville sediments exposed in fault trenches about 2 m deep at Baileys Lake (Fig. 1), with the primary purpose of determining the faulting history. Of interest in the context of this chapter is that no lacustrine sediments were found stratigraphically above sediments of Gilbert-episode age, higher than ~1286 m (4220 ft). Although it is known that Great Salt Lake rose to 1289 m (4230 ft) after the Gilbert episode (Oviatt et al. 2015) (see discussion below), that lake transgression did not leave a stratigraphic record at the Baileys Lake site.

Oviatt et al. (2015)

A double lacustrine barrier at Locomotive Springs at 1289 m (4230 ft) formed sometime after about 11,000 yr BP (Oviatt et al. 2015). The double barriers at this location may have been deposited very rapidly when Great Salt Lake rose from some lower elevation. The barriers contain gravel- and sand-size fragments of hard calcium carbonate, many in the form of flat or curved irregular plates (referred to informally as "chips"), similar in appearance to chips that form in the shallow offshore sediments of modern Great Salt Lake. Clasts of basalt are uncommon in the deposits; the primary source of basalt clasts is the basaltic bedrock, to which the barriers are connected on their western ends (Fig. 4 in Oviatt et al. 2015). If spit progradations were the dominant process of barrier-spit formation, basaltic clasts would dominate the deposit. Instead, the barrier composition suggests the material came from the lake bottom offshore. Therefore, the double barrier was formed not by longshore transport of sediment and spit progradation, but by sediment swept up onto the beach face and barrier crests by waves at the margin of the transgressing lake. Although no information is available on the length of time necessary for the barriers to form, it is possible they formed very quickly, the accumulation rate depending on the number of storms that occurred in months or years.

A limit can be placed on the age of double-barrier formation. Charcoal in dark-gray sandy mud beneath the barrier deposits has an age of about 11,000 yr BP (Fig. 4, Oviatt et al. 2015). Once charcoal has formed and is buried, it can persist for millions of years (Herring 1985), and therefore the barriers could be just slightly younger than, or much younger than, the charcoal, and the precise age of the barriers is not known.

From the end of the Gilbert episode (~11,500 yr BP) until about 10,000 yr BP, sedimentation differed in Great Salt Lake from that of most of postGilbert time (Oviatt et al. 2015). Flat, finely laminated mud that contains brine-shrimp cysts (egg cases) was deposited at that time, in contrast to the ripple-laminated, brine-shrimp fecal-pellet mud that was deposited over most of the lake floor after 10,000 yr BP. Great Salt Lake had to have been low to be salty enough to support brine shrimp, but the undisturbed laminations, showing that wave energy did not reach the bottom of the shallow lake, suggest that the lake was stratified at this time. Oviatt et al. (2015) suggested that enhanced spring discharge on the floor of the Bonneville basin may have led to a freshwater cap on the lake, and thus to stratification.

Thompson et al. (2016)

Thompson et al. (2016) summarized some of the work on cores taken from Great Salt Lake during the 1990s. The primary emphasis in that chapter was on the pollen in Lake Bonneville and Great Salt Lake sediments. Thompson et al. (2016) did not find evidence for extended periods during post-Bonneville time when Great Salt Lake was extremely high or extremely low. See below the discussion of the oxygen-isotope and total inorganic carbon (TIC) records from core GSL96+.

Relationship Between Holocene Paleoclimate and Great Salt Lake

Several lines of evidence suggest that there were major changes in climate in the eastern Great Basin during the Holocene (Currey and James 1982; Grayson 2011; Steponitis et al. 2015; Thompson et al. 2016). The overall record of paleoclimatic variability and change is complicated, with perhaps different changes occurring across space and elevation. This is a multiproxy story, with different sources of data having variable sensitivities to diverse aspects of climate variability and change, different response times, and different resilience to change. The most secure interpretation may be that the early and middle Holocenes were warmer and/or drier than the late Holocene (and/or that enhanced summer insolation during these periods had direct effects on the indicators).

Although climate is a direct control on the water budget of closed-basin lakes, it is unclear how the interpreted regional paleoclimatic records relate to changes in the elevation of Great Salt Lake. Despite decades of a widely held belief in the structure and timing of Holocene climate changes in western North America (much of which has been the basis of the interpreted record of Great Salt Lake elevations), no clear-cut way exists to relate these complex climate changes to what was happening in the lake. It is possible that average lake levels were slightly higher during relatively wet and cool times during the Holocene and that average lake levels were slightly lower during relatively dry and warm times, but those generalizations are poorly supported by data from shorelines and sediments. As the historical record of lake-level changes shows (Fig. 3), the maximum lake level in any particular year is strongly influenced by whatever the maximum lake level was during the previous year or set of years. The effect of this on the geologic record of prehistoric lake-level changes is unknown. Based on our decades of work on these issues, some of which is summarized in this chapter, we think that whether or not Holocene climate change is well-documented in the Great Salt Lake drainage basin, it is not a good idea to infer lake level from knowledge of climate change.

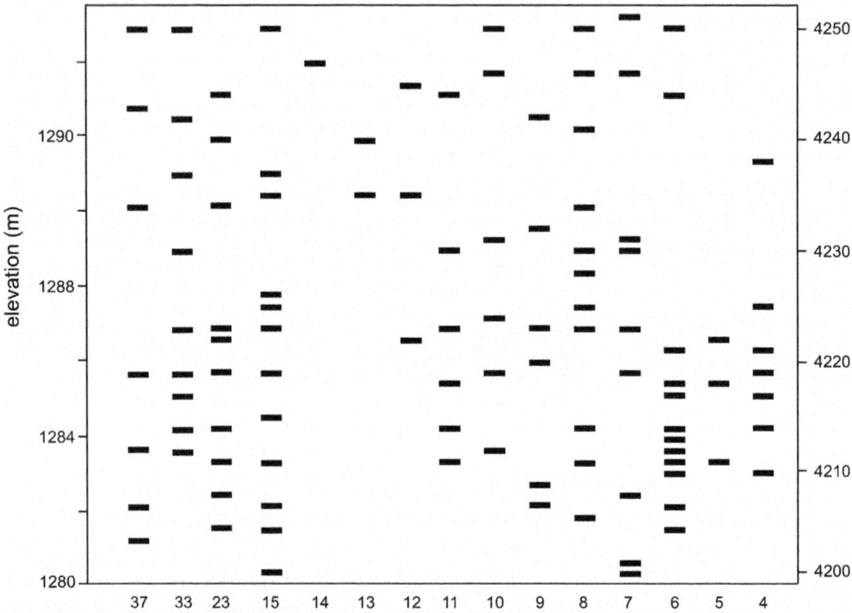

Fig. 6 Elevations of gravel barriers at sites around Great Salt Lake. Data from Murchison (1989). Numbers refer to locations shown on the map in Fig. 1; locations are named in Table 3 (after Murchison 1989). This figure is similar to Fig. 3 of Oviatt et al. (2015). Note that no single shoreline can be traced everywhere around Great Salt Lake at a constant elevation, including at or near the elevation of the "Holocene high" of Currey (1990) (1286.7 m; 4221 ft)

Early Twenty-first Century Interpretations of Great Salt Lake History

Shorelines

Shorelines around Great Salt Lake have provided ambiguous interpretations of post-Bonneville lake-level changes. It is not only difficult to date the shorelines but also they cannot be traced or mapped at consistent elevations around the lake. Figure 6 shows a plot of the elevations of major gravel barriers at a number of localities around Great Salt Lake (see Fig. 1; Table 3); the lack of consistency around the lake is obvious. To our knowledge, no Great Salt Lake shorelines are mappable for more than a few kilometers.

Sediments

The geochemistry of sediments in core GSL96+ (Thompson et al. 2016) is useful for deciphering some of the post-Lake Bonneville history of Great Salt Lake (Fig. 7). In general, the percentage of total inorganic carbon (TIC), a measure of the

Table 3 List of locations numbered in Figs. 1 and 6 (information from Murchison 1989)

Number on the map (Fig. 1)	Location name
4	Promontory east
5	Promontory Point
6	Rozel Point
7	Horseshoe Bay
8	Coyote Bay
9	Desolation Bay
10	Black Mountain Bay
11	Windmill Bay
12	Peplin barrier
13	Hogup barrier
14	Big Wash barrier
15	Fingerpoint spit
23	Stansbury Island west
33	Unicorn Point, Antelope Island
37	White Rock Bay, Antelope Island

percentage of carbonate minerals (calcite, aragonite, and dolomite) in the lacustrine mud, increases with a higher concentration of dissolved solids in the lake water, which is inversely correlated with lake volume and level (the percentage of calcium carbonate in the sample can be estimated by multiplying the percentage of TIC by 8.3). $\delta^{18}O$ in the carbonate minerals in the mud is heavier when lake level is lower (Grey and Bennett 1972), partly due to isotopic fractionation during evaporation (Jones et al. 2009). Using these generalizations, TIC and $\delta^{18}O$ in GSL96+[4] (Fig. 7) can be interpreted as indicating generally when periods of relatively high-lake levels and relatively low lake levels occurred during post-Bonneville time. Great Salt Lake may have experienced higher levels between about 8000 and 6000 yr BP (~550–400 cm), and between 3000 and 2000 yr BP (~300–200 cm); it may have experienced generally lower levels between about 11,500 and 8000 yr BP (~700–600 cm) and between 6000 and 3000 yr BP (~400–300 cm).

Evidence of the Gilbert episode (very low values of TIC), when Great Salt Lake attained elevations higher than during any other post-Bonneville transgression, occurs in the core at 659 cm. The Gilbert-episode lake was the highest and presumably freshest high-lake event in Great Salt Lake history and the TIC record seems to show this, but the event is not apparent in the $\delta^{18}O$ record. The cause of this discrepancy is unknown but may be related to the probable rapidity of the rise during the Gilbert-episode.

The TIC curve, which is based on analytical data from contiguous 1-cm samples, shows a few spikes (that is, lower values), which may represent short-lived episodes of high lakes at ~10,300, ~7600, and ~2200 yr (600, 530, and 230 cm, respectively,

[4]TIC was measured at Kansas State University using a modified Chittick apparatus, following procedures described in Machette (1986). Oxygen isotopes in samples of carbonate mud were analyzed by An Liu at the University of Minnesota and David Dettman at the University of Arizona. Prior to isotopic analyses, grains of sand size or larger were removed from the mud.

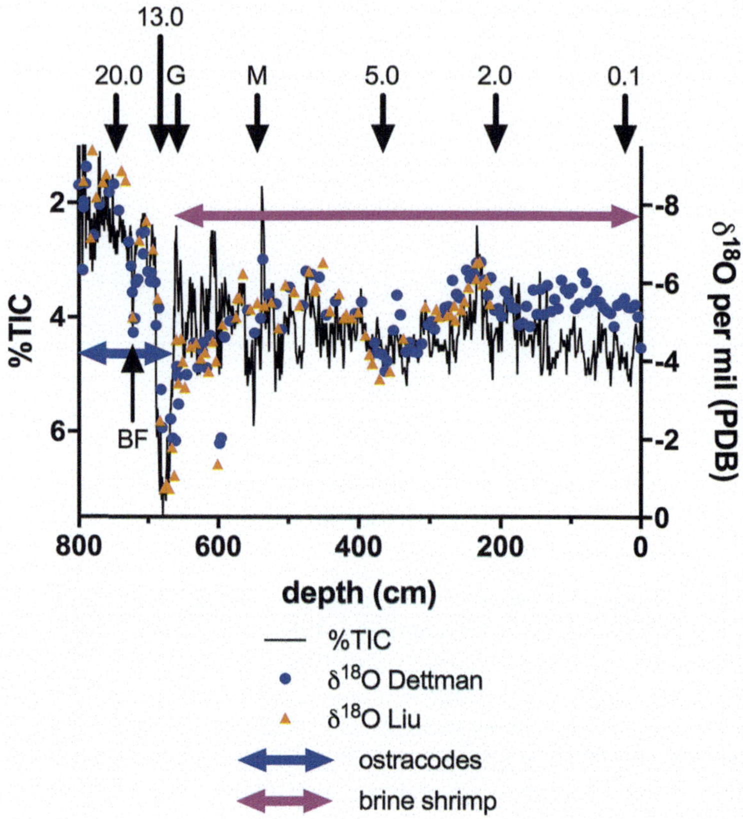

Fig. 7 Total inorganic carbon (%TIC) and oxygen isotopes ($\delta^{18}O$ per mil_{PDB}) in lacustrine sediments in cores from Great Salt Lake (GSL96+; Fig. 1) (Thompson et al. 2016). Neither the $\delta^{18}O$ data nor the TIC curve is a lake-level curve, but both can be interpreted as indicating relative lake-level changes. If reworking of older sediment is an active process, both sets of data may be "smoothed" by an unknown amount. Where %TIC is low, lake water would have been relatively diluted and lake level was relatively high; and where %TIC is high, the water would have been relatively concentrated in dissolved solids and lake level was relatively low. Relatively light isotopes represent periods of low-salinity water and higher lake levels, and heavier isotopes represent high salinity, greater evaporation, and lower lake levels. Note that the vertical scales are inverted so that shifts in the values of TIC and $\delta^{18}O$ mimic change in lake level, but the relationships between TIC or$\delta^{18}O$ and lake level are unlikely to be one-to-one. Lake Bonneville reached its highest level immediately prior to the Bonneville flood (BF) (at approximately 18,000 yr BP). In GSL96+, samples were taken contiguously at a spacing of 1 cm between the top of the core and a depth of about 590 cm, and every 0.5 cm below that. TIC results are available for each of these contiguous samples so the data points are connected with a solid line. Oxygen-isotope results are not connected by a solid line because they are not from contiguous samples. Ages are given for a few depths in the core (marked by black arrows); numbers are approximate ages in thousands of years BP based on age-model results (Thompson et al. 2016); M Mazama ash (7630 yr BP; Zdanowicz et al. 1999), G peak of Gilbert episode (~11,600 yr BP; Oviatt 2014), BF Bonneville flood (~18,000 yr BP; Oviatt 2015); see Thompson et al. (2016) for more information on the chronology of the core

not including the Gilbert episode). In GSL96+, the long-term sedimentation rate for the middle to late Holocene is 0.7 mm/yr, and the spikes in TIC are represented by one to six samples, so the spikes represent time periods between less than a decade and 85 years (it is unknown how much of these time-span estimates are influenced by sediment reworking). Within the spikes, TIC varies by 1–2 percent. The spike centered near 10,300 yr (600 cm) could have formed at the time when the double barrier at Locomotive Springs (1289 m; 4230 ft) was deposited, although we do not know of direct evidence for this (two other spikes, at 530 and 230 cm, are also possible correlatives with the Locomotive Springs double barrier because they are younger than 11,000 yr BP). In the case of the TIC spike at 600 cm (~10,300 yr), the oxygen isotopes do not support the high-lake interpretation. $\delta^{18}O$ becomes lighter in the TIC spikes at 530 and 230 cm (~7600 and ~2200 yr BP) (Fig. 7).

It is important to realize that Fig. 7 is not a lake-level curve, but it is useful for formulating hypotheses about periods when Great Salt Lake may have experienced relatively high and low levels. The key word here is "relatively" because Great Salt Lake probably did not rise or fall beyond the elevation limits shown in Fig. 8, with rare exceptions, such as when it reached the double barrier at Locomotive Springs. Distinctive sedimentological records of persistent high and low lakes, if they ever existed, are not preserved. Extreme highs and lows were probably short-lived (that is, they probably lasted for only a few years; based on analogy with modern lake behavior, year-to-year variation should be expected). Major unconformities that might indicate prolonged lake-level lows, and thick intervals of sediment clearly deposited in deep and fresh or brackish water (characterized by high percentages of calcite instead of aragonite, by very light oxygen isotopes, and/or possibly with ostracodes instead of brine shrimp), have not been observed.

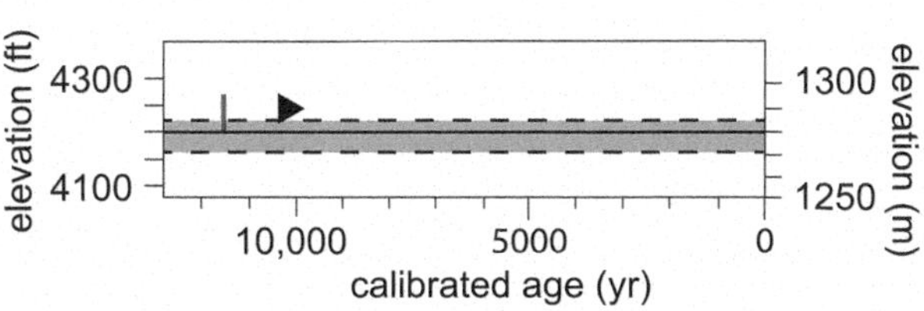

Fig. 8 Summary of current knowledge of Great Salt Lake fluctuations. The precise history of lake-level changes is not known. The gray bar bounded by dashed lines represents the elevation envelope within which most fluctuations have occurred. The upper elevation limit is placed at 1285 m (4217 ft), the elevation of the Desert threshold; the lower elevation limit is placed at 1271 m (4170 ft), the elevation of the modern floor of Great Salt Lake in Gilbert Bay. The horizontal solid black line is drawn at an elevation of 1280 m (4200 ft), the approximate modern average elevation of Great Salt Lake. Superelevation of the lake surface by winds, which is variable from place to place, is not included in the figure. Shorelines shown in Fig. 6 higher than 1285 m (4217 ft) have not been dated and are likely to be related to regressive-phase Lake Bonneville, or may have gravel cores that predate Lake Bonneville. Shoreline evidence of post-Bonneville Great Salt Lake ranges between 2 m above to 0.2 m below the still-water level. The two known exceptions to the elevation limits marked by the dashed lines were the Gilbert-episode lake and the rise to ~1289 m (marked by the black arrow)

The Currey Zone

Shorelines do exist around Great Salt Lake at elevations near 1285 m (4217 ft), both higher and lower than this elevation, but they are not continuous and mappable. Nevertheless, the narrow elevation interval between the historical high (1284 m; 4212 ft) and the Desert threshold (1285 m; 4217 ft) (these are still-water elevations) is of interest because most of the high-lake events during post-Bonneville time probably culminated in this zone. In this chapter, we refer to the zone as the "Currey zone" in honor of Don Currey because he and his students expended so much effort in studying it in various ways. The upper elevation limit of the Currey zone is taken to be the Desert threshold because that seems to have been an important control on late Holocene lake rises due to the expansion and stabilization process.

Summary of Great Salt Lake Fluctuations

Previous reconstructions used smooth solid lines to show interpreted lake-level changes during post-Bonneville time (e.g., Fig. 4c, d). Regarding the lake-level rise referred to as the "late-prehistoric high" (Fig. 4c), Currey et al. (1984) thought Great Salt Lake might have "… persisted near that level for at least 30 years." However, as we emphasize in this chapter, it is more likely that, as a closed-basin lake, fluctuations of Great Salt Lake were frequent and rapid, similar to those observed during the past 150 yr when lake-level records have been maintained (Fig. 3). If it were possible to completely and accurately reconstruct the behavior of Great Salt Lake, a detailed record would appear "jagged" rather than smooth. Since the late 1800s, Great Salt Lake has spent most of its time at relatively low elevations (probably near its historical average, 1280 m; 4200 ft), punctuated by a few short-lived high-lake and low-lake events. This historical pattern is probably typical of the lake's long-term behavior, which would promote almost continuous reworking (and smoothing) of both the shoreline and sediment records.

Great Salt Lake fluctuations are summarized in Fig. 8. Of note in regard to this diagram is that most lake-level rises (that is, still-water elevations) were below 1285 m (4217 ft), and most lake-level falls were above 1271 m (4170 ft); these elevations are shown in the figure as dashed lines that bound a gray-shaded area. The lower limit of 1271 m (4270 ft) is the approximate elevation of the modern lake floor in Gilbert Bay (the south arm) –we choose this elevation because no evidence has been observed of Great Salt Lake reaching elevations the same as or lower than this elevation. Any exceptions to these observations, either highs or lows, were short-lived and did not leave a recognizable sedimentary record. The upper limit of 1285 m (4217 ft) is the elevation of the Desert threshold.

Some general observations about Great Salt Lake are (1) no geomorphic or stratigraphic evidence has been observed of Great Salt Lake in the Great Salt Lake Desert (except for shorezone deposits of the Gilbert episode lake; Oviatt et al.

2018), even though we know Great Salt Lake rose to 1289 m, almost 4 m higher than the Desert threshold, sometime after 11,000 yr BP; (2) As defined here, the Currey zone, between about 1284 m (4212 ft) and 1285 m (4217 ft), represents the vertical range of the upper limits of late Holocene lake-level variations. The upper elevation limit of shorelines in the Currey zone varies from place to place because of spatial variability in coastal processes; and (3) Something similar to our modern view of Great Salt Lake, a shallow, hypersaline puddle that covers a huge area (about 4000 km^2), was probably pervasive throughout post-Bonneville time, with short-lived events of high- and low-lake levels punctuating the general trend.

Future Work and Conclusions

Continued research is needed to precisely decipher Great Salt Lake history. Some future tasks that have potential to clarify and build on what is known are: (1) to search for a location in Great Salt Lake where a high-resolution sedimentary record of lake-level changes during post-Bonneville time is preserved – such a record may not exist in Gilbert Bay (the south arm, south of the railroad causeway), but Gunnison Bay (the north arm) may not have been investigated with this objective in mind; (2) to develop a quantitative estimate of the relationship between lake level and sediment geochemistry – such a quantitative approach might allow geochemical and sedimentological studies to provide better estimates of the prehistorical behavior of Great Salt Lake; (3) to quantify the rate of reworking of older sediment, its variability, and its effect on the geochemistry and ages of Great Salt Lake bottom muds; (4) to develop procedures for mapping shorelines, or perhaps groups of shorelines; (5) to search for ways to date shorelines reliably and reproducibly; and (6) to test the hypothesis that isostatic rebound was complete prior to the Gilbert episode. Other areas of potential future research would include investigations of the effects of nonclimatic processes on lake level, such as tectonics, volcanism, and nonclimatic hydrologic changes. New research will use technology and perspectives that have evolved from older work. Some new technology, tools, and research perspectives that may be promising in future work include lidar, cosmogenic dating, luminescence dating, and studies of the diverse organisms that live in or depend on the lake.

In reviewing previous publications and in conducting our own work on Great Salt Lake, we have realized that researchers assumed a smooth record of century to millennial scale changes in lake level with clearly delimited episodes of higher and lower lake levels that are correlated with regional paleoclimate trends. The hydrology and geomorphology of Great Salt Lake and its basin mitigate against slow and continuous changes. Instead, high-frequency "labile" fluctuations, such as those seen in the historical period, are more likely. We suggest an approach to Great Salt Lake reconstructions that is more evidence based than has been employed in previous efforts.

Acknowledgments Although not everyone we have interacted with about Great Salt Lake can be named here, we acknowledge the help and inspiration of the following people: Ted Arnow, Rob Baskin, Bonnie Baxter, Don Currey, David Dettman, Dave Dinter, Holly Godsey, Don Grayson, Wally Gwynn, Paul Jewell, Kerry Kelts, Blair Jones, An Liu, Don Mabey, David Madsen, Jack McGeehin, Dave Miller, Vicki Pedone, and Doyle Stephens. Much of what we learned about Great Salt Lake in the 1980s and 1990s was inspired by the professionals from the Utah Geological Survey and the U.S. Geological Survey who slogged it out on the lake. We are very grateful to Marith Reheis and William Elliott for their helpful review comments and suggestions.

References

Arnow, T., & Stephens, D. (1990). *Hydrologic characteristics of the Great Salt Lake, Utah: 1847–1986*. U.S. Geological Survey Water-Supply Paper 2332, 32 p.

Atwood, G. (1986). From the Director's Desk. *Utah Geological and Mineral Survey, Survey Notes, 20*(2), 15–16.

Atwood, G. (1994). Geomorphology applied to flooding problems of closed-basin lakes ... specifically Great Salt Lake, Utah. *Geomorphology, 10*, 197–219.

Atwood, G. (2006). *Shoreline superelevation: Evidence of coastal processes of Great Salt Lake* (231 p). Utah: Utah Geological Survey Miscellaneous Publication 06-9.

Atwood, G., & Mabey, D. R. (2000). Shorelines of Antelope Island as evidence of fluctuations of the level of Great Salt Lake. In J. K. King & G. C. Willis (Eds.), *The geology of Antelope Island, Davis County* (pp. 85–97). Utah: Utah Geological Survey Miscellaneous Publication 00-1.

Atwood, G., Wambeam, T. J., & Anderson, N. J. (2016). The present as a key to the past: Paleoshoreline correlation insights from Great Salt Lake. In C. G. Oviatt & J. F. Shroder (Eds.), Lake Bonneville: A scientific update (Developments in Earth Surface Processes 20) (pp. 1–27). Amsterdam: Elsevier.

Bacon, C. R. (1983). Eruptive history of Mount Mazama and Crater Lake caldera, Cascade Range, USA. *Journal of Volcanology and Geothermal Research, 18*, 57–115.

Balch, D. P., Cohen, A. S., Schnurrenberger, D. W., Haskell, B. J., Valero Garces, B. L., Beck, J. W., Cheng, H., & Edwards, R. L. (2005). Ecosystem and paleohydrological response to Quaternary climate change in the Bonneville basin, Utah. *Palaeogeography, Palaeoclimatology, Palaeoecology, 221*, 99–122.

Baskin, R. L. (2014). *Occurrence and spatial distribution of microbial bioherms in Great Salt Lake, Utah* (Ph.D. thesis). University of Utah, Salt Lake City, UT, 190 p.

Baxter, B. K. (2018). *Great Salt Lake microbiology: A historical perspective*. International Microbiology.published online 04 June 2018. https://doi.org/10.1007/s10123-018-0008-z.

Baxter, B. K., & Butler, J. K. (Eds.). (2020). *Great Salt Lake biology: A terminal lake in a time of change* (527 p). Cham: Springer.

Baxter, B., Litchfield, C., Sowers, K., Griffith, J., Dassarma, P., & Dassarma, S. (2005). Microbial diversity of Great Salt Lake. In N. Gunde-Cimerman, A. Oren, & A. Plemenitaš (Eds.), *Adaptation to life at high-salt concentrations in Archaea, Bacteria, and Eukarya: Cellular origin, life in extreme habitats and astrobiology* (pp. 9–25). Netherlands: Springer.

Baxter, B., Atwood, G., Frantz, C., VandenBerg, M., & Butler, J. (2018). Day 1 —Antelope Island and Rozel Point. In *Great Salt Lake microbialites past and present. Field trip guidebook*. American Association of Petroleum Geologists, Annual Convention and Exhibition, May 18–19, 2018, p. 5–52.

Bills, B. G., Currey, D. R., & Marshall, G. A. (1994). Viscosity estimates for the crust and upper mantle from patterns of lacustrine shoreline deformation in the eastern Great Basin. *Journal of Geophysical Research, 99*(B11), 22,059–22,086.

Bouton, A., Vennin, E., Boulle, J., Pace, A., Bourillot, R., Thomazo, C., Brayard, A., Désaubliaux, G., Goslar, T., Yokoyama, Y., Dupraz, C., & Visscher, P. T. (2016). Linking the distribution of microbial deposits from the Great Salt Lake (Utah, USA) to tectonic and climatic processes. *Biogeosciences, 13*, 5511–5526.

Bouton, A., Venin, E., Amiotte-Suchet, P., Thomazo, C., Sizun, J.-P., Virgone, A., Gaucher, E. C., & Visscher, P. T. (2019). Prediction of the calcium carbonate budget in a sedimentary basin: A "source-to-sink" approach applied to Great Salt Lake, Utah, USA. *Basin Research, 32*, 1005. https://doi.org/10.1111/bre.12412.

Bowen, G. J., Nielson, K. E., & Eglinton, T. I. (2019). Multi-substrate radiocarbon data constrain detrital and reservoir effects in Holocene sediments of the Great Salt Lake, Utah. *Radiocarbon, 61*, 905. https://doi.org/10.1017/RDC.2019.62.

Broughton, J. M. (2000). The Homestead Cave ichthyofaunal. In D. B. Madsen (Ed.), *Late Quaternary paleoecology in the Bonneville basin* (Utah Geological Survey Bulletin 130) (pp. 103–121).

Broughton, J. M., & Smith, G. R. (2016). The fishes of Lake Bonneville: Implications for drainage history, biogeography, and lake levels. In C. G. Oviatt & J. F. Shroder (Eds.), Lake Bonneville: A scientific update (Developments in Earth Surface Processes 20) (pp. 292–351). Amsterdam: Elsevier.

Broughton, J. M., Madsen, D. B., & Quade, J. (2000). Fish remains from Homestead Cave and lake levels of the past 13,000 years in the Bonneville basin. *Quaternary Research, 53*, 392–401.

Chen, C. Y., & Maloof, A. C. (2017). Revisiting the deformed high shoreline of Lake Bonneville. *Quaternary Science Reviews, 159*, 169–189.

Clark, D. L., Oviatt, C. G., & Dinter, D. A. (2017) Interim geologic map of the Tooele 30' × 60' quadrangle, Tooele, Salt Lake, and Davis Counties, Utah. *Utah Geological Survey Open-File Report 669DM*, 43 p, scale 1:62,500.

Colman, S. M., Kelts, K. R., & Dinter, D. A. (2002). Depositional history and neotectonics in Great Salt Lake, Utah, from high-resolution seismic stratigraphy. *Sedimentary Geology, 148*, 61–78.

Crittenden, M. D. (1963). New data on the isostatic deformation of Lake Bonneville.U.S. *Geological Survey Professional Paper, 454-E*, E1–E31.

Currey, D. R. (1977). Late quaternary landforms of Great Salt Lake state park and vicinity, Davis County, Utah. In D. C. Greer (Ed.) *Perceptions of Utah: A field guide*. Prepared for the 1977 National Meetings of the Association of American Geographers, Salt Lake City, UT. Appendix, 85–89.

Currey, D. R. (1980). Coastal geomorphology of Great Salt Lake and vicinity. In J. W. Gwynn (Ed.), *Great Salt Lake: A scientific, historical, and economic overview* (Utah Geological and Mineral Survey Bulletin 116) (pp. 69–82).

Currey, D. R. (1982). Lake Bonneville: Selected features of relevance to neotectonic analysis. *U.S. Geological Survey Open File Report 82-1070*, 31 p.

Currey, D. R. (1987). Great Salt Lake levels: Holocene geomorphic development and hydrographic history. In *Third Annual Landsat Workshop*, NASA Laboratory for Terrestrial Physics, 127–132.

Currey, D. R. (1990). Quaternary paleolakes in the evolution of semidesert basins, with special emphasis on Lake Bonneville and the Great Basin, U.S.A. *Palaeogeography, Palaeoclimatology, Palaeoecology, 76*, 189–214.

Currey, D. R., & James, S. R. (1982). Paleoenvironments of the northeastern Great Basin and northeastern basin rim region: A review of geological and biological evidence. In D. B. Madsen & J. F. O'Connell (Eds.), *Man and environment in the Great Basin* (pp. 27–52). Washington, DC: Society for American Archaeology Papers no. 2.

Currey, D. R., Atwood, G., & Mabey, D. R. (1984) *Major levels of Great Salt Lake and Lake Bonneville: Utah Geological and Mineral Survey Map 73*, scale 1:750,000.

Currey, D. R., Berry, M. S., Douglass, G. E., Merola, J. A., Murchison, S. B., & Ridd, M. K. (1988). The highest Holocene stage of Great Salt Lake, Utah. *Geological Society of America Abstracts with Programs, 20*(6), 411.

DeRose, R. J., Wang, S.-Y., Buckley, B. M., & Bekker, M. F. (2014). Tree-ring reconstruction of the level of Great Salt Lake, USA. *The Holocene, 24*(7), 805–813.

Dinter, D. A., & Pechmann, J. C. (2004). Segmentation and Holocene displacement history of the Great Salt Lake fault, Utah. In W. R. Lund (Ed.) *Proceedings volume. Basin and Range Province Seismic hazards summit II*. Western States Seismic Policy Council, Reno-Sparks, Nevada, May 16-19, 2004, 1–5.

Dinter, D. A., & Pechmann, J. C. (2014). *Paleoseismology of the Promontory segment, east Great Salt Lake fault*. Report for U.S. Geological Survey Award Number 02HQGR0105, Department of Geology and Geophysics, University of Utah, Salt Lake City, UT.

Eardley, A. J. (1962). *Gypsum dunes and evaporite history of the Great Salt Lake Desert*. Utah Geological and Mineralogical Survey, Special Studies no. 2, 27 p.

Eardley, A. J., Gvosdetsky, V., & Marsell, R. E. (1957). Hydrology of Lake Bonneville and sediments and soils of its basin. *Geological Society of America Bulletin, 68*, 1141–1202.

Eardley, A. J., Shuey, R. T., Gvosdetsky, V., Nash, W. P., Picard, M. D., Grey, D. C., & Kukla, G. J. (1973). Lake cycles in the Bonneville basin, Utah. *Geological Society of America Bulletin, 84*, 211–216.

Gilbert, G. K. (1890). *Lake Bonneville*. U.S. Geological Survey Monograph 1, 438 p.

Grayson, D. K. (2011). *The Great Basin: A natural prehistory, revised and expanded edition* (418 p). Berkeley: University of California Press.

Grey, D. C., & Bennett, R. (1972). A preliminary limnological history of Great Salt Lake. In *The Great Salt Lake and Utah's water resources*. Proceedings First Annual Conference, Utah Section of the American Water Resources Association, Salt Lake City, Nov. 30, 1972, 3–18.

Hallet, D. J., Hills, L. V., & Clague, J. J. (1997). New accelerator mass spectrometry radiocarbon ages for the Mazama tephra layer from Kootenay National Park, British Columbia, Canada. *Canadian Journal of Earth Sciences, 34*, 1202–1209.

Herring, J. R. (1985). Charcoal flux into sediments of the north Pacific Ocean: The Cenozoic record of burning. In E. T. Sundquist & W. S. Broecker (Eds.), *The carbon cycle and atmospheric CO2: Natural variations from Archean to present* (American Geophysical Union Geophysical Monograph Series 32) (pp. 419–442). Washington, DC: American Geophysical Union.

Homewood, P., Mettraux, M., Schaegis, J.-C., VandenBerg, M., Atwood, G., & Foubert, A. (2018). Lakeside carbonates: Travertine and microbialites. In *Great Salt Lake microbialites past and present*. American Association of Petroleum Geologists, Annual Convention and Exhibition, May 18–19, 2018. Field trip guidebook, Day 2 —Lakeside, 53–138.

Hubbs, C. L., & Miller, R. R. (1948). The Zoological Evidence. In *The Great Basin, with emphasis on glacial and Postglacial Times*. Bulletin of the University of Utah 38, Biological Series 10(7), 17–144.

Hylland, M. D., DuRoss, C. B., McDonald, G. N., Olig, S. S., Oviatt, C. G., Mahan, S. A., Crone, A. J., & Personius, S. F. (2012). Basin-floor Lake Bonneville stratigraphic section as revealed in paleoseismic trenches at the Baileys Lake site, West Valley fault zone, Utah. In M. D. Hylland, & K. M. Harty (Eds.), *Selected topics in engineering and environmental geology in Utah*: Utah Geological Association Publication 41, 175–193.

Hylland, M. D, DuRoss, C. B., McDonald, G. N., Olig, S. S., Oviatt, C. G., Mahan, S. A., Crone, A. J., & Personius, S. F. (2014). Late Quaternary paleoseismology of the West Valley fault zone—Insights from the Baileys Lake trench site. In C. B. DuRoss, & M. D, Hylland (Eds.) *Evaluating surface faulting chronologies of graben-bounding faults in Salt Lake Valley, Utah—New paleoseismic data from the Salt Lake City segment of the Wasatch fault zone and the West Valley fault zone*. Paleoseismology of Utah, volume 24. Utah Geological Survey Special Studies 149, 45–76.

Jones, B. F., Naftz, D. L., Spencer, R. J., & Oviatt, C. G. (2009). Geochemical evolution of Great Salt Lake, Utah, USA. *Aquatic Geochemistry, 15*, 95–121.

Kaufman, D. S., Forman, S. L., & Bright, J. (2001). Age of the Cutler Dam Alloformation (late Pleistocene), Bonneville Basin, Utah. *Quaternary Research, 56*, 322–334.

Kay, P. A., & Diaz, H. F. (Eds.) (1985). *Problems of and prospects for predicting Great Salt Lake levels*. Papers from a conference held in Salt Lake City, Utah, March 26–28, 1985, University of Utah.

Lall, U., & Mann, M. (1995). The Great Salt Lake: A barometer of low-frequency climatic variability. *Water Resources Research, 31*, 2503–2515.

Lall, U., Moon, Y. I., Kwon, H. H., & Bosworth, K. (2006). Locally weighted polynomial regression: Parameter choice and application to forecasts of the Great Salt Lake. *Water Resources Research, 42*, W05422.

Lindsay, M. R., Anderson, C., Fox, N., Scofield, G., Allen, J., Anderson, E., Bueter, L., Poudel, S., Sutherland, K., Munson-McGee, J. H., Van Nostrand, J. D., Zhou, J., Spear, J. R., Baxter, B. K., Lageson, D. R., & Boyd, E. S. (2017). Microbialite response to an anthropogenic salinity gradient in Great Salt Lake, Utah. *Geobiology, 15*, 131–145.

Mabey, D. R. (1986). Notes on the historic high level of Great Salt Lake. *Utah Geological and Mineral Survey, Survey Notes* 20(2), 13–15.

Machette, M. (1986). Calcium and magnesium carbonates. In M. J. Singer & P. Janitzky (Eds.), *Field and laboratory procedures used in a soil chronosequence study* (U.S. Geological Survey Bulletin 1648) (pp. 30–33).

Madsen, D. B. (2000). *Late Quaternary paleoecology in the Bonneville Basin*. Utah Geological Survey Bulletin 130, 190 p.

Madsen, D. B., & O'Connell, J. F. (Eds.) (1982). *Man and environment in the Great Basin*. Society for American Archaeology Papers no. 2, Washington, DC

Madsen, D. B., Oviatt, C. G., Young, D. C., & Page, D. (2015). Old River Bed delta geomorphology and chronology. In D. B. Madsen, D. N. Schmitt, & D. Page (Eds.), *The Paleoarchaic occupation of the Old River Bed delta* (University of Utah Anthropological Papers Number 128) (pp. 30–60).

McKenzie, J. A., & Eberli, G. P. (1985). Late Holocene lake-level fluctuations of the Great Salt Lake (Utah) as deduced form oxygen-isotope and carbonate contents of cored sediments. In P. A. Kay, & H. F. Diaz (Eds.), *Problems of and prospects for predicting Great Salt Lake levels*. Papers from a conference held in Salt Lake City, Utah, March 26–28, 1985, University of Utah, 25–39.

McKenzie, J. A., & Eberli, G. P. (1987). Indications for abrupt Holocene climatic change: Late Holocene oxygen isotope stratigraphy of the Great Salt Lake, Utah. In W. H. Berger & L. D. Labeyrei (Eds.), *Abrupt climate change* (pp. 127–136). Washington, DC: Reidel Publishing Company.

McMahon, S., VanSmeerdijk, A., & Mcllroy, D. (2016).The origin and occurrence of subaqueous sedimentary cracks. In A. T. Brasier, D. Mcllroy, & N. McLoughlin (Eds.) *Earth system evolution and early life: A celebration of the work of Martin Brasier*. Geological Society, London, Special Publications, 448, https://doi.org/10.1144/SP448.15

Mehringer, P. J., Jr., Nash, W. P., & Fuller, R. H. (1971). A Holocene volcanic as from northwestern Utah. *Utah Academy Proceedings 48*(1), 46–51.

Mehringer, P. J. (1977). Great Basin late Quaternary environments and chronology. In D. D. Fowler (Ed.) *Models and Great Basin prehistory: A symposium*. University of Nevada. Desert Research Institute Publications in Social Science12, 113–167.

Mehringer, P. J. (1985). Late-Quaternary pollen records from the interior Pacific Northwest and northern Great Basin of the United States. In V. M. Bryant & R. G. Holloway (Eds.), *Pollen records of late-Quaternary North American sediments* (pp. 167–189). Dallas: American Association of Stratigraphic Palynologists Foundation (AAPG).

Merola, J. A., Currey, D. R., & Ridd, M. K. (1989). Thematic mapper laser profile resolution of Holocene lake limit, Great Salt Lake Desert, Utah. *Remote Sensing Environment, 28*, 233–244.

Miller, D. M. (2016). The Provo Shoreline of Lake Bonneville. In C. G. Oviatt & J. F. Shroder (Eds.), *Lake Bonneville: A scientific update* (Developments in Earth Surface Processes 20) (pp. 127–144). Amersham: Elsevier.

Miller, D. M., Oviatt, C. G., Dudash, S. L., & McGeehin, J. P. (2005). Late Holocene highstands of Great Salt Lake at Locomotive Springs, Utah. *Geological Society of America Abstracts with Programs, 37*(7), 335.

Miller, D. M., Oviatt, C. G., & McGeehin, J. P. (2013). Stratigraphy and chronology of Provo shoreline deposits and lake-level implications, late Pleistocene Lake Bonneville, eastern Great Basin, USA. *Boreas, 42*, 342–361.

Mohammed, I. N., & Tarboton, D. G. (2012). An examination of the sensitivity of the Great Salt Lake to changes in inputs. *Water Resources Research, 48*, W11511.

Moscardelli, L., Dooley, T., Dunlap, D., Jackson, M., & Wood, L. (2012). Deep-water polygonal fault systems as terrestrial analogs for large-scale Martian polygonal terrains. *Geological Society of America, 22*(8), 4–9.

Murchison, S. B. (1989). *Fluctuation history of Great Salt Lake, Utah, during the last 13,000 years* (Ph.D. dissertation). University of Utah, 137 p.

Murchison, S. B., & Mulvey, W. E. (2000). Late pleistocene and heolocene shoreline stratigraphy on antelope island, Davis county, Utah. In J. K. King, & G. C. Willis (Eds.) *Geology of Antelope Island. Salt Lake City*, Utah Geological Survey Miscellaneous Publication 001, 77–83.

Newell, D. L., Jensen, J. L., Frantz, C. M., & Vanden Berg, M. D. (2017). Great Salt Lake (Utah) Microbialite δ13C, δ18O, and δ15N record fluctuations in lake biogeochemistry since the late Pleistocene. *Geochemistry, Geophysics, Geosystems, 18*, 3631–3645.

Oliver, W., Fuller, C., Naftz, D. L., Johnson, W. P., & Diaz, X. (2009). Estimating selenium removal by sedimentation from the Great Salt Lake, Utah. *Applied Geochemistry, 24*, 936–949.

Oviatt, C. G. (1997). Lake Bonneville fluctuations and global climate change. *Geology, 25*, 155–158.

Oviatt, C. G. (2014). *The Gilbert episode in the Great Salt Lake basin, UT*. Geological Survey Miscellaneous Publication14–3, 20 p.

Oviatt, C. G. (2015). Chronology of Lake Bonneville, 30,000 to 10,000 yr B.P. *Quaternary Science Reviews, 110*, 166–171.

Oviatt, C. G., & Jewell, P. W. (2016). The Bonneville shoreline: Reconsidering Gilbert's interpretation. In C. G. Oviatt & J. F. Shroder (Eds.), Lake Bonneville: A scientific update (Developments in Earth Surface Processes 20) (pp. 88–104). Amersham: Elsevier.

Oviatt, C. G., McCoy, W. D., & Reider, R. G. (1987). Evidence for a shallow early or middle Wisconsin lake in the Bonneville basin Utah. *Quaternary Research, 27*, 248–262.

Oviatt, C. G., Currey, D. R., & Miller, D. M. (1990). Age and paleoclimatic significance of the Stansbury shoreline of Lake Bonneville, northeastern Great Basin. *Quaternary Research, 33*, 291–305.

Oviatt, C. G., Thompson, R. S., Kaufman, D. S., Bright, J., & Forester, R. M. (1999). Reinterpretation of the Burmester core, Bonneville basin, Utah. *Quaternary Research, 52*, 180–184.

Oviatt, C. G., Miller, D. M., McGeehin, J. P., Zachary, C., & Mahan, S. (2005). The Younger Dryas phase of Great Salt Lake, Utah, USA. *Palaeogeography, Palaeoclimatology, Palaeoecology, 219*(3–4), 263–284.

Oviatt, C. G., Madsen, D. B., Miller, D. M., Thompson, R. S., & McGeehin, J. P. (2015). Early Holocene Great Salt Lake, USA. *Quaternary Research, 84*, 57–68.

Oviatt, C. G., Pigati, J. S., Madsen, D. B., Rhode, D. E., & Bright, J. (2018). *Juke Box trench: A valuable archive of late Pleistocene and Holocene stratigraphy in the Bonneville basin*. Utah: Utah Geological Survey Miscellaneous Publication18-1, 26 p.

Pace, A., Bourillot, R., Bouton, A., Vennin, E., Galaup, S., Bundeleva, I., Patrier, P., Dupraz, C., Thomazo, C., Sansjofre, P., Yokoyama, Y., Franceschi, M., Anguy, Y., Pigot, L., Virgone, A., & Visscher, P. T. (2016). Microbial and diagenetic steps leading to the mineralisation of Great Salt Lake microbialites. *Scientific Reports, 6*. https://doi.org/10.1038/srep31495.

Paradis, O. P., Corsetti, F. A., Bardsley, A., Hammond, D. E., Xu, X., Stamps, B. W., Stevenson, B. S., Walker, J., & Berelson, W. M. (2017). Microbialite textures and chemical signatures in continental settings: Forging the link between the modern and ancient. *Geological Society of America Abstracts with Programs, 49*(6).

Pederson, J. L., Janecke, S. U., Reheis, M. C., Kaufman, D. S., & Oaks, R. Q., Jr. (2016). The Bear River's history and diversion: Constraints, unsolved problems, and implications for the Lake Bonneville record. In C. G. Oviatt & J. F. Shroder (Eds.), Lake Bonneville: A scientific update (Developments in Earth Surface Processes 20) (pp. 28–59). Amsterdam: Elsevier.

Pedone, V. A. (2002). Oxygen-isotope composition of Great Salt Lake, 1979 to 1996. In W. Gwynn (Ed.), *Great Salt Lake: An overview of change* (Special publication of the Utah Department of Natural Resources) (pp. 121–126). Salt Lake City: Utah Geological Survey.

Reimer, P. J., Bard, E., Bayliss, A., Beck, J. W., Blackwell, P. G., Ramsey, C. B., Grootes, P. M., Guilderson, T. P., Haflidason, H., Hajdas, I., Hatté, C., Heaton, T. J., Hoffmann, D. L., Hogg, A. G., Hughen, K. A., Kaiser, K. F., Kromer, B., Manning, S. W., Niu, M., Reimer, R. W.,

Richards, D. A., Scott, E. M., Southon, J. R., Staff, R. A., Turney, C. S. M., & van der Plicht, J. (2013). IntCal13 and Marine13 Radiocarbon Age Calibration Curves 0–50,000 Years cal BP. *Radiocarbon, 55*, 1869–1887.

Ross, D. S. (1973). *Holocene fluctuations of Great Salt Lake, with special reference to evidence from the eastern shore* (M.S. thesis). University of Utah, 110 p.

Rudy, R. C. (1973). *Holocene fluctuations of Great Salt Lake, with special reference to evidence from the western shore* (M.S. thesis). University of Utah, 76 p.

Simms, S. R., & Stuart, M. E. (2002). Ancient American Indian life in the Great Salt Lake wetlands: Archaeological and biological evidence. In W. Gwynn (Ed.), *Great Salt Lake: An overview of change* (Special publication of the Utah Department of Natural Resources) (pp. 71–83). Utah Geological Survey.

Sorenson, E. D., Hoven, H. M., & Neil, J. (2020). Great Salt Lake shorebirds, their habitats and food base. In B. K. Baxter & J. K. Butler (Eds.), *Great Salt Lake biology: A terminal lake in a time of change* (pp. 263–309). Cham: Springer.

Spencer, R. J., Baedecker, M. J., Eugster, H. P., Forester, R. M., Goldhaber, M. B., Jones, B. F., Kelts, K., McKenzie, J., Madsen, D. B., Rettig, S. L., Rubin, M., & Bowser, C. J. (1984). Great Salt Lake and precursors, Utah: The last 30,000 years. *Contributions to Mineralogy and Petrology, 86*, 321–334.

Steponitis, E., Andrews, A., McGee, D., Quade, J., Hsieh, Y., Broecker, W. S., Shuman, B. N., Burns, S. J., & Cheng, H. (2015). Mid-Holocene drying of the U.S. Great Basin recorded in Nevada speleothems. *Quaternary Science Reviews, 127*, 1–12.

Stokes, W. L. (1966). Introduction. In W. L. Stokes (Ed.), *Guidebook to the geology of Utah* (pp. 1–2). Utah: Geological Society, Guidebook number 20.

Street-Perrott, E. A., & Harrison, S. P. (1985). Lake levels and climate reconstruction. In A. D. Hecht (Ed.), *Paleoclimate Analysis and Modeling* (pp. 291–340). New York: Wiley.

Stuiver, M., & Reimer, P. J. (1993). Extended ^{14}C database and revised CALIB radiocarbon calibration program. *Radiocarbon, 35*, 215–230; CALIB7.0: http://calib.qub.ac.uk/calib/calib.html.

Thompson, R. S., Oviatt, C. G., Honke, J. S., & McGeehin, J. P. (2016). Late Quaternary changes in lakes, vegetation, and climate in the Bonneville basin reconstructed from sediment cores from Great Salt Lake. In C. G. Oviatt & J. F. Shroder (Eds.), Lake Bonneville: A scientific update (Developments in Earth Surface Processes 20) (pp. 221–291). Amsterdam: Elsevier.

Vanden Berg, M. D. (2019). Domes, rings, ridges, and polygons: Characteristics of microbialites from Utah's Great Salt Lake. *The Sedimentary Record, 17*, 4–10.

Venin, E., Bouton, A., Bourillot, R., Pace, A. L., Roche, A., Brayard, A., Thomazo, C., Virgone, A., Gaucher, E. D., Desaubliaux, G., & Vissher, P. T. (2018). The lacustrine microbial carbonate factory of the successive Lake Bonneville and Great Salt Lake, Utah, USA. *Sedimentology, 66*, 165. https://doi.org/10.1111/sed.12499.

Walker, M., Johnsen, S., Rasmussen, S. O., Popp, T., Steffensen, J.-P., Gibbard, P., Hoek, W., Lowe, J., Andrews, J., Björck, S., Cwynar, L. C., Hugen, K., Kershaw, P., Kromer, B., Litt, T., Lowe, D. J., Nakagawa, T., Newnham, R., & Schwander, J. (2009). Formal definition and dating of the GSSP (Global Stratotype Section and Point) for the base of the Holocene using the Greenland NGRIP ice core, and selected auxiliary records. *Journal of Quaternary Science, 24*(1), 3–17.

Walker, M. J. C., Berkelhammer, M., Björck, S., Cwynar, L. C., Fisher, D. A., Long, A. J., Lowe, J. J., Newnham, R. M., Rasmussen, S. O., & Weiss, H. (2012). Formal subdivision of the Holocene Series/Epoch: A Discussion Paper by a working group of INTIMATE (Integration of ice-core, marine and terrestrial records) and the Subcommission on Quaternary Stratigraphy (International Commission on Stratigraphy). *Journal of Quaternary Science, 27*(7), 649–659.

Wurtsbaugh, W. A., Miller, C., Null, S. E., DeRose, R. J., Wilcock, P., Hahnenberger, M., Howe, F., & Moore, J. (2017). Decline of the world's saline lakes. *Nature Geoscience, 10*, 816–821. https://doi.org/10.1038/NGEO3052.

Zdanowicz, C. M., Zielinski, G. A., & Germani, M. S. (1999). Mount Mazama eruption: Calendrical age verified and atmospheric impact assessed. *Geology, 27*(7), 621–624.

What's New About the Old Bonneville Basin? Fresh Insights About the Modern Limnogeology of Great Salt Lake

Kathleen Nicoll

Abstract Located in the Great Basin Desert within the North American continental interior, the Great Salt Lake (GSL) is a remnant of freshwater Palaeolake Bonneville, and today is the fourth largest perennial and closed basin lake in the world. More than a century of study on the Pleistocene megalake and its modern hypersaline environment (latitude 40.7° to 41.7°N, longitude 111.9° to 113.1°W) has significantly contributed to what we know about lake systems, sediments, and Quaternary climate change. Ongoing geolimnological work on GSL is lively and diverse; there is much progress, this review paper highlights select key themes of interest. Some recent studies aim to quantify hypersaline lake water properties and hydrodynamics in context of human activities (e.g., pressures from transportation, mining, development), as science-informed monitoring and managing GSL waters has become a greater priority for extractive industries, resource managers, environmentalists, and regulating officials. Copious research leverages new technologies and methods to identify and describe microbes living in the GSL and its brine, including halophiles, halobacteria, fungi, and viruses. The GSL's self-sustaining microbial ecology drives the high productivity of the lake, which has cascading global impacts for millions of migratory birds that stop at this oasis to feed. Furthermore, the question of what extremophiles live in each niche of the modern saline ecosystem in relation to salts and carbonate organosedimentary sedimentary systems is broadly relevant for understanding the deep rock record, the nature of early Earth evolution, and for astrobiological prospection as we aim to find life on other planets.

Keywords Great Salt Lake · Utah · Lake Bonneville · Great Basin · Paleoclimate · Holocene · Pleistocene

K. Nicoll (✉)
Department of Geography, University of Utah, Salt Lake City, UT, USA

M. R. Rosen et al. (eds.), *Limnogeology: Progress, Challenges and Opportunities*, Syntheses in Limnogeology, https://doi.org/10.1007/978-3-030-66576-0_9

Introduction

Clearly visible from space, the modern hypersaline Great Salt Lake (GSL) is the defining hydrological feature and a key ecosystem in the internally drained Great Basin in North America's largest desert (Grayson 2011). Since initial geographical and topographic surveys of the area were conducted by John Frémont (1845) and Captain Howard Stansbury (1855), the significance of the GSL watershed has been studied and documented. The geologist and geomorphologist G.K. Gilbert recognized the GSL as a relict lake within a larger lake basin, which he named Bonneville. Gilbert published a 438-page monograph describing evidence for regional lake cycles or fluctuations in water level as the freshwater Palaeolake Bonneville expanded during the Pleistocene, then transitioned to become the current hypersaline GSL (Gilbert 1890).

During the Last Glacial Maximum (LGM) ~23,000 to 19,000 BP, the GSL precursor Palaeolake Bonneville was a freshwater megalake that covered ~50,000 sq. km in western Utah and eastern Nevada (Fig. 1). At its highstand, Bonneville was the largest and deepest pluvial palaeolake in North America, comparable in morphology to the modern Lake Michigan. More than a century of research on Bonneville has focused on its world-class geomorphic and sedimentary records. Its deposits and fossils have informed interpretations of Quaternary hydroclimatic and terrestrial landscape change (e.g., Louderback and Rhode 2009; Lyle et al. 2012; Skorko et al. 2012; Reheis et al. 2014; Bekker et al. 2014; Hahnenberger and Nicoll 2014; Nicoll et al. 2014; Nicoll and Keen-Zebert 2016; Dunham et al. 2020, to name some recent contributions). Its environmental facies include evaporites, laminated marls, basinal muds, protodolomites, tufas, microbial mats, mounds, stromatolites, shorelines, spits, bay mouth bars, deltas, gullies, outburst channels, and playa lake features, including salt flats, patterned ground, and aeolian deposits (Figs. 2 and 3).

Given its lake level records dating back to 1848 (Fig. 4), abundant radiometric age control and scores of publications, it can be argued that the Bonneville Basin is the best-studied lake system in North America, if not the world. While studies continue on traditional themes of palaeolake reconstruction from lake levels (e.g., Oviatt et al. 2015; Oviatt and Shroder 2016), this chapter discusses some recent emerging research themes related to the GSL saline setting, and hemispheric-to-global importance. New methodologies – including modeling, molecular approaches to microbiology and community genomics – are being applied to better understand the modern GSL's limnology, biota and modern sedimentary characteristics, and the impact of ongoing climate change and human activities on its hydrodynamics and ecosystematics.

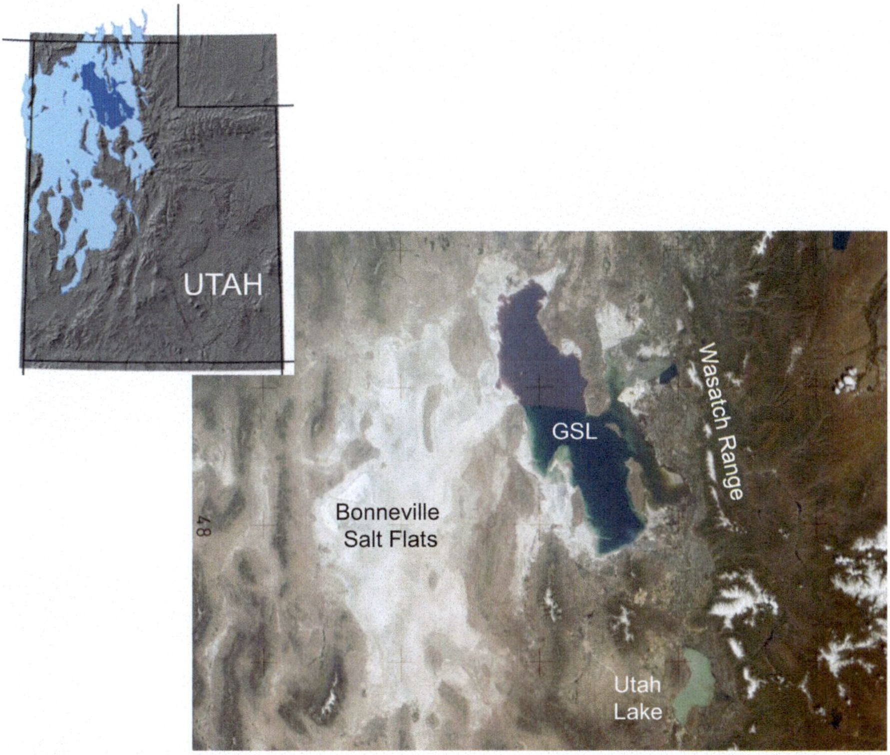

Fig. 1 In Utah, Palaeolake Bonneville during its Pleistocene highstand (light blue) and the modern hypersaline Great Salt Lake (darker blue). Annotated Skylab 2 Photo taken in 1973 by astronauts with a 70 mm camera shows regional features, including Utah Lake, a freshwater remnant of Palaeolake Bonneville. The "line" in the GSL marked by a color difference is due to the railroad causeway that restricts mixing of the waters (Gwynn 2002; Cannon and Cannon 2002); the North Arm is NaCl-saturated and can only support reddish halophytic bacteria. The North Arm contains 107–108 microbial cells ml − 1, a density that gives rise to the opacity and distinctive red color of these brines (Oren 2009). The rest of the lake receives freshwater input from melting snowpack and supports algae. (Orbital image via NASA: https://earthobservatory.nasa.gov/images/81112/view-of-the-great-salt-lake-from-skylab)

GSL Hydroclimatological Attributes

The modern GSL and its associated Bonneville Salt Flats cover portions of northern Utah, southern Idaho, and western Wyoming; this is the fourth largest terminal lake on Earth, and a premier example of a closed (i.e., endorheic), NaCl-dominant saline system. The GSL is an underfilled lake (sensu Carroll and Bohacs 1999); its water level and volume are nontidal and fluctuate as a balance of inflows and outflows. Evaporation is the dominant outflow, and depends on lake area and salinity, both of which are affected by the lake volume. The average lake volume is a balance of direct precipitation and streamflow inflows and removals by evaporation (Arnow

Fig. 2 Select sediments of Bonneville Basin and the GSL. (**a**) Pleistocene laminated marl deposits at Stockton (m section is now eroded); inset box depicted in (**b**) Stratigraphy close up of in situ fossils. (**c**) Stromatolite mounds exposed during 1964 GSL lowstand (UGS). (**d**) Carbonate cemented gravels deposited along wave-cut shoreline. (**e**) Shorelines on hillside in background; debris at 1987 highstand wrack line. (**f**) Spring carbonate microbial mound (tufa); modern salts along the GSL shoreline

and Stephens 1990). Because the GSL is quite shallow with average depth ~ 4–6 m, the water surface area is quite variable, ranging from 3000 to 6000 km^2.

The average GSL water residence time is ~5 years, which is also the time scale implied by the historical range of volume changes (active volume), compared with

Fig. 3 (**a**) Modern hypersaline GSL in salt works area east of Stansbury Island – the distinctive reddish water color is due to halophytic bacteria present in the water. Note the multiple Bonneville Shorelines that appear as ledges in the background. (**b**) Lake carbonates include oolite sands and carbonate encrusted cobbles at the shores of the GSL. (**c**) Orange-colored halophilic algae *Dunaliella salina*, NaCl (halite) salt, and brine fly carcasses along the GSL shore. (**d**) Dunes made of lake-derived oolites on Stansbury Island are stable enough to support sagebrush and a piñon pine-juniper forest

mean inflows (Mohammed and Tarboton 2011). The main variables include precipitation, evaporation, streamflow, groundwater, withdrawals, and return flow. GSL levels fluctuate over time scales of 5–20 years, and longer, and are influenced by meteorological variables (Moon et al. 2008). Over shorter timescales, there is inter-annual fluctuation of ~0.5 m (1–2 ft), with rising levels during winter and spring meltwater runoff (November to June), then falling during summer, when evaporation is high, and inflows are low (July to October). The annual lake low is typically in November. As the GSL volume fluctuates, salts are diluted or concentrated in the lake water, which affects surface salinity, which ranges by location from 5% to 28% (as reviewed by Mohammed and Tarboton 2011, 2012).

One GSL mass balance model (GSLMBM) used an elasticity measure, a ratio of the variability of streamflow, precipitation, evaporation, area, and salinity to the variability in historic volume changes, in order to examine the sensitivity of GSL volume to inputs and various factors internal to its system dynamics (Mohammed

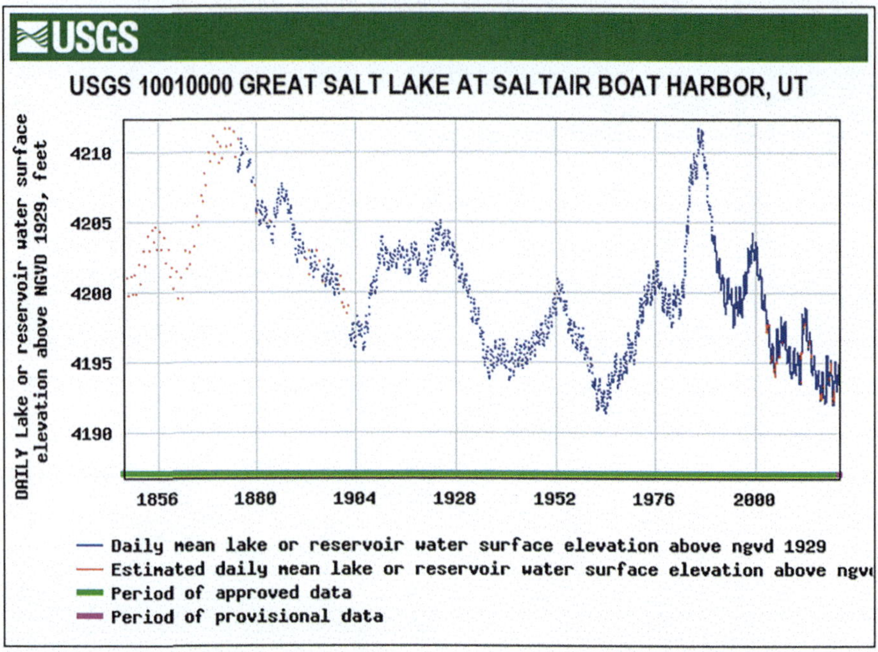

Fig. 4 GSL water surface elevation at USGS 10010000 Great Salt Lake at Saltair Boat Harbor, Utah. Latitude 40°43′53, Longitude 112°12′46" NAD27 Datum of gage: 4186.80 feet above NGVD29. The earliest level data point available is from 18 October 1847. USGS lake level measurements were first made in 1875; lake level values prior to this date were estimates based on observer reports. (Source: https://waterdata.usgs.gov/nwis/dv?cb_62614=on&format=gif_default&site_no=10010000&referred_module=sw&period=&begin_date=1847-10-18&end_date=2020-06-24)

and Tarboton 2012). These models demonstrated that the relations between these physical processes are often nonlinear, and the resulting complexities limit precise forecasts of the GSL level. High levels of uncertainty in modeling the water in the GSL underscores the need to better quantify the interactions between lake level, volume, area, evaporation, and salinity in response to driving inputs of precipitation, streamflow, and climate.

Models and hydro-forecasts are further complicated because GSL is compartmentalized by a railway causeway, creating two "Arms" – a North Arm, known as Gunnison Bay, and a South Arm, also called Gilbert Bay (Cannon and Cannon 2002; Gwynn 2012). Gilbert Bay receives 95% of the total freshwater input to the GSL; Gunnison Bay is in the west desert and receives only ~5% (Loving et al. 2000). The water salinity ranges from 50 to 170 ppt in Gilbert Bay and 150 to 270 ppt in Gunnison Bay (Utah Division of Forestry, Fire and State Lands 2012).

The railroad causeway-built structure significantly limits the exchange of water between the North and South Arms of the GSL, creating chemical stratification and a brine (Madison 1970; Gwynn 2002; Loving et al. 2000). Saltier, denser water from the North arm underflows into the south, creating an anoxic, sulfide-rich deep brine layer. Because this brine is used by local extractive industries, several water and salt balance models have been developed for the GSL (e.g., Gwynn 2002; Hahl and Handy 1969; Loving et al. 2000; Madison 1970; Waddell and Bolke 1973; Waddell and Fields 1977; Wold et al. 1997). These models have changed due to recent changes to the causeway design (Union Pacific Railroad 2013; see review in Freeman 2014). At a local scale, mineral extractive industries separate the lake waters by earthworks, salterns and diversions to intensify evaporation to precipitate mineral salts and recover metals.

The Dynamic Salt and Brine

Due to compartmentalization by built structures and the causeway cutting off the North Arm of the lake, GSL water salinity reaches near saturation at 150 $g \cdot L^{-1}$, up to ~10 times the average salinity of seawater. Due to this high salinity, trace element activity and total dissolved loads in the GSL are difficult to characterize using standard geochemical parameters (e.g., Domagalski et al. 1989; Shope and Angeroth 2015). Additionally, the GSL is monomicitic with a water column that is complicated by the presence of a chemocline that persists over annual to decadal time scales. There are periodic stratification displacement events due to wind (seiches) and/or exposure of the lake beds due to seasonal variations of water input and drought (Beisner et al. 2009). Below the chemocline of the GSL, there is a deep brine layer (referred to as the DBL, a monimolimnion) that accumulates high concentrations of nutrients, organic matter, total mercury (Hg; 59 ng L^{-1}), and methylmercury (MeHg; 33 ng L^{-1}) (Wurtsbaugh and Berry 1990; Jones and Wurtsbaugh 2014).

The GSL DBL has high salinity, ranging from 16.5% to 22.9%, and is anoxic; above the chemocline, the upper brine layer (UBL) has lower salinity (12.6% to 14.7%) and is oxic (Beisner et al. 2009). The dynamics of stratification are strongly impacted by hydroclimate, including weather variables such as temperature and precipitation, which may be reflected in lake inputs. Wind associated with storm frontal passages can force seiches that mix the water column, funnel currents through inlets and engineered structures, and initiate flow reversals over periods of 12 to 24 h, which complicates the lake hydrodynamics and its chemical assessments (Hahnenberger and Nicoll 2012; Freeman 2014).

In late 2013, the earthen railroad causeway restricting flow between the GSL North and South Arms was restructured. This offered Yang et al. (2020) an opportunity to observe the effects of engineered and hydrologic forcings on GSL density stratification, aquatic chemistry, and methylmercury. When the flow mixed North Arm and South Arm water, the meromictic endorheic GSL destratified, and the mixing effectively eliminated the elevated methylmercury (40–90 ng L − 1) in deep waters. When the causeway opened and North-to-South Arm flow was reestablished in 2016, Yang et al. (2020) assessed the geochemical conditions and drivers of density stratification, which support mercury methylation at depth in the GSL. Both the flow across the causeway reopening and natural hydrologic inputs (e.g., snowmelt) were observed to initiate meromixis (stratification). Returning to meromictic conditions drove anoxia at depth, and increased sulfide and methylmercury concentrations. However, shallow brine oxygen concentrations were lower under destratified conditions, as compared to meromictic conditions, and brief periods of anoxia and elevated sulfide and methylmercury occurred at depth even under destratified conditions, demonstrating the oxygen demand by underlying sediment organic matter. In deep water, mercury demethylation was observed only under destratified conditions when oxygen reached bottom waters. These observations of Yang et al. (2020) demonstrate that the deep brine layer acts as a cap that (1) prevents oxygen from the overlying mixolimnion GSL from coming into contact with sediment organic matter and (2) drives methylmercury accumulation in deep waters under meromictic (stratified) conditions.

Stratification displacement of the DBL may involve trace element movement within the water column due to changes in redox potential. Lab simulations conducted by Beisner et al. (2009) observed sediment resuspension over daily, weekly, and monthly time scales to understand the effect of stratification disruption that cause anoxic bottom sediments to be placed in contact with oxic water, and measured the associated effect of trace element desorption and/(or) dissolution. Only a small percentage (1%) of selenium (Se) associated with anoxic bottom sediments is periodically solubilized into the UBL, where it potentially can be incorporated into the biota utilizing the oxic part of the GSL water column.

Trace element movement across sediment–water interfaces and within the water column has implications for the conceptual model of mass balance, load, and cycling in the open waters of GSL (Naftz et al., 2011). Selenium mass balance is important because of its bioaccumulative toxic nature (Diaz et al. 2009; Ohlendorf et al. 2009). In the GSL, selenium is generally thought to originate from abiotic component inputs to the water and sediment and move "up" through the lower food chain where it is increasingly concentrated in tissues of the higher food web components like birds. Ohlendorf et al. (2009) synthesized results of various studies that assessed concentrations and effects of selenium in five species of GSL birds; measured selenium concentrations in water, seston (i.e., organisms and nonliving matter swimming or floating in the water), brine shrimp (*Artemia* sp.), and brine flies (*Ephydra* sp.); measured selenium loads entering GSL; and measured flux of selenium from atmosphere, water to sediment, and the food web. These scientific studies have informed a comprehensive conceptual model for GSL (Naftz et al., 2008) for establishing the site-specific standard for selenium as mandated by the Environmental Protection Agency (EPA).

Extreme Biota Are Ecologically Integral

The hypersalinity and complex, extreme geochemistry control what can live in the GSL. Overall, the biosystematics of highly diverse freshwater lakes are better known than those in saline systems (Javor 1989; Jellison et al., 2008). Emerging work on hypersaline habitats worldwide provides important contexts for studying organisms that are adapted to extreme environments (Torrentera and Dodson 2004). Salt lakes and marsh wetlands typically support a more limited number and variety of life forms and have comparatively simple community structure and food webs (Litchfield and Gillevet 2002; Baxter et al. 2005; Benison and Bowen 2006). Developing research on the GSL continues to illuminate its productivity and detail how salinity and nutrient availability influence the biodiversity within its waters and shorelines (for a review, see Baxter 2018) as well as the gene expression of halotolerant bacteria that dominate the North Arm (Almeida-Dalmet and Baxter 2020).

Life in the GSL has been a topic of interest for more than a century (e.g., Daines 1910; Frederick 1924, to name a couple early workers who described algae and bacteria), but attempts to fully identify and classify GSL microbiota were limited by available technologies. Biomass-rich microbial communities were known to be active in the GSL, even when NaCl concentrations approach saturation (Post 1977, 1980). Modern genomic techniques have helped resolve the microbial consortia present in the GSL and demonstrated the presence of many specialized life-forms in the brine, including fungi and viruses (Weimer et al. 2009 and see reviews by Baxter 2018; Baxter and Butler 2020). More than 100 genera of *Archaea* and *Bacteria* have been isolated from the NaCl-saturated North Arm of the GSL. A diversity of microorganisms, halophiles – salt-tolerant species with adaptations – and extremophiles – organisms that can tolerate anoxia and high radiation – have been identified in GSL by numerous research groups, including Tsai et al. (1995), Wainø et al. (2000), Ingvorsen and Brandt (2002), Kjeldsen et al. (2007), Roney et al. (2009), Beer et al. 2010; Møller et al. (2010); Pugin et al. (2012); Meuser et al., 2013; Barnes and Wurtsbaugh (2015); Boyd et al. (2017); Tazi et al. (2014); and Almeida-Dalmet and Baxter (2020).

Halotolerant organisms in GSL include *Dunaliella salina* (among other *Dunaliella* species) and *Artemia franciscana* (GBIF, 2019). *Artemia*, or brine shrimp, is the only species of macrozooplankton present in the GSL. *Artemia* is a specialized crustacean branchiopod (anostracan) that is able to live under harsh conditions, such as extreme salinity, temperature, frequent desiccation, anoxia, or predation (Dodson and Frey 2001). *Artemia* is globally dispersed and ubiquitous in hypersaline lakes, ephemeral desert ponds, and in coastal lagoons, salterns, pools, saltmarshes, and various temporary habitats.

The microbes present in the GSL are diverse in terms of their phylogeny and ecophysiological functions, and form sustainable ecosystems. Throughout the GSL, primary producers can support heterotrophic, and other trophic, networks without requiring outside inputs. Moreover, halophiles within the GSL's NaCl-saturated consortium have evolved; various culture-independent studies indicate the presence of genes that are collectively capable of completing the biogeochemical cycles that can enable the independent functioning of the microbial ecosystem (e.g., Allen and Banfield 2005; Jakobsen et al. 2006; Baumgartner et al. 2009; Lee et al. 2018; Baxter and Butler 2020).

The ecological interactions of halophilic microbes found in saline lake habitats like the GSL are not limited to other microorganisms; they extend to other trophic levels, and far out beyond the immediate lake locality. As an illustrative example, the microbial community at GSL and other inland salt lakes support a global and diverse avian ecology. *Dunaliella* and other microbial biomass support high densities of *Artemia* (brine shrimp) in salt marshes, salterns, and brine lakes (Clegg and Trotman 2002). The *Artemia* in turn sustains populations of aquatic insects and birds (Torrentera and Dodson 2004) including gulls (e.g., *Larus sp.*) and waders (American avocet, *Recurvirostra Americana*). The GSL feeds residents with a wide home range (e.g, white pelican *Pelecanus erythrorhynchos*) as well as migrating visitors (e.g., common goldeneye duck *Bucephala clangula* and Wilson's phalarope, *Phalaropus tricolor*); more than >35 million birds stop at GSL as they travel along the interhemispheric flyway seasonally (Aldrich and Paul 2002; Roberts 2013). In this manner, the GSL "buffet" is supported by microbial halophiles, which ultimately impact biological systems located far away from the hypersaline lake habitats. Once thought of as self-contained "Dead Lakes" devoid of life, the ecosystems found within many types of NaCl-saturated lakes are increasingly understood as fertile, life-enabling biomes with inherent value, because they are critically connected to the sustainability of avian lifeways around the globe (Marden et al. 2020).

Understanding the versatility of *Artemia* in context of their keystone role in ecology and trophic interactions as well as bioaccumulation of toxins is paramount (Jones and Wurtsbaugh 2014; Boyd et al. 2017; Ogata et al. 2017). Organisms living in saline lakes are often subject to changing aquatic conditions, fluctuating salinities, and even desiccation due to natural climate variability, drought, and inflow diversions (Wurtsbaugh and Berry 1990; Wurtsbaugh and Gliwicz 2001; Belovsky et al. 2011). Microcosm experiments demonstrate the strong influence of salinity on species richness and evenness of phytoplankton communities and the entire GSL ecosystem; this demonstrates that salt lakes require careful management to maintain appropriate salinities for life. Larson and Belovsky (2013) experimented in 12 L container habitats using organisms from the GSL to determine how salinities ranging from 10 to 275 $g{\cdot}L^{-1}$ influenced the ecosystem. Their results define the salinity tolerances of the major organisms living in the GSL: After 30 days, brine shrimp (*Artemia franciscana*) were nearly absent in salinities of 10 $g{\cdot}L^{-1}$ (where fish can survive) and >225 $g{\cdot}L^{-1}$. The biomass assay correlated with salinity; as salinities increased from 75 to 225 $g{\cdot}L^{-1}$, the final biomasses decreased 60% and their total biomass decreased fourfold. The species compositions present were also correlated with salinity; copepod and rotifer biomasses were negligible at salinities >50 $g{\cdot}L^{-1}$ and the brine fly (*Ephydra gracilis*) final biomass decreased 45% as salinity increased from 50 to 250 $g{\cdot}L^{-1}$. When *Artemia* and other grazers were abundantly present, phytoplankton chlorophyll levels were ~4.0μg·L^{-1}, but at higher salinities when grazing rates declined, phytoplankton chlorophyll increased to 130μg·L^{-1}. Mean periphyton chlorophyll levels showed the reverse correlation. Denitrification decreased total N concentrations during the experiments, resulting in final N:P ratios indicative of algal nitrogen limitation (Barnes and Wurtsbaugh, 2015).

Interesting work by Maszczyk and Wurtsbaugh (2017) showed how *Artemia* brine shrimp grazing activity and fecal production increase sedimentation to the GSL DBL. Experiments showed that the presence of *Artemia* affects dissolved nutrient concentrations: planktonic *Artemia* significantly decreased chlorophyll, total nitrogen, and total phosphorus within the mixolimnion (95 cm) and increased particulate matter collected in sediment traps by 28–90%. Maszczyk and Wurtsbaugh (2017) found that the largest increase of "sedimenting material" was at the top of chemocline, but only in the absence of *Artemia*. When it was present, the largest increase of collected matter was in the bottom traps. The presence of *Artemia* significantly decreased the molar TN:TP ratio in the collected sedimenting material, suggesting nitrogen-deficient fecal material. The experiments by Maszczyk and Wurtsbaugh (2017) demonstrated the importance of *Artemia* grazing for increasing material flux from the mixolimnion to the bottom, and determining the stoichiometry of accumulated sediment material. These results further counter the notion of interpreting saline lake core stratigraphies as "dead" lake systems, and emphasize the need to re-evaluate GSL basinal sedimentation rates with material flux as a function of *Artemia* macrozooplankton recruitment and palaeoactivity.

Of Birds, Biostromes, Mats, and Mounds

The GSL supports one of the largest concentrations of birds on Earth. Research indicates that 338 bird species congregate in the wetlands, flats, and waterways associated with the GSL, ~200 of which species are migratory species (Sorenson and Hoven 2020). All of these birds have diets that depend on the local food web, which is linked to two keystone invertebrates: brine fly larvae and brine shrimp – the only invertebrates that can survive in the hypersaline open water in the GSL (Conover and Bell 2020). Because surface water and biota from Great Salt Lake (GSL) contain some of the highest documented concentrations of total mercury (THg) and methylmercury (MeHg) in the United States, the Utah Department of Health and Utah Division of Wildlife Resources has issued tissue mercury (Hg) consumption advisories for several species of birds that consume the GSL brine shrimp (Peterson and Gustin 2008; Boyd et al. 2017). Several studies have now shown how the local food web is linked with carbonate biostromes and microbial mats, which are estimated to cover ~260–700 km^2 of the lake littoral zone (Wurtsbaugh et al. 2011; Kanik et al. 2020). Biosedimentary structures cover ~10% of the GSL's shallow perimeter and in the nearshore where wave activity occurs (Lindsay et al. 2017, 2019) and ooids form (Ingalls et al. 2020). Biostromes are also thought to exist across ~20% of the total modern lake bottom (Lindsay et al. 2020).

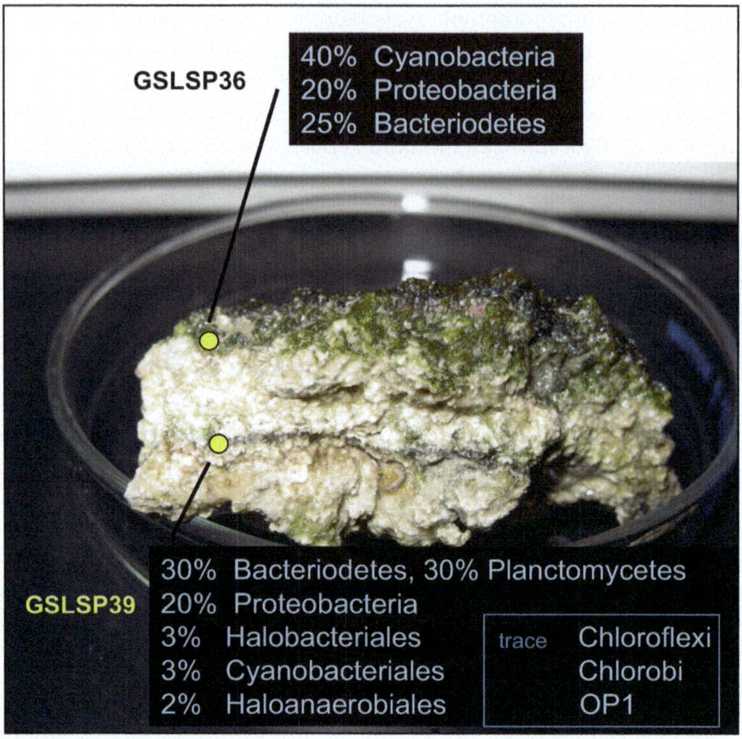

Fig. 5 Microbial mat layers sampled from the GSL with overview of metagenomic results (Beer et al. 2010)

The terrestrial biosedimentary facies in the GSL were first described by Eardley (1938); the area remains an important locale for oolite studies. The organosedimentary structures and microbial mats have a complex classification and nomenclature that is often debated, but includes microbial communities called biostromes, bioherms, mats, stromatolites, microbially induced-sedimentary-structures (MISS) and microbialites (e.g., Carozzi 1962; Halley 1976; Noffke 2008; Bosak et al. 2013; Vanden Berg et al. 2015; Vennin et al. 2019; Kanik et al. 2020). Notably, these organosedimentary features in the GSL have been incorrectly referred to as "Great Salt Lake Coral," which is erroneous. The buildups of calcium and magnesium carbonate deposits are typically cm to m in scale, and are produced by microbial communities, including alternating layers of cyanobacteria and anaerobic microbes (Fig. 5).

Wurtsbaugh (2009) and Wurtsbaugh et al. (2011) assessed whether GSL biostromes are a potential vector for the bioconcentration of high mercury and selenium levels observed within the lake, which exceeds the EPA aquatic life standards. Their study identified the periphyton community of the GSL biostromes as >99%

colonial cyanobacteria. Periphyton chlorophyll levels averaged 900 mg m^{-2} or nine times that of the lake's phytoplankton. Across the GSL, estimates of chlorophyll suggest that periphyton production is about 30% of that of the phytoplankton. Measured brine fly (*Ephydra gracilis*) larval densities on the biostromes increased from 7000 m^{-2} in June to 20,000 m^{-2} in December. In October, pupation and adult emergence halted and larvae of various instars overwintered at temperatures <5 °C. Mean total dissolved and dissolved methyl mercury concentrations in water were 5.0 and 1.2 ηg L^{-1}. Total mercury concentrations in the periphyton, fly larvae, pupae, and adults suggest that bioconcentration is only moderate in the short food web and through fly developmental stages.

Despite the limited bioconcentration observed in the invertebrates, common goldeneye ducks (*Bucephala clangula*) that feed primarily on GSL brine fly larvae had excessively high muscle tissue concentrations near 8000 ηg Hg g^{-1} dry weight (Wurtsbaugh et al. 2011). Selenium concentrations measured in periphyton, brine fly larvae, and goldeneye liver tissue were high (1700, 1200, and 24,000 ηg g^{-1}, respectively) and Hg:Se molar ratios were <1.0 in all tissues, suggesting that the high mercury concentration in the ducks might be partially detoxified by combining with selenium. Wurtsbaugh et al. (2011) concluded that high mercury levels in the Great Salt Lake are routed through the biostrome community, ending up in invertebrate prey that may provide health risks upon consumption by birds. Benthic biostrome studies by Boyd et al. (2017) demonstrated that (1) rates of methyl mercury production in GSL sediments inversely correlate with salinity; (2) numbers of sediment sulfate reducing bacteria (SRB) correlate with methyl mercury production; and (3) salinity constrains SRB activity and the availability of mercury for methylation. More work needs to be done on the SRB within the GSL to understand these biogeochemical dynamics.

Recent metagenomic sequencing of microbialite-associated mats sampled from various locations in the GSL found that their microbiota were taxonomically similar, despite morphological variability (Kanik et al. 2020). The GSL mats analyzed were dominated by the cyanobacterium *Euhalothece* and several heterotrophic bacteria. Subdominant populations of the SRB *Thiohalocapsa* present in the GSL microbial mats harbor the Calvin cycle and nitrogenase, suggesting they supply fixed carbon and nitrogen to heterotrophic bacteria. These findings from the modern hypersaline GSL demonstrate the ability to integrate light into energy metabolism as a fundamental adaptation enabling the development and sustainability of mats. Kanik et al. (2020) report that the unexpected abundance and diversity in the mechanisms of harvesting light energy observed in GSL mat populations has likely minimized niche overlap among its cohabiting taxa; this may provide mechanisms to increase energy yield and osmotic balance during salt stress, thereby enhancing fitness. Together, these physiological benefits promote the formation of robust communities in mats which, in turn, influence the formation of morphologically diverse microbialite organosedimentary structures that can be preserved in the carbonate sediment rock record.

Analogs for the Early, the Extreme, and the Astrobiological "Out There"

Published research demonstrates that GSL-dwelling extremophiles are very versatile; some prefer harsh conditions of salinity and there are many with specific adaptations for tolerating heat, cold, acidity, pressure, and high radiation. Some can even persist through anaerobic conditions or becoming encapsulated in carbonate mats and/or salt crystals. As such, the GSL offers microbial and environmental analogs for extreme habitats on early Earth, Mars, and beyond.

Hypersaline brines like those of the GSL have persisted as a stable habitat throughout Earth's deep geologic history (Warren 2010, 2016). Earth's earliest evidence of life is from organosedimentary structures such as the microbialites preserved in 3.481-billion-year-old (Ga) rocks in the Pilbara of Western Australia (Noffke et al. 2013). Life may have initially evolved in extreme aqueous environments, such as hydrothermal vents and acid-brine pools. Various lines of evidence support the hypothesis that terrestrial life may have inhabited, and may evolved from, brines with a high NaCl, KCl, and/or $MgCl_2$ concentration (e.g., Dundas 1998; Mancinelli 2005a; Stevenson et al. 2015).

The evolution of early cells on Earth was closely linked to water activity, which remains among the key determinants for extraterrestrial habitability (Grant 2004). NASA's directive to "follow the water" and Earth analog studies have shaped approaches to finding habitats suitable for life on Mars and beyond (Mancinelli 2005b; Chan et al. 2011; Nicoll and Finkelstein 2014). Given their role on Early Earth, microbial mat and carbonate-evaporite environments are among the preferred terrestrial models for astrobiological research and prospection (Hoehler 2001; Noffke 2008; Foster and Mobberley 2010; Horgan et al. 2020). There is strong evidence for acid brines formerly existing on Mars (Squyres et al. 2004; Benison and Bowen 2006), and saline minerals found on Mars include chlorides (likely halite) and hydrated calcium sulphates (perhaps gypsum) in lake deposits and nearby lake-derived aeolian deposits (Glotch et al. 2010), all of which have GSL analogs (Chan et al. 2011).

The inferred acid-brine of past surface waters on Mars have led some to speculate that conditions would be practically inhospitable for life (Tosca et al. 2008; Rummel et al. 2014). Some environments that hold potential for supporting "life" on Mars include perchlorate soils at the Phoenix Lander site, which are known to support Archaea on Earth (Hecht et al. 2009; Catling et al. 2010). Recent radar evidence from the Mars Express spacecraft suggests that there may be a 20 km-wide salty liquid lake present below glacier ice in the Planum Australe region of Mars (Orosei et al. 2018). Such environments may be able to support halophiles and other extremophiles (see review by Lee et al. 2018), which refocuses the value of GSL as a direct analog.

Conclusions: GSL Is Much More than a "Mini-Ocean"

Exciting, emerging research on the modern geolimnology of the hypersaline GSL clearly demonstrates its value as a natural laboratory for generating fresh ideas about broad global ecological connections and looking into the deeper geologic past. Applying new molecular and genomic techniques reveals that the harsh GSL ecosystem is not a dead lake – it is actually quite diverse, and has an interesting microbiology that has adapted to its extreme and variable conditions, and has even influenced its basin sedimentation. The GSL's simple community structure at the local scale is profoundly important because it is connected to a larger avian ecology that is global. We can leverage new understandings about modern GSL productivity, food web dynamics, geochemical cycling, and sedimentation as we assess ancient saline lake sediments.

Approaching GSL from an evolutionary standpoint, its brine is an accessible analog for an ancient primordial soup. The setting presents a modern opportunity to study the evolution and adaptations of extremophiles and modern microbial mat communities that have created organosedimentary structures. What we can observe in the brine and salty shores of this GSL has direct analogs with fossil stromatolites and other records of early Earth. Observing hydroclimate dynamics, limnochemistry, sedimentation styles, and ecosystems in shallow underfilled lakes like the modern GSL can spark new insights and perspectives, and yield profound implications for unraveling deep time – both as analogs of the past (e.g., Green River Formation) and for finding life on other planets, like Mars and other planets with harsh chemical environments (Nicoll and Finkelstein 2014, b).

What is the future of research in the GSL? It evolves with new developmental techniques (Baxter and Butler 2020). It is centrally important for scientists to better understand the GSL, and to help lead its management as we face global warming and persistent megadrought conditions. At this juncture of the Anthropocene, we must secure the welfare of a healthy desert watershed system for its dependent biota (Conover and Bell 2020). The GSL is a critical resource for many flora and fauna that are not as charismatic as its bird life, but there is increasing pressure from the activities of people, pollution, and profiteers (Bedford 2009). The GSL provides a multibillion dollar annual economy (Bioeconomics Inc. 2012; Null and Wurtsbaugh 2020) – it provides extractive salts and metals, *Artemia* brine shrimp that are essential for aquaculture, and its waters are the fuel of lake-effect snows delivered to the Wasatch Range as "the Greatest Snow on Earth" – according to Utah's slogan. Hopefully, preserving the natural ecological status of the GSL remains a top priority; it is a lake that is worth its salt.

Acknowledgments This paper is written with thanks to, and admiration for Beth G.-K. The author thanks the two reviewers for their helpful comments and feedback. Special thanks to the editors for convening this volume – and for support of friends Michael, Dave, and Lisa.

References Cited

Aldrich, T. W., & Paul, D. S. (2002). Avian ecology of Great Salt Lake. *Great Salt Lake: An overview of change*. Special publication of the Utah Department of Natural Resources, Salt Lake City, Utah, pp. 343–374.

Allen, E. E., & Banfield, J. F. (2005). Community genomics in microbial ecology and evolution. *Nature Reviews Microbiology, 3*, 489–498.

Almeida-Dalmet, S., & Baxter, B. K. (2020). Unexpected complexity at salinity saturation: Microbial diversity of the North Arm of the Great Salt Lake. In B. K. Baxter & J. K. Butler (Eds.), *Great Salt Lake biology: A terminal lake in a time of change* (pp. 119–144). Springer. https://doi.org/10.1007/978-3-030-40352-2_5.

Arnow, T., & Stephens, D. W. (1990). Hydrologic characteristics of the Great Salt Lake, Utah, 1847–1986. *U.S. Geological Survey Water-Supply Paper*, 2332, pp. 1–32.

Barnes, B. D., & Wurtsbaugh, W. A. (2015). The effects of salinity on plankton and benthic communities in the Great Salt Lake, Utah, USA: A microcosm experiment. *Canadian Journal of Fisheries and Aquatic Sciences, 72*(6), 807–817. https://doi.org/10.1139/cjfas-2014-0396.

Baumgartner, L. K., Dupraz, C., Buckley, D. H., Spear, J. R., Pace, N. R., & Visscher, P. T. (2009). Microbial species richness and metabolic activities in hypersaline microbial mats: Insight into biosignature formation through lithification. *Astrobiology, 9*, 861–874.

Baxter, B. K. (2018). Great Salt Lake microbiology: a historical perspective. *International Microbiology, 21*(3), 79–95. https://doi.org/10.1007/s10123-018-0008-z. Epub 2018 Jun 4.PMID: 30810951.

Baxter, B. K., & Butler, J. K. (2020). *Great Salt Lake biology: A terminal Lake in a time of change* (514 p). Cham, Springer Dordrecht. Switzerland

Baxter, B. K., Litchfield, C. D., Sowers, K., Griffith, J. D., Dassarma, P. A., & Dassarma, S. (2005). Microbial diversity of Great Salt Lake. In N. Gunde-Cimerman, A. Oren, & A. Plemenitaš (Eds.), *Adaptation to life at high salt concentrations in Archaea, Bacteria, and Eukarya* (pp. 9–35). Dordrecht, the Netherlands: Springer.

Bedford, D. (2009). The Great Salt Lake: America's Aral Sea? *Environment, 51*, 8–19. https://doi.org/10.3200/ENVT.51.5.8-21.

Beer, L. L., Pepe-Raney, C., Zvenigorodsky, N., Nicoll, K., Meuser, J. M., Posewitz, M. C., Ghirardi, M., Baxter, B. K., & Spear, J. R. (2010). Mats, microbialites and mud: biodiversity of Great Salt Lake beyond pink water. *Poster at Joint Genome Institute, User's Meeting*, March 24–26, Walnut Creek, CA. http://www.jgi.doe.gov/meetings/usermeeting/2010/JGI-UM5-abstracts.pdf

Beisner, K., Naftz, D. L., Johnson, W. P., & Diaz, X. (2009). Selenium and trace element mobility affected by periodic displacement of stratification in the Great Salt Lake, Utah. *Science of the Total Environment, 407*, 5263–5273.

Bekker, M. F., DeRose, R. J., Buckley, B. M., Kjelgren, R. K., & Gill, N. S. (2014). A 576-year Weber River streamflow reconstruction from tree rings for water resource risk assessment in the Wasatch front, Utah. *Journal of the American Water Resources Association (JAWRA)*, 1–11. https://doi.org/10.1111/jawr.12191.

Belovsky, G. E., Stephens, D., Perschon, C., Birdsey, P., Paul, D., Naftz, D., Baskin, R., Larson, C., Mellison, C., Luft, J., Mosley, R., Mahon, H., Van Leuwen, J., & Allen, D. V. (2011). The Great Salt Lake Ecosystem (Utah, USA): Long term data and a structural equation approach. *Ecosphere, 2*(33), 31–40.

Benison, K. C., & Bowen, B. B. (2006). Acid saline lake systems give clues about past environments and the search for life on Mars. *Icarus, 183*, 225–229.

Bioeconomics, Inc. (2012). Economic Significance of the Great Salt Lake to the State of Utah. Report Prepared for the State of Utah Great Salt Lake Advisory Council. DWQ-2012-006864, 50 p.

Bosak, T., Knoll, A. H., & Petrof, A. P. (2013). The meaning of stromatolites. *Annual Review of Earth and Planetary Sciences, 41*, 21–44. https://doi.org/10.1146/annurev-earth-042711-105327.

Boyd, E. S., Yu, R.-Q., Barkay, T., Hamilton, T. L., Baxter, B. K., Naftz, D. L., & Marvin-DiPasquale, M. (2017). Effect of salinity on mercury methylating benthic microbes and their activities in Great Salt Lake, Utah. *Science of the Total Environment, 581–582*, 495–506. https://doi.org/10.1016/j.scitotenv.2016.12.157.

Cannon, J. S., & Cannon, M. A. (2002). The Southern Pacific railroad trestle – past and present. In J. W. Gwynn (Ed.), *Great Salt Lake, an overview of change* (pp. 283–294). Salt Lake City: Special Publication of the Utah Department of Natural Resources Salt Lake City, Utah.

Carozzi, A. V. (1962). Observations on algal biostromes in the Great Salt Lake, Utah. *Journal of Geology, 70*, 246–252.

Carroll, A. R., & Bohacs, K. M. (1999). Stratigraphic classification of ancient lakes: Balancing tectonic and climatic controls. *Geology, 27*, 99–102.

Catling, D. C., Claire, M. W., Zahnle, K. J., Quinn, R. C., Clark, B. C., Hecht, M. H., & Kounaves, S. (2010). Atmospheric origins of perchlorate on Mars and in the Atacama. *Journal of Geophysical Research Planets, 115*, E00E11. https://doi.org/10.1029/2009JE003425.

Chan, M. A., Nicoll, K., Ormö, J., Okubo, C., & Komatsu, G. (2011). Utah's geologic and geomorphic analogs to Mars—An overview for planetary exploration. In W. B. Garry & J. E. Bleacher (Eds.), *Analogs for planetary exploration: Geological Society of America Special Paper 483* (pp. 349–375). https://doi.org/10.1130/2011.2483(22).

Clegg, J. S., & Trotman, C. (2002). Physiological and biochemical aspects of *Artemia* ecology. In T. J. Abatzopoulos, J. A. Beardmore, J. S. Clegg, P. Sorgeloos (Eds.), *Artemia basic and applied biology* (pp. 129–170). Dordrecht, Springer: Netherlands.

Conover, M. R., & Bell, M. E. (2020). Importance of Great Salt Lake to pelagic birds: Eared grebes, phalaropes, gulls, ducks and white pelicans. In B. K. Baxter & J. K. Butler (Eds.), *Great Salt Lake biology: A terminal Lake in a time of change* (pp. 263–309). Springer. https://doi.org/10.1007/978-3-030-40352-2_8.

Daines, L. L. (1910). *Physiological experiments on some algae of Great Salt Lake*. M.A. thesis, Department of Botany, University of Utah, 14 p.

Diaz, X., Johnson, W. P., & Naftz, D. L. (2009). Selenium mass balance in the Great Salt Lake, Utah. *Science of the Total Environment, 407*, 2333–2341.

Dodson, S., & Frey, D. (2001). Cladoceran and other branchiopoda. In J. H. Thorp & A. P. Covich (Eds.), *Ecology and classification of North American freshwater invertebrates* (pp. 723–776). San Diego, Academic Press. Ch. 20.

Domagalski, J. L., Orem, W. H., & Eugster, H. P. (1989). Organic geochemistry and brine composition in Great Salt, Mono, and Walker Lakes. *Geochimica et Cosmochimica Acta, 53*, 2857–2872.

Dundas, I. (1998). Was the environment for primordial life hypersaline? *Extremophiles, 2*, 375–377.

Dunham, E. C., Fones, E. M., Fang Yihang, L. M. R., Steuer, C., Fox, N., Willis, M., Walsh, A., Colman, D. R., Baxter, B. K., Lageson, D., Mogk, D., Rupke, A., Huifang, X., & Boyd, E. S. (2020). An ecological perspective on dolomite formation in Great Salt Lake, Utah. *Frontiers in Earth Science, 8*, 24. https://doi.org/10.3389/feart.2020.00024. https://www.frontiersin.org/article/10.3389/feart.2020.00024.

Eardley, A. J. (1938). Sediments of Great Salt Lake, Utah. *AAPG Bulletin, 22*, 1305–1411.

Foster, J. S., & Mobberley, J. M. (2010). Past, present, and future: Microbial mats as models for astrobiological research. In J. Seckbach & A. Oren (Eds.), *Cellular origin, life in extreme habitats and astrobiology: Microbial Mats: Modern and ancient microorganisms in stratified systems* (pp. 563–582). Dordrecht, Springer.

Frederick, E. (1924). *On the bacterial flora of Great Salt Lake and the viability of other microorganisms in Great Salt Lake water*. M.S. thesis, Department of Bacteriology, University of Utah, 73 p.

Freeman, M. L. (2014). *Flow reversal events and statistical modeling of flow dynamics of hypersaline water across a constructed causeway, Great Salt Lake, Utah, USA*. MS Thesis, University of Utah, 76 p. https://collections.lib.utah.edu/details?id=196660&facet_setname_s=%22ir_etd%22&q=saltair

Frémont, J. C. (1845). *Report of the exploring expedition to the Rocky Mountains in the year 1842 and to Oregon and North California in the years 1843–44: printed by order of the Senate of the United States* (Vol. 174). Washington: Gales & Seaton.

GBIF Secretariat (2019). GBIF Backbone Taxonomy. Checklist dataset https://doi.org/10.15468/39omei accessed via GBIF.org on 2021-01-11.

Gilbert, G. K. (1890). Lake Bonneville: U.S. Geological Survey, Monograph 1, 248 p.

Glotch, T. D., Bandfield, J. L., Tornabene, L. L., et al. (2010). Distribution and formation of chlorides and phyllosilicates in Terra Sirenum, Mars. *Geophysical Research Letters, 37*, L16202.

Grant, W. D. (2004). Life at low water activity. *Philosophical Transactions of the Royal Society B, 359*, 1249–1267.

Grayson, D. K. (2011). *The Great Basin: A natural prehistory* (432 p). Berkeley: University of California Press.

Gwynn, J. W. (2002). The railroads proximate to Great Salt Lake, Utah. In J. W. Gwynn (Ed.), *Great Salt Lake, an overview of change* (pp. 273–281). Salt Lake City: Special Publication of the Utah Department of Natural Resources.

Gwynn, J. W. (2012). *A lake divided—a history of the Southern Pacific Railroad causeway and its effect on Great Salt Lake, Utah.* http://geology.utah.gov/utahgeo/gsl/lakedivided.htm.

Hahl, D. C., & Handy, A. H. (1969). Chemical and physical variations of the brine, Great Salt Lake, Utah, 1963-1966. *Utah Geological Survey Water Resource Bulletin, 12*, 1–33.

Hahnenberger, M., & Nicoll, K. (2012). Meteorological characteristics of dust storm events in the eastern Great Basin of Utah, U.S.A. *Atmospheric Environment, 60*, 601–612. https://doi.org/10.1016/j.atmosenv.2012.06.029.

Hahnenberger, M., & Nicoll, K. (2014). Geomorphic and land cover identification of dust sources in the eastern Great Basin of Utah, U.S.A. *Geomorphology, 204*, 657–672. https://doi.org/10.1016/j.geomorph.2013.09.013.

Halley, R. B. (1976). Textural variation within Great Salt Lake algal mounds. *Developments in Sedimentology, 20*, 435–445.

Hecht, M. H., Kounaves, S. P., Quinn, R. C., West, S. J., Young, S. M. M., Ming, D. W., Catling, D. C., Clark, B. C., Boynton, W. V., Hoffman, J., Deflores, L. P., Gospodinova, K., Kapit, J., & Smith, P. H. (2009). Detection of perchlorate and the soluble chemistry of Martian soil at the Phoenix lander site. *Science, 325*, 64–67. https://doi.org/10.1126/science.1172466pmid:19574385.

Hoehler, T. M. (2001). The role of microbial mats in the early Earth. *Nature, 412*, 324–327.

Horgan, B. H. N., Anderson, R. B., Dromart, G., Amador, E. S., & Rice, M. S. (2020). The mineral diversity of Jezero crater: Evidence for possible lacustrine carbonates on Mars. *Icarus, 339*, 113526. https://doi.org/10.1016/j.icarus.2019.113526.

Ingalls, M., Frantz, C. M., Snell, K. E., & Trower, E. J. (2020). Carbonate facies-specific stable isotope data record climate, hydrology, and microbial communities in Great Salt Lake, UT. *Geobiology, 00*, 1–28. https://doi.org/10.1111/gbi.12386.

Ingvorsen, K., & Brandt, K. K. (2002). Anaerobic microbiology and sulfur cycling in hypersaline sediments with special reference to Great Salt Lake. In J. W. Gwynn (Ed.), *Great salt Lake, an overview of change* (pp. 387–400). Salt Lake City: Special Publication of the Utah Department of Natural Resources.

Jakobsen, T. F., Kjeldsen, K. U., & Ingvorsen, K. (2006). Desulfohalobium utahense sp. nov., a moderately halophilic, sulfate-reducing bacterium isolated from Great Salt Lake. *Journal of Systematic and Evolutionary Microbiology, 56*, 2063–2069.

Javor, B. (1989). *Hypersaline environments. Microbiology and biogeochemistry*. Berlin: Springer.

Jones, E. F., & Wurtsbaugh, W. A. (2014). The Great Salt Lake's monimolimnion and its importance for mercury bioaccumulation in brine shrimp (*Artemia franciscana*). *Limnology and Oceanography, 59*, 141–155.

Jellison, R., Williams, W. D., Timms, B., Alcocer, J., Aladin, N. V. (2008). Salt lakes: values, threats, and future. Pages 94–110 in N. V. C. Polunin, editor. Aquatic ecosystems: trends and global prospects. Cambridge University Press, Cambridge, United Kingdom. p. 94–110. https://doi.org/10.1017/CBO9780511751790.010

Kanik, M., Munro-Ehrlich, M., Fernandes-Martins, M. C., Payne, D., Gianoulias, K., Keller, L., Kubacki, A., Lindsay, M. R., Baxter, B. K., Vanden Berg, M. D., Colman, D. R., & Boyd, E. S. (2020). Unexpected abundance and diversity of phototrophs in Mats from morphologically variable Microbialites in Great Salt Lake, Utah. *Applied and Environmental Microbiology, 86*(10), e00165-20. https://doi.org/10.1128/AEM.00165-20. Print 2020 May 5 (embargoed). PMID: 32198176.

Kjeldsen, K. U., Loy, A., Jakobsen, T. F., Thomsen, T. R., Wagner, M., & Ingvorsen, K. (2007). Diversity of sulfate-reducing bacteria from an extreme hypersaline sediment, Great Salt Lake (Utah). *FEMS Microbiology Ecology, 60*, 287–298.

Larson, C. A., & Belovsky, G. E. (2013). Salinity and nutrients influence species richness and evenness of phytoplankton communities in microcosm experiments from Great Salt Lake, Utah, USA. *Journal of Plankton Research, 35*, 1154–1166.

Lee, C. J. D., McMullan, P. E., O'Kane, C. J., Stevenson, A., Santos, I. C., Roy, C., Ghosh, W., Mancinelli, R. L., Mormile, M. R., McMullan, G., Banciu, H. L., Fares, M. A., Benison, K. C., Oren, A., Dyall-Smith, M. L., & Hallsworth, J. E. (2018). NaCl-saturated brines are thermodynamically moderate, rather than extreme, microbial habitats. *FEMS Microbiology Reviews, 42*, 672–693. https://doi.org/10.1093/femsre/fuy026.

Lindsay, M. R., Anderson, C., Fox, N., Scofield, G., Allen, J., Anderson, E., Bueter, L., Poudel, S., Sutherland, K., Munson-McGee, J. H., Van Nostrand, J. D., Zhou, J., Spear, J. R., Baxter, B. K., Lageson, D. R., & Boyd, E. S. (2017). Microbialite response to an anthropogenic salinity gradient in Great Salt Lake, Utah. *Geobiology, 15*(1), 131–145. https://doi.org/10.1111/gbi.12201. Epub 2016 Jul 14. PMID: 27418462.

Lindsay, M. R., Johnston, R. E., Baxter, B. K., & Boyd, E. S. (2019). Effects of salinity on microbialite-associated production in Great Salt Lake, Utah. *Ecology, 100*(3), e02611. https://doi.org/10.1002/ecy.2611. Epub 2019 Feb 21. PMID: 30636291.

Lindsay, M. R., Dunham, E. C., & Boyd, E. S. (2020). Microbialites of Great Salt Lakes. In B. K. Baxter & J. K. Butler (Eds.), *Great salt Lake biology: A terminal Lake in a time of change* (pp. 87–118). Springer. https://doi.org/10.1007/978-3-030-40352-2_4.

Litchfield, C. D., & Gillevet, P. M. (2002). Microbial diversity and complexity in hypersaline environments. *Journal of Industrial Microbiology & Biotechnology, 28*, 48–56.

Louderback, L. A., & Rhode, D. E. (2009). 15,000 years of vegetation change in the Bonneville basin: The blue Lake pollen record. *Quaternary Science Reviews, 28*(3–4), 308–326., ISSN 0277-3791. https://doi.org/10.1016/j.quascirev.2008.09.027.

Loving, B. L., Waddell, K. M., & Miller, C. W. (2000). Water and salt balance of Great Salt Lake, Utah, and simulation of water and salt movement through the causeway, 1987–1998. U.S. Geological Survey Water-Resources Investigations Report. (2000–4221).

Lyle, M., et al. (2012). Out of the tropics: The Pacific, Great Basin lakes, and late Pleistocene water cycle in the Western United States. *Science, 337*, 1629–1633.

Madison, R. J. (1970). Effects of a causeway on the chemistry of the brine in Great Salt Lake, Utah. *Utah Geological Survey Water-Resource Bulletin, 14*, 52.

Mancinelli, R. L. (2005a). Microbial life in brines, evaporites and saline sediments: The search for life on Mars. In T. Tokano (Ed.), *Water on Mars and life Berlin* (pp. 277–298). Germany: Springer.

Mancinelli, R. L. (2005b). Halophiles: A terrestrial analog for life in brines on Mars. In N. Gunde-Cimerman, A. Plemenitaš, & A. Oren (Eds.), *Adaptation to life at high salt concentrations in archaea, Bacteria and Eukarya. Volume 9 in the series on cellular origins, life in extreme habitats and astrobiology (COLE)* (pp. 137–149). Dordrecht, the Netherlands: Kluwer Academic Publishers.

Marden, B.; Brown, P.; and Bosteels, T. 2020. Great salt Lake Artemia: Ecosystem functions and services with a global reach. in Baxter, B.K. and Butler, J.K.. Great Salt Lake biology: A terminal Lake in a time of change. Springer 175–237. https://doi.org/10.1007/978-3-030-40352-2_7.

Maszczyk, P., & Wurtsbaugh, W. (2017). Brine shrimp grazing and fecal production increase sedimentation to the deep brine layer (monimolimnion) of Great Salt Lake, Utah. *Hydrobiologia, 802*, 7–22. https://doi.org/10.1007/s10750-017-3235-y.

Meuser, J. E., Baxter, B. K., Spear, J. R., Peters, J. W., Posewitz, M. C., & Boyd, E. S. (2013). Contrasting patterns of community assembly in the stratified water column of Great Salt Lake, Utah. *Microbial Ecology, 66*, 268–280.

Mohammed, I. N., & Tarboton, D. G. (2011). On the interaction between bathymetry and climate in the system dynamics and preferred levels of the Great Salt Lake. *Water Resources Research, 47*, W02525. https://doi.org/10.1029/2010WR009561.

Mohammed, I. N., & Tarboton, D. G. (2012). An examination of the sensitivity of the Great Salt Lake to changes in inputs. *Water Resources Research, 48*(11). https://doi.org/10.1029/2012WR011908.

Møller, M. F., Kjeldsen, K. U., & Ingvorsen, K. (2010). Marinimicrobium haloxylanilyticum sp. nov., a new moderately halophilic, polysaccharide degrading bacterium isolated from Great Salt Lake, Utah. *Antonie Van Leeuwenhoek, 98*, 553–565.

Moon, Y. I., Lall, U., & Kwon, H.-H. (2008). Non-parametric short-term forecasts of the Great Salt Lake using atmospheric indices. *Journal of Climatology, 28*, 361–370. https://doi.org/10.1002/joc.1533.

Naftz, D. L., Johnson, W. P., Freeman, M., Beisner, K., & Diaz, X. (2008). *Estimation of selenium loads entering the South Arm Of Great Salt Lake, Utah* (pp. 1–50). Reston: US Geological Survey.

Naftz, D. L., Millero, F. J., Jones, B. F., & Green, W. R. (2011). An equation of state for hypersaline water in Great Salt Lake, Utah, USA. *Aquatic Geochemistry, 17*, 809–820.

Nicoll, K., & Finkelstein, D. B. (2014). Saline Lakes...A logical step in exploring habitability of "the final frontier". *PALAIOS, 29*, 231–232. https://doi.org/10.2110/pal.2014.s0614.

Nicoll, K., & Keen-Zebert, A. (2016). Initial chronological determinations at an archaic site discovered near Stockton, Utah. *Journal of Archaeological Science: Reports, 6*, 418–423. https://doi.org/10.1016/jasrep.2016.02.026.

Nicoll, K., Yentsch, A., Rood, R., & Jones, K. P. (2014). Site formation and archaic geoarchaeology along the Jordan River, Great Salt Lake Valley, Utah USA. *Quaternary International, 342*, 214–225. https://doi.org/10.1016/j.quaint.2013.08.044.

Noffke, N. (2008). Turbulent lifestyle: Microbial mats on Earth's sandy beaches – Today and 3 billion years ago. *GSA Today, 18*, 4–9.

Noffke, N., Christian, D., Wacey, D., & Hazen, R. M. (2013). Microbially induced sedimentary structures recording an ancient ecosystem in the *ca.* 3.48 billion-year-old dresser formation, Pilbara, Western Australia. *Astrobiology, 13*, 103–1124. https://doi.org/10.1089/ast.2013.1030.

Null, S. E., & Wurtsbaugh, W. A. (2020). Water development, consumptive water uses and Great Salt Lake, pp. 1–21. In B. K. Baxter & J. K. Butler (Eds.), *Great Salt Lake biology: A terminal Lake in a time of change* (pp. 263–309). Springer. https://doi.org/10.1007/978-3-030-40352-2_1.

Ogata, E. M., Wurtsbaugh, W. A., Smith, T. N., & Durham, S. L. (2017). Bioassay analysis of nutrient and *Artemia franciscana* effects on trophic interactions in the Great Salt Lake, USA. *Hydrobiologia, 788*, 1–16.

Ohlendorf, H. M., DenBleyker, J., Moellmer, W. O., & Miller, T. (2009). Development of a site-specific standard for selenium in open waters of Great Salt Lake, Utah. *Natural Resources and Environmental Issues, 15*, Article 4. Available at: https://digitalcommons.usu.edu/nrei/vol15/iss1/4

Oren, A. (2009). Microbial diversity and microbial abundance in salt-saturated brines: Why are the waters of hypersaline lakes red? *Natural Resources and Environmental Issues, 15*, 247–255.

Orosei, R., Lauro, S. E., Pettinelli, E., Cicchetti, A., Coradini, M., Cosciotti, B., Di Paolo, F., Flamini, E., Mattei, E., Pajola, M., Soldovieri, F., Cartacci, M., Cassenti, F., Frigeri, A., Giuppi, S., Martufi, R., Masdea, A., Mitri, G., Nenna, C., Noschese, R., Restano, M., & Seu, R. (2018). Radar evidence of subglacial liquid water on Mars. *Science, 361*, 490–493. https://doi.org/10.1126/science.aar7268.

Oviatt, C. G., & Shroder, J. (2016). Lake Bonneville: A scientific update. *Developments in Earth Surface Processes* (Vol. 20). Edited by Charles G. Oviatt, John F. Shroder. Amsterdam/New York: Elsevier, Pages 1–659

Oviatt, C. G., Madsen, D. B., Miller, D. M., Thompson, R. S., & McGeehin, J. P. (2015). Early Holocene Great Salt Lake, USA. *Quaternary Research, 84*, 57–68.

Peterson, C., & Gustin, M. (2008). Mercury in the air, water and biota at the Great Salt Lake (Utah, USA). *Science of the Total Environment, 405*, 255–268.
Post, F. J. (1977). Microbial ecology of Great Salt Lake. *Microbial Ecology, 3*, 143–165.
Post, F. J. (1980) Biology of the north arm. In: Gwynn JW (ed) Great Salt Lake: a scientific, historical and economic overview. Utah Geological and Mineral Survey, Salt Lake City, pp 314–321.
Pugin, B., Blamey, J. M., Baxter, B. K., & Wiegel, J. (2012). Amphibacillus cookii sp. nov., a facultatively aerobic, spore-forming, moderately halophilic, alkalithermotolerant bacterium. *International Journal of Systematic and Evolutionary Microbiology, 62*, 2090–2096.
Reheis, M. C., Adams, K. D., Oviatt, C. G., Bacon, S. N. (2014). Pluvial lakes in the Great Basin of the western United States: a view from the outcrop. Quaternary Science Reviews 97, 33–57. https://doi.org/10.1016/j.quascirev.2014.04.012
Roberts, A. J. (2013). Avian diets in a saline ecosystem: Great Salt Lake, Utah, USA. *Human–Wildlife Interactions, 7*, 158–168. https://digitalcommons.usu.edu/hwi/vol7/iss1/15.
Roney, H. C., Booth, G. M., & Cox, P. A. (2009). Competitive exclusion of cyanobacterial species in the Great Salt Lake. *Extremophiles, 13*, 355–361.
Rummel, J. D., Beaty, D. W., Jones, M. A., et al. (2014). A new analysis of Mars "special regions": Findings of the second MEPAG Special Regions Science Analysis Group (SR-SAG2). *Astrobiology, 14*, 887–968.
Shope, C. L., & Angeroth, C. E. (2015). Calculating salt loads to Great Salt Lake and the associated uncertainties for water year 2013; updating a 48 year old standard. *Science of the Total Environment, 536*, 391–405. https://doi.org/10.1016/j.scitotenv.2015.07.015.
Skorko, K. W., Jewell, P. W., & Nicoll, K. (2012). Fluvial response to an historic low stand of the Great Salt Lake, Utah. *Earth Surface Processes and Landforms, 37*, 143–156. http://onlinelibrary.wiley.com/doi/10.1002/esp.2226/abstract.
Sorenson, E. D., & Hoven, H. M. (2020). Great Salt Lake shorebirds, their habitats, and food base. In B. K. Baxter & J. K. Butler (Eds.), *Great Salt Lake biology: A terminal Lake in a time of change* (pp. 263–309). Springer. https://doi.org/10.1007/978-3-030-40352-2_9.
Squyres, S. W., Grotzinger, J. P., Arvidson, R. E., et al. (2004). *In situ* evidence for an ancient aqueous environment at Meridiani Planum, Mars. *Science, 306*, 1709–1714.
Stansbury, H. (1855). *Exploration of the valley of the Great Salt Lake: Including a reconnaissance of a new route through the Rocky Mountains*. Philadelphia: Lippincott, Gramabo & Co.
Stevenson, A., Burkhardt, J., Cockell, C. S., et al. (2015). Multiplication of microbes below 0.690 water activity: Implications for terrestrial and extraterrestrial life. *Environmental Microbiology, 17*, 257–277.
Tazi, L., Breakwell, D. P., Harker, A. R., & Crandall, K. A. (2014). Life in extreme environments: Microbial diversity in Great Salt Lake, Utah. *Extremophiles, 18*, 525–535.
Torrentera, L., & Dodson, S. I. (2004). Ecology of the brine shrimp *Artemia* in the Yucatan, Mexico, salterns. *Journal of Plankton Research, 26*, 617–624.
Tosca, N. J., Knoll, A. H., & McLennan, S. M. (2008). Water activity and the challenge for life on early Mars. *Science, 353*, 1204–1207.
Tsai, C. R., Garcia, J. L., Patel, B. K., Cayol, J. L., Baresi, L., & Mah, R. A. (1995). *Haloanaerobium alcaliphilum* sp. nov., an anaerobic moderate halophile from the sediments of Great Salt Lake, Utah. *International Journal of Systematic Bacteriology, 45*, 301–307.
Union Pacific Railroad. (2013). USACE notification re imminent failure of east culvert. Union Pacific Railroad. http://www.waterquality.utah.gov/PublicNotices/docs/2013/UPRRCauseway/UPRRLtr21Oct2013USACENotificationreImminentFailureofEastCulvertFinal20131021 2.pdf
Utah Division of Forestry, Fire and State Lands. (2012). Draft Final Great Salt Lake Comprehensive Management Plan. 305.
Vanden Berg, M. D., Chidsey, T. C., Jr., Eby, D. E., & Kelln, W. (2015). Characterization of microbialites in Bridger Bay, Antelope Island, Great Salt Lake, Utah. In M. R. Rosen, A. Cohen, M. Kirby, E. Gierlowski-Kordesch, S. Starratt, B. L. Valero Garcés, & J. Varekamp (Eds.), *Sixth International Limnogeology Congress—Abstract Volume*, Reno, Nevada, June 15–19, 2015: U.S. Geological Survey Open-File Report 2015–1092, pp. 216–217, https://doi.org/10.3133/ofr20151092

Vennin, E., Bouton, A., Bourillot, R., Pace, A., Roche, A., Brayard, A., Thomazo, C., Virgone, A., Gaucher, E. C., Desaubliaux, G., & Visscher, P. T. (2019). The lacustrine microbial carbonate factory of the successive Lake Bonneville and Great Salt Lake, Utah, USA. *Sedimentology, 66*, 165–204. https://doi.org/10.1111/sed.12499.

Waddell, K. M., & Bolke, E. L. (1973). The effects of restricted circulation on the salt balance of Great Salt Lake, Utah. *Utah Geological Survey Water-Resource Bulletin, 18*, 54.

Waddell, K. M., & Fields, F. K. (1977). Model for evaluating the effect of dikes on the water and salt balance of Great Salt Lake, UT. *Utah Geological Survey Water-Resource Bulletin, 21*, 54.

Wainø, M., Tindall, B. J., & Ingvorsen, K. (2000). *Halorhabdus utahensis* gen. nov., sp. nov., an aerobic, extremely halophilic member of the Archaea from Great Salt Lake, Utah. *International Journal of Systematic and Evolutionary Microbiology, 50*, 183–190.

Warren, J. K. (2010). Evaporites through time: Tectonic, climatic and eustatic controls in marine and nonmarine deposits. *Earth Science Reviews, 98*, 217–268.

Warren, J. K. (2016). *Evaporites: A geological compendium* (2nd ed., 1813 pp). Berlin: Springer.

Weimer, B. C., Rompato, G., Parnell, J., et al. (2009). Microbial biodiversity of Great Salt Lake, Utah. *Natural Resources and Environmental Issues, 15*, 15–22.

Wold, S. R., Thomas, B. E., & Waddell, K. M. (1997). Water and salt balance of the Great Salt Lake, Utah, and simulation of water and salt movement through the causeway. *U.S. Geological Survey Water-Supply Paper, 2450*, 64.

Wurtsbaugh, W. A. (2009). Biostromes, brine flies, birds and the bioaccumulation of selenium in great salt Lake, Utah. *Natural Resources & Environment, 15*, 2.

Wurtsbaugh, W. A., & Berry, T. S. (1990). Cascading effects of decreased salinity on the plankton chemistry and physics of the Great Salt Lake (Utah). *Canadian Journal of Fisheries and Aquatic Science, 47*(1), 100–109.

Wurtsbaugh, W. A., & Gliwicz, Z. M. (2001). Limnological control of brine shrimp population dynamics and cyst production in the Great Salt Lake, Utah. *Hydrobiologia, 466*, 119–132.

Wurtsbaugh, W. A., Gardberg, J., & Izdepski, C. (2011). Biostrome communities and mercury and selenium bioaccumulation in the Great Salt Lake (Utah, USA). *Science of the Total Environment, 409*, 4425–4434.

Yang, S., Johnson, W. P., Black, F., Rowland, R., Rumsey, C., & Piskadlo, A. (2020). Response of density stratification, aquatic chemistry, and methylmercury to engineered and hydrologic forcings in an endorheic Lake (Great Salt Lake, U.S.A.). *Limnology and Oceanography, 65*, 915–926. https://doi.org/10.1002/lno.11358.

Middle Holocene Hydrologic Changes Catalyzed by River Avulsion in Big Soda Lake, Nevada, USA

Michael R. Rosen, Liam Reidy, Scott Starratt, and Susan R. H. Zimmerman

Abstract Big Soda Lake is a 63 m deep, 1.6 km^2 maar lake in the Great Basin of Nevada, USA. Water level in the lake is controlled by groundwater inputs from the surrounding aquifer and the only surface water input is rainfall, which is negligible. A core taken in 2010 records an 8.75 m depositional history of the lake. A radiocarbon date on fossil pollen from 8.4 m below the sediment water interface (BSWI) of 14,740 (+1120/−825) cal yr BP suggests that the core may cover the latest Pleistocene and Holocene depositional history of the lake. Stable isotope values of oxygen and carbon ($\delta^{18}O$ and $\delta^{13}C$) on authigenic calcite, diatom assemblages, and sedimentary structures all show consistent hydrological change from initially saline water at the bottom of the core to fresh/brackish water at about 6 m BWSI, back to saline water at 4.3 m. At 4.3 m depth, the bedding and color of the core change abruptly, and the stable- isotope and diatom assemblages indicate a consistently hypersaline lake until near the top of the core, when fresh water entered the lake due to irrigation and canal building in the twentieth century. The stable isotopes of the calcite abruptly change from inversely varying isotopic compositions below 4.3 m depth to covarying above. This break between relatively fresh and saline conditions in the lake occurs during the middle Holocene, although the exact timing of the transition is unknown due to variability in the ^{14}C age determinations. The cause for

M. R. Rosen (✉)
U.S. Geological Survey, California Water Science Center, Carson City, NV, USA
e-mail: mrosen@usgs.gov

L. Reidy
Department of Geography, University of California, Berkeley, Berkeley, CA, USA

S. Starratt
U.S. Geological Survey, Geology, Minerals, Energy, and Geophysics Science Center, Menlo Park, CA, USA
e-mail: sstarratt@usgs.gov

S. R. H. Zimmerman
Center for Accelerator Mass Spectrometry, Lawrence Livermore National Laboratory, Livermore, CA, USA
e-mail: zimmerman17@llnl.gov

M. R. Rosen et al. (eds.), *Limnogeology: Progress, Challenges and Opportunities*, Syntheses in Limnogeology, https://doi.org/10.1007/978-3-030-66576-0_10

such an abrupt change is difficult to explain through climate shifts, as evidence suggests climate in the Great Basin was different from what the Big Soda Lake record indicates in the Early Holocene. It is hypothesized that the Walker River flowed to the Carson River basin before 5600 cal yr BP, with water either flowing directly into the lake or raising the groundwater table sufficiently to freshen Big Soda Lake. The initial increase in salinity likely was caused by decreased flow of the Walker River due to Middle Holocene aridity. The lake level lowered slowly, and more saline conditions prevailed until 4.3 m depth when water from the Walker River stopped flowing into the Carson River basin. Above 4.3 m depth, diatom and isotopic evidence indicates that the lake became consistently saline. The isotopic and diatom assemblage transitions observed in Big Soda Lake sediment are not consistent with climate reconstructions and demonstrate that hydrologic shifts in a basin can be an important driver of change regardless of climatic conditions. However, climate shifts may also play a role in the hydrologic changes by supplying more or less water to river courses that may induce river avulsion.

Keywords Hydrology · Big Soda Lake · Walker Lake · Mono Lake · Paleoclimate · Paleohydrology · River avulsion

Introduction

Changes in the hydrologic conditions of lakes unrelated to climate are seldom recognized in the sedimentary record of paleolakes and may be uncommon. Some examples do exist, however, particularly for oxbow lakes (Reinfelds and Bishop 1998; Brooks and Medioli 2003) where meandering rivers avulse frequently, and ice-dammed glacial lakes, where expanding or melting ice can either impound or drain lakes more quickly than the climatic variations that are causing the ice changes (Gilbert and Desloges 1987; Valero Garcés et al. 2007), and catastrophic flooding from glacial outbursts that affects other lakes (Hanson and Clague 2016). Abundant evidence also exists for the non-climatic lowering of Lake Bonneville at ~18 ka, when the lake overtopped a sedimentary sill at Red Rock Pass and eroded ~125 m down to a bedrock sill (Miller et al. 2013; Oviatt 2018).

Abrupt nonclimatic hydrological changes in the arid west of the United States have also been documented for Late Holocene changes in the course of the Walker River, based on the appearance of lakes in the Carson Sink (Adams 2003; Adams and Rhodes 2019). These changes greatly affected the lake levels in Walker Lake, due to the diversion of the Walker River from its drainage to Walker Lake into the Carson River drainage. Adams (2003) demonstrated that avulsions of the Walker River occurred in the late Holocene, at around 1400 cal yr BP and again sometime between 915 and 650 cal yr BP, creating lakes at Carson Lake that attained elevations of 1198 m and 1204 m above sea level, respectively (Fig. 1). Adams and Rhodes (2019) documented five such avulsions to the Carson drainage between

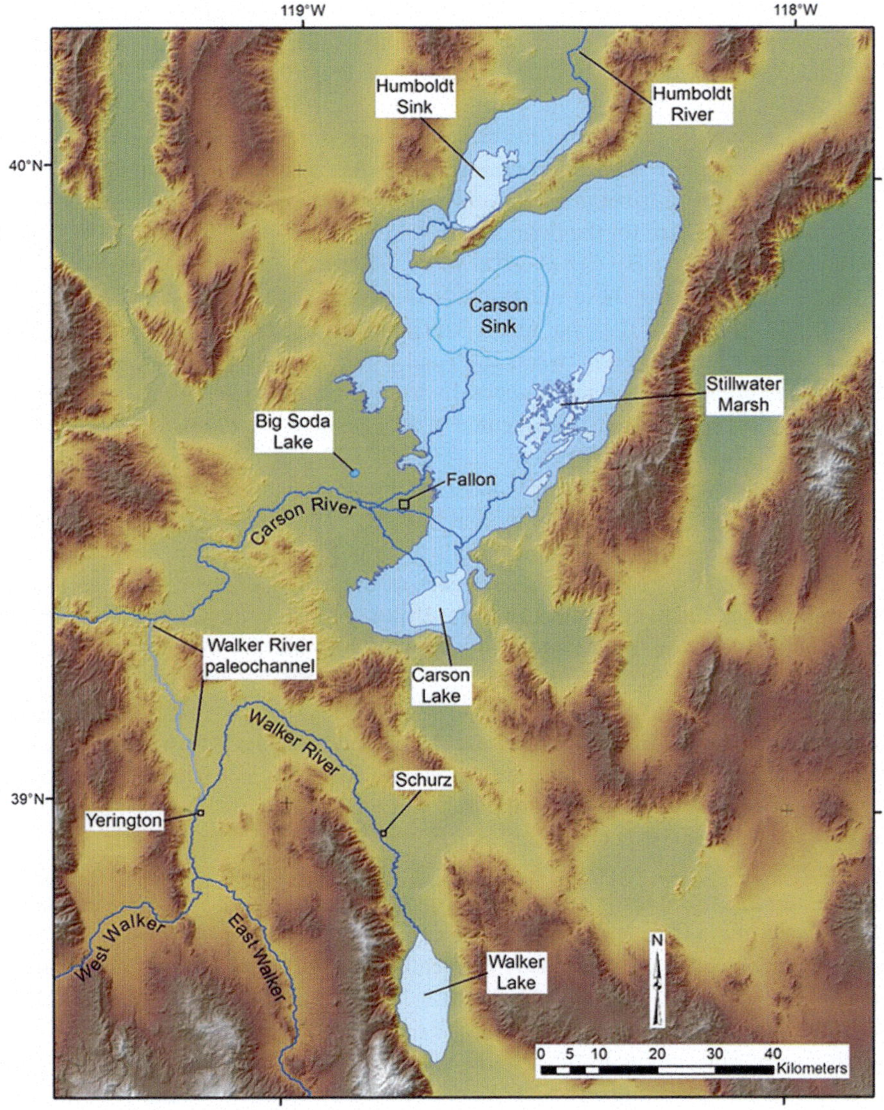

Fig. 1 Map of the Carson River basin showing Holocene lakes determined by Adams and Rhodes (2019). Walker River paleochannel that contributes additional water to the Carson River is identified north of Yerington. Big Soda Lake is located just north and west of Fallon. (Modified from Adams and Rhodes (2019))

2700 and 660 cal yr BP and estimated from tree ring records that each may have lasted 25–50 years. In addition, a roughly 1200 m elevation shoreline at Carson Lake has been shown to be approximately Younger Dryas in age, based on the work of Currey (1988, 1990), Benson et al. (1992), and Adams (2003), indicating the

Carson River basin had relatively more water entering into it than before. It has also been suggested by Bradbury et al. (1989) that the Walker River flowed to the Carson Sink during the Early and Middle Holocene, based on microfossil evidence in cores from Walker Lake. However, the core recovery, sampling resolution, and chronology for the longer of those records are quite poor, leaving doubt about the timing and duration of the hypothesized drainage switches. Additional arguments for low Walker Lake levels attributable to diversion of the Walker River are presented from a better-preserved core for the periods >6100 to 5585 cal yr BP (>5300–4800 yr BP) and 2780–2105 cal yr BP (2700–2100 yr BP) (Bradbury et al. 1989; Benson et al. 1991). The younger of these corresponds to two brief diversions documented by Adams and Rhodes (2019), but also corresponds to the driest part of the regional Late Holocene Dry Period between 2800 and 1850 cal yr BP (Mensing et al. 2013). The climate of the older period is generally less well-defined in regional records and up to this point has not been documented in the Carson River drainage (Adams et al. 2008; Adams and Rhodes 2019).

Periods of drier conditions in Walker Lake due to changes in the course of the Walker River should be mirrored by non-climatic shifts to wetter conditions in the Carson River drainage from the additional river flow, as shown by Adams and Rhodes (2019). However, the consequences of the avulsions have not been identified in Big Soda Lake, the only permanent Holocene lake in the Carson River basin. Moreover, except for the Younger Dryas paleoshoreline at ~1204 m elevation at Carson Lake (Currey 1988, 1990; Benson et al. 1992; Adams 2003), no other extensive late Pleistocene to Middle Holocene shorelines have been found in the Carson River basin. Indeed, Adams (2003) and Adams and Rhodes (2019) demonstrated that the additional flow of water from the Walker River alone, calculated from modern river discharge volumes, into the Carson River would not have been enough on its own to produce the lakes observed in Carson Lake and Carson Sink, and concluded that increased river flows caused by regional climate change would have been required to provide sufficient water into the basin to produce the large lakes that formed in the Late Holocene.

The Carson River itself has also historically changed course before the construction of Lahontan Reservoir in 1905. Russell (1885) and Bell and House (2010) provide evidence that the Carson River at times terminated in Carson Lake to the south, and at other times in the Carson Sink or Stillwater Marsh to the north. The alluvial plain surrounding Big Soda Lake is currently at an elevation between 1220 and 1215 m (Rush 1972), too high for the Late Holocene lakes at 1204 m and 1198 m to reach. However, the elevation of water in Big Soda Lake in 1881–1882 measured by I.C. Russell was near 1198 m elevation (Russell 1885; Rush 1972), which was either just below or close to the elevation of the Late Holocene high stands. Big Soda Lake is fed by groundwater and would not necessarily need a large, deep lake in Carson Lake to provide water that would increase the water level in Big Soda Lake. A raised water table caused either by Walker River input or movement of the Carson River channel closer to Big Soda Lake might be sufficient to add enough water to freshen Big Soda Lake, similar to the twentieth century rise in lake level caused by the increase in leakage from canals and irrigation into Big Soda and

Little Soda lakes caused by the construction of the Newlands Irrigation System (Rush 1972; Kharaka et al. 1984; Rosen et al. 2004).

This paper uses data from an 8.75 m sediment core taken from Big Soda lake in 2010 to provide evidence that an abrupt change observed in the Big Soda Lake sediment record appears to coincide with a Middle Holocene avulsion of the Walker River from the Carson River basin back to the Walker River basin as postulated by Bradbury et al. (1989) and Benson et al. (1991). One final piece of indirect evidence for diversion of the Walker River away from the Walker Lake basin is provided by the calculation by Broecker and Walton (1959) of the time required to accumulate the observed chloride concentrations of several Great Basin lakes, including Walker. The very low concentration of chloride in modern day Walker Lake suggested to those authors that Walker Lake must have desiccated completely at some time during the Holocene, possibly due to the diversion of the Walker River to the Carson Sink. We hypothesize that the abrupt changes observed in the biological and chemical components of Big Soda Lake sediments were caused by this avulsion and that Big Soda Lake provides additional evidence to support a hydrologic change, in addition to climate change that affected lake levels and the sedimentary records of both Big Soda and Walker Lakes.

Description of Big Soda Lake

Big Soda Lake (39°31′25.83″N, 118°52′42.03″W; WGS 1984 datum) is one of two poorly dated Holocene groundwater-fed maar lakes (along with Little Soda Lake) located in the Great Basin near Fallon, Nevada, USA. The age of the eruption that formed the lake has been hypothesized to be between 6000 and 7000 years ago, based on the lack of Holocene wave-cut terraces on Big Soda Lake crater walls (Stark et al. 1980; Cousens et al. 2012), but definitive dating of the eruption(s) has not been successful due to a complete lack of radiogenic argon in the basalt bombs (Cousens et al. 2012).

Big Soda Lake is currently a permanently meromictic lake with water level controlled by groundwater inputs from the surrounding aquifer. The lake was 63 m deep (1215.9 m above mean sea level) with an area of approximately 1.6 km^2 in 1971 (Rush 1972; Kharaka et al. 1984). Direct precipitation (91 mm/year) onto the lake and runoff from the surrounding catchment are small (Rush 1972). Prior to the diversion of Truckee River water to Lahontan Reservoir for irrigation in 1907, Big Soda Lake water level was about 18 m lower (elevation close to 1198 above sea level) than present, and the lake was fully mixed and hypersaline (Russell 1885). Leaking irrigation canals surrounding the lake since 1905 have contributed additional water, raising the lake level over a relatively short period from 1905 to 1930 (Rush 1972) that allowed the formation of 3- to 4-m-high tufa mounds (Rosen et al. 2004) and stratified the lake (Hutchinson 1937; Koenig et al. 1971; Kimmel et al. 1978; Kharaka et al. 1984). Although saline and evaporative, the lake is not completely closed as groundwater leaves the lake to the northeast, based on chloride mass-balance calculations (Hutchinson 1937; Rush 1972).

Previous Work at Big Soda Lake

Big Soda Lake was first studied in the 1880s by I.C. Russell as part of the King Surveys of the 40th parallel (King 1877; Russell 1885), and the limnology and microbiology of the lake are well understood (Hutchinson 1937; Cloern et al. 1983; Kharaka et al. 1984; Oremland et al. 1988; Hernandez-Maldonado et al. 2016). Recent work on the formation of modern groundwater-fed tufa deposits and inhibitions of carbonate precipitation has also contributed to the understanding of the limitations and dynamics of carbonate precipitation in the lake (Rosen et al. 2004; Reddy and Leenheer 2011). Reidy (2013) examined short (2 m length) cores and provides a relatively detailed history of the lake for the past 1600 years, but prior to this study no work on the entire Holocene depositional history of the lake has been undertaken.

Methods

A multidisciplinary study of an 8.75 m core taken from Big Soda Lake in 2010 included examination of the sedimentology, biology, and isotope chemistry of the core, as well as examination of seismic profiles taken after the coring was completed.

Coring

The vessel and coring operations were conducted by personnel from the University of Minnesota National Lacustrine Core Repository (LacCore) R/V KRKII, with the help of members of the investigative team.The R/V KRKII is a large floating platform consisting of two 5.5 m long skiff boats bolted together with aluminum deck plates and beams. Cores were cut into 1.5 m sections or smaller and shipped to the University of Minnesota LacCore core storage facility for subsequent analyses and subsampling. Two locations were successfully cored. The TOPGUN-SODA10-2A-1G and TOPGUN-SODA10- 2A-1K, drives 1–7, (8.87 m long) and TOPGUN-SODA10-3A-1G and TOPGUN-SODA10-3A-1K, drives 1–7, (9.25 m long) were extracted from near the center of Big Soda Lake in 2010 (Fig. 2) at water depths of 63 m using a Kullenberg corer (cores labeled with K). A gravity-core arm (cores labeled with G) was also used for collecting less-disturbed sediments from the sediment-water interface down to about 1 m penetration. However, for this analysis and for consistency, all samples were taken from TOPGUN-SODA10-2A-K,

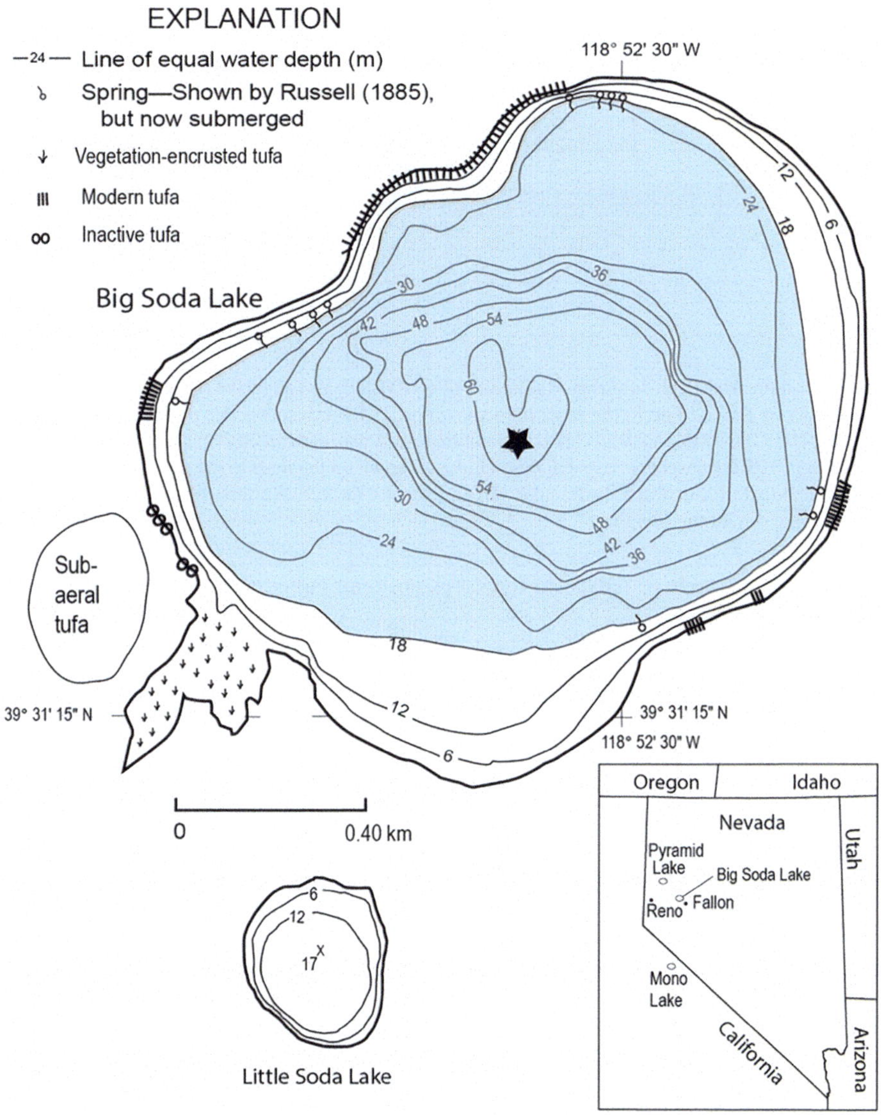

Fig. 2 Map of Big Soda and Little Soda l Lakes showing the water level in 1885 compared to the water level in 2003. Black star represents the location of core TOPGUN-SODA10-2A-1K. Original map is from Rush (1972). Datum is lake surface at stage of 1215.9 m above mean sea level for Big Soda Lake and 1216.1 m above mean sea level for Little Soda Lake. Historic water level of Big Soda Lake, prior to rise of surrounding water table in 1907, is shown as shaded area. (Adapted from map shown in (Russell 1885). Modified from Rosen et al. (2004))

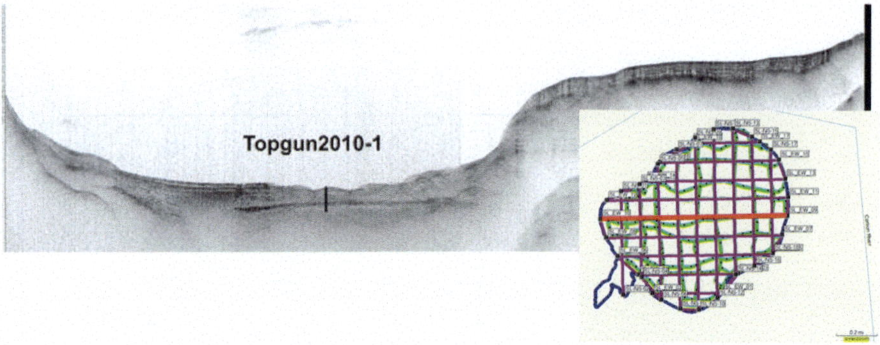

Fig. 3 Compressed High-Intensity Radiated Pulse (CHIRP) image of the sediments at cross section shown by the red line in the inset map showing CHIRP transects. The placement of the core location TOPGUN-SODA10-2A-1K is approximate and was not directly on the transect line. The grid of the CHIRP transects across Big Soda Lake is shown on the inset to the right. Pink lines were the hypothetical transects across the lake. Green lines are the actual course of the boat due to high winds on the day of the imaging

drives 1–7. The archive halves of the cores are maintained at LacCore and are kept under refrigeration.

Seismic Profiling

Seismic profiles (Fig. 3) of the lake bottom were conducted using a Compressed High-Intensity Radiated Pulse (CHIRP) subbottom profiler towed by an 8.5 m long Carolina Skiff. Profiles were conducted after coring was completed due to the availability of the CHIRP equipment and were not used to identify coring sites. The lake was traversed in a north-south and east-west grid on an extremely windy day, and tracklines were as straight as possible given the wind (Fig. 3). The use of the CHIRP data in this paper is to illustrate the transition from laminated to non-laminated sediments in the basin center. The CHIRP profiles are not used to locate depths of transitions and have not been used to create a three-dimensional model of sediment deposition in the lake.

Core Description, Sampling, and Grain-Size Analysis

All cores were split, photographed, and described at LacCore within 6 weeks of coring and further descriptions were completed at LacCore in January 2011. Detailed examination of bedding, sedimentary structures, grain size, and mineralogy using smear slides was conducted in January 2011. Examination of selected samples by X-ray diffraction (XRD) was conducted at the USGS, Reston offices.

Subsamples for radiocarbon analysis, stable isotopes, X-ray diffraction, and diatom examination were collected at various depth intervals depending on the analysis.

Detrital grain-size analysis of the samples from 8.75 to 2.13 m below the sediment-water interface (BSWI) was determined on approximately 0.1–0.5 cm^3 of sediment at 1 cm contiguous intervals. Samples were boiled in DI water and pre-treated with at least 30 mL of 30% H_2O_2 to remove fine-grained organic matter. Calcium carbonate was also removed using 10–20 mL 1 N HCl followed by centrifuging the sample. Biogenic silica was removed using 10–20 mL of 1 N NaOH prior to grain-size analysis. Samples were split, if necessary, to achieve an obscuration of 8–14%. All samples were run on a Malvern Mastersizer 2000 laser diffraction grain-size analyzer coupled to a Hydro 2000G. At the beginning of each measurement day, a tuff standard (TS2) with a known distribution between 1.0 and 16.0 μm (avg. 4.54 ± 0.07; n = 3194 [as of 6/27/09]) was measured twice and compared to past measurements to assess the equipment's accuracy and repeatability. Thereafter, TS2 was run every 10 samples to verify analytical repeatability and stability and once at the end of the day's analyses for a final assessment. The TS2 results were compared to values obtained by measuring known Malvern standards as an additional measure of stability. The measurement principle used is the Mie Scattering principle (de Boer et al. 1987). Sample measurement time was 30 seconds with 30,000 measurement snaps per single sample aliquot averaged per 10,000 snaps. The final three measurements (30,000 measurements/10,000 snaps = three time-averaged measurements) were compared for internal consistency per sample. All data are reported as volume percent and divided into 10 grain-size intervals as well as d(0.1) -10% of the volume distribution is below this value, d(0.5) -volume median diameter, d(0.9) 90% –of the volume distribution is below this value, %clay, %silt, %sand, and mode.

Radiocarbon Dating

The chronology for the core was established with radiocarbon determinations on pine pollen concentrates separated using procedures of Faegri and Iversen (1989), followed by purification with flow cytometry. Samples that yielded a minimum of 10,000 pollen grains were obtained using a COPAS Select large particle flow-cytometry machine in the Flow Cytometry Core Facility at the University of Indiana using procedures developed by Byrne et al. (2003). The purified pollen samples were graphitized and measured by accelerator mass spectrometry at the Center for Accelerator Mass Spectrometry (CAMS) at Lawrence Livermore National Laboratory. The resulting ^{14}C ages were converted to calibrated years before present (hereafter cal yr BP) using the computer program CALIB 4.4 and the IntCal98 curve (Stuiver et al. 1998). The median calibrated ages are reported along with the 2-σ error range (Table 1). A total of 20 radiocarbon dates were obtained from pollen concentrates on core SODA10-2A-1K. In addition, Reidy (2013) used three ^{14}C dates from pollen separates and two independently dated ash layers from an independently collected 1.5 m core to provide age control at the top of the SODA10-2A-1K

Table 1 Big Soda Lake calibrated ^{14}C ages in yr BP. BSWI is below sediment-water interface

LLNL-CAMS #	Sample designation	^{14}C age	±	Core depth (cm BSWI)	cal yr BP median probability	−Error	+Error
				0	50	0	0
152527	SODA10-210 cm	1570	70	210	1465	145	145
152528	SODA10-250 cm	3720	70	250	4070	200	220
155642	SODA10-299 cm	3860	90	299	4275	285	245
152529	SODA10-315 cm	4590	110	315	5265	370	315
152530	SODA10-335 cm	5500	110	335	6295	300	210
155643	SODA10-375 cm	4115	40	375	4645	125	175
152531	SODA10-400 cm	5560	130	400	6360	350	295
152532	SODA10-434 cm	4990	130	434	5740	410	260
155644	SODA10-438 cm	4750	80	438	5480	170	160
155645	SODA10-450 cm	6120	110	450	7000	260	260
155646	SODA10-475 cm	5470	130	475	6250	320	250
155647	SODA10-485 cm	4930	100	485	5680	210	320
155648	SODA10-500 cm	5330	90	500	6110	180	175
155649	SODA10-520 cm	4725	50	520	5465	145	120
155650	SODA10-540 cm	4740	120	540	5460	405	270
155651	SODA10-560 cm	4640	80	560	5380	325	205
155652	SODA10-575 cm	4790	45	575	5520	190	80
155653	SODA10-700 cm	4670	100	700	5400	350	200
155654	SODA10-730 cm	6430	120	730	7345	315	225
155655	SODA10-840 cm	12,570	230	840	14,740	825	1120

All ages from pollen concentrates with >10,000 pollen grains. LLNL-CAMS is the Lawrence Livermore National Laboratory – Center for Accelerator Mass Spectrometry

sediment record. The age control from Reidy (2013) was transferred to the SODA10-2A-1K core examined here using distinctively colored laminae.

Stable Isotopes

Carbon and oxygen isotopic analyses on bulk carbonate were performed on more than 747 dried, salt-free samples (washed in distilled water) taken at approximately 1 cm intervals, throughout the length of the core. At some intervals below 4.3 m BSWI, larger gaps are present because there was insufficient carbonate to produce an isotopic value. Above 4.3 m BSWI 437 samples (including replicates) were analyzed and below 4.3 m 310 samples (including replicates) were analyzed. Samples were reacted with phosphoric acid at 550 °C to release CO_2, using an autoprep device attached to a Micromass Optima mass spectrometer in the Earth and Planetary

Science Department, University of California, Berkeley. Isotopic results are reported from the laboratory relative to the Vienna PeeDee Belemite (VPDB) standard. Long-term external precisions for $\delta^{13}C$ and $\delta^{18}O$ analyses are ± 0.05‰ and ± 0.07‰, respectively. Replicate analyses of core samples were performed on 1% of the samples with an average difference of 0.3‰ VPDB for carbon and 0.75‰ VPDB for oxygen. Values for both oxygen and carbon isotopes were then converted to fractionation-corrected values of oxygen and carbon that reflect the isotopic composition of the water from which the carbonate precipitated using equations from O'Neil et al. (1969) for oxygen and Deines et al. (1974) for carbon (see Rosen et al. 2004 and Rosen 2020, for details). Experimental fractionation factors for monohydrocalcite are almost identical to those for calcite at 25 °C (Jiménez-López et al. 2001), so the fractionation factor for calcite has been used in calculating the isotopic water composition for those samples containing monohydrocalcite (mostly tufa samples). In this process, oxygen-isotope values are converted to Vienna Standard Mean Ocean Water (VSMOW). Correcting to the isotopic composition of the water from which the carbonate precipitated allows comparison to tufa isotopic values reported in Rosen et al. (2004) and allows for an estimate of water isotopic values that can be used for further work on this record. Correlation coefficients using Pearson's r (Pearson correlation coefficient) were determined using OriginLab 2020 software.

Diatoms

A total of 40 (~1 cm^3) sediment samples for diatom studies were collected. Twenty-seven samples were analyzed for diatoms at various intervals above 4.3 m in the core and 13 samples at various intervals were measured below 4.3 m. Seven samples were collected between 4.0 and 4.35 m BSWI across the transition from laminated to non-laminated sediments. Samples were processed using 30% hydrogen peroxide, 36% hydrochloric acid, and 70% nitric acid, (Battarbee 1986) to remove carbonate and organic matter, and 5% sodium pyrophosphate to deflocculate the remaining clay. Approximately 50 μL of the resulting suspension was dried on a 22 × 30 mm coverslip and permanently mounted using Naphrax®. At least 500 frustules were enumerated following the method of Schrader and Gersonde (1978) using a Leica DM LB2 microscope with Nomarski optics at magnifications of 630X and 1000X. Diatom identifications and ecological interpretations were based on Lowe (1974), Gasse (1986), Krammer and Lange-Bertalot (1986, 1988, 1991a, b), Lange-Bertalot and Krammer (1987, 1989), Cumming and others (1995), Round and Bukhtiyarova (1996), Krammer (1997a, b, 2000, 2002), and Lange-Bertalot (2001).

X-Ray Diffraction Analyses

For X-ray diffraction (XRD) analyses conducted at the USGS Reston, Virginia, bulk sediment was dried to obtain a homogeneous, dry powder that passed easily through a 60-mesh sieve. Whole samples were mounted with random orientation of the crystallites by the methods of McCreery (1949) and Zen and Hammerstrom (1975) and then analyzed with XRD to characterize the most prominent minerals. Clay minerals were separated from the whole sediment with grain-size segregation methods taken from Blatt et al. (1980) and Jackson (2005). X-ray diffraction analysis was performed on a Panalytical Xpert X-ray generator equipped with two automated goniometers. The generator produces CuKα radiation spectrally isolated through each goniometer with graphite crystal monochromators. The sollar slits were fixed at 1° and the receiving slits were 1/4°. A typical scan was 2–66° 2θ at a scan rate of 1° 2θ per minute. Settings of 45 kV and 25 mA were used. Percentage error of mineral identifications is ±10%.

All data for this publication is available in Rosen (2020).

Results

Core Description

From the bottom of the core to 7.8 m BSWI, the sediment is indistinctly bedded gray silty mud, and above this the sediment is a massive gray silty mud to 6.6 m BSWI (Fig. 4). Between 6.6 and 5.8 m BSWI, the core has tilted bedding and stringers of coarse sand and gravel that include potentially rotated blocks of mud. Above 5.8 m BSWI, there is more massive gray mud, but there is little to indicate disturbance or movement of the beds other than the massive nature of the interval. Above 4.3 m BSWI, the sediment changes abruptly from gray unlaminated sediment to brown and multicolored sediment that contains millimeter- to centimeter-scale laminations in several places. The area just above 4.3 BSWI may be disturbed, and from 4.3 to 4.0 m BSWI, there are intervals of broken diatom crusts. There is a clearly disturbed zone from about 1.5 to 2.3 m BSWI marked by broken layering and tilted sediment. Disturbed sediments are an issue and make establishing a reliable stratigraphy and chronology difficult. The distinct change at 4.3 m BSWI in sediment type, isotopic composition, and diatom assemblages signals a significant shift in the lake system at this point in time.

The interval of TOPGUN-SODA10-2A-1K from 8.87 to 7.75 m BSWI (Fig. 4) contains between 15% and 25% clay and 10–20% sand. Sand and silt content mirror each other throughout the core (Fig. 5). Sand content increases upward to about 6.59 m BSWI and then decreases sharply to a few percent until 6.50 m BSWI. From 6.50 to 2 m BSWI (where measurements stopped), the percent sand increases relatively consistently from <5% to more than 35% near 2 m BSWI (Fig. 5). However,

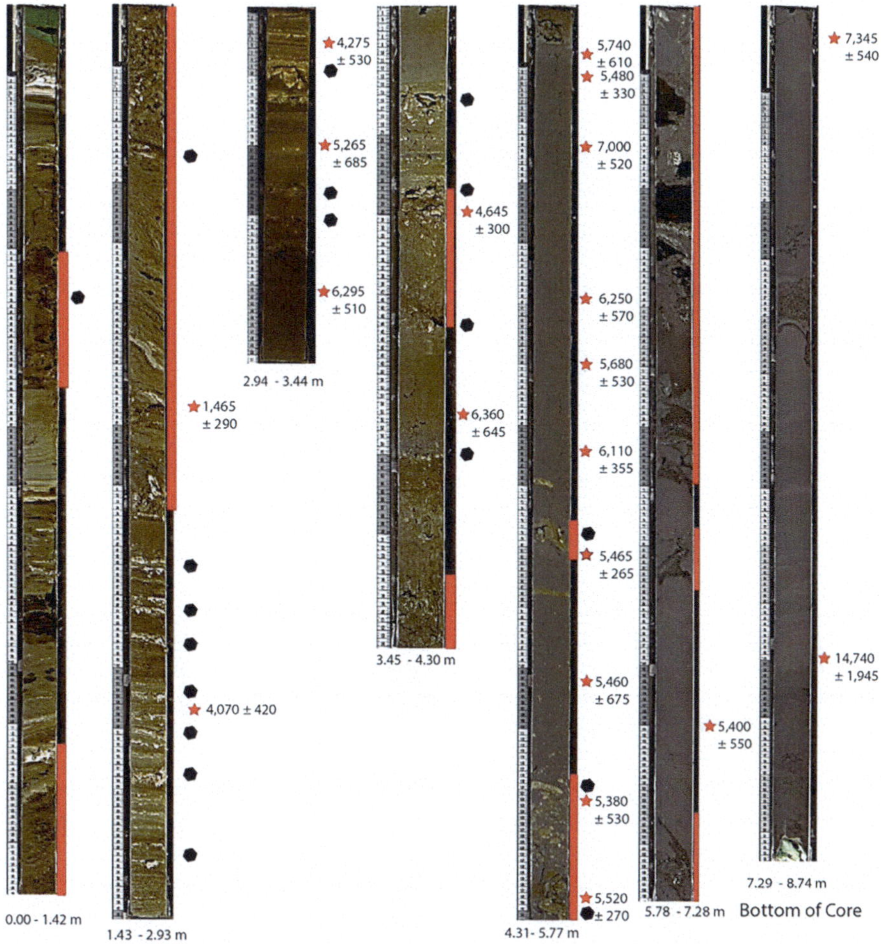

Fig. 4 Photo of SODA10-2A-1K showing the top of each core length at the top of the page. The top of the core is at the top left. Each gray and white block is 10 cm. Transition from brackish to saline is interpreted to be at 4.3 m. Black polygons are areas where diatom crusts are present. Red stars are pollen date locations with dates in cal yr BP listed beside them. Red rectangles are areas of the core that appear to be disturbed. Depths are below the sediment-water interface

two sharp spikes in sand content at 4.94 and 4.03 m BSWI may represent the sandy base of small debris flows. Clay content is mostly <20%, with the exception of the disturbed interval between 6.5 and 6.0 m, where the sediments consist of up to 40% clay. There is no correlation between grain-size and isotopic composition or diatom assemblage.

The mineralogy of the clays (Rosen 2020) is consistent with those derived from mafic volcanic rocks and is dominated by 17–18 Å smectite, with minor amounts of illite and kaolinite. There is a tendency for there to be slightly more illite, kaolinite, and 22–25 Å smectite below 4.3 m BSWI, based on semiquantitative analysis, but

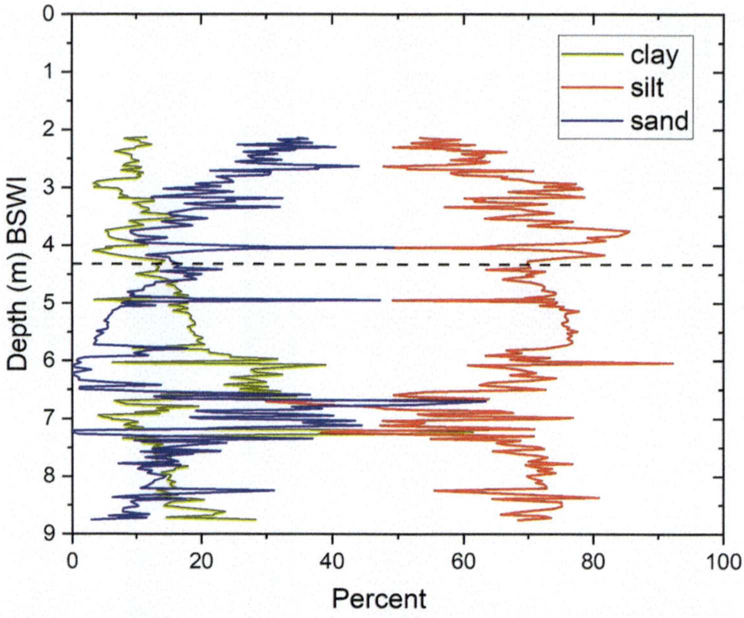

Fig. 5 Graph of sand, silt, and clay content in the core determined by laser diffraction grain-size analyzer. There is no discernable change in grain-size across the 4.3 m BSWI transition from brackish to saline conditions in the lake. Analyses above 2 m BSWI were not done on this core. BSWI = below the sediment-water interface

this is not consistent and there are not enough data to be certain of this trend. Calcite, which is present throughout the core, can vary from trace concentrations below 4.3 m to more than 50% of the sediment in the upper 20 cm of the core. Examination of the calcite using smear slides indicates that the calcite is authigenic and forms euhedral crystals (Fig. 6). There is considerable amorphous silica present in the cores, which is derived from the diagenesis of diatom frustules and occasional tephra layers in the upper 2 m of the core. Other minor and trace minerals that are present include detrital quartz, plagioclase, and potassium feldspars, chlorite, serpentine, pyrite, and mica.

Radiocarbon Dating of the Core

The radiocarbon results (Table 1) obtained from the pine pollen grains clearly do not follow the law of superposition, particularly below the distinctive sedimentary boundary at 4.3 m (Table 1, Fig. 7). The radiocarbon date for the bottom-most pollen sample in the core is 14,740 (+1120/−825) cal yr BP and is in a section that contains indistinct horizontal bedding and appears to be undisturbed. Between this level (8.4 m) and the 4.3 m boundary, eight of the 12 radiocarbon measurements

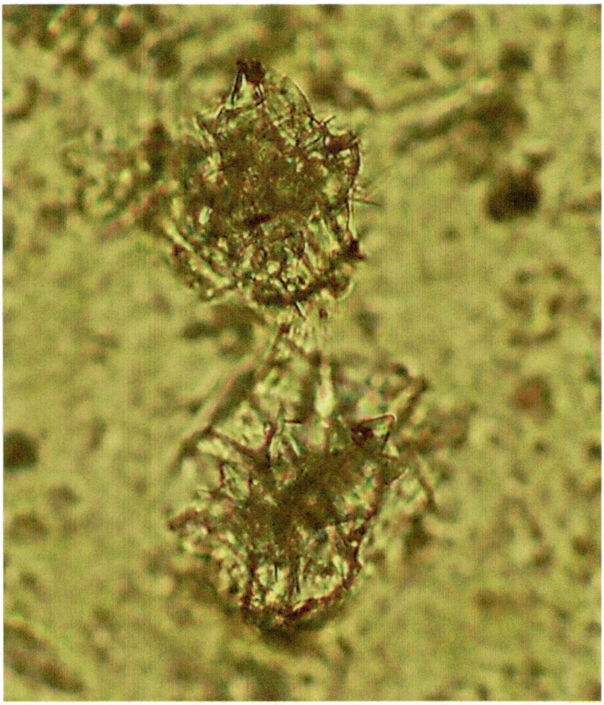

Fig. 6 Photomicrograph of a smear slide of authigenic calcite crystals in aggregates from the core. Crystal aggregates are approximately 10 μm in diameter

cluster between 5740 and 5265 cal yr BP. The other four are between 7345 and 6110 cal yr BP but are also out of stratigraphic order. Above 4.3 m, three of seven radiocarbon measurements fall within sediments identified as being disturbed. The four remaining measurements are in sediment that is not visibly disturbed and fall roughly in stratigraphic order. However, the lower two measurements (3.35 and 3.15 m) are statistically identical to the clusters of dates below the boundary. One possible explanation is that there may have been large instantaneous slumping of material in the lake that redeposited sediment in the bottom of the deep basin. Based on the significant changes in color, texture, bedding, and other sediment characteristics (see below) at 4.3 m depth, the most likely scenario seems to be a large-scale slump at that time, but smaller slumps or other mixing events after that time cannot be ruled out with the data in hand. However, it is not likely that the pollen grains represent sediment reworked from the base of the core to above 4.3 m BSWI, because distinct changes in isotopic composition and diatom assemblages occur below 4.3 m BSWI. That is, if sediment of similar age as that above 4.3 m BSWI was reworked and mixed with sediment below, the isotopic and diatom assemblages would also be mixed (see below). At this time, it is unclear why the pollen dates in the section between about 7.4 m and 3.2 m BSWI give approximately the same age.

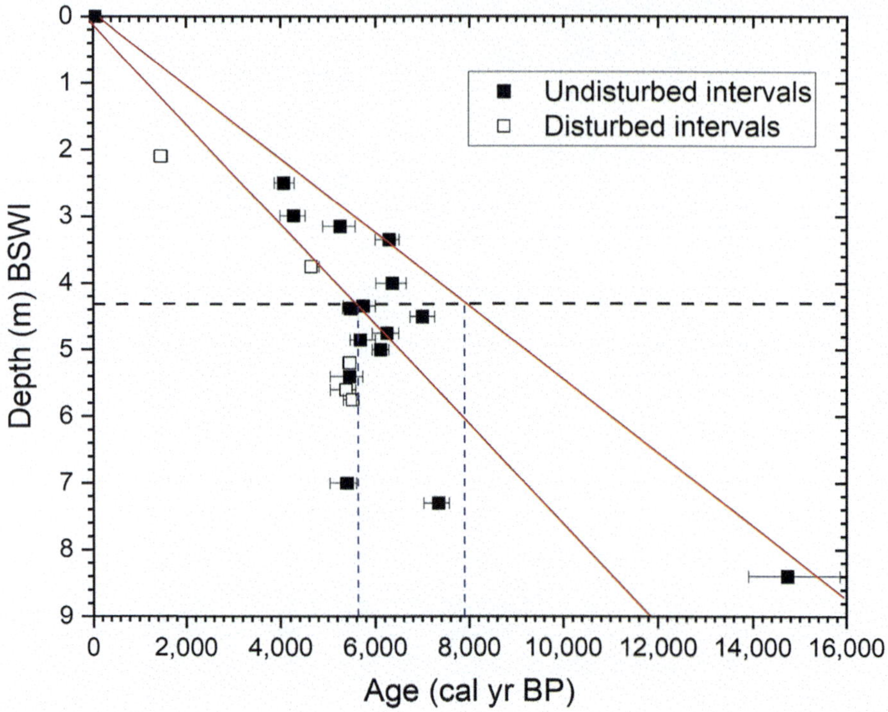

Fig. 7 Carbon-14 age dates from pollen separates using a flow cytometer. Each date represents >10,000 pollen grains. Squares in black are from areas of the core that are laminated and/or show little to no obvious disturbance of the sediment. The five open squares are in areas of the core that are disturbed, with little to no retained horizontal lamination. Black (long-dashed line) line is the transition in the character of the sediment at 4.3 m BSWI. The red lines are boundaries on most of the ages that are chronologically coherent, and the blue (short-dashed line) lines represent the range of dates that are likely at 4.3 m BSWI (7900–5600 cal yr BP). BSWI = below the sediment-water interface

Stable Isotopes

The $\delta^{18}O$ composition of the bulk authigenic carbonate ranges from −14.72 to +0.36‰ VSMOW and the $\delta^{13}C$ ranges from −4.30 to +3.36‰ VPDB (Fig. 8). The distribution of this range is not consistent throughout the core. Below 4.3 m BSWI, the average $\delta^{18}O$ value is −8.1 ± 2.6‰ VSMOW, whereas the average value above 4.3 m BSWI is −3.50 ± 1.3‰. There is little difference in range of $\delta^{13}C$ above or below 4.3 m BSWI, and the average values are not significantly different. However, oxygen and carbon isotopic values inversely covary below 4.3 m BSWI but abruptly switch to positively covarying above 4.3 m BSWI (Fig. 8). From the base of the core to about 5.8 m BSWI, the oxygen isotopic values become more negative to a minimum of −15‰. Above 5.8 m BSWI, the oxygen isotope values become

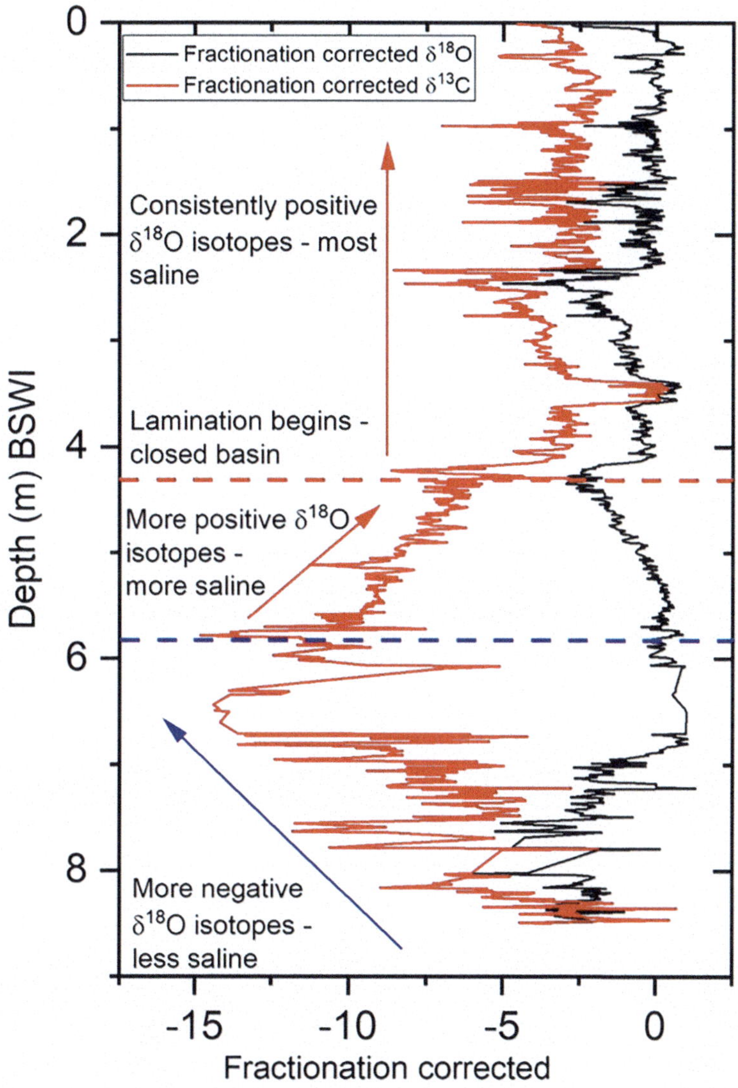

Fig. 8 Fast fourier transform-smoothed changes in fractionation-corrected $\delta^{18}O$ (red line) and $\delta^{13}C$ (black line) of calcite with depth in the core. Blue-dashed line indicates the most negative (freshest) depth of the lake at 5.8 m below the sediment-water interface. The oxygen-isotope composition then becomes more positive as it transitions to saline lake with laminated sediments (red-dashed line) at 4.3 m BSWI. Above 4.3 m depth, the lake remains consistently saline, with some excursions to less-saline lake water. Although never approaching the more negative values below 4.3 m BSWI. BSWI = below the sediment-water interface

progressively more positive until they are close to −5‰. Above 4.3 m BSWI, the oxygen isotopic values stay mostly between −5‰ and 0‰.

While slumping in parts of the core, particularly below 4.3 m BSWI, may obscure some of the trends in isotopic values, the transition from inversely covarying to positively covarying is difficult to explain by disturbance of the sediments (Figs. 4 and 8) because the oxygen and carbon isotopic variations occur in the same carbonate grains that are independent of where the sediment is located in the core. If the carbonate grains above and below 4.3 m BSWI were mixed with carbonate grains below 4.3 m BSWI, the covarying or inversely covarying nature of the carbonates above and below the break at 4.3 BSWI would not be discernable. Overall, the isotopic values follow trends that do not indicate that slumping is obscuring large portions of the record.

Diatoms

There are occasional hard orange crusts about 5–10 mm thick that occur throughout the core down to 5.8 m BSWI. The crusts (Fig. 4) are composed of almost single species of diatoms, although the core in general has a diverse diatom assemblage (Fig. 9a, b). A total of 118 species and 36 varieties of diatoms were observed (Rosen 2020). Diatom preservation was variable, ranging from excellent to good with little evidence of corrosion to very poor, where much of the fine structure has been removed through dissolution. The decrease in quality of preservation is due to dissolution, with limited evidence of physical breakage of frustules. Most identified species prefer alkaline waters, but upon death, the amorphous silica frustules may be subject to dissolution in high pH (10) water of the lake.

The diatoms in Big Soda Lake sediments may be grouped into those with a preference for "fresh" water (<3‰ total dissolved solids [TDS]) and "saline" (>5‰

Fig. 9 (continued) *Chaetoceros muelleri* Lemmermann, 1898, (3) *Cyclotella quillensis* L.W. Bailey 1922, (4) *Discostella stelligera* (Cleve and Grunow) Houk and Klee, 2004, (5) *Lindavia ocellata* (Pantocsek) Nakov et al., 2015, (6) *Cyclotella meneghiniana* Kützing 1844, (7) *Stephanodiscus carsonensis* Grunow in Schneider, 1878 [*S. carconensis*], (8) *Stephanodiscus excentricus* Hustedt, 1952, (9) *Stephanodiscus niagarae* Ehrenberg, 1845, (10) *Stephanodiscus parvus* Stoermer and Håkansson, 1984, (11) *Cocconeis placentula* Ehrenberg, 1838, (12) *Epithemia sorex* Kützing, 1844, (13) *Epithemia turgida* (Ehrenberg) Kützing, 1844, (14) *Fragilaria mesolepta* Rabenhorst, 1861, (15) *Fragilaria vaucheriae* (Kützing) Peterson, 1938, (16) *Rhopalodia gibberula* (Ehrenberg) O. Müller, 1895, (17) *Rhopalodia gibba* (Ehrenberg) O. Müller, 1895. Scale bar equals 10 μm. (**b**) Diatom species common in the upper section (above 4.3 m core depth) of core TOPGUN-SODA10 2A-K1. (1) *Anomoeoneis costata* (Kützing) Hustedt, 1959, (2) *Anomoeoneis sphaerophora* var. *minor* Kociolek and Herbst, 1992, (3) *Halamphora coffeaeformis* (C. Aagardh) Levkov, 2009, (4) *Navicula crucialis* Frenguelli, 1929, (5) *Nitzschia frustulum* Hustedt, 1945, (6) *Nitzschia reimerii* Kociolek and Herbst, 1992, (7) *Nitzschia latens* Hustedt, 1849, (8) *Nitzschia monoensis* Kociolek and Herbst, 1992, (9) *Nitzschia palea* (Kützing) W. Smith, 1856, (10) *Nitzschia pusilla* Grunow, 1862. Scale bar equals 10 μm

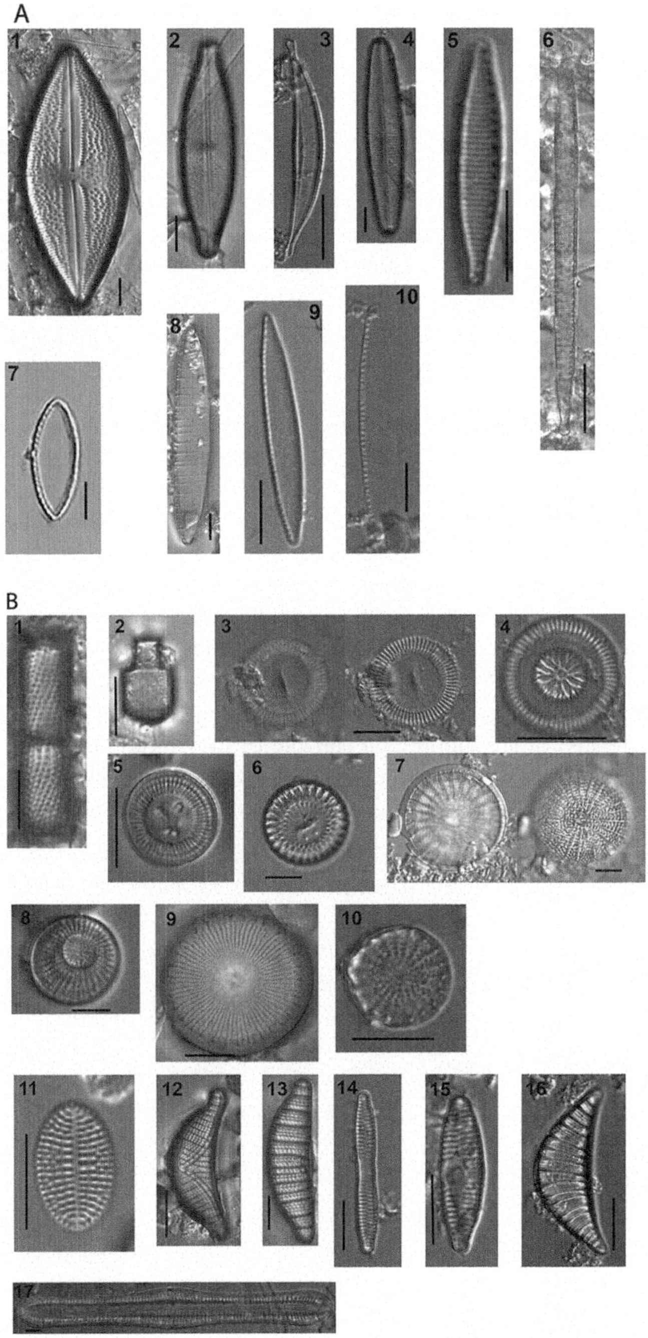

Fig. 9 (**a**) Diatom species common in the lower section (below 4.3 m core depth) of core TOPGUN-SODA10 2A-K1. (1) *Aulacoseira subarctica* (O. Müller) Haworth, 1990, (2)

TDS) (Bradbury 1997). For many species, these are poorly constrained categories, as numerous "freshwater" species with presumed optima below 1‰ TDS are found in waters that are several times more saline. Conversely "saline" species are found in fresh water, as well as waters that contain an order of magnitude higher concentration of dissolved solids than the 5‰ minimum for saline species (Fritz 1990; Fritz et al. 1993). Nutrient concentrations can affect the ability of some species to survive at salinities outside of their presumed ranges (Saros and Fritz 2000). *Rhopalodia* and *Epithemia* (DeYoe et al. 1992) and some *Nitzschia* (Kilham et al. 1986) are able to live in low-nitrogen environments due to the presence of cyanobacterial endosymbionts. Benthic diatoms may also congregate around freshwater springs that flow into more saline lake water, such as in the center of modern tufa mounds that occur at Big Soda Lake (Rosen et al. 2004), which complicates some aspects of the interpretation.

The physical habitat occupied by different diatom species is controlled by the availability of light and nutrients, and a range of other physical and chemical factors which determine the composition of a diatom assemblage. Planktic taxa float in the water column or are suspended by wind-driven turbulence. In some cases, the "planktic" nature of a species depends on the size and depth of the lake in which it is found. Benthic species are present in a number of forms. Semiattached diatoms are lightly attached to a substrate, such as aquatic macrophytes (epiphytic species) or rocks (epilithic species). These often form long chains which in turbulent conditions can be suspended in the water column and become part of the sediment in the deep, aphotic part of a lake. Other benthic taxa are attached by stalks or mucilage pads (adnate forms) or are motile and capable of moving through the mud. Some species exist in multiple environments, depending on seasonal conditions. With this information in mind, the diatom taxa are separated into four ecological groups:

Freshwater planktic: *Aulacoseira* (primarily *A. granulata*), *Discostella pseudostelligera, Lindavia bodanica, L. ocellata, Stephanodiscus* (except *S. excentricus*), *Tabellaria flocculosa.*

Freshwater benthic: *Cocconeis placentula, Epithemia, Fragilaria, Gomphonema, Navicula* (except *N. crucialis*), *Nitzschia frustulum, N. palea, Pinnularia, Pseudostaurosira brevistriata, Staurosira, Staurosirella, Synedra.*

Saline planktic: *Chaetoceros muelleri, Cyclotella* (*C. quillensis*, C. *meneghiniana, C. striata*).

Saline benthic: *Anomoeoneis* (*A. costata, A. sphaerophora* varieties), *Campylodiscus clypeus, Craticula halophila, Halamphora coffeaeformis, Navicula crucialis, Nitzschia latens, N. monoensis, N. pusilla, N. reimerii, Surirella.*

Variations in freshwater planktic species may reflect specific seasonal conditions. *Stephanodiscus* species are generally the dominant species in spring and fall blooms (Kilham et al. 1986, 1996; Bradbury 1988). In the Owens Lake system, California, USA, these blooms are associated with increased snowmelt

(Bradbury 1997). *Stephanodiscus niagarae* blooms in fresher conditions in Walker Lake as well (Bradbury 1987; Bradbury et al. 1989). *Aulacoseira granulata* is more common in warm-season, high-light conditions that are coincident with abundant nutrients. This species is common in shallow, temperate lakes with sufficient turbulence to keep the heavy frustules suspended in the photic zones (Kilham et al. 1986). In Owens Lake, *Aulacoseira granulata* indicates the presence of freshwater into the summer and fall (Bradbury 1997). *Aulacoseira* does not appear in the published Pyramid Lake and Walker Lake records (Bradbury 1987; Bradbury et al. 1989; Galat et al. 1981). *Lindavia ocellata* is present in oligotrophic and ultraoligotrophic alpine temperate lakes including those in the Lake Tahoe basin (Winder et al. 2009; Johnson et al. 2018). *Lindavia bodanica* occupies lakes with similar conditions. The presence of these species in Owens Lake in the latest Pleistocene indicates cold and moist conditions.

Chaetoceros muelleri is common in brackish and saline lakes (Rushforth and Johansen 1986; Fritz 1990; Fritz et al. 1993) and appears in large blooms, along with *Cyclotella quillensis* when the salinity in Walker Lake reaches about 5‰ TDS (Bradbury 1987; Bradbury et al. 1989). *Cyclotella meneghiniana* is present in brackish and saline lakes in Africa (Gasse et al. 1997; Hecky and Kilham 1973; Gasse 2002; Owen et al. 2004, 2008, 2012). This species is present in Walker Lake but is not as abundant as *C. quillensis* (Bradbury 1987; Bradbury et al. 1989). *Stephanodiscus excentricus* increases in abundance during periods of increased salinity.

As with the stable isotopes, the diatom-species assemblages are distinctly different below 4.3 m BSWI and above (Fig. 10a, b). Below 4.3 m, the diatom species *Lindavia ocellata* is most abundant (Rosen 2020), but *Chaetoceros muelleri* and *Stephanodiscus* spp. are also common (Fig. 10a). The diversity of the diatom assemblage below 4.3 m is also much greater than above 4.3 m. Most of the species present in abundance below 4.3 m disappear above 4.3 m BSWI, although some species such as *Chaetoceros muelleri* reappear in the upper 0.1 m of the core, after the recent stratification of the lake, caused by irrigation in the area after 1907 which allowed fresh water to enter the lake and raised the water level ~18 m. This also freshened the upper 30 m of lake water so that the TDS of the water above 30 m (the position of the chemocline) is <26,000 mg/L and the TDS below the chemocline is >85,000 mg/L (Kharaka et al. 1984).

The diatom assemblage above 4.3 m BSWI is characterized by diatoms that dominate for short intervals and then decline (Fig. 10b). From just below 4.3–2.85 m *Nitzschia palea* dominates, from 2.85 to 2.25 m *Nitzschia pusilla* dominates, from 2.25 to 1.0 m *Nitzschia latens* dominates, and above 1.0 m *Navicula crucialis* dominates. Species diversity in the upper 4.3 m is lower than below 4.3 m. Before 1885, Big Soda Lake had a TDS concentration of about 100,000 mg/L (Russell 1885) and was well-mixed. This high TDS content is similar to that found in modern hypersaline Mono Lake water (Domagalski et al. 1990).

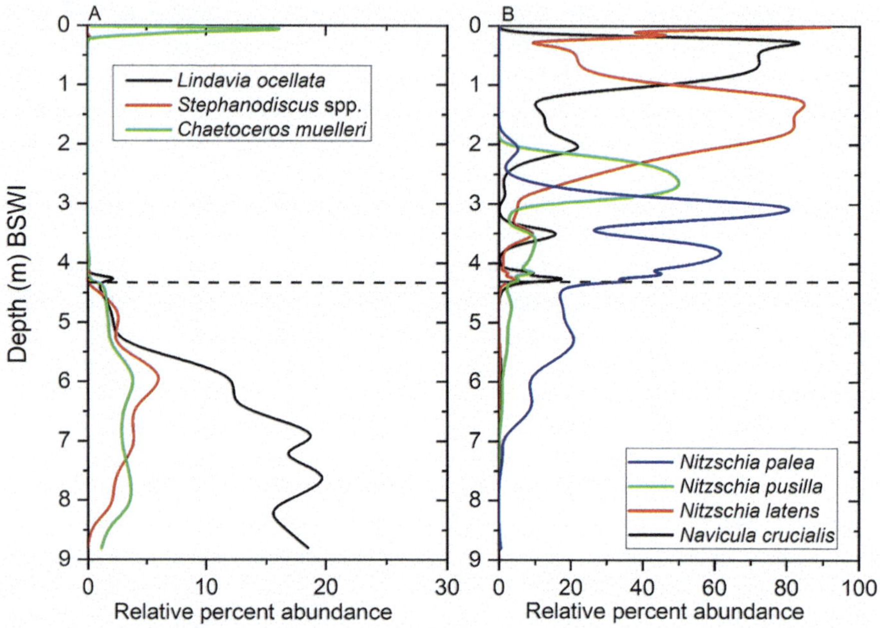

Fig. 10 Diatom abundances through the core: (**a**) abundant diatoms below 4.3 m BSWI. (**b**) abundant diatoms above 4.3 m BSWI. BSWI = below the sediment-water interface

Discussion

Stable Isotopes

The oxygen isotopic composition of the authigenic carbonate above 4.3 m BSWI is on average 5‰ more positive than deeper in the core. As more evaporation occurs in a closed basin, both oxygen and carbon become more positive and as freshwater is brought into the basin, both values become more negative. In addition, the oxygen- and carbon-isotopic compositions of individual samples above 4.3 m BSWI positively covary as is expected in a closed-basin evaporative lake (Rosen et al. 1988, 1989; Kelts and Talbot 1990; Talbot 1990). Below 4.3 m depth, the oxygen and carbon isotopic compositions inversely covary.

Different processes have been used to explain inverse covariation of stable isotopes in lacustrine carbonates. Valero Garcés et al. (1995) showed that inversely covarying oxygen- and carbon- isotopic values occurred when Medicine Lake, Minnesota, USA, became stably stratified (meromixis). Stratification caused productivity of pelagic bacterial communities at the chemocline, which consumed ^{12}C leaving the aragonite precipitated to have heavier $\delta^{13}C$ values. Valero Garcés et al. (1995) also found red laminations in these inversely covarying intervals indicating the large presence of bacterial plate communities. Li et al. (2000) also suggested

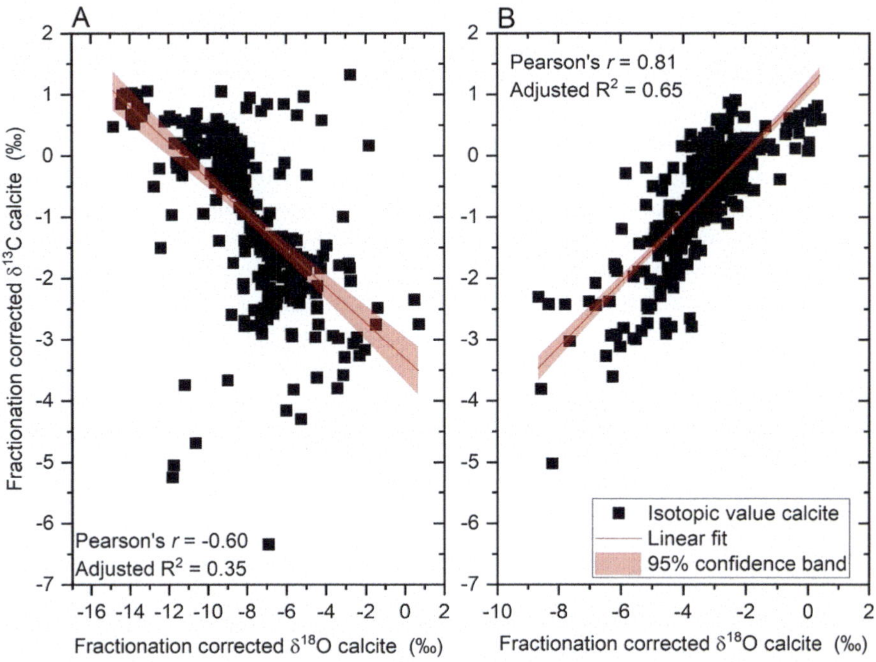

Fig. 11 Correlations of $\delta^{18}O$ and $\delta^{13}C$ of: (**a**) authigenic calcite below 4.3 m BSWI and (**b**) authigenic calcite above 4.3 m BSWI

higher productivity in saline Owens Lake to explain inversely covarying oxygen- and carbon-isotopic compositions of carbonate in Late Holocene samples. The stable meromixis was caused by increasing freshwater into the hypolimnion, decreasing the $\delta^{18}O$ values at the same time. Although Owens Lake was not stratified at this time, increased freshwater input was given as the reason for the decrease in the oxygen isotopic composition. In contrast, Last et al. (1998) explained inversely covarying sections of the basal section of the Clearwater Lake cores, Saskatchewan, Canada, as the product of an open hydrologic system of nonevolved (non-evaporated) waters. However, no limnologic explanation was given for why open hydrologic conditions in the lake would lead to inversely covariant oxygen- and carbon-isotopic values. In all of these explanations, the decrease in the oxygen-isotopic ratios was caused by an increase in freshwater input into saline lakes.

At Big Soda Lake, the statistically significant ($p < 0.001$) but weak (adjusted $R^2 = 0.35$) inverse correlation of oxygen- and carbon-isotopic values (Fig. 11a) is associated with the mostly homogenous gray sediments below 4.3 m. There is very little to no bedding in these units and no evidence of high productivity at a chemocline that would produce bacterial plate communities. In contrast, fine laminations and red and orange color in the sediment are present above 4.3 m depth (Fig. 4) when the oxygen- and carbon-isotopic values (Fig. 11b) positively covary ($p < 0.001$, adjusted $R^2 = 0.65$). Therefore, it is more likely that the inversely covarying trend of the oxygen and carbon isotopes is due to more open water conditions, which is consistent with the hypothesis that more fresh water was being brought in due to the presence of Walker River water in the Carson River basin.

Another possibility is a general increase in wetness in the region. However, the abrupt switch (within a few centimeters of sediment) from inversely covarying to positively covarying oxygen and carbon isotopes persists through the rest of the record (Fig. 8), indicating a sudden change that would be unlikely to be caused by gradually shifting climate. Coupled with the distinctive, abrupt change in sediment type and diatom community at 4.3 m (see below), it is difficult to assign this abrupt transition to climate. It is most likely that the shift from inversely covarying to positively covarying isotopic values occurred when the lake became sufficiently saline to act as a closed basin lake, and that the sudden isotopic change was caused by evaporation greatly exceeding groundwater input to the lake, caused by a combination of a drier climate and the avulsion of the Walker River back to the Walker basin.

Diatoms

The diatom assemblage below 4.3 m BSWI is characterized by a brackish-water community similar in composition to diatoms that were found in Walker Lake before it became more saline in the last 20 years due to freshwater diversions for irrigated agriculture (Fig. 10a). The Walker Lake diatom assemblage represents an average TDS of 10,000 mg/L and suggests an oxygenated water column (Fig. 10a). Above 4.3 m BSWI, the diatom community becomes indicative of diatom communities in Mono Lake, which has TDS between 80,000 and 100,000 mg/L (Fig. 10b). Individual abundances of diatoms appear to quickly come and go in the upper part of the core, possibly indicating the difficult conditions of the saline water body, and possibly climatic controls on individual species.

The change that takes place in the diatom assemblage at a core depth of about 4.3 m reflects a change in salinity as well as a change in the amount of shallow-water environment for benthic species. Below 4.3 m BSWI in core, the dominant species in the surface water represent conditions that are found during wetter parts of the Holocene in Walker Lake (Bradbury 1987; Bradbury et al. 1989). The planktic assemblage in Walker Lake is dominated by *Lindavia ocellata*, *Stephanodiscus niagarae*, and *S. rotula* (not present in Big Soda Lake). The remainder of the assemblage contains benthic species which reflect a freshwater to slightly brackish environment. The dominant benthic species is *Nitzschia palea*, which is found in waters of low conductivity.

As reflected in the stable isotopic composition of the calcite, the more saline diatom community becomes more dominant at about 5.8 m BSWI and the community transitions to saline by 4.3 m BSWI. Above 4.3 m BSWI, the diatom assemblage changes to reflect conditions that are similar to those found in present-day Mono Lake. The relative dominance of the assemblage by these species (*Navicula crucialis, Nitzschia latens, N. palea,* and *N. pusilla*) suggests that the lake level was relatively stable, which is consistent with a relatively stable isotopic composition. The decrease in abundance between 3.25 and 2.00 m may reflect drier climatic conditions, although the stable isotopic composition of the carbonate only reflects more saline conditions in the upper 0.50 m of this section. The sequence in these species also reflects an increase in salinity and an increase in alkalinity (Gasse 1986; Kociolek and Herbst 1992; Fritz et al. 1993; Owen et al. 2004). The recurrence of the brackish water species *Chaetoceros muelleri* at the top of the core follows the freshening of the upper 30 m of the water column to around 25,000 mg/L TDS after leaking canal and irrigation water raised the water level in the lake after 1905.

Age of Brackish to Saline Water Transition

Big Soda Lake became fresher from the base of the core to about 5.8 m BSWI and then progressively became more saline (Fig. 9), but constraining this age is difficult because of the irregularity of the ages in this core. If the date at 8.4 m faithfully records the beginning of sedimentation in the lake, the freshening in the lower part of the core may be related to the Younger Dryas, which several locations in the eastern Sierra Nevada indicate was significantly wetter (Briggs et al. 2005; Adams et al. 2008; Ali 2018; Adams and Rhodes 2019; Zimmerman et al. 2020). The transition from dark-gray and massive to olive-gray, well-bedded sediment, from inversely covarying to positively covarying isotopic values, and to the most saline diatom assemblages all occur at 4.3 m BSWI. Although difficult to interpret, the radiocarbon measurements suggest that the transition likely falls during the Middle Holocene, before about 4000 years ago.

Bradbury et al. (1989) examined a poorly preserved 50 m core taken in Walker Lake and postulated that the lack of microfossils between 26 and 16 m indicated that Walker Lake was essentially a playa between 20,320 (+2305/−2340) and 4465 (+820/−745) cal yr BP (16,800 ± 1000 and 4000 ± 280 cal yr BP). It was proposed that the Walker River broke through an alluvial barrier (Fig. 1) and changed direction to the north and entered the Carson River during that time. Microfossils and several other lines of evidence from a more complete but shorter core also show evidence of rapid filling of Walker Lake at about 5600 cal yr BP (Bradbury et al. 1989; Benson et al. 1991), supporting the inference that the Walker River was diverted to the Carson River drainage before that time. Geomorphic and stratigraphic evidence indicates that avulsion of the Walker River to the Carson River drainage also occurred several times in the Late Holocene (Adams 2003), so this is a recurring dynamic of the Walker River. The switch of Walker River flow back to Walker Lake at around 5600 years ago is within the possible age distribution of the distinct lithologic transition documented in Big Soda Lake from brackish water to saline water at 4.3 m BSWI. The isotopic and diatom transitions to more saline water likely occurred earlier and may be a result of increasing aridity in the region, reducing the volume of Walker River water. The complete switch of Walker River water back to the Walker basin may have occurred near the 5600 cal yr BP date indicated by Benson et al. (1991). Given the uncertainty of the dates measured in Big Soda Lake, it is feasible that the transition at Big Soda Lake was caused by the avulsion of the Walker River. There is some climatic reasoning to suggest flow from the Walker River was high at around 9000 cal yr BP, because Mono and Pyramid Lakes were higher than their later Holocene levels at this time (Adams and Rhodes 2019; Zimmerman et al. 2020), but gradually lowered during the middle Holocene arid period when Walker River flow would have decreased. However, closed basin conditions at Big Soda Lake are not apparent until after the 4.3 m BSWI transition. The event at 4.3 m depth is definitely older than more recent avulsions described by Adams and Rhodes (2019) at around 2700 cal yr BP and is most likely related to the older event proposed by Benson et al. (1991). New undisturbed cores from Big Soda Lake are needed to pinpoint the timing of this transition.

Age of Big Soda Lake

The age of the eruption of Big and Little Soda Lakes has not been determined from analysis of the erupted rocks found around the crater, although several attempts have been made (Cousens et al. 2012). Other informal studies have suggested that the eruption is quite young, less than 6000 and perhaps as recent as 1500 years ago, because there are no Holocene shorelines on the outside of the crater (dePolo et al. 2018). However, the lack of shorelines is a tenuous argument for a young eruption given the crater walls are composed of mostly unconsolidated sediments and may not have preserved any wave-cut terraces. There has also been substantial modern use of the land surrounding the lake that could have obliterated any shoreline benches. Furthermore, these suggested ages appear to be ruled out by the radiocarbon measurements on the SODA10-2A-K core reported here.

The ^{14}C measurement from a fossil pollen concentrate in the deepest sediments of Big Soda Lake at 8.4 m depth indicates the lake may be late Pleistocene in age (14,740 ± 1945 cal yr BP). Bathymetric mapping by Rush (1972) and our recent CHIRP profiling indicate that a shallower (30 m deep) broad explosion event occurred first, which likely raised the crater walls around the lake. A subsequent, more-focused, deeper, explosion created the ~65 m hole in the center of the lake (Fig. 3) after the initial explosion. The deep part of the lake had to have formed after the shallow 30 m shelf, as there is no way to create the shallow shelf without destroying or ejecting the deep basin center sediments if the deep basin had occurred first. The deep center portion of the lake has steep sides and shows evidence of slumping, whereas the shallow event appears relatively undisturbed and has laminated/bedded sediments throughout (Fig. 3).

It is possible that the deep eruption occurred much later than the ^{14}C date of 14,740 (+1120/−825) cal yr BP (but no younger than about 7000 cal yr BP because of dates higher in the core) but was simply filled with Lake Lahontan late Pleistocene sediments derived from the crater wall. However, it is difficult to envision laminated sediments being preserved on the relatively flat shelf at 30 m depth and having

Pleistocene sediments accumulate only in the basin center. If the deep eruption occurred after the shallow eruption as discussed above, Pleistocene sediment would have to cross the relatively flat, shallow shelf without being deposited on it or disturbing it. This seems unlikely; therefore, it is most likely that the Pleistocene sediments at the bottom of the core in the center of the lake were deposited there and not transported there. It is also unlikely that the lake level remained below the upper 30 m shelf during the Early Holocene because there was more fresh water entering the basin at that time, and during more arid times during the Middle and Late Holocene, the lake level was higher than the 30 m shelf. Overall, the timing of the two eruption events is unknown, but the late Pleistocene age of the sediment in the deep crater, suggests a late Pleistocene (~15,000 cal yr BP) age of the basin. It is possible that the two events were closely spaced in time and occurred during the same eruptive sequence.

The steep sides of the center explosion crater allow the likelihood of slumping (Fig. 3), although the first meter of sediment deposited in the basin center does not appear to be disturbed due to preserved coarse bedding. If this 14,740 cal year BP date is correct, it is likely that deposition began when the Walker River was flowing into the Carson River basin and Pleistocene Lake Lahontan was likely present in the basin. Russell (1885) suggested that the interlayered volcanic and lake sediments in the crater walls indicate that the eruption occurred beneath Lake Lahontan, which supports the 14,740 cal year BP age of the deepest sediment. On the other hand, given the overlap between this oldest age and the termination of the final Lahontan highstand at 15.8–15.15 ka (Adams and Wesnousky 1998), it would be interesting to test whether the crater-forming eruptions may have been related to unloading of the regional crust as Lake Lahontan disappeared.

Slumping in the basin may have occurred at two possible times. Slumping could have occurred as the diversion of the Walker River to the Carson River basin began. Rising groundwater levels may have caused instability of the sediments in the steep-sided central basin. Alternatively, when water levels dropped after the return of the Walker River to the Walker River drainage, slumping may have occurred because of the lack of support for the steep-sided walls of the deep inner basin. Slumping at the end of the diversion is more consistent with the ages found in these sediments. Diatoms with higher-salinity affinity and more positive stable-isotope values indicate that the lake started to become more saline around 9000 cal yr BP. At around 8000 cal yr BP, the regional climate became more arid (Bacon et al. 2006; Negrini et al. 2006; Miller et al. 2010; Grayson 2011; Kirby et al. 2012, 2015; Zimmerman et al. 2020) and the isotopic signature in Big Soda Lake appears to be consistent with this, even though Walker River water may still have been flowing into the Carson River basin. When the Walker River returned to the Walker basin around 5600 cal yr BP (Benson et al. 1991), groundwater levels may have fallen further and Big Soda Lake became permanently saline. The amount of groundwater decline after the Walker River water stopped flowing into the Carson River basin is unknown. While groundwater levels may have dropped below the level of the basin center resulting in desiccation, it is unlikely given that there is no evidence that the lakebed was ever dry. Moreover, water levels in the lake in 1880, the lowest historical levels recorded, were still about 20 m above the elevation of the lowest part of the basin.

Conclusions

While the stratigraphy and radiocarbon measurements in the SODA10-2A-K core are problematic, the timing of both the incursion of fresh water from the Walker River into the Carson River basin and its return to the Walker River basin, based on the dates of Benson et al. (1991) and Bradbury et al. (1989), is close to the range of dates obtained for the transition of fresh/brackish water to saline conditions in Big Soda Lake. More saline conditions are seen after 8000 cal yr BP as the Walker River los flow as the region became more arid. Sometime between 6000 and 4000 years ago (4.3 m BSWI), the lake became permanently saline when the Walker River avulsed back to the Walker River Valley. Subsequent diversions of the Walker River into the Carson basin shown by Adams (2003) around 1400 and sometime between 915 and 650 cal yr BP may correspond to more limited negative excursions in the isotopic record observed by Reidy (2013) in other short Big Soda Lake cores, but these events were likely more subdued if less water was available than at the end of the Pleistocene and early Holocene. This study shows that the combined use of stable isotopes of precipitated carbonate, diatoms, ^{14}C dating of concentrated pollen grains, and sedimentology can show that changes in the hydrochemistry of this saline lake can be attributed dominantly to hydrologic processes rather than exclusively climate change during the Middle Holocene. Although slumping of sediments complicates the age profile, the change in the hydrochemistry can be constrained to before the deepest date obtained in the laminated sediments because there is little to no mixing of the saline sediments with the fresher water sediments below.

Acknowledgments The authors would like to thank the anonymous book reviewers whose comments were extremely helpful in clarifying our arguments. Ken Adams, Desert Research Institute, provided a revised version of a figure he created for Adams and Rhodes (2019) that he kindly allowed us to use in the manuscript; we greatly appreciate this generosity. We would like to thank Roger Byrne (deceased) and Lynn Ingram, University of California, Berkeley for advice and support for LR, and for providing stable isotope data for this work. Many thanks to LacCore (University of Minnesota), in particular, Amy Myrbo, Kristina Brady, and Anders Noren for providing coring services and for enduring some extreme difficulties and expenses in obtaining the cores, and for shipping emergency equipment as needed. Thanks to Kurt Fritz (USGS) for help in the field during coring and to Kip Allander (USGS) for obtaining the seismic data. Many thanks also to C. Hassel of Indiana University for her enthusiastic collaboration on the development of flow cytometry separation of pollen over the years. Funding for the coring was provided as a grant to the University of California, Berkeley (LR) by the USGS Geosciences and Environmental Change Science Center, with particular thanks to Marith Reheis (USGS emeritus) for facilitating the funding. The radiocarbon dates were funded by LLNL Laboratory Directed Research and Development (LDRD) grant 09-ERI-003 to SZ; this is LLNL-JRNL-805156. Any use of trade, firm, or product names is for descriptive purposes only and does not imply endorsement by the U.S. Government.

References

Adams, K. D. (2003). Age and paleoclimatic significance of late Holocene lakes in the Carson Sink, NV, USA. *Quaternary Research, 60*, 294–306. https://doi.org/10.1017/qua.2018.151.

Adams, K. D., & Rhodes, E. J. (2019). Late Holocene paleohydrology of Walker Lake and the Carson Sink in the western Great Basin, Nevada, USA. *Quaternary Research, 92*, 165–182. https://doi.org/10.1017/qua.2018.151.

Adams, K. D., & Wesnousky, S. (1998). Shoreline processes and the age of the Lake Lahontan highstand in the Jessup Embayment, Nevada. *Geological Society of America Bulletin, 110*, 1318–1332.

Adams, K. D., Goebel, T., Graf, K., Smith, G. M., Camp, A. J., Briggs, R. W., & Rhode, D. (2008). Late Pleistocene and Early Holocene lake-level fluctuations in the Lahontan Basin, Nevada: Implications for the distribution of archaeological sites. *Geoarchaeology, 23*, 608–643.

Ali, G. A. H. (2018). *Late Glacial and Deglacial Fluctuations of Mono Lake, California* (Ph.D. thesis). Columbia University, 227 p.

Bacon, S. N., Burke, R. M., Pezzopane, S. K., & Jayko, A. S. (2006). Last glacial maximum and Holocene lake levels of Owens Lake, eastern California, USA. *Quaternary Science Reviews, 25*, 1264–1282. https://doi.org/10.1016/j.quascirev.2005.10.014.

Battarbee, R. A. (1986). Diatom analysis. In B. E. Berglund (Ed.), *Handbook of Holocene palaeoecology and palaeohydrology* (pp. 527–570). New York: Wiley.

Bell, J. W., & House, P. K. (2010). *Geologic map of the Grimes Point quadrangle, Churchill County, Nevada*. Nevada Bureau of Mines and Geology Map 173, 1:24,000 scale, 24 p. text.

Benson, L. V., Meyers, P. A., & Spencer, R. J. (1991). Change in the size of Walker Lake during the past 5000 years. *Palaeogeography, Palaeoclimatology, Palaeoecology, 81*, 189–214. https://doi.org/10.1016/0031-0182(91)90147-J.

Benson, L. V., Currey, D. R., Lao, Y., & Hostetler, S. W. (1992). Lake-size variations in the Lahontan and Bonneville basins between 13,000 and 9000 ^{14}C yr B.P. *Palaeogeography, Palaeoclimatolology, Palaeoecology, 95*, 19–32.

Blatt, H., Middleton, G., & Murray, R. (1980). *Origin of sedimentary rocks* (2nd ed., 782 p). Englewood Cliffs: Prentice-Hall.

Bradbury, J. P. (1987). Late Holocene diatom paleolimnology of Walker Lake, Nevada. *Archiv fur Hydrobiologie, Supplement 79, Monographische Beitrage, 1*, 1–27.

Bradbury, J. P. (1988). A climatic-limnologic model of diatom succession for paleolimnological interpretation of varved sediments at Elk Lake Minnesota. *Journal of Paleolimnology, 1*, 115–131.

Bradbury, J. P. (1997). A diatom record of climate and hydrology from the past 200 ka from Owens Lake, California with comparison to other Great Basin records. *Quaternary Science Reviews, 16*, 203–219.

Bradbury, J. P., Forester, R. M., & Thompson, R. S. (1989). Late Quaternary paleolimnology of Walker Lake, Nevada. *Journal of Paleolimnology, 1*, 249–267.

Briggs, R. W., Wesnousky, S. G., & Adams, K. D. (2005). Late Pleistocene and Late Holocene lake highstands in the Pyramid Lake subbasin of Lake Lahontan, Nevada, USA. *Quaternary Research, 64*, 257–263.

Broecker, W. S., & Walton, A. F. (1959). Re-evaluation of the salt chronology of several Great Basin Lakes. *Bulletin of the Geological Society of America, 70*, 601–618.

Brooks, G. R., & Medioli, B. E. (2003). Deposits and cutoff ages of Horseshoe and Marion oxbow lakes, Red River, Manitoba. *Géographie Physique et Quaternaire, 57*, 151–158.

Byrne, R., Park, J., Ingram, L., & Hung, T. (2003). *Cytometric sorting of Pinaceae pollen and its implications for radiocarbon dating and stable isotope analyses*. Presented at 20th annual Pacific climate workshop, Asilomar, CA, April 6–9, 2003.

Cloern, J. E., Cole, B. E., & Oremland, R. E. (1983). Seasonal changes in the chemistry and biology of a meromictic lake (Big Soda Lake, Nevada, U.S.A.). *Hydrobiologia, 105*, 195–206.

Cousens, B., Henry, C. D., & Gupta, V. (2012). Distinct mantle sources for Pliocene–Quaternary volcanism beneath the modern Sierra Nevada and adjacent Great Basin, northern California and western Nevada, USA. *Geosphere, 8*, 562–580. https://doi.org/10.1130/GES00741.1.
Cumming, B. F., Wilson, S. E., Hall, R. J., & Smol, J. P. (1995). Diatoms from British Columbia (Canada) lakes and their relationship to salinity, nutrients and other limnological variables. *Bibliotheca Diatomologica, 31*, 1–207.
Currey, D. R. (1988). Isochronism of final Pleistocene shallow lakes in the Great Salt Lake and Carson Desert regions of the Great Basin. In *Program and abstracts of the tenth biennial meeting* (p. 117). American Quaternary Association.
Currey, D. R. (1990). Quaternary palaeolakes in the evolution of semidesert basins, with special emphasis on Lake Bonneville and the Great Basin, U.S.A. *Palaeogeography, Palaeoclimatology, Palaeoecology, 76*, 189–214.
de Boer, G. B. J., de Weerd, C., Thoenes, D., & Goossens, H. W. J. (1987). Laser diffraction spectrometry: Fraunhofer diffraction versus Mie scattering. *Particle & Particle Systems Characterization, 4*, 14–19. https://doi.org/10.1002/ppsc.19870040104.
Deines, P., Langmuir, D., & Harmon, R. S. (1974). Stable carbon isotope ratios and the existence of a gas phase in the evolution of carbonate ground waters. *Geochimica et Cosmochimica Acta, 38*, 1147–1164.
dePolo, C. M., Henry, C. D., Zuza, A. V., Micander, R., & Faulds, J. E. (2018). *Sparkling or still? A tour of geology from Soda Lakes to Stillwater Marsh, Nevada*. Guide for the earth science week field trip, October 13, 2018. Nevada Bureau of Mines and Geology Educational Series E-63, 25 p. http://pubs.nbmg.unr.edu/Sparkling-or-still-p/e063.htm
DeYoe, H. R., Lowe, R. L., & Marks, J. C. (1992). Effects of nitrogen and phosphorus on the endosymbiont load of *Rhopalodia gibba* and *Epithemia turgida* (Bacillariophyceae). *Journal of Phycology, 28*, 773–777.
Domagalski, J. I., Eugster, H. P., & Jones, B. F. (1990). Trace metal geochemistry of Walker, Mono, and Great Salt Lakes. In R. J. Spencer & F.-M. Chou (Eds.), *Fluid-mineral interactions: A tribute to H. P. Eugster. The Geochemical Society, Special Publication No. 2* (pp. 315–353). San Antonio: The Geochemical Society.
Faegri, K., & Iversen, J. (1989). *Textbook of pollen analysis* (328 p). New York: Wiley.
Fritz, S. C. (1990). Twentieth-century salinity and water-level fluctuations in Devils Lake, North Dakota: Test of a diatom-based transfer function. *Limnology and Oceanography, 35*, 1771–1781.
Fritz, S. C., Juggins, S., & Battarbee, R. W. (1993). Diatom assemblages and ionic characterization of lakes of the northern Great Plains, North America: A tool for reconstructing past salinity and climate fluctuations. *Canadian Journal of Fisheries and Aquatic Sciences, 50*, 1844–1856.
Galat, D. L., Lider, E. L., Vigg, S., & Robinson, S. R. (1981). Limnology of a large, deep, North American terminal lake, Pyramid Lake, Nevada. *Hydrobiologia, 82*, 281–317.
Gasse, F. (1986). East African diatoms: Taxonomy, ecological distribution. *Bibliotheca Diatomologica, 11*, 1–201.
Gasse, F. (2002). Diatom-inferred salinity and carbonate oxygen isotopes in Holocene waterbodies of the western Sahara and Sahel (Africa). *Quaternary Science Reviews, 21*, 737–767.
Gasse, F., Barker, P., Gell, P. A., Fritz, S. C., & Chalie, F. (1997). Diatom-inferred salinity in palaeolakes: An indirect tracer of climate change. *Quaternary Science Reviews, 16*, 547–563.
Gilbert, R., & Desloges, J. R. (1987). Sediments of ice-dammed, self-draining Ape Lake, British Columbia. *Canadian Journal of Earth Sciences, 24*, 1735–1747.
Grayson, D. K. (2011). *The Great Basin: A natural history* (418 p). Berkeley: University of California Press.
Hanson, M. A., & Clague, J. J. (2016). Record of glacial Lake Missoula floods in glacial Lake Columbia, Washington. *Quaternary Science Reviews, 133*, 62–76. https://doi.org/10.1016/j.quascirev.2015.12.009.
Hecky, R. E., & Kilham, P. (1973). Diatoms in alkaline, saline lakes: Ecology and geochemical implications. *Limnology and Oceanography, 18*, 53–71.

Hernandez-Maldonado, J., Sanchez-Sedillo, B., Stoneburner, B., Miller, L., McCann, S., Rosen, M. R., Oremland, R. S., & Saltikov, C. W. (2016). The genetic basis of anoxygenic photosynthetic arsenite oxidation. *Environmental Microbiology*. https://doi.org/10.1111/1462-2920.13509.

Hutchinson, G. E. (1937). A contribution to the limnology of arid regions, primarily founded on observations made in the Lahontan Basin. *Transactions of the Connecticut Academy of Arts and Sciences, 33*, 47–132.

Jackson, M. L. (2005). *Soil chemical analysis: Advanced course* (revision of 2nd ed.). Parallel Press, University of Wisconsin, Madison, 895 p.

Jiménez-López, C., Caballero, E., Huertas, F. J., & Romanek, C. S. (2001). Chemical, mineralogical and isotope behavior, and phase transformation during precipitation of calcium carbonate minerals from intermediate ionic solution at 25 °C. *Geochimica et Cosmochimica Acta, 65*, 3219–3231.

Johnson, B. E., Noble, P. J., Heyvaert, A. C., Chandra, S., & Karlin, R. (2018). Anthropogenic and climatic influences on the diatom flora within the Fallen Leaf Lake watershed, Lake Tahoe Basin, California over the last millennium. *Journal of Paleolimnology, 59*, 159–173.

Kelts, K., & Talbot, M. (1990). Lacustrine carbonates as geochemical archives of environmental change and biotic/abiotic interactions. In M. M. Tilzer & C. Serruya (Eds.), *Large lakes: Ecological structure and function* (pp. 288–315). Berlin: Springer–Verlag.

Kharaka, Y. K., Robinson, S. W., Law, L. M., & Carothers, W. W. (1984). Hydrogeochemistry of Big Soda Lake, Nevada: An alkaline meromictic desert lake. *Geochimica et Cosmochimica Acta, 48*, 823–835.

Kilham, P., Kilham, S. S., & Hecky, R. E. (1986). Hypothesized resource relationships among African planktonic diatoms. *Limnology and Oceanography, 31*, 1169–1181.

Kilham, S. S., Theriot, E. C., & Fritz, S. C. (1996). Linking planktonic diatoms and climate change in the large lakes of the Yellowstone ecosystems using resource theory. *Limnology and Oceanography, 41*, 1052–1062.

Kimmel, B. L., Gersberg, R. M., Paulson, L. J., Axler, R. P., & Goldman, C. R. (1978). Recent changes in the meromictic status of Big Soda Lake, Nevada. *Limnology and Oceanography, 23*, 1021–1025.

King, C. (1877). *United States geological exploration of the 40th parallel*. U.S. Army Engineer Department Professional Paper 18, 7 volumes.

Kirby, M. E., Zimmerman, S. R. H., Patterson, W. P., & Rivera, J. J. (2012). A 9170-year record of decadal-to-multi-centennial scale pluvial episodes from the coastal Southwest United States: A role for atmospheric rivers. *Quaternary Science Reviews, 46*, 57–65. https://doi.org/10.1016/j.quascirev.2012.05.008.

Kirby, M. E., Knell, E. J., Anderson, W. T., Lachniet, M. S., Palermo, J., Eeg, H., Lucero, R., Murrieta, R., Arevalo, A., Silveira, E., & Hiner, C. A. (2015). Evidence for insolation and Pacific forcing of late glacial through Holocene climate in the Central Mojave Desert (Silver Lake, CA). *Quaternary Research, 84*, 174–186.

Kociolek, J. P., & Herbst, D. B. (1992). Taxonomy and distribution of benthic diatoms from Mono Lake, California, U.S.A. *Transactions of the American Microscopical Society, 111*, 338–355.

Koenig, E. R., Baker, J. R., Paulson, L. J., & Tew, R. W. (1971). Limnological status of Big Soda Lake, Nevada, October 1970. *Great Basin Naturalist, 31*, 106–108. Available at: https://scholarsarchive.byu.edu/cgi/viewcontent.cgi?article=1688&context=gbn

Krammer, K. (1997a). Die cymbelloiden diatomeen. Eine monographie der weltweit bekannten Taxa. Teil 1. Allgemeines und Encyonema part. *Bibliotheca Diatomologica, 36*, 1–382.

Krammer, K. (1997b). Die cymbelloiden diatomeen. Eine monographie der weltweit bekannten Taxa. Teil 2. Encyonema part., Encyonopsis, und Cymbellopsis. *Bibliotheca Diatomologica, 37*, 1–469.

Krammer, K. (2000). *Diatoms of Europe. Diatoms of the European inland waters and comparable habitats. Volume 1: The genus Pinnularia* (703 p). Ruggell: A.R.G. Gantner Verlag K.G.

Krammer, K. (2002). *Diatoms of Europe. Diatoms of the European inland waters and comparable habitats. Volume 3: Cymbella* (584 p). Ruggell: A.R.G. Gantner Verlag K.G.

Krammer, K., & Lange-Bertalot, H. (1986). Bacillariophyceae. 1. Teil: Naviculaceae. In H. Ettl, J. Gerloff, H. Heynig, & D. Mollenhauer (Eds.), *Süsswasser flora von Mitteleuropa, Band 2/1* (876 p). Stuttgart, New York: Gustav Fischer Verlag.

Krammer, K., & Lange-Bertalot, H. (1988). Bacillariophyceae. 2. Teil: Bacillariaceae, Epithemiaceae, Surirellaceae. In H. Ettl, J. Gerloff, H. Heynig, & D. Mollenhauer (Eds.), *Süsswasserflora von Mitteleuropa, Band 2/2* (611 p). Jena: Gustav Fischer Verlag.

Krammer, K., & Lange-Bertalot, H. (1991a). Bacillariophyceae. 3. Teil: Centrales, Fragilariaceae, Eunotiaceae. In H. Ettl, J. Gerloff, H. Heynig, & D. Mollenhauer (Eds.), *Süsswasserflora von Mitteleuropa, Band 2/3* (576 p). Jena: Gustav Fischer Verlag.

Krammer, K., & Lange-Bertalot, H. (1991b). Bacillariophyceae. 4. Teil: Achnanthaceae, Kritische Ergänzungen zu Navicula (Lineolatae) und Gomphonema, Gesamtliteraturverzeichnis Teil 1-4. In H. Ettl, G. Gärtner, J. Gerloff, H. Heynig, & D. Mollenhauer (Eds.), *Süsswasserflora von Mitteleuropa, Band 2/4* (437 p). Jena: Gustav Fischer Verlag.

Lange-Bertalot, H. (2001). *Diatoms of Europe: Diatoms of the European inland waters and comparable habitats. Volume 2: Navicula sensu strict, 10 Genera Separated from Navicula sensu lato. Frustulia* (526 p). Ruggell: A.R.G. Gantner Verlag K.G.

Lange-Bertalot, H., & Krammer, K. (1987). Bacillariaceae, Epithemiaceae, Surirellaceae. *Bibliotheca Diatomologica, 15*, 1–289.

Lange-Bertalot, H., & Krammer, K. (1989). Achnanthes eine Monographie der Gattungen. *Bibliotheca Diatomologica, 18*, 1–393.

Last, W. M., Vance, R. E., Wilson, S., & Smol, J. P. (1998). A multi-proxy limnologic record of rapid early-Holocene hydrologic change on the northern Great Plains, southwestern Saskatchewan, Canada. *The Holocene, 8*, 503–520.

Li, H. C., Stott, L. D., Bischoff, J. L., Ku, T. L., & Lund, S. P. (2000). Climate variability in East-Central California during the past 1000 years reflected by high-resolution geochemical and isotopic records from Owens Lake sediments. *Quaternary Research, 54*, 189–197.

Lowe, R. L. (1974). *Environmental requirements and pollution tolerance of freshwater diatoms: U.S. Environmental Protection Agency Report EPA-670/4-74-005* (340 p). Cincinnati: National Environmental Research Center.

McCreery, G. L. (1949). Improved mount for powder specimens used on the Geiger Counter X-ray spectrometer. *Journal of American Ceramic Society, 32*, 141–146.

Mensing, S. A., Sharpe, S. E., Tunno, I., Sada, D. W., Thomas J. M., Starratt, S., Smith, J. (2013). The Late Holocene Dry Period: multiproxy evidence for an extended drought between 2800 and 1850 cal yr BP across the central Great Basin, USA. *Quaternary Science Reviews 78*, 266–282

Miller, D. M., Schmidt, K. M., Mahan, S. A., McGeehin, J. P., Owen, L. A., Barron, J. A., Lehmkuhl, F., & Löhrer, R. (2010). Holocene landscape response to seasonality of storms in the Mojave Desert. *Quaternary International, 215*, 45–61. https://doi.org/10.1016/j.quaint.2009.10.001.

Miller, D. M., Oviatt, C. G., & McGeehin, J. P. (2013). Stratigraphy and chronology of Provo shoreline deposits and lake-level implications, Late Pleistocene Lake Bonneville, eastern Great Basin, USA. *Boreas, 42*, 342–361. https://doi.org/10.1111/j.1502-3885.2012.00297.x.

Negrini, R. M., Wigand, P. E., Draucker, S., Gobalet, K., Gardner, J. K., Sutton, M. Q., & Yohe, R. M. (2006). The Rambla highstand shoreline and the Holocene lake-level history of Tulare Lake, California, USA. *Quaternary Science Reviews, 25*, 1599–1618. https://doi.org/10.1016/j.quascirev.2005.11.014.

O'Neil, J. R., Clayton, R. N., & Mayeda, T. K. (1969). Oxygen isotope fractionation in divalent metal carbonates. *The Journal of Chemical Physics, 51*, 5547–5558.

Oremland, R. S., Cloern, J. E., Sofer, Z., Smith, R. L., Culbertson, C. W., Zehr, J., Miller, L., Cole, B., Harvey, R., Iversen, N., Klug, M., Des Marais, D. J., & Rau, G. (1988). Microbial and biogeochemical processes in Big Soda Lake, Nevada. In A. J. Fleet et al. (Eds.), *Lacustrine petroleum source rocks. Geological Society Special Publication No. 40* (pp. 59–75). London: Geological Society.

Oviatt, C. G. (2018). Geomorphic controls on sedimentation in Pleistocene Lake Bonneville, eastern Great Basin. In S. W. Starratt & M. R. Rosen (Eds.), *From saline to freshwater: The diver-*

sity of Western Lakes in space and time. Geological Society of America Special Paper 536. https://doi.org/10.1130/2018.2536(04).
Owen, R. B., Renaut, R. W., Hover, V. C., Ashley, G. M., & Muasya, A. M. (2004). Swamps, springs and diatoms: Wetlands of the semi-arid Bogoria-Baringo Rift, Kenya. *Hydrobiologia, 518*, 59–78.
Owen, R. B., Renaut, R. W., & Jones, B. (2008). Geothermal diatoms: A comparative study of floras in hot spring systems of Iceland, New Zealand, and Kenya. *Hydrobiologia, 610*, 175–192.
Owen, R. B., Lee, R. K. L., & Renaut, R. (2012). Early Pleistocene lacustrine sedimentation and diatom stratigraphy at Munya wa Gicheru, southern Kenya Rift Valley. *Palaeogeography, Palaeoclimatology, Palaeoecology, 331–332*, 60–74.
Reddy, M. M., & Leenheer, J. (2011). Calcite growth – rate inhibition by fulvic acids isolated from Big Soda Lake, Nevada, USA, the Suwannee River, Georgia, USA and by polycarboxylic acids. *Annals of Environmental Science, 5*, 41–53.
Reidy, L. M. (2013). *Lake sediments as evidence of natural and human-induced environmental change from California and Nevada* (Ph.D. dissertation). University of California, Berkeley, 99 p.
Reinfelds, I., & Bishop, P. (1998). Palaeohydrology, palaeodischarges and palaeochannel dimensions: Research strategies for meandering alluvial rivers. In G. Benito, V. R. Baker, & K. J. Gregory (Eds.), *Palaeohydrology and Environmental Change* (pp. 27, 353 p–42). Chichester: Wiley.
Rosen, M. R. (2020). *Mineralogic, grain-size, biologic, and stable isotopic analyses of core TOPGUN-SODA10 2A-K from Big Soda Lake, Nevada, USA*. U.S. Geological Survey data release. https://doi.org/10.5066/P9INH1ID
Rosen, M. R., Miser, D. E., & Warren, J. K. (1988). Sedimentology, mineralogy, and isotopic analysis of Pellet Lake, Coorong Region, South Australia. *Sedimentology, 35*, 105–122.
Rosen, M. R., Miser, D. E., Starcher, M. A., & Warren, J. K. (1989). Formation of dolomite in the Coorong Region, South Australia. *Geochimica et Cosmochimica Acta, 53*, 661–669.
Rosen, M. R., Arehart, G. B., & Lico, M. S. (2004). Exceptionally fast growth of <100-year-old tufa, Big Soda Lake, Nevada: implications for using tufa as a paleoclimate proxy. *Geology, 32*, 409–412.
Round, F., & Bukhtiyarova, L. (1996). Four new genera based on Achnanthes (Achnanthidium), together with a redefinition of *Achnanthidium*. *Diatom Research, 11*, 345–361.
Rush, F. E. (1972). *Hydrologic reconnaissance of Big and Little Soda Lakes, Churchill County, Nevada*. Carson City, Nevada Department of Conservation and Natural Resources, Division of Water Resources, Report 11, 3 p. Available at: http://images.water.nv.gov/images/publications/Information%20series/11.pdf
Rushforth, S. R., & Johansen, J. R. (1986). The inland *Chaetoceros* (Bacillariophyceae) species of North America. *Journal of Phycology, 22*, 441–448.
Russell, I. C. (1885). *Geological history of Lake Lahontan, a Quaternary lake of northwestern Nevada* (288 p). Washington, DC: U.S. Geological Survey.
Saros, J. E., & Fritz, S. C. (2000). Nutrients as a link between ionic concentration/composition and diatom distributions in saline lakes. *Journal of Paleolimnology, 23*, 449–453.
Schrader, H., & Gersonde, R. (1978). Diatoms and silicoflagellates. *Utrecht Micropaleontology Bulletin, 17*, 129–176.
Stark, M., Wilt, M., Haught, J. R., & Goldstein, N. (1980). *Controlled-source electromagnetic survey at Soda Lakes geothermal area, Nevada*. Lawrence Berkeley Laboratory Report LBL11221, 13 p.
Stuiver, M., Reimer, P. J., Bard, E., Beck, J. W., Burr, G. S., Hugen, K. A., Kromer, B., McCormac, G., van der Plicht, J., & Spurk, M. (1998). IntCal98 radiocarbon age calibration, 24,000–0 cal BP. *Radiocarbon, 40*, 1041–1083.
Talbot, M. (1990). A review of the paleohydrological interpretation of carbon and oxygen isotopic ratios in primary lacustrine carbonates. *Chemical Geology (Isotope Geoscience Section), 80*, 261–279.

Valero Garcés, B. L., Kelts, K., & Ito, E. (1995). Oxygen and carbon isotope trends and sedimentological evolution of a meromictic and saline lacustrine system: the Holocene Medicine Lake basin, North American Great Plains, USA. *Palaeogeography, Palaeoclimatology, Palaeoecology, 117*, 253–278.

Valero Garcés, B. L., Moreno, A., Morellón, M., Gonzáles-Sampériz, P., Corella, P., & Rico, M. (2007). From Carboniferous to Quaternary: a space and time lake transect from the Pyrenees to the Ebro Basin. In C. Arenas, A. M. Alonso-Zarza, & F. Colombo (Eds.), *Geo-Guías, Lacustrine field trips to the lacustrine deposits of the northeast of Spain* (pp. 111–152). Salamanca: Sociedad Geológica de España.

Winder, M., Reuter, J. E., & Schladow, S. G. (2009). Lake warming favours small-sized planktonic diatom species. *Proceedings of the Royal Society, Series B, 276*, 427–435.

Zen, E.-A., & Hammerstrom, J. G. (1975). Quantitative determination of dawsonite in Green River Shale by powder-sample X-ray diffraction: effect of grinding. *United States Geological Survey Journal of Research, 31*, 21–30.

Zimmerman, S. R. H., Hemming, S. R., & Starratt, S. (2020). Holocene sedimentary architecture and paleoclimate variability at Mono Lake, CA. In S. W. Starratt & M. R. Rosen (Eds.), *GSA Special Paper 536, From saline to freshwater: The diversity of Western Lakes in space and time*. https://doi.org/10.1130/SPE536.

Diatom Record of Holocene Moisture Variability in the San Bernardino Mountains, California, USA

Scott W. Starratt, Matthew E. Kirby, and Katherine Glover

Abstract Lower Bear Lake, in the San Bernardino Mountains, contains a Holocene paleohydrology record for southern California. The diatom and sediment geochemistry record indicates that the region experienced a wet Early Holocene followed by a gradual decrease in precipitation, which was punctuated by four strong and five weak pluvial episodes. The Lower Bear Lake record is compared with that of Silver Lake, a Mojave River terminal lake with headwaters in the San Bernardino Mountains, which exhibited several pluvial events at roughly the same time. The comparison is extended to records in relative proximity to Bear Lake (Dry Lake, Lake Elsinore, and San Joaquin marsh) and to two lakes with headwaters in the Sierra Nevada (Tulare Lake and Owens Lake). All exhibit a wet Early and early Middle Holocene wet interval and gradual drying through the remainder of the Holocene but differ in the expression of the pluvial episodes observed at Lower Bear Lake. The pluvial episodes are likely the result of changes in the storm track that affects the frequency and magnitude of winter storms in the area. These episodes are controlled by complex oceanic and atmospheric interactions and may be the result of the synchronous interaction of several teleconnections.

Keywords Lower Bear Lake · California · Holocene paleoclimate · Paleohydrology · Diatoms

---

Supplementary Information The online version of this chapter (https://doi.org/10.1007/978-3-030-66576-0_11) contains supplementary material, which is available to authorized users.

S. W. Starratt (✉)
U.S. Geological Survey, Geology, Minerals, Energy, and Geophysics Science Center, Menlo Park, CA, USA
e-mail: sstarrat@usgs.gov

M. E. Kirby
Department of Geological Sciences, California State University-Fullerton, Fullerton, CA, USA

K. Glover
Sawyer Environmental Research Building, Climate Change Institute, Orono, ME, USA

M. R. Rosen et al. (eds.), *Limnogeology: Progress, Challenges and Opportunities*, Syntheses in Limnogeology, https://doi.org/10.1007/978-3-030-66576-0_11

Introduction

California is a place of hydroclimatic extremes, both wetness (e.g., the A.D. 1861–1862 floods that drowned much of the Sacramento and San Joaquin valleys) and dryness (e.g., the 1990 multiple-year and recent four-year once-in-millennium droughts). Both extremes are related to the amount of winter precipitation the state receives and where that precipitation falls, on an annual basis. Climate model projections suggest an uncertain future for California, perhaps wetter or drier or some other combination of frequency versus changes in intensity (Overpeck et al. 2013; Stevenson et al. 2015; Kim et al. 2018). Whatever the scenario, paleoperspectives are required to understand the breadth of possible outcomes and their forcings. Over the past few decades, a wealth of paleoclimatic archives have helped to develop this paleoperspective including: lake-level studies (Lindström 1990; Stine 1990; Negrini et al. 2006; Kleppe et al. 2011; Morgan and Pomerleau 2012), speleothems (Oster et al. 2009, 2017; McCabe-Glynn et al. 2013), tree rings (Hughes and Brown 1992; Graumlich 1993; Hughes and Graumlich 1996; Hughes and Funkhouser 1998; Meko et al. 2001; MacDonald et al. 2008), lacustrine sediment and tufa isotopes (Benson et al. 2002), pollen and charcoal (Davis et al. 1985; Davis 1992, 1999; Brunelle and Anderson 2003; Anderson et al. 2010; Dingemans et al. 2014), and microfossils (Starratt 2004; Malamud-Roam et al. 2006). In addition to terrestrial sites, marine records reveal evidence for terrestrial climate variation in the form of vegetation reconstructions (Heusser et al. 2015), flood histories (Inman and Jenkins 1999; Hendy et al. 2015), and sea surface temperature (Fisler and Hendy 2008); the latter play an important role in modulating winter storm track trajectories and storm frequency (Kim et al. 2018, 2019; Toste et al. 2018). Despite the wealth of data, there remain many questions yet to be answered and regions yet to be explored. The southwestern part of California is one such region where additional Holocene records are required to develop fully the spatiotemporal patterns of Holocene hydroclimates and their forcings.

Kirby et al. (2004, 2005, 2006, 2007, 2010, 2012, 2013), Bird and Kirby (2006), and Bird et al. (2010), using a number of proxies, have produced late Pleistocene and Holocene environmental histories for several lakes in the region. Using a sediment core collected from Lower Bear Lake in the San Bernardino Mountains, Kirby et al. (2012) provided a record of Holocene precipitation variability in the watershed that serves as the water source for many of the ephemeral lakes of the Mojave Desert. Using loss-on-ignition (LOI; 550 °C and 950 °C), molar C:N, grain size, and biological proxies (gastropods and ostracods), Kirby et al. (2012) identified five major and four minor intervals (pluvial periods) representing increased precipitation and/or runoff, spanning ~9170 years. The primary purpose of this paper is to provide more detailed information on the extent and characteristics of the Holocene pluvial episodes in coastal southwest California (Lower Bear Lake) using changes in the diatom flora. Chemical variables indirectly affected by climate change such as nutrient load (Wolfe et al. 2001; Sickman et al. 2013), pH (Bigler and Hall 2002; Wolfe 2002; Schmidt et al. 2004; Sickman et al. 2013), and ionic concentration (Rühland et al. 2003) also affect productivity and ecosystem structure. Here, we use the ecological specificity of diatoms to characterize individual pluvial episodes and provide a clearer picture of broader-scale changes over the Holocene in southern California.

Previous Work

In the context of coastal southwestern California, a number of studies have reconstructed the hydrology of late Pleistocene and Holocene using lacustrine sediments (e.g., Kirby et al. 2004, 2007, 2010, 2013, 2014, 2015; Bird and Kirby 2006; Bird et al. 2010; Dingemans et al. 2014; Feakins et al. 2014; Heusser et al. 2015). Specifically, for Lower Bear Lake, Kirby et al. (2012) used a multi-proxy methodology to identify nine pluvial episodes representing increased precipitation and/or runoff, spanning about 9200 years. These episodes are inferred using (1) increased runoff and transportation of terrestrial biomass during wetter conditions explained by high C:N ratios and greater abundance of total organic matter (LOI 550 °C); (2) low gastropod counts that may indicate reduced habitat and increased transportation distance from littoral zone to profundal zone where degradation of organic matter decreased the pH which reduced preservation; and (3) low levels of total carbonate (LOI 950 °C) suggesting increased carbonate solubility resulting from a cooler spring-summer epilimnion (depending on timing of snow melt) and potentially more corrosive bottom waters. In this paper, we re-examine this record using diatoms and an updated chronology to put the Bear Lake record into a regional context that is consistent with other paleoclimate records.

Study Site and Climatology

Lower Bear Lake (2080 masl) is located in the San Bernardino Mountains of the east-west trending Transverse Ranges in southern California, about 160 km northeast of Los Angeles (Fig. 1). The highest point is San Gorgonio Mountain (3502 m). The Lower Bear Lake watershed is underlain by Paleozoic carbonate rocks and quartzite, Mesozoic plutonic and metamorphic rocks, and Tertiary sedimentary rocks. In much of the basin, the basement rocks are overlain by Quaternary alluvium (Leidy 2003; Flint and Martin 2012). The mountains were formed about 11 million years ago through transpression associated with the San Andreas Fault and are still rising. Many local rivers, including the Mojave River, originate in the range. The higher elevations held glaciers during the Pleistocene (Owen et al. 2003).

The term Lower Bear Lake was used in an 1857 public land survey for a small lake in the Big Bear Valley (Kirby et al. 2012). The Lower Bear Lake was covered with water to a depth of 10–11 m (Fig. 2) when the valley was filled following the construction of Big Bear Dam in 1884, creating Big Bear Reservoir (Leidy 2003).

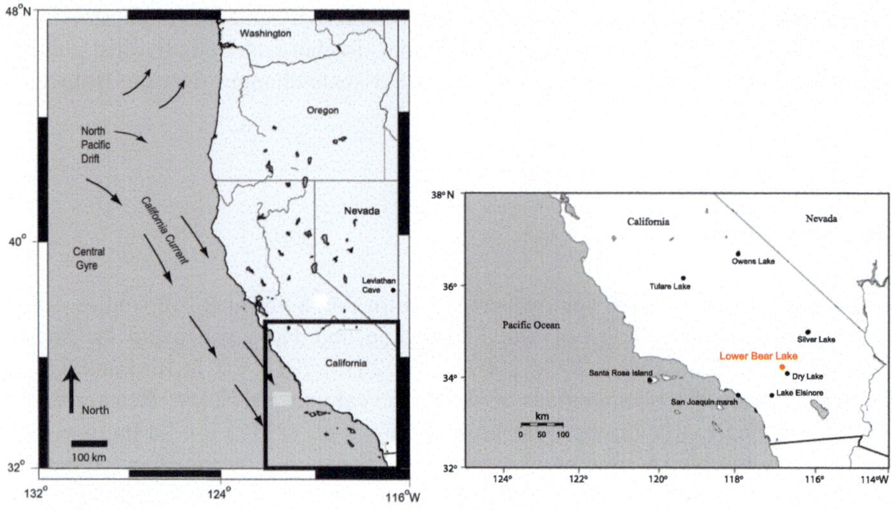

Fig. 1 Location of Lower Bear Lake and other sites discussed. (Modified from Kirby et al. (2012))

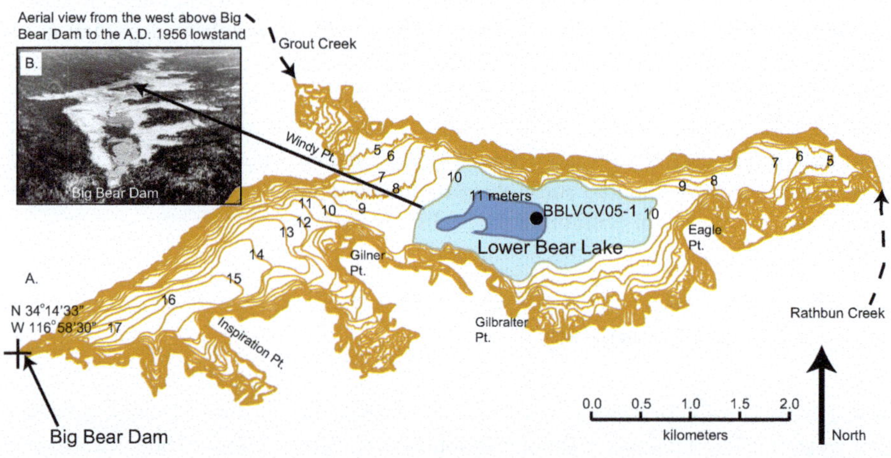

Fig. 2 Bathymetric map of Lower Bear Lake and coring site. (Modified from Kirby et al. (2012)). Numbers indicate depth below surface of Big Bear Reservoir (2055 m). Note area east and west of coring site with slope of less than 1 m/km, indicating the large ecospace available for colonization with a small change in depth

Modern bathymetry (Fig. 2) reveals a depression near the center of the reservoir in the original location of Lower Bear Lake. Survey reports from 1857 to 1878 show that Lower Bear Lake was full, with an outlet at the west end of the lake (Leidy 2003). During the mid-1950s drought, the level of Big Bear Reservoir decreased to the point that only the central depression contained water (Leidy 2003). Kirby et al. (2012) have suggested that the lake was probably less than 3 m deep in the nineteenth century and could have been more that 8 m deep in the Early Holocene. In

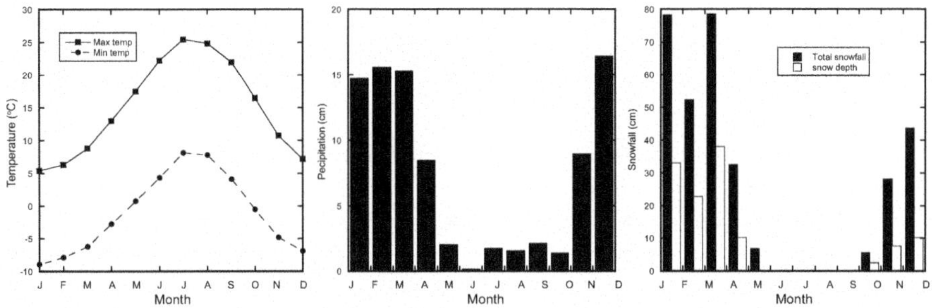

Fig. 3 Temperature, precipitation, and snowfall record measures at Big Bear Dam (WRCC)

addition to permanent and ephemeral creeks flowing directly into the lake, surface runoff, snowmelt, and groundwater also contribute to the hydrologic budget of the lake.

The highly seasonal coastal southern California Mediterranean climate is strongly connected to large-scale variations in atmospheric and oceanographic circulation in the northeast Pacific Ocean. Precipitation is greatest between November and March (Fig. 3) when a relatively weak North Pacific high-pressure system is displaced by a southward-migrating subpolar low-pressure system to migrate to a position off the California coast (Pyke 1972; Dailey et al. 1993; Hickey 1993; Haston 1997). Strong regional precipitation events are generated by the southern branch of the polar jet stream drawing warm, moist subtropical air into the region (Browning and Pardoe 1973; Dettinger 2011). Such events may be more strongly associated with El Niño events (Andrews et al. 2004). Maximum snowfall occurs between January and March. Snow accumulates between October and April, reaching maximum depth in March (Fig. 2). Kirby et al. (2006, 2010, 2012, 2014) have shown that regional lake levels are strongly tied to total winter precipitation, although local conditions may affect the timing and magnitude of these changes. Orographic conditions and occasional monsoonal moisture from the Gulf of California contribute to annual precipitation between July and October (Adams and Comrie 1997; Comrie and Glenn 1998; Metcalfe et al. 2015). The remnants of dissipating eastern tropical cyclones may also be an important source of moisture during this time (Ritchie et al. 2010). Based on modern climatology, it is unlikely that summer and early fall precipitation have contributed more than about 15% of annual precipitation during the Holocene (Bird and Kirby 2006; Hereford et al. 2006; Kirby et al. 2007, 2014, 2015).

The strongest climate forcing along the southern California margin is the interaction between Pacific Ocean sea surface temperatures (SST) and regional atmospheric circulation (Trenbreth and Hurrell 1994; Barron et al. 2012). A weakening and southward shift in the position of the North Pacific high-pressure system directs winter storm systems southward and intermittently into southern California (Cayan and Peterson 1989). Namias et al. (1988) showed that warmer than average SST in the northeast Pacific Ocean favors a southerly migration of the polar jet stream, which increases the number of storm systems crossing the southern California coast, including those that entrain subtropical moisture from the central Pacific (Fye et al. 2004; Dettinger 2007, 2011). During the summer, the North Pacific high-pressure system forces storm tracks to the north and the region receives little or no precipitation associated with jet stream-entrained cyclonic systems (Barlow et al. 2001).

The El Niño-Southern Oscillation (Redmond and Koch 1991; Haston and Michaelsen 1994; Barron and Anderson 2011) and Pacific Decadal Oscillation (PDO; Mantua and Hare 2002) also affect the amount of winter precipitation across the region. El Niño and a positive PDO generally combine to generate larger amounts of winter precipitation, due, in part, to the changing strengths of the California Current and Equatorial Countercurrent which result in the elevation of surface temperature by 7–10 °C (Hood 1993). The relationship between these large-scale processes and atmospheric rivers in not well understood (Woodhouse et al. 2005; Cook et al. 2011; Dettinger 2011).

Methods

In 2005, a single-drive, 4.50-m-long sediment core (BBLVC05-1) was collected from Lower Bear Lake. Kirby et al. (2012) provide a detailed discussion of the sampling and preparation of samples for physical properties, grain size, and sediment geochemistry analyses. This study utilizes the loss-on-ignition (LOI; 550 °C and 950 °C) and elemental C and N (C:N ratio) data from the previous study.

Thirty-three samples were collected for AMS ^{14}C dating (31 discrete materials [e.g., seeds, wood, charcoal, etc.] and two on bulk organic matter) from the upper 405 cm of the core. The samples were analyzed at either the Lawrence Livermore National Laboratory Center for Accelerator Mass Spectrometry or the University of California, Irvine Keck Carbon Cycle AMS Laboratory.

Diatom analysis was not part of the original sampling protocol but was conducted following the development of the age model and completion of the geochemical analyses. Sampling was focused on the pluvial episodes (PE), which resulted in majority of the samples being collected from those intervals. In some cases, sampling was limited by the lack of core material in sections that had been previously sampled for other proxies. One hundred five samples were processed for diatom analysis using 30% hydrogen peroxide, 37% hydrochloric acid, and 70% nitric acid (Battarbee 1986) to remove carbonate and organic matter and 5% sodium pyrophosphate to deflocculate the remaining clay. The samples were then washed repeatedly with di-ionized water until neutralized. Approximately 50 μl of the resulting suspension was dried on a 22 × 30 mm coverslip and permanently mounted using Naphrax®. Five hundred frustules per sample were enumerated following the method of Schrader and Gersonde (1978) using a Leica DM LB2 microscope with Nomarski differential interference contrast (DIC) optics at magnifications of 630× and 1000×. Diatom identification was based on Gasse (1986), Krammer and

Lange-Bertalot (1986, 1988, 1991a, b), Lange-Bertalot and Krammer (1987, 1989), Cumming et al. (1995), Round and Bukhtiyarova (1996), Krammer (1997a, b, 2000, 2002), Lange-Bertalot (2001), and Cantonati et al. (2017). Data for some taxa were combined because similar or gradational morphological characteristics prevented species differentiation in some samples.

Results and Interpretation

Age Model

Kirby et al. (2012) obtained 33 AMS dates from core BBLVC05-1, of which 23 were used to construct an age model. Ten dates were rejected for various reasons including coupled discrete-bulk samples with different ages, coupled discrete ages of samples from samples from different materials, stratigraphic inversion, and obvious contamination or reworking. A detailed explanation for the rejected dates is in Kirby et al. (2012) Table 1. The dates were calibrated using CALIB ver. 6.0.0 and IntCal09.14c (Stuiver and Reimer 1993; Reimer et al. 2009).

A new age model (Fig. 4) was generated using Bacon 2.2 (Blaauw 2010; Blaauw and Christen 2011; Blaauw and Heegaard 2014) using the same 23 dates plus an additional date (32) from the interval 354–355 cm. The dates were calibrated using IntCal13 (Reimer et al. 2013). The biggest difference between the new model (Supplementary Table S1) and that presented in Kirby et al. (2012) is the 300–400-year age deviation in the Early Holocene. The timing of PE-V, the earliest and longest lasting of the pluvial events in the Lower Bear Lake record (Kirby et al. 2012), starts about 300 years earlier.

The age model is also extrapolated to the top and bottom of the core. The extrapolation to the bottom of the core is not an issue in this study because the sediments in the section of the core below 385 cm are devoid of diatoms. The age at the top (0–25 cm) of the core varied but indicated that it is about 250 years younger than the ages obtained using the model in Kirby et al. (2012).

General Trends and Proxy Interpretation

Geochemical Properties

Kirby et al. (2012) have discussed the physical (magnetic susceptibility, sediment grain size), geochemical (percent organic carbon, percent carbonate, molar C:N), and biological proxies (number of ostracods and gastropods per gram of sediment) in some detail. The discussion here will focus on the geochemical properties as they express the greatest amount of variability in the record and only for the upper 400 cm of the core.

Table 1 Pluvial episode characteristics

PE-I (251–152 cal yr B.P.)
Interpretation: Changes in organic matter, $CaCO_3$, and molar C:N indicate a return to conditions similar to those in PE-II
Geochemistry:
Organic matter – 60% average; decreased to 44% at top of interval
$CaCO_3$ – Decreased to 6% through interval
Molar C:N ratio – Increased from 7 to 10 at top of interval
Diatoms (% relative abundance): No sediment available for analysis
Inter-pluvial II-I
Interpretation: Increase in subsaline species may reflect summer increase in TDS and fresher conditions during wetter winters
Geochemistry:
Organic matter – Stepped decrease from 60% in lower half to 38% in upper half of interval
$CaCO_3$ – Stepped increase to 17% in lower half to high of 44% in upper half of interval; more variability in upper half of interval
Molar C:N ratio – Decrease from 9 at bottom to average of 7 in upper half of interval
Diatoms (% relative abundance):
Freshwater – *Aulacoseira* average 5% with greater variability than previous interval; *Fragilaria* varies between 0% and 10%, including up to 4% planktic *Fragilaria*; small Fragilarioids average 3%
Subhaline – *Cocconeis placentula* varies from 6% to 32%; *Craticula* reaches peak of 30% at 650 cal yr B.P.; *Anomoeoneis sphaerophora* average 2%
Epiphytic – *Amphora libyca* varies from 2% to 9%; *Gomphonema* varies from 5% to 9%; *Cymbella* varies from 0% to 4%
Heterotrophic – *Nitzschia amphibia/frustulum* varies between 0% and 11%; average 6%; *Epithemia* varies from 3% to 25%; average 14%; *Rhopalodia gibba* range 0–2%
PE-II (834–738 cal yr B.P.)
Interpretation: Reduction of TDS concentration to that of the previous inter-pluvial episode is reflected in the abrupt decrease in $CaCO_3$ and subsaline species and the increase in freshwater taxa; some aquatic macrophytes around the margin of the lake
Geochemistry:
Organic matter – Rapid increase to average of 72% for interval
$CaCO_3$ – Rapid decrease to average of 6% for interval; more variability than organic matter
Molar C:N ratio – 9.5 average
Diatoms (% relative abundance):
Freshwater – *Aulacoseira* average 5%; *Fragilaria* varies between 0% and 15% in upper part of interval, including up to 8% planktic *Fragilaria*; small Fragilarioids increases to 8% at top of interval
Subhaline – *Cocconeis placentula* gradual increase from 6% to 20%; *Craticula* gradual decrease from 15% to 4%; *Anomoeoneis sphaerophora* 7% at bottom rapidly decreases to average of 2%
Epiphytic – *Amphora libyca* 9% at bottom and top of interval and 5% in upper part of interval; *Gomphonema* 10% at bottom decreases to 3% in middle and rises to 12% at top; *Cymbella* increases to 7% near bottom decreasing to 1% near top of interval

(continued)

Table 1 (continued)

Heterotrophic – *Nitzschia amphibia/frustulum* average 3%; *Epithemia* 20% at base decreasing to 6% at top of interval
Inter-pluvial IIa-II
Interpretation: Continues the trend present in PE-IIa
Geochemistry:
Organic matter – Decrease to 28% near bottom; 47% average for remainder of interval
$CaCO_3$ – Rapid increase from 22% to 50% at near base; rapidly returns to 22% average for remainder of interval
Molar C:N ratio – 8 average
Diatoms (% relative abundance):
Freshwater – *Aulacoseira* average increases to 5% near top of interval; benthic *Fragilaria* decreases to 2% in middle of interval; increases to 12% near top of interval, including 5% planktic *Fragilaria*
Subhaline – *Cocconeis placentula* decreases to 15% in upper part of interval and increases to 30% at top; *Craticula* ranges from 2% to 10%; *Anomoeoneis sphaerophora* reaches peak of 5% in middle of interval
Epiphytic – *Amphora libyca* 2% average; *Gomphonema* decreases from 13% in middle to 7% near top of interval; *Cymbella* increases to 3% at top of interval
Heterotrophic – *Nitzschia amphibia/frustulum* average 9%; *Epithemia* 12% through most of interval with peak of 19% at top of interval
PE-IIa (1749–1562 cal yr B.P.)
Interpretation: Continues the trend present in inter-pluvial III-IIa; increase in subsaline species suggests that lake has reached the point where at least seasonal evaporation is increasing the TDS to the point that the higher concentration is expressed in the composition of the diatom flora
Geochemistry:
Organic matter – 44% average
$CaCO_3$ – 22% at bottom decreasing to 9% in the middle; 23% at top of interval
Molar C:N ratio – 9 average
Diatoms (% relative abundance):
Freshwater – *Aulacoseira* average 3%; benthic *Fragilaria* ranges from 2% to 10%
Subhaline – *Cocconeis placentula* ranges from 16% to 33%; *Craticula* average 7%; *Anomoeoneis sphaerophora* average 2%
Epiphytic – *Amphora libyca* 2% average; *Gomphonema* increases from 4% to 9%; *Cymbella* decreases from 2% to 0%
Heterotrophic – *Nitzschia amphibia/frustulum* varies between 8% and 20%; *Epithemia* 17% at base decreasing to 4% in middle and increasing to 15% at top of interval; *Rhopalodia gibba* peak at 1%
Inter-pluvial III-IIa
Interpretation: Lake decreases in size, with aquatic macrophytes supporting an increased abundance of epiphytic species; no longer deep enough to support *Aulacoseira* in large numbers
Geochemistry:
Organic matter – Increases from 22% near bottom to 46%
$CaCO_3$ – Decrease from 44% near bottom of interval to 20%

(continued)

Table 1 (continued)

	Molar C:N ratio – Decreases from 7.5 to 5.5 near bottom; increases to 9 at top of interval
	Diatoms (% relative abundance):
	Freshwater – *Aulacoseira* decreases to 0% through most of interval; *Fragilaria* decreases to 0% at base; benthic *Fragilaria* increases to 6% near top of interval
	Subhaline – *Cocconeis placentula* peak of 40% in middle of interval; *Craticula* average 7%; *Anomoeoneis sphaerophora* average 1%
	Epiphytic – *Amphora libyca* decreases to 2% average; *Gomphonema* 6% average; *Cymbella* increases to 2% average at top of interval
	Heterotrophic – *Nitzschia amphibia/frustulum* increases rapidly from 8% to 24%, decreases to 8% at top of interval; *Epithemia* decreases rapidly from 21% to 10% at base; average 12%
PE-III (3329–2928 cal yr B.P.)	
	Interpretation: Small lake with fluctuating changes in the stability of the water column based on changes in the abundance of *Aulacoseira* (mixed, high nutrient levels) and heterotrophs (stable, low nutrient levels)
	Geochemistry:
	Organic matter – 46% in bottom, decreasing to 14% in middle; returning to 36% at top of interval
	$CaCO_3$ – Decreased to 4% in middle; increasing to 28% at top of interval
	Molar C:N ratio – C:N: 11 average
	Diatoms (% relative abundance):
	Freshwater – Highest *Aulacoseira* pulse at 55%; planktic *Fragilaria* 4% at top of interval
	Subhaline – *Cocconeis placentula* average 3%; peak of 19% at top of interval; *Craticula* average 5%; *Anomoeoneis sphaerophora* average 2%; range 0–4%
	Epiphytic – *Amphora libyca* decreases to 2% average in most of interval; increasing to 17% at top; *Gomphonema* decreases from 8% to 0% at top of interval
	Heterotrophic – *Nitzschia amphibia/frustulum* 8% at base increasing to 40% in middle and returning to 7% at top of interval; *Epithemia* 18% at base decreasing to 7% at top of interval
Inter-pluvial IIIa-III	
	Interpretation: Small lake surrounded by aquatic epiphytes at beginning and end of the interval; decreases in size in the middle of the interval
	Geochemistry:
	Organic matter – 30% at bottom decreasing to 18% in the lower part; 40% at top of interval
	$CaCO_3$ – 30% at bottom increasing to 50% in the lower part; decreasing to 25% at top of interval
	Molar C:N ratio – 7 average
	Diatoms (% relative abundance):
	Freshwater – *Aulacoseira* increases from 8% to 11% at top of interval; benthic *Fragilaria* increases to 4% near top and decreases to 2%
	Subhaline – *Cocconeis placentula* peak of 13% near top of interval; *Craticula* average 5%; *Anomoeoneis sphaerophora* average 2%
	Epiphytic – *Amphora libyca* decreases to 3% in upper part to 23% at top of interval; *Gomphonema* increases to 13% in upper part and decreases to 8% at top of interval
	Heterotrophic – *Nitzschia amphibia/frustulum* decreases at 21% to 7% at top of interval; *Epithemia* average 12% decreases to 7% at top of interval

(continued)

Table 1 (continued)

PE-IIIa (4048–3697 cal yr B.P.)
Interpretation: Slight increase in lake size with increase in all types of benthic species (epiphytic, epipsammic, epipelic)
Geochemistry:
Organic matter – 29% average
$CaCO_3$ – 38% at base decreasing to 30% at top of interval
Molar C:N ratio – 6.5 at bottom to 8.5 at top of interval
Diatoms (% relative abundance):
Freshwater – *Aulacoseira* 11% peak; benthic *Fragilaria* declines from 5% to 1%; small Fragilarioids declines from 4% to 0%
Subhaline – *Cocconeis placentula* increases from 5% to 9% at top of interval; *Craticula* average 6%; *Anomoeoneis sphaerophora* average 2%
Epiphytic – *Amphora libyca* increases from 3% to 17% at top of interval; *Gomphonema* ranges from 5% to 11%; *Cymbella* 2% decreases to 0% at bottom
Heterotrophic – *Nitzschia amphibia/frustulum* average 19%; *Epithemia* average 8%
Inter-pluvial IIIb-IIIa
Interpretation: Small lake with aquatic macrophytes around margin
Geochemistry:
Organic matter: – 3% average; more variability than PE-IIIb
$CaCO_3$ – Varies from 35 to 56%; 38% at top of interval
Molar C:N ratio – 7 average
Diatoms (% relative abundance):
Freshwater – *Aulacoseira* reaches 26% before decreasing to 0% at top of interval; benthic *Fragilaria* average 3% in upper part of interval; small Fragilarioids average 2% in upper part of interval
Subhaline – *Cocconeis placentula* and *Craticula* increase to 5% at top of interval; *Anomoeoneis sphaerophora* increases to average of 1% at top of interval
Epiphytic – *Amphora libyca* average 4%; *Gomphonema* 6% to 9% at top of interval; *Cymbella* average 1%
Heterotrophic – *Nitzschia amphibia/frustulum* increases from 6% in the middle to 20% at top of interval; *Epithemia* varies from 20% to 8% to 18% in upper part of interval
PE-IIIb (4744–4434 cal yr B.P.)
Interpretation: Minor change in lake size
Geochemistry:
Organic matter – 24% average
$CaCO_3$ – 36% average
Molar C:N ratio – 7 average
Diatoms (% relative abundance): No sediment available for analysis
Inter-pluvial IIIc-IIIb
Interpretation: Small lake with little terrestrial organic matter input; variable water column stability
Geochemistry:
Organic matter – 22% average
$CaCO_3$ – 54% high; 44% at top of interval

(continued)

Table 1 (continued)

Molar C:N ratio – 6 average
Diatoms (% relative abundance):
Freshwater – *Aulacoseira* 19%
Subhaline – *Cocconeis placentula* average 3%; *Craticula* average 3%
Epiphytic – *Amphora libyca* increases from 0% to 3%; *Gomphonema* increases from 3% to 6%; *Cymbella* 1%
Heterotrophic – *Nitzschia amphibia/frustulum* 21%; *Epithemia* decreases from 26% to 11%; *Rhopalodia gibba* 3% (highest value in core)
PE-IIIc (5506–5175 cal yr B.P.)
Interpretation: Small nitrogen-limited lake
Geochemistry:
Organic matter – 38% high
$CaCO_3$ – 21% low
Molar C:N ratio – 6 at bottom and top of interval; high of 10
Diatoms (% relative abundance):
Freshwater – *Aulacoseira* decreases from 18% to 4%; benthic *Fragilaria* less than 3%; small Fragilarioids decrease to 0%
Subhaline – *Cocconeis placentula* average 2%; *Craticula* average 1%; *Anomoeoneis sphaerophora* average 1%
Epiphytic – *Amphora libyca* decreases from 3% to 0%; *Gomphonema* decreases from 10% to 3%
Heterotrophic – *Nitzschia amphibia/frustulum* 13% increasing to 21%; *Epithemia* 10% increasing to 27%; *Rhopalodia gibba* increasing from 1.5% to 3%
Inter-pluvial IV-IIIc
Interpretation: Small lake surrounded by aquatic macrophytes, becoming more abundant up core
Geochemistry:
Organic matter – Decreases from 30% to 15% average
$CaCO_3$ – Increases from 28% to 58%; 40% at top of interval
Molar C:N ratio – Average 6
Diatoms (% relative abundance):
Freshwater – *Aulacoseira* peak at 25%; planktic *Fragilaria* maximum 2%; smaller Fragilarioids 0–6%
Subhaline – *Cocconeis placentula* peak of 10% in middle of interval; *Craticula* average 2%; *Anomoeoneis sphaerophora* average 1%
Epiphytic – *Amphora libyca* average 2%; *Gomphonema* increases from 1% at bottom to 10% at top of interval
Heterotrophic – *Nitzschia amphibia/frustulum* decreases from 34% to 18%; *Epithemia* decreases from 22% to 9%; *Rhopalodia gibba* high of 1.5% at top of interval
PE-IV (6939–6410 cal yr B.P.)
Interpretation: Lake remains relatively small; higher abundance of *Aulacoseira* in the earlier part of the pluvial episode suggests more overturn; the increase in heterotrophs indicates a decrease in nutrients which may be associated with stronger seasonal stability
Geochemistry:
Organic matter – Base and top of interval 28%; rises to 70% in upper part

(continued)

Table 1 (continued)

$CaCO_3$ - Average 5%
Molar C:N ratio – Increases from 9 to 16 at top of interval
Diatoms (% relative abundance):
Freshwater – *Aulacoseira* peak at 35%; planktic *Fragilaria* peak at 6%; small Fragilarioids account for maximum of 6%
Subhaline – *Cocconeis placentula* decreasing trend; average 2%; *Craticula* average 3%; *Anomoeoneis sphaerophora* increases to 4% in middle of interval
Epiphytic – *Amphora libyca* average 9% in middle and decreases to 1% at top of interval; *Gomphonema* 11% at bottom decreasing to 1% at top of interval
Heterotrophic – *Nitzschia amphibia/frustulum* 7% at base; upper interval varies from 0% to 32%; *Epithemia* varies from 0% to 22%
Inter-pluvial V–IV
Interpretation: Rapid decrease in lake size; algae become more important organic carbon source; increase in molar C:N ratio, decrease in *Aulacoseira* and increase in epiphytes and various attached benthic species indicate that aquatic macrophytes increase in abundance up core
Geochemistry:
Organic matter – Decreases to 16% in lower part of interval; average 30%
$CaCO_3$ – Lower part of interval to 50%; upper part of interval to 37%
Molar C:N ratio – Increases from 8 at base to 12 at top of interval
Diatoms (% relative abundance):
Freshwater – Three samples vary between 10% and 30% *Aulacoseira*; planktic *Fragilaria* 22% near base; smaller Fragilarioid peak at 40% in sample above *Fragilaria* peak
Subhaline – *Cocconeis placentula* 1–5%; *Craticula* and *Anomoeoneis sphaerophora* 1%
Epiphytic – *Amphora libyca* increases to 7% in middle of interval; *Gomphonema* increases to 3% in middle of interval
Heterotrophic – *Nitzschia amphibia/frustulum* average 6%; *Epithemia* average 5%; *Rhopalodia gibba* 1.5% at bottom decreasing to 0% at top of interval
PE-V (9,250–8586 cal yr B.P.)
Interpretation: Lake reaches maximum Holocene extent, flooding extensive shelf (maximum small Fragilarioid abundance) to a depth of 1–2 m which extends more than 2 km east and west of the coring site; increased organic matter is largely terrestrial in nature; variations in organic matter reflect changes in relative amount of detrital sediment transported into the basin; decreases nitrogen at top of PE-V; salinity pulse in middle of episode is not reflected in $CaCO_3$ values, suggesting that it is transient, and any carbonate deposited is quickly dissolved
Geochemistry:
Organic matter – Rises to 70%; decreases to 20% at top of PE-V
$CaCO_3$ – 2–11%
Molar C:N ratio – Increases from 10 to 14; average 12.5
Diatoms (% relative abundance):
Freshwater – *Aulacoseira* 25% at base and 39% at top of interval; varies around 10% for remainder of interval; benthic *Fragilaria* average 2%; planktic *Fragilaria* reaches 5% in upper part of interval; small Fragilarioids reach greatest abundance in core; two pulses with maxima at 73% and 54%

(continued)

Table 1 (continued)

Subhaline – *Cocconeis placentula* reaches peak of 13% in lower part of interval and declines to 3% at top; *Craticula* average 1% except for spike of 16% at 8950 cal yr B.P.; *Anomoeoneis sphaerophora* average 1% except for spike of 11% at 8950 cal yr B.P.
Epiphytic – *Amphora libyca* average 1%; peak 7%; *Gomphonema* average 2%
Heterotrophic – *Nitzschia amphibia/frustulum* 16% at base decreases to average of 4% in lower half increasing to 35% at top of interval; *Epithemia* varies around average of 12%
Pre-PE-V
Interpretation: Small lake filling the deepest part of the central depression and occasionally flooding a portion of the surrounding shallow margin; large portion of organic matter is transported from surrounding area; some aquatic macrophytes around margin of lake; nitrogen-limited at times
Geochemistry:
Organic matter – Increases from 3% to 50% at top of interval
$CaCO_3$ – 2–8%
Molar C:N ratio – 10–12
Diatoms (% relative abundance):
Freshwater – *Aulacoseira* fluctuated between 0% and 30%; planktic *Fragilaria* 2% of lowest sample; remainder are benthic species; small Fragilarioids 23% of lowest sample; decreased to 0% in most remaining samples in interval
Subhaline – *Cocconeis placentula* increases from 3% to 11%; *Craticula* and *Anomoeoneis sphaerophora* 1% average
Epiphytic – *Amphora libyca* average 2%; *Gomphonema* average 4%; peak 13% at top of interval
Heterotrophic – *Nitzschia amphibia/frustulum* 7% to 30%; *Epithemia* 17% increasing to 36% at top of interval; *Rhopalodia gibba* varies around average of 1.5%

The percent total organic matter, as determined by 550 °C LOI (Fig. 5), measures a combination of algal productivity in the lake and the terrestrial material transported into the lake from the surrounding watershed. In Lower Bear Lake total organic matter varied from 4% to more than 70%. Kirby et al. (2012) used LOI 950 °C values as a proxy for carbonate abundance, where pluvial episodes are represented by low values that result either from decreased production or increased dissolution in colder water. Carbonate precipitation is enhanced during drier periods and reduced during pluvial episodes. The increased organic matter deposited during pluvial episodes lowers the pH, further enhancing the dissolution of carbonate in the sediments. The variability in LOI 950 °C is greater than that observed in the LOI 550 °C record.

Molar C:N values are indicative of the relative amounts of algal and terrestrial organic matter in the sediment. Lacustrine algae, including diatoms, vary between 4 and 10, while values for terrestrial plants are above 20 (Meyers and Ishiwatari 1993).

Characteristics of the Diatom Record

The fossil diatom flora in Lower Bear Lake is comprised of 213 species and varieties from 42 genera. Diatom preservation was variable, ranging from excellent to good, with little evidence of corrosion, to very poor, where much of the fine

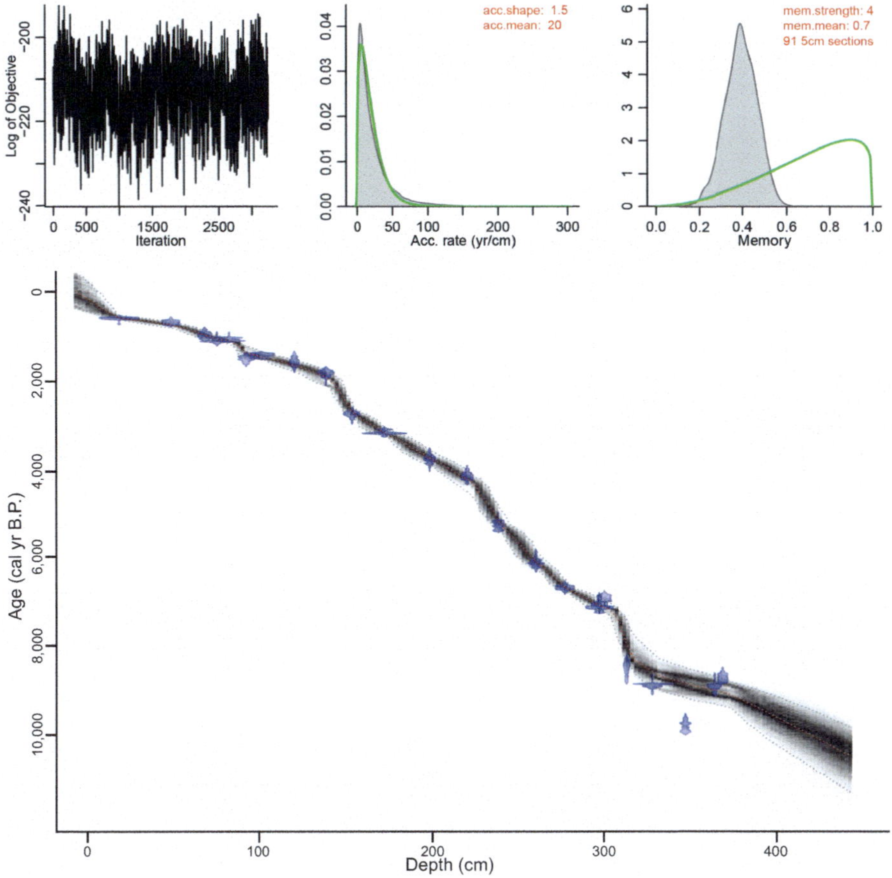

Fig. 4 Age model for core BBLV05-1

structure was removed through dissolution and physical breakage. Assemblages fluctuate between those dominated by freshwater, alkaliphilic taxa and those containing a greater abundance of species characteristic of saline or high-conductivity waters (hereafter referred to as saline taxa). Environmental preferences for most of the species enumerated in this study are listed in Supplementary Table S2. Although the record is dominated by freshwater species, those preferring more saline conditions are always present in low numbers and gradually increase above 150 cm. The lake appears to have been mesotrophic throughout the Holocene.

The diatoms may be grouped into those with a preference for "fresh" water (<3‰ total dissolved solids TDS) and "saline" (>5‰ TDS) (Bradbury 1997). For many species these are generalized categories as numerous "freshwater" species that have optima below 1‰ TDS are found in waters that have several times more salinity. Conversely "saline" species can be found in freshwater, as well as waters that contain an order of magnitude-higher concentration of dissolved solids than the 5‰ minimum for saline species (Fritz 1990; Fritz et al. 1993). Nutrient levels can affect the ability of some species to survive at the salinity extremes of their normal

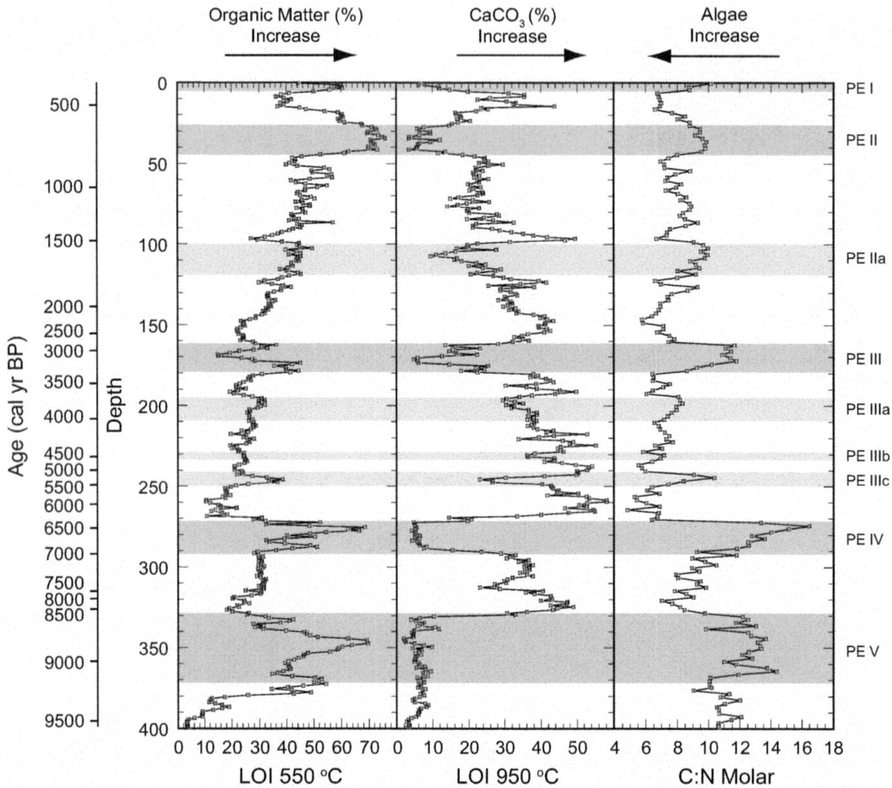

Fig. 5 Loss-on-ignition and molar C:N records. (Data and pluvial episodes after Kirby et al. (2012))

ranges (Saros and Fritz 2000). *Rhopalodia* and *Epithemia* (DeYoe et al. 1992) and some *Nitzschia* (Kilham et al. 1986) can live in low-nitrogen environments due to the presence of cyanobacterial endosymbionts.

The physical habitat occupied by different diatom species affects the availability of light, nutrients, and a range of other physical and chemical factors, which determine the composition of a diatom assemblage. Kirby et al. (2012) suggested that the Holocene depth of Lower Bear Lake was 8 m, which is probably shallow enough that light would penetrate to the bottom except during short-lived windy conditions. Planktic taxa float in the water column or are suspended by wind-driven turbulence. In some cases, the "planktic" nature of a species depends on the size and depth of the lake in which it is found. Benthic species are present in a number of forms. Tychoplanktic species often begin as long chains attached to the bottom in shallow water and break apart and float into deep water suspended by wind-driven convection. Semi-attached diatoms are lightly attached to a substrate, such as aquatic macrophytes (epiphytic species) or rocks (epilithic species). These often form long

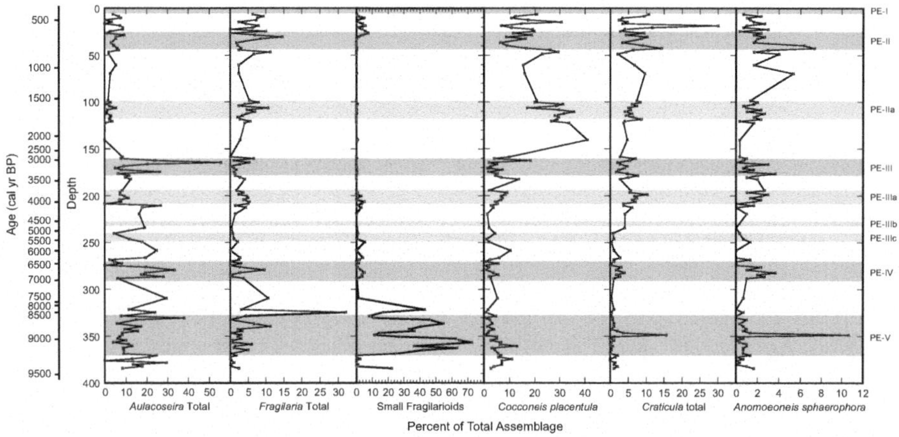

Fig. 6 Abundance of important freshwater and subsaline diatom taxa. (Pluvial episodes after Kirby et al. (2012))

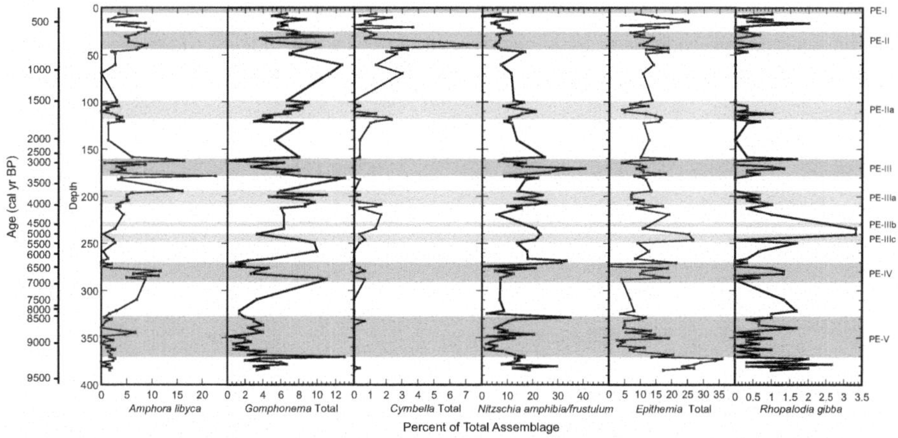

Fig. 7 Abundance of important taxa: epiphytic and heterotrophic diatom taxa. (Pluvial episodes after Kirby et al. (2012))

chains which in turbulent conditions can be suspended in the water column and become part of the sediment in the deep, aphotic part of a lake. Other benthic taxa are attached by stalks or mucilage pads (adnate forms) or are motile and capable of moving through the mud. Some species may exist in multiple environments, depending on seasonal conditions. Using this information, the diatom taxa (Supplementary Table S2, Figs. 6 and 7) are separated into four ecological groups:

Freshwater species (Fig. 6) – *Aulacoseira* spp. (planktic; primarily *A. italica*, *A. ambigua*; also *A. crenulata*, *A. islandica*, *A. lirata*, *A. subarctica*, *A. valida*),

Fragilaria spp. (planktic; primarily *F. nanana*, *F. tenera*; also *F. crotonensis*, *F. radians*), *Fragilaria* spp. (benthic; primarily *F. capucina*, *F. vaucheriae*; also *F. capucina* var. *gracilis*), and small Fragilarioids (tychoplanktic; primarily *Pseudostaurosira brevistriata*, *Staurosira construens* var. *construens*, *S. construens* var. *venter*, *Staurosirella neopinnata*; also *P. brevistriata* var. *inflata*, *P. subsalina*, *S. construens* var. *binodis*, *S. construens* var. *triundulata*, *S. elliptica*, *Staurosirella lapponica*, *S. martyi*)

Saline species (Fig. 6) – *Cocconeis placentula*, *Craticula* spp. (*C. cuspidata*, *C. halophila*) and *Anomoeoneis sphaerophora* (includes *A. costata*)

Epiphytic species (Fig. 7) – *Amphora libyca*, *Cymbella* spp. (*C. cistula*, *C. cymbiformis*, *Encyonema gracile*, *E. minutum*, *Encyonopsis cesatii*), and *Gomphonema* spp. (*G. acuminatum*, *G. affine*, *G. affine* var. *insigne*, *G. angustatum*, *G. angustum*, *G. clavatum*, *G. gracile*, *G. olivaceum*, *G. parvulum*, *G. subclavatum*, *G. subtile*, *G. truncatum*, *G. truncatum* var. *capitatum*, *G. ventricosum*)

Heterotrophic species (Fig. 7) – *Epithemia* spp. (primarily *E. turgida*; also *E. adnata*, *E. argus*, *E. hyndmanii*, *E. turgida* var. *granulata*), *Nitzschia amphibia*, *Nitzschia frustulum*, *Rhopalodia gibba*

The majority of the species not included in these groups comprise less than 1% of the assemblage in fewer than three samples. In other cases, the environmental preferences are not known or are overly broad, being listed only as "fresh" or "marine."

The changes in relative abundance of these four ecological groups and trends in geochemical proxies are described for the nine pluvial episodes identified by Kirby et al. (2012) and the inter-pluvial intervals in Table 1.

Discussion

Lower Bear Lake history

Kirby et al. (2012) identified five major and four minor pluvial episodes (Table 1, Supplementary Table S1). The five major pluvial episodes are identified as intervals of increased precipitation based on the near absence of gastropods, near-zero values for percent carbonate (LOI 950 °C), peaks in percent organic matter (LOI 550 °C), and peaks in the molar C:N ratios. The minor episodes reflect similar changes as the major episodes but on a smaller and sometimes less well-defined scale.

The present study builds on these initial results with a revised age model and a detailed analysis of the diatom flora of the lake. The combined sediment proxy (percent organic matter, percent carbonate, and molar C/N) and diatom data suggest a wet Early and early Middle Holocene, drier later half of the Middle Holocene, and a return to wetter conditions in the Late Holocene .

For the first 2800 years of the record (PE-V, PE-IV, and the intervening period), the trends in total organic matter and carbonate are similar. In PE-V and PE-IV, the longest pluvial episodes, organic matter varies by more than 40%, with the highest values reached in the upper half of both episodes. Carbonate values are consistently low in both pluvial intervals; there is also an abrupt drop of 20% at about 7300 cal yr B.P. Changes in the molar C:N values show that these are not a simple precipitation-evaporation relationship. Overall, the molar C:N values in the pluvial episodes indicate a mixture of algal and terrestrial organic matter; the values are stable in PE-V and trend upward toward a terrestrial source in PE-IV. The C:N values show a largely algal source during the inter-pluvial period. The stable carbonate values and limited variations in the C:N in PE-V suggest that the changes in organic matter are probably a function of dilution by sediments transported into the lake.

Kirby et al. (2012) have indicated that the pre-dam depth of Lower Bear Lake was about 3 m and the Early Holocene depth could have been >8 m. The bathymetric map (Fig. 2) suggests that the lake had a deep central basin surrounded by an extended, shallow shelf, particularly to the east, where it extends more than a kilometer from the coring site while only gaining a meter in elevation. Stone and Fritz (2004) have shown that the relative distribution of shallow marginal areas to deep basins plays a significant role in the understanding of lake-level changes. The decrease in organic matter between 9200 and 9000 cal yr B.P. corresponds to the greatest relative abundance of small Fragilarioids suggesting that the lake level rose, flooding the shelf space to the east of the core site.

Fritz et al. (1993) and Finkelstein and Gajewski (2007, 2008) have shown that due to high growth rates, this group may be over-represented in the sediment record, but the bathymetry of Lower Bear Lake and the sediment proxy records support this interpretation. It is likely that Lower Bear Lake reached its greatest Holocene extent during this time.

The lower and upper boundaries of the inter-pluvial interval are marked by evidence of abrupt changes in moisture availability. Organic matter values are well below those in the pluvial episodes, and carbonate values are more than triple those of the pluvial episodes. The molar C:N values indicate an algal carbon source with a gradual increase of terrestrial input up core, suggesting a gradual increase in moisture into PE-IV. Lower carbonate values in the upper part of the interval support this conclusion. The abrupt decrease in carbonate and increase in molar C:N ratio may be evidence of the 8.2 ka event during which the California coast experienced more North Pacific storms (Oster et al. 2017). An increase in *Aulacoseira* and planktic *Fragilaria* show that lake level rose at that time.

The organic matter and carbonate patterns in PE-IV follow a pattern similar to that in PE-V. The molar C:N ratio, however, shows a gradual change from algal-dominated carbon to a more mixed algal-terrestrial value. The absence of small Fragilarioids in PE-IV suggests that the level of the lake was not high enough to flood a large area of the shelf surrounding the deep central basin. The increase in epiphytic and heterotrophic species indicates more oligotrophic conditions with at least some aquatic macrophytes along the margin of the lake at the top of the interval. This is corroborated by the molar C:N ratio and presence of pollen from aquatic vascular plants.

The interval between the top of PE-IV and the bottom of PE-III, ~6400–3400, corresponds to a "mid-Holocene arid interval" (Kirby et al. 2015). Organic matter values are lower and carbonate values are as much as five times higher than in the Early Holocene. Molar C:N values are consistently dominated by algal carbon. This relative stability is punctuated by small pluvial episodes (PE-IIIc, PE-IIIb, and PE-IIIa) which all express small increases in organic matter, decreases in carbonate, and molar C:N ratios that shift toward a terrestrial carbon source. There are no strong trends in the diatom assemblages.

The last 3400 years, beginning with PE-III, indicate an overall increase in precipitation. Organic matter gradually increases in abundance during most of the Late Holocene, with notable increases during the pluvial episodes, in particular, PE-II, which corresponds to the wet interval in the Medieval Climate Anomaly that is present at a number of sites in western North America (Rodysill et al. 2018). The carbonate trend until about 1500 cal yr B.P. reflects an increase in precipitation, but about 1400 cal yr B.P. there is little overall change and the variability is low which suggests that the increase in organic matter at this time may be due to a decrease in sediment transport into the lake. The molar C:N ratio is relatively stable. The diatom response indicates an initial deepening with a shift from epiphytic diatoms to *Aulacoseira*, which is followed by a decrease in available nitrogen and another in *Aulacoseira* to a relative abundance of over 50%, a pattern which appears only in PE-III. This change may be related to a change in stratification. Kilham et al. (1986) have suggested that the high abundance of *Nitzschia fonticola*, an obligate heterotroph, in association with the blue-green alga *Microcystis aeruginosa* is an indication of hydrodynamically stable conditions. High numbers of *Aulacoseira* spp. are generally associated with wind-driven convective mixing (Kilham 1990; Kilham et al. 1996) and that there is a relationship between nutrient availability and light level (Stoermer et al. 1981; Tilman 1981). It should, however, be noted that the relationships discussed are based on research conducted in large lakes and may not be applicable in this case. At the end of PE-III, planktic diatoms largely disappear from the record and are replaced by a variety of benthic taxa filling a variety of habitat and nutrient-controlled niches.

Climatic Forcing

There are three predominant multi-millennial-scale forcings that probably influence Holocene climate in the coastal southwest United States: (1) during the Early Holocene, the diminishing, yet still present Laurentide and Cordilleran Ice Sheets (Dyke 2004) and for the entire Holocene, (2) Milankovitch forcing via changes in seasonal insolation values (Bird and Kirby 2006; Kirby et al. 2007, 2010, 2014, 2015; Lachniet et al. 2014, 2020), and (3) Pacific sea surface temperatures (Namias et al. 1988; Kirby et al. 2010, 2014, 2015; Barron and Anderson 2011; Barron et al. 2012; McCabe-Glynn et al. 2013; Oster et al. 2017).

Beginning in the late Pleistocene and continuing into the Early Holocene, the remnant North American ice sheets influenced atmospheric circulation by modulating the north-south latitudinal thermal gradients and likely displacing the mean average position of the circumpolar vortex south of its present position. This likely resulted in a higher frequency of winter storms crossing the central and southern California coast (Kirby et al. 2005; Wong et al. 2016; Oster et al. 2017).

Insolation drives climate broadly through the strength of seasonality (Harrison et al. 2003, 2014; Metcalfe et al. 2015) and by modulating the position of the Intertropical convergence zone (ITCZ) and polar jet stream (Barton and Ellis 2009; Chiang 2009; Ellis and Barton 2012). The position of the ITCZ and polar jet stream, in turn, influences the strength and position of storm tracks (Lau 1988; Sawada et al. 2004; Chiang and Fang 2010; Gan and Wu 2013). Kirby et al. (2005, 2007, 2010, 2015) and Bird and Kirby (2006) suggested a relationship between long-term changes in seasonal insolation and Holocene climate change across the coastal southwest United States; their research is based on sediment cores from Lake Elsinore, Silver Lake, and Dry Lake.

Insolation at the beginning of the Lower Bear Lake record was about 7% higher in the summer and 7% lower in the winter than modern (Fig. 8). Spring and fall values are near the mean modern insolation values, at about 6000 cal yr B.P.; the summer and winter values are near the mean and spring and fall are, respectively, at their minimum (~ −3.5%) and maximum (~ +3.5%). Greater summer insolation in the southwestern United States increases the strength and extent of the North American Monsoon (NAM; Metcalfe et al. 2015). In southern California the NAM is expressed as local convective thunderstorms (Tubbs 1972; Mitchell et al. 2002; Liu et al. 2003). Higher summer insolation would likely result in an increase in the number of tropical cyclones and depressions that make landfall or follow a track that brings them close enough to southern California to increase precipitation (Corbosiero et al. 2009; Ritchie et al. 2010). Early Holocene summer precipitation values were higher than modern in response to higher summer insolation (Kutzbach 1981; Kutzbach and Guetter 1986; COHMAP 1988; Diffenbaugh and Sloan 2004). Notably in the modern system, summer precipitation makes no difference in the annual hydrologic budget of the coastal southwest United States. Likewise, the impact of summer precipitation to the Early Holocene's hydrologic budget is debatable without fingerprinting relative source contributions. However, as proposed by Kirby et al. (2005, 2007, 2010, 2015) and Bird and Kirby (2006), enhanced summer precipitation may have helped to sustain, but not create, regional lakes in combination with winter precipitation. Lower winter insolation during the Early Holocene caused the southward migration of the polar jet stream, increasing the frequency of winter storms across the southwestern United States (Kirby et al. 2005, 2007, 2010, 2015; Bird and Kirby 2006; Chiang and Fang 2010; Ellis and Barton 2012; Oster et al. 2017). This increase in winter precipitation, the dominant contributor to the regions' annual hydrologic budget, helps to explain the regional evidence for wetter conditions, more runoff, groundwater input, and/or higher lake levels. Coupled with enhanced summer precipitation, the Early Holocene is the wettest extended period across the study region over the past 11,700 years.

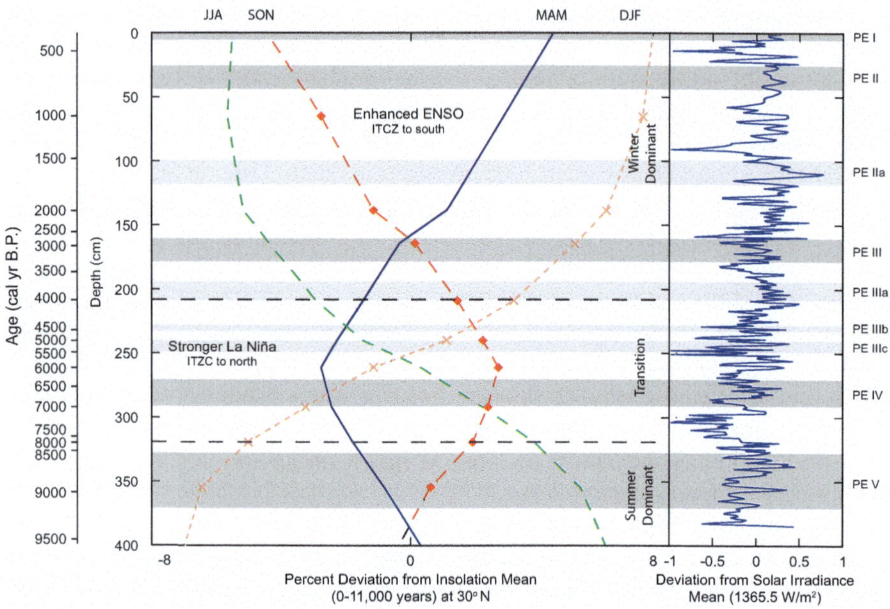

Fig. 8 Insolation deviation from 0 to 11,000 year at 30° N compared to modern. (Modified from Berger and Loutre (1991)). Deviation of solar irradiance from mean (1365.5 W/m^2). (Data from Steinhilber et al. (2012))

Sea Surface Temperatures

Tropical and North Pacific sea surface temperatures (SST) are important controls of western US paleohydroclimate on timescales ranging from sub-decadal to millennial (Namias et al. 1988; Ely et al. 1994, Barron et al. 2003, 2005, 2012; Barron and Bukry 2007; Kirby et al. 2010, 2014; Barron and Anderson 2011; Lachniet et al. 2014, 2020). Higher North Pacific and Tropical SST generally generate wetter winters along the California coast. Precipitation increases if the warmer SST are present in conjunction with El Niño and/or warm Pacific Decadal Oscillation (PDO) conditions (Namias et al. 1988; Andrews et al. 2004; Cayan et al. 1999; Dai and Wigley 2000; Gedalof et al. 2002; Castello and Shelton 2004; Seager et al. 2005; D'Orgeville and Peltier 2009; Cook et al. 2011; Wang et al. 2012, 2013; An and Choi 2013; Dai 2013; Gan and Wu 2013; McCabe-Glynn et al. 2013; Carré et al. 2015).

The Early Holocene wet interval, including PE-V at Lower Bear Lake, corresponds with a period of higher SST (warm-phase PDO) at ODP Site 1019 (Barron et al. 2003) and a weaker tropical Pacific SST gradient (El Niño conditions) (Koutavas and Joanides 2012). Conditions were similar during the Late Holocene, although the SST at ODP 1019 was not at high, a period during which conditions were becoming drier at Lower Bear Lake, but there were a series of weaker (relative to PE-V) pluvial episodes.

Although the Middle Holocene aridity is not as strongly expressed at Lower Bear Lake as it is at the Mojave Desert terminal lakes, it is a period of the highest percent carbonate values and is when diatoms favoring brackish water and/or higher TDS begin to increase in abundance. During this time SST along the northern California coast reach their lowest values and the tropical Pacific SST gradient anomaly is the strongest (La Niña conditions). Kirby et al. (2015) have suggested that southern California is strongly affected by eastern Pacific ENSO (Karamperidou et al. 2015) variability and that this may in part be responsible for the Middle Holocene aridity at Silver Lake.

The Late Holocene is marked by more frequent and possibly stronger El Niño events (Moy et al. 2002; Conroy et al. 2008) which may have been responsible for the increased moisture in southern California. This corresponds to the long-term decrease in summer and increase in winter insolation which would favor more El Niño events (Clement et al. 1999). The trend in insolation may be the reason that Lower Bear Lake levels do not return to Early Holocene levels.

Regional Comparisons

In order to assess the regional extent of the pluvial episodes, several Holocene paleoclimate records were evaluated. Comparison sites include local lakes (Dry Lake and Lake Elsinore), Silver Lake (terminal lake for the Mojave River which has its headwaters in the San Bernardino Mountains), Tulare Lake (Central Valley), Owens Lake (eastern Sierra Nevada rain shadow), three coastal sites influences by marine conditions (San Joaquin marsh and Abalone Rocks marsh and Soledad Pond on Santa Rosa Island), and Leviathan Cave in southern Nevada. The summaries for these sites are in Table 2.

While the overall trends in precipitation observed in the Lower Bear Lake record were present in the other records, there are inconsistencies in timing and magnitude of the trends. The variations in timing may be real or, at least in part, due to issues associated with age control. Differences in magnitude may be due to differences in the ability of individual proxies to express subtle changes in precipitation and evaporation and/or local conditions. While all of the comparison sites showed some evidence of pluvial episodes, no site contained evidence of all the events identified by Kirby et al. (2012). As noted above, there are a number of reasons why an event may not be present in the record at a number of sites. It is also possible that if atmospheric rivers are responsible for the pluvial episodes recorded in Lower Bear Lake, as proposed by Kirby et al. (2012), these events reflect a short-term synergistic relationship between local and regional conditions.

Table 2 Regional comparison of evidence for pluvial episodes in lacustrine, marsh, and speleothem records from southern California and Nevada

	Regional comparisons						Regional comparisons	
Site	Reference	Latitude	Longitude	Elevation (m)	Period covered (cal yr B.P.)	Proxy	Proxy trends	Pluvial episodes
Dry Lake	Bird and Kirby (2006) and Bird et al. (2010)	34° 07′ 30″	116° 49′ 41″	2763	9000-A.D. 1860	% sand, MS, %TOM, charcoal, ground-penetrating radar	Early Holocene highstand interrupted by 8.2 ka event continues until 5550; low lake level for remainder of the record	High %TOM due to increased precipitation from 9000 to ~8500 which corresponds to PE-V
Lake Elsinore	Kirby et al. (2004, 2005, 2007, 2010)	33° 39′ 30″	117° 21′ 00″	393	19,250-0	MS, TOM, carbonate, total inorganic P, %Al, % sand, $\delta^{18}O$, $\delta^{13}C$	Wet Early and early Middle Holocene; highly variable Middle and Late Holocene due to ENSO	Peaks correspond to all pluvial episodes (except PE-II), but numerous other equivalent peaks are present in TOM; Late Holocene MS, $\delta^{18}O$ peaks include PE-III, PE-IIa, PE-II; % sand includes parts of PE-V, PE-IV, PE-III
Leviathan Cave	Lachniet et al. (2014, 2020)	37° 49′ 53″	115° 36′ 26″	2420	16,000-0	$\delta^{18}O$, $\delta^{13}C$	Early Holocene warmer moisture(?), drier(?); early Middle Holocene following 8.2 ka event to 6500 colder moisture(?), wetter(?); Middle Holocene warmer moisture(?), drier; Late Holocene colder moisture(?), wetter(?)	PE-V corresponds to driest(?) interval; PE-IV, PE-III, PE-IIa to wet(?) peaks

	Regional comparisons						Regional comparisons	
Site	Reference	Latitude	Longitude	Elevation (m)	Period covered (cal yr B.P.)	Proxy	Proxy trends	Pluvial episodes
Owens Lake	Bacon et al. (2006, 2018)	36° 27′ 00″	118° 00′ 00″	1095	27,000-0	Sedimentology, TIC, geomorphology, shoreline mapping	Wet Early and early Middle Holocene, dry Middle Holocene; wet Late Holocene (not as wet as Early Holocene)	PE-V, PE-III(?), PE-II; Bacon et al. (2018) identified a number of short wetter intervals, none of which appear to match those in Lower Bear Lake
San Joaquin marsh	Davis (1992)	33° 39′ 34″	117° 51′ 11″	3	7000-0 uncalibrated	Pollen	Wetter early Middle Holocene followed by drying trend in remainder of Holocene (apparent drying may be due to saltwater intrusion)	Peaks in composite correspond to PE-IV and PE-III
Santa Rosa Island (two sites)	Anderson et al. (2010)	33° 57′ 55″; 33° 57′ 20″	120° 05′ 50″; 119° 05′ 45″	275 m; 0 m	12,000-0; 7000-0	MS, pollen, diatoms	Progressive drying in Early Holocene and early Middle Holocene; increased moisture in the Late Holocene	No evidence of pluvial episodes
Silver Lake	Kirby et al. (2015)	35° 22′ 30″	116° 08′ 15″	280	14,800-0	MS, TOM, carbonate, C:N, grainsize, ostracod abundance	Wet Early Holocene; arid Middle Holocene; wet (ephemeral lake) Late Holocene	Pluvial episodes not present; may have been smoothed in terminal lake record

(continued)

Table 2 (continued)

	Regional comparisons						Regional comparisons	
Site	Reference	Latitude	Longitude	Elevation (m)	Period covered (cal yr B.P.)	Proxy	Proxy trends	Pluvial episodes
Tulare Lake	Davis (1999), Negrini et al. (2006), and Blunt and Negrini (2015)	36° 00′ 00″	119° 45′ 00″	56	11,600-0	Sedimentology, TOC, TIC, pollen	Wetter Early Holocene prior to ~7500; dry Middle Holocene; moist neopluvial ~4500–2800 followed by drought (corresponds to Late Holocene Dry Period (Mensing et al. 2013); wet 1900–1100	PE-V, PE-IV, PE-III, PE-I; several Middle Holocene highstands do not correspond to PE-IIIc, PE-IIIb, PE-IIIa

Conclusion

The Holocene sediment proxy and diatom evidence from Lower Bear Lake, in the San Bernardino Mountains near the southern California coast, record an insolation-controlled drying trend and identify several major pluvial episodes. Although the revised age model adjusts the timing of these intervals, the results confirm the original conclusions of Kirby et al. (2012). The changes in the diatom flora provide a more detailed qualitative understanding of the variations in lake size during the multi-centennial pluvial episodes. The diatom flora exhibits a small and gradual trend toward slightly brackish water or waters with a high level of dissolved solids. This trend is punctuated by the several pluvial episodes of varying extent. Small Fragilarioid taxa dominate the flora during PE-V, the longest of the pluvial episodes, which indicates that the lake reached its greatest extent during this time. The relative abundance of *Aulacoseira* spp. from the end of PE-V through the end of PE-III shows that the lake occupied a deep central basin and rarely extended onto the surrounding shelf. During this time, taxa that tolerate more saline conditions, such as *Craticula* spp., *Anomoeoneis sphaerophora*, and *Halamphora coffeaeformis*, increase in abundance. The abundance of epiphytic species also increases during this time suggesting an increase in aquatic macrophytes around the margin of the lake. Species with cyanobacterial endosymbionts are in greater abundance from the end of PE-IV to PE-III, during the middle Holocene arid interval. These taxa do not respond consistently to the pluvial episodes. The analysis of several lake records across southern California supports the conclusion that the lake level changes observed in Lower Bear Lake are synoptic in nature and not the result of microclimatic variability.

The timing of the pluvial episodes does not correspond consistently to traditional sources of moisture such as the ENSO and PDO and based on conditions present in the early twentieth century. Kirby et al. (2012) proposed that atmospheric rivers provided extra moisture during the pluvial episodes. However, it is presently unclear as to the oceanic and atmospheric conditions necessary for the formation of atmospheric rivers (Dettinger 2011). The key to understanding these episodes of excess moisture may lie in an understanding of the composite effects of multiple teleconnections.

These results generally agree with and build on those of Enzel et al. (1989, 1992), Enzel (1992), Ely et al. (1993, 1994), Enzel and Wells (1997), Wells et al. (2003), and Kirby et al. (2015) in an attempt to explain the complex interaction of atmospheric and marine processes that control the highly variable precipitation in coastal California and the Mojave Desert to the east. The projected increase in global temperatures over the coming decades and the increase in moisture suggested by climate modeling make the understanding of these processes and the hazards associated with them increasingly important.

Acknowledgments The authors thank John Barron (USGS) for his comments on several drafts of this manuscript, Michael Rosen for the USGS review, and Jeffrey Stone and an anonymous reviewer for their insightful and helpful comments. We thank the editor, Michael Rosen, for his patience. This research was supported by the USGS Land Change Science Program.

References

Adams, D. K., & Comrie, A. C. (1997). The North American monsoon. *Bulletin of the American Meteorological Society, 78*(10), 2197–2213.

An, S.-I., & Choi, J. (2013). Mid-Holocene tropical Pacific climate state, annual cycle, and ENSO in PMIP2 and PMIP3. *Climate Dynamics, 43*(3–4), 957–970. https://doi.org/10.1007/s00382-013-1880-z.

Anderson, R. S., Starratt, S. W., Jass, R. M. B., & Pinter, N. (2010). Fire and vegetation history on Santa Rosa Island, Channel Islands, and long-term environmental change in southern California. *Journal of Quaternary Science, 25*(5), 782–797.

Andrews, E. D., Antweiler, R. C., Neiman, P. J., & Ralph, F. M. (2004). Influence of ENSO on flood frequency along the California coast. *Journal of Climate, 17*, 337–348.

Bacon, S. N., Burke, R. M., Pezzopane, S. K., & Jayko, A. S. (2006). Last glacial maximum and Holocene lake levels of Owens Lake, eastern California, USA. *Quaternary Science Reviews, 25*, 1264–1282.

Bacon, S. N., Lancaster, N., Stine, S., Rhodes, E. J., & McCarley Holder, G. A. (2018). A continuous 4000-year lake-level record of Owens Lake, south-central Sierra Nevada, California, USA. *Quaternary Research, 90*, 276–302. https://doi.org/10.1017/qua.2018.50.

Barlow, M., Nigam, S., & Berbery, E. H. (2001). ENSO, Pacific decadal variability, and U.S. summertime precipitation, drought, and stream flow. *Journal of Climate, 14*(9), 2105–2128.

Barron, J. A., & Anderson, L. (2011). Enhanced late Holocene ENSO/PDO expression along the margins of the eastern North Pacific. *Quaternary International, 235*, 1–10.

Barron, J. A., & Bukry, D. (2007). Solar forcing of the Gulf of California during the past two thousand years suggested by diatoms and silicoflagellates. *Marine Micropaleontology, 62*, 115–139.

Barron, J. A., Heusser, L. A., Herbert, T., & Lyle, M. (2003). High-resolution climatic evolution of coastal northern California during the past 16,000 years. *Paleoceanography, 18*(1). https://doi.org/10.1029/2002PA000768.

Barron, J. A., Bukry, D., & Dean, W. E. (2005). Paleoceanographic history of the Guaymas Basin, Gulf of California, during the past 15,000 years, based on diatoms, silicoflagellates, and biogenic sediments. *Marine Micropaleontology, 56*, 81–102.

Barron, J. A., Metcalfe, S. E., & Addison, J. A. (2012). Response of the North American monsoon to regional changes in ocean surface temperature. *Paleoceanography, 27*. https://doi.org/10.1029/2011PA002235.

Barton, N. P., & Ellis, A. W. (2009). Variability in wintertime position and strength of the North Pacific jet stream as represented by re-analysis data. *International Journal of Climatology, 29*, 851–862.

Battarbee, R. A. (1986). Diatom analysis. In B. E. Berglund (Ed.), *Handbook of Holocene palaeoecology and palaeohydrology* (pp. 527–570). New York: Wiley.

Benson, L., Kashgarian, M., Rye, R., Lund, S., Paillet, F., Smoot, J., Kester, C., Mensing, S., Meko, D., & Lindstrom, S. (2002). Holocene multidecadal and multicentennial droughts affecting Northern California and Nevada. *Quaternary Science Reviews, 21*, 659–682.

Berger, A., & Loutre, M. F. (1991). Insolation values for the climate of the last 10 million years. *Quaternary Science Reviews, 10*, 297–317.

Bigler, C., & Hall, R. I. (2002). Diatoms as indicators of climatic and limnological change in Swedish Lapland: A 100-lake calibration set and its validation for paleoecological reconstructions. *Journal of Paleolimnology, 27*, 97–115.

Bird, B. W., & Kirby, M. E. (2006). An alpine lacustrine record of early Holocene North American monsoon dynamics from Dry Lake, southern California (USA). *Journal of Paleolimnology, 35*, 179–192.

Bird, B. W., Kirby, M. E., Howat, I. M., & Tulaczyk, S. (2010). Geophysical evidence for Holocene lake-level change in southern California (Dry Lake). *Boreas, 39*(1), 131–144. https://doi.org/10.1111/j.1502-3885.2009.00114.

Blaauw, M. (2010). Methods and code for 'classical' age-modeling of radiocarbon sequences. *Quaternary Geochronology, 5*, 512–518.

Blaauw, M., & Christen, J. A. (2011). Flexible paleoclimate age-depth models using an autoregressive gamma process. *Bayesian Analysis, 6*(3), 457–474.

Blaauw, M., & Heegaard, E. (2014). Estimation of age-depth relationships. In H. J. B. Birks, A. F. Lotter, S. Juggins, & J. P. Smol (Eds.), *Tracking environmental change using lake sediments: data handling and numerical techniques: Developments in paleoenvironmental research* (Vol. 5, pp. 379–414). Dordrecht: Springer Netherland.

Blunt, A. B., & Negrini, R. M. (2015). Lake levels for the past 19,000 years from the TL05-4 cores, Tulare Lake, California, USA. *Geophysical and geochemical proxies: Quaternary International, 387*, 122–130.

Bradbury, J. P. (1997). A diatom record of climate and hydrology for the past 200 ka from Owens Lake, California with comparison to other Great Basin lake records. *Quaternary Science Research, 16*, 203–219.

Browning, K. A., & Pardoe, C. W. (1973). Structure of low-level jetstreams ahead of mid-latitude cold fronts. *Quarterly Journal of the Royal Meteorological Society, 99*(422), 619–638.

Brunelle, A., & Anderson, R. S. (2003). Sedimentary charcoal as an indicator of late-Holocene drought in the Sierra Nevada, California, and its relevance to the future. *Holocene, 13*, 21–28.

Cantonati, M., Kelly, M. G., & Lange-Bertalot, H. (2017). *Freshwater benthic diatoms of Central Europe: Over 800 common species used in ecological assessment* (942 p). Schmitten-Oberreifenberg: Koeltz Botanical Books.

Carré, M., Sachs, J. P., Purea, S., Schauer, A. J., Braconnot, P., Falcón, R. A., Julien, M., & Lavallée, D. (2015). Holocene history of ENSO variance and asymmetry in the eastern tropical Pacific. *Science, 345*, 1045–1048.

Castello, A. F., & Shelton, M. L. (2004). Winter precipitation on the US Pacific coast and El Niño-Southern Oscillation events. *International Journal of Climatology, 24*(4), 481–497.

Cayan, D. R., & Peterson, D. H. (1989). The influence of North Pacific atmospheric circulation on streamflow in the west. *American Geophysical Union Geophysical Monographs, 55*, 375–397.

Cayan, D. R., Redmond, K. T., and Riddle, R. G. (1999). ENSO and hydrologic extremes in the western United States: Journal of Climate, v. 12, no. 9, p. 2881–2893.

Chiang, J. C. H. (2009). The tropics in paleoclimate. *Annual Review of Earth and Planetary Sciences, 37*, 263–297.

Chiang, J. C. H., & Fang, Y. (2010). Was the North Pacific wintertime climate less stormy during the mid-Holocene? *Journal of Climate, 23*, 4025–4037.

Clement, A. C., Seager, R., & Cane, M. A. (1999). Orbital controls on the El Niño/Southern Oscillation and the tropical climate. *Paleoceanography, 15*, 731–737.

COHMAP Members. (1988). Climatic changes of the last 18,000 years: Observations and model simulations. *Science, 241*, 1043–1052.

Comrie, A. C., & Glenn, E. C. (1998). Principal components-based regionalization of precipitation regimes across the southwest United States and northern Mexico, with an application to monsoon precipitation variability. *Climate Research, 10*, 201–210.

Conroy, J. L., Overpeck, J. T., Cole, J. E., Shanahan, T. M., & Steinitz-Kannan, M. (2008). Holocene changes in eastern tropical Pacific climate inferred from a Galápagos lake sediment record. *Quaternary Science Reviews, 27*, 1166–1180.

Cook, B. I., Seager, R., & Miller, R. L. (2011). On the causes and dynamics of the early Twentieth-Century North American pluvial. *Journal of Climate, 24*, 5043–5060.

Corbosiero, K. L., Dickinson, M. J., & Bosart, M. J. (2009). The contribution of eastern North Pacific tropical cyclones to the rainfall climatology of the southwest United States. *Monthly Weather Review, 137*, 2415–2435.

Cumming, B. F., Wilson, S. E., Hall, R. J., & Smol, J. P. (1995). Diatoms from British Columbia (Canada) lakes and their relationship to salinity, nutrients and other limnological variables. *Bibliotheca Diatomologica, 31*, 1–207.

Dai, A. (2013). The influence of the inter-decadal Pacific oscillation on US precipitation during 1923–2010. *Climate Dynamics, 15*(7), 1062–1071.

Dai, A., & Wigley, T. M. L. (2000). Global patterns of ENSO-induced precipitation. *Geophysical Research Letters, 27*, 1283–1286.

Dailey, M. D., Anderson, J. W., Reish, D. J., & Gorsline, D. S. (1993). The Southern California Bight: Background and setting. In M. D. Dailey, D. J. Reish, & J. W. Anderson (Eds.), *Ecology of the Southern California Bight: A synthesis and interpretation* (pp. 1–18). Berkeley: University of California Press.

Davis, O. K. (1992). Rapid climatic change in coastal southern California inferred from pollen analysis of San Joaquin marsh. *Quaternary Research, 37*, 89–100.

Davis, O. K. (1999). Pollen analysis of Tulare Lake, California: Great Basin-like vegetation in central California during the full-glacial and early Holocene. *Review of Palaeobotany and Palynology, 107*, 249–257.

Davis, O. K., Anderson, R. S., Fall, P. L., O'Rourke, M. K., & Thompson, R. S. (1985). Palynological evidence for early Holocene aridity in the southern Sierra Nevada, California. *Quaternary Research, 24*, 322–332.

Dettinger, M. (2007). Fifty-two years of Pineapple-Express storms across the west coast of North America. In S. W. Starratt, P. Cornelius, & J. G. Joelson Jr. (Eds.), *Proceedings of the twenty-first annual Pacific climate workshop, technical report 77 of the interagency ecological program for the San Francisco Estuary* (pp. 41–58).

Dettinger, M. (2011). Climate change, atmospheric rivers, and floods in California – A multimodel analysis of storm frequency and magnitude changes. *Journal of the American Water Resources Association, 47*(3), 514–523.

DeYoe, H. R., Lowe, R. L., & Marks, J. C. (1992). Effects of nitrogen and phosphorus on the endosymbiont load of *Rhopalodia gibba* and *Epithemia turgida* (Bacillariophyceae). *Journal of Phycology, 28*, 773–777.

Diffenbaugh, N. S., & Sloan, L. C. (2004). Mid-Holocene orbital forcing of regional-scale climate: A case study of western North America using a high-resolution RCM. *Journal of Climate, 17*, 2927–2937.

Dingemans, R., Mensing, S. A., Feakins, S. J., Kirby, M. E., & Zimmerman, S. R. H. (2014). 3000 years of environmental change at Zaca Lake, California, USA. *Frontiers of Ecology and Evolution, 2*, art. 34. https://doi.org/10.3389/fevo.2014.00034.

D'Orgeville, M., & Peltier, W. R. (2009). Implications of both statistical equilibrium and global warming simulations with CCSM3. Part 1: On the decadal variability in the North Pacific basin. *Journal of Climate, 22*, 5277–5297.

Dyke, A. S. (2004). An outline of North American deglaciation with emphasis on central and northern Canada. In J. Ehlers & P. L. Gibbard (Eds.), *Developments in quaternary sciences, quaternary glaciations–extent and chronology, part II* (pp. 373–424). Amsterdam: Elsevier.

Ellis, A. W., & Barton, N. P. (2012). Characterizing the North Pacific jet stream for understanding historical variability in western United States winter precipitation. *Physical Geography, 33*, 105–128.

Ely, L. L., Enzel, Y., Baker, V. R., & Cayan, D. R. (1993). A 5000-year record of extreme floods and climate change in the Southwestern United States. *Science, 262*, 410–412.

Ely, L. L., Enzel, Y., & Cayan, D. R. (1994). Anomalous North Pacific atmospheric circulation and large winter floods in the southwestern United States. *Journal of Climate, 7*, 977–987.

Enzel, Y. (1992). Flood frequency of the Mojave River and the formation of late Holocene playa lakes, southern California, USA. *The Holocene, 2*(1), 11–18.

Enzel, Y., & Wells, S. G. (1997). Extracting Holocene paleohydrology and paleoclimatology information from modern extreme flood events: An example from southern California. *Geomorphology, 19*, 203–226.

Enzel, Y., Cayan, D. R., Anderson, R. Y., & Wells, S. G. (1989). Atmospheric circulation during Holocene lake stands in the Mojave Desert: Evidence of regional climate change. *Nature, 341*, 44–47.

Enzel, Y., Brown, W. J., Anderson, R. Y., McFadden, L. D., & Wells, S. G. (1992). Short-duration Holocene lakes in the Mojave River drainage basin, southern California. *Quaternary Research, 38*, 60–73.

Feakins, S. J., Kirby, M. E., Cheetham, M. L., Ibarra, Y., & Zimmerman, S. R. H. (2014). Fluctuation in leaf wax D/H ratio from a southern California lake records significant variability in isotopes in precipitation during the late Holocene. *Organic Geochemistry, 66*, 48–59.

Finkelstein, S. A., & Gajewski, K. (2007). A palaeolimnological record of diatom-community dynamics and late-Holocene climatic changes from Prescott Island, Nunavut, central Canadian Arctic. *The Holocene, 17*(6), 803–812.

Finkelstein, S. A., & Gajewski, K. (2008). Responses of Fragilarioid-dominated diatom assemblages in a small Arctic lake to Holocene climatic changes, Russell Island, Nunavut, Canada. *Journal of Paleolimnology, 40*, 1079–1095.

Fisler, J., & Hendy, I. L. (2008). California current system response to late Holocene climate cooling in southern California. *Geophysical Research Letters, 35*, L09702. https://doi.org/10.1029/2008GL033902.

Flint, L. E., Martin, P., with contributions by Brandt, J., Christensen, A. H., Flint, A. L., Flint, L. E., Hevesi, J. A., Jachens, R., Kulongoski, J. T., Martin, P., & Sneed, M. (2012). *Geohydrology of Big Bear Valley, California: Phase 1 – geologic framework, recharge, and preliminary assessment of the source of age of groundwater*. U.S. Geological Survey Scientific Investigations Report 2012–5100, 112 p.

Fritz, S. C. (1990). Twentieth-century salinity and water-level fluctuations in Devils Lake, North Dakota: Test of a diatom-based transfer function. *Limnology and Oceanography, 35*(8), 1771–1781.

Fritz, S. C., Juggins, S., & Battarbee, R. W. (1993). Diatom assemblages and ionic characterization of lakes of the northern Great Plains, North America: A tool for reconstructing past salinity and climate fluctuations. *Canadian Journal of Fisheries and Aquatic Sciences, 50*, 1844–1856.

Fye, F. K., Stahle, D. W., & Cook, E. R. (2004). Twentieth-Century sea surface temperature patterns in the Pacific over decadal moisture regimes over the United States. *Earth Interactions, 8*(22), 1–22.

Gan, B., & Wu, L. (2013). Seasonal and long-term coupling between wintertime storm tracks and sea surface temperature in the North Pacific. *Journal of Climate, 26*, 6123–6136.

Gasse, F. (1986). East African diatoms: Taxonomy, ecological distribution. *Bibliotheca Diatomologica, 11*, 1–201.

Gedalof, Z., Mantua, N. J., & Peterson, D. L. (2002). A multi-century perspective of variability in the Pacific Decadal Oscillation: New insights from tree rings and coral. *Geophysical Research Letters, 29*, 2204. https://doi.org/10.1029/2002GL015824.

Graumlich, L. J. (1993). A 1000-yr record of temperature and precipitation in the Sierra Nevada. *Quaternary Research, 39*, 249–255.

Harrison, S. P., Kutzbach, J. E., Liu, Z., Bartlein, P. J., Otto-Bliesner, B., Muhs, D., Prentice, I. C., & Thompson, R. S. (2003). Mid-Holocene climates of the Americas: A dynamical response to changed seasonality. *Climate Dynamics, 20*, 663–688.

Harrison, S. P., Bartlein, P. J., Brewer, S., Prentice, I. C., Boyd, M., Hessler, L., Holmgren, K., Izumi, K., & Willis, K. (2014). Climate model benchmarking with glacial and mid-Holocene climates. *Climate Dynamics, 43*, 671–688.

Haston, L. (1997). Spatial and temporal variability of southern California precipitation over the last 400 yr and relationships to atmospheric circulation patterns. *Journal of Climate, 10*, 1836–1852.
Haston, L., & Michaelsen, J. (1994). Long-term central California precipitation variability and relationships to El Niño-Southern Oscillation. *Journal of Climate, 7*, 1373–1387.
Hendy, I. L., Napier, T. J., & Schimmelmann, A. (2015). From extreme rainfall to drought: 250 years of annually resolved sediment deposition in Santa Barbara Basin, California. *Quaternary International, 387*, 3–12.
Hereford, R., Webb, R. H., & Logpre, C. I. (2006). Precipitation history and ecosystem response to multidecadal precipitation variability in the Mojave Desert region, 1893–2001. *Journal of Arid Environments, 67*, 13–34.
Heusser, L. E., Hendy, I. L., & Barron, J. A. (2015). Vegetation response to southern California drought during the Medieval Climate Anomaly and early Little Ice Age (AD 800–1600). *Quaternary International, 387*, 23–35.
Hickey, B. M. (1993). Physical oceanography. In M. D. Dailey, D. J. Reish, & J. W. Anderson (Eds.), *Ecology of the Southern California Bight: A synthesis and interpretation* (pp. 19–70). Berkeley: University of California Press.
Hood, D. W. (1993). Ecosystem interrelationships. In M. D. Dailey, D. J. Reish, & J. W. Anderson (Eds.), *Ecology of the Southern California Bight: A synthesis and interpretation* (pp. 782–835). Berkeley: University of California Press.
Hughes, M. K., & Brown, P. M. (1992). Drought frequency in central California since 101 B.C. recorded in giant sequoia tree rings. *Climate Dynamics, 6*, 101–197.
Hughes, M. K., and Graumlich, L. J. (1996). Multimillennial dendroclimatic records from the western United States: in Bradley, R. S., Jones, P. D., and Jouzel, J., eds., Climatic Variations and Forcing Mechanisms of the last 2000 Years, NATO Advanced Studies Workshop Series. Springer-Verlag, New York, p. 109–124.
Hughes, M. K., & Funkhouser, G. (1998). Extremes of moisture availability reconstructed from tree rings for recent millennia in the Great Basin of western North America. In M. Benniston & J. L. Innes (Eds.), *The impacts of climate variability on forests. Lecture notes in earth science* (Vol. 74, pp. 99–107). Berlin, New York: Springer.
Inman, D. L., & Jenkins, S. A. (1999). Climate change and the episodicity of sediment flux of small California rivers. *Journal of Geology, 107*, 251–270.
Karamperidou, C., Di Nezio, P. M., Timmermann, A., Jin, R. F., & Cobb, K. M. (2015). The response of ENSO flavors to mid-Holocene climate: Implications for proxy interpretation. *Paleoceanography, 30*, 527–547.
Kilham, P. (1990). Ecology of *Melosira* species in the great lakes of Africa. In M. M. Tilzer & C. Serruya (Eds.), *Large Lakes: Ecological structure and function* (pp. 414–427). Berlin: Springer Verlag.
Kilham, P., Kilham, S. S., & Hecky, R. E. (1986). Hypothesized resource relationships among African planktonic diatoms. *Limnology and Oceanography, 31*(6), 1169–1181.
Kilham, S. S., Theriot, E. C., & Fritz, S. C. (1996). Linking planktonic diatoms and climate change in the large lakes of the Yellowstone ecosystems using resource theory. *Limnology and Oceanography, 41*(5), 1052–1062.
Kim, J., Guan, B., Waliser, D. E., Ferraro, R. D., Case, J. L., Iguchi, T., Kemp, E., Putman, W., Wand, W., Wu, D., & Tian, B. (2018). Winter precipitation characteristics in western US related to atmospheric river landfalls: Observations and model evaluations. *Climate Dynamics, 50*, 231–248.
Kim, H.-M., Zhou, Y., & Alexander, M. A. (2019). Changes in atmospheric rivers and moisture transport over the northeast Pacific and western North America in response to ENSO diversity. *Climate Dynamics, 52*, 7375–7388.
Kirby, M. E., Poulsen, C. J., Lund, S. P., Patterson, W. P., Reidy, L., & Hammond, D. E. (2004). Late Holocene lake level dynamics inferred from magnetic susceptibility and stable oxygen isotope data: Lake Elsinore, southern California (USA). *Journal of Paleolimnology, 31*, 275–293.

Kirby, M. E., Lund, S. P., & Poulsen, C. J. (2005). Hydrologic variability and the onset of modern El Niño-Southern Oscillation: A 19,250-year record from Lake Elsinore, southern California. *Journal of Quaternary Science, 20*(3), 239–254.

Kirby, M. E., Lund, S. P., & Bird, B. W. (2006). Mid-Wisconsin sediment record from Baldwin Lake reveals hemispheric climate dynamics (southern CA, USA). *Palaeogeography, Palaeoclimatology, Palaeoecology, 241*, 267–283.

Kirby, M. E., Lund, S. P., Anderson, M. A., & Bird, B. W. (2007). Insolation forcing of Holocene climate change in southern California: A sediment study from Lake Elsinore. *Journal of Paleolimnology, 38*, 395–417.

Kirby, M. E., Lund, S. P., Patterson, W. P., Anderson, M. A., Bird, B. W., Ivanovici, L., Monarrez, P., & Nielsen, S. (2010). A Holocene record of Pacific Decadal Oscillation (PDO)-related hydrologic variability in Southern California (Lake Elsinore, CA). *Journal of Paleolimnology, 44*, 819–839.

Kirby, M. E., Zimmerman, S. R. H., Patterson, W. P., & Rivera, J. J. (2012). A 9170-year record of decadal-to-multi-centennial scale pluvial episodes from the coastal southwest United States: A roll for atmospheric rivers? *Quaternary Science Reviews, 46*, 57–65.

Kirby, M. E., Feakins, S. J., Bonuso, N., Fantozzi, J. M., & Hiner, C. A. (2013). Latest Pleistocene to Holocene hydroclimates from Lake Elsinore, California. *Quaternary Science Reviews, 76*, 1–15.

Kirby, M. E., Feakins, S. J., Hiner, C. A., Fantozzi, J., Zimmerman, S. R. H., Dingemans, T., & Mensing, S. A. (2014). Tropical Pacific forcing of late Holocene hydrologic variability in the coastal southwest United States. *Quaternary Science Reviews, 102*, 27–38.

Kirby, M. E., Knell, E. J., Anderson, W. T., Lachniet, M. S., Palermo, J., Eeg, H., Lucero, R., Murrieta, R., Arevalo, A., Silveira, E., & Hiner, C. A. (2015). Evidence for insolation and Pacific forcing of late glacial through Holocene climate in the central Mojave Desert (Silver Lake, CA). *Quaternary Research, 84*, 174–186.

Kleppe, J. A., Brothers, D. S., Kent, G. M., Biondi, F., Jensen, S., & Driscoll, N. W. (2011). Duration and severity of Medieval drought in the Lake Tahoe Basin. *Quaternary Science Reviews, 30*, 3269–3279.

Koutavas, A., & Joanides, S. (2012). El Niño-Southern Oscillation extremes in the Holocene and last glacial maximum. *Paleoceanography, 27*, PA4208. https://doi.org/10.1029/2012PA002378.

Krammer, K. (1997a). Die cymbelloiden diatomeen. Eine monographie der weltweit bekannten Taxa. Teil 1. Allgemeines und Encyonema part. *Bibliotheca Diatomologica, 36*, 1–382.

Krammer, K. (1997b). Die cymbelloiden diatomeen. Eine monographie der weltweit bekannten Taxa. Teil 2. *Encyonema* part., *Encyonopsis*, und *Cymbellopsis*. *Bibliotheca Diatomologica, 37*, 1–469.

Krammer, K. (2000). *Diatoms of Europe. Diatoms of the European inland waters and comparable habitats. Volume 1: The genus Pinnularia* (703 p). Ruggell: A.R.G. Gantner Verlag K.G.

Krammer, K. (2002). *Diatoms of Europe. Diatoms of the European inland waters and comparable habitats. Volume 3: Cymbella* (584 p). Ruggell: A.R.G. Gantner Verlag K.G.

Krammer, K., & Lange-Bertalot, H. (1986). Bacillariophyceae. 1. Teil: Naviculaceae. In H. Ettl, J. Gerloff, H. Heynig, & D. Mollenhauer (Eds.), *Süsswasser flora von Mitteleuropa, Band 2/1* (876 p). Stuttgart, New York: Gustav Fischer Verlag.

Krammer, K., & Lange-Bertalot, H. (1988). Bacillariophyceae. 2. Teil: Bacillariaceae, Epithemiaceae, Surirellaceae. In H. Ettl, J. Gerloff, H. Heynig, & D. Mollenhauer (Eds.), *Süsswasserflora von Mitteleuropa, Band 2/2* (611 p). Jena: Gustav Fischer Verlag.

Krammer, K., & Lange-Bertalot, H. (1991a). Bacillariophyceae. 3. Teil: Centrales, Fragilariaceae, Eunotiaceae. In H. Ettl, J. Gerloff, H. Heynig, & D. Mollenhauer (Eds.), *Süsswasserflora von Mitteleuropa, Band 2/3* (576 p). Jena: Gustav Fischer Verlag.

Krammer, K., & Lange-Bertalot, H. (1991b). Bacillariophyceae. 4. Teil: Achnanthaceae, Kritische Ergänzungen zu *Navicula* (Lineolatae) und *Gomphonema*, Gesamtliteraturverzeichnis Teil 1–4. In H. Ettl, G. Gärtner, J. Gerloff, H. Heynig, & D. Mollenhauer (Eds.), *Süsswasserflora von Mitteleuropa, Band 2/4* (437 p). Jena: Gustav Fischer Verlag.

Kutzbach, J. E. (1981). Monsoon climate of the early Holocene: Climatic experiment using the earth's orbital parameters for 9000 years ago. *Science, 234*, 59–61.
Kutzbach, J. E., & Guetter, P. J. (1986). The influence of changing orbital parameters and surface boundary conditions on climate simulations for the past 18,000 years. *Journal of Atmospheric Science, 43*, 1726–1759.
Lachniet, M. S., Denniston, R. F., Asmerom, Y., & Polyak, V. J. (2014). Orbital control of western North America atmospheric circulation and climate over two glacial cycles. *Nature Communications, 5*, 3805. https://doi.org/10.1038/ncomms4805.
Lachniet, M. S., Asmerom, Y., Polyak, V. P., and Denniston, R. F. (2020). Great Basin paleoclimate and aridity linked to Arctic warming and tropical Pacific sea surface termperatures: Paleoceanography and Paleoclimatology, v. 35, e2019PA003785. https://doi.org/10.1029/2019PA003785.
Lange-Bertalot, H. (2001). *Diatoms of Europe: Diatoms of the European inland waters and comparable habitats. Volume 2: Navicula sensu strict, 10 Genera Separated from Navicula sensu lato. Frustulia* (526 p). Ruggell: A.R.G. Gantner Verlag K.G.
Lange-Bertalot, H., & Krammer, K. (1987). Bacillariaceae, Epithemiaceae, Surirellaceae. *Bibliotheca Diatomologica, 15*, 1–289.
Lange-Bertalot, H., & Krammer, K. (1989). *Achnanthes* eine Monographie der Gattungen. *Bibliotheca Diatomologica, 18*, 1–393.
Lau, N.-C. (1988). Variability of the observed midlatitude storm tracks in relation to low-frequency changes in the circulating pattern. *Journal of Atmospheric Science, 45*, 2718–2743.
Leidy, R. (2003). *Prehistoric and historic environmental conditions in Bear Valley, San Bernardino County, California* (48 p). Sacramento: EIP Associates.
Lindström, S. (1990). Submerged tree stumps as indicators of mid-Holocene aridity in the Lake Tahoe region. *Journal of California and Great Basin anthropology, 12*, 146–157.
Liu, A., Shields, C., Otto-Bliesner, B., Kutzbach, J., & Li, L. (2003). Coupled climate simulation of the evolution of global monsoons in the Holocene. *Journal of Climate, 16*, 2472–2490.
MacDonald, G. M., Kremenetski, K. V., & Hildago, H. G. (2008). Southern California and the perfect drought: Simultaneous prolonged drought in southern California and the Sacramento and Colorado River systems. *Quaternary International, 188*, 11–23.
Malamud-Roam, F. P., Ingram, B. L., Hughes, M., & Florsheim, J. L. (2006). Holocene paleoclimate records from a large California estuarine system and its watershed region: Linking watershed climate and bay conditions. *Quaternary Science Reviews, 25*, 1570–1598.
Mantua, N. J., & Hare, S. R. (2002). The Pacific decadal oscillation. *Journal of Oceanography, 58*(1), 35–44.
McCabe-Glynn, S., Johnson, K. R., Strong, C., Berkelhammer, M., Sinha, A., Cheng, H., & Edwards, R. L. (2013). Variable North Pacific influence on drought in southwestern North America since AD 854. *Nature Geoscience, 6*, 617–621. https://doi.org/10.1038/NGEO1862.
Meko, D., Therrel, M., Baisan, C., & Hughes, M. (2001). Sacramento River flow reconstructed to A.D. 869 from tree rings. *Journal of the American Water Resources Association, 37*, 1029–1039.
Mensing, S. A., Sharpe, S. E., Tunno, I., Sada, D. W., Thomas, J. M., Starratt, S. W., & Smith, J. (2013). The Late Holocene Dry Period: Multiproxy evidence for an extended drought between 2800 and 1850 cal yr BP across the central Great Basin, USA. *Quaternary Science Reviews, 78*, 266–282.
Metcalfe, S. E., Barron, J. A., & Davies, S. J. (2015). The Holocene history of the North American Monsoon: "Known knowns" and "known unknowns" in understanding its spatial and temporal complexity. *Quaternary Science Research, 120*, 1–27.
Meyers, P. A., & Ishiwatari, R. (1993). Lacustrine organic geochemistry – an overview of indicators of organic matter sources and diagenesis in lake sediments. *Organic Geochemistry, 20*(7), 867–900.
Mitchell, D. L., Redmond, K., Ivanova, D., Rabin, R., & Brown, T. J. (2002). Gulf of California sea surface temperatures and the North American Monsoon: Mechanistic implications from observations. *Journal of Climate, 15*, 2261–2281.

Morgan, C., & Pomerleau, M. M. (2012). New evidence for extreme and persistent terminal medieval drought in California's Sierra Nevada. *Journal of Paleolimnology, 47*, 707–713.
Moy, C. M., Selzer, G. O., Rodbell, D. R., & Anderson, D. A. (2002). Variability of El Niño/Southern Oscillation activity at millennial timescales during the Holocene Epoch. *Nature, 420*, 162–165.
Namias, J., Yuan, X., & Cayan, D. R. (1988). Persistence of North Pacific sea surface temperature and atmospheric flow patterns. *Journal of Climate, 1*(7), 682–703.
Negrini, R. M., Wigand, P. E., Draucker, S., Gobalet, K., Gardner, J. K., Sutton, M. A., & Yoheli, R. M. (2006). The Rambla highstand shoreline and the Holocene lake-level history of Tulare Lake, California, USA. *Quaternary Science Reviews, 25*(13–14), 1599–1618.
Oster, J. L., Montañez, I. P., Sharp, W. D., & Cooper, K. M. (2009). Late Pleistocene California droughts during deglaciation and Arctic warming. *Earth and Planetary Science Letters, 288*, 434–443.
Oster, J. L., Sharp, W. D., Covey, A. K., Gibson, J., Rogers, B., & Mix, H. (2017). Climate response ot the 8.2 ka event in coastal California. *Scientific Reports, 7*(3886), 1–9. https://doi.org/10.1038/s41598-017-04215-5.
Overpeck, J., Garfin, G., Jardine, A., Busch, D. E., Cayan, D., Dettinger, M., Fleischman, E., Gershunov, A., MacDonald, G., Redmond, K. T., Travis, W. R., & Udall, B. (2013). Summary for decision makers. In G. Garfin, A. Jardine, R. Merideth, M. Black, & S. LeRoy (Eds.), *Assesssment of climate change in the Southwest United States* (pp. 1–20). Washington, DC: Island Press/Center for Resources Economics.
Owen, L. A., Finkel, R. C., Minnich, R. A., & Perez, A. E. (2003). Extreme southwestern margin of late Quaternary glaciation in North America: Timing and controls. *Geology, 31*, 729–732.
Pyke, C. B. (1972). *Some meteorological aspects of the seasonal distribution of precipitation in the western United States and Baja California.* University of California Water Resources Center, Report No. 139, University of California Water Resources Center, 209 pp.
Redmond, K. T., & Koch, R. W. (1991). Surface climate and streamflow variability in the western United States and their relationship to large-scale circulation indices. *Water Resources Research, 27*, 2381–2399.
Reimer, P. J., Baillie, M. G. L., Bard, E., Bayliss, A., Beck, J. W., Blackwell, P. G., Ramsey, C. B., Buck, C. E., Burr, G. S., Edwards, R. L., Friedrich, M., Grootes, P. M., Guilderson, T. P., Hajdas, I., Heaton, T. J., Hogg, A. G., Hughen, K. A., Kaiser, K. F., Kromer, B., McCormac, F. G., Manning, S. W., Reimer, R. W., Richards, D. A., Southon, J. R., Talamo, S., Turney, C. S. M., van der Plicht, J., & Weyhenmeyer, C. E. (2009). INTCAL 2009 dataset. *Radiocarbon, 51*, 1111–1150.
Reimer, P. J., Bard, E., Bayliss, A., Beck, J. W., Blackwell, P. G., Bronk Ramsey, C., Grootes, P. M., Guilderson, T. P., Haflidason, H., Hajdas, I., Hatté, C., Heaton, T. J., Hoffmann, D. L., Hogg, A. G., Hughen, K. A., Kaiser, K. F., Kromer, B., Manning, S. W., Niu, M., Reimer, R. W., Richards, D. A., Scott, E. M., Southon, J. R., Staff, R. A., Turney, C. S. M., & van der Plicht, J. (2013). IntCal13 and Marine13 radiocarbon age calibration curves 0–50,000 years cal BP. *Radiocarbon, 55*(4), 1869–1887.
Ritchie, E. A., Wood, K. M., Gutzler, D. S., & White, R. S. (2010). The influence of eastern Pacific tropical cyclone remnants on the southwestern United States. *Monthly Weather Review, 139*(1), 192–210.
Rodysill, J. R., Anderson, L., Cronin, T. M., Jones, M. C., Thompson, R. S., Wahl, D. B., Willard, D. A., Addison, J. A., Alder, J. R., Anderson, K. H., Anderson, L., Barron, J. A., Bernhardt, C. E., Hostetler, S. W., Kehrwald, N. M., Khan, N. S., Richey, J. N., Starratt, S. W., Strickland, L. E., Toomey, M. R., Treat, C. C., & Wingard, G. L. (2018). A North American Hydroclimate Synthesis (NAHS) of the common era. *Global and Planetary Change, 162*, 175–198. https://doi.org/10.1016/j.gloplacha.2017.12.025.
Round, F., & Bukhtiyarova, L. (1996). Four new genera based on *Achnanthes* (*Achnanthidium*), together with a redefinition of *Achnanthidium*. *Diatom Research, 11*, 345–361.

Rühland, K., Priesnitz, A., & Smol, J. P. (2003). Paleolimnological evidence from diatoms for recent environmental changes in 50 lakes across Canadian Arctic treeline. *Arctic, Antarctic, and Alpine Research, 35*, 110–123.
Saros, J. E., & Fritz, S. C. (2000). Changes in the growth rates of saline-lake diatoms in response to variation in salinity brine type and nitrogen form. *Journal of Plankton Research, 22*(6), 1071–1083.
Sawada, M., Viau, A. E., Vettoretti, G., Peltier, W. R., & Gajewski, K. (2004). Comparison of North American pollen-based temperature and global lake-status with CCCma AGCM2 output at 6 ka. *Quaternary Science Review, 23*, 225–244.
Schmidt, R., Kamenik, C., Kaiblinger, C., & Hetzel, M. (2004). Tracking Holocene environmental changes in an alpine lake sediment core: Application of regional diatom calibration, geochemistry, and pollen. *Journal of Paleolimnology, 32*, 177–196.
Schrader, H., & Gersonde, R. (1978). Diatoms and silicoflagellates. *Utrecht Micropaleontology Bulletin, 17*, 129–176.
Seager, R., Harnik, N., Robinson, W., Kushnir, Y., Ting, M., Huang, H. P., & Velez, J. (2005). Mechanisms of ENSO-forcing of hemispherically symmetric precipitation variability. *Quarterly Journal of the Royal Meteorological Society, 131*, 1501–1527.
Sickman, J. O., Bennett, D. M., Lucero, D. M., Whitmore, T. J., & Kenny, W. F. (2013). Diatom-inference models for acid neutralizing capacity and nitrate based on 41 calibration lakes in the Sierra Nevada, California, USA. *Journal of Paleolimnology*. https://doi.org/10.1007/s10933-013-9711-0.
Starratt, S. W. (2004). Diatoms as indicators of late Holocene fresh water flow variation in the San Francisco Bay estuary, central California, USA. In M. Poulin (Ed.), *Seventeenth international diatom symposium* (pp. 371–397). Bristol: Biopress Ltd.
Steinhilber, F., Abru, J. A., Beer, J., Brunner, I., Christi, M., Fischer, H., Heikkilä, U., Kubik, P. W., Mann, M., McCracken, K. G., Miller, H., Miyahara, H., Oerter, H., & Wilhelms, F. (2012). 9,400 years of cosmic radiation and solar activity from ice cores and tree rings. *Proceedings of the National Academy of Sciences, 109*(16), 5967–5971.
Stevenson, S., Timmermann, A., Chikamoto, Y., Langford, S., & DeNezio, P. (2015). Stochastically generates North American megadroughts. *Journal of Climate, 28*(5), 1865–1880.
Stine, S. (1990). Late Holocene fluctuations of Mono Lake, eastern California. *Palaeogeography, Palaeoclimatology, and Palaeoecology, 78*, 333–381.
Stone, F. R., & Fritz, S. C. (2004). Three-dimensional modeling of lacustrine diatom habitat areas: Improving paleolimnological interpretation of planktic: Benthic ratios. *Limnology and Oceanography, 49*(5), 1540–1548.
Stoermer, E. F., Kreis, Jr., R. G., and Sicko-Goad, A. (1981). A systematic, quantitative, and ecological comparison of Melosira islandica O. Mull. with M. granulata (Ehr.) Ralfs from the Laurentian Great Lakes: Journal of Great Lakes Research, v. 7, p. 345–356.
Stuiver, M., & Reimer, P. J. (1993). Extended ^{14}C data base and revised CALIB 3.0 ^{14}C age calibration program. *Radiocarbon, 35*, 215–230.
Tilman, D. (1981). Tests of resource competition theory using four species of Lake Michigan algae. *Ecology, 62*, 802–815.
Toste, R., de Freitas, L. P., & Landau, L. (2018). Changes in the North Pacific current divergence and California Current transport based on HadGRM2-ES CMIP5 projections to the end of the century. *Deep-Sea Research, Part II, 169–170*, art. 104641.
Trenbreth, K. B., & Hurrell, J. W. (1994). Decadal atmosphere-ocean variations in the Pacific. *Climate Dynamics, 9*, 303–319.
Tubbs, A. M. (1972). Summer thunderstorms over southern California. *Monthly Weather Review, 100*(11), 799–807.
Wang, T., Ottera, O. H., Gao, Y., & Wang, H. (2012). The response of the North Pacific decadal variability to strong tropical volcanic eruptions. *Climate Dynamics, 39*, 2917–2936.
Wang, F., Liu, Z., and Notaro, M. (2013). Extracting the dominant SST modes impacting North America's observed climate: Journal of Climate, v. 26, p. 5434–5452.

Wells, S. G., Brown, J. B., Enzel, Y., Anderson, R. Y., & McFadden, L. D. (2003). Late Quaternary geology and paleohydrology of pluvial lake Mojave, southern California: in Enzel, Late Quaternary geology and paleohydrology of pluvial lake Mojave, southern California. In Y. Enzel, S. G. Wells, & N. Lancaster (Eds.), *Paleoenvironments and Paleohydrology of the Mojave and Southern Great Basin deserts. Geological Society of America Special Paper 368* (pp. 79–115). Boulder: Geological Society of America.

Wolfe, A. P. (2002). Climate modulates the acidity of Arctic lakes on millennial time scales. *Geology, 30*, 215–218.

Wolfe, A. P., Baron, J. S., & Cornett, R. J. (2001). Anthropogenic nitrogen deposition induces rapid ecological changes in alpine lakes of the Colorado Front Range (USA). *Journal of Paleolimnology, 25*, 1–7.

Wong, C. L., Potter, G. L., Montañez, I. P., Otto-Bliesner, B. L., Behling, P., & Oster, J. L. (2016). Evolution of moisture transport to the western U.S. during the last deglaciation. Geophysical Research Letters, 43, 3468–3477. https://doi.org/10.1002/2016GL068389.

Woodhouse, C. A., Kunkel, K. E., Easterling, D. R., & Cook, E. R. (2005). The twentieth-century pluvial in the western United States. *Geophysical Research Letters, 32*, L07701. https://doi.org/10.1029/2005GL022413.

A 12,000 Year Diatom-Based Paleoenvironmental Record from Lago De Zirahuén, Mexico

Isabel Israde-Alcántara, C. G. Vázquez, Sarah Davies, Ben Aston, and Margarita Caballero Miranda

Abstract The volcanic dammed Zirahuén Lake is one of the least eutrophic lakes in the state of Michoacán in the central Mexican highlands. Diatom-based records from two cores – one littoral in a water column of 15 m deep and one from the center of the ca. 40 m-deep lake – provide a sequence of environmental change spanning the last c. 12 ky. Diatom assemblages during the late Glacial are representative of shallow water or marsh-type environments, indicating lower lake levels in the lake margin. Fluctuating conditions occurred in the early Holocene, as evidenced by diatom assemblages in the central core, while an unconformity was observed in the littoral core between 12.1 and 7.2 ka BP associated with the damming of the lake as product of the development of the Magueyera volcano. The cores recorded a lake deepening and more stable lake levels between ca. 8.8 and 2 ka BP, with the planktonic *Aulacoseira ambigua* dominating the diatom assemblages. The lake

Supplementary Information The online version of this chapter (https://doi.org/10.1007/978-3-030-66576-0_12) contains supplementary material, which is available to authorized users.

I. Israde-Alcántara (✉)
Instituto de Investigaciones en Ciencias de la Tierra, Universidad Michoacana de San Nicolás de Hidalgo, Morelia, Mexico
e-mail: isabel.israde@umich.mx

C. G. Vázquez
Escuela Nacional de Estudios Superiores, Unidad Morelia, Universidad Nacional Autónoma de México (UNAM), Mexico City, Mexico
e-mail: gvazquez@enesmorelia.unam.mx

S. Davies
Department of Geography and Earth Sciences, Aberystwyth University, Aberystwyth, UK
e-mail: sjd@aber.ac.uk

B. Aston
Institute of Geography and Earth Science, Aberystwyth University, Aberystwyth, UK

M. C. Miranda
Institute of Geophysics at the UNAM (Universidad Nacional Autónoma de México), Coyoacán, Mexico
e-mail: UNAM.maga@geofisica.unam.mx

M. R. Rosen et al. (eds.), *Limnogeology: Progress, Challenges and Opportunities*, Syntheses in Limnogeology, https://doi.org/10.1007/978-3-030-66576-0_12

deepening is associated with the last two La Magueyera lava flows that caused the final damming of the lake. These volcanic events increased hydrothermal activity and eutrophic conditions in the lake and favored increased silica availability for the *Urosolenia eriensis* blooms, with almost monospecific diatoms at c. 10.2 ka BP and again between 6.8 and 6.6 ka BP. Anthropogenic influence on the lake ecosystem is evident from c. 3.6 cal ka BP as increased catchment erosion. Significant changes from the mid-eighteenth century associated mainly with colonial copper smelting, deforestation, and recent intensification of land use in the catchment for avocado production are responsible for lake eutrophication.

Keywords Volcanic dammed lake · Zirahuén · Late Pleistocene · Holocene · Climate change

Introduction

The state of Michoacán in the highlands of western-central Mexico contains numerous Neogene-Quaternary lakes formed by volcanic and tectonic activity (Fig. 1). These basins are ideal for the investigation of climatic and paleoenvironmental change in the region over the Late Pleistocene and Holocene, providing the focus for some of the first paleolimnological research in Mexico (e.g., Hutchinson et al. 1956). Over the last few decades, multiple paleolimnological investigations of lakes in the highlands of Michoacán have revealed complex patterns of environmental change related to climate, volcanic, and human activity (Metcalfe 1995; Davies et al. 2004; Vázquez-Castro et al. 2008; Israde-Alcántara et al. 2010; Lozano-García et al. 2013). Deciphering a clear climatic signal has been problematic, particularly during the last 3–4 thousand years when human activity has intensified in the region.

Paleolimnological records indicate sedentary agriculture was established in the highlands of Michoacan from ca. 3.5 cal ka BP, based on palynological evidence from the Pátzcuaro lake (Watts and Bradbury 1982) located 12 km at NE of Zirahuén Lake (Fig. 2a, b) The archaeological record has provided evidence of Pre-Classic settlements (2000 BCE to 250 CE; 3950–1700 cal ka BP), with dense, urban settlements since the Classic (250 CE to 900 CE; 1700–1050 cal ka BP). The region formed the heartland of the Post-Classic (900 CE to 1521 CE; 1521–429 cal ka BP) Purepecha culture which had its maximum development before the arrival of Spaniards to the region in 1521 CE (Fisher et al. 2003; Pollard 2000).

The first paleolimnological record from Lake Zirahuén focused on a series of short cores covering the last ~1000 years (Davies et al. 2004). These provided evidence of responses to changing human activity in the basin during the Post-Classic Purepecha settlement and through Colonial expansion and settlement, particularly from the eighteenth century. A rapid response to recent land-use intensification was also identified (Davies et al. 2005). Two longer cores were retrieved in 2003 (Fig. 2c,

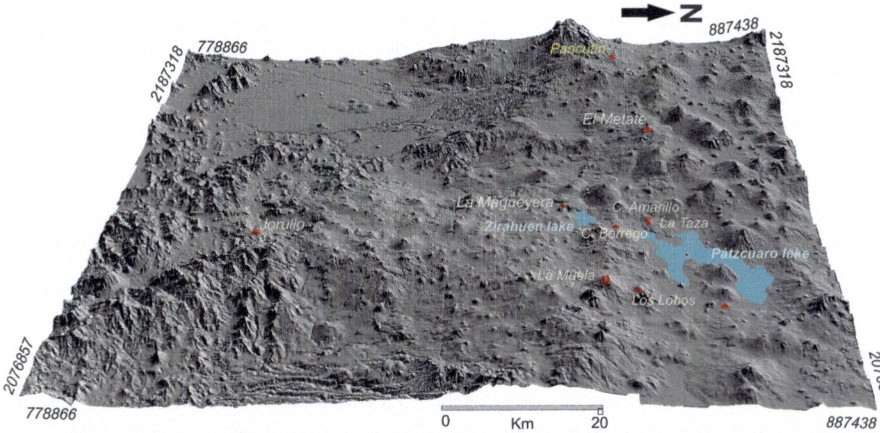

Fig. 1 (**a**) Location of Zirahuén Lake in the TMVB (Trans Mexican Volcanic Belt within the monogenetical Michoacán Guanajuato volcanic field). Elevation model showing the volcanic highlands surrounding the study area, with lakes Pátzcuaro and Zirahuén and a high elevation area between the lakes

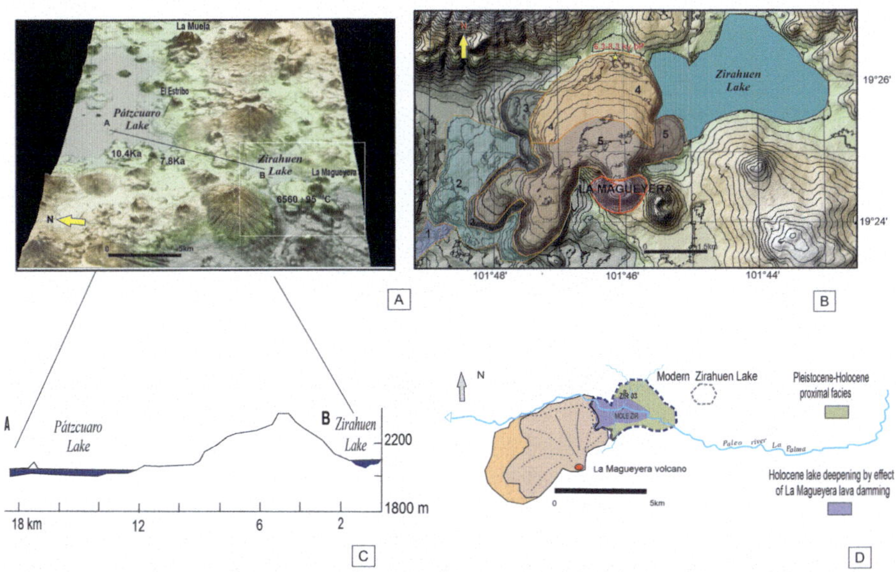

Fig. 2 (**a**) Altitudinal relationship between Lake Patzcuaro and Lake Zirahuén, observed in section A–B, A being the southern shores of Lake Patzcuaro. (**b**) La Magueyera volcano and its relationship with Lake Zirahuén and the La Palma River, as well as the location of MOLE ZIR 03 central core and ZIR 03-01 littoral core. The arrow shows the direction of the paleo river flow toward the Balsas basin; (**c**) main stages of the dome construction of the La Magueyera volcano. Lava flows 1, 2, and 3 are the oldest flowing toward the Balsas basin. Lava flows 4 and 5 emissions dammed the lake

d). The first, 6.61 m in length, was retrieved in approximately 12 m of water in littoral northern sector (ZIR 03 01). It spans the last ca. 17 cal ka, with a hiatus observed between 12.1 and 7.2 cal ka BP (Ortega et al. 2010). The second (MOLE ZIR 03), obtained from the deepest part of the lake in the central north sector, in 36.5 m of water, provides a continuous record of the last 11.6 cal ka BP.

A detailed stratigraphic correlation between the two cores, age model and sedimentary evolution of the lake, was presented by Vázquez et al. (2010). Ortega et al. (2010) focused on the paleoenvironmental record of the last 17 cal ka based on diatom, geochemical, and physical properties of the littoral core, ZIR 03 01. The palynological and microcharcoal records from MOLE ZIR 03 and ZIR 03 01 were reported by Lozano-García et al. (2013), identifying the multiple influences of climate and anthropogenic disturbance on the landscape, vegetation, and fire regime.

Here, we present the diatom record from MOLE ZIR 03, building on previous paleolimnological investigations at Lago de Zirahuén. The new data provide insights into aquatic responses to volcanic activity and climate change during the early Holocene which are not captured in the littoral core (Lozano-García et al. 2013; Ortega et al. 2010). More recent anthropogenic impacts on the aquatic ecology can be placed within the longer term context of the evolution of the lake through the Holocene.

Site Description

Lago de Zirahuén is located on the Trans Mexican Volcanic Belt (TMVB), Michoacán state, Mexico (19° 26′ N, 101° 45′ W), at an altitude of 2075 m asl at the limit of the watershed between the Lerma and the Balsas basins (Fig. 1) (Israde-Alcántara et al. 2011). The drainage basin has an area of 260.8 km^2 (López 1982) and the lake area is 9.86 km^2 (Vázquez 2012). The lava dammed lake is hydrologically closed with a principal stream, the Rio de la Palma, flowing into the lake from the east (Fig. 2c). The lake has a maximum depth of 42 m, but may be two meters lower during dry periods. It is one of the most dilute bodies of water in central Mexico, with an average electrical conductivity of 75.0 μS/cm with Ca-Mg-HCO_3 as the dominant ions (Armienta et al. 2008; Tavera and Martínez-Almeida 2005). The surface water pH of lake water varies, with recorded values of 8.4 in 1997 (Davies et al. 2002) and 6.5 in 2013. Lago de Zirahuén is a tropical oligo-mesotrophic lake and one of the least eutrophic lakes in Michoacán state. However, in the 1980s, a trend toward nutrient enrichment was observed in the recent sedimentary record (Davies et al. 2004, 2005) which was attributed to an increase in agricultural and tourist activities within the basin, in combination with the lack of water treatment facilities, increasing the concentration of nutrients. Since the summer of 2002, *Mycrocystis aeruginosa* has been observed in the north and south sectors of the lake. Analysis from June 2013 indicates a lake clearly in the process of eutrophication, with a Secchi disc transparency of 2.26 and an oxygen concentration of 5.70 mg/L^{-1} compared with 2010 measures with Secchi disc transparency of 3.70 m and oxygen concentration of 7.82 mg/L^{-1}, respectively (Bernal-Brooks et al. 2016). Toward the Rio de La Palma, the oxygen concentration decreases to 1.82 ± 1.17 mg/L (Gomez-Tagle et al. 2013). Lago de Zirahuén is a monomictic lake, thermally stratified with a December–January mixing regime (Tavera and Martínez-Almeida 2005).

The basin is covered by numerous lava emissions and pyroclastic materials (Fig. 1) extruded over the last 8 million years (Osorio-Ocampo et al. 2018). The continued emissions of those volcanic materials in the region have noticeably changed the configuration of the basin during the Neogene-Quaternary. Like many sites in the region, the principal volcanic products observed are Quaternary basaltic-andesitic lava flows forming large semi-shield volcanoes (Cerro Cumburinos to the NW and Cerro Zirahuén to the NE). Based on volcanic and geomorphologic relations (Fig. 2c), it has been suggested that La Magueyera Lava Flows (LMLF), to the SW, were produced around 6560 ± 950 BP (Ortega et al. 2010) modifying the morphology of the lacustrine basin, contributing to the change of lake level. Nevertheless, through a detailed morphological analysis, five lava flows have been identified (Fig. 2c). The dated lava flow (Ortega et al. 2010) belongs to the fourth lava flow event. The age of the lake is not yet known and could originate in the initiation of La Magueyera lava flow in the area, damming the perennial Rio de la Palma River during the late Pleistocene–Early Holocene.

Around the lake there is a thick layer of paleosols which are the main sources of detritus for the lake. One of them is a reddish Luvisol with the characteristic coloration of the region (Fig .3). To the SW, covering the LMLF there is a more recent ochric andosol of yellow to brown color (Vázquez 2012). To the N and SE of the lake the landscape has some flat areas that have been used for agriculture; the continued deforestation has caused marked erosion which is intensified during the rainy months. The dominant vegetation in the upper elevations of the basin (3300–2800 m) is fir forest, replaced at lower elevations by oak and pine forest (Lozano-García et al. 2013). The forest at lower levels has been increasingly replaced with the expansion of avocado cultivation. Areas below 2400 m asl are largely covered with grassland and scrub vegetation (Tavera and Martínez-Almeida 2005).

The climate of the region is temperate subhumid (Cw_2), with annual mean temperature of 16 °C. The annual rainfall is between 800 and 1200 mm with a mean of 891 mm (Servicio Meteorológico Nacional). The climate is markedly seasonal with a concentration of rainfall in summer between June and October with the arrival of humidity from the Gulf of Mexico and Pacific Ocean, controlled by variations in atmospheric circulation of the High Pressure Subtropical Cells and the Intertropical Convergence Zone. These changes in effective moisture are reflected in Lago de Zirahuén through changes in lake level and chemical composition and aquatic ecology.

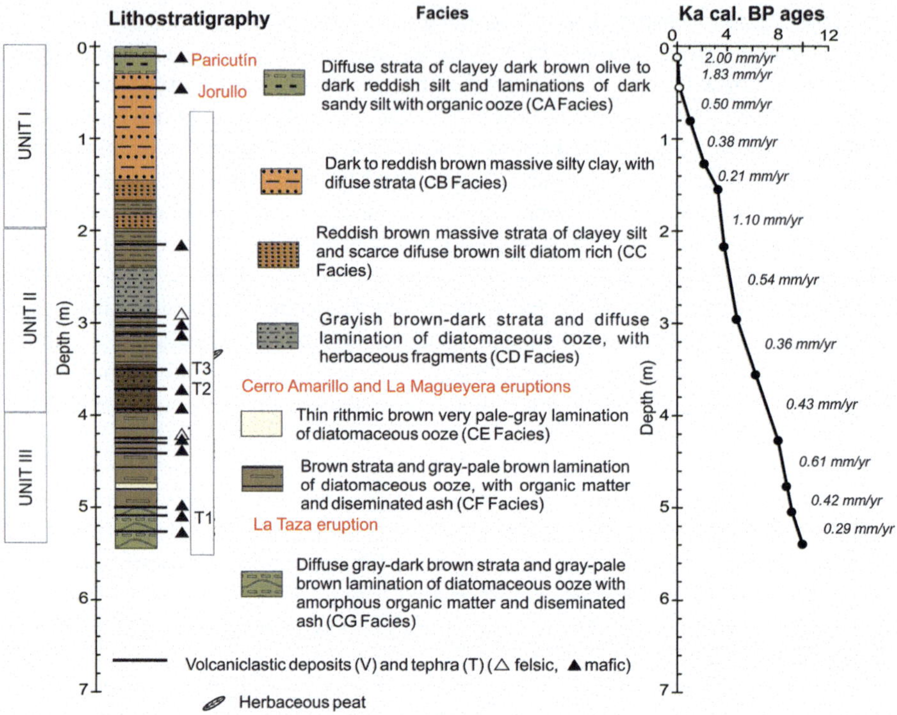

Fig. 3 Stratigraphic column of MOLE ZIR 03 core record showing the main facies along the core. Diagram showing age model (ka) using IntCal20. Black circles are the midpoint of 2 σ calibrated ^{14}C and open circles are the known tephra ages of Paricutin (TP) and Jorullo (TJ). Sedimentation rates are shown in mm/year. Other evidence of nearby volcanic activity around the lake are La Taza, La Magueyera, and Cerro Amarillo

Methods

During the Mesoamerican Lakes Expedition (MOLE) in October 2003, three cores were collected in the deepest central zone of Zirahuén Lake at ca. 36.5 m below the water surface using a modified Kullenberg coring system (Kelts et al. 1986; Kullenberg 1947) (Fig. 2c), obtaining a central sequence of 5.4 m length. The polycarbonate core tubes were sectioned longitudinally, photographed, and lithostratigraphy was described following standard protocols (Schnurrenberger et al. 2003). The sediment was packed into weakly diamagnetic resealable plastic cubes (8 cm^3) for magnetic analyses (251 samples were taken continuously). Mass-specific low-field magnetic susceptibility (χ_{lf}) was measured in a Bartington system. To correct for the presence of paramagnetic minerals, ferrimagnetic susceptibility (χ_f) was calculated by subtracting the paramagnetic contribution (χ_p) estimated from the high field slope of the hysteresis loops. Total inorganic carbon (TIC) was measured on 94 freeze-dried samples on a UIC 5030 coulometer, and total carbon (TC) in a UIC 5020. Total organic carbon (TOC) was obtained from the difference between TC and TIC for each sample (Myrbo 2004). In order to complement the study, we made use of scanning electron microscopy observations (SEM) and X-ray diffraction in a Phillips 1130/96 diffractometer (Vargas et al. 2007).

Diatom samples were taken at 5 cm intervals through the entire sequence. The samples were dried and processed for analysis in 1 cm^3 volumes through digestion with 10% hydrochloric acid and 30% hydrogen peroxide to eliminate carbonates and organic matter. Once the diatoms were cleaned, 200 μl aliquots were taken and dried on a cover slip for mounting with Naphrax. Observations were made at 1000× magnification with an Olympus Bymax 50 microscope; taxa were identified using the works of Krammer and Lange Bertalot (1986, 1988, 1991a, b), Rumrich et al. (2000), Rott et al. (2006), Bradbury (2000), and reference to the standard floras of Geology Department at the INICIT (Instituto de Investigaciones en Ciencias de la Tierra) and previous studies on the lake (Ortega et al. 2010). A minimum of 400 valves was counted for each sample and 200 valves in samples of scarce diatom content. Percentages in samples, variation in abundance through time, and preferred habitat of diatoms were graphically represented by means of Tilia Graph software (Grimm 1987, 1992). Total diatom abundance is expressed in valves per gram of dry sediment. Only diatoms with an abundance >5% are shown in diagrams (Figs. 4 and 5).

The age model for the central core was provided by 10 AMS ^{14}C dates determined on bulk organic sediment and pollen extracts (Table 1). Radiocarbon dates were calibrated to calendar years BP, with the northern hemisphere terrestrial calibration curve (IntCal 20, Reimer et al. 2020) using the midpoint of the 2σ probability interval of calendar ages. The age model also incorporates two tephras of known origin, Jorullo at 0.43 m depth (1759–1764 CE) and Paricutin (1943–1945 CE) at 0.2 m depth (Vázquez et al. 2010).

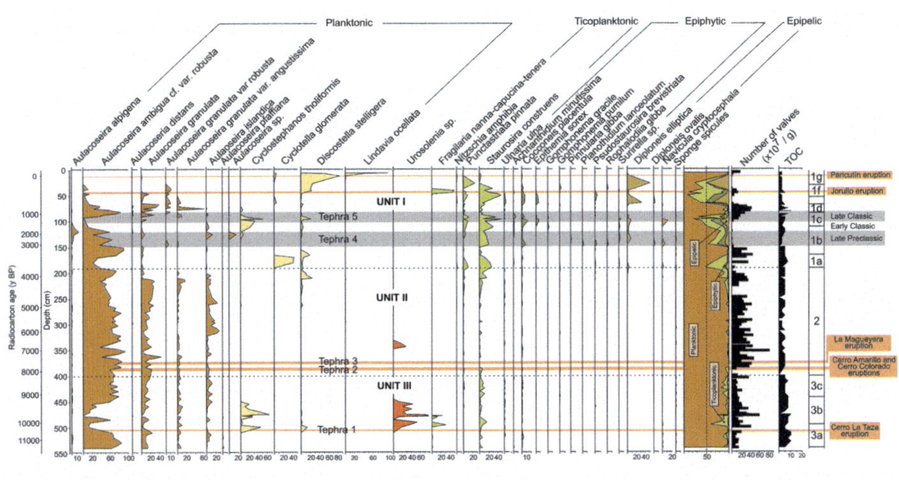

Fig. 4 Diatom record from MOLEZIR 03 sequence showing the main zones: selected taxa as percentages of total counts, and total counts grouped by ecological habitat, plotted versus core depth. Model age is in cal ka BP

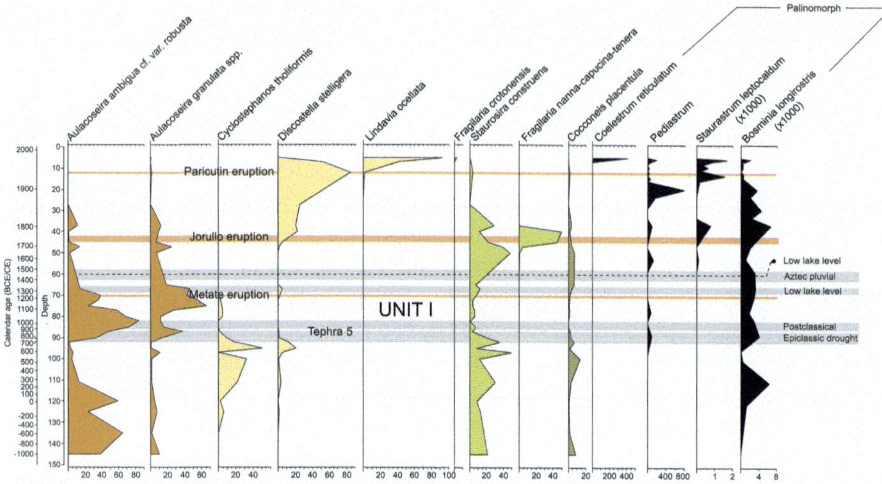

Fig. 5 Detailed diatom and palinomorphs record of the past 2000 years and the associated historical events during this time: Epiclassic drought, Postclassic, Aztec pluvial, El Metate, Jorullo, and Paricutin Eruptions

Table 1 AMS ^{14}C dates for the MOLE ZIR 03 sequence

Lab. Code	Depth (m)	AMS ^{14}C yr BP	2 δ range (cal yr BP)	Mean value
Beta – 221711	0.80	990 ± 40	793–956	874 BP
Beta - 221712	1.25	2090 ± 40	1943–2289	2116 BP
Beta - 223997	1.55	3230 ± 40	3379–3559	3469 BP
SUERC – 8106	2.15	3732 ± 26	3984–4152	4068 BP
SUERC – 8103	2.97	4725 ± 27	5326–5577	5451 BP
SUERC – 8102	3.56	6235 ± 38	7010–7254	7139 BP
SUERC – 8104	4.27	8014 ± 48	8651–9014	8832 BP
Beta – 223999	4.76	8710 ± 40	9545–9887	9716 BP
Beta – 223998	5.03	9130 ± 40	10221–10410	10350 BP
Beta – 221711	5.39	10020 ± 100	11242–11871	11556 BP
La Taza		9300 ± 40	10302–10647	10475 BP
Cerro Amarillo		6970 ± 40	7694–7924	7709 BP
La Magueyera		6560 ± 950	5593–9661	7627 BP
El Metate			1250 CE	
Jorullo			1759–1764 CE	
Paricutin			1943–1945 CE	

Calibration based on IntCal 20 (Reimer et al. 2020) program, calculated between consecutive data. *TP* Paricutín Tephra (1943 CE), *TJ* Jorullo Tephra (1762 CE), and other nearby eruptions

Results

^{14}C Dating and Stratigraphy

The stratigraphy of MOLE ZIR 03 has been described in detail in Vázquez et al. (2010). To summarize, it is composed of three main units (Fig. 3): Unit III (5.40–3.98 m, 11.5–8.1 ka BP) is characterized by laminated gray diatom oozes with disseminated organic matter and six volcanic ashes, with a mean sedimentation rate of 0.44 mm/yr; Unit II (3.98–1.98 m, 8.1–3.91 ka BP) is characterized by diffuse laminations of black organic-rich diatom ooze and seven volcanic ashes, with a mean sedimentation rate of 0.57 mm/yr; and Unit 1 (1.98–0 m, 3.91 ka BP-2003 CE) composed mainly of clastic sediments of black-reddish silty clay and silty sand, with few organic and diatomaceous layers and a mean sedimentation rate of 2.0 mm/yr. Unit 1 contains the historic tephras of Jorullo and Paricutin volcanoes (Fig. 1). Sedimentation rates calculated from the revised age model indicate a range of 0.21–2.00 mm/yr through the c. 11.5 kyr record.

Magnetic Susceptibility and TOC

Magnetic susceptibility (Fig. 6) shows a clear relationship to the stratigraphy of the core. The bottom and middle of the sequence have a very low concentration of magnetic minerals with little variability. Unit III has the lowest values with $\chi_f \approx 0.02 \times 10^{-6}$ m^3/kg on average, and for Unit II intermediate values $\chi_f \approx 0.1 \times 10^{-6}$ m^3/kg. The highest magnetic concentration is present in Unit I with the maximum values of $\chi_f \approx 5 \times 10^{-6}$ m^3/kg. In all cases χ_f was high when tephras occur. The TOC values (Fig. 4) show downcore variations clearly related to stratigraphy too (Fig. 6). TOC is highest in Unit III (highest value at the base 12.36 wt%). TOC shows a clearly decreasing trend from Unit III to I with a stable trend in Unit II, and some high values at the top of the core. TIC is interpreted as a measure of carbonate precipitation in the lake sediments as no detrital carbonates enter in the lake. Higher values would therefore be related to changes in ionic concentration during drier conditions (Eugster and Hardie 1978). TIC values are very low, ranging from 0% to 0.05%, with very low amplitude variations, and are not discussed further.

Diatom Assemblages

The diatom record from MOLE ZIR 03 (Figs. 4 and 5) is almost complete, with diatoms well preserved throughout except between 0.65 and 0.52 m. A total of 84 diatom taxa were identified.

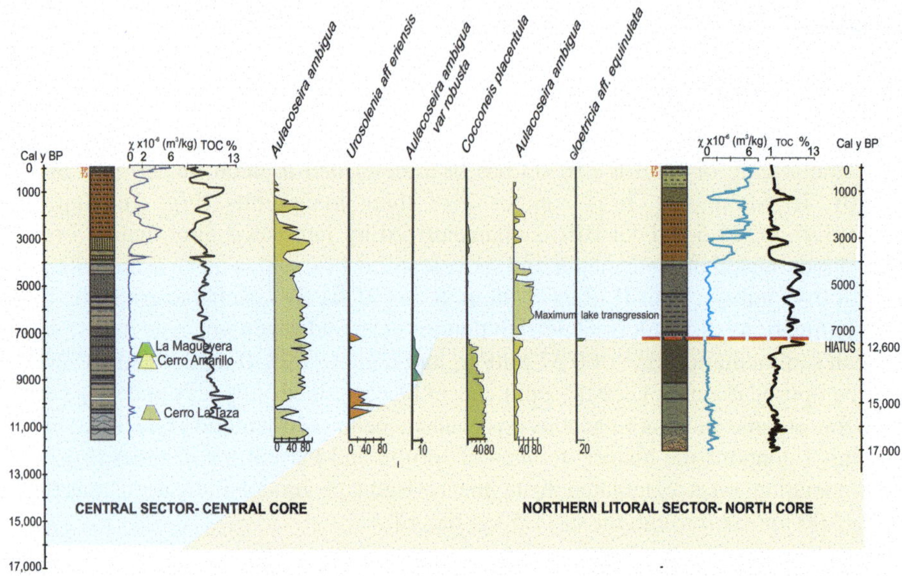

Fig. 6 Stratigraphic units. Correlation between the central coring site, MOLEZIR 03 core, and the northern littoral site, ZIR03 01 core based on lithology, magnetic susceptibility, TOC (Total Organic Carbon), and selected diatom taxa. Maximum lake stages and associated volcanism in relation to main organisms are shown in blue. The red dotted line highlights the discordance in the littoral core

The base of the core to 3.98 m depth, corresponding to stratigraphic Unit III is split into three subzones. The lower part (Subzone 3c, 5.40–4.99 m; 11.6–10.26 ka BP) marks the start of the Holocene and is dominated by *Aulacoseira ambigua* cf. var. *robusta* and *A. granulata* (over 80% of total species), followed by *Staurosira construens* and *Cocconeis placentula*. Diatom abundance is 20 × 10^7 vgds (valves per gram of dried sediment). Above this, in subzone 3b (4.99–4.49 m; 10.26–9.22 ka BP), a significant decline in the percentage of *Aulacoseira* spp. is observed, which is substituted by the planktonic *Urosolenia* aff. *eriensis*, which increases to over 80% of the assemblage at 4.8 m. Diatom abundance increases to 41 × 10^7 vgds through subzone 3b. The appearance of *Urosolenia* is accompanied by *Cyclostephanus tholiformis*, which reaches values of over 60%. The group *Fragilaria nanana*, *F. capucina*, *F. tenera* is present at up to 30% of the total diatom count above 4.95 m in depth. *Staurosira construens* also appears in small numbers, coinciding with clear laminations in the core, mainly between 4.78 and 4.70 m. The upper part of the zone (Subzone 3a, 4.49–3.98 m; 9.22–8.20 ka BP) is characterized by a return to an assemblage dominated by *Aulacoseira ambigua* cf. var. *robusta*, along with *A. granulata* and *A. islandica*. *Staurosira construens* is present at c. 10–15% of the count, with *Cocconeis placentula* and *Ulnaria ulna* also present in small amounts.

In Unit II (zone 2, 3.98–1.98 m; 8.20–3.92 ka BP), the diatom abundance increases relative to zone 3, with values between 20 and 40 × 10^7 vpgds and a short-lived peak of 80 × 10^7 vpgds at 3.5 m. The assemblage continues to be dominated by *Aulacoseira ambigua* cf. var. *robusta*, and *Aulacoseira granulata, Aulacoseira islandica* is present at lower abundances but does increase from 3.25 m to 2.5 m. *Urosolenia* makes a brief reappearance between 3.43 and 3.37 m at c. 30% of the count. Overall, species diversity is low.

Unit I, (Fig. 5) representing the uppermost 1.98 m of core, is split into seven subzones. It is characterized by the reduced dominance of *Aulacoseira ambigua* cf. var. *robusta* and increased species diversity, with greater contribution of tycho-planktonic, epiphytic, and epipelic taxa. Subzone 1a (1.98–1.70 m; 3.92–3.63 ka BP) sees an abrupt decrease in *Aulacoseira ambigua* var. *robusta* abundance and increase of *Cyclotella glomerata* in association with *Staurosira construens*. Between 1.70 and 1.11 m (Subzone 1b, 3.63–1.69 ka BP) the diatom assemblage *Aulacoseira* spp. again, reaching their maximum abundance in the diatom record (92% at 1.55 m). *Staurosira construens, Punctastriata pinnata,* and *Pseudostaurosira brevistriata* increase in this interval along with small numbers of *Diploneis elliptica, Cocconeis placentula, Planothidium lanceolatum, and Pinnularia* gibba are present. In the next interval (Subzone 1c, 1.11–0.80 m; 1.69–0.88 ka BP), *Aulacoseira* spp. almost disappear, with a corresponding increase in *Cylostephanos tholiformis*. The overall diatom abundance declines in subzone 1c, which also sees an increasing proportion of small fragilaroid species, *Cocconeis placentula* and *Epithemia sorex*. Between 0.80 and 0.65 m (Subzone 1d, 0.88–0.58 ka BP), *Aulacoseira granulata, A. granulata v. angustissima,* and *A. ambigua* reappear with an increase in the number of vpgs. The interval 0.65–0.52 m (Subzone 1e, 0.58–0.33 ka BP) has the lowest diatom abundance of the record; it is characterized by diatoms of robust, highly silicified diatoms such as *Diploneis* and *Aulacoseira ambigua* cf. var. *robusta*. From 0.52 to 0.37 m (Subzone 1f, 330 y −145 y BP), the *Fragilaria nanna, F. capucina,* and *F. tenera* group, together with *Diploneis elliptica,* form a distinctive assemblage. The upper part of Unit I (Subzone 1 g, 0.37–0 m; c. 145 y BP–2003 CE) is characterized by a major change in the diatom flora, with high percentages of *Discostella stelligera* and *D. pseudostelligera.* These taxa are rapidly replaced by *Lindavia ocellata* from 10 cm (1943 AD) to 5 cm depth (1982 AD), dominate the assemblage. In the last 5 cm of the sedimentary record (1982 CE) appear *Fragilaria crotonensis,* a characteristic cultural eutrophication diatom taxa (Reavie and Cai 2019).

Interpretation and Discussion

Paleoenvironmental Interpretation

At the base of the core to ~2 m depth (ca. 11.6–4 ka BP), low magnetic susceptibility values suggest minimal catchment erosion into the lake (Fig. 6). Occasional peaks in magnetic susceptibility are associated with tephra layers as described by

Vázquez et al. (2010). The highest % TOC values are found in the basal sediments, up to 4 m depth (11.6 ka BP – 8.2 ka BP), indicating a productive aquatic environment as catchment sources do not appear to be significant based on the magnetic signal. The diatom assemblage through this section (Unit III) is formed by the planktonic taxa *Aulacoseira ambigua* cf. var. *robusta* and *A. granulata*, particularly in subzones 3a and 3c. *A. ambigua* var. *robusta* was also found in Late Pleistocene sediments of nearby Lake Patzcuaro. These are heavily silicified species that require substantial water depth and deep circulation to remain buoyant (Bradbury 2000) and an abundant silica supply (Kilham et al. 1986). The presence of *A. granulata* also indicates nutrient-rich lake waters. Subzone 3b shows a change in diatom flora, *Aulacoseira* reduces its abundance and instead pioneer diatom *Urosolenia* cf. *eriensis* becomes abundant more so between 10.2 and 9.3 cal ka BP where it is associated with *Cyclostephanos tholiformis* suggests increased nutrients. Significant numbers of *Urosolenia* cf. *eriensis* and *Cyclostephanos tholiformis* are common in eutrophic lakes in North America (Stoermer et al. 1987) and Brazil Tremarin et al. (2015); *Urosolenia* is a delicate, lightly silicified diatom and its exceptional preservation within this section of core is notable. Its presence may suggest a higher degree of stratification in the water column as it commonly occurs in the late-summer autumn plankton bloom of moderately stratified temperate lakes (Rott et al. 2006). In this interval, the needle-shaped diatoms planktic group (*Fragilaria nanna*, *F. capucina, F. tenera*) appears for the first time by ca. 10 cal ka BP. Fragilaroid taxa commonly occur in association with volcanic ash deposition in lake sediments (T1), in response to increased nutrient inputs (Telford et al. 2004). These nutrient-rich conditions in the lake were possibly a consequence of the mafic tephra associated to the Cerro La Taza eruption. This diatom assemblage gives an indication of the status of the lake after the eruption, with a stratified water column in a warm climate. A productive aquatic environment is also consistent with the relatively high TOC values found in the basal sediments (11.6 cal ka BP–8.2 cal ka BP). The palynomorphs record in subzone 3b showed a *Pinus*-dominated landscape with low percentages of spores of *Pediastrum boryanum* var. *longirostrus* and *Botryococcus* (Lozano-García et al. 2013). The renewed dominance of *Aulacoseira* spp. in subzone 3c suggests a return to deeper circulation in the water column.

The continued dominance of *A. ambigua* cf. var *robusta* and *A. granulata* and other Aulacoseira species through Unit II, between ca. 8.2 and 3.8 cal ka BP, indicates relatively stable lake conditions, with only minor fluctuations in the assemblage through this period. A relatively deep mixing regime would have been required to enable these heavily silicified taxa to remain the major component of the diatom assemblage throughout this 4000 year period. The lack of epiphytic and periphytic forms throughout Unit II suggest minimal input of littoral material to the center of the lake, again supporting more stable conditions. A brief reappearance of *Urosolenia* cf. *eriensis* between 6.6 and 6.8 cal yBP suggests increased silica supply as this coincides with the highest diatom concentrations in the core (80×10^7 vgds). This occurs directly above a mafic tephra, which relates to the La Magueyera eruption dated to 6.9–7.1 cal. ka BP (Vázquez et al. 2010) and a mean of 7.6 cal y BP (intCal 20) and is likely a direct response to nutrients inputs and increased ionic concentration from this volcanic source. After this eruption, visible peat and the highest TOC percentages are recorded.

Above 1.98 m to the top of the core, the magnetic susceptibility values increase significantly, with several large peaks (Fig. 6). While a small peak is associated with the deposition of a tephra layer at 1.98 m, these high values are more sustained and are associated with the influx of reddish brown soil from the surrounding slopes (Vázquez et al. 2010). The sedimentation rate through Unit I is substantially higher than the rest of the core. The increase in the proportion of epiphytic and other periphytic diatoms and more variable diatom assemblages support the interpretation of episodes of increased catchment erosion.

Subzone 1a (3.92–3.63 cal ka BP) shows an abrupt decrease of *Aulacoseira ambigua* var. *robusta* and an increase of *Cyclotella glomerata* in association with *Staurosira construens,* that indicates an increase in nutrient concentration but with a decrease in mixing. Subzone 1b (3.63–1.69 cal ka BP) sees a return to the dominance of *Aulacoseira ambigua* cf. var *robusta*. Above this, the increased importance of tychoplanktonic species *Staurosira construens* and *Pseudostaurosira brevistriata* along with periphytic taxa (e.g., *Planothidium, Pinnularia*), suggest increased inputs of material from the littoral zone, related to catchment disturbance and/or lower lake level. The trend of fluctuating conditions continues into subzone 1c (1.69–0.88 cal ka BP) as *Cyclostephanos tholiformis* dominates the planktonic taxa, suggesting lower mixing and nutrient enrichment. The associated increase in epiphytic forms represented by *Cocconeis placentula* and *Epithemia sorex* indicates an expansion of shoreline habitat suitable for macrophytes, possibly due to lower lake levels. A return to the *Aulacoseira* spp. dominated assemblage in subzone 1d (0.88–0.58 cal ka BP) suggests more turbid, well-mixed lake waters and an increase in lake level as epiphytic taxa decline. Increased catchment erosion is evident above this in subzone 1e (0.58–0.33 cal ka BP) from a further peak in magnetic susceptibility. The poor preservation coinciding with increase in catchment inputs indicates that sedimentation may have been more rapid, diluting the diatom component of the sediments. The occurrence of the Jorullo tephra (1764–1759 CE) at the boundary between subzones 1e and 1f at 43 cm depth marks the onset of a significant change in the diatom assemblage, with a short-lived spike in planktic *Fragilaria tenera-nanna-capucina* group. This is likely a response to the deposition of the tephra and consequent increase in nutrient availability in the lake waters. Above this, the increase in *Discostella stelligera* becomes the dominant component of the plankton, and was previously interpreted as being associated with increased anthropogenic catchment disturbance (Davies et al. 2002).

At the end of this phase in subzone 1f (0.33–0.24 cal ka BP) (ca 1830 CE–1880 CE), an increase in epipelic species (*Diploneis elliptica*) is observed, which may indicate a decrease in the level of water in the lake. Subzone 1g, from 37 cm to the top of the core (since ca. 1810 CE), indicates further catchment disturbance, with a peak in magnetic susceptibility and poor diatom preservation. The Paricutin tephra (1943 CE, at 20 cm depth) contributes to this signal. The rapid switch from *D. stelligera* to *L. ocellata* in the uppermost sediments in this core at 10 cm depth was also reported by Davies et al. (2004) in short cores and interpreted as a response to intensification of agriculture since the 1980s. *Lindavia ocellata* is a common taxon in Lake Tacambaro, south of Lake Zirahuén and appears in the last 10 cm of the record (Sanchez 2007; Caballero et al. 2016). Its co-occurrence with *Fragilaria crotonensis*, a commonly used indicator species for mesotrophic conditions and the onset of eutrophication, supports this interpretation. The presence of epiphytic, epipelic, and aerophilous diatom species suggest the expansion of marginal habitat.

Correlation of MOLE ZIR 03 and ZIR03-1

Correlation of diatom evidence from MOLE ZIR 03 with the previously published 17 ka, 6.61 m record from the littoral zone (ZIR 03-1; Ortega et al. 2010; Vázquez et al. 2010; Lozano-García et al. 2013) provides an opportunity to examine the consistency between the diatom records from the two sequences. The correlation between ZIR 03-1 and MOLE ZIR 03 was established by Vázquez et al. (2010) and is based on the presence of five common tephra layers in the two cores. The oldest of the five common tephras is dated between 7.1 and 6.9 cal ka BP (Vázquez et al. 2010). This stratigraphic framework which links the cores over the last ca. 7 cal ka BP is supported by the magnetic susceptibility and pollen records from both cores (Lozano-García et al. 2013). The diatom assemblages recorded in MOLE ZIR 03 between ca. 7 cal ka BP and the base of the sequence at ca. 11.6 cal ka BP is not represented in ZIR 03-1. This covers a key time period as the North American Monsoon was becoming re-established over Mexico during the early Holocene (Metcalfe et al. 2015). Furthermore, there is evidence for significant volcanic activity at this time which may have impacted lake development. The diatom record therefore may provide insights into aquatic responses to these local and regional factors.

Volcanic Impacts on the Paleoenvironmental Record

In central Mexican lacustrine basins, paleoenvironmental records are influenced by episodes of volcanic activity (Telford et al. 2004). Results from tephrastratigraphy from nearby Lake Pátzcuaro shows a peak in activity from monogenetic volcanoes around the area between 15 and 10 ka BP (Newton et al. 2005). More recent research shows that this activity persisted into the early Holocene with the development of two monogenetic volcanoes close to Zirahuén: Cerro la Taza which erupted at 9300 ± 40 uncalibrated ^{14}C years BP (10.4 cal ka BP) and Cerro Amarillo dated at 6970 ± 40 uncalibrated ^{14}C years BP (7.8 cal ka BP), Osorio-Ocampo et al. (2018) was the source of the more recent emitted fall deposits this period (Fig. 1).

A direct influence on the morphology of Lago de Zirahuén was caused by eruptions of La Magueyera volcano (Fig. 2c) with lava flows produced in five stages. Only the second to last Magueyera lava (4) has been dated, and was erupted between

6.3 and 8.3 cal ka BP (Ortega et al. 2002). 7.6 cal ka BP (median probability IntCal 20).

The impacts of the early Holocene volcanic activity are recorded in specific changes in diatom assemblages as well as other proxies. In the littoral core (northern lake sector) (Figs. 2c and 4), the diatom record from 17 cal ka BP to 10 cal ka BP indicates that Zirahuén was a shallow lake with the epiphytic species *Cocconeis placentula* as the dominant diatom at abundances greater than 80%. This suggests that the earlier eruptions that created the La Magueyera volcano (lava 1 and lava 2, Fig. 2c) were not sizeable enough to produce a high enough lava to dam the lake. Lavas 3 and 4 finally dammed Lake Zirahuén resulting in the rise in lake level needed to produce a rapid change, from the shallow, aquatic environment with epiphytic diatoms *Cocconeis,* to a planktonic assemblage-dominated *Aulacoseira* ambigua cf. var. robusta (Fig. 4). As outlined earlier, this species and the others it occurs in association will require a deeply circulating water column to keep it buoyant (Bradbury 2000).

Maximum lake transgression is indicated by the appearance of *Urosolenia-Cyclostephanos* assemblage along with *Botryococcus* in the palynological record (Lozano-García et al. 2013). The peaks in *Urosolenia* are also associated with a decline in *Isoetes* between 10.2 cal ka BP and 9.6 cal ka BP, coincident with the end of the La Taza volcano eruption (Figs. 1 and 6) in correspondence with high valve abundance and an inverse relation with the lowest magnetic susceptibility values. The *Urosolenia* peaks are associated with a sharp decrease in *Aulacoseira* and *Isoetes* (Lozano-García et al. 2013) between 10.2 cal ka BP and 9.6 cal ka BP, suggesting a decrease in turbulence and an increase in water depth. This period is associated with the end of the La Taza volcanic eruption (Fig. 1).

In the littoral ZIR 03-01 core, an unconformity is marked by a high peak in magnetic susceptibility between 12 ka cal BP (Late Pleistocene) and 7.2 cal ka BP (Fig. 6). Prior to this, the presence of *Isoetes* that together with the presence of *Cocconeis* indicates a shallow lake of less than 18 meters (Lozano-García et al. 2013), compared with the present lake conditions of 42 m maximum depth and an area of 9.86 km^2. If we subtract an 18 m regression, during the late Pleistocene the lake could have had 24 m of maximum depth and be more restricted to the southern sector. The flood plain could be expanded by the sediment discharged from the La Palma River three kilometers inside the lake in the eastern delta of La Palma River (Fig. 2d).

A second lake transgression and final damming of the lake could be related to lava 4, dated between 8.3 BP and 6.3 ka BP (Ortega et al. 2002) when *Gloetrichia equinulata* (8.9 cal ka BP to ca. 6.8 cal ka BP) appears in the record (Lozano-García et al. 2013). This change is followed by a further *Urosolenia* peak. During this time another nearby volcanic center, Cerro Amarillo, also produces an eruption between 7.8 cal ka BP and 7.7 cal ka BP. Both the eruptions of Cerro Amarillo and La Magueyera contributed to high charcoal peaks, high percent carbon, and the disappearance of *Pinus* and *Quercus*, suggesting a period of significant disturbance in the basin.

In this volcanic context, the hydrothermal influence is a possible factor in influencing the lake chemical and thermal conditions and possibly produced the *Urosolenia* and *Cyclostephanus tholiformis* assemblages. However, nutrient enrichment and changes in lake configuration and water column dynamics could also explain the observed changes. *Cyclostephanus tholiformis* is common in eutrophic lakes and ponds (Davies et al. 2005) and in warm water lakes in North America (Stoermer et al. 1987). At Lake Kivu, an African volcano tectonic lake, *Urosolenia,* appears as a result of a rise in lake level during the Late Pleistocene as a result of subaquatic springs that maintained the thermohaline stratification of the lake (Ross et al. 2015). As at Kivu Lake, previous phreatomagmatic activity could have enhanced hot water infiltration via groundwater into the lake as hydrothermal pulses produce well-established lake stratification and diatom blooms (Sarmiento et al. 2006). At Zirahuén, a volcanic complex divides the Pátzcuaro basin (2042 masl) from that of Lake Zirahuén (2082 masl), this eruptive activity pulse could produce a rapid change in the lake conditions (probably associated with hydrothermal discharge into this alkaline volcanic lake).

The potential impact of more recent tephra deposition from the Jorullo and Paricutin eruptions on the diatom record from Lago de Zirahuén was investigated by Telford et al. (2004), who identified a very limited response to these historical eruptions. The high levels of catchment erosion over the last few hundred years coupled with anthropogenic eutrophication have likely been the stronger influence on the diatom record over this more recent time period (Telford et al. 2004), masking any minor impacts of tephra deposition.

Regional Paleoclimatic Record

Paleoenvironmental studies in northern Michoacán cover the last ca. 50 ka BP, with Late Pleistocene records from Pátzcuaro (Bradbury 2000; Metcalfe et al. 2007) and Cuitzeo (Israde-Alcántara et al. 2010) indicating contrasting conditions to the Holocene. Between 47 ka BP and 18 ka BP, Pátzcuaro was a more extensive and deeper and colder lake than at present (Bradbury 2000), while Zacapu (Metcalfe 1992) and Cuitzeo (Israde-Alcántara et al. 2010) were deep lakes before 28 ka BP, both lakes having conditions of more stability and depth than at present. During this time period, the landscape was also modified by either volcanic or tectonic activities (Pooja et al. 2016).

Younger Dryas to Early Holocene

From the interval between 13.5 and 12 ka BP, Lakes Pátzcuaro, Cuitzeo (Israde-Alcántara et al. 2012), and Parangueo (Domínguez-Vázquez et al. 2018) experienced wet conditions and high water levels, with a shallowing from 12 to 10 ka

BP. Generally dry conditions at the onset of the Holocene are observed in records across Central Mexico (Metcalfe et al. 2015). During the early Holocene in Lago de Zirahuén, a deep lake was established at the central core site but the littoral core records a hiatus. This could be indicative of lower lake levels than present but the hiatus is more likely related to the catchment disturbance linked to volcanic activity and subsequent establishment of the current lake configuration.

From 10.2 ka BP to 9.5 ka BP, the MOLE ZIR 03 core high percentages of *Urosolenia and Botryococcus* together with the highest TOC percentages (Fig. 3) and an *Isoetes* decrease suggests a deep, fresh, and nutrient-rich lake. The littoral core records a hiatus, which could be interpreted as low water level at that location, but is more likely related to the catchment disturbance linked to Holocene volcanism that led to the establishment of the present lake configuration.

The volcanic activity in the area at Pátzcuaro Lake 8 km NE from Zirahuén Lake continued from ca. 7 ka BP to 8.4 ka BP with the construction of Cerro Amarillo (7.7 ka BP) and emissions 4 and 5 of La Magueyera (8.3 cal ka BP to 6.3 cal ka BP) (Ortega et al. 2010) (Fig. 2c) with the emission of ash fall deposits. This caused an increase in the charcoal and magnetic susceptibility signal. As a result, *Pinus* and *Quercus* disappeared in the basin, replaced by a succession of atypical communities with the presence of *Gloetricchia equinulata* and *Tilia* (Lozano-García et al. 2013) and *Urosolenia*-*Cyclostephanus toliformis* (Fig. 4).

The pollen spectra at Zirahuén at 8.2 ka cal BP shows a sharp decrease in *Alnus* abundance that has been associated with a cold and dry episode caused by a glacial advance (Heinrich event). The nearly global expression of climate conditions at 8.2 cal ka BP indicates a short-lived cold and dry episode promoted by the reduction on the northward heat transport (Walker et al. 2019).

Toward 9 ka BP a decrease in water level occurred in Lake Pátzcuaro. Records of diatoms, pollen, and mineralogy indicate still wetter than present conditions that prevailed just a little after 4.5 ka BP (Bradbury 2000; Metcalfe et al. 2007). In Zacapu, the records of several paleoenvironmental indicators suggest a geography dominated by shallow marshes and somewhat deeper ponds between 9 ka BP and 7 ka BP. The diatom record from Zirahuén (between 11.6 ka BP and 8.2 ka BP) indicates fluctuating conditions during the lower Holocene, but the driest conditions occurred during the late Holocene too. Also in Zacapu, Xelhuantzi Lopez (1994) reported a temperate humid or subhumid environment between ca. 8.1 and ca. 6.7 ka BP. This is coincident with an inundation of the shore at Zirahuén between 7.3 and 7.1 ka BP.

Middle Holocene (7–4 ka BP)

In the period between 7 ka BP and 4 ka BP, the Zirahuén diatom record indicates a period of increased stability in the lake that is represented by the high abundance of diatoms characteristic of deep, freshwater lakes. During that period, the central MOLE ZIR 03 core displays diffuse lamination. In Lake Zacapu, 38 km north of

Zirahuén Lake, the diatom record between 6.9 ka BP and 5.8 ka BP (Metcalfe 1994) reveals a period of semidry temperate climate and after this at ca. 4.5 ka BP, Zacapu became an alkaline marshland, Metcalfe (1994) recorded the poor preservation of diatom valves as well as discontinuities in the diatom record, which following recent studies of Garduño et al. (personal communication) and Soria et al. (2020), we interpret as being associated with significant seismic activity. In the neighboring region of Cuitzeo, diatom associations between 8 ka BP and 4 ka BP suggest the establishment of more saline conditions. The deep, fresh, and stable lake through the mid-Holocene in Zirahuén is consistent with other records across the region that indicates generally wetter conditions and an enhanced summer monsoon (Metcalfe et al. 2015).

Late Holocene (4 ka BP to Present)

Toward the top of the Zirahuén cores (ca. 3.5 cal ka BP), significant changes are seen in the diatom preservation, abundance of diatoms, increasing diversity, and a noticeable increase in tychoplanktonic (*Staurosira*) and periphytic (*Cocconeis*, *Diploneis*) species (in both cores). Fluctuations of well-preserved and abundant diatoms were observed intercalated between the cycles of detrital input to the lake, marking periods of stability in contrast to unstable periods characterized by the common presence of detritus, phytoliths, sponge spicules, and fragmented diatoms such as *Surirella* and *Campylodiscus*, species indicating periods of increased ionic concentration and becoming gradually scarce until their disappearance from the record.

Beginning at about 2 ka, the appearance of *Diploneis eliptica* in the diatom record of Zirahuén together with other epiphytic life forms such as *Gomphonema*, *Epithemia*, *Planothidium*, and *Cocconeis*, and the abrupt oscillations in the abundance of *Aulacoseira* spp. The more diverse assemblages with increased inputs of periphytic taxa are coincident with such fluctuations in combination with high values of magnetic susceptibility represent episodes of catchment erosion in the basin. A general trend toward drier conditions was observed in the pollen record by Lozano-García et al. (2013). A drier and more variable climate suggests a weakening of the summer monsoon over the Late Holocene, with drier conditions associated with an increasing influence of the El Niño-Southern Oscillation (ENSO) mechanism in west central Mexico (Lozano-García et al. 2013; Metcalfe et al. 2015).

During the archaeological period between the Late Formative (0 a 100 CE) and the Terminal Formative (300 CE), corresponding to the top of subzone 1f of core MOLE ZIR 03, the presence of *Aulacoseira* spp. and the scarcity of epiphytic forms in association with high magnetic susceptibility values suggest humid conditions. During that period, the first appearance of *Zea mays* pollen recorded in the Pátzcuaro region was observed (Watts and Bradbury 1982) and in Zirahuén, *Zea mays* and *Teocinte* pollen (varieties of maize pollen) appear at ca. 3 ka (Lozano-García et al. 2013) indicating human occupation and cultivation within the basin. Due to the influence of anthropogenic activity, interpretation of finer scale climatic variability is tentative. The generally drier climate evident from the paleoenvironmental record may have made the catchment more sensitive to erosion from anthropogenic disturbance.

The better-documented period is the Late Classic period between 300 and 900 CE, distinguished by the building of numerous ceremonial centers in central Mexico. In the state of Michoacán, the most relevant of such centers were located in Cuitzeo and Zacapu (Pollard 2000). During the period of 290 to c. 900 CE, in MOLE ZIR 03, sediments become darker and the diatom record provides evidence of a lake rich in nutrients and with fluctuating conditions.

Between ca. 900 and 1270 CE a slightly more humid condition than before is recorded by an increase in the frequency of *Aulacoseira ambigua*, although the values of magnetic susceptibility decrease significantly indicating the absence of high erosion rates in the basin. The increase in *Aulacoseira granulata* var *angustissima* around 1100 CE may indicate slightly increased ionic concentration and drier conditions. This period is coincident with the Medieval Climate Anomaly, during which wet conditions have been reported in Mexico from the eastern Yucatan Peninsula to the west at Santa Maria del Oro, Nayarit (Vázquez-Castro et al. 2008). Records from across the region indicate the possibility of a two-part MCA (Metcalfe et al. 2015).

In several sectors of E-W central Mexico between 19° and 20° of latitude, there is also evidence of an increase in water levels of lakes like Pátzcuaro around 750–1000 CE (Fisher et al. 2003), Zacapu at 760–1070 CE (Metcalfe 1995), southern Guanajuato (Metcalfe et al. 1989), and Zempoala at 650–850 CE (Almeida-Leñero et al. 2005).

Our record also confirms the decrease in the lake's water level and the widely fluctuating conditions with high peaks in the erosion rate between 1368 and 1620 CE, a period during which diatoms are absent from the sediments and the magnetic susceptibility values are high, indicating an interval of significant erosion. Between 1250 and 1550 CE, the pollen record is dominated by the cypress family (*Cupressaceae*) and toward 1600 CE the pollen of *Acacia* spp. and *Agave* spp. become characteristic of this interval.

In Lake Zirahuén and its surrounding area, besides the presence of *Teocinte* pollen in the core record there is no evidence of human occupation, even though historical documents between ca. 1460 and 1650 CE indicate a significant decrease in human population in the region associated with viral and bacterial epidemics introduced from the Old World to which native Americans had had no previous exposure. This made them more vulnerable to infectious diseases such as smallpox and measles causing high mortality and the abandonment of cultivated fields. Lake Pátzcuaro showed the highest water levels, during 1522 CE, a level that has not been surpassed since the Spanish Conquest in 1521 CE. During the colonial period land disputes were frequent during drought conditions, particularly where land was exposed due to lake level lowering (Endfield and O'Hara 1999).

The climatic record of the eighteenth and nineteenth century is characterized by numerous drought episodes recorded in historical documents. Over this time, we do see evidence for lake level lowering, with increased presence of lake marginal diatom species, but the record is complicated by the increasing intensity of human impacts within the basin during the expansion of Colonial activities (Davies et al. 2004). The onset of this increased activity follows the deposition of the Jorullo tephra (1759–1764 CE). Diatom preservation is poor throughout as catchment inputs increase.

The concluding part of the environmental history of Lake Zirahuén comprises in the upper 0.37 m of the sediment core extending from the years 1805 to 2003 CE. Three types of events occurred during that period, the 1845 and 1858 CE earthquakes, the eruption of Paricutín volcano in 1943 CE, and the unequivocal signal of the development of the Spanish metallurgic industry for the processing of copper. Copper processing effluent was carried via the La Palma River to Lake Zirahuén, generating a peak in copper concentrations in the lake's sediments (Davies et al. 2004). In that interval of the sediment core, there is an increase in the presence of *Cyclotella stelligera* and *Cyclotella ocellata* together with *Fragilaria crotonensis*, which are a species indicative of eutrophication of the lake.

Conclusions

This chapter refines the age model and presents the diatom record that can be used to reconstruct the paleoclimate of Lake Zirahuén from 12 ka BP to the present. The data presented illustrate the following:

- Lake Zirahuén was dammed in early-middle Holocene as product of the Magueyera volcano lava flows located in the south sector of the lake.
- During the stages of eruptions and lava flows from La Magueyera volcano, five lava flows were produced, the first three lavas (lavas 1–3) were erupted some time before 12 ka BP, and lava flows 4 and 5 were erupted between ca. 8.3 ka and 6.3 ka BP. Lava flows 4 and 5 produced the maximum transgression in the lake leading to the deepening of the lake and the substitution of periphytic littoral diatoms assemblages by planktonic diatoms.
- The regional climatic system in this high volcanic plateau show dry conditions associated with a weak monsoon during the late Pleistocene-Early Holocene.
- From Early Holocene to 3.5 ka BP, the Zirahuén catchment shows the lowest erosion rates in the basin due to low volcanic activity and high lake levels. This was a period of stable lake conditions corresponding to a strengthening of the monsoon system.
- During the late Holocene, drier conditions than present prevailed from ca. 3.5 ka BP and the trend toward drier conditions has been observed at local and regional levels and was related to an increase in ENSO activity.
- The last 3 ka BP of Lake Zirahuén shows evidence of the combined impacts of drier conditions and human activity, periodic droughts, and constant fluctuations in regional volcanism.

- The past 900 years show a clear trend of decreasing water levels in the lake. Nonetheless, a peak of the planktonic diatom *Aulacoseira ambigua* between 800 and 1400 CE suggests an improvement of the environmental conditions, consistent with the Medieval Climate Anomaly that occurred between 950 and 1350 CE.

The volcanic activity in this sector of the Michoacán-Guanajuato volcanic field has been intense in the Pleistocene and along the Holocene, volcanism together with the climatic signal controls lake levels behavior in Zirahuén Lake, with increasing influence of human activity during the last 3000 years.

Acknowledgments The core was extracted as part of the Mesoamerican Lakes Expedition (MOLE) with the support of the Limnological Research Center (LRC), University of Minnesota, USA, and is part of the G28528-T CONACYT group project and internal projects of Universidad Michoacana de San Nicolás de Hidalgo (UMSNH), and Universidad Nacional Autónoma de Mexico (UNAM). Fieldwork was supported by Aberystwyth University and the University of Nottingham. Thanks to Beatriz Ortega, Socorro Lozano-García, and Alejandro Rodríguez for the field work and research support. Radiocarbon dating was provided through UK NERC and organic geochemical analyses through the UK NERC Isotope Geosciences Facilities. Gabriel Vázquez Castro acknowledges receipt of a PhD studentship from Conacyt (163229) and from UNAM. PAPIIT (in 113408). Ben Aston acknowledges receipt of a PhD studentship from Aberystwyth University. We are grateful to Sarah Metcalfe and Amy Myrbo for the external reviews of the early versions of this manuscript. Many thanks to the two anonymous reviewers and the editors (Michael Rosen and David Finkelstein) for their comments to improve the quality of the manuscript. Thanks to Elizabeth Gierlowski-Kordesch who always encouraged the Hispanic language community to show their research.

The first author dedicates this work to her beloved husband Victor Hugo Garduño, thanking him for all the discussions about the volcanic landscape evolution in the area.

References

Almeida-Leñero, L., Hooghiemstra, H., Cleef, A., & Geel, B. (2005). Holocene climatic and environmental change from pollen records of lakes Zempoala and Quila, central Mexican highlands. *Review of Palaeobotany and Palynology, 136*(1–2), 63–92.

Armienta, M., Vilaclara, G., De la Cruz-Reyna, S., Ramos, S., Ceniceros, N., Cruz, O., Aguayo, A., & Arcega-Cabrera, F. (2008). Water chemistry of lakes related to active and inactive Mexican volcanoes. *Journal of Volcanology and Geothermal Research, 178*, 249–258.

Bernal-Brooks, F., Sanchez-Chavez, J., Bravo Inclán, L., Hernández Morales, R., Martínez Cano, A. K., Lind, O. T., & Dávalos-Lind, L. (2016). The algal growth limiting-nutrient of lakes located at Mexico's Mesa central. *Journal of Limnology, 75*, 169–178. https://doi.org/10.4081/jlimnol.2016.1439.

Bradbury, J. (2000). Limnologic history of Lago de Pátzcuaro, Michoacán, Mexico for the past 48,000 years: impacts of climate and man. *Palaeogeography, Palaeoclimatology, Palaeoecology, 163*, 65–95.

Caballero, M., Vazquez, G., Ortega, B., Favila, M., & Lozano-Garcia, S. (2016). Responses to a warming trend and "El Niño" events in a tropical lake in western Mexico. *Aquatic Sciences, 78*, 591–604.

Davies, S. J., Metcalfe, S. E., Caballero, M. E., & Juggins, S. (2002). Developing diatom based transfer functions for Central Mexican lakes. *Hydrobiologia, 467*, 199–213.

Davies, S., Metcalfe, S., MacKenzie, A., Newton, A., Endfield, G., & Farmer, J. (2004). Environmental changes in the Zirahuén basin, Michoacán, Mexico, during the last 1000 years. *Journal of Paleolimnology, 31*, 77–98.

Davies, S., Metcalfe, S., Bernal-Brooks, F., Chacón-Torres, A., Farmer, J., Mackenzie, A., & Newton, A. (2005). Lake sediments record sensitivity of two hydrologically closed upland Lakes in Mexico to human impact. *Ambio, 34*(6), 470–475.

Domínguez-Vázquez, G., Osuna-Vallejo, V., Castro-López, V., Lsrade-Alcántara, I., & Bischoff, J. A. (2018). Changes in vegetation structure during the Pleistocene–Holocene transition in Guanajuato, central Mexico. *Vegetation History and Archaeobotany, 28*(1), 81–91. https://doi.org/10.1007/s00334-018-0685-8. Publisher Name Springer Berlin Heidelberg Print ISSN 0939-6314. Online ISSN 1617-6278.

Endfield, G. H., & O'Hara, S. L. (1999). Degradation, drought and dissent: An environmental history of colonial Michoacán, west central Mexico. *Annals of the Association of American Geographers, 89*, 402–419.

Eugster, H., & Hardie, L. (1978). Saline lakes. In A. Lerman (Ed.), *Lakes: chemistry, geology and physics* (pp. 237–293). New York: Springer Verlag.

Fisher, C., Pollard, H., Israde, I., Garduño, V., & Banerjee, S. (2003). A reexamination of human-induced environmental change within the Lake Pátzcuaro Basin, Michoacán, Mexico. *Proceedings of the National Academy of Sciences of the USA, 100*(8), 4957–4962.

Gomez-Tagle, C. A., Madrigal, G. J., Israde-Alcántara, I., Hernández, M. R., Ortega Murillo, M. R., & Alvarado, V. R. (2013). Evaluación rápida del Lago de Zirahuén. Universidad Michoacana de San Nicolás de Hidalgo. *Scientific Report.*

Grimm, E. C. (1987). CONISS: A FORTRAN 77 program for stratigraphically constrained cluster analysis by the method of incremental sum of squares. *Computers & Geosciences, 13*, 13–35.

Grimm, E. C. (1992). *TILIA 2 and Tilia graph.* Illinois: Illinois State Museum.

Hutchinson, E. G., Patrick, R., & Deewey, E. S. (1956). Sediment of Lake Patzcuaro, Michoacán, Mexico. *Geological Society of America Bulletin, 67*, 1491–1504.

Israde-Alcántara, I., Velázquez Durán, R., Lozano García, M. S., Garduño-Monroy, V. H., Bischoff, J., & Domínguez Vázquez, G. (2010). Evolución paleolimnologica del lago de Cuitzeo, Michoacán. *Boletín de la Sociedad Geológica Mexicana, 62*(3), 345–357. ISSN: 1405-3322.

Israde-Alcántara, I., Miller, W., Garduño Monroy, V. H., & Barron, J. (2011). Estratigrafía y marco geodinámico de las Cuencas lacustres del Centro de México In: B. Ortega, & M. Caballero (Eds.), Escenarios de Cambio Climático. Registros del Cuaternario en América Latina. UNAM, 73–90.

Israde-Alcántara, I., Bischoff, J., Domínguez-Vázquez, G., Li, H.-C., McGeehin, J., DeCarli, P., Bunch, T., Wittke, J., Weaver, J., Xie, S., Firestone, R., West, A., Kennett, J., Mercer, C., Kinzie, C., & Wolbach, W. (2012). Evidence from Central Mexico supporting the Younger Dryas Impact Hypothesis. *PNAS, 109*(13), E738–E747. https://doi.org/10.1073/pnas.1110614109. ISSN: 0027-8424.

Kelts, K., Briegel, U., Ghilardi, K., & Hsü, K. (1986). The limnogeology-ETH coring system. *Schweizerische Zeitschrift für Hydrologie, 48*(1), 104–115.

Kilham, P., Kilham, S. S., & Hecka, R. E. (1986). Hypothesized resource relationships among African planktonic diatoms. *Limnology and Oceanography, 31*, 1169–1181.

Krammer, K., & Lange-Bertalot, H. (1986). *Süsswasserflora von Mitteleuropa. Bacillariophyceae 1 Gustav*. New York: Fischer Verlag Stuttgart.

Krammer, K., & Lange-Bertalot, H. (1988). *Süsswasserflora von Mitteleuropa. Bacillariophyceae. 2. Teil: Epithemiaceae, Bacillariaceae, Surirellaceae* (Vol. 2/2). Stuttgart: Gustav Fischer.

Krammer, K., & Lange-Bertalot, H. (1991a). *Süsswasserflora von Mitteleuropa. Bacillariophyceae. 3. Teil: Centrales; Fragilariaceae, Eunotiaceae* (Vol. 2/3). Stuttgart: Gustav Fischer.

Krammer, K., & Lange-Bertalot, H. (1991b). *Süsswasserflora von Mitteleuropa. Bacillariophyceae. 4. Teil: Achnanthaceae* (Vol. 2/4). Stuttgart: Gustav Fischer.

Kullenberg, B. (1947). The piston core sampler. *Svenska Hydrografisk-Biologiska Komm. Skr. ser. 3 Hydrografi, 1*(2), 1–46.

López, S. (1982). *Dinámica hidrológica del Lago Zirahuén. Informe de Servicio Social. (Inédito), Depto. de Zootecnia. Div. de Ciencias Biológicas y de la Salud.* México, D.F.: Universidad Autónoma Metropolitana Iztapalapa.

Lozano-García, S., Torres-Rodríguez, E., Ortega, B., Vázquez, G., & Caballero, M. (2013). Ecosystem responses to climate and disturbances in western central Mexico during the late Pleistocene and Holocene. *Palaeogeography, Palaeoclimatology, Palaeoecology, 370*, 184–195.

Metcalfe S. E. Street-Perrott F. A, Brown R. B., Hales P. E., Perrot R. A., Steininger F. M. (1989). Late Holocene human impact on lake basins in central Mexico. *Geoarchaeology, 4*(2), 119–141.

Metcalfe, S. E. (1992). *Changing environments of the Zacapu Basin central Mexico: a diatom-based history spanning the last 30,000 years*. Research paper 38, School of Geography, Oxford, p. 38.

Metcalfe S. E. Street-Perrot F. A., O´Hara S. L., Hales P. E., and Perrott R. A. (1994). The Paleolimnological record of environmental change: examples from the arid frontier of Mesoamerica. In: Millington A. C. and Pye K., eds.. Environmental Change in Drylands:Biogeographical and geomorphological perspectives. Wiley, London. Pp.131–145.

Metcalfe, S. (1995). Holocene environmental change in the Zacapu basin, Mexico: A diatom based record. *The Holocene, 5*, 196–208.

Metcalfe, S. E., Davies, S. J., Brasisby, J. D., Leng, M. J., Newton, A. J., Terrett, N. L., & O´Hara, S. L. (2007). Long and short-term change in the Pátzcuaro Basin, central Mexico. *Palaeogeography Palaeoclimatology Palaeoecology, 247*, 272–295.

Metcalfe, S., Barron, A. J., & Davies, J. S. (2015). The Holocene History of the North American Monsoon: 'known, knowns' and 'known unknowns' in understanding its spatial and temporal complexity. *Quaternary Science Reviews, 120*, 1–27.

Myrbo, A. (2004). *Limnological Research Center Core Facility, Analityc procedures: Carbon Coulometry* compilation date October 22, http://lrc.geo.umn.edu/laccore/assets/pdf/sops/coulometry.pdf

Newton, A., Metcalfe, E. S., Davies, S., Cook, G., Barker, P., & Telford, J. (2005). Late Quaternary volcanic record from lakes of Michoacán, central Mexico. *Quaternary Science Reviews, 24*(1–2), 91–104.

Ortega, B., Caballero, C., Lozano, S., Israde, I., & Vilaclara, G. (2002). 52,000 years of environmental history in Zacapu basin, Michoacán, Mexico: The magnetic record. *Earth and Planetary Science Letters, 202*, 663–675.

Ortega, B., Vazquez, G., Caballero, M., Israde, I., Lozano-García, S., Schaaf, P., & Torres, E. (2010). Late Pleistocene: Holocene record of environmental changes in Lake Zirahuén, central Mexico. *Journal of Paleolimnology, 44*, 745–760.

Osorio-Ocampo, S., Macias, J. L., Cardona-Melchor, S., Sosa-Ceballos, G., Garcia-Sanchez, L., Perton, M., Pola, A., Garduno-Monroy, V. H., Layer, P. W., Perton, M., Benowitz, J., & Benowitz, J. (2018). The eruptive history of the Patzcuaro lake area in the Michoacán-Guanajuato volcanic field, central Mexico: Field mapping. C-14 and 40Ar/39 ar geochronology. *Journal of Volcanology and Geothermal Research, 358*, 307–328.

Pollard, H. (2000). Tarascan and their Ancestors: Prehistory of Michoacán. In *Greater mesoamerica: the archaeology of west and northwest Mexico.* Gorenstei University of Utah Press, Salt Lake.

Pooja, K., Siebe, C., Guilbaud, M. N., & Salinas, S. (2016). Geological and environmental controls on the change of eruptive style (phreatomagmatic to Strombolian-effusive) of Late Pleistocene El Caracol tuff cone and its comparison with adjacent volcanoes around the Zacapu basin (Michoacán, México). *Journal of Volcanology and Geothermal Research, 318*, 114–133.

Reavie, D. E., & Cai, M. (2019). Consideration of species-specific diatom indicators of anthropogenic stress in the Great Lakes. *PLoS One, 14*(5), e0210927. https://doi.org/10.1371/journal.pone.0210927.

Reimer, P. J., Reimer, R. W., Austin, W. E. N., Austin, W. E. N., Bard, E., Capano, M., Bayliss, A., Blackwell, P. G., Heaton, T. J., Bronk Ramsey, C., Butzin, M., Kohler, P., Cheng, H., Cheng, H., Edwards, R. L., Edwards, R. L., Friedrich, M., Grootes, P. M., Guilderson, T. P., Guilderson,

T. P., Hajdas, I., Wacker, L., Sookdeo, A., Hogg, A. G., Hughen, K. A., Kromer, B., Manning, S. W., Muscheler, R., Palmer, J. G., Turney, C. S. M., Sookdeo, A., Pearson, C., Van Der Plicht, J., Richards, D. A., Scott, E. M., Southon, J. R., Fahrni, S. M., Adolphi, F., Buntgen, U., Buntgen, U., Reinig, F., Buntgen, U., Buntgen, U., Fahrni, S. M., Fogtmann-Schulz, A., Kudsk, S., Friedrich, R., Miyake, F., Olsen, J., Sakamoto, M., & Talamo, S. (2020). The Intcal20 Northern Hemisphere radiocarbon age calibration curve (0–55 cal kBP). *Radiocarbon, 62*(4), 725–757. https://doi.org/10.1017/RDC.2020.41.

Ross, K. A., Gashugi E., Gafasi A., Wuest A., Schmid M, Alvarez I., (2015) Characterisation of the Subaquatic Groundwater Discharge That Maintains the Permanent Stratification within Lake Kivu; East Africa. PLOS ONE 10(3):e0121217.

Rott, E., Kling, H., & McGregor, G. (2006). Studies on the diatom *Urosolenia* Round & Crawford (Rhizosoleniophycideade). Part 1. New re-classified species from subtropical and tropical freshwaters. *Diatom Research, 21*(1), 105–124.

Rumrich, U., Lange-Bertalot, H., & Rumrich, M. (2000). *Diatomeen der Anden. Von Venezuela bis Patagonia/Tierra de Fuego. Iconografia diatomologica* (Vol. 9). Koeltz Scientific Books.

Sánchez Gonzalez M. J. (2007). Evolución paleoambiental del lago Crater de Tacambaro, Michoacán. Tesis de licenciatura.. Facultad de Biología. Universidad Michoacana de San Nicolás de Hidalgo. 83 p..

Sarmiento, H., Isumbisho, M., & Descy, J. P. (2006). Phytoplankton ecology of Lake Kivu (eastern Africa). *Journal of Plankton Research, 28*(9), 815–829.

Schnurrenberger, D., Russell, J., & Kelts, K. (2003). Classification of lacustrine sediments based on sedimentary components. *Journal of Paleolimnology, 29*, 141–154.

Sigala I, Caballero M, Correa-Metrio A, Lozano-Garcia S, Vazquez G, Perez L, Zawisza E (2017). Basic limnology of 30 continental waterbodies of the Transmexican Volcanic Belt across climatic and environmental gradients. *Boletin de la Sociedad Geologica Mexicana 69*(2), 313–370.

Stoermer, E. F., Kociolek, J. P., Schelske, C. L., & Conley, D. J. (1987). Quantitative analysis of siliceous microfossils in the sediments of Lake Erie's central basin. *Diatom Research, 2*, 113–134.

Tavera, R., & Martínez-Almeida, V. (2005). Atelomixis as a possible driving force in the phytoplankton composition of Zirahuén, a warm-monomictic tropical lake. *Hidrobiologia, 533*, 199–208.

Telford, R., Barker, P., Metcalfe, S., & Newton, A. (2004). Lacustrine responses to tephra deposition, Examples from Mexico. *Quaternary Science Reviews, 23*, 2337–2353.

Tremarin, P. E., Gomes, F. E., Majewski, A. V., & Veiga, L. T. (2015). *Acanthoceras* and *Urosolenia* species (Diatomeae) in subtropical reservoirs from South Brazil: Ultrastructure, distribution and autoecology. *Biota Neotropica, 15*(1).

Vargas, M., Beltrána, H., Vázquez-Labastida, E., Linares-López, C., & Salmón, M. (2007). Synthesis and characterization of montmorillonite clays with modulable porosity induced with acids and superacids. *Journal of Materials Research, 22*(3), 788–800.

Vázquez, C. G. (2012) *Magnetismo ambiental de los últimos 17,000 años en el lago Zirahuén* (Michoacán México. Ph. D. Disertation thesis). Universidad Nacionaql Autonoma de Mexico, p. 202.

Vázquez, G., Ortega, B., Davies, S., & Aston, B. (2010). Registro sedimentario de los últimos ca. 17000 años del lago de Zirahuén, Michoacán, México. *Boletín de la Sociedad Geológica Mexicana, 62*(3), 325–343.

Vázquez-Castro, G., Ortega-Guerrero, B., Rodríguez, A., Caballero, M., & Lozano-García, S. (2008). Mineralogía magnética como indicador de sequía en los sedimentos lacustres de los últimos ca. 2600 años de Santa María del Oro occidente de México. *Revista Mexicana de Ciencias Geológicas, 25*(1), 21–38.

Walker, M., Gibbard, P., Head, M. J., Berkelhammer, M., Bjorck, S., Bjorck, S., Cheng, H., Cwynar, L. C., Fisher, D., Gkinis, V., Long, A., Lowe, J., Newnham, R., Rasmussen, S. O., & Weiss, H. (2019). Formal subdivision of the holocene series/epoch: A summary. *Journal of the Geological Society of India, 93*, 135–141.

Watts, W., & Bradbury, J. (1982). Palaeoecological studies at Lake Pátzcuaro on the west-central Mexican plateau and at Chalco at the basin of Mexico. *Quaternary Research, 17*, 56–70.

Xelhuantzi Lopez, M. S. (1994). Estudio palinológico de cuatro sitios ubicados en la Cuenca de Zacapu: fondo de ciénega, contacto Lomas-ciénega, pantano interno y Loma Alta, en Petrequin, P. (ed) 8000 años de la cuenca de Zacapu, evolución de los paisajes y primeros desmontes: Morelia. *Michoacan, Cuadernos de Estudios Michoacán, 6*, 81–93.

Sedimentary Record of the Zacapu Basin, Michoacán, México, and Implications for P'urhépecha Culture During the Preclassic and Postclassic Periods

Diana C. Soria-Caballero, Víctor Hugo Garduño-Monroy, Isabel Israde-Alcántara, Ángel G. Figueroa-Soto, M. Gabriela Gómez-Vasconcelos, and Nathalie Fagel

Abstract The analysis of syn-sedimentary deformations is a helpful tool for paleoenvironmental reconstructions; these structures are better preserved in fluvio-lacustrine sequences. In the Zacapu basin, lacustrine sedimentation has been active since late Pliocene until modern times. During the Holocene, the lake evolved into a wetland that was the home of the first P'urhépecha settlements in Michoacán, México. Since the beginning, the Zacapu basin was affected by volcanic activity from the Michoacán-Guanajuato Volcanic Field and seismicity from the Morelia-Acambay Fault System. The effects of these geologic events were recorded in the basin sediments and in the pictorial records of the P'urhépecha people. According to archeological studies, P'urhépechas settled in the Zacapu basin around 150 BC (2100 y BP) (Loma Alta phase) occupying the site for at least 1700 years until the

Supplementary Information The online version of this chapter (https://doi.org/10.1007/978-3-030-66576-0_13) contains supplementary material, which is available to authorized users.

D. C. Soria-Caballero (✉)
Escuela Nacional de Estudios Superiores unidad Morelia, Universidad Nacional Autónoma de México, Morelia, Michoacán, Mexico

V. H. Garduño-Monroy (Deceased) · I. Israde-Alcántara (✉)
Instituto de Investigaciones en Ciencias de la Tierra, Edificio U. Ciudad Universitaria. Universidad Michoacana de San Nicolás de Hidalgo, Morelia, Michoacán, Mexico

Á. G. Figueroa-Soto · M. G. Gómez-Vasconcelos
Cátedra CONACYT en Instituto de Investigaciones en Ciencias de la Tierra, Edificio U. Ciudad Universitaria. Universidad Michoacana de San Nicolás de Hidalgo, Morelia, Michoacán, Mexico

N. Fagel
Université de Liège, Département de Géologie, U.R. Argiles, Géochimie et Environnements sédimentaires (AGEs), Liège, Belgium

M. R. Rosen et al. (eds.), *Limnogeology: Progress, Challenges and Opportunities*, Syntheses in Limnogeology, https://doi.org/10.1007/978-3-030-66576-0_13

Milpillas phase (also known as Late Uricho phase). In this time span, the population may have witnessed the volcanic activity of the Capáxtiro volcano (between 193 BC (2143 y BP) and 84 BC (2034 y BP) and the Malpaís Prieto volcano (AD 826 - 962 or 1124 - 988 y BP), but also the tectonic activity of nearby faults (i.e., Villa Jiménez fault). In this work, we analyze the syn-sedimentary deformations exposed in the sedimentary record of trenches at the uplifted area of Las Lomas, and identify the presence of liquefaction structures, folding, microfaulting, and slumps, which have characteristics that suggest that they are related to seismic activity. We propose the occurrence of at least four earthquakes with moderate to high magnitude (MI ≥ 5) in the basin based in the deformations. The oldest syn-sedimentary deformations recorded in the lacustrine sequence are liquefied dikes and flames dated around 40 ka. Meanwhile, the youngest syn-sedimentary deformation are microfaulted layers which contain fragments of P'urhépecha pottery dated in the Milpillas/Late Uricho phase (1350 AD). Because of the time span when the volcanic and seismic events occurred, we discuss their relation with the human activities of the early settlements in the Zacapu basin.

Keywords Sedimentary deformations · P'urhépecha culture · Zacapu basin · Holocenic volcanoes · geologic risk · human–landscape interaction

Introduction

The analysis of sedimentary deformations, either syn-depositional or postdepositional, is a helpful tool for paleoenvironmental reconstructions. These structures form in unconsolidated materials, particularly in fluvio-lacustrine sequences composed of alternate clayey and sandy layers (Horváth et al. 2005). The origin of such deformations could be related to very different processes, like climatic change (e.g., Vanneste et al. 1999; Horváth et al. 2005), lithostatic load processes (e.g., Kuhn 2005), gravitational instability or brittle deformation (e.g., Neuwerth et al. 2006), fast sedimentation, tectonic events, or a combination of the previous (e.g., Tuttle et al. 2002).

Syn-sedimentary deformation typically refers to a restricted stratigraphic level, delimited above and underneath by undeformed layers (Phratiyal & Sharma 2009). When syn-sedimentary deformations appear due to tectonic activity, they are known as *seismites* (Seilacher 1969). Seismites differ from structures generated by lithological load or storms because they have a maximum thickness of 1 m, are confined to a single layer with an extensive spatial continuity, delimited by horizontal strata at the bottom and top, and usually located close to known active faults (Alfaro 2006; Schillizi et al. 2010).

Seismites can be related to earthquakes with magnitudes $M \geq 5$, because lower magnitude events form small and sparse structures that are difficult to identify (Ambraseys 1988; Obermeier 1996; Tuttle et al. 2002). The usefulness of the seismites relies in that they allow to identify secondary effects on hidden seismogenic faults and add information about ancient earthquakes in a given region (Obermeier et al. 1987; Amick & Gelinas 1991; Tuttle & Schweig 1995; Hibsch et al. 1997; Vanneste et al. 1999; Takahama et al. 2000; Rodríguez-Pascua et al. 2010).

Tectonic sedimentary deformations are classified as: (a) Microfaulting; (b) Distorted sedimentation, such as convolute and contorted strata, microfolds, macrofolds, and slumps; and (c) Intrusions of sediments and dewatering structures, such as dikes, pillars, clay diapirs, mud lenses, and flame and pillow structures. The last are the most common (Hibsch et al. 1997; Alfaro et al. 2001; Neuwerth et al. 2006; Koç Taşgin & Türkmen 2009; Phratiyal & Sharma 2009).

In this paper, we analyze the sedimentary record of the Zacapu basin, located in the NW of Michoacán, México, to identify syn-sedimentary deformations that could be related to ancient seismicity in the area, outlining their relation with the P'urhépecha settlements in the region.

Study Area

The Zacapu basin (19°51'N, 101°40'W, Fig. 1a) is a volcano–tectonic intra-arc basin formed during the Late Pliocene-Pleistocene by a NW-SE extensional tectonic regime, and evolved under the influence of the Michoacán-Guanajuato Volcanic Field (MGVF) and the Morelia-Acambay Fault System (MAFS) (Demant et al. 1992). The MGVF, located in the Trans-Mexican Volcanic Belt, is a region of monogenetic volcanism with quaternary ages and more than three thousand volcanic vents, most of which are cinder cones, spatters, shield volcanoes, domes, maars, and only three stratovolcanoes (Hasenaka and Carmichael 1987; Connor 1987). The MAFS is an area of ~200 km length and ~50 km width, that comprises E-W-trending fault segments with prehistorical, historical, and modern cortical seismicity; this system has been active since Miocene (Szynkaruk et al. 2004; Israde et al. 2005; Norini et al. 2010; Rodríguez-Pascua et al. 2010).

In its early state, the Zacapu basin was connected with the Cuitzeo basin (Fig. 1b), one of the biggest basins of México (Alvarez 1972; Israde-Alcántara 1997; Moncayo et al. 2001; Antaramián & Correa 2003), but during the Quaternary they were separated due to the reactivation of the MAFS, where the faults changed from a normal to oblique displacement (normal–sinistral strike slip) (Ego & Ansan 2002; Szynkaruk et al. 2004; Mennella 2011), and the volcanic activity in the MGVF intensifies (Pérez-López et al. 2011; Hernández Jiménez et al. 2014).

The Zacapu basin has an area of 335 km^2 and used to enclose a lake that gradually became shallower until it turned into a wetland. The modern landscape shows a wide valley surrounded by volcanoes reaching heights up to 3365 masl (Figs. 1b and 2). The relative basement of the basin is composed of andesitic lava flows and ignimbritic sequences from early Miocene polygenetic volcanoes (17–19 Ma, Demant et al. 1992; Israde-Alcántara & Garduño-Monroy 1999; Israde-Alcántara

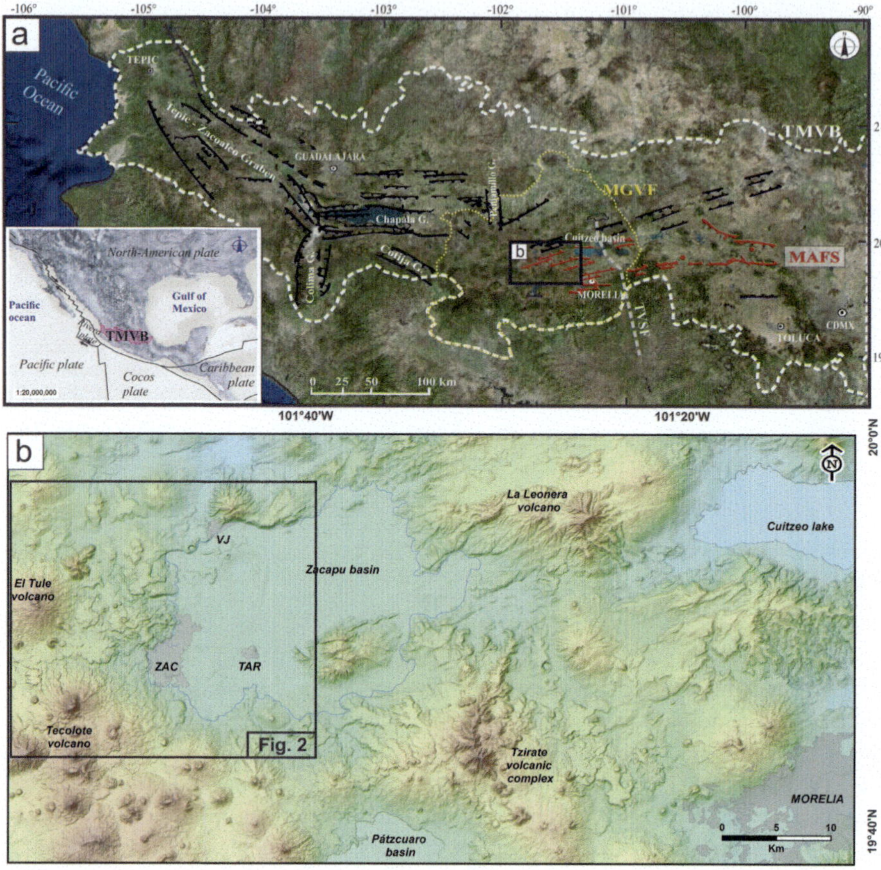

Fig. 1 (**a**) Location of the Trans-Mexican Volcanic Belt (TMVB), México, and the study area: the Zacapu basin, between the Michoacán-Guanajuato Volcanic Field (MGVF) and the Morelia Acambay fault system (MAFS). (**b**) Zacapu basin general view, we can observe its closeness to the Cuitzeo basin and prominent volcanic vents that enclosed and separate the basin, such as La Leonera volcano and Tzirate volcanic complex. Abbreviations: ZAC Zacapu city, TAR Tarejero volcano and town, VJ Villa Jiménez town

et al. 2011; Reyes-Guzmán et al. 2018). The basement is covered by Pliocene lacustrine sediments (that continued to accumulate until historical times) and volcanic materials of Pleistocene and Holocene ages belonging to the Tres Cerritos-Capáxtiro-Malpaís Prieto volcanic complex (Fig. 2). The Pleistocene lavas enclosed the basin to the northwest, which were cut by normal faults of the MAFS, and developed morphologies of triangular facets. At the western edge, the faulting is expressed by lineaments of volcanic cones emplaced through tectonic fractures (Le Corvec et al. 2013).

In central México, lacustrine basins have been studied following three approaches: anthropological (e.g., Metcalfe et al. 1989; Street-Perrott et al. 1989; Arnauld et al.

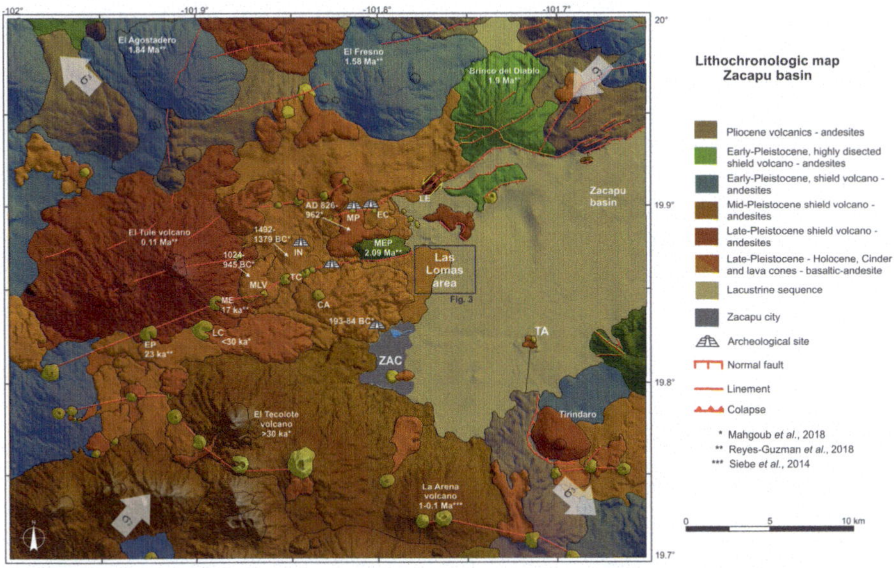

Fig. 2 Lithochronologic map of Zacapu basin. We can observe the main Quaternary volcanic vents, the lacustrine area, and archeological sites. White arrows indicate σ_2 and σ_3. The ages were obtained from previous works (indicated with *). Abbreviations: CA Capáxtiro volcano, EC El Caracol volcano, EP El Pueblito volcano, IN Infiernillo lava flow, LC Las Cabras volcano, LE Los Espinos maar, ME El Molcajete de Eréndira volcano, MEP Mesa el Pinal lava flow, MLV Malpaís Las Víboras lava flow, MP Malpaís Prieto lava flow, TA Tarejero volcano, TC Tres Cerritos volcanoes, ZAC Zacapu city

1993; O'Hara et al. 1993; Fisher et al. 2003; Davies et al. 2011; Mahgoub et al. 2018), paleoclimatic/paleoenvironmental (e.g., Watts & Bradbury 1982; Metcalfe et al. 1994; Xelhuantzi 1994; Ortega et al. 2002; Davies et al. 2004; Park et al. 2010; Vázquez et al. 2010; Lozano-García & Sosa-Nájera 2011; Ortega & Caballero 2011; Vilaclara et al. 2011), and volcano-tectonic (e.g., Rodríguez-Pascua et al. 2004, 2010, Landgridge et al. 2013; Siebe et al. 2012, 2014). Volcano-tectonic studies indicate that lacustrine basins in central Mexico are predisposed to sedimentary deformations related to tectonic activity (folding, liquefaction structures, etc.), and provide a valuable record of Quaternary changes.

According to the *Relación de Michoacán* chronicle (De Alcalá 2010) and the works of Michelet et al. (1989), the Zacapu basin was the site of the first P'urhépechsa settlements, with four phases of maximum ocscupation: Loma Alta (Loma Alta 1: 100 BC (or 2050 y BP)–0; Loma Alta 2: 0–350 AD; Loma Alta 3: 350–550 AD; and Jarácuaro interphase: 550–600 AD), Lupe (Early Lupe: 600–700 AD; Late Lupe: 700–850 AD; and La Joya interphase: 850–900 AD), Palacio (900–1200 AD) and Milpillas (1200–1450 AD); in the work of Pollard (2015), the Palacio and Milpillas phases together correspond to the Uricho phase (early and late, respectively) . According to Metcalfe et al., (1994) and Ortega et al. (2002), the human impact on the lake began around 4000 cal y BP. Before that, the water level passed from high lacustrine levels (before ~40 ka) to fluctuant lacustrine levels, and later developed fragmented wetland environments.

In the nineteenth century, the shallow lake was drained for agricultural purposes, causing a lake regression and leaving a wide plain behind (Petrequin 1994; Metcalfe et al. 1994). Human activities modified the basin, which was also modified by volcanic and tectonic activity. The "Lienzo de Michoacán" — an ancient painted cloth — shows the relationship between earthquakes and sacrifices by representing a collapsing pyramid (Fig. 3), the stairs of which are stained with the blood of recent sacrifices (Garduño-Monroy & Rodríguez-Pascua 2014).

Fig. 3 *Lienzo de Michoacán*, an ancient Mexican codex published for the first time in 1540 in the *Relación de Michoacan* chronicle by Fray Jerónimo de Alcalá. This codex depicts a destructed pyramid (perhaps by effect of an earthquake), blood falling down the stairs (referring to a recent sacrifice), and the inhabitants debating the relation with the premonitions observed in the sky (comet)

The Zacapu basin has an instrumental seismicity record reported by the National Seismological Service (http://www.ssn.mx) related to subduction processes and cortical activity.

Methods

Thematic maps (scale 1:50,000) and digital elevation models (15 m resolution), obtained from the Mexican Geological Service (SGM by its acronym in Spanish) and the National Institute of Statistics and Geography (INEGI by its acronym in Spanish), were used to delimitate the geological and morphological units, as well as the fault traces in the basin.

A microtopographic survey was carried out on the area of Las Lomas, located to the west of the basin, where there is more evidence of P'urhépecha settlements. We used a high-precision Leyca GPS and gathered 4822 points to generate a contour map with 50 cm equidistance, which facilitated the description of the geometry of the area.

Thirteen trenches were excavated and their sedimentary sequences recorded (Figs. 4 and 5). The trenches where distributed preferentially at the NE of the Las Lomas area, over a ~3 km^2 area. For each trench, the stratigraphic sequence was

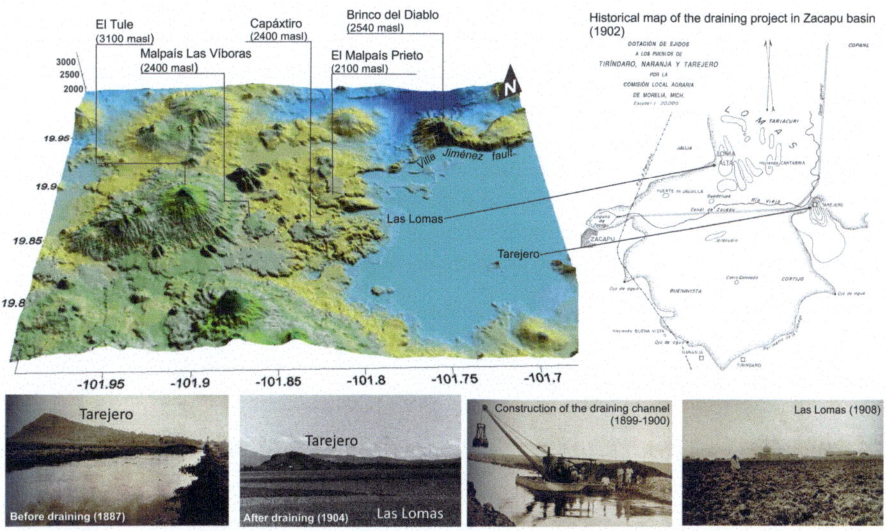

Fig. 4 Zacapu basin DEM, indicating the main volcanic vents with elevations. Las Lomas area in the DEM and in the historical plan for draining the basin (around 1902 by the Local Agrarian Commission). At the bottom a series of photos of the draining process in the area, from left to right: view from Las Lomas before desiccation, view from Las Lomas after desiccation, construction of the draining channel, and material added to Las Lomas with agricultural purposes. (Photos from Espinoza Jiménez & Baltazar Caballero, 2014)

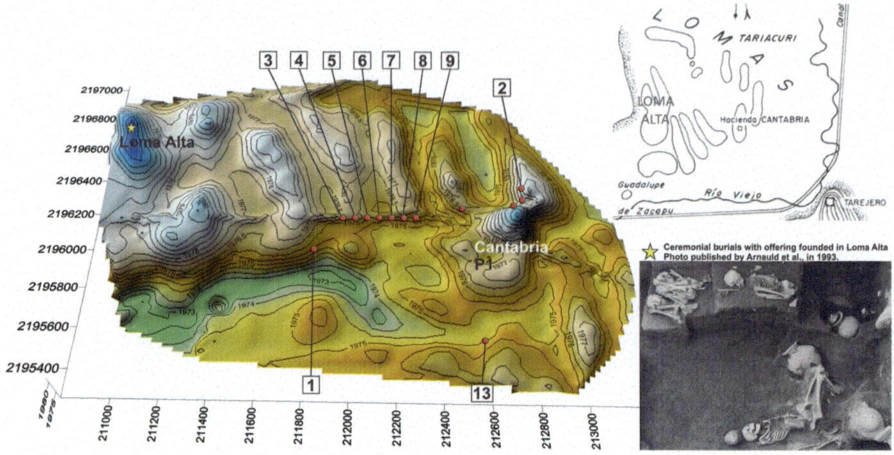

Fig. 5 Detailed DEM obtained from the microtopographic survey in Las Lomas area and comparison with the morphology drawing in the historical map (upper right). Loma Alta and Cantabria (P1) are the principal P'urhépecha settlements, where ceremonial burials were found (bottom right). The location of trenches is indicated with red dots and the ones recorded in Fig. 5 are indicated with numbers

recorded in photos and sketches of the observed contacts and sediment characteristics (i.e., composition, organic content, grain size, etc.). Individual sequences were correlated using guide layers. We carried out an analysis of the syn-sedimentary structures, recording their geometries, sizes, and spatial distribution. Identification of these structures was performed following the method described in Audemard et al. (2011).

The trenches were sampled at every change of layer and analyzed for grain size, mineralogy, and microfossil contents. In addition, charcoal fragments were collected, dated by radiocarbon, and calibrated with INTCAL-20 (August 2020. See online data supplement). The samples analyzed by granulometry required six sieves (from Φ - 6 to Φ 5), and size distributions were graphed.

Results and Discussion

Zacapu Basin Landscape and Las Lomas Geometry

The landscape in the Zacapu area reflects volcanic and tectonic activity coexisting with sedimentary processes. The morphologic analysis reflects the interaction between the volcanic vents and its deposits, the faults mapped in this work, and the sedimentary sequences of the basin.

At the northeast, we observe a group of polygenetic andesitic volcanoes and scoria cones of Miocene to Pliocene ages stand out, affected by ENE-WSW, E-W, and NE-SW faults (Fig. 2). The ENE-WSW faults belong to the MAFS, have normal movement dipping to the SSE, and forming scarps up to 120 m high.

The Tiríndaro volcanic complex, observed at the south of the area, is formed by lavas and breccias from small-shield volcanoes with andesitic and basaltic compositions that reach heights of 3365 masl. This volcanic complex has a Pleistocene age and a morphology that resembles a caldera with a moderately developed drainage, where the volcanic deposits are aligned along two major trends: E-W and NE-SW.

At the west and northwest of the basin, we identify a group of scoria cones and domes resulted from a major episode of effusive volcanism that closed the basin to the north (Fig. 4). These monogenetic vents and their lava flows are aligned in NE-SW and ENE-WSW direction, highlighting the emplacement of the Infiernillo (1492 – 1379 BC or 3442 – 3329 y BP), Malpaís Las Víboras (1024 – 945 BC or 2974 – 2895 y BP), Capáxtiro (193 – 84 BC or 2143 – 2034 y BP), and Malpaís Prieto (AD 826-962 or 1124 – 988 y BP) volcanoes (Mahgoub et al. 2018). Also in this group, Los Espinos maar was emplaced and deformed by the Villa Jiménez fault causing a NE-SW elongated crater (Fig. 2). According to Münn et al. (2006), the elongation can be used as an indicator of the direction of maximum shear (σ_1) and allows us to calculate the maximum extension (σ_3, Fig. 2). Accordingly, in the Zacapu basin the maximum extension has a NW-SE direction. This is consistent with the extensional tectonic regime proposed by Demant et al. (1992).

In the mapped area, the basin is a wide valley with elevations around 1970 masl, which contained the lacustrine sediments until 2200 cal y BP when the lake evolved to a wetland (Petrequin 1994). Later, between 1887 and 1908, the area was drained for agricultural purposes (Fig. 4, Ávila 1985), causing deflation (Ortega et al. 2002). The west portion of this area known as Las Lomas was the site of the human settlements of a sedentary agrarian society which uses it as funeral and ritual spaces (Fig. 5; Arnauld et al. 1993; Villanueva 2019).

The microtopographic survey shows that Las Lomas is an uplifted area composed by a series of mounds of between 2 and 10 m height, limited south by a depression of the ancient drainage of Lake Zacapu (Fig. 5). Las Lomas shows two types of mounds: elongated and circular. The elongated mounds are dominant and considered of natural formation, they have gentle slopes and trend north–south: the major axis ranges from 200 to 600 m in length, while the minor axis ranges from 50 to 200 m. The circular mounds have steep slopes and are the result of human activities in the area, particularly the addition of material to raise the base of settlements according to water-level fluctuations. The volume of added material ranges from 5×10^5 to 6×10^7 m^3; these mounds contain the oldest P'urhépecha burials of the archeological sites of Loma Alta and P1 CEMCA (Arnauld et al. 1993; Ortega et al. 2002).

Although the origin of Las Lomas is uncertain, we can suggest three different scenarios according to the morphology of the area, the direction of its major axis, and the sedimentary deformation recorded in its stratigraphy. These scenarios are:

I A large-scale folding event, occurring along a NNW-SSE direction due to the propagation of seismic waves, which uplifted and deformed the unconsolidated sediments.

II Seismically induced lateral spreading, where the syn-sedimentary deformation is related to seismic waves and liquefaction of the underlying strata; although this type of deformation has been described in coastal zones (e.g., Guarnieri et al. 2008), the granulometry and high plasticity of the Las Lomas sediments make them prone to deformation (e.g., Limon earthquake (Ms = 7.5) in Panamá, Camacho and Viquez 1992).

III Non-seismically induced lateral spreading, produced by the load of the Holocene volcanoes, where a thin ductile layer of sediments underlying the volcanoes is required in order to causes volcanic vents to spread laterally along the basal layer, creating protuberances and uplifting the sediments near the volcano (van Wyk de Vries and Matela 1998; Münn et al. 2006). This type of scenario has been reported in the El Estribo volcano, southern of Pátzcuaro lake, where nearby lacustrine sediments were lifted and folded, and a fault was reactivated (Israde et al. 2005; Garduño-Monroy et al. 2009; Garduño-Monroy et al. 2011).

The characteristics of the Zacapu basin make it difficult to favor one scenario over the other, because here the Pliocene lacustrine sediments could behave as a ductile layer under the Holocenic volcanoes and lava flows (such as the Tres Cerritos–Capáxtiro–Malpaís Prieto volcanic complex), and the faults have documented evidence of recent activity.

For the first scenario, the main objection is that the uplifted sediments are restricted in a small area (Las Lomas) instead of been widely distributed in the basin as expected. The second scenario has not been reported in lacustrine areas, but in predominantly sandy sediments. Finally, for the third scenario we could expect severely deformed sediments closer to the lava flows, but we need more trench work in the area to confirm it. In any case, geophysical studies are required in order to reliably suggest an origin for the Las Lomas area.

Stratigraphy of the Las Lomas Area

The Las Lomas area shows deformed and reworked sediments. We analyzed 13 trenches and identified four co-relatable units in the trenches: a sequence of lacustrine sediments covered by a carbonated layer, followed by an erosional unconformity of volcanic ash layer, and finally covered by wetland sediments and a modern agricultural soil (Fig. 6). Each unit has the following characteristics:

A. *Lacustrine sediments:* a ~3 m-thick sequence composed by alternate clayey and diatomitic layers interbedded with ash-fall deposits. This sequence is laminar at the base and severely deformed at the top. The age obtained for this unit is 41258–40776 cal yr BP obtained from charcoal fragments in organic-rich sedi-

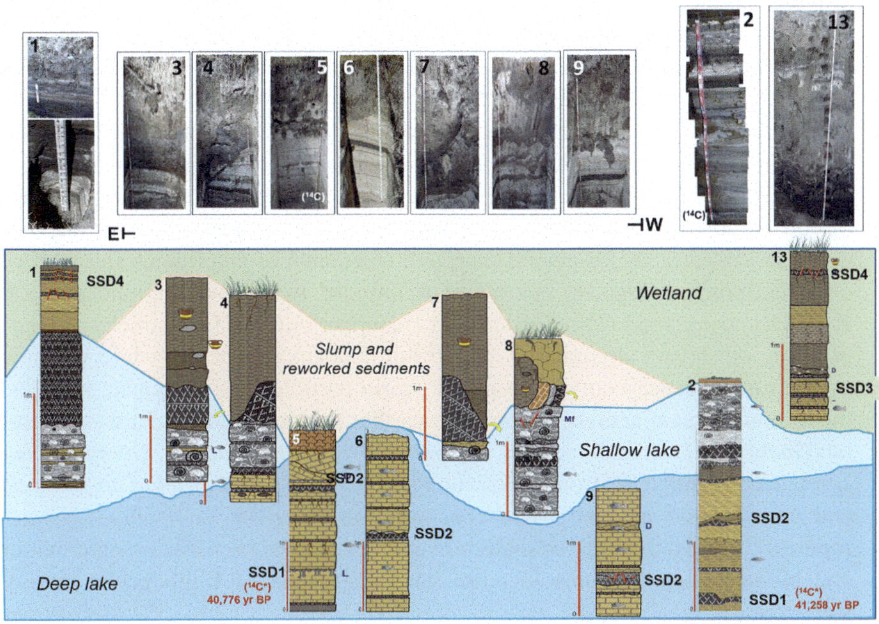

Fig. 6 Sedimentary record of the 10 selected trenches and correlation between guide layers across the stratigraphy. Transition from a deep lake to a shallow lake, and finally a wetland is recorded. Layers with syn-sedimentary deformations (SSD) are indicated in each column

ment layers in trenches 2 and 5, at 350 cm and 190 cm depth, respectively (Fig. 6). The clayey layers contain fragments of plants, animals (scales and fish bones), coal, and oxidized scoria particles. The volcanic ash layers range from 3 to 10 cm thick, and are composed of glass, scoria, and altered pumices. The analysis of clays shows the presence of halloysite and montmorillonite. The diatomite analyses show species associated with deep-water, planktonic habitats such as *Stephanodiscus niagarae, S. minutulus, Fragilaria tenera,* and *Ulnaria ulna.*

B. *Carbonated unit:* a 90 cm-thick sequence composed mainly (80%) by carbonate shells of pulmonate gastropods, bivalves, and ostracods mixed with clay (metahalloysite), silt, volcanic ash, diatomitic sediments, charcoal, and fragments of fish. Ostracod species belong to four families (*Candonidae, Cyprididae, Darwinulidae,* and *Limnocytheridae*), that prefer shallow waters with aquatic vegetation, but can also be drought resistant. The most abundant gastropods correspond to genus *Physa,* associated to perennial water bodies. The diatomological analysis shows species that prefer turbid, slightly acid, and fluctuating waters (*Aulacoseira granulata* var. *Angustissima, Eunotia implicata, E. circumborealis,* and *Hanztchia amphioxys*).

C. *Volcanic ash-fall unit:* a 90 cm-thick layer of black volcanic ash with high contents of glass, scoria, and fragments of fish bones. This unit was related to the effusive activity of the Capáxtiro volcano, with an age of 200 to 80 BC (or 2144 to 2030 y BP) (Mahgoub et al. 2018).
D. *Wetland sediments:* composed by two units: (A) At the bottom a 1 m-thick sequence of silty-clay (montmorillonite) layers intercalated with ash-fall deposits (2 to 5 cm-thick) in lenticular geometries; this sequence contains circular bioturbation structures with 7 cm of diameter (filled with silts and gravel), oxidized scoria particles, dispersed charcoal, fragments of fish (bones and scales), and diatomite aggregates, composed mainly of two tycoplanctonic species: *Fragilaria tenera* and *Ulnaria ulna.* (B) The upper unit is a 1.1 m-thick sequence of sandy-silt layers alternated with ash-fall deposits (black and white volcanic ash). This sequence contains oxidized scoria particles and carbonate nodules; the diatomological analyses show species that prefer shallow, acid water habitats (including *Eunotia implicata, E. circumborealis, E. bilunaris, Gomphonema spp.*), accompanied by drought-resistant species (such as *Hanztchia amphioxys*) and species that indicate ionic concentration (such as *Anomonoeis sphaerophora*); toward the top, the diatomite content decreases, while organic matter and bioturbation structures increase, also fragments of P'urhépecha pottery were found.

Paleolimnological Conditions of the Basin

According to the stratigraphic record and proxies, we can infer four environmental stages in the Zacapu basin (Fig. 6): (1) A deep lake >40 ka BP, characterized by fine-grained sediments (clay and silt), an abundance of argillaceous minerals, and deep-water species of diatoms; (2) A shallow lake characterized by the change between two mineral phases (halloysite to metahalloysite), with the presence of carbonate nodules. In this environment, populations of ostracods, gastropods, and diatoms are common; all these species prefer basic, turbid waters with high content of dissolved salts. The transition to this stage could be explained by an increase in the rate of evaporation, a decrease in the rate of precipitation, or an increase of the sediment input to the lake, as observed by Ortega et al. (2002), with shallow conditions prevailing since 15 ka BP; (3) A wetland where laminar clayish silt layers were deposited and human activities became frequent; and (4) The final stage has been interpreted as part of historical sediments, because the stratigraphic units are mixed by human activities, organic matter contents increase, and fragments of P'urhépecha pottery dated from the Milpillas/Uricho phase, ancient burials, and other archeological remains are present (Figs. 5 and 6).

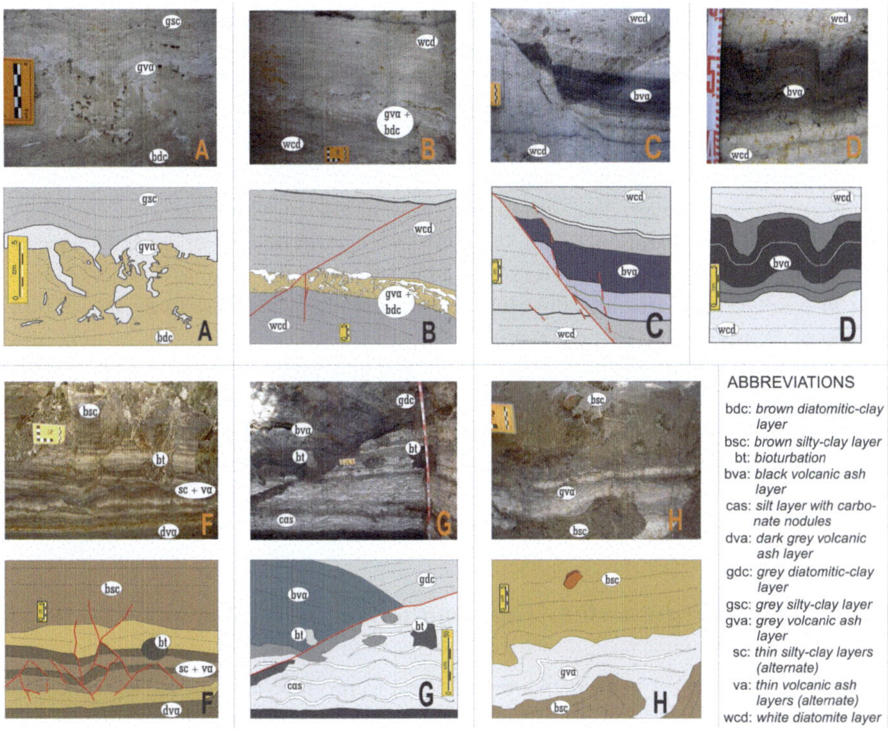

Fig. 7 Sedimentary deformations identified in the stratigraphic record of the trenches. Photo and interpretative sketch. Abbreviations refer to the layers identified in the field. (**a**, **b**) corresponds to SSD1, (**c**, **d**) corresponds to SSD2, (**f**) to SSD3, (**h**) to SSD4, and (**g**) to the slump zone

Record of Sedimentary Deformations in the Las Lomas Area

The sedimentary units previously described show different types of sedimentary alterations, which were classified according to their geometry and size (Fig. 7).

In the lacustrine sediments unit, we can observe a cross-bedding layer and also a liquefied layer, a folded layer, a deformed layer, and the presence of a normal fault plane which causes a 10°-rotation of the unit in the hanging wall (Fig. 7a–d). In the liquefied layer, the presence of a small dike of gray volcanic ash intruding a 10 cm-thick diatomitic layer stands out (Fig. 7a, b). The folded layer refers to symmetrical folds formed in a 12 cm-thick, black volcanic ash layer, delimited at the base and at the top by horizontal diatomitic layers; these folds have wavelengths of ~10 cm, rising up to 4 cm from the normal stratification (Fig. 7d). The deformed layer shows the loss of the original lamination, where a 5 cm-thick silt layer shows ripples and is covered by an 8 cm-thick horizontal clay layer.

The volcanic ash-fall unit unconformity overlays the carbonated unit forming a slump structure, particularly visible at the mounds sides. In turn, the carbonated unit shows current ripples (Fig. 7g). Slumps might have occurred due to destabilization of the sediments related to the uplift in the Las Lomas area, when the basin had shallow waters, causing the hiatus in the sedimentary record reported by other authors, or even abnormalities in its age due to reworking (Xelhuatzin 1994; Arnauld et al. 1997; Enzel et al. 2000; Ortega et al. 2002; Rodríguez-Pascua et al. 2010). Also, the slumps show a "tray effect" (Schillizi et al. 2010) where the involved sedimentary units slide over but retain their original lamination and sequence (a volcanic ash-fall layer, a carbonated layer, and a diatomitic-clay sequence with a liquefied layer).

In the wetland sediments unit, we observe a liquefied layer and widespread microfaulted layers. The first refers to a 10 cm-thick gray volcanic ash layer liquefied in a silt layer, forming convoluted stratification, where the majority of fragments of P'urhépecha pottery were found (Fig. 7h). Microfaulting occurs in a 30 cm-thick alternate sequence of brown silt and clay layers, causing 12 cm of vertical displacement of an 8 cm-thick gray volcanic ash layer. Microfaulted layers are delimited at the base by a horizontal, 10 cm-thick black volcanic ash layer, and at the top by a laminated sequence of clayey-silt layers (Fig. 7f). Load structures are visible at the top of this unit.

Origin of the Sedimentary Deformations

The sediments of the Las Lomas area are ideal to produce and appreciate deformation structures due to the contrast of black volcanic ash deposits in alternation with white clays and laminar silty-clay layers.

In the sedimentary record, we can differentiate two groups of structures: one of primary sedimentary structures (cross bedding and ripples), and the others, possibly related to sedimentation changes due to volcanic and tectonic activity in the basin (e.g., liquefaction structures, folding, microfaulting, and slumps). The second group has characteristics such as widespread spatial distribution, confinement to a single layer delimited by undeformed layers, association to fault planes, and liquefaction geometries that could suggest a tectonic origin, following a scenario where the vibration affected unconsolidated saturated sediments in the basin. For example, landslides in subaquatic conditions can be observed in turbiditic deposits, where a steep-slope topography allows the gravity to attract unconsolidated sediments. Notably, Las Lomas has subaquatic landslides despite its gentle-slope topography. These deposits were found along the flanks of the mounds and were composed of brecciated deposits and a mixture of young and old sediments. Slumps on smooth slopes may result from volcanic or tectonic deformation in the basin.

If we assume that structures of group two have a tectonic origin, we can use the considerations of Audemard et al. (2011) and group them as seismites (specifically type-A seismites), affecting sediments as old as 40 ka and as young as historical ages (sediments with P'urhépecha pottery). From the stratigraphic record we can deduce at least four ground-shaking events, produced either by tremors or earthquakes, which were recorded in the sediments. We suggest that the last two sedimentation disturbance events happened between 150 BC (2100 cal y BP) (Loma Alta 1 phase) and 1350 AD (Milpillas/Late Uricho phase).

Type-A seismites allow to indirectly estimate the magnitudes of the events that produced them. Different authors suggest that liquefaction structures are formed during earthquakes with a minimum magnitude of M ≥ 5, but we must also consider their proximity to historical volcanoes and active faults from the MAFS, which, in the case of Zacapu are really close (e.g., Allen 1986; Obermeier 1996; Lafuente et al. 2008; Sola 2010; Wetzler et al. 2010). As we mentioned earlier, the Zacapu basin has instrumental seismicity recorded at least since 1970 (http://www.ssn.mx); from that year and until 2017, the maximum frequency of earthquakes concentrates between 0 and <40 km depth, and the maximum magnitude recorded has been M 4.4 (Fig. 8), this indicate that Zacapu is a seismically active region where bigger earthquakes have occurred in the past and may occur in the future.

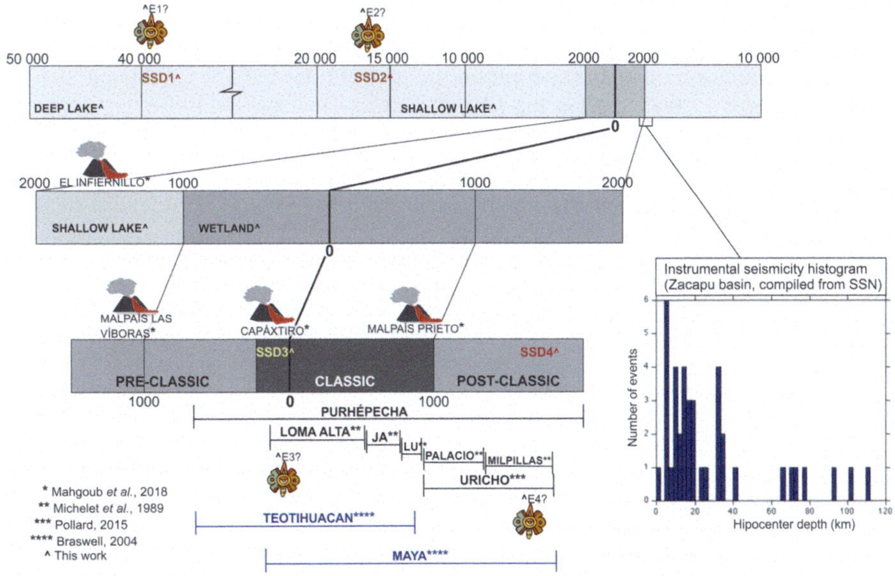

Fig. 8 Geological events and development of the P'urhépecha culture in the Zacapu basin and comparison with other indigenous cultures in México. Volcanoes refer to historical activity reported by Mahgoub et al. (2018). The abbreviations E 1? to E4? refer to a probable earthquake (*ollin* symbol) recorded by the sediments in the basin through syn-sedimentary deformations (SSD 1-4). Instrumental seismicity from 1970 to 2017 period compiled from the SSN catalog (www.ssn.mx). The dates for the occupation phases belong to the works of Michelet et al. (1989), Braswell (2004), and Pollard (2015). Abbreviations: JA Jarácuaro phase, LU Lupe phase

Coexistence of Human Settlements and Major Volcanic and Tectonic Events in the Zacapu Basin

According to the archeological studies in the Zacapu basin, the P'urhépecha people settled in the area since the late Preclassic until Postclassic Period, occupying the site for at least 1700 years. The beginning of these settlements, the Loma Alta phase (100 BC –150 BC or 2050 – 2100 cal y BP), coincides with the emplacement and activity of the Capáxtiro volcano and the uplift of Las Lomas (Fig. 8). During the Loma Alta phase, the evidence of density and size of settlements around the Zacapu and Cuitzeo basins suggest a population of 5000–8000 people (Pollard 2015).

At the beginning of the Uricho phase (or Palacio phase, Postclassic Period), the Malpaís Prieto volcano had intense pyroclastic activity, spreading a thick ash-fall layer over the basin, particularly over the Las Lomas area; during this phase, around 900 AD, the massive population on the malpaís begins (Migeon 1998; Villanueva 2019). The fragments of P'urhépecha pottery recovered from the liquefied and microfaulted layers of the wetland sequence belong to the Milpillas phase (or Late Uricho phase), suggested that a tectonic event occurred after the Malpaís Prieto activity (Fig. 8). At the time of the Late Uricho phase, a major settlement expansion was documented over the basins of Zacapu, Pátzcuaro, and Cuitzeo, reaching an estimated population of 48,000 people in the area (Pollard 2015; De Alcalá 2010), of which at least 12,000 lived in the Milpillas site in the Zacapu basin, surrounded by other small settlements in the Malpaís Prieto and Infiernillo sites (Michelet et al. 2012). These people could be affected by the volcanic and tectonic events referred. The tectonic event occurred during the Milpillas phase could correspond to that which is represented in the "Lienzo de Michoacán" (see Fig. 3), and also could be the event recorded in the island of Jarácuaro, SW of Pátzcuaro, where the residents moved two large stone blocks and placed them on the rupture surfaces caused by an earthquake as an offering to their gods (see Garduño-Monroy & Rodríguez-Pascua 2014).

Conclusive Remarks

An earthquake can modify the dynamics and morphology of a basin; therefore, to understand past tectonic and volcano-tectonic activity is fundamental to carry out a paleoenvironmental analysis supported by coseismic stratigraphy. The aim of coseismic stratigraphy is to approach the multidisciplinary studies in lacustrine basins evolved under the influence of active faults. In these scenarios, syngenetic sedimentary structures record the effects of a seismic event ($M > 5$), forming slumps, liquefaction structures, microfaulting, folding and sedimentation alterations such as uplift, hiatus, and age reversals in the dated material. Therefore, the accurate interpretation of the sedimentary record requires widespread trenches and sediment cores, distributed and collected around the basin.

In the Zacapu basin, stratigraphic and microtopographic data suggest that the lacustrine sequence was lifted up and likely deformed due to lateral spreading related to the tectonic activity in the basin. This process allowed the formation of the Las Lomas area, an important site for the development of P'urhépecha culture (Tricart 1992; Petrequin 1994). Las Lomas' stratigraphic sequence helps us understand and differentiate the secondary effects of tectonic events from climatic and anthropogenic effects on the basin.

The presence of sedimentary deformations with A-type seismites characteristics suggests the occurrence of four tectonic events, possibly moderate magnitudes earthquakes or volcanic tremors. The oldest of these events happened at least 40,000 years ago, and the youngest episode likely occurred during the Milpillas/Late Uricho phase of the P'urhépecha settlement affecting a large population in the area. These results suggest that seismic activity in the Zacapu region may have been underestimated.

The P'urhépecha culture was affected by the volcanic and tectonic activity in the Zacapu basin. They witnessed at least two volcanic eruptions and possibly two earthquakes, and even made offerings and dedicated pictorial records to these events. The Capáxtiro activity coincides with the start of the Loma Alta phase, at the end of the Preclassic period. There is significant evidence that the formation of the Las Lomas area was important for the success of the early P'urhépecha settlements.

Scenarios similar to that of the Zacapu basin are common in central México, where Quaternary volcanism, active faults, and detailed stratigraphic records coexist. The study of these scenarios is crucial to understand tectonic and volcanic processes along the Chapala-Tula fault system.

Research has so far focused on the effects of subduction earthquakes; however, cortical fault systems could be more devastating due to their shallowness and the dense urban populations of central México. For example, the effects of the Puebla earthquake occurred on September 19, 2017, showed that more seismic studies are required for better urban planning.

Acknowledgments The authors dedicated this work to the beloved memory of Victor Hugo Garduño Monroy who offered all his knowledge with great enthusiasm throughout his scientific career and specially to expand the paleoseismological research in Mexico. Funding to support this work came from the SEP-CONACYT-CB- 5222009-01-134151 project entitled "Tectonic, Paleoseismologic and Archeoseismologic study in Holocene to recent lakes of the Trans-Mexican Volcanic Belt and the Jalisco block." We thank Dr. Antonio Pola Villaseñor and Dr. Agnes Samper for their comments to improve this work.

References

Alfaro, P. (2006). Estructuras sedimentarias de deformación interpretadas como sismitas en el Mioceno Superior (Turoliense) de la cuenca de Granada (Cordillera Bética). *Geogaceta, 40*, 255–258. ISSN: 0213683X.

Alfaro, P., Delgado, J., Estévez, A., & López-Casado, C. (2001). Paleoliquefaction in the Bajo Segura basin (Eastern Betic Cordillera). *Acta Geológica Hispánica, 36*(3–4), 233–244.

Allen, J. R. L. (1986). Earthquake magnitude-frequency, epicentral distance, and soft- sediments deformation in sedimentary basins. *Sedimentary Geology, 46*, 67–75.
Álvarez, J. (1972). Ictiología Michoacana V. Origen y distribución de la Ictiofauna Dulceacuícola de Michoacán. *Anales de la Escuela Nacional de Ciencias Biológicas de México, 19*, 155–161.
Ambraseys, N. N. (1988). Engineering seismology. *Earthquake Engineering and Structural Dynamics, 17*(1), 1–105.
Amick, D., & Gelinas, R. (1991). The search for evidence of large prehistoric earthquakes along the Atlantic seaboard. *Science, 251*(4994), 655–658.
Antaramián, H. E., & Correa, G. (2003). Fisiografía. In G. Correa (Ed.), *Chapter II: Geografía Física. Atlas Geográfico del estado de Michoacán. UMSNH Colmich. SEP, Michoacán EDDISA S.A* (Vol. 2, pp. 1–454).
Arnauld, M. C., Carot, P., & Fauvet-Berthelot, M. F. (1993). Arqueología de las Lomas en la cuenca lacustre de Zacapu, Michoacán, México. *Cuadernos de Estudios Michoacanos 5,* México. Vol. 13, Centre D'Etudes Mexicaines Et Centramericaines. 1–233.
Arnauld, M. C., Metcalfe, S. E., & Petrequin, P. (1997). Holocene climatic change in Lake Zacapu basin, Michoacán: Synthesis of results. *Quaternary International, 43/44.*, Elsevier Science Ltd, 173–179.
Audemard, M., F.A, Michetti, A. M., & McCalpin, J. P. (2011). Geological criteria for evaluating seismicity revisited. Forty years of Paleoseismic investigations and the natural records of past earthquakes. *The Geological Society of America.* Special Paper, *479*, 1–21.
Ávila, J. N. G. (1985). La desecación de la ciénaga de Zacapu: orígenes y consecuencias. *Tzintzun: Revista de Estudios Históricos, 6*, 26–37.
Braswell, G. E. (2004). *The Maya and Teotihuacan: Reinterpreting early classic interaction* (430 pp). Austin: University of Texas Press.
Camacho, E., & Víquez, V. (1992). Historical seismicity of the North Panama Deformed Belt. Inst. Geoci., Universidad de Panamá [Inf. interno]. 50 pp.
Connor, C. B. (1987). Structure of the Michoacan - Guanajuato Volcanic Field Mexico. Journal of Volcanology and Geothermal Research, 33, 191 - 200
Davies, S. J., Metcalfe, S. E., MacKenzie, A. B., Newton, A. J., Endfield, G. H., & Farmer, J. G. (2004). Environmental changes in the Zirahuen Basin, Michoacan, Mexico, during the last 1000 years. *Journal of Paleolimnology, 31*, 77–98.
Davies, S. J., Metcalfe, S. E., & Israde-Alcántara, I. (2011). Reconstrucción de las interacciones entre el ser humano, el clima y el ambiente en los lagos de Michoacán. In M. Caballero & B. Ortega (Eds.), *Escenarios de cambio climático: registros del Cuaternario en América Latina I* (pp. 115–136). México: UNAM Press.
De Alcalá, J. (2010). *Relación de Michoacán* (314 pp). México: El Colegio de Michoacán.
Demant, A., Labat, J. N., Michelet, D., & Tricart, J. (1992). El Proyecto Michoacán 1983 –1987. Medio ambiente e Introducción a los trabajos arqueológicos. *Cuadernos de Estudios Michoacanos, 4*, CEMCA. 1–197.
Ego, F., & Ansan, V. (2002). Why is the central Trans-Mexican volcanic belt (102–99° W) in transtensive deformation? *Tectonophysics, 359*(1), 189–208.
Enzel, Y., Kadan, G., & Eyal, Y. (2000). Holocene earthquakes inferred from a fan-delta sequence in the Dead Sea graben. *Quaternary Research, 53*(1), 34–48.
Espinoza Jiménez J. A., & Baltazar Caballero, A. (2014). La desecación de los lagos en México en la época porfirista. Agua, Cultura y Sociedad en México. Napoleón Guzmán. In: http://www.purepecha.mx/threads/5742-La-desecaci%C3%B3n-de-los-lagos-en-M%C3%A9xico-en-la-%C3%A9poca-porfirista. Accessed on 14 Nov 2014.
Fisher, C. T., Pollard, H. P., Israde Alcántara, I., Garduño Monroy, V. H., & Banerjee, S. (2003). A reexamination of human induced environmental change within the Lake Pátzcuaro Basin, Michoacán, México. *Proceedings of the National Academy of Sciences., 100*(8), 4957–4962.
Garduño-Monroy, V. H., & Rodríguez-Pascua, M. A. (2014). Sismos, Ofrendas y Sacrificios-humanos. *Revista Saber Más.* Num. 13. In: https://www.sabermas.umich.mx/secciones/articulos/192-sismos-ofrendas-y-sacrificios-humanos.html. Accessed on 14 Nov 2014.

Garduño-Monroy, V. H., Chávez-Hernández, J., Aguirre-González, J., Vázquez-Rosas, R., Mijares Arellano, H., Israde-Alcántara, I., Hernández-Madrigal, V.M., Rodríguez-Pascua, M.A., & Pérez López, R. (2009). Zonificación de los periodos naturales de oscilación superficial en la ciudad de Pátzcuaro, Mich., México, con base en microtremores y estudios de paleosismología. *Revista Mexicana de Ciencias Geológicas, 26*(3), 623–637.

Garduño-Monroy, V. H., Israde-Alcántara, I., Rodríguez-Pascua, M. A., & Hernández- Madrigal, V.M. & Ortiz-Hurtado H.J. (2011). Eventos sísmicos de tiempos prehistóricos e históricos en cuencas lacustres: ejemplo, la cuenca de Pátzcuaro, Michoacán, México. In M. Caballero & B. Ortega (Eds.), *Escenarios de cambio climático: registros del Cuaternario en América Latina I* (pp. 91–114). México: UNAM Press.

Guarnieri, P., Pirrotta, C., Barbano, M. S., De Martini, P. M., Pantosti, D., Gerardi, F., & Smedile, A. (2008). Paleoseismic investigation of historical liquefactions along the Ionian coast of Sicily. *Journal of Earthquake Engineering, 13*(1), 68–79.

Hasenaka, T., & Carmichael, I.S.E. (1987). The cinder cones of Michoacán-Guanajuato, Central Mexico: petrology and chemistry. Journal of Petrology, 28 (2), 241–269.

Hernández Jiménez, A., Guilbaud, M. N., & Siebe, C. (2014). Complex eruption of Late Pleistocene Las Cabras scoria cone (Michoacán, México): From magma mingling to cone breaching. *IAVCEI - 5IMC. Abstracts, PS3-1.9*, 165–166.

Hibsch, C., Alvarado, A., Yepes, H., Perez, V. H., & Sébrier, M. (1997). Holocene liquefaction and soft-sediment deformation in Quito (Ecuador): A Paleoseismic history recorded in lacustrine sediments. *Journal of Geodynamics, 24*(1–4), 259–280.

Horváth, Z., Michéli, E., Mindszenty, A., & Berényi-Üveges, J. (2005). Soft-sediment deformation structures in Late Miocene – Pleistocene sediments on the pediment of the Mátra Hills (Visonta, Atkár, Verseg): Cryoturbation, load structures or seismites? *Tectonophysics, 410*(1–4), 81–95.

Israde-Alcántara, I., & Garduño – Monroy, V. H. (1999). Lacustrine record in a volcanic intra arc setting: The evolution of the late Neogene Cuitzeo basin system (central – Western Mexico, Michoacan). *Palaeo, 151*, 209–227.

Israde-Alcántara, I. (1997). Neogene diatoms of Cuitzeo Lake, central sector of the Trans-Mexican Volcanic Belt and their relationship with the volcano-tectonic evolution. *Quaternary International, 43*, 137–143.

Israde-Alcántara, I., Garduño-Monroy, V. H., Fisher, C. T., Pollard, H. P., & Rodríguez- Pascua, M. A. (2005). Lake level change, climate, and the impact of natural events: The role of seismic and volcanic events in the formation of the Lake Pátzcuaro Basin, Michoacán Mexico. *Quaternary International, 135*, 35–46.

Israde-Alcántara, I., Miller, W., Garduño-Monroy, V. H., & Barron, J. (2011). Estratigrafia y marco geodinámico de las cuencas lacustres del centro de México. In M. Caballero & B. Ortega (Eds.), *Escenarios de cambio climático: registros del Cuaternario en América Latina I* (pp. 73–90). México, DF: UNAM Press.

Koç Taşgin, C., & Türkmen, ˙I. (2009). Analysis of softsediment deformation structures in Neogene fluvio-lacustrine deposits of Çaybaği Formation, Eastern Turkey. *Sedimentary Geology, 218*(1–4), 16–30. https://doi.org/10.1016/j.sedgeo.2009.04.009.

Kuhn, G. G. (2005). Paleoseismic features as indicators of earthquakes hazards in North Coastal, San Diego County, California, USA. *Engineering Geology, 80*, 115–150.

Lafuente, P., Rodríguez-Pascua, M. A., Simón, J. L., Arlegui, L. E., & Liesa, C. L. (2008). Sismitas en depósitos Pliocenos y Pleistocenos de la Fosa de Teruel. *Revista de la Sociedad Geológica de España, 21*(3–4), 133–149.

Langridge, R. M., Persaud, M., Zúñiga, F. R., Aguirre-Díaz, G. J., Villamor, P., & Lacan, P. (2013). Preliminary paleoseismic results from the Pastores fault and its role in the seismic hazard of the Acambay graben, Trans-Mexican Volcanic Belt, Mexico. *Revista Mexicana de Ciencias Geológicas, 30*(3), 463–481.

Le Corvec, N., Spörli, K. B., Rowland, J., & Lindsay, J. (2013). Spatial distribution and alignments of volcanic centers: Clues to the formation of monogenetic volcanic fields. *Earth-Science Reviews, 124*, 96–114.

Lozano-García, S., & Sosa-Nájera, S. (2011). Dinámica de la vegetación en la cuenca de México durante el último ciclo glacial/interglacial. In M. Caballero & B. Ortega (Eds.), *Escenarios de cambio climático: registros del Cuaternario en América Latina I* (pp. 239–254). México: UNAM, Press.

Mahgoub, A. N., Reyes-Guzmán, N., Böhnel, H., Siebe, C., Pereira, G., & Dorison, A. (2018). Paleomagnetic constraints on the ages of the Holocene Malpaís de Zacapu lava flow eruptions, Michoacán (México): Implications for archeology and volcanic hazards. *The Holocene, 28*(2), 229–245.

Mennella, L. (2011). *Sismotectónica del sector occidental del Sistema Morelia-Acambay, México, a partir del análisis de poblaciones de fallas*. Master Thesis. UMSNH, Morelia, Mexico. 1–189.

Metcalfe, S. E., Brown, R. B., Hales, P. E., Perrott, R. A., Steininger, F. M., & Strett – Perrott, F. A. (1989). Late Holocene human impact on Lake Basins in Central Mexico. *Geoarchaeology an International Journal, 4*(2), 119–141.

Metcalfe, S. E., Street-Perrott, F. A., O'Hara, S. L., Hales, P. E., & Perrott, R. A. (1994). The Paleolimnological record of environmental change: Examples from the Arid Frontier of Mesoamerica. In A. C. Millington & K. Pye (Eds.), *Environmental change in drylands biogeographical and geomorphological perspectives. 131–145.Metcalfe, S.E. Changing environments of the Zacapu basin, Central Mexico: A diatom based history spanning the last 30 000 years*. Hull: School of Geography and Earth Resources, University of Hull.

Michelet, D., Arnauld, M. C., & Fauvet-Berthelot, M. F. (1989). El Proyecto del CEMCA en Michoacán, Etapa I: Un Balance. *Trace*. Centro de Estudios Mexicanos y Centroamericanos, *16*, 70–87.

Michelet, D., Forest, M., Bortot, S., & Darras, V. (2012). Almacenar en El Malpaís de Zacapu, Centro-Norte de Michoacan (1250–1450 d. C.). Séverine and Michelet; Dominique and Darras; Véronique. *Almacenamiento prehispánico. Del Norte de México al Altiplano central,* CEMCA, Universidad de San Luis Potosí, ARCHAM, Université de Paris 1. 129–140.

Migeon, G. (1998). El Poblamiento del Malpaís de Zacapu y de sus Alrededores, del Clásico al Posclásico", Génesis, Culturas y Espacios en Michoacán, Veronique Darras (coord.), Centro de Estudios Mexicanos y Centroamericanos, 35–45.

Moncayo Estrada, R., Israde Alcántara, I., & Garduño Monroy, V. H. (2001). La Cherahuita, Hubbsina turneri De Buen (1941) (Pices: Goodeidae). Origen, distribución y su uso en la regionalización de la cuenca del Lerma. *Hidrobiológica, 11*(1), 1–18.

Münn, S., Walter, T. R., & Klügel, A. (2006). Gravitational spreading controls rift zones and flank instability on El Hierro, Canary Islands. *Geological Magazine, 143*(3), 257–268.

Neuwerth, R., Suter, F., Guzman, C. A., & Gorin, G. E. (2006). Soft-sediment deformation in a tectonically active area: The Plio-Pleistocene Zarzal formation in the Cauca Valley (Western Colombia). *Sedimentary Geology, 186*, 67–88.

Norini, G., Capra, L., Borselli, L., Zuniga, F. R., Solari, L., & Sarocchi, D. (2010). Large scale landslides triggered by Quaternary tectonics in the Acambay graben, Mexico. *Earth Surface Processes and Landforms, 35*, 1445–1455.

O'Hara, S. L., Alayne, S. P. F., & Burt, T. P. (1993). Accelerated soil erosion around a Mexican highland lake caused by prehispanic agriculture. *Letters to Nature, 362*, 43–46.

Obermeier, S. F. (1996). Use of liquefaction-induced features for paleoseismic analysis: An overview of how seismic liquefaction features can be distinguished from other features and how their regional distribution and properties of source sediment can be used to infer the location and strength of Holocene paleoearthquakes. *Engineering Geology, 44*, 1–76.

Obermeier, S. F., Weems, R. E., & Jacobson, R. B. (1987). Earthquake-induced liquefaction features in the coastal South Carolina region. *U.S. Geological Survey Professional Paper, 1536*, 27.

Ortega, B., & Caballero, M. (2011). Registros lacustres del centro de México: una aproximación a los paleoambientes de los últimos 50 000 años. In M. Caballero & Ortega (Eds.), *Escenarios de cambio climático: registros del Cuaternario en América Latina I* (pp. 163–182). México: UNAM Press.

Ortega, B., Caballero, C., Lozano, S., Israde, I., & Vilaclara, G. (2002). 52 000 years of environmental history in Zacapu basin, Michoacan, Mexico: The magnetic record. *Earth and Planetary Science Letters, 202*, 663–675.

Park, J., Byrne, R., Böhnel, H., Molina-Garza, R., & Conserva, M. (2010). Holocene climate change and human impact, Central Mexico: A record based on maar lake pollen and sediment chemistry. *Quaternary Science Reviews, 29*, 618–632.

Pérez-López, R., Legrand, D., Garduño-Monroy, V. H., Rodríguez-Pascua, M. A., & Giner-Robles, J.L. (2011). Scaling laws of the size-distribution of monogenetic volcanoes within the Michoacán-Guanajuato Volcanic Field (Mexico). *Journal of Volcanology and Geothermal Research, 201*(1–4), 65–72.

Petrequin, P. (1994). 8000 años de la Cuenca de Zacapu. Evolución de los paisajes y primeros desmontes. *Cuadernos de Estudios Michoacanos, 6, CEMCA*, 1–144.

Phratiyal, B., & Sharma, A. (2009). Soft-sediment deformation structures in the Late Quaternary sediments of Ladakh: Evidence for multiple phases of seismic tremors in the North western Himalaya region. *Journal of Asian Earth Sciencies, 34*, 761–770.

Pollard, H. P. (2015). In 1522 the Tarascan king ruled over a domain of more than 75,000 km2 in the west-central highlands of Mexico, including the modern state of Michoacán. The Tarascan Empire was at that time the second largest in Mesoamerica and was ethnically dominated by a population the Spaniards called Tarascos, who spoke a language known as Tarasco or Purépecha. The. *From Tribute to Communal Sovereignty: The Tarascan and Caxcan Territories in Transition,* 92.

Reyes-Guzmán, N., Siebe, C., Chevrel, M. O., Guilbaud, M. N., Salinas, S., & Layer, P. (2018). Geology and radiometric dating of quaternary monogenetic volcanism in the western Zacapu lacustrine basin (Michoacán, México): Implications for archeology and future hazard evaluations. *Bulletin of Volcanology, 80*(18), 1–20.

Rodríguez-Pascua, M. A., Garduño-Monroy, V. H., & Israde-Alcántara, I. (2004). Evidencias paleosísmicas en fallas activas y sedimentos deformados del Lago de Pátzcuaro Jarácuaro, Michoacán, México. *Geotermas, 6*(3), 151–154.

Rodríguez-Pascua, M. A., Garduño-Monroy, V. H., Israde-Alcántara, I., & Pérez-López, R. (2010). Estimation of the paleoepicentral area from the spatial gradient of deformation in lacustrine seismites (Tierras Blancas Basin, Mexico). *Quaternary International, 219*(1), 66–78.

Schillizzi, R., Luna, L., & Falco, J. I. (2010). Estructuras de deformación (¿sismitas?) en la formación Río Negro, Provincia de Río Negro, Argentina. *Latin American Journal of Sedimentology and Basin Analysis, 17*(1), 17–32.

Seilacher, A. (1969). Fault - graded beds interpreted as seismites. *Sedimentology, 13*(1–2), 155–159.

Siebe, C., Guilbaud, M. N., Salinas, S., & Chédeville-Monzo, C. (2012). *Eruption of Alberca de los Espinos tuff cone causes transgression of Zacapu lake ca. 25,000 yr BP in Michoacán, México*. IAVCEI – CMV/CVS – IAS 4IMC conference. Auckland, New Zealand, pp. 74–75.

Siebe, C., Guilbaud, M. N., Salinas, S., Kshirsagar, P., Chevrel, M. O., De la Fuente, J. R., Hernández-Jiménez, A., & Godínez, L. (2014, 13–17 Nov). Monogenetic volcanism of the Michoacán-Guanajuato volcanic field: Maar craters of the Zacapu basin and domes, shields, and scoria cones of the Tarascan highland (Paracho-Paricutin region). In *Field guide, pre-meeting fieldtrip for the 5th international Maar conference (5IMC-IAVCEI)*, Querétaro, 33 pp.

Sola, F. (2010). Caracterización sedimentológica de un nivel guía y su posible origen sísmico, Corredor de Almanzora, Almería. *Geogaceta, 49*, 31–34.

Street-Perrott, F. A., Perrott, R. A., & Harkness, D. D. (1989). Anthropogenic soil erosion around Lake Pátzcuaro, Michoacán, México, during the Preclassic and Late Postclassic – Hispanic periods. *American Antiquity, 54*(4), 759–765.

Szynkaruk, E., Garduño-Monroy, V. H., & Bocco, G. (2004). Active fault systems and tectono-topographic configuration of the central Trans-Mexican Volcanic Belt. *Geomorphology, 61*, 111–126.

Takahama, N., Otsuka, T., & Brahmantyo, B. (2000). A new phenomenon in ancient liquefaction – The draw-in process, its final stage. *Sedimentary Geology, 135*, 157–165.

Tricart, J. (1992). La Cuenca Lacustre de Zacapu: Un acercamiento geomorfológico. En: *El Proyecto Michoacán 1983–1987. Medio ambiente e Introducción a los trabajos arqueológicos. CEMCA* 4. 115–197.

Tuttle, M. P., & Schweig, E. C. (1995). Archeological and pedological evidence for large prehistoric earthquakes in the New Madrid seismic zone, Central United States. *Geology, 23*(3), 253–256.

Tuttle, M. P., Dryer-Williams, K., & Barstow, N. L. (2002). Paleoliquefaction study of the Clarendon – Linden fault system, western New York State. *Tectonophysics, 353*, 263–286.

Van Wyk de Vries, B., & Matela, R. (1998). Styles of volcano-induced deformation: Numerical models of substratum flexure, spreading and extrusion. *Journal of Volcanology and Geothermal Research, 81*, 1–18.

Vanneste, K., Meghraoui, M., & Camelbeeck, T. (1999). Late quaternary earthquake-related soft-sediment deformation along the Belgian portion of the Felbiss Fault, Lower Rhine Gaben system. *Tectonophysics, 309*, 57–79.

Vázquez, G., Ortega, B., Davies, S. J., & Aston, B. J. (2010). Registro sedimentario de los últimos ca. 17000 años del lago de Zirahuén, Michoacán, México. *Boletin de la Sociedad Geológica Mexicana, 62*(3), 325–343.

Vilaclara, G., Silva-Romo, G., Cuna, E., Mendoza, C., & Robledo, R. (2011). Las cuencas lacustres neógenas de Tlaxcala como indicadoras de un pasado más húmedo en el trópico norteamericano. In M. Caballero & B. Ortega (Eds.), *Escenarios de cambio climático: registros del Cuaternario en América Latina I* (pp. 367–382). México: UNAM Press.

Villanueva, E. F. (2019). Algunas manifestaciones gráfico-rupestres de la Ciénaga de Zacapu, Michoacán. In *Las manifestaciones rupestres en México: técnica, iconografía y paisaje* (pp. 190–209). Enredars.

Watts, A. W., & Bradbury, J. L. (1982). Paleoecological studies at Lake Pátzcuaro on the West-Central Mexican Plateau and at Chalco in the Basin of Mexico. *Quaternary Research, 17*, 56–70.

Xelhuantzi López, M. S. (1994). Estudio palinológico de cuatro sitios ubicados en la cuenca de Zacapu: Fondo de la ciénega, Contacto loma – ciénega, Pantano interno y Loma alta. *Cuadernos de Estudios Michoacanos, CEMCA, 6*, 81–94.

Wetzler, N., Marco, S., & Heifetz, E. (2010). Quantitative analysis of seismogenic shear-induced turbulence in lake sediments. *Geology, 38*(4), 303–306.

Stratigraphy and Sedimentology of the Upper Pleistocene to Holocene Lake Chalco Drill Cores (Mexico Basin)

Blas Valero-Garcés, Mona Stockhecke, Socorro Lozano-García, Beatriz Ortega, Margarita Caballero, Peter Fawcett, Josef P. Werne, Erik Brown, Susana Sosa Najera, Kristin Pearthree, David McGee, Alastair G. E. Hodgetts, and Rodrigo Martínez

Abstract The Basin of Mexico is a high elevation (2240 m asl), large (9540 km^2), tectonic endorheic basin developed in the central-eastern Trans-Mexican Volcanic Belt. In 2016, the ICDP MexiDrill project recovered a total of 1152 m of sediments from a maximum depth of 520 m in Lake Chalco, in the SW of the Basin of Mexico. The upper 309.15 m (composite sequence, mc) are composed of fine-grained lacustrine sediments alternating with discrete visible tephra layers, and the lower 200 m are primarily volcaniclastic facies and basaltic lavas with some intercalated fluvial and alluvial facies. Initial lacustrine deposition started in the Chalco Basin, after the deposition of a thick volcaniclastic sequence, and continued until the lake

B. Valero-Garcés (✉)
Instituto Pirenaico de Ecología - Consejo Superior de Investigaciones Científicas (IPE- CSIC), Zaragoza, Spain
e-mail: blas@ipe.csic.es

M. Stockhecke (✉)
Large Lakes Observatory and Department of Earth & Environmental Sciences, University of Minnesota Duluth, Duluth, MN, USA

Eawag, Swiss Federal Institute of Aquatic Science and Technology, Surface Waters – Research and Management, Kastanienbaum, Switzerland
e-mail: mona.stockhecke@eawag.ch

S. Lozano-García · S. S. Najera · R. Martínez
Instituto de Geología, Universidad Nacional Autónoma de Mexico, Ciudad Universitaria, Coyoacán, Mexico

B. Ortega · M. Caballero
Instituto de Geofísica, Universidad Nacional Autónoma de Mexico, Ciudad Universitaria, Coyoacán, Mexico

P. Fawcett · K. Pearthree
Department of Earth & Planetary Sciences, University of New Mexico, Albuquerque, NM, USA

M. R. Rosen et al. (eds.), *Limnogeology: Progress, Challenges and Opportunities*, Syntheses in Limnogeology, https://doi.org/10.1007/978-3-030-66576-0_14

was drained in recent centuries. Lake Chalco has remained as a shallow lake until the present day.

Five main lithotypes throughout the core have been defined: (1) Organic, with (i) organic-rich silty clay, (ii) sapropelic silty clay, and (iii) peat; (2) Diatomaceous, with laminated, banded, and mottled diatomaceous silty clay; (3) Calcareous, with banded or massive carbonate silty clay and ostracod-rich layers; (4) Clastic, with silty clay, sand, and gravel, both clast and matrix-supported; and (5) Volcanic, as (i) volcaniclastic layers (tephras and clastic deposits) and (ii) lavas. Eighteen lithological units have been defined based on lithotypes and magnetic susceptibility values. Laminated diatomaceous facies occurred during phases with deeper depositional environments. Carbonate deposition marked more alkaline, shallower phases. Peat and organic-rich silts deposited during periods with better development of wetlands in the basin. The preliminary chronological framework suggests the lake sequence spans at least the last 367 ky. More saline diatom assemblages in diatomaceous facies and the occurrence of carbonate lithotypes and peat layers suggest relatively lower lake levels during interglacials; higher lake levels during glacial stages are indicated by more frequent freshwater diatom assemblages and the occurrence of finely laminated diatomaceous facies. The detailed stratigraphy and facies descriptions from the MexiDrill cores provide the basis for subsequent development of chronology and paleoenviromental studies of the Mexico basin record. Furthermore, the depositional model for Chalco is applicable to lakes developed in active tectonic and volcanic settings.

Keywords ICDP MexiDrill · Chalco Basin · Mexico Basin · Pleistocene · Holocene · volcanism · lake

Introduction

The pioneering work of Elizabeth Gierlowski and Kerry Kelts in the 1990s produced a global record of lake basins including some Quaternary examples (Gierlowski - Kordesch and Kelts 1994, 2001). During the last decades of the twentieth century and under the umbrella of the International Continental Scientific

J. P. Werne
Department of Geology & Environmental Science, University of Pittsburgh, Pittsburgh, PA, USA

E. Brown
Large Lakes Observatory and Department of Earth & Environmental Sciences, University of Minnesota Duluth, Duluth, MN, USA

D. McGee
Department of Earth, Atmospheric & Planetary Sciences, Massachusetts Institute of Technology, Cambridge, MA, USA

A. G. E. Hodgetts
School of Geography, Earth and Environmental Sciences, University of Birmingham, Edgbaston, Birmingham, UK

Drilling Program (ICDP, https://www.icdp-online.org/home/), a number of long Quaternary lacustrine sequences extending beyond the last glacial cycle have been made available worldwide: Lake Chad (Schuster et al. 2009), Qinghai Lake (Colman et al. 2007; An et al. 2012), Lake Baikal (Lake Baikal Paleoclimate Project Members 1992), Dead Sea (Stein et al. 2011), Bosumtwi (Koeberl et al. 2005), Lake Malawi (Scholz et al. 2007), Potrok Aike (Zolitschka et al. 2013), Lake Van (Litt et al. 2012), El'gygytgyn (Melles et al. 2012), Lake Ohrid (Wagner et al. 2014), and Towuti (Russell et al. 2016) among others. In the Americas, several lacustrine records have provided the basis for detailed reconstructions of the climate and environmental evolution in the tropical regions: e.g., Lake Titicaca (Fritz et al. 2007), Petén Itzá (Hodell et al. 2006), Chalco (Brown et al. 2012), and Junin (Rodbell and Abbott 2012). These long sequences span several glacial cycles and contain a detailed history of the origin and evolution of the lake basins, and provide reconstructions of regional vegetation, climate, and environmental change (Fig. 1a).

In this contribution, we summarize the stratigraphy of Lake Chalco in the Basin of Mexico. Investigations of lake sediments of the Valley of Mexico extend back to the 1950s (see references in Bradbury 1989) and most of the work at Lake Chalco was undertaken by researchers at the Universidad Nacional Autónoma de Mexico (UNAM) since the 1990s (see detailed references in Lozano-Garcia et al. 1993; Lozano-Garcia and Ortega-Guerrero 1994, 1998,; Caballero and Ortega Guerrero 1998; Caballero et al. 1999). These studies focused on relatively short (<25 m) cores spanning the last 40 kyr, utilized sedimentary facies, magnetic mineralogy, pollen, and diatom data and transfer functions to reconstruct paleoenvironmental conditions. They showed broadly that lake salinity was higher prior to 27 ka, decreased

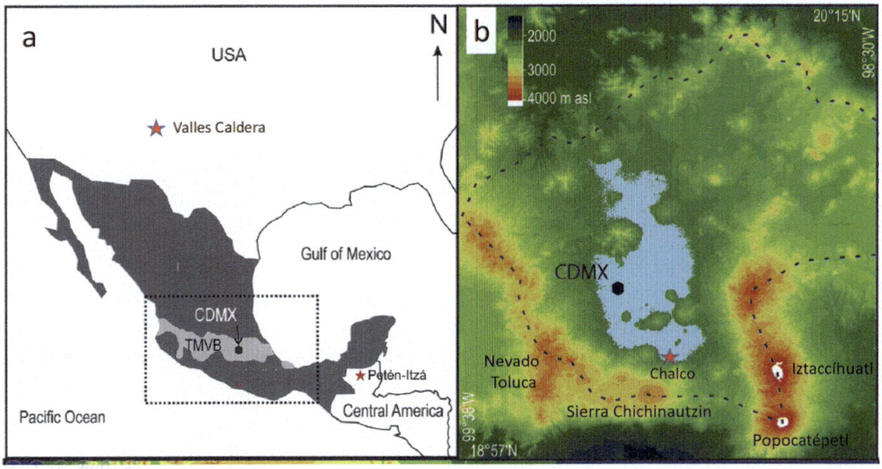

Fig. 1 (**a**) The Trans-Mexican Volcanic Belt (TMVB, light gray area), with the location of Mexico City (CDMX), and some long lacustrine records from the tropics and subtropics of North and Central America (Valles Caldera, New Mexico, USA; Lake Chalco (close to CDMX), Mexico); Petén Itza (Guatemala) (from Caballero et al. 2019). (**b**) Map of the Basin of Mexico showing maximum extent of Lake Chalco during the Pleistocene, Mexico City, and major volcanic centers within the Basin of Mexico and drill site location (red star)

during the Last Glacial Maximum and deglaciation, and increased again during the Early Holocene as temperatures increased. Several drought episodes occurred in Central Mexico during the Late Holocene and, even though human impact limits the interpretation of many records in the region, the late Classic drought (AD 600–1100) seems to have been the most important (Metcalfe et al. 2010). Longer cores from the Chalco Basin recovered in 2008 (>125 m) reached to the last interglacial and show a complex stratigraphy with numerous interbedded volcaniclastic deposits, abundant diatom-rich and laminated facies, and some carbonate-rich intervals (Avendaño 2017; Ortega et al. 2015; Lozano-García et al. 2015). In 2016, the complete lacustrine sequence of the Chalco Basin (>500 m thick) was recovered as part of the ICDP-UNAM-NSF-funded project MexiDrill-Chalco (Brown et al. 2019). In the following sections, we describe the stratigraphy and main lithotypes of the Chalco lacustrine sequence and explore some of the paleoclimate, paleoenvironmental, and basin development implications of this unique sedimentary record.

Basin of Mexico

The Trans-Mexican Volcanic Belt (TMVB) is the largest Neogene volcanic continental arc in North America (160,000 km^2 in extension, 1000 km in length) (Fig. 1) (Ferrari et al. 2012), and is built upon Cretaceous and Cenozoic magmatic provinces (Lugo-Hubp 1984; Lugo-Hubp et al. 1994) The Basin of Mexico is surrounded entirely by volcanic ranges: Sierra de las Cruces to the west includes the oldest geologic formations dated between 3.7 and 0.7 Ma (Delgado-Granados and Martin del Pozo 1993), Sierra de Pachuca and Tezontlalpan to the north (de Cserna et al. 1988), Sierra Nevada to the west (Macías et al. 2012), and Sierra Chichinautzin to the south (Martin del Pozzo 1982; Siebe et al. 2005). Several isolated volcanic structures and clusters are present on the plain, such as Sierra de Guadalupe (García-Palomo et al. 2006) and Sierra de Santa Catarina (Lugo-Hubp et al. 1994; Jaimes-Viera et al. 2018) (Fig. 1).

The Sierra Nevada consists of a chain of andesitic-dacitic stratovolcanoes in a 45 km long N–S alignment, extending from Tláloc in the North (4151 m asl.) to Popocatépetl in the South (5454 m asl.). Iztaccíhuatl (5285 m asl) sits North of Popocatépetl and is composed of a number of coalescing and superimposed edifices that began cone construction ca. 1.8 Ma ago but is now believed to be extinct (Nixon 1989; Macías et al. 2012). Popocatépetl is known to have built upon the remnants of at least three previous edifices (Siebe et al. 1995) with its oldest eruptive products dated since the last magnetic reversal ca. 750,000 years ago. Gravitational collapses, followed by subsequent rebuilding events, have been instrumental in the growth of Popocatépetl to the stratovolcano we see today. The most recent collapse has been dated between 24,000 and 29,000 cal yr BP. During the last 29,000 years, at least eight large Plinian eruptions have occurred in the volcanoes of Sierra Nevada, with seven from Popocatépetl. The Plinian fall deposit of Popocatépetl named "Pómez Tutti Frutti" (Siebe et al. 1997) erupted ca. 17,000 cal yr BP, and is currently the best stratigraphic marker in the Basin of Mexico (Sosa-Ceballos et al. 2012). It consists of juvenile dacitic pumice and polylithologic lithic componentry making it an easily recognizable fall (Siebe et al. 1995). Since December 21, 1994, activity at Popocatépetl began again, with frequent magmatic eruptions and paroxysms occurring from January 9, 2005. Regularly, tephra and gas are emitted and are visible from the surrounding municipalities. Dome growth and ballistics also pose a significant hazard to those who live around the volcano resulting in an exclusion zones of up to 12 km being enforced by the authorities.

The Sierra Chichinautzin extends from the southern end of the Sierra Nevada to the Nevado de Toluca volcano in the Lerma basin (west of the Basin of Mexico). It is a monogenetic volcanic field with more than 220 volcanic vents consisting of scoria cones and shield volcanoes. Products from these smaller scale volcanoes include lavas and tephra deposits covering an area of ca. 2500 km^2 (Bloomfield 1975; Martin del Pozzo 1982; Lugo Hubp 1984). The compositions of these products are more basic than the nearby polygenetic volcanoes, and generally erupt as andesites with subordinate basalts and dacites defining a calc-alkaline series (e.g. Siebe et al. 2005). The oldest products are 1.2 Ma (Arce et al. 2013a), and the youngest cone is Xitle volcano of an age of 1570 cal yr BP (Siebe 2000).

Among the ranges within the Basin of Mexico, the Sierra de Santa Catarina is prominent and divides the southern Xochimilco and Chalco subbasins from the northern Texcoco subbasin. This range is composed of seven volcanoes aligned E–W, which include cones and a maar with associated lavas and domes of andesitic composition (Lugo-Hubp et al. 1994; Arce et al. 2015). Lugo-Hubp et al. (1994) suggested a Holocene age of Santa Catarina structures based on their relatively young relief; however, $^{40}Ar/^{39}Ar$ dating of its products gives an age between 132,000 and 23,000 years (Jaimes-Viera et al. 2018).

The Nevado de Toluca stratovolcano (80 km to the SW of Mexico City) has had at least one eruption which has dispersed tephra across the Basin of Mexico. This eruption, producing the Upper Toluca Pumice, was Plinian and occurred ca. 12,300 yr BP (Arce et al. 2003). These deposits have been identified and correlated across the Basin of Mexico displaying thicknesses between 12 and 30 cm (Siebe et al. 1999; Ortega-Guerrero et al. 2015).

The Basin of Mexico is a tectonic depression bound by a series of fault systems. The Sierra Chichinautzin is a topographic high of a horst oriented E–W with the Xochimilco and Xicomulco normal fault system as its northern limit and the La Pera normal fault system as its southern limit (Siebe et al. 2004; Alaniz-Álvarez and Nieto-Samaniego 2005; García-Palomo et al. 2008). These extensional faults have served as preferred pathways for frequent, small batches of ascending magma that have successively added altitude to the Sierra Chichinautzin (Siebe et al. 2004). The Sierra de las Cruces was constructed over a N–S alignment and is constituted by three blocks limited by faults oriented E–W and NE–SW (García-Palomo et al. 2008). The existence of NW–SE-oriented normal faults has been inferred from deep wells in Mexico City in the inner part of the basin, crossing most of the metropolitan area. The limestone substrate has been located at different depths at Mixhuca and Tulyehualco wells suggesting the existence of an asymmetric graben limited by faults dipping to the SE and NW, respectively (Pérez-Cruz 1988). The closure of the ancient southward drainage of the Basin of Mexico by the Sierra Chichinautzin caused the development of a large lacustrine system that spans over the last 1.2 Ma and reached a maximum area of 1690 km^2 at an elevation of 2258 m asl (Ruiz-Angulo and López-Espinoza 2015). Continued subsidence since then has allowed accumulation of a thick sequence (~300 m) of Quaternary lacustrine sediment.

During the Quaternary, Popocatépetl and Iztaccíhuatl supported glaciers; however, a glacial chronology is only available for Iztaccíhuatl (White 1962, 1987), where six advances have been dated by ^{36}Cl between 205,000 and 1000 years ago (Vazquez-Selem and Heine 2011).

Lake Chalco

The lakes in the central Basin of Mexico were drained after the Spanish conquest for flood control, agriculture, and urban development. Documentary evidence shows that during the sixteenth century, prior to these modifications, the lake system covered an area of 1500 km^2 (Ezcurra 1990). Mexico City is built upon these lake beds. Presently, the plain of Lake Chalco is an agricultural region located in the southern subbasin of the Basin of Mexico in which the last remnant of the lake was drained during the late nineteenth and early twentieth centuries. The modern lake at Chalco covers an area less than 40 km^2 (Fig. 1b). It is a small alkaline marsh with an average depth of 3 m with turbid carbonate/bicarbonate, sodium-rich water (Lozano-García et al. 1993; Caballero et al. 2019). Although Lake Chalco falls within the administrative limits of Mexico City, it represents one of the largest remaining blocks of uninhabited land within the megalopolis and thus was identified as a target location for scientific drilling.

Surrounding vegetation has been severely affected by several thousand years of human occupation; the Basin of Mexico is one of the earliest human settlement sites in the Americas, with continuous agricultural activity since the Middle Holocene (Lozano-Garcia et al. 1993; Bhattacharya, and Byrne 2016). The lower elevation parts of the basin are used for agriculture and the foothills are covered by introduced grass and crops. *Pinus* spp. (pine) and *Quercus* spp. (oak) forests are present only in some of the highest areas. Halophytic vegetation currently dominates some of the shallow areas.

Mexico City depends on groundwater for >70% of its water supply. Since the 1950s, much of Mexico City, including the Chalco subbasin, has experienced accelerated ground subsidence with accompanying damage to urban infrastructure as a result of groundwater extraction from the regional aquifer that underlies the thick lacustrine aquitard on which the city is built. By 1991, total subsidence had reached 8 m with subsidence rates exceeding 350 mm/yr (Cabral-Cano et al. 2010). In the Chalco subbasin, these rates exceed 200 mm/yr and have favored the formation of a new lake on top of old lacustrine sediments. The extreme groundwater extraction rates and subsequent high subsidence rates pose a major challenge for a sustainable long-term water supply.

The Valley of Mexico is characterized by a high-altitude subtropical climate, with little temperature variability during the annual cycle and a rainy season from May to October, following the northward Intertropical Convergence Zone ITCZ migration during the boreal summer, although occasional polar fronts can bring precipitation during winter (Mosiño Aleman and Gracia 1974).

Coring the Chalco Basin

(a) *Geophysical surveys*

The subsurface stratigraphy of the Basin of Mexico is known from wells drilled in the metropolitan area to 2–3 km depth (Oviedo de León 1970; de Cserna et al. 1988; Pérez-Cruz 1988; Enciso-De la Vega 1992; Arce et al. 2013b, 2015). The basement rocks are Lower Cretaceous limestones and calcareous conglomerates and anhydrites, most likely of Eocene age. Overlying these deposits are volcanic rocks, which include mostly lavas and less abundant pyroclastic deposits, ages varying from Oligocene to Quaternary. Lacustrine sediments are found interlayered with lavas, breccias, conglomerates, and volcaniclastics. The older lacustrine sediments are present in most wells at depths between 750 and 300 m, although in Copilco-I well a ca. 100 m-thick lacustrine sequence was reported at 1600 m depth (Pérez-Cruz 1988).

A number of geophysical surveys were undertaken prior to the ICDP drilling campaign to characterize the sedimentary infilling of the basin (Brown et al. 2012, 2019). In the Chalco subbasin, near-surface conditions resulting from intensive agricultural development of the area complicate meaningful data collection with conventional seismic reflection surveys. Information from Bouguer gravity anomalies and passive seismic H/V spectral ratio methods, coupled with data from nearby water well logs, indicate the presence of a ca. 300 m "lacustrine sediment" package underlain by ca. 150 to 200 m of "granular material" (Ortiz-Zamora and Ortega-Guerrero 2007). This pattern has been interpreted as a graben bound to the north by the Sierra Santa Catarina and Cerro de la Estrella and to the south by a horst-shaped block composed of the Teuhtli and Xitle volcanoes (Campos-Enriquez et al. 2003). Geophysical surveys (Brown et al. 2012, 2019) provided a map of sediment thickness that was calibrated with measurements of resonance frequencies at the Santa Catarina water wells where the depth to the base of the lacustrine sediment package was known.

(b) *Previous coring campaigns and the ICDP MexiDrill*

In 2008 and 2011, six overlapping cores were drilled to different depths near the depocenter of ancient Lake Chalco: CHA08-II (1–27 m), CHA08-III (1–90 m), CHA08-IV (85–122 m), CHA08-V (29–72 m), CHA08-VI (70.8–85 and 106–122.4 m), and CHA11-VII (18 m). The cores CHA08 III to VI were drilled with a Shelby corer in 1.10 m sections with inner 10 cm diameter PVC tubes. The cores CHA08-II and CHA11-VII were drilled with a modified Livingstone piston corer in 1 and 2 m long sections.

The ICDP MexiDrill project goal was to recover the complete lake sequence of the Chalco Basin. During 6 weeks of continuous work in March–April 2016, the MexiDrill field campaign recovered more than 1000 m of drill core using wireline diamond coring techniques in four holes at a site identified by a suite of techniques as described above as the location of the thickest lacustrine sequence (Lozano-Garcia et al. 2017). The drillers, through rotational drilling techniques, reached depths of 420 m in core 1A, 310 m in 1B, and 520 m in 1C. Core 1D used a piston corer with percussion to enhance recovery of the upper soft sediments. A total of 1152 m of core sediments was recovered reaching a maximum depth of 520 m. Recovery ranged from 88% to 92% in the three cores (Fig. 2).

The upper 300 m of core 1A is composed of fine-grained lacustrine sediments alternating with discrete tephra layers, and the lower sections (ca. 300–422 m) are composed of volcaniclastics and lavas. Full downhole geophysical logs were collected from hole 1A. Cores 1B and 1C recovered the upper lacustrine sequence (313 m), and 1C continued through the volcaniclastic-lava unit reaching poorly sorted, coarse clastic deposits (ca. 512–520 m). Triple core recovery with offset core intervals in the upper 300 m lacustrine unit provides sufficient core overlap to create a composite sequence and continuous paleoenvironmental and volcanological reconstructions. In total, the project drilled 1262 m (including reamed sections) and recovered 1065 m of cores. (Fig. 2). The composite sequence is 554 mc (composite depth in meters) long including 52 m (9.4%) of gaps, mainly occurring within the lower coarse-grained volcaniclastic deposits, which were difficult to recover.

Lacustrine Facies

A number of clastic, carbonate, biogenic, chemical, and volcanic facies were defined in the Chalco 2008 cores (Herrera-Hernandez 2011). Clastic facies were composed of clay or silt with variable amounts of organic matter, varied colors (black brown, reddish), and with laminated, banded, and massive textures. Discrete occurrences of struvite (hydrated magnesium-ammonium phosphate) were identified at several levels. Biogenic facies included thin layers of diatomaceous ooze, thick layers of diatomaceous, organic-rich mud, and about 15 layers, between 5–10 cm thick, of ostracod-rich intervals. These composed more than 80% ostracod carapaces (ostracod *hash*). In the lower part of the sequence (between 113 and 122 m), up to 11 layers about 3–5 cm thick of carbonate mud mostly composed of endogenic carbonates (Mg-calcite, dolomite, and siderite) occurred. A number of tephra layers (lapilli and ash) were described and identified.

In the 2016 ICDP cores, lithotypes were defined based on lithologies and textures determined by macroscopic and microscope (smear slide) observations (Schnurrenberger et al. 2003; Table 1 and Fig. 3). In Table 1, the main features of the lithotypes are summarized and their associated depositional processes and environments interpreted. Figure 3 shows core images of selected intervals illustrating the main lithotypes identified in the MexiDrill cores.

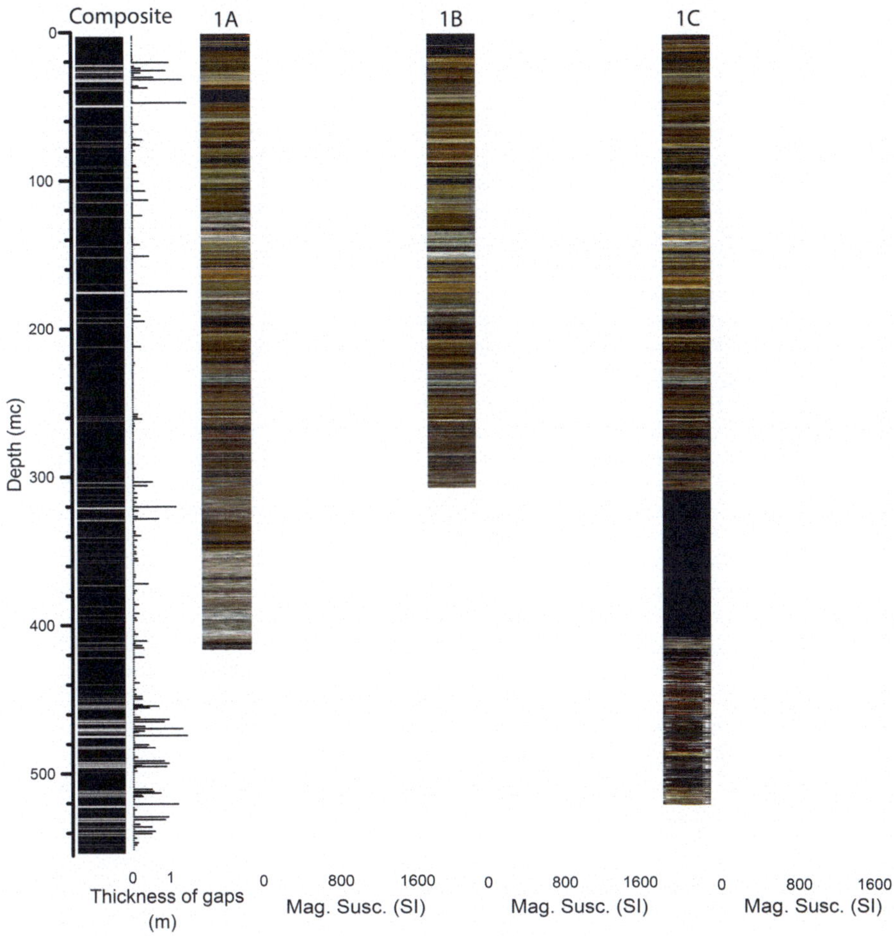

Fig. 2 Composite record from the Chalco drill cores and the ICDP MexiDrill cores including magnetic susceptibility and stacked color core images of the sections for site 1 cores (holes A, B, and C). White colored gaps of the composite record in the vertical column are given in thickness (m) by the bar chart

Total organic carbon (TOC) and inorganic carbon (TIC) analyses of selected samples are consistent with the three main types of lacustrine lithotypes: carbonate, organic, and diatomaceous (Fig. 4). The five main groups of lithotypes are:

- Organic. This includes: (i) organic-rich silty clay, (ii) sapropelic silty clay, and (iii) peat. The TOC range is quite large (up to 45%) and the TIC is below 1% in most samples.
- Diatomaceous. These are diatomaceous silty clays and include (i) laminated, (ii) banded, and (iii) "mottled" lithotypes. TIC content is generally lower than 1% and TOC reaches up to 4%.

Table 1 Chalco Basin lithotypes defined in the MexiDrill sequence: description, total thickness (in meter and percentage), and depositional processes and environments

Lithotypes	Description	Thickness	Depositional processes and environments
Organic-rich (O)	Organic-rich silty clay, banded or mottled in layers up to several dm. Organic matter includes sponge spicules, phytoliths, diatoms as well as terrestrial and amorphous aquatic OM. Lithotypes range from peaty to organic-rich silty clays. Some intervals contain abundant terrestrial OM clasts and variable amount of volcanic glass.	18 m 3.3%	Swamp/wetland to relatively shallow (and small) eutrophic lacustrine environment
Diatomaceous laminated (Dl)	Finely laminated diatomaceous silty clay with couples of light beige and dark brown laminae, light laminae (1–2 mm) contain almost only diatoms, dark laminae (2–5 mm) contain more clastic components. Thickness ranges from few cm to several dm.	44 m 7.8%	Relative deep and distal fresh-water lake undergoing strong likely seasonal changes and a at least partly anoxic water column Light laminae reflect deposition during diatom blooms Dark laminae reflect baseline deposition in the distal areas of the lake
Diatomaceous banded (Db)	Thin to medium bedded diatomaceous silty clay intercalated often by gray siliciclastic-rich 2–5 mm thin clay beds	80 m 14.4%	Distal, deep lake environments dominated by fine clastic and diatom deposition. Gray layers deposited during flood events transporting material from littoral areas and/or watershed.
Diatomaceous mottled (Dm)	Diatom-rich clayey silt with speckle of beige diatom clasts in a relatively homogenous matrix	1 m 0.2%	Distal but shallower environments with redox fluctuations responsible for mottling textures.
Carbonated banded (Cb)	Dm-thick beds composed of autochthonous carbonate minerals (Aragonite, Calcite), ostracod shells, few diatoms with etched surfaces and minor organics remains and clastic minerals; reddish intervals contain coarser carbonate grains, light gray intervals are Ostracod and Phacotus rich; dark gray intervals have more fine-grained (micritic) carbonate grains	57 m 10.4%	Shallower alkaline/ brackish environment dominated by carbonate precipitation and high biogenic productivity with fine carbonate clastic input from littoral areas

(continued)

Table 1 (continued)

Lithotypes	Description	Thickness	Depositional processes and environments
Carbonated massive (Co)	Mm to cm thick laminae composed mostly of biogenic or autochthonous carbonate material. Some dominated by ostracod shells and Phacotus (few layers almost purely of ostracod shells). Some are cemented if directly overlaying a v-layer	8 m 1.4%	Rapid carbonate deposition by endogenic processes (carbonate minerals) or biogenic accumulation (ostracods) in a highly productive system with limited detrital input
Coquina (C)	Bivalves, gastropods, and ostracods within a diatomaceous-silty clay matrix	0.6 m 0.1%	Biogenic accumulation in littoral areas
Silty clay (Sc)	Medium to thick beds of light gray silty clay dominated by silicate minerals and with variable diatom and carbonate minerals	18 m 3.2%	Distal, deep lake environments dominated by fine clastic deposition and with low bioproductivity
Sand (S)	Fine to coarse sand (carbonate, volcanic clasts) with silty clay matrix	1.4 m 0.2%	Alluvial (or close to alluvial-influenced lacustrine littoral) high-energy environment during a relatively low lake stand
Gravel (G)	Clast-supported, coarse sand matrix with rounded heterolithic pebbles, in massive, graded, and laminated layers		Alluvial /Fluvial, high-energy environment during relatively low lake stands
Tephra (T) and Lapilli (L)	Fine ash to lapilli pumice beds with sharp boundaries to over- and underlying silty clays		Reworked or fallout volcanic deposits from a volcanic source in vicinity to the lake
Volcaniclastic (V)	Coarse deposits (from gravel to sand) composed only of volcanic rock fragments and volcanic material matrix. Mostly matrix-supported; some clast-supported; massive, banded, and laminated structures. Dm to meter thick layers.	207.1 m 37.4%	Primary coarse volcaniclastic deposits and reworked volcanic materials by slope-failure, alluvial, mass wasting, and fluvial processes
Basalt (B)	Black to reddish vesicular Basalts	68.4 m 12.4%	Lava flows

- Calcareous. These are carbonate-bearing silty clay and include (i) banded and (ii) massive lithotypes. A unique lithotype is an ostracod and mollusk-rich hash or coquina at the base of the lacustrine sequence. Carbonate content covers a wide range, from 1% to almost 100% in some thin layers.
- Clastic. These include silty clay.
- Volcanic. These occur as (i) volcaniclastic layers as relatively thin tephra layers (ash and lapilli) and clastic deposits and (ii) lavas).

Fig. 3 (**a**) Core images of main lithotypes; vertical scale in centimeters: Organic-rich (O, MEXI-CHA16-1C-133Y-1 117.5–127.5 cm, MEXI-CHA16-1A-76Y-1 32–42 cm); Diatomaceous laminated (Dl; CHA08-VI-18 70–80 cm, MEXI-CHA16-1C-140Y-1 99–109 cm, MEXI-CHA16-1B-147Y-1 8–18 cm); Diatomaceous banded (Db, MEXI-CHA16-1A-169Y-1 83–93 cm, MEXI-CHA16-1C-167Y-1 110–120 cm); Diatomaceous mottled (Dm, MEXI-CHA16-1D-10C-2 30–40 cm); Carbonated banded (Cb, MEXI-CHA16-1D-11C-13 90–100 cm, MEXI-CHA16-1C-32Y-1 94–104 cm); Carbonated massive (Co, MEXI-CHA16-1C-174Y-1

Fig. 3 (continued) 132–142 cm); Coquina (C, MEXI-CHA16-1A-175Y-1 25–47 cm); Silty clay (Sc, MEXI-CHA16-1A-27Y-1 56–66 cm, MEXI-CHA16-1A-54Y-1 66–76 cm); Struvite (MEXI-CHA16-1C-18Y-1). (**b**): Core images of main lithotypes; vertical scale in centimeters Volcaniclastic (V, MEXI-CHA16-1C-178Y-1 49–59 cm, MEXI-CHA16-1C-182Y-1 13–23 cm), MEXI-CHA16-1A-269Y-1 93–103 cm, MEXI-CHA16-1A-271Y-1 105–115 cm, MEXI-CHA16-1A-175Y-1 54–76 cm, MEXI-CHA16-1D-8C-2-A 52–62 cm, MEXI-CHA16-1B-181Y-1 70–80 cm and MEXI-CHA16-1B-177Y-1 50–60 cm); Basalt (B, MEXI-CHA16-1C-231Y-1 50–60 cm, MEXI-CHA16-1C-217Y-3 10–20 cm, MEXI-CHA16-1C-270Y-1-A 50–60 cm)

Along the sequence, these main lithotypes show associations at several scales. At a decimeter scale, alternation of sapropelic, diatomaceous, organic (peat), and carbonate facies are common. Lithotypes group in thicker (multimeter scale) intervals that define lithostratigraphic units (see next section) and illustrate dynamic changes of the lacustrine basin encompassing a mosaic of depositional environments: shallow alkaline carbonate-producing lakes, deeper lakes with diatomaceous

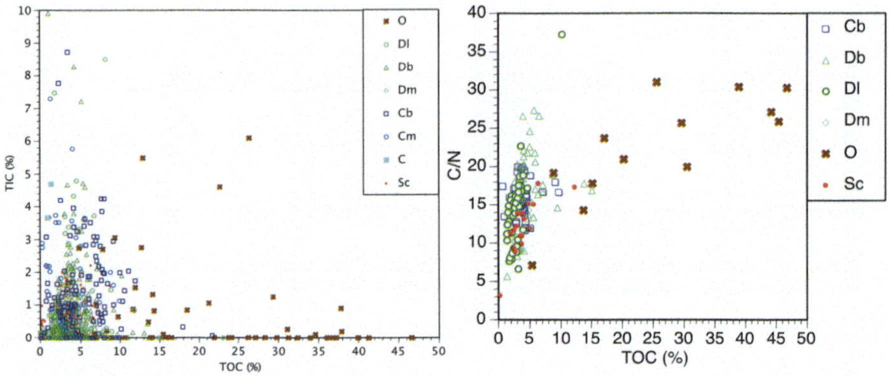

Fig. 4 (**a**) Total Inorganic Carbon (TIC) contents plotted against Total Organic Carbon (TOC) contents of main lithotypes: carbonate (blue boxes), organic (brown open circles), and diatomaceous (pink crosses) clayey silt. (**b**) TOC plotted versus C/N for the main lithotypes: carbonate (blue boxes), organic (brown open circles), and diatomaceous (pink crosses) clayey silt

deposition, wetlands with peat accumulation, and lakes dominated by organic deposition (sapropelic muds).

Stratigraphy of the Chalco Basin

Previous Stratigraphies Based on five shorter sediment cores, an initial stratigraphy was defined by Ortega-Guerrero (1992) for the upper 26 m of the sequence. The top unit 1 (0–1.8 m) is composed of brown silt with altered lapilli fragments, and more abundant ostracod and plant remains toward the base. Unit 2 (1.8–2.5 m) includes brown silt with abundant plant remains and ostracods. Unit 3 (2.5–3 m) includes diatomaceous facies with intercalated tephra intervals. Unit 4 (3–8 m) is composed of dark brown to black peat with abundant tephra layers. Brown silt dominates the lower units (5 to 7): with more plant fragments in Unit 5 (8–11 m), more gastropod shells in Unit 6 (11–16 m), and more ostracod levels and abundant lapilli layers in Unit 7 (16–26 m). The "Gran Ceniza Basáltica" at 21 m depth provides an important chronostratigraphic marker for the sequence (older than 31 ka, Ortega – Guerrero et al. 2015, 2017, 2018). Other key tephra layers are identified, including: Pómez Ocre (ca. 5 ^{14}C kyr); "Pómez Toluca Superior" or "sal y pimienta" (12,520 +/− 135 ^{14}C yr, Arce et al. 2003); "Pómez con Andesita" (ca. 14 ka) and "Pómez Tutti Frutti" (17,600 cal ka BP), Sosa Cebalos et al. 2012).

After the 2008 drill campaign, a new stratigraphy was developed for the upper 125 m with five main units (Herrera 2011): Unit 1 (0–8 m) groups organic-rich clay and silt and diatomaceous facies with several tephra layers. Unit 2 (8–68 m) is dominated by yellowish brown clayey silt (massive, banded, laminated) with more mottled intervals (subunit 2A (8–40 m)) and more frequent intercalation of gray silty clay layers (subunit 2B, 40–68 m). Unit 3 (68–90 m) contains olive brown and olive clay-silt with some thick intervals of reddish-brown silts and the presence of struvite (Fig. 3a) at some levels. Unit 4 (90–104 m) is characterized by the occurrence of olive brown clay-silt with some diatom ooze layers. Unit 5 (104–124 m) is defined by the abundance of laminated facies with alternation of carbonate-rich layers, diatomaceous facies, brown, gray clayey silt, and olive brown clayey silt.

A. *ICDP MexiDrill lithostratigraphy*

The 2016 ICDP MexiDrill drilling campaign obtained four long drill cores in the Chalco Basin, reaching up to 553 mc (composite depth) in the volcanic basement, and recovering the entire lake sequence (upper 295 mc). Magnetic susceptibility values supported correlation of the drill cores (Fig. 2). Cores were split and imaged at LacCore/CSDCO (University of Minnesota) and further correlation was achieved by visual stratigraphy using the CoreWall software. A 554 mc long spliced sequence was generated including core section from both the 2008 and 2016 projects. Eighteen lithological units have been defined (Table 2) based on lithotypes and magnetic susceptibility values (Fig. 5). The upper five units correspond to those defined in 2008 cores. The depth boundaries are those from the spliced composite core. Furthermore, a composite corrected depth scale (mcc) was developed to reflect relatively constant background lacustrine sedimentation by collapsing volcanic layer depths to zero thickness to account for instantaneous event deposition. Similar approaches have been used within other lake drilling records facing frequent volcaniclastic input (Deino et al. 2019; Stockhecke et al. 2014). The composite-corrected sequence totals in 241.68 mcc (Fig. 6).

Unit 1 (0–15 m) groups organic and diatom-rich lithotypes with relatively low clastic mineral content and includes a number of volcanic tephra that are well-correlated in the basin. It corresponds to Units 1 to 4 as defined by Ortega-Guerrero (1992) and Unit 1 as defined in the 2008 cores. A conspicuous black diatomaceous interval includes the Pómez Toluca Superior (also known as "Sal y Pimienta") tephra. The laminated and silty clay and dark brown massive and banded silty clay include some peat layers, and the Tutti Frutti Tephra is present in this unit as well. The environments associated with these lithotypes are relatively shallow lacustrine settings dominated by organic and diatomaceous sedimentation.

Unit 2 (15–90 m) is characterized by carbonate lithotypes. It broadly correlates with Unit 2 in the 2008 cores. The upper unit 2A (15–53.5 m) corresponds to Units 5 through 7 (Ortega-Guerrero 1992) and is dominated by banded calcareous lithotypes. Subunit 2B (53.5–90 m) contains more grayish- and greenish-brown calcareous clayey silt and the terrigenous content is higher with common fining upward sequences. Subunit 2A groups light brown/yellowish banded diatomaceous and calcareous silty clay with plant fragments, gastropod shells, and some ostracod-rich levels and mottled intervals. It also includes the "Gran Ceniza Basáltica" tephra. Facies in Unit 2 have textures indicative of shallow lake levels with occurrence of soft clasts, subaerial cracks, and mottling. Periods of higher terrigenous content suggest higher run-off or discrete flooding episodes; more alkaline conditions were conducive to periods with more abundant ostracods. Frequent cm-thick Fe-stained intervals suggest shifting in redox conditions during early diagenesis.

Table 2 Main lithostratigraphic units (as shown in Fig. 5) defined in the composite MexiDrill sequence using corrected composite depth (mc). The thicknesses of the main units in core 1A are also indicated

Unit	Dominant lithotypes	Base of unit (mc) Splice composite	Thickness (core1A)
1	Organic and diatomaceous (laminated, massive)	15	13.02
2A	Diatomaceous and calcareous (banded)	53.5	32.36
2B	Calcareous and diatomaceous (banded), more gray silt toward the base	90	30.39
3	Diatomaceous (laminated), calcareous and diatomaceous (banded)	103	11.92
4	Diatomaceous (laminated)	150	43.42
5	Volcaniclastic (mottled)	164	10.68
6	Calcareous, volcaniclastic, diatomaceous, and organic	174	9.51
7	Diatomaceous (laminated)	190	11.99
8	Calcareous	204.5	13.56
9	Diatomaceous (laminated)	210.8	5.6
10	Volcaniclastic (with some intercalated lake sediments)	218	18.53
11	Organic, calcareous, and diatomaceous	236.6	9.73
12	Diatomaceous (laminated) and organic	256.5	18.36
13	Volcaniclastic	269	14.87
14A	Volcaniclastic and diatomaceous		5.76
14B	Calcareous/Diatomaceous (laminated)		6.59
14C	Diatomaceous (laminated) and organics	292	10.54
15	Volcaniclastic, clastic, and diatomaceous	317	22.6
16	Volcaniclastic	447	128.48
17	Basalt lava	541	92.04
18	Volcaniclastic	555	12.15

Fig. 5 The complete synthetic lithostratigraphy of the spliced sequence of MexiDrill including 2008 and 2016 cores on composite depth (mc) on the right and the magnification into the acustrine sequence covering the upper 295 mc on the left. Preliminary age constraints all transferred on the composite depth consist of correlated previously ^{14}C-dated tephras (in gray text; the "Great Basaltic Ash" (older than 34 ka) (Ortega Guerrero et al. 2015, 2017, 2018), the "Pómez Toluca Superior (PTS)" dated as 12,520 +/− 135 14C yr BP (Arce et al. 2003), the "Pómez Tutti Frutti (PTF)" as 17,670 cal ka BP (Sosa-Ceballos et al. 2012) and "White pumice (WP)" as 27,800 cal BP (Siebe et al. 2017)). Radiocarbon dates of pollen (in green text, Ortega-Guerrero et al. 2018; Lozano-García et al. 2015), a U/Th date of zircon grain selected from tephra (in red text, Ortega Guerrero et al. 2017; Torres-Rodriguez et al. 2018), and a new U/Th age from lacustrine micritic carbonate (in blue text)

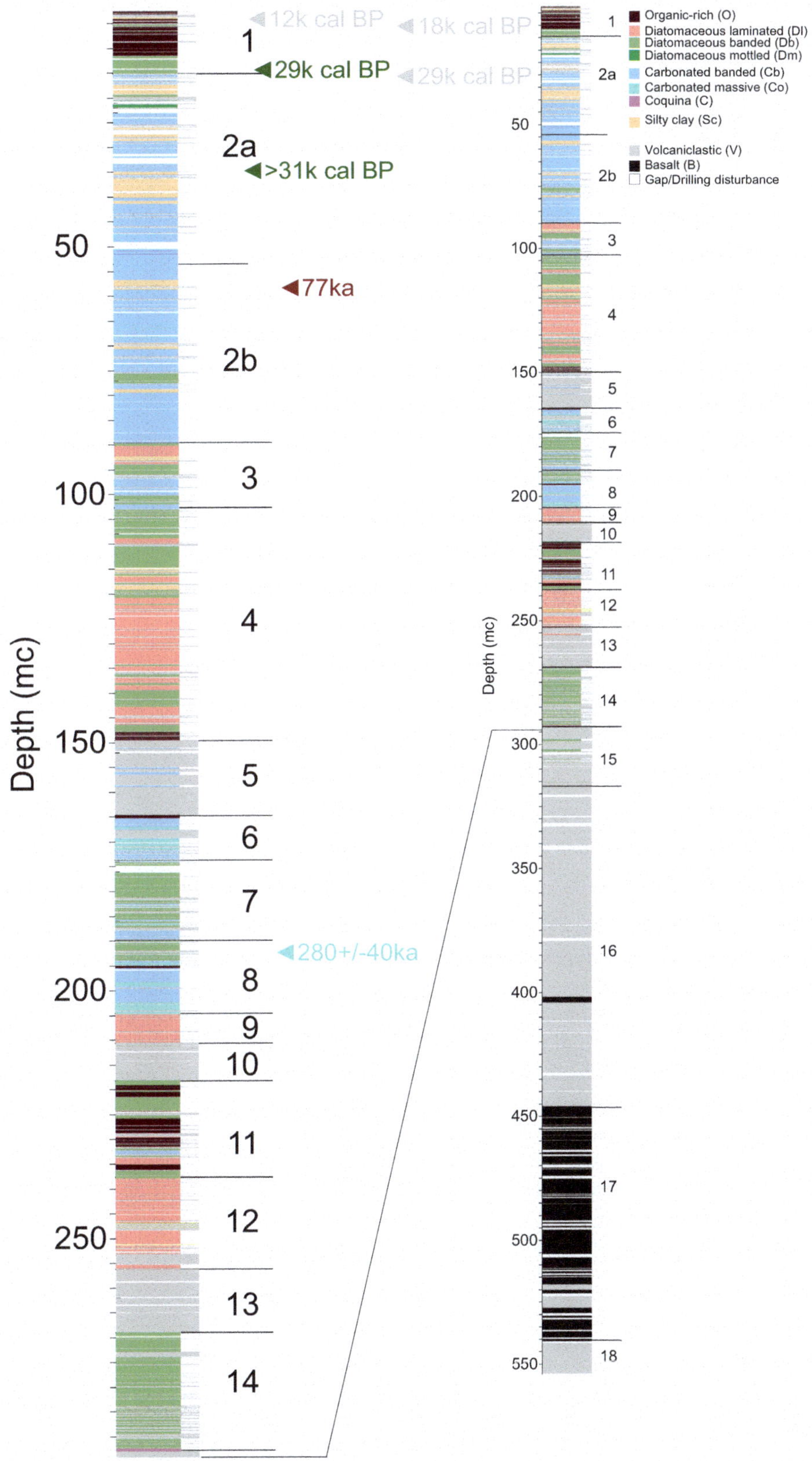

Depth (mc)
12k cal BP
18k cal BP
29k cal BP
29k cal BP
>31k cal BP
77ka
280+/-40ka
1
2a
2b
3
4
5
6
7
8
9
10
11
12
13
14
15
16
17
18
Organic-rich (O)
Diatomaceous laminated (Dl)
Diatomaceous banded (Db)
Diatomaceous mottled (Dm)
Carbonated banded (Cb)
Carbonated massive (Co)
Coquina (C)
Silty clay (Sc)
Volcaniclastic (V)
Basalt (B)
Gap/Drilling disturbance

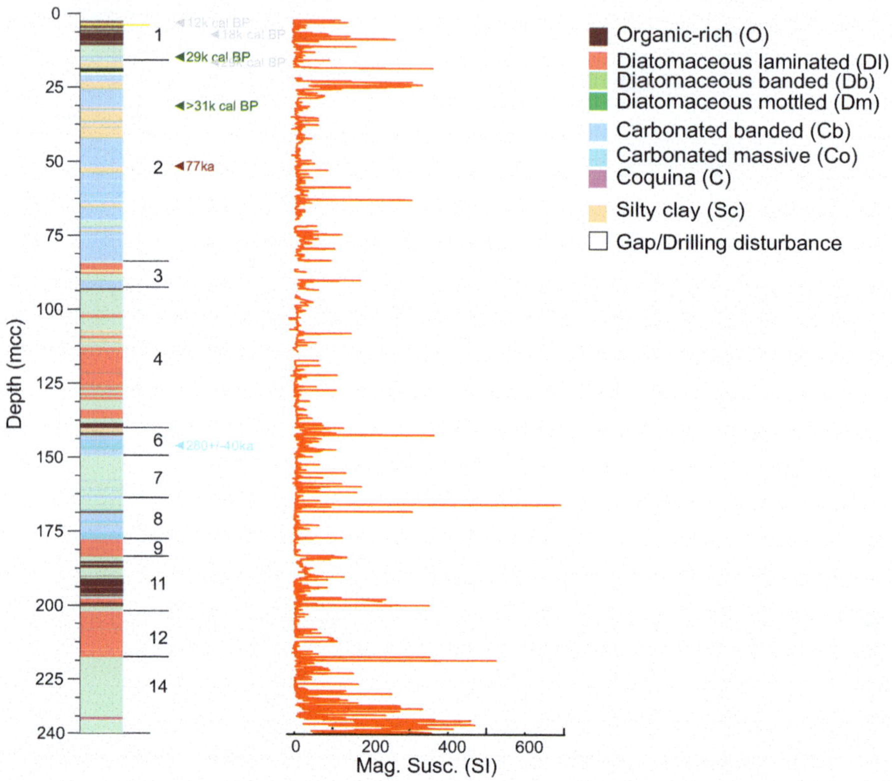

Fig. 6 The lithostratigraphy of the MexiDrill record on the composite corrected depth (mcc). Volcaniclastic layers have been collapsed in this version to show only the lacustrine sediments. Legend is shown in Fig. 5

Unit 3 (90–103 m) is characterized by the occurrence of carbonate and diatomaceous laminated lithothypes, with some reddish-brown carbonate silts. This unit presents several organic-rich levels with struvite occurrences that have been interpreted as guano accumulation during hyper-productivity phases (Pi et al. 2010). Reddish intervals have higher MS values and represent oxidized facies, likely during early diagenesis. The large facies variability (saline and freshwater diatoms, mottling, periods of high clastic input) points to significant hydrological changes during relatively low lake levels.

Unit 4 (103–150 m) is defined by the occurrence of alternations of finely laminated diatomaceous, carbonate, banded diatomaceous, and organic-rich lithotypes. Finely laminated facies are indicative of deeper lake environments and more frequent suboxic conditions. Carbonate deposition reflects periods with higher alkalinity, but still during relatively high lake levels as suggested by the finely laminated facies.

Unit 5 (150–164 m) is a volcaniclastic unit, composed of massive volcanic material with abundant secondary textures such as nodules, cracks, and staining. This may suggest subaerial deposition and/or reworking by surface processes of pyroclastic deposits, e.g., ignimbrites or lahars. The unit is topped by a tephra layer.

Unit 6 (164–174 m) is characterized by the occurrence of calcareous lithotypes (variegated, banded to laminated). They alternate with massive brown silty clays at the top and with laminated diatomaceous and more organic-rich intervals with secondary Fe-staining toward the base. A thick volcanic interval occurs in the middle of the unit.

Unit 7 (174–190 m) groups laminated to banded greenish diatomaceous silty clay and *Unit 8 (190–204.5 m)* contains carbonate-rich, massive, and banded lithotypes, with some reddish and brownish intervals.

Unit 9 (204.5–211 m) is composed of greenish laminated diatomaceous silty clay with more dark brown intervals toward the base.

Units 6 to 9 are arranged in alternating carbonate and diatomaceous units, as Units 6 and 8 are dominated by calcareous lithotypes, while Units 7 and 9 mostly include diatomaceous ones.

Unit 10 (211–220 m) is a thick volcaniclastic unit that includes two lake sediment intervals (organic-rich and diatomaceous silty clay).

Unit 11 (224–236.5 m) and *Unit 12 (236.5–256.5 m)* are composed of alternating organic-rich lithotypes (peaty silty clay) and diatomaceous (dark brown, massive diatomaceous silty clay) with some reddish banded carbonate and silty clay. In Unit 12, diatomaceous silty clay is dominant (with some greenish/brownish, banded/laminated intervals). Banded, brownish intervals are more frequent toward the base, then grading into greenish and better laminated.

Unit 13 (256.5–269 m) is a volcaniclastic unit with intercalated diatomaceous layers and tephra at the bottom.

Unit 14 (269–292 m) includes diatomaceous lithotypes with tephra layers, more abundant at the top, brownish/reddish, banded carbonated and laminated diatomaceous lithotypes, and dark brown diatomaceous silty clay intervals in the middle and laminated diatomaceous silty clays and oozes at the bottom At the base, a unique decimeter-thick ostracod and mollusk-rich layer (coquina) is found on top of a white pumice tephra, with overlying ash.

Unit 15 (292–317 m) contains interbedded m-thick fine volcaniclastic intervals, coarse clastic lithotypes deposited by mass transport, alluvial and fluvial processes, and dm-thick laminated brownish diatomaceous silty clay. The top volcaniclastic layer is a conspicuous white, laminated, pumice-rich layer.

Units 16 to 18 are the basal volcanic and volcaniclastic units.

Unit 16 (317–447 m) is a thick volcaniclastic unit composed of m-thick subunits, displaying polylithologic clasts among a matrix of varying coarseness.

Unit 17 (447–541.7 m) is composed of lava with laminated, massive, vesicular textures.

Unit 18 (541.7–551 m) is a volcaniclastic unit, with distinct clastic horizons. Largely, the deposit appears to be fining with stratigraphic height.

Evolution of the Chalco Basin

(a) *Depositional evolution*

The lower 200 m of the Chalco sequence are composed of volcaniclastic deposits and lavas. The volcaniclastic deposits of basal unit 18 could be associated with a period of volcanic activity at the end of the volcanic-tectonic phase responsible for the formation of the modern Chalco Basin. Emplacement of nearly 100 m of lava with vesicular textures (unit 17) followed this early episode of volcanic activity. Additional volcanic events occurred leading to the deposition of a thick interval (almost 130 m) of volcaniclastic facies (unit 16) that continued with the infilling of the Chalco Basin. Thick, massive layers of coarse to fine matrix-supported clasts were deposited as alluvial activity dismantled either existing volcanic structures or deposits associated with major evolution events from nearby polygenetic volcanoes.

Unit 15 represents the onset of the infilling of the newly formed Chalco Basin by lacustrine depositional processes. This interval has been described and interpreted in detail by Martinez-Abarca (2019) and Martínez-Abarca et al. (2021). During this period, input of volcaniclastic material continued. Later on, increased efficiency of erosion is indicated by the occurrence of rounder clasts and clast-supported textures in clast-rich sections. Smaller sizes and fine lamination textures in the finer-grained layers indicate an increased influence from a hydrologic regime. A fluvial-lacustrine setting is interpreted as the depositional environment for the upper part of unit 15, followed by the final development of the Chalco Lake (Martínez-Abarca et al. 2021). The lacustrine phase in the Chalco Basin started at the top of unit 15 with deposition of banded diatomaceous silty clay intercalated with fluvial deposits. Diatom assemblages indicate freshwater conditions in this newly formed lake (Martínez-Abarca et al. 2021). The intercalation of lacustrine and alluvial facies suggests that the alluvial and fluvial drainage networks carved the volcanic watersheds and connected to the lake environment. A relatively large number of tephras interbedded with the lake sediments suggest a high frequency of explosive eruptions occurred which transported deposits to the Chalco region. The occurrence of a unique ostracod, mollusk, and gastropod coquina (Unit 14C) indicates periods of high bioproductivity in freshwater with the presence of hard substrates on the lake floor. A stepwise development with early shallow lake stages have been described in other volcanic basins (Pueyo et al. 2011; Martínez-Abarca et al. 2021). Higher lake levels and deeper environments led to deposition of laminated diatomaceous lithotypes (unit 14 B and A). Unit 13 demonstrates volcanic activity was key producer of material here, leading to the deposition of more volcaniclastic deposits. Lake deposition continued with laminated diatomaceous and organic facies (Units 12 and 11) interrupted by deposition of another thin volcaniclastic interval (Unit 10). Fine-grained lake deposition continued with alternating diatomaceous (Units 9 and 7) and carbonate (Units 8 and 6) dominated phases. After emplacement of another volcaniclastic interval (Unit 5), the lake deepened and laminated carbonate, diatomaceous, and organic lithotypes (Unit 4) were deposited indicative of more anoxic, alkaline, and productive conditions. Deposition of banded to massive facies dominated Unit 3, suggesting relatively low lake levels. The top of this unit marks the last deposition of finely laminated diatomaceous facies.

Unit 2 is dominated by carbonate lithotypes and some silty facies which demonstrate high magnetic susceptibility. During deposition of Unit 2, lake levels continued to be relatively shallow and the lake experienced periods of higher alkalinity, conducive to precipitation of carbonates and deposition of some layers with abundant ostracods and gastropods. Terrigenous content progressively increased toward the top as indicated by higher magnetic susceptibility values and higher clastic content. The presence of intense mottling and root structures suggests multiple episodes of extremely low lake levels with subaerial exposure of the sediment at this site.

Unit 1 represents deposition in a lake dominated by sedimentation of organic-rich clay, silt, and diatomaceous facies. During this period, biogenic activity dominated over detrital watershed input.

(b) *Paleoenvironmental, paleovolcanism, and paleoclimate implications*

A preliminary chronological framework for the Chalco sequence has been established using a variety of dating techniques in several cores (Herrera 2011; Torres-Rodriguez et al. 2015, 2018; Martínez-Abarca 2019; Caballero et al. 2019) including:

- tephra layers: the Pómez Ocre (ca. 5 ^{14}C kyr), the Upper Toluca Pumice (UTP, 12,520 +/− 135 ^{14}C yr) (Arce et al. 2003), the Tutti Fruti Pumice (TFP, 17,670 cal ka BP) (Sosa-Ceballos et al. 2012), the "Great Basaltic Ash" (older than 34 ka), the "Pómez con Andesita" (ca. 14 ka) (Ortega Guerrero et al. 2015, 2017, 2018), and the White pumice (WP, 27,800 cal BP) (Siebe et al. 2017).
- Radiocarbon ages in pollen extracts from samples of the upper units (Ortega Guerrero et al. 2017; Torres-Rodriguez et al. 2018).
- U/Th date (76.74 ka) obtained from zircons collected from an ash layer at 69.84 mcc (75.28 mc in the MexiDrill-Chalco composite sequence) (Ortega Guerrero et al. 2017; Torres-Rodriguez et al. 2018).
- U/Th date obtained from a micritic carbonate sample containing few ostracods from 192.63 mc (165.73 mcc) (CHA16-1C_107Y-1 5.5 to 7.5 cm) yielding an age of 280 +/− 40 ka.

Extrapolation of sedimentation rates based on these preliminary chronological constraints suggests that onset of lacustrine deposition (top of Unit 15, 240.34 mcc, 309.15 mc) occurred between 364 and 505 ka, suggesting that the MexiDrill record will extend back to MIS 10 or MIS 13. This basal date was estimated by linear extrapolation of the two U/Th dates (zircon at 69.85 mcc and carbonate at 165.73 mcc) and considering the 40 ka error for the carbonate U/Th age. Ongoing work using 40Ar/39Ar dating of tephra and more radiocarbon dates in upper cores will refine this chronology. More saline diatom assemblages in diatomaceous facies and the occurrence of carbonate lithotypes and organic-rich layers suggest more arid conditions during interglacials, whereas higher humidity during glacial stages is indicated by freshwater diatom assemblages and the occurrence of finely laminated diatomaceous facies.

This preliminary view of glacial–interglacial conditions is consistent with the mid-Pleistocene sequence from Valles Caldera, New Mexico (Fawcett et al. 2011). In that sequence, the driest conditions occurred during the warmest phases of interglacials, with a significant reduction in summer precipitation, possibly in response to a poleward migration of the subtropical dry zone. In the South American tropics, the Lake Titicaca sequence showed a distinctive pattern during the last 400 ka of clastic, glacial-related facies deposited during wetter periods associated with glacier advances and relatively high lake level stages, and more carbonate-rich facies deposited in shallower conditions during warmer and relatively less humid interglacials (Fritz et al. 2007). In the shallow Lake Junín system, depositional variability during the last 700 ka is characterized by alternating intervals of carbonate lake-wetland facies (interglacial periods) and clastic-facies (glacial) (Cheng et al. 2020).

For the last glacial cycle, the age model is more robust. A number of studies in Lake Chalco indicate that the lake had high salinity prior to 27 ka BP and during the Last Glacial Maximum (LGM, 23 to 17 ka cal BP), and then deepened during the deglaciation, between 15 and 11 ka BP. During the Early Holocene, the lake became shallower and saline. These results suggest that lake chemistry is highly controlled by evaporation, with low evaporation periods, such as the LGM, showing lower salinity than intervals with high evaporation (end of MIS 3 and early MIS 1) which are saline. However, at Chalco, the LGM (27–15 ka cal BP) shows lower lake level conditions compared to the late glacial, consistent with lower fluxes of terrigenous material into the Cariaco Basin (Peterson et al. 2000).

In contrast, the results from the scientific drilling in Lake Peten Itza (ICDP PISDP project, Yucatan Peninsula, Guatemala) (Hodell et al. 2006) showed wet conditions during the Last Glacial Maximum, a transition to drier conditions during the late Glacial (including the Younger Dryas), and a generally wet Holocene, illustrating an opposite trend at the orbital scale. Interestingly, during the last 40,000 years at a millennial scale, the Chalco record has been interpreted to show a pattern of cold and dry conditions during colder northern hemisphere periods (stadials), comparable to other tropical northern hemisphere sites, further supporting the idea that there is a strong coupling between tropical and North Atlantic regions (Caballero et al. 2019).

Volcanic activity has played a significant role in the origin and evolution of Lake Chalco. Multiple eruptions from a variety of source volcanoes are present within the record. Early volcanic output (Units 18 to 15, Fig. 5) was responsible for the deposition of over 200 m of volcanic and volcaniclastic material that originated and partially filled the basin. From Unit 13, over 16 m of primary and reworked tephra was deposited during a mostly stable lacustrine setting. From Unit 10, over 19 m of volcanic deposits occur within the core and from Unit 5, there is ca. 11 m of volcanically derived deposits. From Unit 2, volcaniclastic input had ceased, but many discrete tephra layers 1–50 cm can be identified; two of which are from well-documented eruptions from Popocatépetl and Nevado de Toluca volcanoes. Uplift of the Santa Catarina Sierra and the Chichinautzin Sierra volcanoes have mostly likely contributed to higher sediment input from these watersheds. Ortega-Guerrero et al. (2015) suggest that the monogenetic volcano Teuhtli, in the southwestern part of the Chalco lacustrine plain, is the origin of the so-called Gran Ceniza Basáltica tephra (dated older than 34 ka, Ortega Guerrero et al. 2015, 2017, 2018). Volcanic input occurred during the last 23 ka from Santa Catarina to the north, Xico to the east, and Teuhtli volcano to the west and are mostly likely sources of some of these tephras. Ongoing work on the tephrastratigraphy and volcanological history of the Mexico City region aims to address the volcanic contributors to the Basin of Mexico and their associated hazard implications for the region today.

Conclusions

The cores from the Basin of Mexico provide a continuous lacustrine record from tropical North America for the Upper Pleistocene and the early to mid-Holocene, including several glacial cycles and volcanic events. The >500 m long MexiDrill sequence illustrates the complex relationship between hydroclimate and volcanic activity in the evolution of lacustrine basins. The Chalco Basin originated as a result of volcanic activity and was infilled in the early stages by mostly volcaniclastic facies and lavas reworked by fluvial and alluvial processes. The thickness of this basal unit is over 200 m at the coring site and our preliminary chronology suggests a minimal age of 367 ka. Initial lacustrine deposition started in the Chalco Basin after this fluvial–alluvial phase and has continued into historic times. Deposition was dominated by alternating diatomaceous, carbonate, and organic-rich facies on scales of 10s of meters. Laminated diatomaceous facies occurred during phases with relatively deep depositional environments. Carbonate deposition marked more alkaline phases. Peat and organic-rich silt were deposited during shallow phases with development of wetlands in the basin.

The preliminary results of the MexiDrill cores show the high potential of the Chalco sequence to provide detailed information on millennial scale variability, the climate during past interglacials, and the relationships among rates of climate change, ecosystem responses, and biodiversity, as well as volcanic and tectonic evolution as the geochronology improves. The cores hold geologic, paleoenvironmental, and paleobiological information that is directly relevant to the >25 million people living in the Basin of Mexico who face challenges related to climate change, volcanic hazards, seismicity, and hydrology. Knowledge of the complex history of past environments can aid in developing policies to mitigate regional consequences of a warming world.

Acknowledgments The MexiDrill Field campaign was funded by ICDP, US National Science Foundation, and UNAM DGAPA-PAPIIT IV100215. M. S. gratefully acknowledges support from the Swiss National Science Foundation grant P300P2 158501.

References

Alaniz-Álvarez, S. A., & Nieto-Samaniego, A. F. (2005). El sistema de fallas Taxco-San Miguel de Allende y la Faja Volcánica Transmexicana, dos fronteras tectónicas del centro de Mexico activas durante el Cenozoico. *Boletín de la Sociedad Geológica Mexicana, 57*(1), 65–82.

An, Z., Colman, S. M., Zhou, W., Li, X., Brown, E. T., Jull, A. J. T., Cai, Y., Huang, Y., Lu, X., Chang, H., Song, Y., Sun, Y., Xu, H., Liu, W., Jin, Z., Liu, X., Cheng, P., Liu, Y., Ai, L., Li, X., Liu, X., Yan, L., Shi, Z., Wang, X., Wu, F., Qiang, X., Dong, J., Lu, F., & Xu, X. (2012). Interplay between the Westerlies and Asian monsoon recorded in Lake Qinghai sediments since 32 ka. *Nature Scientific Reports, 2*, 1–6.

Arce, J. L., Macías, J. L., & Vázquez-Selem, L. (2003). The 10.5 ka Plinian eruption of Nevado de Toluca volcano, Mexico: Stratigraphy and hazard implications. *Geological Society of America Bulletin, 115*, 230–248.

Arce, J. L., Layer, P. W., Lassiter, J. C., Benowitz, J. A., Macias, J. L., & Ramírez-Espinosa, J. (2013a). 40Ar/39Ar dating, geochemistry, and isotopic analyses of the Quaternary Chichinautzin volcanic field, south of Mexico City: Implications for timing, eruption rate, and distribution of volcanism. *Bulletin of Volcanology, 75*, 774.

Arce, J. L., Layer, P. W., Morales-Casique, E., Benowitz, J. A., Rangel, E., & Escolero, O. (2013b). New constraints on the subsurface geology of the Mexico City Basin: The San Lorenzo Tezonco deep well, on the basis of 40Ar/39Ar geochronology and whole-rock chemistry. *Journal of Volcanology and Geothermal Research, 266*, 34–49.

Arce, J. L., Layer, P., Martínez, I., Salinas, J. I., Macías-Romo, M. D. C., Morales-Casique, E., & Lenhardt, N. (2015). Geología and estratigrafía del pozo profundo San Lorenzo Tezonco and de sus alrededores, sur de la Cuenca de Mexico. *Boletín de la Sociedad Geológica Mexicana, 67*(2), 123–143.

Avendaño-Villeda, D. (2017). *Reconstrucción paleolimnológica en el registro del lago de Chalco durante la transición del Estadio Isotópico Marino 6 a 5 (MIS 6 a MIS 5)*, Mexico, Universidad Nacional Autónoma de Mexico, Mexico.

Bhattacharya, T., & Byrne, R. (2016). Late Holocene anthropogenic and climatic influences on the regional vegetation of Mexico's Cuenca Oriental. *Global and Planetary Change, 138*, 56–69.

Bloomfield, K. (1975). A late-quaternary monogenetic volcano field in Central Mexico. *Geologische Rundschau, 64*(1), 476–497.

Bradbury, J. P. (1989). Late Quaternary lacustrine paleoenvironments in cuenca de Mexico. *Quaternary Science Reviews, 8*, 75–100.

Brown, E. T., Werne, J. P., Lozano-García, S., Caballero, M., Ortega-Guerrero, B., Cabral-Cano, E., Valero-Garces, B. L., Schwalb, A., & Arciniega-Ceballos, A. (2012). Scientific drilling in the Basin of Mexico to evaluate climate history, hydrological resources, and seismic and volcanic hazards. *Scientific Drilling, 14*, 72–75.

Brown, E. T., Caballero, M., Fawcett, J. P. T., Lozano-García, S., Ortega, B., Pérez, L., Schwalb, A., Smith, V., Steinman, B. A., Stockhecke, M., Valero-Garcés, B., Watt, S., Wattrus, N. J., Werne, J. P., Wonik, T. H., Myrbo, A. E., Noren, A., O'Grady, R., Schnurrenburger, D., & The MexiDrill Team. (2019). Scientific drilling of Lake Chalco, Basin of Mexico (Mexidrill). *Scientific Drilling, 26*, 1–15. https://doi.org/10.5194/sd-26-1-2019.

Caballero, M., & Ortega-Guerrero, B. (1998). Lake levels since about 40,000 years ago at Lake Chalco, near Mexico City. *Quaternary Research, 50*, 69–79.

Caballero, M., Lozano, S., Ortega, B., Urrutia, J., & Macias, J. L. (1999). Environmental characteristics of Lake Tecocomulco, northern Basin of Mexico, for the last 50,000 years. *Journal of Paleolimnology, 22*, 399–411.

Caballero, M., Lozano-García, S., Ortega-Guerrero, B., & Correa-Metrio, A. (2019). Quantitative estimates of orbital and millennial scale climatic variability in Central Mexico during the last ~40,000 years. *Quaternary Science Reviews, 205*, 62–75.

Cabral-Cano, E., Osmanoglu, B., Dixon, T., Wdowinski, S., DeMets, C., Cigna, F., & Díaz-Molina, O. (2010). Subsidence and fault hazard maps using PSI and permanent GPS networks in Central Mexico. *International Association of Hydrological Sciences, Publication Series, 339*, 255–259.

Campos-Enríquez, J. O., Alatriste-Vilchis, D. R., Huizar-Álvarez, R., Marines-Campos, R., & Alatorre-Zamora, M. A. (2003). Subsurface structure of the Tecocomulco sub-basin (northeastern Mexico Basin), and its relationship to regional tectonics. *Geofísica Internacional, 42*(1), 3–24.

Chen, Ch., McGee, D., Woods, A., Pérez, L., Hatfield, R., Edwards, L. ,Cheng, H., Valero-Garcés, B., Lehmann, S., Stoner, J., Schwalb, A., Tal, I., Seltzer, G., Tapia, P., Abbott, M., Rodbell, D. (2020). U-Th dating of lake sediments: Lessons from the 700 ka sediment record of Lake Junín, Peru. *Quaternary Science Reviews, 244*, https://doi.org/10.1016/j.quascirev.2020.106422.

Colman, S. M., Yu, S.-Y., An, Z., Shen, J., & Henderson, A. C. G. (2007). Late-Cenozoic climate changes in China's western interior: A review of research on Lake Qinghai and comparison with other records. *Quaternary Science Reviews, 26*, 2281–2300.

De Cserna, Z., de la Fuente-Duch, M., Palacios-Nieto, M., Triay, L., Mitre-Salazar, L. M., & Mota-Palomino, R. (1988). Estructura geológica, gravimetría, sismicidad y relaciones neotectónicas regionales de la Cuenca de Mexico: Mexico, D.F., Universidad Nacional Autónoma de Mexico, Instituto de Geología, Boletín 104, 71 p.

Deino, A. L., Dommain, R., Keller, C. B., Potts, R., Behrensmeyer, A. K., Beverly, E. J., King, J., Heil, C. W., Stockhecke, M., Brown, E. T., Moerman, J., de Menocal, P., & Olorgesailie Drilling Project Scientific Team. (2019). Chronostratigraphic model of a high-resolution drill core record of the past million years from the Koora Basin, South Kenya Rift: Overcoming the difficulties of variable sedimentation rate and hiatuses. *Quaternary Science Reviews, 215*, 213–231.

Delgado-Granados, H., & Martín del Pozo, A. L. (1993). Pliocene to Holocene volcanic geology at the junction of Las Cruces, Chichinahutzin and Ajusco ranges southwest of Mexico City. *Geofísica Internacional, 34*, 341–351.

Enciso-De la Vega, S. (1992). Propuesta de nomenclatura estratigráfica para la cuenca de Mexico: Mexico, Universidad Nacional Autónoma de Mexico, Instituto de Geología. *Revista, 10*(1), 26–36.

Ezcurra, E. (1990). The Basin of Mexico. In B. L. Turner (Ed.), *The earth as transformed by human action, global and regional changes in the biosphere over the past 300 years* (pp. 577–588). New York: Cambridge University Press.

Fawcett, P. J., Werne, J. P., Anderson, R. S., Heikoop, J. M., Brown, E. T., Berke, M. A., Smith, S., Goff, F., Donohoo-Hurley, L. L., Cisneros-Dozal, L. M., Schouten, S., Sinninghe Damsté, J. S., Huang, Y., Toney, J., Fessenden, J., Wolde, G. G., Atudorei, V., Geissman, J. W., & Allen, C. D. (2011). Extended Megadroughts in the Southwestern United States during Pleistocene Interglacials. *Nature, 470*, 518–521. https://doi.org/10.1038/nature09839.

Ferrari, L., Orozco-Esquivel, T., Manea, V., & Manea, M. (2012). The dynamic history of the Trans-Mexican Volcanic Belt and the Mexico subduction zone. *Tectonophysics, 522–523*, 122–149.

Fritz, S., Baker, P., Seltzer, G., Ballantyne, A., Tapia, P., Cheng, H., & Edwards, L. (2007). Quaternary glaciation and hydrologic variation in the South American tropics as reconstructed from the Lake Titicaca drilling project. *Quaternary Research, 68*, 410–420.

García-Palomo, A., Carlos-Valerio, V., López-Miguel, C., Galván-García, A., & Concha-Dimas, A. (2006). Landslide inventory map of Guadalupe range, north of the Mexico Basin. *Boletín de la Sociedad Geológica Mexicana, 58*(2), 195–204.

García-Palomo, A., Zamorano, J. J., López-Miguel, C., Galván-García, A., Carlos-Valerio, V., Ortega, R., & Macías, J. L. (2008). El arreglo morfoestructural de la Sierra de Las Cruces, Mexico central. *Revista Mexicana de Ciencias Geológicas, 25*(1), 158–178.

Gierlowski-Kordesch, E., & Kelts, K. (Eds.). (1994). *Global geological record of Lake Basins* (World and regional geology series, 427 pp) (Vol. I. IGCP Project 324). Cambridge University Press.

Gierlowski-Kordesch, E., & Kelts, K. (2001). Lake Basins through space and time. *American Association of Petroleum Geologists Study Geology, 46*, 648 p.

Herrera-Hernández, D. (2011). *Estratigrafía y análisis de facies de los sedimentos lacustres del Cuaternario tardío en la cuenca de Chalco*. Mexico: Universidad Nacional Autónoma de Mexico, Ciudad de Mexico.

Hodell, D., Anselmetti, F., Brenner, M., Ariztegui, D., & the PISDP Science Party. (2006). The Lake Petén Itzá scientific drilling project. *Scientific Drilling, 3*, 25–29.

Jaimes-Viera, C., Martin del Pozzo, A. L., Layer, P. W., Benowitz, J. A., & Nieto-Torres, A. (2018). Timing the evolution of a monogenetic volcanic field: Sierra Chichinautzin, Central Mexico. *Journal of Volcanology and Geothermal Research, 356*, 225–242.

Koeberl, C., Peck, J., King, J., Milkereit, B., Overpeck, J., & Scholz, C. (2005). The ICDP lake Bosumtwi drilling project: A first report. *Scientific Drilling, 1*, 23–27.

Lake Baikal Paleoclimate Project Members. (1992). Initial results of U.S.-Soviet Paleoclimate Study of Lake Baikal. *Eos, Transactions, American Geophysical Union, 73*(43), 457–462.

Litt, T., Anselmetti, F. S., Baumgarten, H., Beer, J., Cagatay, N., Cukur, D., Damci, E., Glombitza, C., Haug, G., Heumann, G., Kallmeyer, J., Kipfer, R., Krastel, S., Kwiecien, O., Meydan, A. F., Orcen, S., Pickarski, N., Randlett, M. E., Schmincke, H.-U., Schubert, C. J., Sturm, M., Sumita, M., Stockhecke, M., Tomonaga, Y., Vigliotti, L., Wonik, T., & the PALEOVAN Scientific Team. (2012). 500,000 years of environmental history in Eastern Anatolia: The PALEOVAN drilling project. *Scientific Drilling, 14*, 18–29.

Lozano-García, M. S., & Ortega-Guerrero, B. (1994). Palynological and magnetic susceptibility records of Chalco Lake, Central Mexico. *Palaeogeography, Palaeoclimatology, Palaeoecology, 109*, 177–191.

Lozano-García, M. S., & Ortega-Guerrero, B. (1998). Late quaternary environmental changes of the central part of the Basin of Mexico; correlation between Texcoco and Chalco Basins. *Review of Paleobotany and Palynology, 99*, 77–93.

Lozano-García, S., Ortega-Guerrero, B., Caballero-Miranda, M., & Urrutia-Fucugauchi, J. (1993). Late Pleistocene and Holocene paleoenvironments of the Chalco lake, Central Mexico. *Quaternary Research, 40*, 332–342.

Lozano-García, S., Ortega, B., Roy, P. D., Beramendi-Orosco, L., & Caballero, M. (2015). Climatic variability in the northern part of the American tropics since the latest MIS 3. *Quaternary Research, 85*(2), 261–272.

Lozano-García, S., Brown, E., Ortega, B., Caballero, M., Werne, J., Fawcett, P. J., Schwalb, A., Valero-Garcés, B., Schnurrenberger, D., O'Grady, R, Stockhecke, M., Steinman, B., Cabral-Cano, E., Caballero, C., Sosa-Nájera, S., Soler, A., Pérez, L., Noren, A., Myrbo, A., Bücker, M., Wattrus, N., Arciniega, A., Wonik, T. H., Watt, S., Kumar, D., Acosta, C., Martínez, I., Cossio, R., Ferland, T., & Vergara-Huerta, F. (2017). Perforación profunda en el lago de Chalco: reporte técnico. *Boletín de la Sociedad Geológica Mexicana, 69*(2), 299–311.

Lugo-Hubp, J. (1984). Geomorfología del Sur de la Cuenca de Mexico, Instituto de Geografía, UNAM, Mexico. *Serie Varia, 1*(8), 1–95.

Lugo-Hubp, J., Mooser, F., Pérez-Vega, A., & Zamorano-Orozco, J. (1994). Geomorfología de la Sierra de Santa Catarina, D.F., Mexico. *Revista Mexicana de Ciencias Geológicas, 11*(1), 43–52.

Macías, J. L., Arce, J. L., García-Tenorio, F., Layer, P. W., Rueda, H., Reyes-Agustín, G., López-Pizaña, F., & Avellán, D. (2012). Geology and geochronology of Tlaloc, Telapón, Iztaccíhuatl and Popocatépetl volcanoes, Sierra Nevada, Central Mexico. In J. J. Aranda-Gómez, G. Tolson, & R. S. Molina-Garza (Eds.), The southern Cordillera and Beyond *Geological Society of America, field guide* (Vol. 25, pp. 163–193).

Martín del Pozzo, A. L. M. (1982). Monogenetic vulcanism in sierra Chichinautzin, Mexico. *Bulletin of Volcanology, 45*(1), 9–24.

Martínez Abarca, R. (2019). *Lago de Chalco: registro sedimentario y estratigráfico de sus etapas formativas*. Tesis de Maestria. UNAM, 128 p.

Martínez-Abarca, R., Ortega-Guerrero, B., Lozano-García, S., Caballero, M., Valero-Garcés, B., McGee, D., ... Hodgetts, A. G. E. (2021). Sedimentary stratigraphy of Lake Chalco (Central Mexico) during its formative stages. International Journal of Earth Sciences, 1–21. https://doi.org/10.1007/s00531-020-01964-z

Melles, M., Brigham-Grette, J., Minyuk, P. S., Nowaczyk, N. R., Wennrich, V., DeConto, R. M., Anderson, P. A., Andreev, A. A., Coletti, A., Cook, T. L., Haltia-Hovi, E., Kukkonen, M., Lozhkin, A. V., Rosén, P., Tarasov, P. E., Vogel, H., & Wagner, B. (2012). 2.8 million years of Arctic climate change from Lake El'gygytgyn, NE Russia. *Science, 337*(6092), 315–320.

Metcalfe, S. E., Jones, M. D., Davies, S. J., Noren, A., & MacKenzie, A. (2010). Climate variability over the last two millennia in the North American monsoon region, recorded in laminated lake sediments from Laguna de Juanacatlán, Mexico. *The Holocene, 28*, 1195–1206.

Mosiño-Alemán, P. A., & García, E. (1974). The climate of Mexico. In R. A. H. Bryson & F. K. Hare (Eds.), *Climates of North America* (Vol. 2, pp. 345–405). New York: Elsevier.

Nixon, G. (1989). The geology of Iztaccíhuatl Volcano and adjacent areas of the Sierra Nevada and Valley of Mexico: Geological Society of America Special Paper 219, 58 p.

Ortega-Guerrero, B. (1992). *Paleomagnestismo, magnetoestrtifia y paleoecologia del Cuaternario tardio en el Lago de Chalco, Cuenca de Mexico (Tesis doctoral)*. Universidad Nacional Autónoma de Mexico.

Ortega-Guerrero, B., Lozano García, M., Caballero, M., & Herrera Hernández, D. A. (2015). Historia de la evolución deposicional del lago de Chalco, Mexico, desde el MIS 3. *Boletín de la Sociedad Geológica Mexicana, 67*(2), 185–201.

Ortega-Guerrero, B., Lozano-García, M. S., Herrera-Hernández, D., Caballero, M., Beramendi Orosco, L. E., Bernal, J. P., Torres-Rodríguez, E., & Avendaño-Villeda, D. (2017). Lithostratigraphy and physical properties of lacustrine sediments of the last ca. 150 kyr from Chalco Basin, Central Mexico. *Journal of South American Earth Sciences, 79*, 507–524.

Ortega-Guerrero, B., García, L. C., & Linares-López, C. (2018). Tephrostratigraphy of the late Quaternary record from Lake Chalco, Central Mexico. *Journal of South American Earth Sciences, 81*, 122–140.

Ortiz Zamora, D. D. C., & Ortega Guerrero, M. A. (2007). Origen y evolución de un nuevo lago en la planicie de Chalco: implicaciones de peligro por subsidencia e inundación de áreas urbanas en Valle de Chalco (Estado de Mexico) y Tláhuac (Distrito Federal). *Investigaciones geográficas, 64*, 26–42.

Oviedo de León, A. (1970). El Conglomerado Texcoco y el posible origen de la Cuenca de Mexico. *Revista del Instituto Mexicano del Petróleo, 2*, 5–20.

Pérez-Cruz, G. A. (1988). Estudio sismológico de reflexión del subsuelo de la Ciudad de Mexico: Mexico, D.F., Universidad Nacional Autónoma de Mexico, Tesis de Maestría, División de Estudios de Posgrado de la Facultad de Ingeniería, 83 p.

Peterson, L. C., Haug, G. H., Hughen, K. A., & Rohl, U. (2000). Rapid changes in the hydrologic cycle of the tropical Atlantic during the last glacial. *Science, 290*, 1947–1951.

Pi, T., Lozano-García, S., Caballero-Miranda, M., Ortega-Guerrero, B., & Roy, P. (2010). Discovery and characterization of a struvite layer in the Chalco paleolake, Mexico. *Revista Mexicana de Ciencias Geológicas, 27*(3), 573–580.

Pueyo, J. J., Sáez, A., Giralt, S., Valero-Garcés, B. L., Moreno, A., Bao, R., Schwalb, A., Herrera, C., Klosowska, B., & Taberner, C. (2011). Carbonate and organic matter sedimentation and isotopic signatures in Lake Chungará, Chilean Altiplano, during the last 12.3 kyr. *Palaeogeography, Palaeoclimatology, Palaeoecology, 307*, 339–355.

Ruiz-Angulo, A., & López-Espinoza, E. D. (2015). Estimación de la respuesta térmica de la cuenca lacustre del Valle de Mexico en el siglo XVI: un experimento numérico. *Boletín de la Sociedad Geológica Mexicana, 67*(2), 215–225.

Russell, J. M., Bijaksana, S., Vogel, H., Melles, M., Kallmeyer, J., Ariztegui, D., Crowe, S. A., Fajar, S. J., Hafidz, A., Haffner, D., Hasberg, A. K. M., Ivory, S. J., Kelly, C., King, J. W., Kirana, K. H., Morlock, M. A., Noren, A., O'Grady, R., & Ordonez, L. (2016). The Towuti Drilling Project: Paleoenvironments, Biological Evolution, And Geomicrobiology Of A Tropical Pacific Lake. *Scientific Drilling, 21*, 29–40.

Schnurrenberger, D., Russell, J., & Kelts, K. (2003). Classification of lacustrine sediments based on sedimentary components. *Journal of Paleolimnology, 29*, 141–154.

Scholz, C. A., Johnson, T. C., Cohen, A. S., King, J. W., Peck, J. A., Overpeck, J. T., Talbot, M. R., Brown, E. T., Kalindekafe, L., Amoako, P. Y. O., Lyons, R. P., Shanahan, T. M., Castañeda, I. S., Heil, C. W., Forman, S. L., McHargue, L. R., Beuning, K. R., Gomez, J., & Pierson, J. (2007). East African megadroughts between 135–75 kyr ago and implications for early human history. *Proceedings of the National Academy of Sciences, 104*, 16416–16421.

Schuster, M., Duringer, P., Ghienne, J.-F., Roquin, C., Sepulchre, P., Moussa, A., Lebatard, A.-E., Mackaye, H. T., Likius, A., Vignaud, P., & Brunet, M. (2009). Chad Basin: Paleoenvironments of the Sahara since the late Miocene. *Comptes Rendus Geosciences, 341*, 612–620.

Siebe, C. (2000). Age and archaeological implications of Xitle volcano, southwestern Basin of Mexico City. *Journal of Volcanology and Geothermal Research, 104*(1), 45–64.

Siebe, C., Macias, J. L., Abrams, M., Rodriguez, S., Castro, R., & Delgado, H. (1995). Quaternary explosive volcanism and pyroclastic deposits in East Central Mexico: Implications for future hazards. In *Guidebook of geological excursions: In conjunction with the annual meeting of the Geological Society of America, New Orleans, Louisiana, November 6–9, 1995* (pp. 1–48). Baton Rouge: Louisiana State University. Basin Research Institute. Center for Coastal Energy & Environmental Resources.

Siebe, C., Macías, J., Abrams, M., Rodríguez, S., & Castro, R. (1997). *Catastrophic prehistoric eruptions at Popocatépetl and Quaternary explosive volcanismin the Serdán-Oriental basin, east-central Mexico*. Pre-meeting excursion fieldtrip guidebook No 4, January 12–18: Puerto Vallarta, Mexico, International Association of Volcanology and Chemistry of the Earth's Interior (IAVCEI) General Assembly (88 pp.).

Siebe, C., Schaaf, P., & Urrutia-Fucugauchi, J. (1999). Mammoth bones embedded in a late Pleistocene lahar from Popocatépetl volcano, near Tocuila, Central Mexico. *Geological Society of America Bulletin, 111*, 1550–1562.

Siebe, C., Rodríguez-Lara, V., Schaaf, P., & Abrams, M. (2004). Geochemistry, Sr-Nd isotope composition, and tectonic setting of Holocene Pelado, Guespalapa and Chichinautzin scoria cones, south of Mexico City. *Journal of Volcanology and Geothermal Research, 130*, 197–226.

Siebe, C., Arana-Salinas, L., & Abrams, M. (2005). Geology and radiocarbon ages of Tláloc, Tlacotenco, Cuauhtzin, Hijo del Cuauhtzin, Teuhtli, and Ocusacayo monogenetic volcanoes in the central part of the Sierra Chichinautzin, Mexico. *Journal of Volcanology and Geothermal Research, 141*(3), 225–243.

Siebe, C., Salinas, S., Arana-Salinas, L., Macías, J. L., Gardner, J., y Bonasia, R. (2017). The~ 23,500 y 14C BP White Pumice Plinian eruption and associated debris avalanche and Tochimilco lava flow of Popocatépetl volcano, México. *Journal of Volcanology and Geothermal Research, 333*, 66–95.

Sosa-Ceballos, G., Gardner, J. E., Siebe, C., & Macías, J. L. (2012). A caldera-forming eruption~ 14,100 14 Cyr BP at Popocatepetl volcano, Mexico: Insights from eruption dynamics and magma mixing. *Journal of Volcanology and Geothermal Research, 213*, 27–40.

Stein, M., Ben-Avraham, Z., Goldstein, S., Agnon, A., Ariztegui, D., Brauer, A., Haug, G., Ito, E., & Yasuda, Y. (2011). Deep drilling at the dead sea. *Scientific Drilling, 11*, 46–47.

Stockhecke, M., Kwiecien, O., Vigliotti, L., Anselmetti, F. S., Beer, J., Namik Çağatay, M., Channell, J. E. T., Kipfer, R., Lachner, J., Litt, T., Pickarski, N., & Sturm, M. (2014). Chronostratigraphy of the 600,000 year old continental record of Lake Van (Turkey). *Quaternary Science Reviews, 104*, 8–17.

Torres-Rodríguez, E., Lozano-García, S., Roy, P., Ortega, B., Beramendi-Orosco, L., Correa-Metrio, A., & Caballero, M. (2015). Last glacial droughts and fire regimes in the central Mexican highlands. *Journal of Quaternary Science, 30*, 88–99.

Torres-Rodríguez, E., Lozano-García, S., Caballero-Miranda, M., Ortega-Guerrero, B., Sosa-Nájera, S. and Debajyoti-Roy, P. (2018). Pollen and non-pollen palynomorphs of Lake Chalco as indicators of paleolimnological changes in high-elevation tropical central Mexico since MIS 5. *Journal of Quaternary Science, 33*(8), 945–957.

Vazquez-Selem, L., & Heine, K. (2011). Late quaternary glaciation in Mexico. In J. E. Ehlers, P. L. Gibbard, & P. D. Hughes (Eds.), *Quaternary glaciations-extent and chronology: A closer look* (Vol. 15, pp. 849–861). Amsterdam: Elsevier.

Wagner, B., Wilke, T., Krastel, S., Zanchetta, G., Sulpizio, R., Reicherter, K., Leng, M. J., Grazhdani, A., Trajanovski, S., Levkov, Z., Reed, J., & Wonik, T. (2014). More than one million years of history in Lake Ohrid Cores. *Eos, Transactions American Geophysical Union, 95*(3), 25–26.

White, S. E. (1962). Late Pleistocene glacial sequence for the west side of Iztaccíhuatl, Mexico. *Geological Society of America Bulletin, 73*, 935–958.

White, S. E. (1987). Quaternary glacial stratigraphy and chronology of Mexico. *Quaternary Science Reviews, 5*(1–4), 201–206.

Zolitschka, B., Anselmetti, F., Ariztegui, D., Corbella, H., Francus, P., Lücke, A., Maidana, N., Ohlendorf, C., Schäbitz, F., & Wastegard, S. (2013). Environment and climate of the last 51,000 years—New insights from the Potrok Aike maar lake Sediment Archive Drilling Project (PASADO). *Quaternary Science Reviews, 71*, 1–12.

Submarine Groundwater Discharge as a Catalyst for Eodiagenetic Carbonate Cements Within Marine Sedimentary Basins

Elizabeth H. Gierlowski-Kordesch, Gar W. Rothwell, Ruth A. Stockey, and David B. Finkelstein

Abstract Submarine groundwater discharge (SGD) is the input of freshwater into ocean water and sediments, especially in estuaries, from continental aquifers. Such meteoric water influx can occur at depths up to 3000 m. The mixing of calcium-rich meteoric water with saline marine water in the presence of organic matter can produce early calcite cement in pores. Pervasive to stratabound carbonate cementation in shallow to offshore marine sandstones has been assumed to result from long-term slow diffusion or advective processes. However, results from the sedimentologic and geochemical studies of Cretaceous calcitic concretion-bearing layers in sandstones in a marine coastal setting indicate otherwise. Permineralization of delicate fungi, liverworts, mosses, and macrophytic plant material as well as associated textures and isotopic signatures of carbonate cement within stratabound concretions in the Longarm Formation on Vancouver Island, Canada, indicate high flux flow of calcium-rich meteoric waters during very early diagenesis (eogenesis) to synsedimentary freshwater diagenesis. We propose that calcite cementation from meteoric waters along horizons in estuarine to fully marine sediments can occur through SGD directly via terrestrial aquifer flow through carbonates. Because SGD can be introduced into marine sediments along shallow coastlines as well as at great depths, this groundwater input may be an important geochemical process in the diagenesis of coastal marine sediments, and is key to fossil preservation, including permineral-

Elizabeth H. Gierlowski-Kordesch contributed equally with all other contributors.

E. H. Gierlowski-Kordesch
Department of Geological Sciences, Ohio University, Athens, OH, USA

G. W. Rothwell · R. A. Stockey
Department of Botany and Plant Pathology, Oregon State University, Corvallis, OR, USA
e-mail: rothwelg@science.oregonstate.edu; stockeyr@science.oregonstate.edu

D. B. Finkelstein (✉)
Department of Geosciences, Hobart and William Smith Colleges, Geneva, NY, USA
e-mail: finkelstein@hws.edu

M. R. Rosen et al. (eds.), *Limnogeology: Progress, Challenges and Opportunities*, Syntheses in Limnogeology, https://doi.org/10.1007/978-3-030-66576-0_15

ization of land plants along ancient coastlines and possibly the formation of Carboniferous coal balls.

Keywords Cretaceous · Sandstones · Carbonate cementation · Freshwater aquifers · Longarm Formation

Introduction

Submarine groundwater discharge (SGD) is the direct input of continental freshwater (rarely saline) from aquifers into the ocean bypassing rivers and streams. Post et al. (2013) document offshore groundwater occurrences up to 1020 km offshore, with total dissolved solids of 13 g/L and higher with onshore connections. Taniguchi et al. (2002) and Bratton (2010) point out the misunderstandings and complexities of the term when comparing it to other freshwater discharges. Here we follow the terminology of Taniguchi et al. (2002) and refer to SGD as flow across the sea floor and represents all direct discharge of subsurface fluids across the land-ocean interface. This flow is composed of submarine fresh groundwater discharge and recirculated saline groundwater discharge and can occur at scales from nearshore (up to ~10 m wide), embayment scale (up to ~10 km wide), and shelf scale (average width of ~80 km) (Bratton 2010). Driving forces for SGD include convection, hydraulic head, tidal pumping, and wave set-up (Taniguchi et al. 2002). SGD is a significant source of meteoric waters entering the ocean, perhaps 10% of the total discharge (Miller and Ullman 2004; Zelster et al. 2006), and is not limited to estuarine sediments (Moore 1999; Schwartz 2003). Freshwater can mix with saline water across a large mixing zone from unconfined coastal aquifers and from confined aquifers allowing for seepage as dispersed and focused discharge in shallow or deep waters (Shaban et al. 2005; Bratton 2010; Moore 2010). This seepage occurs on beaches and in coastal wetlands (Ullman et al. 2003; Price et al. 2006; Zapata-Rios and Price 2012), and where there are seasonal changes in water levels and oceanic/freshwater interactions, landward saline intrusion is possible. Theoretical modeling of SGD dynamics shows that the rate of transfer of ions and chemicals from the continental freshwater aquifers to coastal ocean waters can be quite high in comparison to surface water input (Johannes 1980; Li et al. 1999; Rodellas et al. 2015). Nutrient fluxes of dissolved inorganic Nitrogen, Phosphorous, and Silica from SGD were calculated to be up to an order of magnitude greater than those delivered by river inputs (Rodellas et al. 2015). Cai et al. (2003) measured dissolved inorganic carbon in SGD and found concentrations to coastal waters orders of magnitude greater than in river or ocean water. Holmden et al. (2012) found excess calcium in SGD is delivered to coastal waters of Florida Bay (541 mg/L) compared to seawater (450 mg/L). Their finding of a $\delta^{44/40}$Ca gradient in the sediments and waters of Florida Bay supports the idea that local carbon cycling affects epeiric seas in the present and the geological past.

SGD could act as a carbonate source in nearshore to offshore and shelfal marine sands where flow derives from carbonate aquifers. Calcite precipitation can occur within an area where alkaline freshwater mixes with ocean water (Plummer 1975; Wigley and Plummer 1976), especially those associated with organic matter and bacterial abundance (Mitterer and Cunningham Jr. 1985; Coleman 1993; Mozley and Davis 2005). Modern examples of calcite cementation within beach sands (beachrock) down-gradient from a marsh can be found at discrete points along the coast of Louisiana (Roberts and Whelan III 1975; Kocurko 1986), where concretions and subsurface bedding plane "crusts" or beachrock are cemented by Mg-calcite in association with wood fragments, algae, and burrows. These processes also occur farther downslope in the upper-to-middle shoreface of the beach if groundwater from a deeper confined aquifer or fault zone seeps into the ocean affecting only certain levels of the offshore environment (Schmalz 1971; Robb 1990; Shaban et al. 2005). For example, Jørgensen (1976) reports on recent carbonate sedimentation in littoral to submarine sands and gravels along the northern Denmark coastline and documents other occurrences from earlier literature as well. Deeper offshore carbonate cementation in the accretionary prism sediments off the western coast of North America, attributed to "pore water expulsion" from tectonic structural deformation of sediment (see Ritger et al. 1987), may be enhanced by the discharge of a steeply dipping freshwater or mixed saline aquifers at water depths of 800–2100 m. It is known that meteoric (fresh) groundwater is present in surficial and deeper subsurface sediments of the modern shelf along the eastern shore of the United States (Manheim and Paull 1981; Post et al. 2013). The depth of submarine discharge of terrestrial aquifers (meteoric water influx into the ocean) presently has been documented to be up to 3000 m below the ocean surface (Taniguchi et al. 2002).

Stable isotopic composition of carbonate cements is used to document the role of meteoric waters in cementation of marine carbonates (Gross 1964; Allan and Matthews 1977) and concretions (Hudson 1978). Mozley and Burns (1993) note that most cementation in concretions is presumed to be very early and must be linked to shallow burial of marine pore waters. They invoke cementation as a result of mixing of meteoric and marine waters in the near offshore as one way to explain the depleted carbon and oxygen isotopic values in calcite and siderite concretions. Others have observed more negative oxygen isotopic compositions and support precipitation of cements at higher temperatures than shallow burial or changes in water composition during growth of the concretions (after Raiswell and Fisher 2000 and references therein).

Frequent sea level or climate change as well as erosion and subaerial exposure of sands are normally inferred for cement growth, but not proven unequivocally, especially in examples for meteoric influx to supply ion-rich waters to the marine subsurface and to produce the necessary volume of carbonate cement (e.g., Dutton et al. 2000; Al-Ramadan et al. 2005). For example, general diffusive and advective models using surface input of meteoric waters generally explain calcium carbonate cementation in sandstones (see Bjørkum and Walderhaug 1990, 1993; Bjørlykke and Jahren 2012). Calcium sources for these models include limited or undocumented fossil shell populations and associated shales (e.g., Dutta 1992; Park 2009).

Unfortunately, these proposed calcium sources cannot adequately explain the volume of calcium-rich sources needed to produce thick, calcite-cemented discrete horizons (stratabound cementation) or well-aligned, carbonate concretionary bodies lacking textural differences with non-cemented adjacent horizons (see Wilkinson 1993). Such stratabound cementation distributions have been attributed to "some self-organization processes" (Abdel-Wahab and McBride 2001), unidentified "local factors" (Milliken et al. 1998), or called simply "enigmatic" (Al-Ramadan et al. 2005). No detailed models exist for linking the calcium source to the observed volume of carbonate cement. Insightfully, McBride et al. (1995) blame "highly localized hydrologic factors" of pore water flow on stratabound carbonate cement in the Tertiary sandstones of the Apennines (Italy) and McBride et al. (1994) and McBride and Parea (2001) invoke phreatic groundwater flowing oceanward within Italian Pleistocene coastal sands for highly elongate, carbonate concretion growth.

Meteoric carbonate cements, a usual component in many studies of diagenetic evolution in marine basins, may not require a marine influence to experience sea level change and exposure nor proximity to land for precipitation since SGD occurs at depths of ~3000 m below the ocean surface. SGD may enable processes by which meteoric calcite cementation occur in early-to-late diagenesis within marine sands in shallow-to-deep shelf areas without any temperature or sea level changes. Focused SGD discharge can also explain stratabound carbonate cementation of sandstones with similar textures across both uncemented and cemented layers. This chapter focusses on the Cretaceous sandstones of the Apple Bay section of the Longarm Formation equivalent of Vancouver Island (Canada), and suggests SGD is a mechanism for very early to synsedimentary cementation that preserves delicate plant and fungi material within carbonate nodules that are stratabound.

Geologic and Paleontologic Setting

The study area in Vancouver Island is within the Insular Belt of the Canadian Cordillera. This island comprises sediments deposited along an island arc (Jeletzky 1976; Campbell et al. 1991), experiencing only shallow burial (Longstaffe 1994). The studied Apple Bay section is located at the northern end of Vancouver Island (British Columbia, Canada), in the area of Quatsino Sound, which is dominated by approximately four to five major fault blocks (see Fig. 1). Each block contains Mesozoic to early Cenozoic rocks. Faulting patterns have major trends to the northwest to west-north-west and minor faults strike in several directions (north, northeast, northwest, and east-west) producing rock sections over the area that cannot be definitively correlated, even over short distances (Jeletzky 1976). The Apple Bay succession (50° 36′ 21″ N, 127° 39′25″ W) is exposed only for a few hundred meters, delineated by faults (see Fig. 1). This section, proposed to be Valanginian to Barremian (~140–125 Ma) in age by Jeletzky (1970), has been correlated tentatively with the upper Kyuquot Group to Longarm Formation (Haggart 1991, 1996) (see Fig. 2). Stockey and Rothwell (2009) established a Valanginian-Hauterivian

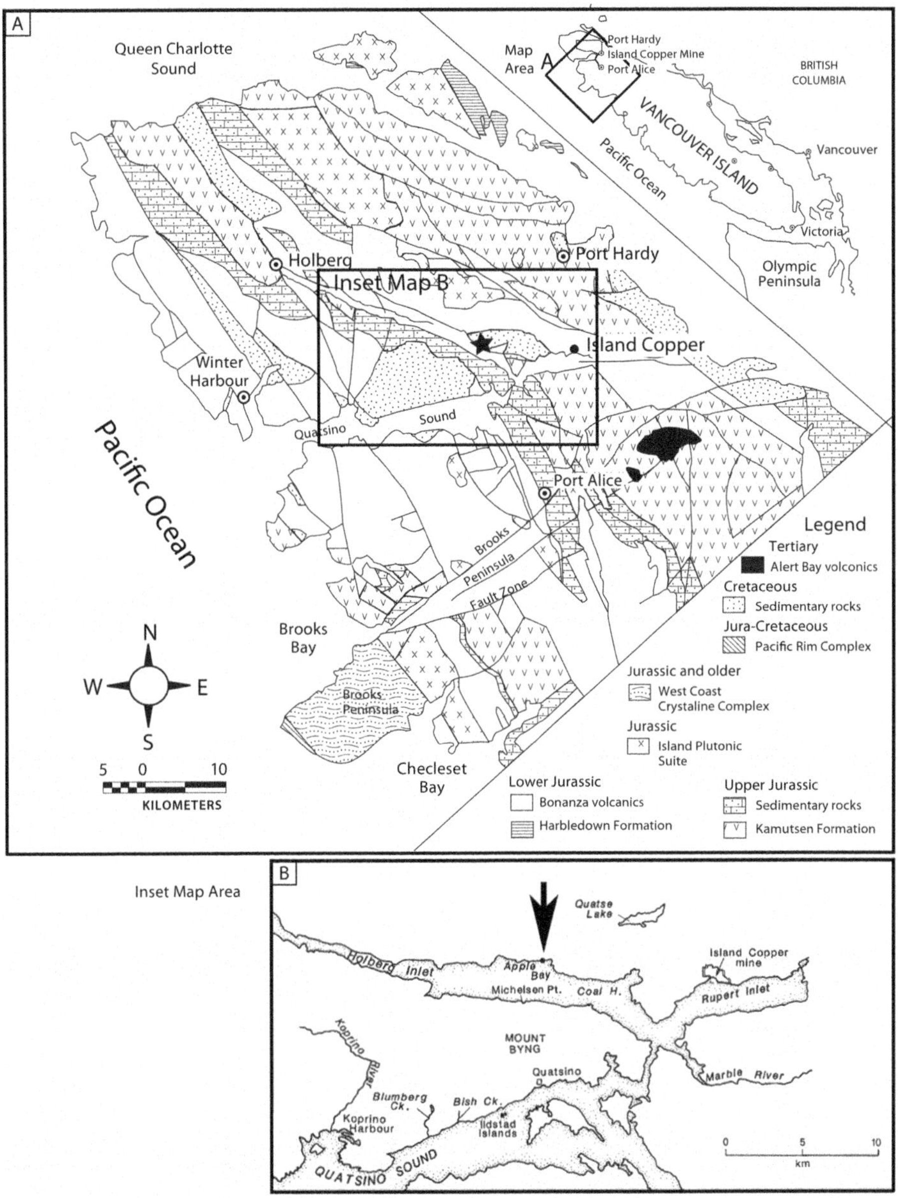

Fig. 1 Map of northern Vancouver Island, Canada, the regional geology of Northern Vancouver Island (**a**) and the inset map (**b**) showing the location of the measured section and fossil locality (star in 1A and arrow in 1B). Regional faulting is responsible for the offset of units. (Modified from Nixon et al. (1994))

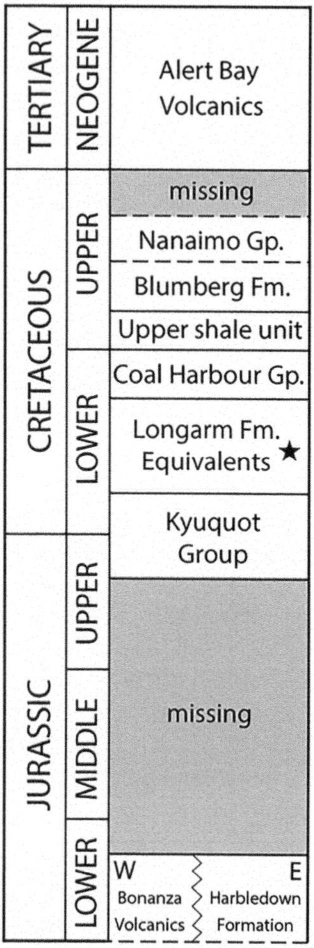

Fig. 2 Stratigraphy of Quatsino Sound, Vancouver Island (From Jeletzky 1976). The star denotes the focus of the study being within the Longarm Formation. (See Fig. 1 for locality)

age for the permineralized flora preserved in the calcareous concretions of Apple Bay limited to discrete horizons within mostly fine- to medium-grained sandstones and siltstones (Smith et al. 2003). The flora is of *Lagerstätten* quality, including fungi, liverworts, mosses, lycopsids, and sphenopsids as well as ferns, conifers, equisetophytes, bennettialeans, seed ferns, and gnetophytes (Little et al. 2006; Rothwell and Stockey 2006; Stockey et al. 2006; Vavrek et al. 2006; Stockey and Wiebe 2008; Stanich et al. 2009; Rothwell et al. 2009; Stockey and Rothwell 2009) (Fig. 3). Elements preserved at Apple Bay include fungal hyphae, pollen, flowering plant seeds and reproductive structures, spores, rhizomes, sori, pinnules, leafy shoots, leaves, and cones. Precipitation of calcite within cells (see Scott and Collinson 2003 for processes) allows for three-dimensional preservation of the most diverse pre-flowering plant flora in the world (Stockey and Rothwell 2006).

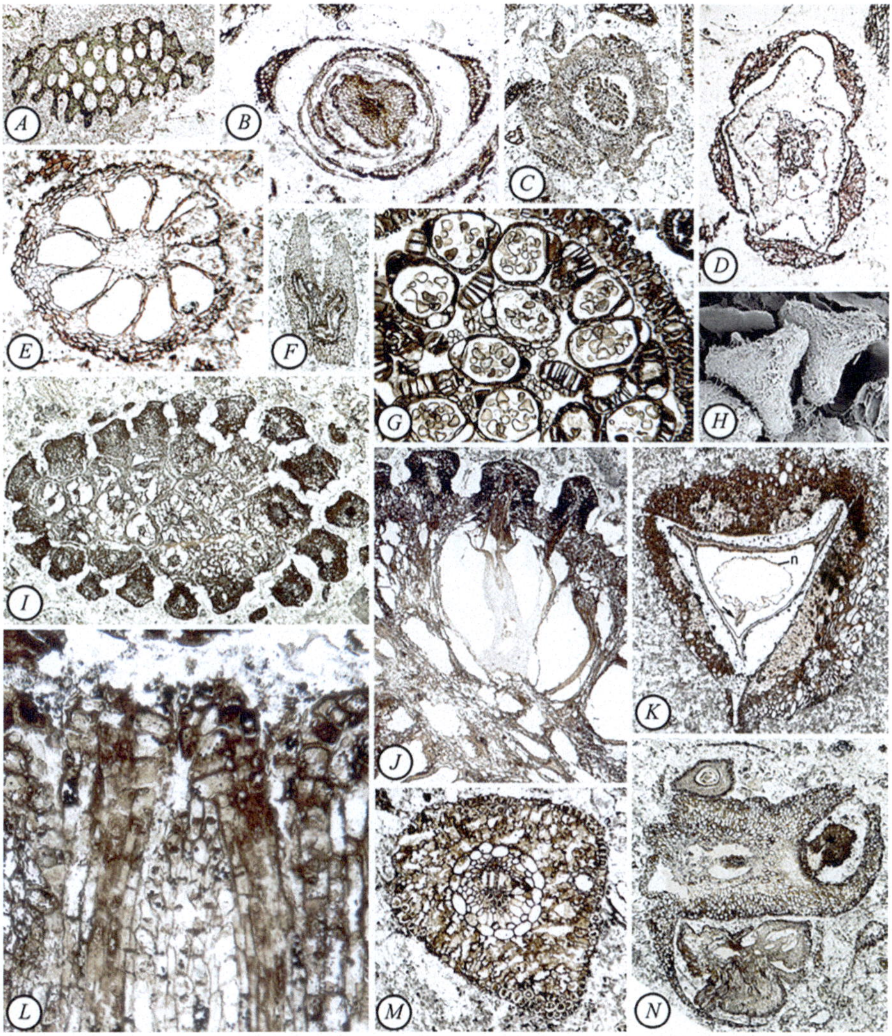

Fig. 3 Representative plants and fungi, showing diversity and quality of cellular preservation from the carbonate concretion horizons of the Longarm Formation at Apple Bay, Vancouver Island. (**a**) Fungal fruiting body (hymenophore), *Quatsinoporites cranhamii* Smith, Currah et Stockey, showing pores in which spores are produced, × 1.0. (**b**) Polytrichaceous moss gametophyte, *Meantoinea alophosioides* Bippus, Stockey, Rothwell et Tomescu, in cross section, × 60. (**c**) Lycopod leafy stem with plectostele (*Lycopodium* sp. or *Diphasiastrum* sp.) in cross section, × 8.0. (**d**) Equisetophyte aerial shoot, *Equisetum haukeanum* Stanich, Rothwell et Stockey in cross section, × 35. (**e**) Marattiaceous fern synangium with spores, *Escapia christensenioides* Rothwell, Millay et Stockey, × 6.0. (**f**) Dennstaedteoid filicalean fern pinna in cross section, × 11. (**g**) Filicalean fern sorus with sporangia and spores, *Cyathea cranhamii* Smith, Rothwell et. Stockey, × 50. (**h**) SEM photo of *Cyathea cranhamii* spores, × 450. (**i**) Bennettitalean seed "cone," *Foxeoidea connatum* Rothwell et Stockey, showing seeds in cross section, × 5.0. (**j**) Embedded seed of *Foxeoidea connatum* in longitudinal section, × 20. (**k**) Uni-ovulate "seed fern" cupule and seed; *Doylea tetrahedrasperma* Stockey et Rothwell (*n* = nucellus of seed, arrow identifies tip of pollen chamber), × 20. (**l**) Longitudinal section of immature ovule of bennettitalean cone *Williamsonia* sp. at stage of meiosis, × 150. (**m**) Pinaceous conifer needle-like leaf of *Midoriphyllum piceoides* Stockey et Wiebe in cross section, × 18. (**n**) Seed cone of gnetalean gymnosperm, *Protoephredites eamesii* Rothwell et Stockey with attached seeds, × 10

Sedimentology and Petrography

The 6.5-m section at Apple Bay, Vancouver Island, is comprised of carbonate-rich, ripple cross-laminated, parallel-laminated, and massive siltstones to fine- and medium-grained sandstones and rare mudstones, with carbonate and iron-rich concretions (Fig. 4). Ripple cross-lamination is wave-dominated with intricately interwoven cross-lamination and less common current-dominated, climbing ripple sets (20–30 cm thick) intercalated with calcareous siltstone to mudstone. Rippled units are erosionally truncated at their upper and lower contacts, and contain rare trace fossils. Plant fragments (compressions) are present throughout the section. Exquisitely preserved, permineralized plant material is present only within carbonate concretions at two different levels (0.2 and 2.5 m). Oysters are present in various layers as broken shells (no hash layers) with only one stratigraphic level (~4.7 m) where probable in situ oysters are preserved. Iron-cemented burrows exist at some stratigraphic levels as well. Mudstones are generally massive in fabric and crop out in thin layers.

The Longarm quartz arenites to subarkoses contain mostly quartz and feldspar grains that are fairly well sorted and rounded with sharp stratal boundaries. Within carbonate concretions, mineral grains and plant debris are supported by calcite cement; no matrix or clay cement is present with textures composed of sand grains with little contact and sand grains with point to rare long contacts (Fig. 5). No evidence for displacive cement growth is present, using criteria of Watts (1978). Permineralized plant remains float within the concretion sand or silt fabric with common bivalve (oyster) fossils (e.g., *Quoiecchia aliciae* Crickmay) (Fig. 6). Outside the carbonate concretions, the surrounding sand to silt fabric is grain-supported with later diagenetic calcite cement replacing plagioclase in rock fragments, clay cement replacing feldspar grains, and carbonized plant debris (Fig. 5).

The Apple Bay succession is interpreted as a middle-to-lower shoreface deposit along a coastal shoreline, perhaps within an estuary, with rare background fine-grained units with bioturbation features. Abundant hydrodynamically light particles, such as delicate plant material from land, intercalated in the sandy to silty sediment suggest rapid deposition (e.g., Kiteley and Field 1984; Pemberton et al. 2001). Early to penecontemporaneous carbonate cementation of the concretions is suggested by the lack of compaction (Curtis and Coleman 1986) as well as the lack of matrix material or clay cement and diagenetic replacement of primary minerals. Also, the nearly perfect, three-dimensional preservation of delicate plant material in coarse sediment also suggests very early to synsedimentary cementation. Preservation of the delicate plant material is evidence for early cementation prior to or contemporaneous with sand deposition. The rest of the sandy deposit underwent shallow burial diagenesis, including replacement of plagioclase by calcite cement in rock fragments (pre-depositional) and alteration of feldspar into clays (post-depositional) but common pristine biotite remained present.

Large thin sections of whole concretions as well as surrounding sediment were studied under cathodoluminescence to determine calcite cementation patterns (after

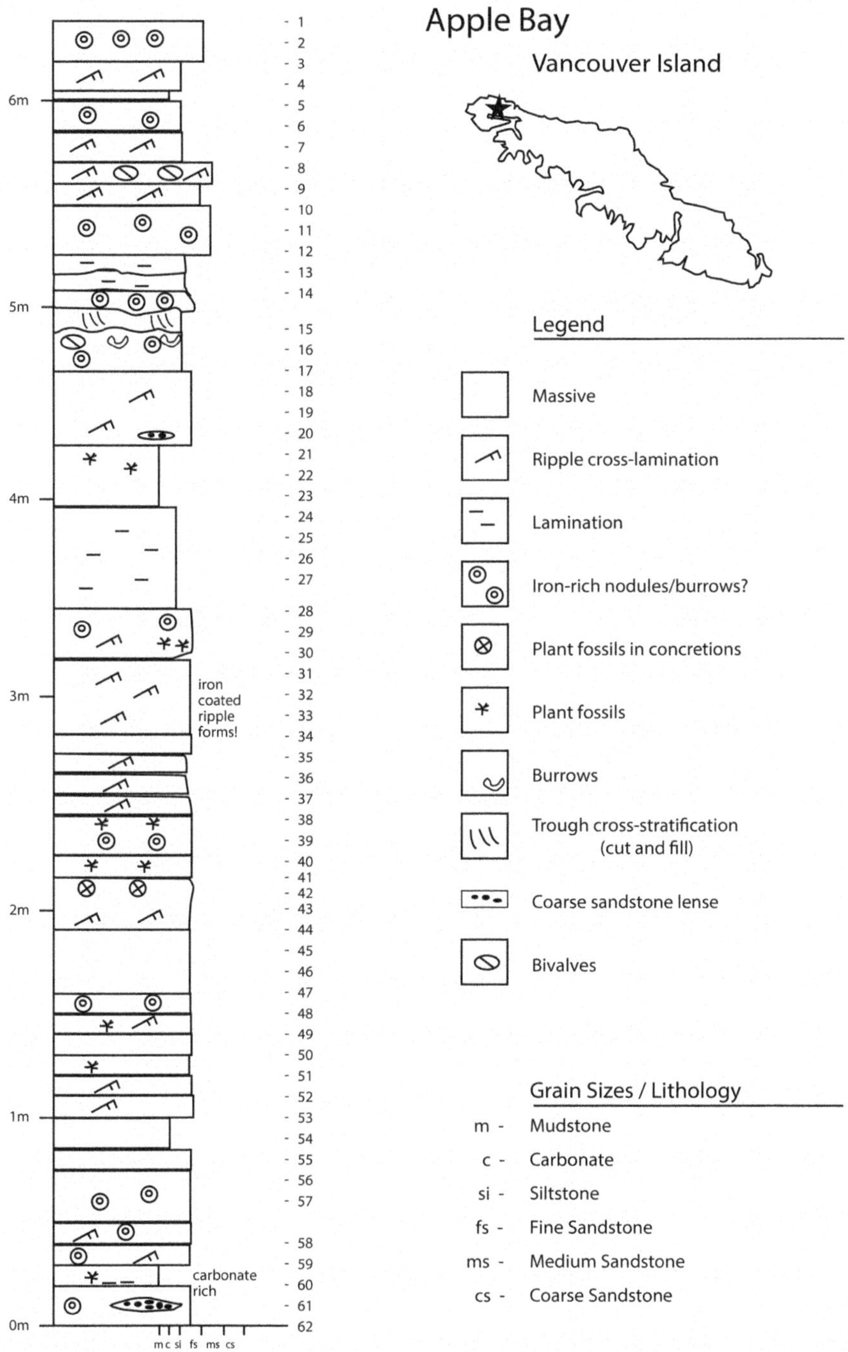

Fig. 4 Sedimentologic log of Cretaceous sandstones of the Longarm Formation equivalent exposed at Apple Bay, northern Vancouver Island, Canada, containing layers of carbonate concretions with a sandstone-mudstone succession. Sample numbers are listed to the right of the sedimentary column

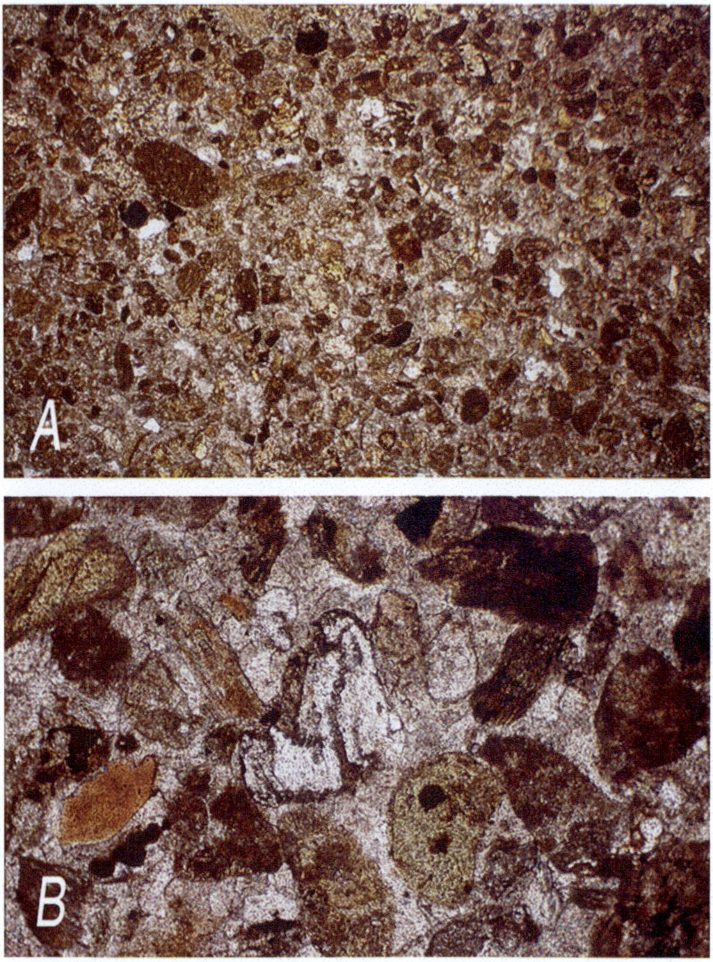

Fig. 5 Petrographic thin section photos under plane polarized light of the fabric of sandstone grains and plant material mixed within two carbonate-cemented concretions. Note grain contacts ranging from no grain contact (cement-supported) to point and long contacts. (**a**) Field of view is 4.8 mm across photo. (**b**) Field of view is 0.9 mm across photo

Machel 2000). Cement of the concretions with sand and coarse silt grains exhibit a generally uniform dull orange luminescence (Fig. 7). The absence of any changes in luminescence and therefore lack of zonation in the cements suggests there is no change in the chemical parameters of the precipitating fluids (e.g., Kantorowicz et al. 1987) as it compositionally relates to trace and rare earth elements acting as luminescence activators, sensitizers and quenchers. Only a single cementation event in concretions is inferred. Petrographic evidence suggests that the concretions were formed through a single cementation event during shallow burial and is consistent with the findings of Longstaffe (1994).

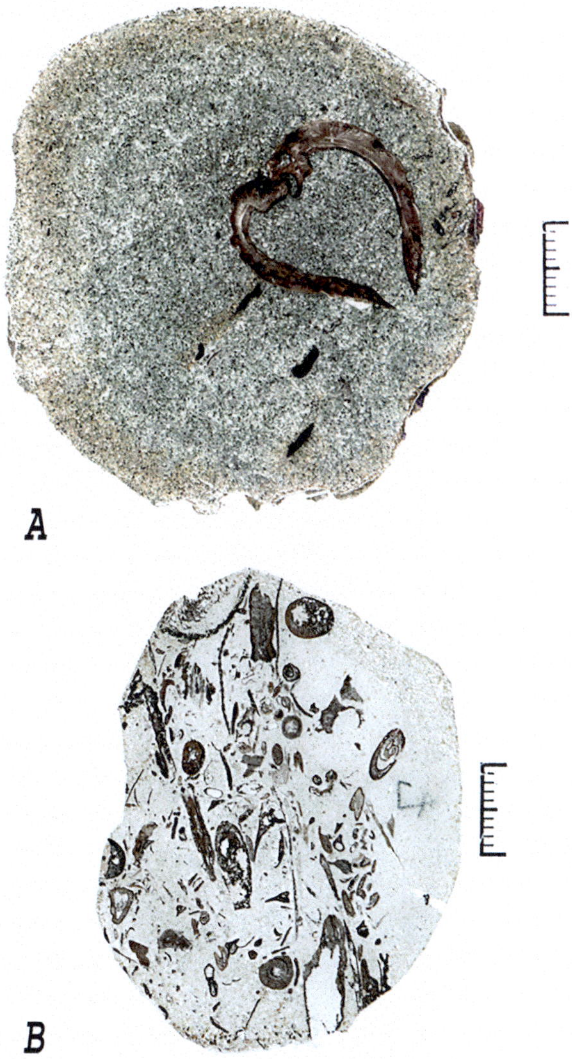

Fig. 6 (**a**) Photo of a cut carbonate concretion of cemented sandstone with a cross-section of a bivalve shell (*Quoiecchia aliciae* Crickmay). Scale in mm. (**b**) Photo of an acetate peel of a cut carbonate concretion exhibiting suspended plant debris embedded in carbonate cement and very fine sandstone to siltstone. Scale in mm

Stable Isotopic Analyses

Concretions were randomly collected across the Apple Bay outcrop for the initial paleobotanical analysis. Sandstone samples were taken at approximately 5–10 cm intervals stratigraphically in the exposed stratigraphic succession to characterize the unit hosting the concretions. Carbon and oxygen isotopic ratios were analyzed from carbonate concretions and carbonate cemented sandstones (see Fig. 4 for position of concretion layers and of samples collected from background sandstone succession). Samples were analyzed on a Thermo Electron Delta Plus mass spectrometer in a

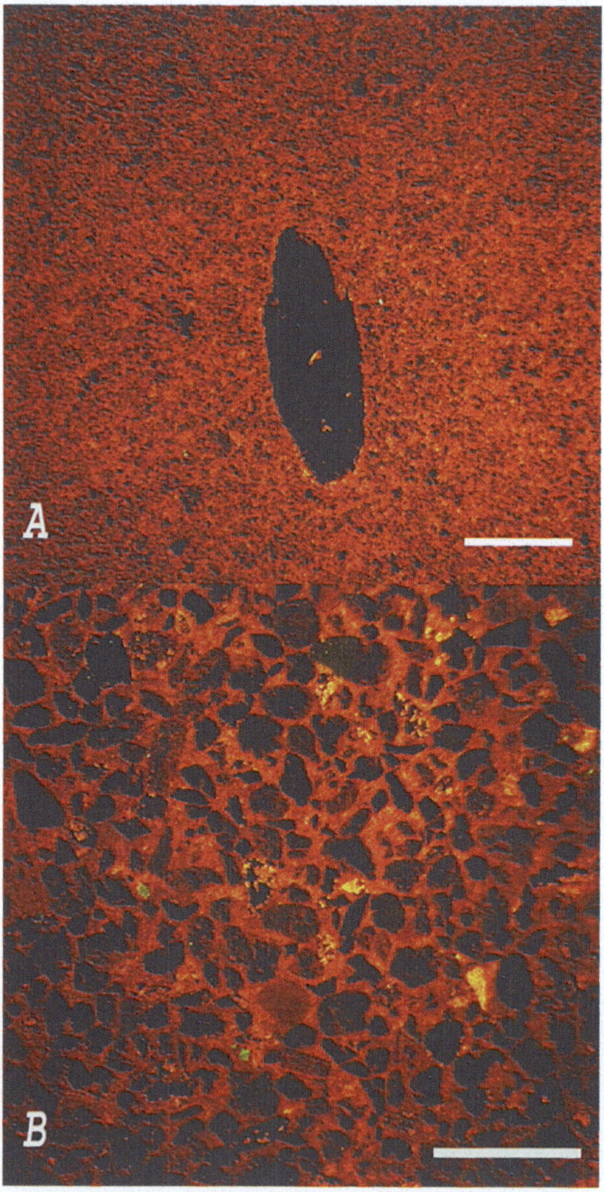

Fig. 7 (**a**) Thin section photo of coarse siltstone in carbonate concretion under cathodoluminescence showing homogeneity of calcite cement through uniform dull orange luminescence. Large plant fragment in the center of photo. Scale bar is 1 mm wide. (**b**) Thin section photo of fine sand grains in a carbonate concretion floating in carbonate cement showing homogeneous dull orange luminescence. Scale bar is 0.8 mm wide

Table 1 Sample descriptions and stable isotopic analyses

Stratigraphic height (m)	Sample number	Description	$\delta^{13}C$ ‰ (VPDB)	$\delta^{18}O$ ‰ (VPDB)
4.87	15	Carbonate cemented sandstone	−3.06	−9.97
4.80	16	Carbonate cemented sandstone	−2.24	−10.39
4.68	17	Carbonate cemented sandstone	−3.31	−10.01
3.83	25	Carbonate cemented sandstone	−6.10	−13.49
3.64	27	Carbonate cemented sandstone	−6.44	−13.16
3.47	28	Carbonate cemented sandstone	−5.83	−12.27
3.18	31	Carbonate cemented sandstone	2.50	−9.20
3.08	32	Carbonate cemented sandstone	2.34	−9.14
2.96	33	Carbonate cemented sandstone	−2.20	−8.87
2.87	34	Carbonate cemented sandstone	−2.85	−10.15
2.77	35	Carbonate cemented sandstone	−2.53	−9.96
2.68	36	Carbonate cemented sandstone	−6.08	−12.82
2.10	42	Carbonate cemented sandstone	−2.19	−11.31
2.02	43	Carbonate cemented sandstone	−8.20	−14.04
1.71	46	Carbonate cemented sandstone	−8.47	−9.29
1.61	47	Carbonate cemented sandstone	−7.05	−12.92
1.40	49	Carbonate cemented sandstone	−6.02	−14.49
1.30	50	Carbonate cemented sandstone	−6.36	−12.75
1.21	51	Carbonate cemented sandstone	2.10	−7.96
1.11	52	Carbonate cemented sandstone	−1.92	−9.56
0.99	53	Carbonate cemented sandstone	−2.47	−11.02
0.89	54	Carbonate cemented sandstone	−6.38	−12.86
0.80	55	Carbonate cemented sandstone	−2.44	−9.16
0.70	56	Carbonate cemented sandstone	−1.99	−8.95
0.60	57	Carbonate cemented sandstone	−5.86	−8.14
0.39	58	Carbonate cemented sandstone	−5.55	−10.97
0.29	59	Carbonate cemented sandstone	−3.96	−9.16
0.19	60	Carbonate cemented sandstone	−5.55	−8.72
0.10	61	Carbonate cemented sandstone	−7.92	−14.75
0.05	62	Carbonate cemented sandstone	−6.41	−10.71
0.00	63	Carbonate cemented sandstone	−0.83	−9.14

Kiel III automated carbonate preparation device for their carbon and oxygen isotopic compositions ($\delta^{13}C$ and $\delta^{18}O$). Samples were reacted with phosphoric acid at 70 °C to liberate CO_2 gas. Values are reported as ‰ relative to the VPDB standard via regular calibration to an internal laboratory standard, HS-1 (−1.25‰ VPDB for $\delta^{13}C$ and −8.47‰ VPDB for $\delta^{18}O$). Precision is better than 0.05‰ VPDB for $\delta^{18}O$ and 0.02‰ VPDB for $\delta^{13}C$ (1σ). Isotopic data of carbonate cemented sandstones with stratigraphic location and carbonate concretions are presented in Tables 1 and 2.

Carbonate cements in the surrounding sandstones range from −8.47 to 2.5‰ for $\delta^{13}C$ VPDB and −7.96 to −14.75‰ VPDB for $\delta^{18}O$, and average −3.98 ± 2.98‰

Table 2 Stable isotopic analyses of concretions

Description	$\delta^{13}C$ ‰ (VPDB)	$\delta^{18}O$ ‰ (VPDB)	Concretion group
Unpaired concretion analyses			
Carbonate concretion a	−3.60	−5.91	I
Carbonate concretion b	−16.43	−3.81	II
Carbonate concretion 3	−15.58	−4.50	II
Carbonate concretion 8	−11.85	−4.29	II
Paired concretion analyses			
Carbonate concretion 1A, outer edge	−9.24	−12.10	
Carbonate concretion 1B, inner core	−15.65	−3.88	II
Carbonate concretion 2A, outer edge	−10.11	−8.68	
Carbonate concretion 2B, inner core	−13.50	−5.96	II
Carbonate concretion 6A outer edge	−2.05	−8.73	
Carbonate concretion 6B, inner core	−2.96	−9.33	I
Carbonate concretion 7A, outer edge	−15.80	−3.79	
Carbonate concretion 7B, inner core	−15.89	−3.64	II

VPDB and −10.82 ± 1.98‰ VPDB, respectively. Concretions are isotopically different than the cement of the surrounding sandstones (Fig. 8). Concretions fall into two groups based on their $\delta^{13}C$ values. Group I carbonate concretions have $\delta^{13}C$ averaging −2.87 ± 0.78‰ VPDB and $\delta^{18}O$ −7.99 ± 1.83‰ VPDB, and Group II carbonate concretions with $\delta^{13}C$ averaging −13.52 ± 2.81‰ VPDB and $\delta^{18}O$ −5.88 ± 3.01‰ VPDB. Overall, the concretions are isotopically lighter in $\delta^{13}C$ and heavier in $\delta^{18}O$ compared to the cements from surrounding sandstones when projected into the stratigraphic section where concretions and plant materials are found. Some concretions have a more complex isotopic cement history from core to margin. There is only one sample representing Group I (Concretion 6) with a core of $\delta^{13}C$ −2.96‰ VPDB and $\delta^{18}O$ 9.33‰ VPDB whereas the margin has $\delta^{13}C$ −2.05‰ VPDB and $\delta^{18}O$ −8.73‰ VPDB. Group II Concretions (1, 2 and 7) have cores averaging $\delta^{13}C$ −15.01 ± 1.32‰ VPDB and $\delta^{18}O$ −4.49 ± 1.28‰ VPDB, and have margins averaging $\delta^{13}C$ −11.72 ± 3.56‰ VPDB and $\delta^{18}O$ −8.19 ± 4.18‰ VPDB. Group II core to margin changes in $\delta^{13}C$ consistently shift from lighter to heavier isotopic value consistent with changing sources of dissolved inorganic carbon possibly related to mixing of groundwaters and marine waters. Core to margin changes in $\delta^{18}O$ are consistent with changing water compositions, as calculated temperatures would be unreasonable for the paleoenvironment and the preservation of the delicate features of the plant materials within the concretions.

Outward growth and changing proportions of fresh groundwater to marine water is suggested by the core to margin changes in $\delta^{13}C$ from light-to-heavy $\delta^{13}C$ values consistent with the breakdown of terrestrial organic matter as sources of alkalinity and $\delta^{18}O$ (mixing of marine and groundwater contributions). In addition, the fine-scale of preservation of plant materials within the concretions coupled with the changing $\delta^{18}O$ from core to margin of the concretions constrain the possible growth models (after Raiswell and Fisher 2000) responsible for precipitation of these

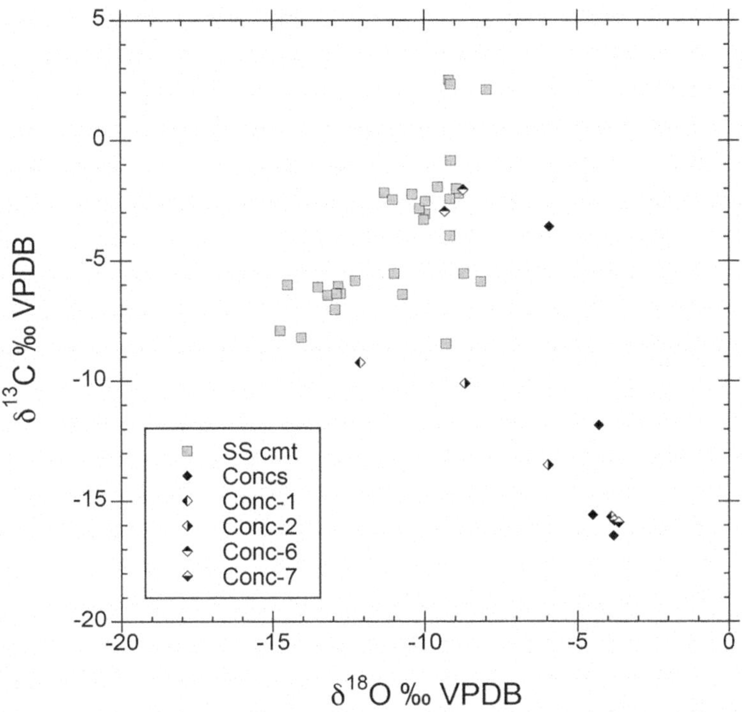

Fig. 8 $\delta^{13}C$ vs. $\delta^{18}O$ of sandstone and concretion samples. Most concretions stand out as light in $\delta^{13}C$ and heavy in $\delta^{18}O$ compared to sandstone cements. Ground waters have $\delta^{13}C_{DIC}$ values that are typically low, generally between −10‰ and −15‰ from plant respiration/decay and CO_2 production in soils. (After Leng and Marshall 2004)

concretions. Localized presence of plant materials (as a source for dissolved inorganic carbon) and mixing of fresh groundwaters and marine waters (to overcome inhibitors of calcite precipitation) provide a reasonable mechanism for precipitating carbonate penecontemporaneously with deposition. This combined mechanism described by Raiswell and Fisher (2000) preserves the delicate plant fossils and allows the cement to support both the mineral grains and plant debris without any petrographic evidence of compaction within these concretions from Apple Bay.

Results and Implications

Stable isotopic analyses confirm the presence of fresh groundwater as an important component in early carbonate cementation of concretions in marine sandstones, especially in western Canadian Cretaceous basins (e.g., Longstaffe and Ayalon 1987; Longstaffe 1994; *this study).* In addition, organic matter and its microbial degradation aid in the carbonate precipitation process (e.g., Mitterer and Cunningham

Jr. 1985; Kantorowicz et al. 1987; Wilkinson 1991, 1993; Coleman 1993; Ferris et al. 1995; Maliva et al. 2000). For example, Mozley and Davis (2005) attribute elongate concretion development in localized areas to favorable water chemistry (i.e., higher alkalinity) resulting from microbially induced degradation of organic matter within down-gradient groundwater flow.

Stratabound carbonate concretions at Apple Bay clearly formed contemporaneously with initial sand deposition in order to preserve delicate plants quickly. Eogenesis of these sandstones is not limited to bio-mediated, diffusive chemical processes, but is also affected by short-term "hydrodynamic" processes resulting in rapid cementation, most probably from groundwater flow (e.g., Longstaffe and Ayalon 1987; McBride et al. 1994, 1995; McBride and Parea 2001). For rapid permineralization and three-dimensional preservation of fragile plants within sands the transport modeling of Li et al. (1999) indicates the important role of meteoric water flushing at "high" rates of flow, perhaps on the order of submarine springs, must be invoked. Modeling of sandstone diagenesis by Wilkinson and Dampier (1990) shows that higher velocities of pore water fluid can decrease the time needed to form concretions. Additional evidence for rapid cementation at Apple Bay is the presence of cement-supported sand and silt grains in the concretions, which indicate cementation occurred before grains completely settled. Displacive growth of cement is not observed. Groundwater flow can also readily explain the alignment and restriction of concretions to bedding planes of many stratabound carbonates and concretions as well as the extrusion of calcium-rich groundwater from continental carbonate aquifers for the precipitation of calcite within sandstones (see Fig. 9 for flow model).

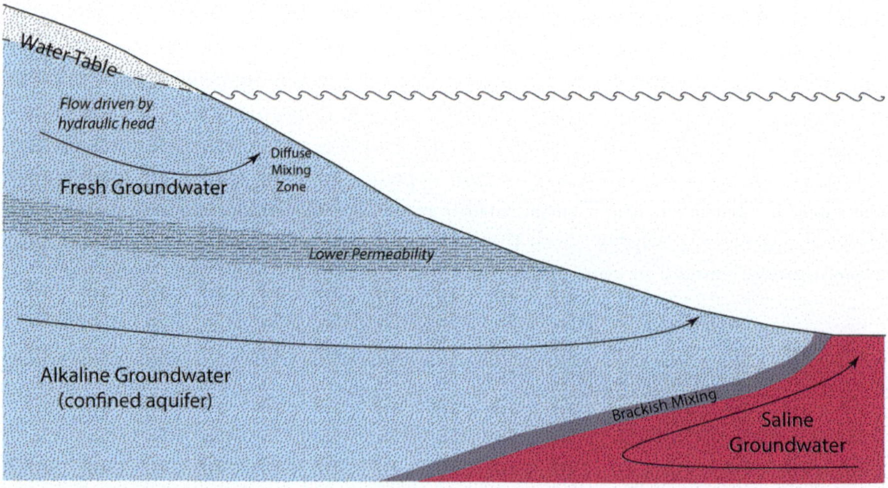

Fig. 9 Generalized diagenetic model of submarine discharge of freshwater aquifers directly into Cretaceous ocean sands at Apple Bay. (Modified from Bratton 2010). High fluid flux at the middle shoreface from input of a confined aquifer caused carbonate cementation as alkaline freshwater mixed with ocean water at the sediment/water interface. Model not to scale

There are other similar sandstones of Cretaceous marine sections in which permineralized paleobotanical material are reported from within calcareous nodules in discrete horizons as well (Nishida and Nishida 1988; Nishida et al. 1992, 1996; Lantz et al. 1999; Ratzel et al. 2001; Rothwell and Stockey 2002). Preservation of such delicate features, such as mosses and liverworts, points to a very early diagenetic cementation of carbonate (eogenesis) to protect these elements before significant degradation could occur through transport and microbial decay. Relatively quick cementation is clear from the exquisite preservation and association with organic matter is crucial for the precipitation process. SGD is also interpreted as the major mechanism in preserving these Cretaceous floral examples as well.

SGD needs to be recognized as an important process in eodiagenesis within marine sedimentary basins through time.

Older diagenetic models for meteoric water diagenesis of marine sediments only consider freshwater input from nearshore coastlines or from connate water preserved in deep water sediments (Bjørlykke 2010). Freshwater intrusion through continental aquifers at various depths along a continental margin, especially the freshwater within continental shelf sediments (Post et al. 2013), can simplify diagenetic models to explain meteoric influx for stratabound carbonate cementation within marine sediments. Freshwater intrusion can help explain the insufficient sources of calcium in host rock or unsubstantiated external sources as well as the inferred sea level fluctuations without substantiating evidence. More work is needed on the carbonate geochemistry associated with regions of SGD and more extensive geochemical mapping of the environments of carbonate-cemented sandstones along ancient marine exposures is critical to better understand eodiagenetic patterns of calcite cementation in marine sandstones through time.

Stratigraphic and lateral mapping coupled with stable isotopic analysis of carbonate-cemented sandstones and carbonate concretions in ancient marine exposures could delineate the extent of paleosignatures of SGD in the nearshore environment and possible driving forces. For example, systematic changes from light-to-heavy $\delta^{13}C$ values in concretions over a stratigraphic interval might be consistent with both an evolving dissolved inorganic carbon pool and chemistries of local aquifers in the source region. Changes in convective mixing of ocean and ground water cannot be ruled out. Constraining appropriate drivers will require a regional perspective and understanding of the paleogeography, paleoclimate, and paleowater chemistry.

SGD processes can help explain the "enigmatic" origins of Carboniferous coal balls.

An interesting parallel to our work may involve the origins of Carboniferous coal balls. Early meteoric calcite cementation can also be identified in Carboniferous coal balls, which are concretions of uncompressed peat cemented by calcium carbonate as well as pyrite, silica, siderite, and dolomite (Scott and Rex 1985; Scott et al. 1996; DeMaris 2000; Raymond et al. 2012; Richter et al. 2014). The peat is attributed to coastal swamps and mires with meteoric carbonate cementation, confirmed by stable isotopic data and occurring during an early stage, preserving over 75% of the flora (Scott and Rex 1985). Marine influence is evidenced by some stable isotopic evidence and associated marine animal remains (Holmes and Scott 1981; Scott et al. 1996). Marine fauna includes bryozoans, brachiopods, conodonts, ostracodes, trilobites, sponges, foraminifera, bivalves, gastropods, cephalopods, and worm tubes (Mamay and Yochelson 1962) preserved in various ways, such as pyritic, baritic, or phosphatic replacement or silica cement infill. Researchers have postulated the source of calcium for the coal ball cement to be from marine waters because of the presence of these marine fauna but have been confounded as to the process of adding such large quantities of seawater to the peat without isotopic evidence for it (Evans and Amos 1961; Scott et al. 1996; Raymond et al. 2012). Groundwater influence has been hinted at in some of the literature above, but interpretations have been counter to hydrodynamic gradients with models suggesting saline water bodies diffusing upward into the freshwater peat areas (see summary in Scott et al. 1996).

Coastal groundwater studies from recent literature suggest saltwater intrusion into freshwater coastal wetlands can occur seasonally, depending on the position of the freshwater/seawater interface according to the Ghyben-Herzberg Principle (Price et al. 2006; Zapata-Rios and Price 2012). The term "coastal groundwater discharge" (CGD) has been coined by Price et al. (2006) to describe the brackish water discharge from a coastline as a result of the mixing of meteoric groundwater with oceanic waters during seawater intrusion. The timing and extent of this brackish CGD is affected by regional precipitation patterns and regional evapotranspiration affecting the recharge patterns of terrestrial aquifers associated with freshwater coastal wetlands as well as seasonal tidal changes and wave activity on the seaward side (Michael et al. 2005; Price et al. 2006). Over geologic time with the influence of tectonic subsidence and uplift, the position of this coastal freshwater/seawater groundwater interface can change over time, allowing fresh groundwater discharge to cement peat within a freshwater swamp, mire, or wetland or sea groundwater or surface water intrusion to bring in marine groundwaters over long or short periods of time.

The meteoric carbonate cementation of the coal balls along the Carboniferous low-gradient coastlines now preserved in the rocks of North America and Europe (Scott et al. 1996) can easily be explained by the flow of calcium-rich fresh groundwater directly into the peat during various stages of preservation and position in the subsurface, from very early to later diagenesis, as the peat deposits were exposed to subsidence or sea level rises. The calcium-rich groundwaters could be derived from terrestrial carbonate aquifers. Siderite could form if the limestone of these terrestrial aquifers contained iron minerals or organic matter. High magnesium calcite, and possibly dolomite, could form from groundwaters percolating through dolostones before reaching the peat deposits (e.g., Siewers and Phillips 2015). Coastal groundwater discharge and dynamics of the freshwater/seawater groundwater interface can be a mechanism to theoretically apply to the early cementation and preservation of coal balls, noting the influence of marine incursions with stable isotopic analyses and marine faunal remains can account for later cementation events. The terrestrial groundwater would act as an unlimited source of the ions, dependent upon the regional bedrock geology, necessary for early cementation of coal balls, and future modeling of these groundwater processes can test the efficacy of this theory.

Acknowledgments This manuscript was originally drafted by BGK with contributions from GWR, RAS, and DBF. All authors designed and performed the research leading to this manuscript, which has evolved from earlier drafts dating back to 2014. This particular version emerged from discussions in 2016 prior to Beth's unfortunate passing. Every attempt has been made to retain Beth's original vision and thought process. Special thanks to Jim Haggart for consultations involving paleontologic identification and David Kidder and Bill Ullman for valuable comments on earlier versions of the manuscript. Martin Kordesch and Ricardo Zanatta aided in the cathodoluminescence analysis. The manuscript was improved by reviewer comments from Christine Chen, Kathleen Nicoll and Russell S. Shapiro. This study was funded by a National Science Foundation grant (EAR0308931) to GWR and a NSERC grant (A6908) to RAS.

References

Abdel-Wahab, A., & McBride, E. F. (2001). Origin of giant calcite-cemented concretions, Temple Member, Qasr El Sagha Formation (Eocene). *Faiyum Depression, Egypt: Journal of Sedimentary Research, 71*, 70–81.

Al-Ramadan, K., Morad, S., Proust, J. N., & Al-Aasm, I. (2005). Distribution of diagenetic alterations in siliciclastic shoreface deposits within a sequence stratigraphic framework: Evidence from the Upper Jurassic Boulonnais. *NW France: Journal of Sedimentary Research, 75*, 943–959.

Allan, J. R., & Matthews, R. K. (1977). Isotope signatures associated with early meteoric diagenesis. *Sedimentology, 29*, 797–817.

Bjørkum, P. A., & Walderhaug, O. (1990). Geometrical arrangement of calcite cementation within shallow marine sandstones. *Earth Science Reviews, 29*, 145–161.

Bjørkum, P. A., & Walderhaug, O. (1993). Isotopic composition of a calcite-cemented layer in the Lower Jurassic Bridport Sands, southern England: Implications for formation of laterally extensive calcite-cemented layers. *Journal of Sedimentary Petrology, 63*, 678–682.

Bjørlykke, K. (2010). *Petroleum geoscience: From sedimentary environments to rock physics*. Berlin: Springer, 508p.

Bjørlykke, K., & Jahren, J. (2012). Open or closed geochemical systems during diagenesis in sedimentary basins: Constraints on mass transfer during diagenesis and the prediction of porosity in sandstone and carbonate reservoirs. *American Association of Petroleum Geologists Bulletin, 96*, 2193–2214.

Bratton, J. F. (2010). The three scales of submarine groundwater flow and discharge across passive continental margins. *The Journal of Geology, 110*, 565–575.

Cai, W., Wang, Y., Krest, J., & Moore, W. S. (2003). The geochemistry of dissolved inorganic carbon in a surficial groundwater aquifer in North Inlet, South Carolina, and the carbon fluxes to coastal ocean. *Geochimica et Cosmochimica Acta, 67*, 631–637.

Campbell, R. B., Dodds, C. J., Yorath, C. J., & Brown, A. S. (1991). Insular belt. In H. Gabrielse & C. J. Yorath (Eds.), *Geology of the Cordilleran Orogen in Canada: Geological Survey of Canada* (Vol. 4 (G-2), pp. 574–581). Sherbrooke, QC: Geology of Canada Series.

Coleman, M. L. (1993). Microbial processes: controls on the shape and composition of carbonate concretions. *Marine Geology, 113*, 127–140.

Curtis, C. D., & Coleman, M. L. (1986). Controls on the precipitation of early diagenetic calcite, dolomite, and siderite concretions in complex depositional sequences. In D. L. Gautier (Ed.), *Roles of organic matter in sediment diagenesis* (Vol. 38, pp. 23–33). Tulsa, Oklahoma: SEPM Special Publication.

DeMaris, P. J. (2000). Formation and distribution of coal balls in the Herrin Coal (Pennsylvania), Franklin County, Illinois Basin, USA. *Journal of the Geological Society, London, 157*, 221–228.
Dutta, P. K. (1992). Climatic influence on diagenesis of fluvial sediments. In K. H. Wolf & G. V. Chillingarian (Eds.), *Diagenesis III, Developments in Sedimentology* (Vol. 47, pp. 191–252). Amsterdam: Elsevier.
Dutton, S. P., Willis, B. J., White, C. D., & Bhattacharya, J. P. (2000). Outcrop characterization of reservoir quality and interwell-scale cement distribution in a tide-influenced delta, Frontier Formation, Wyoming, USA. *Clay Minerals, 35*, 95–105.
Evans, W. E., & Amos, D. H. (1961). An example of the origin of coal-balls. *Proceedings of the Geologist's Association, 72*, 445–454.
Ferris, F. G., Fratton, C. M., Gerits, J. P., Schultze-Lam, S., & Sherwood Lollar, B. (1995). Microbial precipitation of a strontium calcite phase at a groundwater discharge zone near Rock Creek, British Columbia, Canada. *Geomicrobiology Journal, 13*, 57–67.
Gross, M. (1964). Variations in the O^{18}/O^{16} and C^{13}/C^{12} ratios of diagenetically altered limestones in the Bermuda Islands. *Journal of Geology, 72*, 170–194.
Haggart, J. W. (1991). A synthesis of cretaceous stratigraphy, Queen Charlotte Islands, British Columbia. In G. J. Woodsworth (Ed.), *Evolution and hydrocarbon potential of the queen Charlotte Basin* (Geological Survey of Canada, British Columbia, Paper 90-10) (pp. 253–277).
Haggart, J. W. (1996). Stratigraphy and correlation of cretaceous rocks of the northern Insular Belt, western Canada. *Mitteilungen Geologie-Paläontologie Institut, Universität Hamburg, 77*, 67–73.
Holmden, C., Papanastassiou, D. A., Blanchon, P., & Evans, S. (2012). $\delta^{44/40}$Ca variability in shallow water carbonates and the impact of submarine groundwater discharge on Ca-cycling in marine environments. *Geochimica et Cosmochimica Acta, 83*, 179–194.
Holmes, J., & Scott, A. C. (1981). A note on the occurrence of marine animal remains in a Lancashire coal ball. *Geological Magazine, 118*, 307–308.
Hudson, J. D. (1978). Concretions, isotopes, and the diagenetic history of the Oxford Clay (Jurassic) of central England. *Sedimentology, 25*, 339–370.
Jeletzky, J. A. (1970). Cretaceous macrofaunas. In E. W. Bamber, T. E. Bolton, M. J. Copeland, L. M. Cumming, H. Frebold, W. H. Fritz, J. A. Jeletzky, D. C. McGregor, D. J. McLaren, B. S. Norford, A. W. Norris, G. W. Sinclair, E. T. Tozer, & F. J. E. Wagner (Eds.), *Biochronology: Standard of phanerozoic time* (Geological Survey of Canada, Economic Geology Report 1B) (pp. 649–662).
Jeletzky, J. A. (1976). Mesozoic and ?Tertiary rocks of Quatsino Sound, Vancouver Island, British Columbia. *Geological Survey of Canada Bulletin, 242*, 1–243.
Johannes, R. E. (1980). The ecological significance of the submarine discharge of groundwater. *Marine Ecology Progress Series, 3*, 365–373.
Jørgensen, N. O. (1976). Recent high magnesian calcite/aragonite cementation of beach and submarine sediments from Denmark. *Journal of Sedimentary Petrology, 46*, 940–951.
Kantorowicz, J. D., Bryant, I. D., & Dawans, J. M. (1987). Controls on the geometry and distribution of carbonate cements in Jurassic sandstones: Bridgport Sands, southern England and Viking Group, Troll Field, Norway. In J. D. Marshall (Ed.), *Diagenesis of Sedimentary Sequences* (Vol. 36, pp. 103–118). Geological Society (London) Special. Publication.
Kiteley, L. W., & Field, M. E. (1984). Shallow marine depositional environments in the Upper Cretaceous of northern Colorado. In R. W. Tillman & C. T. Siemers (Eds.), *Siliciclastic shelf sediments* (Vol. 34, pp. 179–204). Tulsa, Oklahoma: SEPM Special Publication.
Kocurko, M. J. (1986). Interaction of organic matter and crystallization of high magnesium calcite, south Louisiana. In D. L. Gautier (Ed.), *Roles of organic matter in sediment diagenesis* (Vol. 38, pp. 13–21). SEPM Special Publication, Tulsa, Oklahoma.
Lantz, T. C., Rothwell, G. W., & Stockey, R. A. (1999). *Conantiopteris schuchmanii*, gen. et sp. nov., and the role of fossils in resolving phylogeny of Cyatheaceae s.l. *Journal of Plant Research, 112*, 361–381.

Leng, M. J., & Marshall, J. D. (2004). Paleoclimate interpretation of stable isotope data from lake sediment archives. *Quaternary Science Reviews, 23*, 811–831.
Li, L., Barry, D. A., Stagnitti, F., & Parlange, J.-Y. (1999). Submarine groundwater discharge and associated chemical input to a coastal sea. *Water Resources Research, 35*, 3253–3259.
Little, S. A., Stockey, R. A., & Rothwell, G. W. (2006). *Solenostelopteris skogiae* sp. nov. from the Lower Cretaceous of Vancouver Island. *Journal of Plant Research, 119*, 525–532.
Longstaffe, F. J., & Ayalon, A. (1987). Oxygen-isotope studies of clastic diagenesis in the Lower Cretaceous Viking Formation, Alberta: implications for the role of meteoric water. In J. D. Marshall (Ed.), *Diagenesis of sedimentary sequences* (Geological Society (London) Special Publication) (Vol. 36, pp. 277–296).
Longstaffe, F. J. (1994). Stable isotope constraints on sandstone diagenesis in the western Canada sedimentary basin. In A. Parker & B. W. Sellwood (Eds.), *Quantitative diagenesis: Recent developments and applications to reservoir geology* (pp. 223–274). Dordrecht: Kluwer Academic Publishers.
Machel, H. G. (2000). Application of cathodoluminescence to carbonate diagenesis. In M. Pagel, V. Barbin, P. Blanc, & D. Ohnenstetter (Eds.), *Cathodoluminescence in geosciences* (pp. 271–301). Berlin: Springer.
Maliva, R. G., Missimer, T. M., Leo, K. C., Statom, R. A., Dupraz, C., Lynn, M., & Dickson, J. A. D. (2000). Unusual calcite stromatolites and pisoids from a landfill leachate collection system. *Geology, 28*, 931–934.
Mamay, S. H., & Yochelson, E. L. (1962). *Occurrence and significance of marine animal remains in American coal balls*. U.S. Geological Survey Professional Paper 354-I, 224p. plus plates.
Manheim, F. T., & Paull, C. K. (1981). Patterns of groundwater salinity changes in a deep continental-oceanic transect off the southeastern Atlantic coast of the U.S.A. *Journal of Hydrology, 54*, 95–105.
McBride, E. F., & Parea, G. C. (2001). Origin of highly elongate, calcite-cemented concretions in some Italian coastal beach and dune sands. *Journal of Sedimentary Research, 71*, 82–87.
McBride, E. F., Picard, M. D., & Folk, R. L. (1994). Oriented concretions, Ionian coast, Italy: evidence of groundwater flow direction. *Journal of Sedimentary Research, A64*, 535–540.
McBride, E. F., Milliken, K. L., Cavazza, W., Cibin, U., Fontana, D., Picard, M. D., & Zuffa, G. G. (1995). Heterogeneous distribution of calcite cement at the outcrop scale in Tertiary sandstones, northern Apennines, Italy. *American Association of Petroleum Geologists Bulletin, 79*, 1044–1063.
Michael, H. A., Mulligan, A. E., & Harvey, C. F. (2005). Seasonal oscillations in water exchange between aquifers and the coastal ocean. *Nature, 436*, 1145–1148.
Miller, D. C., & Ullman, W. J. (2004). Ecological consequences of groundwater discharge to Delaware Bay, United States. *Ground Water, 42*, 959–970.
Milliken, K. L., McBride, E. F., Cavazza, W., Cibin, U., Fontana, D., Picard, M. D., & Zuffa, G. G. (1998). Geochemical history of calcite precipitation in Tertiary sandstones, northern Apennines, Italy. In S. Morad (Ed.), *Carbonate Cementation in Sandstones: Distribution Patterns and Geochemical Evolution* (Vol. 26, pp. 213–239). International Association of Sedimentologists Special Publication.
Mitterer, R. M., & Cunningham, R., Jr. (1985). The interaction of natural organic matter with grain surfaces: Implications for calcium carbonate precipitation. In N. Schneidermann & P. M. Harris (Eds.), *Carbonate Cements* (Vol. 36, pp. 17–31). Tulsa, Oklahoma: SEPM Special Publication.
Moore, W. S. (1999). The subterranean estuary: A reaction zone of groundwater and sea water. *Marine Chemistry, 65*, 111–125.
Moore, W. S. (2010). The effect of submarine groundwater discharge on the ocean. *Annual Review of Marine Science, 2*, 59–88.
Mozley, P. S., & Burns, S. J. (1993). Oxygen and carbon isotopic composition of marine carbonate concretions: An overview. *Journal of Sedimentary Petrology, 63*, 73–83.

Mozley, P. S., & Davis, J. M. (2005). Internal structure and mode of growth of elongated calcite concretions: evidence for small-scale microbially induced, chemical heterogeneity in groundwater. *Geological Society of America Bulletin, 117*, 1400–1412.

Nishida, H., & Nishida, M. (1988). *Protomonimia kasai-nakajhongii* gen. et sp. nov.: a permineralized Magnlialean fructification from the mid-Cretaceous of Japan. *Botanical Magazine of Tokyo, 101*, 397–426.

Nishida, M., Ohsawa, T., & Nishida, H. (1992). Structure and affinities of the petrified plants from the Cretaceous of northern Japan and Saghalien VIII. *Parataiwania nihongii* gen. et sp. nov., a taxodiaceous cone from the Upper Cretaceous of Hokkaido. *Journal of Japanese Botany, 67*, 1–9.

Nishida, M., Yoshida, A., & Nishida, H. (1996). *Cretocycas yezonakajimae* gen. et sp. nov., a permineralized cycad petiole from the Upper Cretaceous of Hokkaido. *Journal of Japanese Botany, 71*, 223–230.

Nixon, G. T., Hammack, J. L., Koyanagi, V. M., Payie, G. J., Panteleyev, A., Massey, N. W. D., Hamilton, J. V., & Haggart, J. W. (1994). *Preliminary geology of the Quatsino-Port McNeill Map areas, northern Vancouver Island* (pp. 63–85). Paper – Ministry of Energy, Mines and Petroleum Resources.

Park, A. J. (2009). *Prediction of calcite cementation in sandstones associated with sandstone-shale interaction, Gulf of Mexico* (p. 160). 2009 American Association of Petroleum Geologists Annual Meeting (Denver CO), Abstract Volume,.

Pemberton, S. G., Spila, M., Pulham, A. J., Saunders, T., MacEachern, J. A., Robbins, D., & Sinclair, I. K. (2001). Ichnology and Sedimentology of Shallow to Marginal Marine Systems: Ben Nevis and Avalon Reservoirs, Jeanne d'Arc Basin. *Geological Association of Canada Short Course Notes, 15*, 1–343.

Plummer, L. N. (1975). Mixing of sea water with calcium carbonate groundwater. In W. H. T. Whitten (Ed.), *Quantitative Studies in the Geological Sciences* (Vol. 142, pp. 219–236). Geological Society of America Memoir.

Post, V. E. A., Groen, J., Kooi, H., Person, M., Ge, S., & Edmunds, W. M. (2013). Offshore fresh groundwater reserves as a global phenomenon. *Nature, 504*, 71–78.

Price, R. M., Swart, P. K., & Fourquean, J. W. (2006). Coastal groundwater discharge – An additional source of phosphorus for the oligotrophic wetlands of the Everglades. *Hydrobiologia, 569*, 23–36.

Raiswell, R., & Fisher, Q. J. (2000). Carbonate concretions: a review of growth mechanisms and their influence on chemical and isotopic composition. *Journal of the Geological Society (London), 157*, 239–257.

Ratzel, S. R., Rothwell, G. W., Mapes, G., Mapes, R. H., & Doguzhaeva, L. A. (2001). *Pityostrobus hokodzensis*, a new species of pinaceous cone from the Cretaceous of Russia. *Journal of Paleontology, 75*, 895–900.

Raymond, A., Guillemette, R., Jones, C. P., & Ahr, W. M. (2012). Carbonate petrology and geochemistry of Pennsylvanian coal balls from the Kalo Formation of Iowa. *International Journal of Coal Geology, 94*, 137–149.

Richter, D. K., Heinrich, F., Geske, A., Neuser, R. D., Gies, H., & Immenhauser, A. (2014). First description of Phanerozoic radiaxial fibrous dolomite. *Sedimentary Geology, 304*, 1–10.

Ritger, S., Carson, B., & Suess, E. (1987). Methane-derived authigenic carbonates formed by subduction-induced pore-water expulsion along the Oregon/Washington margin. *Geological Society of America Bulletin, 98*, 147–156.

Robb, J. M. (1990). Groundwater processes in the submarine environment. In C. G. Higgins & D. R. Coates (Eds.), *Groundwater geomorphology: The role of subsurface water in earth-surface processes and landforms* (Vol. 252, pp. 267–281). Geological Society of America Special Paper.

Roberts, H. H., & Whelan, T., III. (1975). Methane-derived carbonate cements in barrier and beach sands of a subtropical delta complex. *Geochimica et Cosmochimica Acta, 39*, 1085–1089.

Rodellas, V., Garcia-Orellana, J., Masque, P., Feldman, M., & Weinstein, Y. (2015). Submarine groundwater discharge as a major source of nutrients to the Mediterranean Sea. *Proceedings of the National Academy of Sciences, 112*, 3927–3930.

Rothwell, G. W., Crepet, W. L., & Stockey, R. A. (2009). Is the anthophyte hypothesis alive and well? New evidence from the reproductive structures of Bennettitales. *American Journal of Botany, 96*, 296–322.

Rothwell, G. W., & Stockey, R. A. (2002). Anatomically preserved *Cycadeoidea* (Cycadeoidaceae) with a re-evaluation of systematic characters for the seed cones of Bennettitales. *American Journal of Botany, 89*, 1447–1458.

Rothwell, G. W., & Stockey, R. A. (2006). Combining characters of Pteridaceae and tree ferns: *Pterisorus radiata* gen. et. sp. nov., a permineralized Lower Cretaceous filicalean fern. *International Journal of Plant Sciences, 167*, 695–701.

Schmalz, R. F. (1971). Formation of beachrock at Eniwetok atoll. In O. P. Bricker (Ed.), *Carbonate Cements* (pp. 17–24). Baltimore: The John Hopkins Press.

Schwartz, M. C. (2003). Significant groundwater input to a coastal plain estuary: Assessment from excess radon: Estuarine. *Coastal and Shelf Science, 56*, 31–42.

Scott, A. C., & Rex, G. (1985). The formation and significance of Carboniferous coal balls. *Philosophical Transactions of the Royal Society of London, Series B, Biological Sciences, 311*, 123–137.

Scott, A. C., Mattey, D. P., & Howard, R. (1996). New data on the formation of carboniferous coal balls. *Review of Palaeobotany and Palynology, 93*, 317–331.

Scott, A. C., & Collinson, M. E. (2003). Non-destructive multiple approaches to interpret the preservation of plant fossils: implications for calcium-rich permineralizations. *Journal of the Geological Society (London), 160*, 857–862.

Shaban, A., Khawlie, M., Abdallah, C., & Faour, G. (2005). Geologic controls on submarine groundwater discharge: Application of remote sensing to North Lebanon: Environmental. *Geology, v., 47*, 512–522.

Siewers, F. D., & Phillips, T. L. (2015). Petrography and microanalysis of Pennsylvanian coal-ball concretions (Herrin Coal, Illinois Basin, USA): Bearing on fossil plant preservation and coal-ball origins. *Sedimentary Geology, 329*, 130–148.

Smith, S. Y., Rothwell, G. W., & Stockey, R. A. (2003). *Cyathea cranhami* sp. nov. (Cyatheaceae) anatomically preserved tree fern sori from the Lower Cretaceous of Vancouver Island, British Columbia. *American Journal of Botany, 90*, 755–760.

Stanich, N. A., Rothwell, G. W., & Stockey, R. A. (2009). Phylogenetic radiation of *Equisetum* (Equisetales) as inferred by Lower Cretaceous species of from British Columbia, Canada. *American Journal of Botany, 96*, 1289–1299.

Stockey, R. A., & Rothwell, G. W. (2006). The last of the pre-angiospermous vegetation: A Lower Cretaceous flora from Apple Bay, Vancouver Island. Presentation No. 45. In *Advances in paleobotany – recognizing the contributions of David L. Dilcher and Jack A. Wolfe on the occasion of their 70th Birthday, Florida Museum of Natural History*. University of Florida.

Stockey, R. A., & Rothwell, G. W. (2009). Distinguishing angiophytes from the earliest angiosperms: A Lower Cretaceous (Valanginian-Hauterivian) fruit-like reproductive structure. *American Journal of Botany, 96*, 323–335.

Stockey, R. A., & Wiebe, N. J. P. (2008). Lower Cretaceous conifers from Apple Bay, Vancouver Island: Picea-like leaves, *Midoriphyllum piceoides* gen. et sp. nov. (Pinaceae). *Botany, 86*, 649–657.

Stockey, R. A., Rothwell, G. W., & Little, S. A. (2006). Relationships among fossil and living Dipteridaceae: Anatomically preserved *Hausmannia* from the Lower Cretaceous of Vancouver Island. *International Journal of Plant Sciences, 167*, 649–663.

Taniguchi, M., Burnett, W. C., Cable, J. E., & Turner, J. V. (2002). Investigation of submarine groundwater discharge. *Hydrological Processes, 16*, 2115–2129.

Ullman, W. J., Chang, B., Miller, D. C., & Madsen, J. A. (2003). Groundwater mixing, nutrient diagenesis, and discharges across a sandy beachface, Cape Henlopen, Delaware (USA). *Estuarine, Coastal, and Shelf Science, 57*, 539–552.

Vavrek, M. J., Stockey, R. A., & Rothwell, G. W. (2006). *Osmunda vancouverense* sp. nov. (Osmundaceae), permineralized fertile frond segments from the Lower Cretaceous of British Columbia, Canada. *International Journal of Plant Sciences, 167*, 631–637.

Watts, N. L. (1978). Displacive calcite: evidence from recent and ancient calcretes. *Geology, 6*, 699–703.

Wigley, T. M. L., & Plummer, L. N. (1976). Mixing of carbonate waters. *Geochimica et Cosmochimica Acta, 40*, 989–995.

Wilkinson, M. (1991). The concretions of the Bearreraig Sandstone Formation: geometry and geochemistry. *Sedimentology, 38*, 899–912.

Wilkinson, M. (1993). Geometrical arrangement of calcite cementation within shallow marine sandstones. *Earth Science Reviews, 34*, 47–51.

Wilkinson, M., & Dampier, M. D. (1990). The rate of growth of sandstone-hosted calcite concretions. *Geochimica et Cosmochimica Acta, 54*, 3391–3399.

Zapata-Rios, X., & Price, R. M. (2012). Estimates of groundwater discharge to a coastal wetland using multiple techniques: Taylor Slough, Everglades. *Hydrogeology Journal, 20*, 1651–1668.

Zelster, I. S., Everett, L. G., & Dzhamalov, R. G. (2006). *Submarine groundwater*. Boca Raton, FL: CRC Press, Taylor and Francis Group, 475p.

Part V
South America

Reconstructing Paleoenvironmental Conditions Through Integration of Paleogeography, Stratigraphy, Sedimentology, Mineralogy and Stable Isotope Data of Lacustrine Carbonates: An Example from Early Middle Triassic Strata of Southwest Gondwana, Cuyana Rift, Argentina

C. A. Benavente, A. C. Mancuso, and K. M. Bohacs

Abstract Reconstructing continental paleoenvironmental conditions (temperature, precipitation, hydrology, and hydrography) is essential for constraining the influences on terrestrial ecosystems, sediment and solute yields to the ocean and carbon cycles, as well as for calibrating numerical paleoclimate models. Making such reconstructions from lacustrine strata, however, is quite challenging because of varying water sources, flow paths and residence times, paleogeographic effects, and eodiagenesis. Indeed, even stratigraphic evidence of fluctuations in lake level cannot be directly interpreted in terms of changing precipitation because of the complex response of lakes to changing input conditioned by upstream and downstream factors. Such challenges can be addressed by thorough integration of paleogeography, stratigraphy, sedimentology, mineralogy, and stable isotope data of lacustrine carbonates within a framework that accounts for the many and convoluted controls on lake systems.

C. A. Benavente (✉)
Geology, FCEN-UNCuyo, Mendoza, Argentina

Instituto Argentino de Nivología, Glaciología y Ciencias Ambientales (IANIGLA), CCT-Mendoza, CONICET, Mendoza, Argentina

A. C. Mancuso
Facultad de Ciencias Exactas y Naturales (FCEN), Universidad Nacional de Cuyo (UNCuyo), Mendoza, Argentina

K. M. Bohacs
KMBohacs Geoconsulting, Houston, TX, USA

M. R. Rosen et al. (eds.), *Limnogeology: Progress, Challenges and Opportunities*, Syntheses in Limnogeology, https://doi.org/10.1007/978-3-030-66576-0_16

We illustrate this approach using data from three lacustrine units from the Cuyana Basin (Argentina) of early Middle Triassic age—the Cerro de las Cabras, Santa Clara Arriba, and Cerro Puntudo formations. They represent sedimentation in carbonate-rich lacustrine systems during a time for which information on continental paleoenvironments is sparse, especially for interior Pangea. Our detailed study of the stable carbon and oxygen isotope composition of the carbonate beds of these paleolakes, integrated with other geological evidence, enabled interpretation of their complex hydrology and paleoclimate conditions. Sedimentological, stratigraphic, and mineralogical data suggest surface water flow ranged from intermittently open to persistently closed, whereas C and O stable isotope values indicate that both Cerro de las Cabras and Cerro Puntudo paleolakes had open groundwater flow. The Santa Clara Arriba paleolake, although intermittently open to surface flow, had long water residence time, based on high correlation of C and O stable isotopes.

This evidence of the hydrographic-hydrologic complexity of these continental interior basins obviates confident quantitative estimates of paleotemperature from oxygen isotopes. The isotope data do, however, indicate significant evaporation and frequently varying precipitation, runoff, and nutrient supply. These data, along with ephemeral-stream strata, subaerial-exposure features, vertisols, and clay-mineral assemblages dominated by smectite/illite, suggest a seasonally varying warm semi-arid to sub-humid paleoclimate. This is in agreement with paleoclimate models for the Triassic that pointed to marked seasonality in the Cuyana Rift Basin.

Keywords Carbonates · $\delta^{13}C$ · $\delta^{18}O$ · Dolomite · Hydrogeochemistry · Lake basin · Pangea · Triassic

Introduction

Continental carbonates in general, and lacustrine carbonates in particular, preserve detailed information on paleohydrochemistry that is very helpful in reconstructing such paleoenvironmental parameters as groundwater hydrology, temperature, and precipitation (Kelts and Talbot 1990). Therefore, the study of carbon and oxygen stable isotope signatures from continental carbonates in an integrated paleogeographic, stratigraphic, and sedimentological context provides a rich data set to supplement the classic paleobotany approach. Although palynoflora can be sensitive indicators of paleoclimate, stable isotope analyses of lacustrine carbonates provide the benefits of being easier to find and sample, allowing characterization of closely spaced vertical samples to detect stratigraphic trends, avoiding taphonomic biases, and having the potential for quantitative estimates of paleotemperature (Benavente and Mancuso 2016). Incorporating stable-isotope data is especially helpful for situations where samples are available only from limited outcrops that do not span the entire depositional profile from alluvial plain to lake center, or when time or resource limitations require an early interpretation based on focused sample sets. Great caution, however, must be used in the selection, characterization, and interpretation of such continental carbonates to yield reliable results (Benavente et al. 2019a).

Challenges to interpreting lacustrine carbonates include varying water sources, flow paths, and residence times linked to varying upstream and downstream conditions, paleogeographic effects, and eodiagenesis. All these factors affect lake level but are not discernable by stratigraphic evidence alone (Bohacs et al. 2000). Reliable interpretations require characterizing these factors using an integrated multiproxy approach (Table 1). As a fundamental part of such approach, the meticulous C and O stable isotope study of the primary signature from lacustrine carbonates provides a unique data set for a lake basin with insights about paleohydrology, including the possibility of discriminating between lacustrine systems dominated by within-lake processes or extra-basin controls and constraining paleoclimate.

Such a robust integrated approach based on fundamental principles is particularly important when interpreting the continental environmental conditions of

Table 1 Integrated multiproxy approach for deep time lake basin paleoenvironmental interpretations

Describe sedimentology, mineralogy, stratigraphy, and paleogeographic setting of lake strata
Acknowledge complexities difficult to model in the lake basin (paleogeography linked to catchment areas and rain shadows, size of the lacustrine basin versus size of catchment, drainage patterns: surface flow versus groundwater flow, among others)
Collect carbonate samples in context and analyze
Examine carbonate samples to locate intervals with primary depositional fabrics: macroscopic fabric
At microfabric scale combine petrography, CL, microprobe, and SEM to determine the primary carbonate
Sample primary carbonate for geochemical analysis with microdrill
Analyze replicates and closely spaced vertical and lateral samples of all facies to estimate temporal, spatial, and facies variability
Collect siliciclastic samples (siltstone and mudstone) and analyze
Examine siliciclastic samples at microscale for primary composition (SEM)
Sample primary siliciclastics for XRD
Determine geological age
Interpret hydrography and hydrology
Characterize lake-basin type from stratigraphy, sedimentology, mineralogy, etc.
Characterize hydrology from C-O covariance, etc.
Interpret paleoclimate
Use geological age and paleogeography (latitude, altitude) for O-isotope boundary conditions
Use absolute range of $\delta^{18}O$ to estimate seasonality of precipitation/temperature if appropriate
Use primary clay assemblages as proxies for climate conditions
Interpret paleoprecipitation/humidity
Compare interpretations with coeval units from same general vicinity to differentiate effects of regional climate from local hydrology
Compare interpretations with paleoclimate models for the same age and region

deep-time spans with significantly different paleogeography and paleoclimate, such as the early Middle Triassic Period, the age of the strata we studied.

During the Triassic, the supercontinent Pangea extended across the equator from about 85°N to 90°S (Frakes 1979; Ziegler et al. 1983; Parrish 1993; Torsvik et al. 2012; Torsvik and Cocks 2013; Holz 2015). This apparently caused extraordinary effects on global climate (Parrish et al. 1986). Models that attempt to explain global atmospheric circulation propose a mega-monsoonal regime during the Triassic linked to highly marked seasonality, a displacement of the arid belts polewards, and a decrease of the precipitation/evaporation rate in the mid-latitude interior of Pangea (Kent and Tauxe 2005; Sellwood and Valdes 2006). The development of highly marked seasonality during the Triassic has been attributed to the almost symmetric position of Pangea over the equator during a time of changing climate mode from warm to hot with high-latitude ice-free areas (Parrish 1993; Sellwood and Valdes 2006; Holz 2015; Retallack 2013).

Globally, large perturbations of carbon and oxygen stable-isotope signatures have been reported from marine sections across the Permian-Triassic boundary and into the Early Triassic (Atudorei and Baud 1997; Krull et al. 2004; Galfetti et al. 2007; Henkes et al. 2018). Stable carbon isotopes are characterized by a negative excursion across the P-T boundary and a gradual trend to less negative values during Early Triassic. Subsequent positive excursions during the early Middle Triassic (Anisian) have been attributed to productivity peaks in the oceans (Atudorei and Baud 1997). Such perturbations have also been recorded in terrestrial ecosystems (Retallack et al. 2011).

The Cuyana Rift Basin in west-central Argentina spanned a broad area during the Triassic, with numerous sub-basins that contained carbonate-rich lacustrine systems (Spalletti 2001) (Fig. 1a). The stratal record of these systems offers an excellent opportunity to study paleohydrology patterns and paleoclimate conditions for the Middle Triassic. In the Cerro Puntudo sub-basin, at the northernmost margin of the rift, the studied deposits belong to the Cerro Puntudo Formation (Fig. 1b) and the Santa Clara sub-basin (Fig. 1c), located south of the Cerro Puntudo sub-basin, presents lacustrine deposits of the Santa Clara Arriba Formation. The Potrerillos sub-basin, located south of the Cerro Puntudo sub-basin, includes the Cerro de las Cabras Formation (Fig. 1d).

The Cerro Puntudo and Cerro de las Cabras formations have been characterized as the record of Ca-rich playa-lake systems that developed during the synrift phase of the Cuyana Basin in Anisian times (Benavente et al. 2015). Interpretations of their paleohydrology have been obtained through stratigraphy, sedimentology, mineralogy, and carbon and oxygen stable isotope studies (Benavente et al. 2015, 2019a, b).

The Santa Clara sub-basin, which is in between these two sub-basins (Fig. 1c), however, is an area that remains fairly unexplored but is of great interest since it also contains thick early Middle Triassic carbonate-rich lacustrine successions. In this contribution, our same multiproxy approach is used to characterize this third sub-basin. These new data are integrated across the region of all three sub-basins to achieve detailed paleoenvironmental reconstructions, including hydrology and paleoclimate inferences for deep-time lacustrine systems.

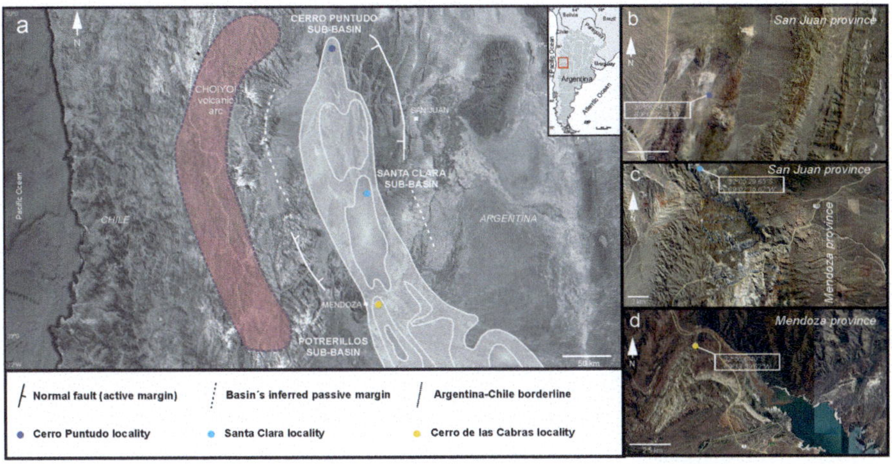

Fig. 1 (**a**) Location map of the Triassic sedimentary basins of the west-central region of Argentina and the location of the master fault bounding the half grabens of the Cuyana Triassic rift. (**b**) Location of the Cerro Puntudo sub-basin, northern area of the rift. (**c**) Location of the Santa Clara sub-basin, latitudinally between Cerro Puntudo (north) and Potrerillos (south). (**d**) Location of the Potrerillos sub-basin at the south of Santa Clara sub-basin

The aims of this study are: (1) to characterize the stable isotopic composition of the lacustrine carbonates of the Santa Clara Arriba Formation; (2) to understand the paleohydrology of this system; (3) to compare its paleohydrology pattern with the hydrologic interpretation of the Cerro de las Cabras and Cerro Puntudo formations (Benavente et al. 2019a, b); and (4) to provide insights into continental paleoclimate conditions for the early Middle Triassic in south-west Gondwana.

Geological Setting

The Cuyana Basin spans 30,000 km^2 across the San Juan, Mendoza, and San Luis provinces of Argentina. The basin formed due to rifting of the south-west margin of Gondwana during the Triassic (Uliana and Biddle 1988). The basin trends NNW-SSE and is formed by several asymmetric half-grabens with the location of the master fault alternating between the western and eastern sides of the rift basin (Legarreta et al. 1992; Kokogian et al. 1993; Barredo et al. 2011). In the northern Cuyana Basin, the master fault is located in the east; and in the southern area of the basin, the master fault is located to the west (Ramos and Kay 1991; Legarreta et al. 1992; Barredo 2005) (Fig. 1a). We studied strata from three sub-basins (from north to south): Cerro Puntudo, Santa Clara, and Potrerillos (Fig. 2). Their ages have been constrained to the early Middle Triassic (Ávila et al. 2006; Benavente et al. 2019b; Teixeira et al. 2018; Mancuso et al. 2010).

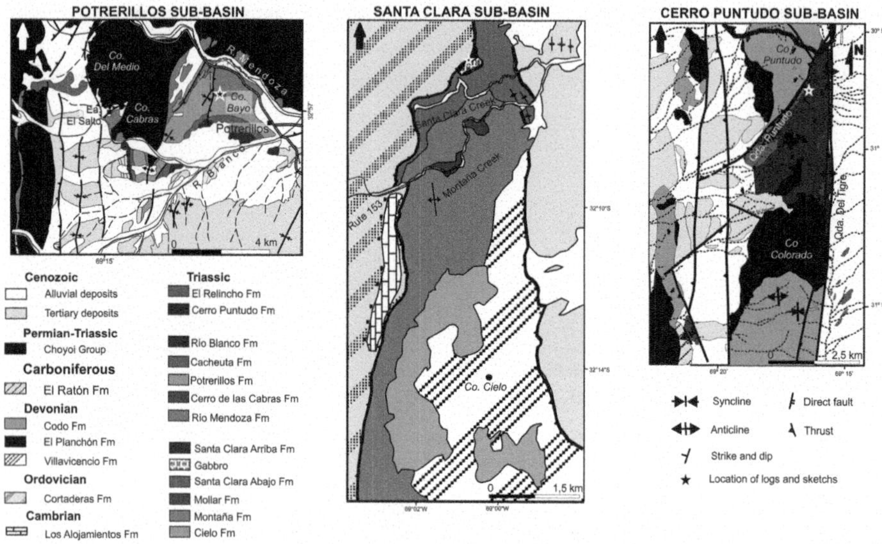

Fig. 2 Geologic map of the Cerro Puntudo, Santa Clara, and Cerro de las Cabras sub-basins of the Cuyana Rift Basin

Paleogeographically, the study area was about 800 to 1000 km from the southwest ocean coast of Pangea (Markello et al. 2007) and occupied paleolatitudes between 44°S and 52°S (calculated using Van Hinsbergen et al. 2015) during the Anisian stage. Paleoaltitudes likely ranged from 500 to 1000 meters above sea level, based on the elevations of analogous modern lake systems in the East Africa Rift (Moernaut et al. 2010). Paleoclimate models indicate that the area had prevailing westerly winds off the proto-Pacific Ocean that generated downwelling along the ocean coast south of about 40°S (Parrish 1993; Sellwood and Valdes 2006; Holz 2015). This wind pattern was due to an atmospheric gyre located just offshore that shifted seasonally by about 10°, between 35°S and 45°S, which could have impacted our study area. All three sub-basins had an extensive topographic barrier to the west (Choiyoi volcanic arc). In addition, the Cerro Puntudo and Santa Clara sub-basins had topographic barriers along their eastern margins due to active rift-shoulder uplift (Benavente et al. 2018) but the Potrerillos sub-basin had its active margin and topographic barrier to the west in addition to the Choiyoi volcanic arc.

All three sub-basins contain carbonate-rich lacustrine strata (Benavente et al. 2012, 2015, 2018) and their basement is the volcanic rocks of the Permian-Triassic Choiyoi Group (Llambías 2001). There also are abundant Paleozoic and Cenozoic sedimentary rocks in their catchment areas (Folguera and Etcheverría 2004) (Fig. 2).

In the southernmost Potrerillos sub-basin (Fig. 1d), the Cerro de las Cabras formation (CC) is an aggradational succession that has been interpreted as an evaporative facies association, diagnostic of underfilled-lake basins (Benavente et al. 2015). In the Santa Clara sub-basin (Fig. 1c), the Santa Clara Arriba formation (SCA) contains thin clayey tuffaceous fine-grained sandstone beds alternating with laminated limestone and mudstone (Harrington 1971) and was previously interpreted as an overfilled-lake basin, based on sedimentology and stratigraphy only (Benavente et al. 2018). Integration of mineralogical and isotopic information, however, suggests this formation records balanced-fill conditions.

In the northernmost Cerro Puntudo sub-basin (Fig. 1b), sedimentological studies of the Cerro Puntudo formation (CP) revealed an alkaline playa-lake system with a succession characterized by an aggradational-progradational stacking pattern of fluctuating profundal lake-facies-association strata that has been interpreted as a balanced-fill lake (Benavente et al. 2015).

Methods

A total of 98 samples from all three units were analyzed (CP: 71; SCA: 12; CC: 15). These samples represent all the carbonates found in these successions. The samples were processed at the Laboratorio de Paleontología del Instituto Argentino de Nivología, Glaciología y Ciencias Ambientales (IANIGLA). Twenty milligrams of powder were extracted from polished slabs of samples under low-magnification binocular microscope with a manual micro-drill (Dremel Multipro 395 JP). Criteria to differentiate primary carbonate from diagenetic carbonate phases included a thorough multiproxy study with petrography, cathodoluminescence, and electron microprobe analysis data (Murphy et al. 2014; Parrish et al. 2018; Henkes et al. 2018; Benavente et al. 2019a). Those analyses led to the identification of three main carbonate phases that were sampled separately: non-luminescent dolomite, non-luminescent micrite, and luminescent spar (Benavente et al. 2019a).

Samples were analyzed on Thermo Fisher Scientific GasBench II, equipped with a PAL autosampler, and coupled to a ConFlow IV interface and a MAT 253 mass spectrometer (Thermo Fisher Scientific) for their carbon and oxygen isotopic compositions ($\delta^{13}C$ and $\delta^{18}O$) at the Stable Isotope Ratio Facility for Environmental Research (SIRFER) University of Utah, Salt Lake City, USA. Carrara marble and LSVEC were used as primary reference materials, and Marble-Std was used as secondary reference material to cross-check the final values. Internal reference material(s) were calibrated against international standards NBS-18 and NBS-19. The oxygen fractionation factor was calculated using the alpha value proposed by Swart et al. (1991). Values are reported as ‰ relative to the VPDB standard. Precision is better than 0.05‰ for $\delta^{18}O$ and 0.02‰ for $\delta^{13}C$. The variability of C and O stable isotope values within beds was tested at random carbonate beds in which microsamples were taken at each lamina. The obtained standard deviation for the measurements was 0.57‰ and therefore each sample was considered representative of its bed. Nevertheless, wherever there was a change in the carbonate texture a new microsample was taken and analyzed.

The mineralogical composition of samples was analyzed with x-ray diffraction (XRD) of the total fraction and finest fraction <2 μm sample fraction using a Panalytical X'PERT Pro with CuK radiation with 40 kV y 40 mA at Laboratorio de

Análisis de Cuencas from the Universidad Nacional de Córdoba (UNC) with 30 g of selected mudstone and siltstone samples; and at the Instituto Tecnológico Minero (INTEMIN) of the Servicio Geológico Minero Argentino (SEGEMAR) of Buenos Aires with an X-ray diffractometer (Philips X'Pert MPD).

Results

Santa Clara Setting, Stratigraphy, and Clastic Sedimentology

The Santa Clara sub-basin (SCA Fm.) (Fig. 1c) is a separate sub-basin of the Cuyana Rift, midway between the Potrerillos (CC Fm.) and Cerro Puntudo (CP Fm.) sub-basins. It has the same structural configuration as the Cerro Puntudo sub-basin: an active margin on the east and flexural margin on the west (Benavente et al. 2018) (Fig. 1a). It contains Triassic continental strata of the El Peñasco Group (Stappenbeck 1910; Stipanicic 1947; Groeber and Stipanicic 1953; Spalletti 2001; Cortés et al. 2003; Spalletti and Zavattieri 2009). This group comprises five units from base to top: (1) Cielo Formation—thickly bedded coarse-grained sandstone and structureless conglomerate with sandstone matrix (Harrington 1971), probably deposited in an alluvial-fan setting. (2) Mollar Formation—fine-grained, stratified to structureless sandstone alternating with organic-carbon-rich black mudstone (Harrington 1971), interpreted as a lacustrine system with broad-scale cycles representing lake expansion and contraction (Spalletti and Zavattieri 2009). (3) Montaña Formation—coarse-grained stratified sandstone interbedded with finely stratified siltstone (Harrington 1971), interpreted as a fluvial system with wood-bearing floodplain (Artabe et al. 2003). (4) Santa Clara Abajo Formation—siltstone, fossiliferous mudstone (fish remains), sandstone, and subordinate conglomerate belonging to a fluvio-deltaic-lacustrine depositional system (Harrington 1971; López-Arbarello and Zavattieri 2008). (5) Santa Clara Arriba Formation—thin clayey tuffaceous fine-grained sandstone alternating with laminated limestone, mudstone, and siltstone (Harrington 1971) deposited in a low-gradient deltaic-lacustrine system (Benavente et al. 2018).

Two facies associations occur in the Santa Clara Arriba: A and B (Table 2 and Fig. 3). Exposure indicators are abundant toward the top of the unit and include mud cracks. Organic matter (OM) preservation is good, biota diversity is low, fauna is represented by fish remains (scales) and abundant invertebrate traces associated with the mud cracks; flora is represented by very fragmented macrophyte remains. These attributes, along with the occurrence of microbial limestones and progradational/ aggradational stacking pattern, were interpreted as a fluvial-lacustrine lake-facies association typical of an overfilled lake basin (Benavente et al. 2018). However, integration of mineralogic and isotopic information presented below suggests this formation records balanced-fill conditions.

Table 2 Facies associations of the Santa Clara Arriba Formation (early Middle Triassic) at the Santa Clara sub-basin, Cuyana Basin, Mendoza, Argentina

Facies association (FA)	Facies	Sedimentary structures	Bed geometry	Vertical and lateral relations	Fossil content	Processes	FA Interpretation
A	*Fossiliferous wavy bedding* **(Sw1)**	Fine- to very fine-grained, well-sorted sandstones alternating with muddy silty layers. Sandy and silty layers are 1 cm thick with horizontal lamination and ripple cross-lamination. Mudcracks with 10 cm wide polygons in muddy silt layers.	Tabular to lenticular, 0.5–2.0 m thick	Underlies facies Fh, Sra and Srs; overlies facies Sra	Abundant vertical tubes (trace fossils or plant stems), plant remains, vertebrate footprints	Tractive flows alternating with subaerial exposure	Distributary channels and ponds in the lower delta plain associated with mouth bars at the delta front
	Current ripple sandstones **(Sra1)**	Medium-grained, moderately sorted sandstones. Sra: asymmetric ripple cross-lamination, 5–7 cm thick ripple forms, amplitude is 10–15 cm. Rip-up clasts and convolute structures.	Tabular, 0.5–1.5 m thick, inclined tops, convex to erosive bases	Underlies and overlies facies Sw1	Rare trace fossils	Tractive flows	
	Wave ripple sandstones **(Srs)**	Alternating very fine-coarse-grained well-sorted sandstone. Srs: symmetric, 2–4 cm thick ripple forms, amplitude is 20 cm.	Tabular, 0.25–3.00 m thick	Underlies facies Sra and overlies facies Sw1	Rare trace fossils	Waves	
	Siltstones **(Fh)**	Horizontally laminated muddy siltstones; laminae are 1 cm thick.	Tabular, 0.1–0.5 m thick	Underlies and overlies facies Sw1	Plant remains	Suspension settle-out	
	Trough cross-stratifed sandstones **(St)**	Medium- to coarse-grained, poorly sorted sandstones with pebbles. St: 2 cm laminae in sets of 10 cm and cosets 50 cm thick; clast lags.	Lenticular, 0.5–1.0 m thick	Underlies facies Sw1 and Fc, overlies facies Sra	———	Channelized tractive flows	

(continued)

Table 2 (continued)

Facies association (FA)	Facies	Sedimentary structures	Bed geometry	Vertical and lateral relations	Fossil content	Processes	FA Interpretation
	Coaly mudstones **(Fc)**	Structureless mudstones rich in OM	Tabular, 0.2 m thick	Underlies facies Sw1 and overlies facies St	———	Suspension settle out	
B	*Finely laminated argillaceous mudstone* **(Fl)**	Finely laminated mudstone: laminae are 1 mm thick, Fe concretions 5–20 cm in diameter, variable OM content associated with pyrite. Clay assemblage composed of illite and smectite.	Tabular 0.05–2.5 m thick	Underlies facies Sw2, Sra and Th; overlies facies Sw2, Sra, and Th	Plant remains, fish scales	Suspension settle- out	Prodelta to offshore lacustrine
	Laminated limestone **(Ll)**	Structureless to faintly laminated carbonate of dominant dolomite composition.	Tabular, 0.05–0.5 m thick	Underlies and overlies facies Fl	———		
	Tuffaceous sandstones **(Th)**	Very fine-grained-sandstone – single layer, containing angular quartz (70%) and K-feldspar (5%) crystals up to 1 mm long and 0.5 mm crystals of opaque minerals in a structureless texture in the upper 85 cm with faint relict horizontal laminae (5 mm thick) of silty sandstone in the lowermost 15 cm.	Tabular 1 m thick	Underlies and overlies facies Fl	———	Ash fall	

Facies association (FA)	Facies	Sedimentary structures	Bed geometry	Vertical and lateral relations	Fossil content	Processes	FA Interpretation
	Bioturbated wavy bedding **(Sw2)**	Fine to very fine-grained, well-sorted sandstones alternating with muddy silt layers. Horizontal lamination and ripple cross-lamination. Sandy and silty layers are 1 cm thick.	Tabular, 0.5–2.0 m thick	Underlies facies Sra and Fl; overlies facies Sra and Fl	Abundant trace fossils only	Tractive flows	
	Current ripple sandstones **(Sra2)**	Medium-grained, moderately sorted sandstones. Sra: asymmetric ripple cross-lamination, 5–7 cm thick ripple forms; amplitude is 10–15 cm	Tabular, 1.5 m thick, inclined upper contacts, convex bases	Underlies facies Sw2 and Fl, overlies facies Sw2 and Fl	———	Tractive flows	

Modified from Benavente et al. (2018)

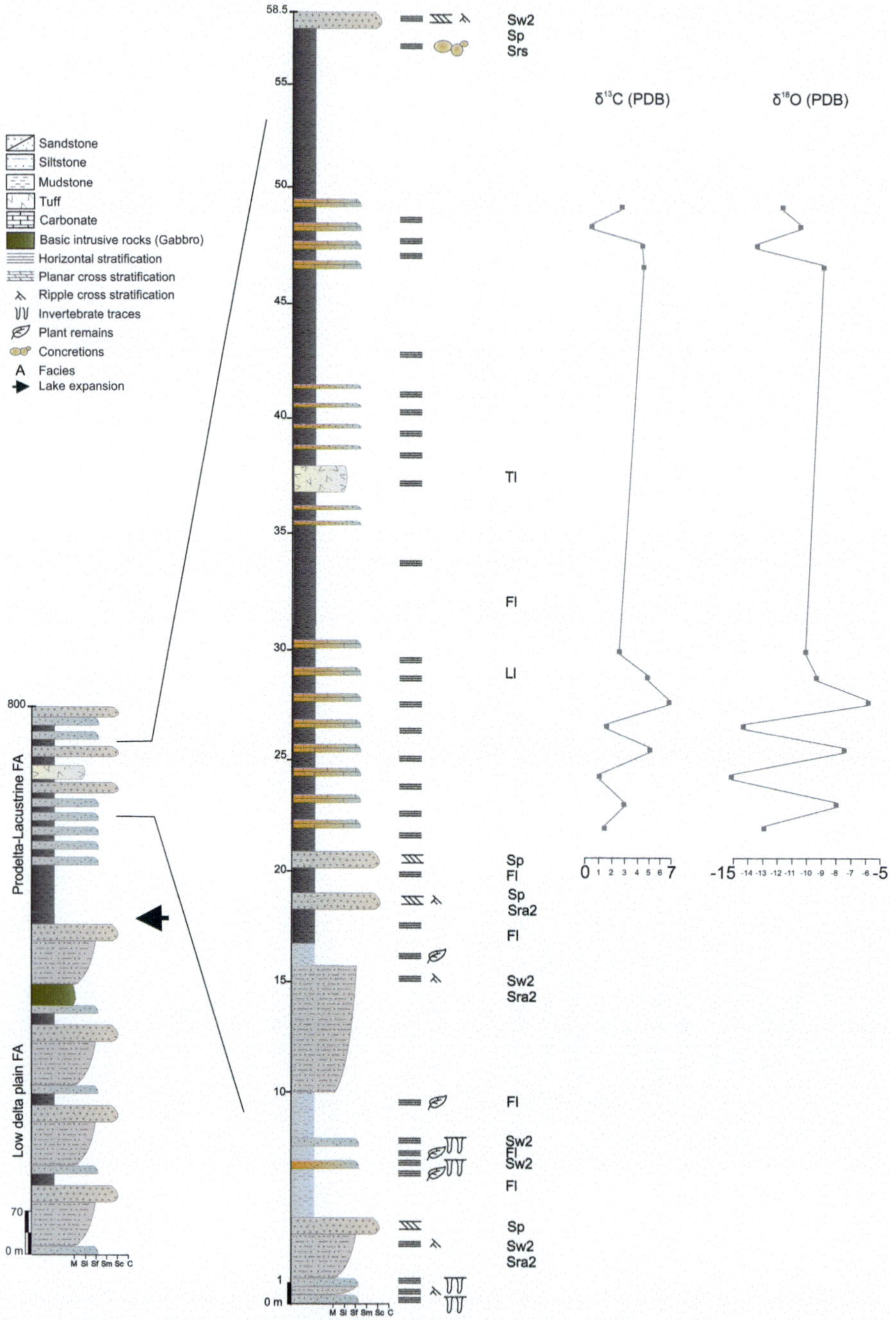

Fig. 3 Stratigraphic succession and $\delta^{18}O$ (x axis) and $\delta^{13}C$ (y axis) trends for the Santa Clara Arriba carbonates, Cuyana Basin continental carbonates, Santa Clara sub-basin, Cuyana Basin

Santa Clara Arriba Carbonate Sedimentology and Petrography, and Clay Mineralogy

Beds of the laminated limestone facies (Ll) of the Santa Clara Arriba Formation (Table 2) are laterally continuous with irregular bases, geometry is tabular, and they are 10–50 cm thick (Figs. 3 and 4a). They have faint-to-distinct horizontal lamination and laminae are 1 mm thick. This facies interfingers with finely laminated mudstones that are grey in the outcrop but black in fresh cut rocks (Fl facies) (Fig. 4a) with abundant OM, pyrite, and fossil remains (Benavente et al. 2018). In thin section, carbonates are mainly dolomite shrub mosaics (Fig. 4b). Dolomite crystals measure 5 μm. There is abundant reddish to brownish OM associated with the dolomite and dark opaque spheroids of 10 μm in diameter with a birefringent nucleus and dense external rim (Fig. 4b). The dolomite is non-luminescent (Fig. 4c).

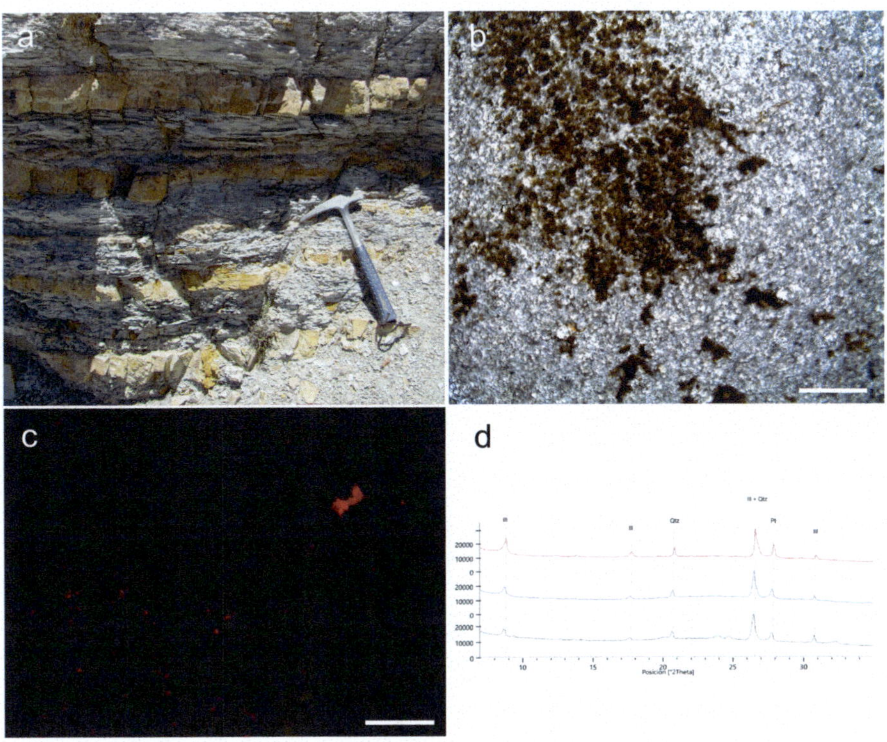

Fig. 4 Field photograph and microphotographs under plane light and cold-cathode illumination of carbonates of the Santa Clara Arriba Fm. (**a**) Tabular laminated limestone facies (Ll) (weathered color: yellow) levels interbedded with the Laminated mudstone facies (Fl) (weathered color: grey). Hammer for scale is 35 cm. (**b**) Microphotograph of the shrub-like microfabric of the dolomite containing brownish opaque OM. Scale is 40 μm. (**c**) Cold-cathode microphotograph showing non-luminescent dolomite. Scale is 25 μm. (**d**) XRD spectra showing narrow and high peak of expandable clays

The Fl facies is composed of finely laminated argillaceous mudstone in tabular beds 0.05–2.5 m thick with 1-mm-thick laminae, Fe concretions 5–20 cm in diameter, variable OM content associated with pyrite, plant remains, and fish scales. Mineralogy of the Ll facies is predominantly non-stoichiometric dolomite, whereas the Fl facies is an expandable-clay-rich assemblage of the smectite group (Fig. 4d and Table 2).

Interpretation: Crinkly to wavy laminae in Ll facies are interpreted as algal biofilms that stabilized the substrate (Noffke et al. 2001). The shrub microfabric of the dolomite suggests a primary origin (Sanz-Montero et al. 2008). Non-luminescent dolomite also attests to primary precipitation. Spheroidal structures interpreted as bacterial bodies are preserved in the dolomite (Armenteros 2010). This would suggest the microorganisms thrived in high-Mg hydrochemical conditions and nucleated precipitation (Sanz-Montero et al. 2008). This facies accumulated in a lake-center setting.

Fl facies was deposited as suspension settle-out plumes in the profundal zone of the paleolake and its OM content corresponds to preserved cocoidal algae remains (Benavente et al. 2018). OM presence along with pyrite and Fe-rich concretions suggests dysoxic to anoxic conditions at the lake bottom and persistent lake stratification, consistent with long water-residence time and closed hydrology (Benavente et al. 2018). The expandable-clay-rich assemblage found for Fl suggests seasonality (Chamley 1989; Deocampo and Ashley 1999).

Santa Clara Arriba Carbon and Oxygen Stable Isotopes

The lake-center carbonates of the Santa Clara Arriba unit have $\delta^{18}O$ values between −15.2 and −5.9‰ (χ average −10.7‰; σ 2.88) and $\delta^{13}C$ values range from +0.3 to +6.8 (χ average 2.8‰; σ 2.07) (Table 3 and Fig. 5). The correlation coefficient is positive and high: $r = 0.9$ ($r < 0.7$). Stratigraphically the C and O values show

Table 3 Stable isotopes $\delta^{13}C$ and $\delta^{18}O$, obtained from different textures of the limestones of the Santa Clara Arriba unit (early Middle Triassic), Santa Clara sub-basin, Cuyana Basin

Carbonates	Texture	$\delta^{13}C$	$\delta^{18}O$
Lake center limestones	Dolomicrite	1.4	−13.1
	Dolomicrite	2.9	−8.1
	Dolomicrite	1.1	−15.2
	Dolomicrite	5.3	−7.5
	Dolomicrite	1.5	−14.3
	Dolomicrite	6.8	−5.9
	Dolomicrite	4.8	−9.4
	Dolomicrite	2.6	−10.2
	Dolomicrite	4.4	−8.9
	Dolomicrite	0.3	−13.4
	Dolomicrite	0.4	−10.5
	Dolomicrite	2.7	−11.7

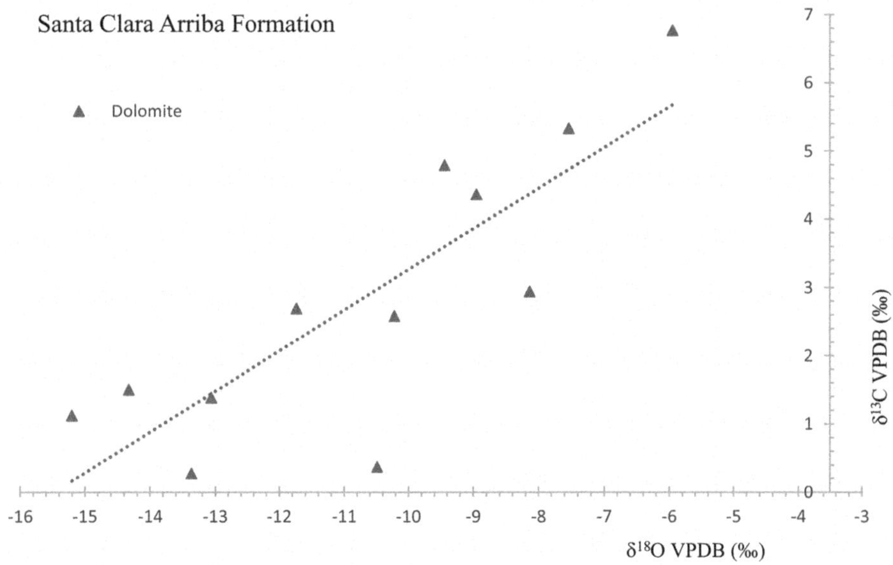

Fig. 5 Cross-plot of $\delta^{18}O$ and $\delta^{13}C$ values for the Santa Clara Arriba Formation (Santa Clara sub-basin). High covariance between $\delta^{18}O$ and $\delta^{13}C$ suggests close hydrology conditions and a long residence time

significant variation through time (up to 5‰) (Fig. 3). Covariation of C and O occurs mostly at the base and middle part of the section with a shift in $\delta^{18}O$ values at approximately 29 m while the $\delta^{13}C$ value remains roughly invariant (Fig. 3). Towards the top, C and O stable isotope fluctuations are opposite (Fig. 3).

Interpretation: the negative signature of $\delta^{18}O$ suggests that Mg concentration was not linked to evaporation of lake waters (Sanz-Montero et al. 2006), but, on the contrary, to a dilution effect by meteoric water input (Benavente et al. 2019a) and biomediation of dolomite precipitation (Sanz-Montero et al. 2008) during long water-residence times. This points to an O signature dominated by within-lake processes.

The positive signature of $\delta^{13}C$ has been linked to concentration of brines and generation of methane (Potter et al. 2004; Leng et al. 2005). Another probable significant factor is the preservation of reduced OM due to the anoxic conditions of the paleolake. This would have been related to concentration of Mg-rich waters linked to dolomite precipitation and to the long water-residence episodes consistent with chemostratified lake waters (although thermal stratification cannot be ruled out). All these factors are also consistent with a C signature also dominated by within-lake-processes.

The strong correlation between O and C stable isotope values and variation through time is indicator of long residence time (Talbot 1990; Alonso-Zarza et al. 2012) which strengthens our interpretation of long residence time of waters in the system that, linked to a high Mg concentration, would have enabled dolomite biomediated precipitation (Sanz-Montero et al. 2006, 2008, 2009).

The covariation of C and O stable isotope values observed at the base and middle of the stratigraphic section and its loss toward the top of the unit most likely indicates a shift in residence time of the lake waters from long to short. This is in agreement with sedimentology and stratigraphy data that show a shift from aggradation to progradation stratal stacking toward the top of the succession.

Paleoenvironmental Insights Into the Cerro Puntudo and Cerro de Las Cabras Formations from Published Data

To place our findings about the Santa Clara Arriba formation into a broader regional context, we summarize in this section the evidence about early Middle Triassic paleoenvironmental conditions in the neighboring sub-basins based on our previous work and other publications.

Cerro Puntudo Setting, Stratigraphy, and Clastic Sedimentology

The Cerro Puntudo formation accumulated at the ramp or platform margin of its sub-basin, a relatively low-accommodation area with wide lateral drainage networks (Fig. 1b). The lake was probably shallow with very low bottom gradient and fed by surface and groundwater (Pla-Pueyo et al. 2009, 2015; Billi 2007)—as attested by the recognition of spring discharge aprons in the rocks (Benavente et al. 2015). From base to top, this unit comprises an overall aggradational–minor-progradational succession up to 90 m thick (Benavente et al. 2015). The lower part contains red conglomerates and sandstones that represent alluvial-fan deposits overlain by alternating dusky red conglomerates and medium-grain to coarse-grain, poorly sorted sandstones (braided fluvial). The middle interval comprises fine- to medium-grain sandstones of moderate red color in narrow, thin channel bodies and blackish red argillaceous mudstone (fluvial system) interbedded with thin limestones. The upper interval contains structureless, yellowish grey limestones, olive grey and light olive grey stromatolitic limestones, fine-grained very dark red sandstones, and argillaceous mudstones with carbonate cement (lacustrine system) that are associated with sparse greyish green tuffs (Benavente et al. 2015, their Figs. 7f and 8). Six facies associations have been defined (Table 4): Alluvial fan (A), Sandflat (B), Muddy sandflat to mudflat (C), Carbonate channel (D), Lake margin carbonates (E) and Lake center carbonates (F).

Sheetflood deposits, soft-sediment deformation, and exposure indicators suggest most overland flow was ephemeral and occurred during flash floods. Exposure indicators are diverse and common: mud drapes, mud chips, microbialite intraclasts,

Table 4 Facies associations and their characteristics defined for the Cerro Puntudo Formation (Anisian) at the Cerro Puntudo sub-basin, Cuyana Basin

Facies associations	Facies	Sedimentary structures fabric	Bed geometry	Vertical and lateral relations	Fossil content	Interpretation
Alluvial fan (A)	Pebbly sandstones **(Ps)**	Very coarse to medium grain supported sandstone of moderate red color (5R 4/6). Clasts up to 8 cm, poor sorting, imbrication. Mottled fabric.	Tabular, 20–40 cm thick, erosive bases	Underlies the stratified intraclastic sandstone facies (b)	Rhizoliths and rhizohaloes	Fluvial system in the distal alluvial fan
	Stratified intraclastic sandstones **(Sp)**	Coarse to medium sandstone, moderate sorting, normal grading, intraclastic lags of quartz and feldspar, very dark reddish brown color (10R 3/4). Sp cosets of 20 cm. Planar cross-bedding and asymmetric ripple cross-lamination.	Lenticular to tabular, 0.5–1.5 m thick	Overlies and grades laterally with the pebbly sandstone facies (a)	Invertebrate traces *Skolithos* and *Scoyenia*	Sheetfloods
Sandflat (B)	Graded sandstones **(Sh)**	Very fine to medium sandstone, normal grading, Sh faint, dark very reddish brown color (10R 3/4). mottled fabric, cracked mud drapes	Tabular, 20–45 cm thick, erosive bases	Overlies and is interstratified with the laminated mudrock facies (e) and underlies the laminated limestone facies (k).	Rhizohaloes and invertebrate traces *Skolithos* and *Scoyenia*	Sheetfloods in a low gradient slope with pedogenesis
	Laminated sandstones **(Sl)**	Very fine to medium sandstone, Sh faint, Sl, millimeter lamination, Sp, inverse grading, convolute structures, mud chips randomly dispersed, moderate brownish red (10R 4/6) to very pale orange (10YR 8/2) colors	Tabular, 15–120 cm thick	Overlies the laminated mudrock facies (e)	Rhizoliths and rhizohaloes	Sheetfloods with pedogenesis

(continued)

Table 4 (continued)

Facies associations	Facies	Sedimentary structures fabric	Bed geometry	Vertical and lateral relations	Fossil content	Interpretation
Mudflat (C)	Laminated argillaceous mudstones **(Lm)**	Fl very thin lamination, Pale red (5R 6/2) to grayish red (10R 4/2) colors. Clay assemblages dominated by illite, smectite and analcime.	Tabular, 10 a 180 cm thick	Underlies the oncolitic wackestone facies (g) and is interstratified with the laminated sandstone facies (d) and laminated limestone facies (k)	Rhizohaloes	Settle out deposition and subsequent paleosol (moderate-good to poor drainage in the external mudflat zone
	Laminated siltstones **(Fl)**	Faint horizontal lamination, lamina 1 cm thick, convolute structures, mottling, mudcrack systems, moderate red (5R 5/4) to grayish red (10R 4/2) colors	Tabular, 7–120 cm thick, erosive bases	Interstratified with the laminated mudrock facies (e)	Rhizohaloes	Settle out in vegetated floodplains with scarce tractive flows
Carbonate channels (D)	Oncolitic wackestones **(Ow)**	Oncolites are (0.1–3 cm diameter), randomly distributed in the units, poor sorting, inverse grading. Light olive gray (5Y 6/1) to light greenish gray (5GY 8/1) colors	Lenticular, 10–30 cm thick	Overlies and underlies the laminated mudrock facies (e)	Rhizohaloes and filamentous algae	Tractive flows in ephemeral channels
	Oncolitic laminated sandstones **(Sol)**	Very fine sandstone, oncolite imbrication (oncolites cannot imbricate) (0.1–1.5 cm diameter), moderate yellow (5Y 7/6) to dark yellowish orange (10 YR 6/6) colors	Tabular, 20 cm thick	Underlies the oncolitic boundstone facies (m)	Filamentous algae	Tractive flows in ephemeral channels
Lake margin carbonates (E)	Mottled carbonate mudstones **(Mc)**	Mottling, mudcrack systems, vertical coloration changes, light bluish gray (5B 7/11) to light greenish gray (5G 8/11) colors	Lenticular, 75 cm thick	Laterally interstratified with the laminated siltstone facies (f) and the nodular wackestone facies (j)	Rhizoliths and rhizohaloes	Palustrine carbonate area with pedogenesis

Table 4 (continued)

Facies associations	Facies	Sedimentary structures fabric	Bed geometry	Vertical and lateral relations	Fossil content	Interpretation
	Nodular wackestones **(Nw)**	Wackestone, with nodules that are ellipsoidal-subrounded (5 cm diameter), light bluish gray (5B 7/11) to light greenish gray (5G 8/11)	Tabular, 0.11–3 m thick	Overlies the laminated sandstone facies (d) and the laminated siltstone facies (f) and grades laterally to the mottled carbonate mudstone facies (i)	Rhizoliths and rhizohaloes	Palustrine carbonate area with pedogenesis
	Laminated carbonates **(Lc)**	Laminated fabric, plane to domal lamination, crenulated laminae (1–2 mm thick), crack systems, mud chips, tepee structures, peloids, spherulites, medium dark gray (N5) to dark gray (N3)	Tabular to lenticular, 0.5–1.05 m thick	Over- and underlies the laminated mudrock facies (e) and grades laterally to the mottled carbonate mudstone facies (i)	Filamentous algae, articulated ostracode valves, vertebrate footprints	Palustrine carbonate area with pedogenesis
Lake center carbonates (F)	Laminated marlstones **(Ml)**	Very fine lamination (0.2 mm), dark blue (5 PB 3/2) to pale reddish violet (5RP 6/2) colors	Tabular, 10 cm thick	Overlies and underlies the laminated mudrock facies (e)	———	Settle out in small ponds of the mudflat
	Oncolitic boundstones **(Ob)**	Matrix-supported boundstones with, oncolites (1–12 cm diameter) medium gray (N5) to dark gray (N3) colors	Lenticular, 10–70 cm thick	Overlies and underlies the laminated mudrock facies (e) and grades laterally to the mottled carbonate mudstone facies (i). Erosive contact with the oncolitic laminated sandstone facies (h)	Filamentous algae, Charophyta, articulated ostracode valves, phytoclasts	Carbonate shallow ponds of the mudflat

Modified from Benavente et al. (2015)

desiccation cracks, tepee structures, rhizobrecciation, and tetrapod tracks. Fluctuating groundwater levels are indicated by vertic paleosols with Fe-depleted (yellow) halos, cracks induced by root activity, rhizoliths, silica nodules, and horizontal calcite-crystal crusts. Evaporites include gypsum and efflorescent calcite-silica crusts. OM preservation is moderate to low, the biota is moderately diverse (cyanobacteria, charophytes, ostracods and tetrapods), and the flora included herbaceous hygrophytes, sphenophytes, and lycopsids (Mancuso 2009). These attributes, along with the overall aggradational/progradational stratal stacking, indicate a fluctuating profundal lake-facies association, suggesting a balanced-fill lake type. Integration of the sedimentological, petrographic, and geochemical data presented below helps refine this interpretation and explains the occurrence of minor evaporites in this formation.

Cerro Puntudo Carbonate Sedimentology, Petrography and Clay Mineralogy, and Geochemistry

Carbonate beds span the mottled carbonate mudstone facies (Mc), nodular wackestone facies (Nw), and laminated carbonate facies (Lc) (that together constitute the facies association E) as well as the oncolitic boundstone facies (Ob) (the facies association F) (Table 4). These carbonates interfinger with the laminated argillaceous mudstone (Ml) and laminated marlstone facies (Lm). In this depositional system, these facies are found from the lake margin to the lake center. In thin section, the nodular limestones comprise homogeneous micrite with minor detrital components and unlined vertical cracks. The nodules in the nodular wackestone facies (Nw) are also formed of micrite. In both facies, the nodules are ellipsoidal or subrounded, coated by silica, and vary in size from 3 to 5 cm. The laminated carbonate facies (Lc) is characterized by planar to domal laminated micrite. Laminae are 1-mm thick, alternating light and dark in color, and are rarely disrupted by cracks. They preserve abundant bioclasts, plant remains, and tubules 100-µm long and 10-µm thick (Fig. 6a). Oncolitic boundstones (Ob) have oncolites that range in size from 1 to 12 cm dispersed in a micrite matrix. The cortices of oncolites contain radially oriented tubules that are 100-µm long and 10-µm thick. The nuclei can be charophyte gyrogonites (Fig. 6b). The micrite in all the limestone facies described is non-luminescent (Fig. 6c). Laminated argillaceous mudstones (Lm) are structureless to faint horizontally laminated. Laminated siltstones (Fl) are argillaceous-siliceous in composition, and structureless to laminated with mottling and convolute structures. Both of these clastic facies are characterized by expandable-clay-rich assemblages of the smectite group (Fig. 6d). The $\delta^{13}C$ values of the Cerro Puntudo carbonates vary from −0.60‰ to −5.77‰ (χ −3.10‰; σ 1.15‰); the $\delta^{18}O$ values range from −5.04‰ to −14.84‰ (χ −12.58‰; σ 1.8‰) (Table 5 and Fig. 7). The correlation coefficient for the variables $\delta^{13}C$ and $\delta^{18}O$ is $r = 0.23$ ($p = 0.05$), very low.

Interpretation: the mottled carbonate limestone (Mc) and nodular wackestone facies (Nw) are interpreted as due to subaqueous primary precipitation of carbonate

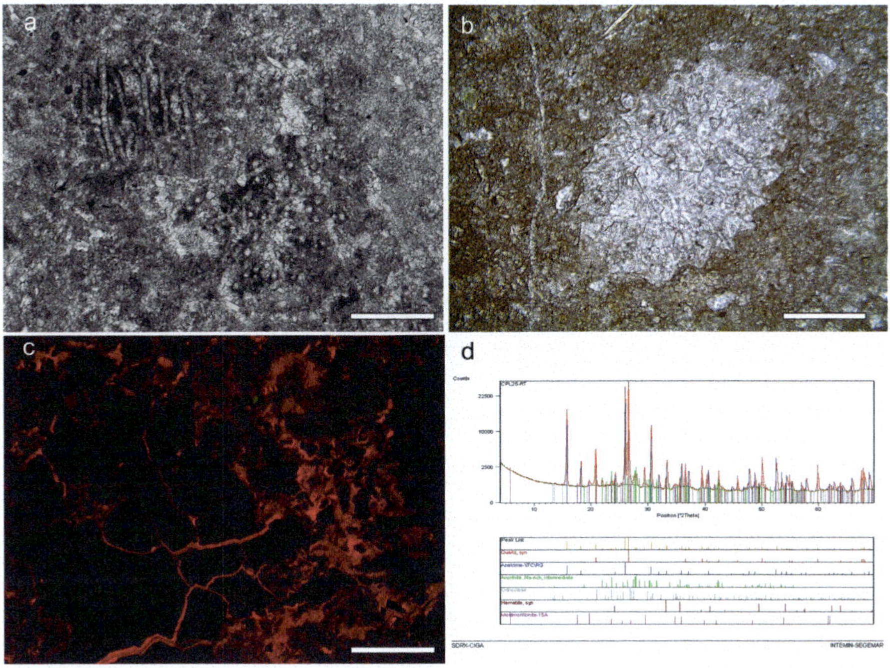

Fig. 6 Microphotographs under plane light and cold-cathode illumination of carbonates of the Cerro Puntudo Fm. (**a**) Microphotograph of micrite with tubules. Scale is 100 μm. (**b**) Microphotograph of a gyrogonite. Scale is 200 μm. (**c**) Cold-cathode microphotograph showing non-luminescent micrite and zoned spar crystals. Scale is 50 μm. (**d**) XRD spectra showing expandable clays (montmorillonite) and zeolites (analcime) with narrow and high peaks in this case

with subsequent subaerial exposure (Alonso-Zarza 2003; Benavente et al. 2015). The laminated carbonate facies (Lc) corresponds to subaqueous precipitation of carbonate but the presence of bioclasts and algae and the less common disruption by cracks suggests a more distal position in the paleolake (Benavente et al. 2015). Both of these facies likely accumulated in a lake-margin setting.

The oncolitic boundstone facies (Ob) represents precipitation of carbonate directly from standing water over charophyte thalli and gyrogonites (Benavente et al. 2015), probably in a lake-center setting characterized by persistent standing water. Non-luminescent micrites and their petrographic character indicate primary precipitation of carbonate (Benavente et al. 2019a).

The very low correlation found between the $\delta^{18}O$ and $\delta^{13}C$ values for both lake-margin and lake-center carbonates points to open hydrology and short residence time (Benavente et al. 2019a). Nevertheless, minor evaporites formed in this system, due to the equilibrium in the leakage ratio, between groundwater input and evaporation (Rosen 1994; Benavente et al. 2019a).

Expandable-clay-rich assemblages of illite-smectite associated are strong evidence for seasonality (Chamley 1989; Deocampo and Ashley 1999). The presence of analcime characterizes alkaline lake waters (Chamley 1989).

Table 5 Stable isotopes $\delta^{13}C$ and $\delta^{18}O$, obtained from different textures of the limestones of the Cerro Puntudo and Cerro de las Cabras formations (Anisian), Cuyana Basin

Cerro Puntudo				Cerro de las Cabras			
Carbonates	Texture	$\delta^{13}C$	$\delta^{18}O$	Carbonates	Texture	$\delta^{13}C$	$\delta^{18}O$
Lake margin	Micrite	−4.03	−12.85	*Lake margin*	Micrite	−5.19	−13.36
		−2.29	−11.95			−5.89	−9.89
		−3.60	−12.19			−5.70	−12.50
		−3.49	−13.36			−5.80	−13.60
		−2.69	−11.96			−6.20	−12.70
		−4.47	−9.08			−5.80	−12.40
		−3.14	−14.84			−5.90	−13.30
		−2.97	−12.14			−6.00	−12.50
		−3.49	−12.87			−5.70	−13.00
		−2.66	−13.73			−6.40	−12.30
		−4.53	−14.53			−6.40	−12.40
		−2.97	−13.62			−6.00	−14.10
		−3.68	−13.75			−6.10	−12.60
						−8.10	−12.30
	Equant spar	−3.64	−14.84				
	Granular spar	−5.89	−18.27				
	Granular spar	−4.94	−13.44				
	Micrite	−2.49	−11.50				
		−2.92	−11.96				
		−4.39	−11.94				
		−3.93	−5.04				
		−3.33	−10.55				
		−3.62	−13.33				
		−3.05	−12.35				
		−3.62	−11.93				
		−3.15	−11.22				
		−1.04	−13.48				
		−2.05	−14.34				
		−1.64	−12.15				
		−0.60	−12.62				
		−0.85	−12.44				
		−0.73	−11.31				
		−2.47	−8.41				
		−1.34	−14.18				
		−4.71	−13.34				
		−5.77	−13.12				
		−5.55	−12.44				
		−5.72	−13.27				
		−5.44	−13.97				
		−3.27	−13.02				
		−2.25	−13.8				
		−2.47	−11.53				
		−3.01	−12.54				
	Granular spar	−2.32	−14.70				
	Equant spar	−5.93	−13.33				
	Granular spar	−4.76	−14.29				

(continued)

Table 5 (continued)

Cerro Puntudo				Cerro de las Cabras			
Carbonates	Texture	$\delta^{13}C$	$\delta^{18}O$	Carbonates	Texture	$\delta^{13}C$	$\delta^{18}O$
	Equant spar	−6.63	−18.13				
Lake center	Micrite	−4.05 −4.21 −1.79 −3.37 −2.05 −3.14 −2.48 −2.68	−13.79 −10.40 −13.57 −5.97 −13.73 −13.98 −12.07 −14.32				
	Oncolitic micrite	−3.17 −3.85 −2.81 −3.37	−13.55 −12.83 −11.95 −13.27				
	Carbonate matrix	−2.20 −2.40 −2.74	−14.02 −13.38 −12.94				
	Oncolite cortex	−3.31 −4.51 −3.77	−12.41 −12.92 −14.48				
	Oncolite nuclei	−3.52 −2.32	−14.19 −12.55				
	Oncolitic micrite	−2.45 −1.95	−12.22 −13.51				
	Granular spar	−3.28	−13.49				
	Granular spar	−3.57	−10.40				
	Granular spar	−2.39	−14.68				

Modified from Benavente et al. (2019a, 2019b)

Cerro de Las Cabras Setting, Stratigraphy, and Clastic Sedimentology

The Cerro de las Cabras formation accumulated in the north end of the Potrerillos sub-basin, near a fault-transfer zone (Fig. 1d). From base to top, this unit comprises an overall fining- and thickening-upward aggradational succession up to 900 m thick that contains conglomerates, pebbly sandstones, and sandstones in channel bodies in the lower interval, and sandstones that interfinger with fine and coarse mudstones and limestones (biogenic and authigenetic) in the upper part (Benavente et al. 2015). Two facies associations occur: sandflat and mudflat (A) and lake margin carbonates (B) (Table 6).

Sheetflood deposits and exposure indicators attest to ephemeral streams with widely varying discharge. Exposure indicators such as microtepee structures, rhizobrecciation, mud drapes, mud cracks, mud chips, and pedogenesis are common. Paleosols are gleyed, mottled dark-reddish brown to greyish red purple with calcitic intraclasts and early calcite spar when in sandstone or interbedded red and violet

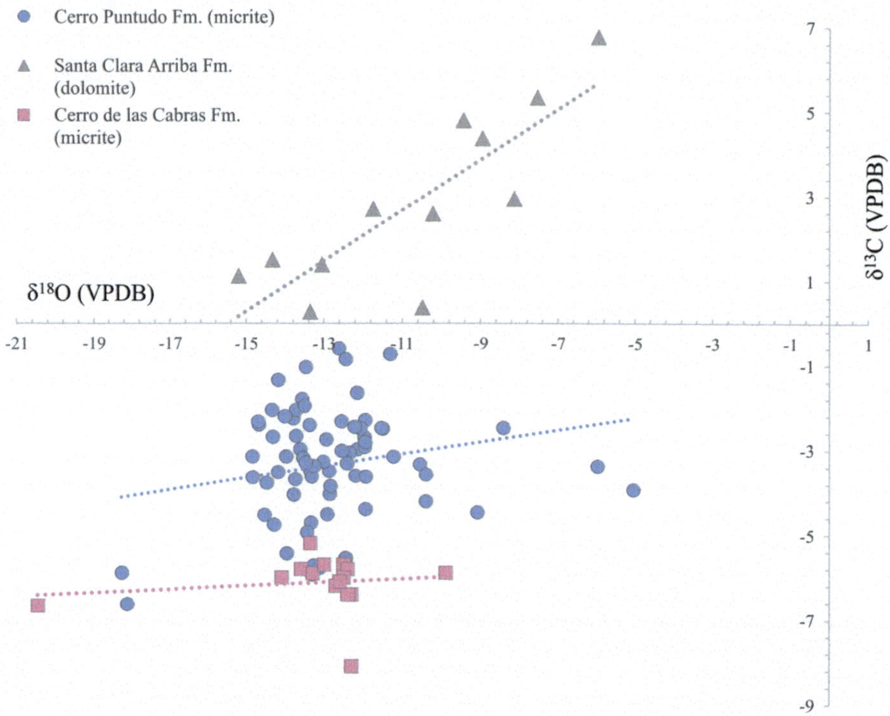

Fig. 7 Cross-plot of $\delta^{18}O$ and $\delta^{13}C$ values for the Cerro Puntudo (Cerro Puntudo sub-basin), Cerro de las Cabras (Potrerillos sub-basin), and Santa Clara Arriba Formation. Lack of covariance between $\delta^{18}O$ and $\delta^{13}C$ indicates open-hydrologic conditions in Cerro Puntudo and Cerro de las Cabras sub-basins, opposite to the high covariance between C and O stable isotope values of the Santa Clara Arriba Formation

halos in a grey matrix and green to orange halos in a red matrix when in mudstone, all indicating significant water-table fluctuations. These paleosols are vertisols with meter-scale horizonation, possible gigals, shrink-swell broken clay layers, and desiccation cracks. OM preservation is low; the biota is low diversity (mostly ostracods). These attributes, along with the occurrence of microbialitic limestones and aggradational stacking pattern overall, indicate an evaporative lake-facies association (Bohacs et al. 2000) characteristic of underfilled lake basins (see Table 6). This unit does, however, lack thick evaporites, a contradiction that the following detailed petrographic and geochemic data help explain.

Cerro de Las Cabras Carbonate Sedimentology, Petrography, Clay Mineralogy, and Geochemistry

The carbonate-rich layers in this formation represent two facies within facies association B: mottled-nodular carbonate (Mc) and laminated carbonate (Lc); they

Table 6 Facies associations and their characteristics defined for the Cerro de las Cabras Formation (Anisian) at the Potrerillos sub-basin, Cuyana Basin

Facies associations	Facies	Sedimentary structures	Bed geometry	Vertical and lateral relations	Fossil content	Interpretation
Sandflat and mudflat (A)	Pebbly sandstones **(Ps)**	Medium sandstone of red violet color (5R 3/4) with pebbles (matrix supported)	Tabular, 90 cm thick, erosive bases	Overlies massive mudstone facies (c)	———	Proximal sheetfloods
	Intraclastic massive sandstones **(Sm)**	Fine-medium sandstones, mottling, red brownish dark color (10R 3/4), carbonate intraclasts	Tabular, 60 cm thick	Interstratified with pebbly sandstone facies (a) and	Rhizoliths and rhizohaloes	Distal sheetfloods with pedogenesis (gleying)
	Structureless argillaceous mudstones **(Fm)**	Mottling, mudcrack systems, grayish brown (5YR 3/2), dark green (5G 3/2) or very dark red (5R 2/6) colors. Clay assemblages are dominated by illite and smectite.	Tabular-lenticular, 20–90 cm thick	Overlies mottled-nodular limestone facies (d) and underlies massive intraclastic sandstone facies (b) and the mottled-nodular limestones (d)	Rhizoliths and rhizohaloes	Paleosols with moderate-good drainage to poorly drained
Lake margin carbonates (B)	Mottled-nodular carbonates **(Mc)**	Carbonate mudstone, micrite, mottling, crack systems, nodules randomly dispersed, yellowish grey color (5Y 8/1)	Lenticular-tabular, 75 cm thick	Underlies massive mudstone facies (c)	Rhizoliths and rhizohaloes	Palustrine ponds
	Laminated carbonates **(Lc)**	Laminated, with crenulated discontinuous micrite laminae (1–3 mm thick), olive gray color (5Y 4/1), mudcrack systems, microtepee structures, peloids	Tabular, 70 cm thick	Overlies and underlies massive mudstones (c)	Filamentous algae, articulated ostracode valves	Stromatolites in carbonate ponds

Modified from Benavente et al. (2015)

interfinger with structureless argillaceous mudstones (Fm) and together characterize the palustrine environment (Benavente et al. 2015) (Table 6). In thin section, the mottled-nodular carbonate facies (Mc) is dominated by homogeneous micrite disrupted by vertical structures that bifurcate downward and are filled by spar, coated by cutans, and with circumgranular cracks (Fig. 8a). The laminated carbonate facies (Lc) contains micrite with horizontal lamination of alternating dark and light color laminae. Laminae are crinkly and 1-mm thick (Fig. 8b). In the micrite, tubular structures 100-μm long and 10-μm wide were identified. This facies also contains voids that are filled by granular and equant spar (Fig. 8b). Micrite in both limestone facies show compaction against spar mosaics and is non-luminescent under cold-cathode illumination (Fig. 8c). The argillaceous structureless mudstone facies (Fm) is characterized by an expandable clay assemblage dominated by smectite-group minerals (Fig. 8d). The Cerro de las Cabras carbonates have $\delta^{13}C$ values between −5.19‰ and −8.10‰ (χ average −6.03‰; σ 0.67‰); and $\delta^{18}O$ values range from −9.89‰ to −14.10‰ (χ −12.64‰; σ 0.55‰) (Table 5 and Fig. 7). The correlation coefficient is very low and negative ($r = -0.18$, $p = 0.22$). The $\delta^{13}C$ values of the Cerro de las Cabras limestones have a narrow range (< 2‰) (Fig. 7). The $\delta^{13}C$ value for spar is −6.67‰ and for $\delta^{18}O$ is −20.46‰ (Table 5 and Fig. 7).

Interpretation: the mottled-nodular limestone facies (d) is interpreted as recording subaqueous primary precipitation of carbonate with subsequent subaerial exposure (Alonso-Zarza 2003; Benavente et al. 2015). The bifurcating vertical structures disrupting the micrite matrix are interpreted as rhizoliths (Huerta and Armenteros 2005; Benavente et al. 2015). This indicates the formation of these limestones in a lake-margin position with vegetation development and subaerial exposure during low water levels (Meléndez et al. 2009; Benavente et al. 2015). The laminated carbonate facies (Lc) is interpreted as stromatolites resulting from the subaqueous bio-induced precipitation of carbonate by cyanobacteria probably represented by the tubules found that resemble filamentous algae (Freytet and Verrecchia 1998; Dupraz et al. 2009; Benavente et al. 2015). These carbonate facies accumulated in lake-margin settings. Non-luminescent micrite and early compaction attests to primary precipitation of carbonate (Benavente et al. 2019a).

Expandable-clay-rich assemblages indicate seasonality (Chamley 1989; Deocampo and Ashley 1999). The lack of correlation between $\delta^{18}O$ and $\delta^{13}C$ values for this Lake-margin carbonates is interpreted as indication of open hydrology and short residence time in a recharge hydrology situation (Benavente et al. 2019a)—which may explain the lack of evaporites (Rosen 1994; Benavente et al. 2019a).

Discussion

There exist several challenges to integrating geochemical data into paleoenvironmental interpretations of lake strata: extra-basinal and paleogeographic factors, water sources and flow paths, and within-lake complexities. In particular, robust interpretations of isotopic analyses require comprehensive assessment of the main

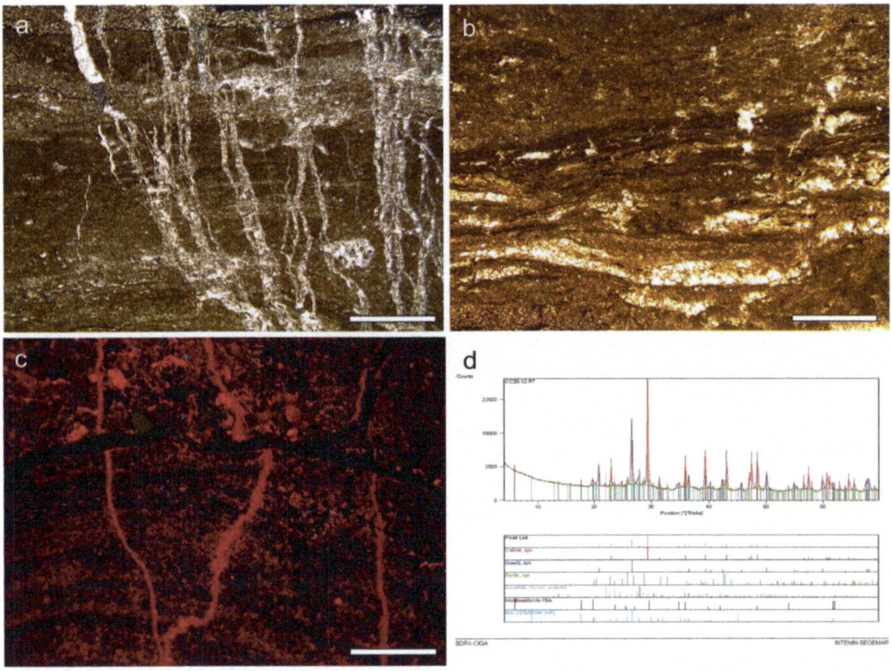

Fig. 8 Microphotographs under plane light and cold-cathode illumination of carbonates of the Cerro de las Cabras Fm. (**a**) Microphotograph of micrite disrupted by vertical structures filled by spar. Scale is 100 μm. (**b**) Microphotograph with detailed of crinkly lamination. Scale is 25 μm. (**c**) Cold-cathode microphotograph showing high luminescence in vertical cracks filled by microspar. Scale is 25 μm. (**d**) XRD spectra showing narrow and high peak characteristic of expandable clays (montmorillonite)

factors of the depositional system and their influences on C and O isotope signatures (Table 1). We interpret that the surface areas of standing water of the Cuyana Basin paleolakes were restricted in size relative to their catchments; thus, these carbonates were particularly sensitive to these factors (e.g., Hammarlund et al. 2003).

Acknowledging Extra-basinal and Paleogeographic Complexities

Extra-basinal and paleogeographic factors can influence the isotopic composition of lake waters in several ways: global climate mode and its connection to precipitation $\delta^{18}O$, and carbon-cycle evolution, paleolatitude, and paleogeographic influence on water composition, paleolatitude and paleoaltitude effects on surface temperature and precipitation, bedrock composition variations across the lake's drainage network, and lateral topographic barriers.

Global-scale land-mass distribution and climate mode influence ocean and meteoric $\delta^{18}O$ water composition through atmospheric and oceanic circulation. In the

Triassic, all these factors were quite different from present day. Calculation of surface ocean water temperature for the Phanerozoic is still being refined based on a new paradigm of changing $\delta^{18}O$ composition (Veizer and Prokoph 2015).

In addition, extra-basinal factors may have influenced the $\delta^{13}C$ values. Multiproxy studies for terrestrial ecosystems reveal that the Triassic was characterized by several Greenhouse crises in which the atmospheric $\delta^{13}C$ fluctuated between extremely negative and positive values (Retallack et al. 2011; Retallack 2013). The extremely low $\delta^{13}C$ values found by Retallack et al. (2011) and Retallack (2013) can be explained by the release of large amounts of marine-sourced CH_4. In particular, the Anisian was bracketed by a negative and positive excursion. In this context, our data would be linked to a transition: recovering from light C release and toward heavy C accumulation in the atmosphere. Significant residence times would have enabled equilibration of lake waters with this fluctuating CO_{2atm} – this would be particularly noticeable in the Santa Clara Arriba paleolake.

The paleoaltitude of a lake and its catchment area affects water composition through lapse-rate influence on surface temperature, precipitation, and evaporation. The estimated paleoaltitude of the Cuyana Rift lakes was between 500 and 1000 m, and the catchment areas may have been many 100 s of meters higher (Franzese et al. 2003). Thus, high-altitude recharge may have been a significant influence on the most negative $\delta^{18}O$ values (e.g., Ostera and Dapeña 2003). This would make a significant difference in lake-water surface temperatures and result in a 5–10 °C warmer record if the paleolatitudinal location were the same but in lower elevation.

A significant lateral topographic barrier most likely bounded these sub-basins and might have exerted strong effects on isotopic compositions (Andres et al. 2006). During early Middle Triassic the Choiyoi volcanic-arc complex developed as a significant topographic high at the south-west margin of Gondwana, to the west of the Cuyana Basin (Ramos and Kay 1991; Jenchen and Rosenfeld 2002; Franzese et al. 2003) (Fig. 1a). This high would have had a distillation effect on winds from the west off the paleo-Pacific Ocean affecting all the studied sub-basins. Another topographic high was located to the east of the Cerro Puntudo and Santa Clara sub-basins, formed by a half-graben escarpment adjacent to the master fault. According to Parrish (1993), this uplift had a strong effect on local paleoclimate, accentuating seasonality. The relatively high elevation setting we hypothesize for these lake basins would implicate lower surface lake water temperatures and a more negative signature for $\delta^{18}O$. For the Potrerillos sub-basin, this effect would have been more accentuated since, in addition to the volcanic arc, the master fault located to the west generated a more prominent barrier than for the northern area of the rift.

Acknowledging Intra-basinal Complexities of Water Sources and Flow Paths

The chemical and isotopic character of lake waters is affected by water source (meteoric, connate, juvenile) and flow paths (direct fall, overland, subsurface). These factors varied among our three sub-basins.

Also relevant is whether precipitation falls or flows directly into the lake or if rainfall is diverted by rift shoulders and drained toward the back of the escarpment (Cohen 1990). A backshed drainage pathway would generate groundwater recharge that may have taken longer to contribute to the lake-pool water. Waters flowing along these two separate pathways would eventually evolve distinct $\delta^{18}O$ values. In particular, the groundwater-dominated pathway would be affected by the presence or absence of a carbonate-rich bedrock composition. Ultimately, such different water inputs would mix in the lake resulting in a convolved signature. These complex hydrologic factors in a lake basin can generate confusion when interpreting the lacustrine stable isotope record in the most parsimonious way.

The most negative $\delta^{18}O$ values from SCA and CP point to meteoric water input and a dilution effect (Arenas et al. 1997; Gibson et al. 2002; Benavente et al. 2015). Spring discharge aprons are reported from CP strata (Benavente et al. 2015). The narrow range of the $\delta^{18}O$ values of CC samples is characteristic of open-hydrology systems with no balance between hydrologic inputs and evaporation whose water residence time was too brief (less than a year) for carbon and oxygen isotope ratios to co-evolve (Talbot 1990; Li and Ku 1997).

Based on the evidence presented herein, the Cerro de las Cabras appears to have been fed mostly by groundwater of meteoric origin with persistently closed hydrography but persistently open hydrology. The Santa Clara Arriba had an intermittently open hydrography but closed hydrology. The Cerro Puntudo was fed by surface and groundwater with an intermittently open hydrography and persistently open hydrology. Thus, each of these systems presents their own particular challenges for paleoenvironmental interpretations of geochemical data.

Acknowledging Within-Lake Complexities

In addition to temperature, within-lake complexities can affect isotopic signatures by evaporation before and during carbonate precipitation, cation concentrations, and kinetic and vital fractionation. Evaporation can change the isotopic value of lake waters from the long-term isotopic value of meteoric waters. This variation is not constant and varies with changing climate and water-residence times. Lake systems with open hydrography and hydrology minimize evaporative fractionation because lake waters have short residence times. Thus, the long-term value of meteoric waters is best recorded in primary carbonates formed in distal lake-center settings not subject to intermittent subaerial exposure.

Both O and C stable isotope signatures of CC and CP lake-margin carbonates and shallow lake-center carbonates were likely affected by evaporative processes (Benavente et al. 2019a), whereas the CP and SCA lake-center carbonates were not. The more marginal settings were subjected to more frequent subaerial exposure and saline-alkaline conditions, and therefore a higher probability that their $\delta^{18}O$ values correspond to disequilibrium precipitation. Thus, the most reliable interpretations for $\delta^{18}O$ would be based on CP and SCA lake-center carbonates.

Fluctuations in the $\delta^{13}C$ values of the Cuyana Basin carbonates may have been due to such within-lake processes as OM decay in a palustrine subenvironment (Anderson and Leng 2004; Liutkus et al. 2005; Benavente et al. 2015), aquatic OM accumulation on the paleolake bottom (Camoin et al. 1997), and/or paleosol C input (CO_{2soil}) (Benavente et al. 2019a). All of these processes can contribute to ^{12}C enrichment.

Integrated Characterization of the Santa Clara Arriba System

The strong correlation between carbon and oxygen isotope values for the Santa Clara Arriba lacustrine limestones found in this study is a strong indication of long water-residence times and closed hydrology (Talbot 1990; Leng and Marshall 2004; Alonso-Zarza et al. 2012). These conditions would also be linked to a stratified water column evidenced by abundant preserved OM with pyrite and concretions in Fl facies (Benavente et al. 2018). Since these carbonates are dolomitic, Mg-rich waters can be hypothesized and chemostratification suspected, which agrees with all parameters mentioned so far for the lake basin. Nevertheless, sedimentology shows a fairly persistent surface-water supply (Benavente et al. 2018).

A feasible explanation is that in this balanced-fill lake basin, episodes of hydrological closure were longer than those of open surface water+sediment input, generating a long residence time. This would be linked to how often lake level reached the spill point (Bohacs et al. 2000). This is in contraposition to the overfilled-lake-basin type that was proposed previously for the Santa Clara Arriba system based on sedimentology and stratigraphy only (Benavente et al. 2018). This shows that C and O stable isotope analysis from primary lacustrine carbonates allows reconstructing hydrology of lake basins and furthermore impacts greatly in the lake-basin-type classification (Benavente et al. 2019a). These data can therefore be considered as reliable proxies for reconstructing paleohydrology of deep-time lacustrine basins.

Paleoclimate Implications

Temperature: Paleotemperatures of other basins have been estimated from $\delta^{18}O$ values of carbonates using an equation based on modern-day relations among latitude, altitude, and $\delta^{18}O$ of meteoric precipitation (Bowen and Wilkinson 2002; Leng and Marshall 2004). For the continental interior lake basin we studied, calculating paleotemperature ranges would require using a range of paleolatitudes and paleo-altitudes. However, given the other complexities of the lacustrine system revealed by our geochemical data and discussed in previous sections (varying water sources, seasonal/monsoonal variations, flow paths, evaporative and vital fractionation), as well as the differences in land-mass distribution, climate mode, and atmospheric and ocean circulation between the modern and the Triassic Period, exactly what such ranges of estimated paleotemperatures would mean is unclear. Further studies need to be done to better constrain issues that strongly affect paleotemperature calculations from C and O stable isotopes from primary lacustrine carbonates such as equilibrium precipitation, temperature transfer functions (Hren and Sheldon 2012), and $\delta^{18}O_w$ of precipitation during the Triassic. Monsoonal climates have also been found to influence such other paleothermometers as leaf physiognomy (e.g., Chen et al. 2019; Wang et al. 2019) and such studies integrated with the multiple proxies we have described would be useful.

So far, for the three sub-basins, warm paleotemperatures were interpreted for Anisian microflora of the Cerro de las Cabras unit (Zavattieri 1991). Within the Cuyana Rift Basin context, paleoenvironmental conditions have been reconstructed as warm as well from paleobotanical studies of an adjacent sub-basin inferred as Middle Triassic (Bodnar et al. 2018; Spalletti et al. 2003).

Humidity: The O stable isotope data from lake-margin carbonates and shallow lake-center carbonates of the Potrerillos and Cerro Puntudo sub-basins suggest an evaporative signature (Benavente et al. 2019a). Nevertheless, it is relevant to consider the paleogeographic context of these two sub-basins. The Cerro Puntudo sub-basin has its master fault toward the east whereas the Potrerillos sub-basin has its master fault to the west (Benavente et al. 2015). This would have generated a significant topographic high blocking Pacific Ocean humidity for the Potrerillos sub-basin. Indeed, the CC system is interpreted as an underfilled recharge lake basin (Benavente et al. 2019b) most likely fed by an aquifer through infiltration. This is opposite to CP that was interpreted as a balanced-fill through flow basin (Benavente et al. 2019b) in which evaporation is evidenced by the development of minor evaporites. This means that CC was probably drier than CP. However, as has been considered, this might have been due to a regional tectonic configuration effect limiting a more general paleoclimate interpretation. However, multiple proxies analyzed for CC and CP like C and O stable isotope data, along with the presence of gypsum and efflorescent calcite-silica crusts, and clay assemblages dominated by illite-smectite support semi-arid climate conditions, at least seasonally.

In contrast, the O stable isotope signature of the Santa Clara Arriba carbonates is interpreted as controlled by within-lake processes and therefore do not provide further general paleoclimate insight. Its negative composition points to dilution by meteoric water input to the system. In this case, the positive signature of $\delta^{13}C$ from Santa Clara Arriba has been linked to generation of methane and concentration of brines.

Seasonality: Since we are constraining deep time lacustrine system, it is relevant to clarify that by seasonality we refer to alternation of wet and dry periods of time with no further implication on the duration of such time spans. Substantial seasonality in precipitation, runoff, and phreatic levels are indicated by sheetflood deposits, soft-sediment deformation, exposure indicators (microtepee structures, rhizobrecciation, mud drapes, desiccation cracks, mud chips, microbialite intraclasts, *Scoyenia* traces, and tetrapod tracks), vertic pedogenesis with meter-scale horizonation, possible gigals, silica nodules, and shrink-swell broken clay layers, gypsum and efflorescent calcite-silica crusts, and low OM preservation and biotic diversity in lake-margin settings. Recurrent episodes or areas of higher soil moisture are evidenced by well-developed vegetation cover that included herbaceous hygrophytes, sphenophytes, and lycopsids.

The $\delta^{18}O$ values of CP samples span a range of 9‰ (Fig. 7) that might be due to seasonal variations of $\delta^{18}O$ of water input, particularly given the relatively small volume of these lakes (e.g., Leng and Marshall 2004). The less negative end member $\delta^{18}O$ values could be explained as changes in several factors such as a different air-mass source of rainfall during a different season which affects the isotope composition of precipitation itself (Leng and Marshall 2004; Leng et al. 2005). Moreover, in modern circulation patterns and moisture transport at latitudes near 30°S, similar to early Middle Triassic paleolatitude for the Cuyana Basin (Van Hinsbergen et al. 2015), annual variability in isotope composition of precipitation is up to approximately 6‰ (Rozanski and Araguás 1995; Bowen and Revenaugh 2003). Alternatively, the less negative $\delta^{18}O$ values could be due to local evaporation processes. This is the more obvious influence, since the carbonates were precipitated in palustrine and shallow lacustrine subenvironments.

C isotope compositions vary by up to +5‰, similar to the +3‰ variation in TDIC noted in modern lakes during cyanophyta annual blooms (Thompson et al. 1997) which could indicate seasonal changes in nutrient supply to the paleolake. Moreover, the overlying very negative $\delta^{18}O$ values (see Fig. 6 in Benavente et al. 2019a) indicate dilution through meteoric water input, which indirectly supports a seasonal nature of rainfall in the sub-basins.

The clay-mineral assemblages of all three units are dominated by expandable clays with smectite-group components along with illite (Figs. 4d, 6d, and 8d). This supports semi-arid and seasonal climate conditions (Chamley 1989; Deocampo and Ashley 1999).

Regional context: All three sub-basins show similar evidence of aridity and seasonality, indicating these characteristics are likely due to regional climate and not local hydrology (Hutchinson 1957; Bohacs et al. 2003). As Hutchinson (1957) wrote: "Only when a considerable group of lakes exhibits more or less parallel fluctuation are we justified in drawing conclusions as to the time and nature of the climatic change causing the fluctuations." Aridity was probably enhanced by a rain-shadow effect of the upwind topographic barrier between the nearest ocean and the Cuyana Rift. Seasonal climates are characteristic of the paleolatitudes of the study area. Nevertheless, the three sub-basins studied are separated by 200 km along a north-south line, which in modern time is enough latitudinal variation as to represent different climate sub-regimes. However, the data presented for early Middle Triassic points to homogeneous climate regime in the paleolatitudinal range studied supporting the hypothesis of a low meridional climate gradient.

Similar paleoenvironmental conditions of warm dry subtropical to temperate strongly seasonal climate with high storminess and large seasonal rainfall (monsoon) were also inferred based on extensive paleobotanical studies in inferred coeval strata of adjacent sub-basins of the Cuyana Rift (Bodnar et al. 2018; Spalletti et al. 2003) supporting our interpretation of a regional paleoclimate signal. A typical warm-temperate climate (dry summers, wet winters) was interpreted for the Patagonia region in the Triassic by the reconstruction of phytogeographic provinces (Scotese et al. 1999; Quattrocchio et al. 2011).

In contrast, a study of microflora of the Cerro de las Cabras unit established an affinity with the Ipswich microflora (Zavattieri 1990), which suggested a warm and humid paleoclimate (Zavattieri 1991). A recent paleofloral study of coeval strata in an adjacent sub-basin interpreted warm to subtropical humid or seasonal paleoclimate (Beltrán et al. 2018). This study, however, was based on dipterid ferns that are known to flourish (and be best preserved) in areas of high groundwater and the authors admit that the humid conditions may have only been local. Therefore, these paleoflora studies reach conflicting interpretations of climatic humidity for the same area. We think this is caused by taphonomic bias (Benavente and Mancuso 2016) and the lack of integrated multiproxy studies of these other strata. Our paleoclimate interpretations are in accord with the modeling of Parrish et al. (1986) who proposed a warm arid regime with moderately high but seasonally restricted rainfall for paleolatitudes from 15 to 45°S—the three sub-basins we studied occupied paleolatitudes around 45°S and probably moved southwards by about 3° during early Middle Triassic, according to the paleolatitude calculator developed by van Hinsbergen et al. (2015).

Conclusions

The complexities of water sources, flow paths, and groundwater hydrology in lacustrine systems make it essential to interpret stable isotope data within a robust stratigraphic-sedimentological context when attempting paleoclimate reconstructions. In the Cuyana Basin, hydrography and hydrology varied substantially among the three sub-basins during the Middle Triassic. Carbonates from the Santa Clara Arriba paleolake have strongly correlated C and O stable isotopes, newly reported herein. This indicates that lake had a closed hydrology probably caused by long periods of inactive surface flow alternating with shorter surface water+sediment supply episodes. This hydrological reconstruction, added to previous sedimentological and stratigraphic studies of the unit, points to a balanced-fill lake-basin type. This is opposite to the previous overfilled-lake type classification based on sedimentology and stratigraphy alone. This confirms that integrated multiproxy dataset analysis using C and O stable isotope primary signatures from lacustrine carbonates can provide reliable detailed paleohydrology reconstructions.

In contrast, the Cerro de las Cabras appeared to have been fed by groundwater flow with persistently closed surface hydrography but open groundwater hydrology, whereas the Cerro Puntudo system was supplied by both surface and groundwater, with intermittently open hydrography and open hydrology.

These water source and flow complexities appear to obviate the quantitative estimation of paleotemperatures from oxygen isotopes. In contrast, in terms of qualitative paleohumidity and seasonality, all three basins show evidence of similar paleoclimate despite varying hydrography and hydrology among them. Regional paleoclimate conditions are interpreted to have been seasonally semi-arid to subhumid throughout the history of these paleolakes in the early Middle Triassic in southwest Gondwana, based on stratal stacking, sedimentary structures, exposure indicators, paleosol type and development, evaporite occurrence, clay-mineral assemblages, and an evaporative O stable isotope signature. This is consistent with global paleoclimate models for the Triassic that indicate substantial seasonality in the Cuyana Rift Basin region.

Following a robust interpretation flow as shown in Table 1 that establishes a strong paleogeographic, stratigraphic, sedimentological, petrographic, and mineralogic context provides a solid basis for interpreting stable isotope data and for avoiding inappropriate applications and misleading conclusions.

Acknowledgments We want to thank Elizabeth Gierlowski-Kordesch for expanding the field of Limnogeology and being the spark that ignited integrated Mesozoic lake basin studies in Argentina. It is our wish to continue honoring her memory carrying on her passion, lake science.

We also thank Editors Sila Pla Pueyo and Michael Rosen and anonymous Reviewers that improved significantly this contribution. Funding was provided by PICT 2013-0805 (ACM) and PICT 2014-0489 (CAB).

References

Alonso-Zarza, A. M. (2003). Palaeoenvironmental significance of palustrine carbonates and calcretes in the geological record. *Earth-Science Reviews, 60*, 261–298.

Alonso-Zarza, A. M., Meléndez, A., Martín-García, R., Herrero, M. J., & Martín-Pérez, A. (2012). Discriminating between tectonism and climate signatures in palustrine deposits: Lessons from the Miocene of the Teruel Graben, NE Spain. *Earth-Science Reviews, 113*, 141–160.

Anderson, N. J., & Leng, M. J. (2004). Increased aridity during the early Holocene in West Greenland inferred from stable isotopes in laminated-lake sediments. *Quaternary Science Reviews, 23*, 841–849.

Andres, M. S., Sumner, D. Y., Reid, R. P., & Swart, P. K. (2006). Isotopic fingerprints of microbial respiration in aragonite from Bahamian stromatolites. *Geology, 34*, 973–976.

Arenas, C., Casanova, J. O.-L., & Pardo, G. (1997). Stable-isotope characterization of the Miocene lacustrine systems of Los Monegros (Ebro Basin, Spain) palaeogeographic and palaeoclimatic implications. *Palaeogeography, Palaeoclimatology, Palaeoecology, 128*, 133–155.

Armenteros, I. (2010). Diagenesis of carbonates in continental settings. In A. M. Alonso-Zarza & L. Tanner (Eds.), *Carbonates in continental settings, geochemistry, diagenesis and applications, developments in sedimentology* (Vol. 62, pp. 61–151). Oxford: Elsevier.

Artabe, A. E., Spalletti, L. A., Bodnar, J., & Morel, E. M. (2003). Estudio paleoxilológico y sedimentológico de la Formación Montaña (Triásico), provincia de Mendoza, Argentina. *Ameghiniana, 46*(1), 141–152.

Atudorei, V., & Baud, A. (1997). Carbon isotope events during the Triassic. *Albertiana, 20*, 45–49.

Ávila, J. N., Chemale, F. J. R., Mallmann, G., Kawashita, K., & Armstrong, R. A. (2006). Combined stratigraphic and isotopic studies of Triassic strata, Cuyo Basin, Argentine Precordillera. *GSA Bulletin, 118*, 1088–1098.

Barredo, S. P. (2005). *Análisis estructural y tectosedimentario de la subcuenca de Rincón Blanco*. Precordillera Occidental, provincia de San Juan. Universidad de Buenos Aires, PhD Dissertation, 325p.

Barredo, S. P., Chemale, F., Marsicano, C., Ávila, J. N., Ottone, E. G., & Ramos, V. A. (2011). Tectono-sequence stratigraphy and U–Pb zircon ages of the Rincón Blanco Depocenter, northern Cuyo Rift, Argentina. *Gondwana Research, 21*, 624–636.

Benavente, C. A., & Mancuso, A. C. (2016). *Lacustrine Triassic paleoenvironments from Argentina: A context to paleobiology* (p. 106). Río Negro, Argentina: 11° Congreso de la Asociación Paleontológica Argentina.

Benavente, C. A., Mancuso, A. C., & Bohacs, K. M. (2019a). Paleohydrogeologic reconstruction of Triassic carbonate paleolakes from stable isotopes: Encompassing two lacustrine models. *Journal of South American Earth Sciences, 95*. https://doi.org/10.1016/j.jsames.2019.102292.

Benavente, C. A., Mancuso, A. C., Bohacs, K. M., & Irmis, R. B. (2019b). Multiproxy approach to paleohydrologic and paleoclimate reconstruction of Triassic lake basins of Argentina. *Geological Society of America Abstracts with Programs, 51*, 5. https://doi.org/10.1130/abs/2019AM-339785.

Benavente, C. A., Zavattieri, A. M., Mancuso, A. C., Abarzúa, F., & Gierlowski-Kordesch, E. H. (2018). Paleolimnology of the Santa Clara Arriba paleolake (Triassic Cuyana rift basin): Integrating sedimentology and palynology. *Journal of Paleolimnology, 59*, 5–20.

Benavente, C. A., Mancuso, A. C., & Cabaleri, N. G. (2012). First occurrence of charophyte algae from a Triassic paleolake in Argentina and their paleoenvironmental context. *Palaeogeography, Palaeoclimatology, Palaeoecology, 363-364*, 172–183.

Benavente, C. A., Mancuso, A. C., Cabaleri, N. G., & Gierlowski-Kordesch, E. H. (2015). Comparison of lacustrine successions and their paleohydrologic implications in the two subbasins of the Triassic Cuyana rift, Argentina. *Sedimentology, 62*, 1771–1813.

Beltrán, M., Bodnar, J., & Coturel, E. P. (2018). *Lycopodites* (Lycopodiidae, Lycopodiales): un nuevo integrante de las floras triásicas de la Argentina. *Revista del Museo Argentino de Ciencias Naturales, 20*(2), 205–216.

Billi, P. (2007). Morphology and sediment dynamics of ephemeral stream terminal distributary systems in the Kobo Basin (northern Welo, Ethiopia). *Geomorphology, 85*, 98–113.

Bodnar, J., Drovandi, J., Morel, E., & Ganuza, D. (2018). Middle Triassic dipterid ferns from west-central Argentina and their relation to palaeoclimatic changes. *Acta Palaeontologica Polonica, 63*. https://doi.org/10.4202/app.00459.2018.

Bohacs, K. M., Carroll, A. R., Neal, J. E., & Mankiewicz, P. J. (2000). Lake-basin type, source potential, and hydrocarbon character: An integrated sequence-stratigraphic—Geochemical framework. In E. Gierlowski-Kordesch & K. Kelts (Eds.), *Lake basins through space and time* (AAPG Studies in Geology) (Vol. 46, pp. 3–37).

Bohacs, K. M., Carroll, A. R., & Neal, J. E. (2003). Lessons from large lake systems – thresholds, nonlinearity, and strange attractors. In M. A. Chan & A. W. Archer (Eds.), *Extreme depositional environments: Mega end members in geologic time: Boulder, Colorado* (Geological Society of America Special Paper) (Vol. 370, pp. 75–90).

Bowen, G. J., & Revenaugh, J. (2003). Interpolating the isotopic composition of modern meteoric precipitation. *Water Resources Research, 39*, 10–1299.

Bowen, G. J., & Wilkinson, B. (2002). Spatial distribution of δ18O in meteoric precipitation. *Geology, 30*(4), 315–318.

Camoin, G., Casanova, J., Rouchy, J. M., Blanc-Valleron, M. M., & Deconinck, J. F. (1997). Environmental controls on perennial and ephemeral carbonate lakes: The central palaeo-Andean Basin of Bolivia during Late Cretaceous to early Tertiary times. *Sedimentary Geology, 113*, 1–26.

Chamley, H. (1989). *Clay sedimentology*. Berlin: Springer, pp. 623.

Chen, W.-Y., Su, T., Jia, L.-B., & Zhou, Z.-K. (2019). The relationship between leaf physiognomy and climate based on a large modern dataset: Implications for palaeoclimate reconstructions in China. *Palaeogeography, Palaeoclimatology, Palaeoecology, 527*, 1–13. https://doi.org/10.1016/j.palaeo.2019.04.022.

Cohen, A. S. (1990). A tectonostratigraphic model for sedimentation in Lake Tanganyika, Africa. In B. Katz (Ed.), *Lacustrine Basin exploration. Case studies and modern analogs* (Vol. 50, pp. 137–150). AAPG Memoir.

Cortés, J. M., González, B. G., & Koukharsky, M. M. L. (2003). *Hoja Geológica 3369-03 Yalguaráz. Provincias de San Juan y Mendoza República Argentina* (p. 95). Subsecretaría de Minería de la Nación, Servicio Geológico Minero Argentino.

Deocampo, D. M., & Ashley, G. M. (1999). Siliceous islands in a carbonate sea: Modern and Pleistocene spring-fed wetlands in Ngorongoro crater and Oldupai* gorge, Tanzania. *Journal of Sedimentary Research, 69*(5), 974–979.

Dupraz, C., Reid, R. P., Braissant, O., Decho, A. W., Norman, R. S., & Visscher, P. T. (2009). Processes of carbonate precipitation in modern microbial mats. *Earth-Science Reviews, 96*, 141–162.

Frakes, L. A. (1979). *Climates throughout geologic time* (p. 304). Amsterdam, New York: Elsevier Scientific.

Galfetti, T., Hochuli, P. A., Brayard, A., Bucher, H., Weissert, H., & Vigran, J. O. (2007). The Smithian/Spathian boundary event: A global climatic change in the wake of the end-Permian biotic crisis. Evidence from palynology, ammonoids and stable isotopes. *Geology, 35*, 291–294.

Gibson, J. J., Prepas, E. E., & Mceachern, P. (2002). Quantitative comparison of lake throughflow, residency, and catchment runoff using stable isotopes: Modelling and results from a regional survey of Boreal lakes. *Journal of Hydrology, 262*, 128–144.

Folguera, A., & Etcheverría, M. (2004). Hoja Geológica 3369-15 Potrerillos, Provincia de Mendoza. Programa Nacional de Cartas Geológicas de la República Argentina, 1:100.000. Servicio Geológico Minero Argentino. *Instituto de Geología y Recursos Minerales, 301*, 1–136. Buenos Aires.

Franzese, J., Spalletti, L. A., Gómez, P. I., & Macdonald, D. (2003). Tectonic and palaeoenvironmental evolution of Mesozoic sedimentary basins along the Argentinian Andes foothills (32° – 54° S.L.). In: *Structure and Development of the Pacific Margin of Gondwana* (Ed. by R.J. Pankhurst, L.A. Spalletti). *Journal of South American Earth Sciences, 16*(1), 81–90.

Freytet, P., & Verrecchia, E. P. (1998). Freshwater organisms that build stromatolites: A synopsis of biocrystallization by prokaryotic and eukaryotic algae. *Sedimentology, 45*, 535–563.

Groeber, P., & Stipanicic, P. N. (1953). Triásico. In: Groeber P, Stipanicic PN, Mingramm (eds.) A.R.G.: Mesozoico. Geografía de la República Argentina. *Sociedad Argentina de Estudios Geográficos GAEA, 2*(1), 13–141.

Hammarlund, D., Björck, S., Buchardt, B., Israelson, C., & Thomsen, C. T. (2003). Rapid hydrological changes during the Holocene revealed by stable isotope records of lacustrine carbonates from Lake Igelsjön, southern Sweden. *Quaternary Science Reviews, 22*, 353–370.

Harrington, H. J. (1971). *Descripción geológica de la Hoja 22c Ramblón. Provincias de Mendoza y San Juan. Boletín n° 114* (p. 81). Argentina: Dirección Nacional de Geología y Minería.

Henkes, G. A., Passey, B. H., Grossman, E. L., Shenton, B. J., Yancey, T. E., & Pérez-Huerta, A. (2018). Temperature evolution and the oxygen isotope composition of Phanerozoic oceans from carbonate clumped isotope thermometry. *Earth and Planetary Science Letters, 490*, 40–50.

Holz, M. (2015). Mesozoic paleogeography and paleoclimates – A discussion of the diverse greenhouse and hothouse conditions of an alien world. *Journal of South American Earth Sciences, 61*, 91–107.

Hren, M. T., & Sheldon, N. D. (2012). Temporal variations in lake water temperature: Paleoenvironmental implications of lake carbonate $\delta^{18}O$ and temperature records. *Earth and Planetary Science Letters, 337–338*, 77–84.

Huerta, P., & Armenteros, I. (2005). Calcrete and palustrine assemblages on a distal alluvial-floodplain: A response to local subsidence (Miocene of the Duero basin, Spain). *Sedimentary Geology, 177*, 253–270.

Hutchinson, G. E. (1957). *A treatise on limnology* (Geography, physics and chemistry) (Vol. 1). New York: Wiley, 1015 p.

Jenchen, U., & Rosenfeld, U. (2002). Continental Triassic in Argentina: Response to tectonic activity. *Journal of South American Earth Sciences, 15*, 461–479.

Kelts, K., & Talbot, M. (1990). Lacustrine carbonates as geochemical archives of environmental change and biotic/abiotic interactions. In M. M. Tilzer & C. Serruya (Eds.), *Large lakes: ecological structure and function* (pp. 290–317). Madison: Science Technical Publications.

Kent, D. V., & Tauxe, L. (2005). Corrected Late Triassic latitudes for continents adjacent to the North Atlantic. *Science, 307*, 240–244.

Kokogian, D. A., Seveso, S. S., & Mosquera, A. (1993). Capítulo I-7 Las secuencias sedimentarias triásicas. In V. Ramos (Ed.), *Relatorio Geología y Recursos Naturales de Mendoza* (pp. 65–78). XII Congreso Geológico Argentino and II Congreso de Exploración de Hidrocarburos.

Krull, E. S., Lehrmann, D. J., Druke, D., Kessel, B., Yu, Y. Y., & Li, R. (2004). Stable carbon isotope stratigraphy across the Permian-Triassic boundary in shallow marine carbonate platforms, Nanpanjiang Basin, South China. *Palaeogeography, Palaeoclimatology, Palaeoecology, 204*, 297–315.

Legarreta, L., Kokogian, D. A., & Dellapé, D. A. (1992). Estructura terciaria de la Cuenca Cuyana: ¿Cuánto de inversión tectónica? *Revista de la Asociación Geológica Argentina, 47*, 83–86.

Leng, M. J., & Marshall, J. D. (2004). Paleoclimate interpretation of stable isotope data from lake sediment archives. *Quaternary Science Reviews, 23*, 811–831.

Leng, M. J., Lamb, A. L., Heaton, T. H. E., Marshall, J. D., Wolfe, B. B., Jones, M. D., Holmes, J. A., & Arrowsmith, A. (2005). Isotopes in lake sediments. In M. J. Leng (Ed.), *Isotopes in palaeoenvironmental research* (pp. 147–184). The Netherlands: Springer.

Li, H. C., & Ku, T. L. (1997). $\delta^{13}C$- $\delta^{18}O$ covariance as a paleohydrological indicator for closed-basin lakes. *Palaeogeography, Palaeoclimatology, Palaeoecology, 133*, 69–80.

Liutkus, C. M., Wright, J. D., Ashley, G. M., & Sikes, N. E. (2005). Paleoenvironmental interpretation of lake-margin deposits using $\delta^{13}C$ and δ ^{18}O results from early Pleistocene carbonate rhizoliths, Olduvai Gorge, Tanzania. *Geology, 33*, 377–380.

Llambías, E. J. (2001). Complejos magmáticos triásicos al norte de los 40° S. In A. Artabe, E. Morel, & A. Zamuner (Eds.), *El Sistema Triásico de la Argentina* (pp. 55–68). La Plata: Fundación Museo de La Plata "Francisco P. Moreno".

López-Arbarello, A., & Zavattieri, A. M. (2008). Systematic revision of *Pseudobeaconia* Bordas, 1944 and *Mendocinichtys* Whitley, 1953 (Actinopterygii "Perleidiformes") from the Triassic of Argentina. *Palaeontology, 51*, 1025–1052.

Mancuso, A. C. (2009). Taphonomic analysis in lacustrine environment: Two very different Triassic lake paleoflora contexts from Western Gondwana. *Sedimentary Geology, 222*, 149–159.

Mancuso, A. C., Chemale, F., Barredo, S., Ávila, J. N., Ottone, E. G., & Marsicano, C. (2010). Age constraints for the northernmost outcrops of the Triassic Cuyana Basin, Argentina. *Journal of South American Earth Sciences, 30*, 97–103.

Markello, J. R., Koepnick, R. B., Waite, L. E., & Collins, J. F. (2007). The carbonate analogs through time (CATT) hypothesis and the global atlas of carbonate fields—A systematic and predictive look at Phanerozoic carbonate systems. *SEPM Special Publication, 89*, 1–31.

Meléndez, N., Liesa, C. L., Soria, A. R., & Meléndez, A. (2009). Lacustrine system evolution during early rifting: El Castellar Formation (Galve subbasin, Central Iberian Chain). *Sedimentary Geology, 222*, 64–77.

Moernaut, J., Verschuren, D., Charlet, F., Kristen, I., Fagot, M., & De Batist, M. (2010). The seismic-stratigraphic record of lake-level fluctuations in Lake Challa: Hydrological stability and change in equatorial East Africa over the last 140 kyr. *Earth and Planetary Science Letters, 290*, 214–223.

Murphy, J. T. J. R., Lowenstein, T. K., & Pietras, J. T. (2014). Preservation of primary lake signatures in alkaline earth carbonates of the Eocene Green River Wilkins Peak-Laney Member transition zone. *Sedimentary Geology, 314*, 75–91.

Noffke, N., Gerdes, G., Klenke, T., & Krumbein, W. E. (2001). Microbially induced sedimentary structures—A new category within the classification of primary sedimentary structures. *Journal of Sedimentary Research, 71*, 649–656.

Ostera, H. A., & Dapeña, C. (2003). *Environmental Isotopes and Geochemistry of Bañado Carilauquen. Mendoza, Argentina*. Short Papers – IV South American Symposium on Isotope Geology, 461–464.

Parrish, J. (1993). Climate of the supercontinent Pangea. *The Journal of Geology, 101*, 215–233.

Parrish, J. T., Hyland, E. G., Chan, M. A., & Hasiotis, S. T. (2018). Stable and clumped isotopes in desert carbonate spring and lake deposits reveal palaeohydrology: A case study of the Lower Jurassic Navajo Sandstone, South-Western USA. *Sedimentology, 66*, 32–52.

Parrish, J. M., Parrish, J. T., & Ziegler, A. M. (1986). Permian-Triassic paleogeography, paleoclimatology, and implications for therapsid distributions. In N. H. Hotton, P. D. Maclean, & E. C. Roth (Eds.), *The biology and ecology of mammal-like reptiles* (pp. 109–132). Washington, DC: Smithsonian Press.

Pla-Pueyo, S., Viseras, C., Candy, I., Soria, J. M., García-García, F., & Schreve, D. (2015). Climatic control on palaeohydrology and cyclical sediment distribution in the Plio-Quaternary deposits of the Guadix Basin (Betic Cordillera, Spain). *Quaternary International, 389*, 56–69.

Pla-Pueyo, S., Gierlowski-Kordesch, E. H., Viseras, C., & Soria, J. M. (2009). Major controls on sedimentation during the evolution of a continental basin: Pliocene–Pleistocene of the Guadix Basin (Betic Cordillera, southern Spain). *Sedimentary Geology, 219*, 97–114.

Potter, J., Siemann, M. G., & Tsypukov, M. (2004). Large-scale carbon isotope fractionation in evaporites and the generation of extremely ^{13}C-enriched methane. *Geology, 32*(6), 533–536.

Quattrocchio, M. E., Volkheimer, W., Borromei, A. M., & Martínez, M. A. (2011). Changes of the palynobiotas in the Mesozoic and Cenozoic of Patagonia: A review. *Biological Journal of the Linnean Society, 103*, 380–396.

Ramos, V. A., & Kay, S. M. (1991). Triassic rifting and associated basalts in the Cuyo Basin, central Argentina. In R. S. Harmon & C. W. Rapela (Eds.), *Andean Magmatism and its Tectonic Setting* (Geological Society of America Special Papers) (Vol. 265, pp. 79–91).

Retallack, G. J. (2013). Permian and Triassic greenhouse crises. *Gondwana Research, 24*, 90–103.

Retallack, G. J., Sheldon, N. D., Carr, P. F., Fanning, M., Thompson, C. A., Williams, M. L., Jones, B. G., & Hutton, A. (2011). Multiple Early Triassic greenhouse crises impeded recovery from Late *Permian* mass extinction. *Palaeogeography, Palaeoclimatology, Palaeoecology, 308*(1–2), 233–251.

Rozanski, K., & Araguás, L. (1995). Spatial and temporal variability of stable isotope composition of precipitation over the South American continent. *Bulletin de l'Institut français d'études andines, 24*(3), 379–390.

Rosen, M. R. (1994). The importance of groundwater in playas: A review of playa classifications and the sedimentology and hydrology of playas. *Geological Society of America Special Paper, 289*, 1–18.

Sanz-Montero, M. E., Rodríguez-Aranda, J. P., & García Del Cura, M. A. (2009). Bioinduced precipitation of barite and celestite in dolomite microbialites Examples from Miocene lacustrine sequences in the Madrid and Duero Basins, Spain. *Sedimentary Geology, 222*, 138–148.

Sanz-Montero, M. E., Rodríguez-Aranda, J. P., & García Del Cura, M. A. (2008). Dolomite–silica stromatolites in Miocene lacustrine deposits from the Duero Basin, Spain: The role of organotemplates in the precipitation of dolomite. *Sedimentology, 55*, 729–750.

Sanz-Montero, M. E., García Del Cura, M. A., & Rodríguez-Aranda, J. P. (2006). Facies dolomíticas de sistemas lacustres miocenos en las cuencas del Duero y de Madrid. Rasgos indicativos de su origen microbiano. *GeoTemas, 9*, 205–208.

Scotese, C. R., Boucot, A. J., & Mckerrow, W. S. (1999). Gondwanan palaeogeography and palaeoclimatology. *Journal of African Earth Sciences, 28*, 99–114.

Sellwood, B. W., & Valdes, P. J. (2006). Mesozoic climates: General circulation models and the rock record. *Sedimentary Geology, 190*, 269–287.
Spalletti, L. A. (2001). Evolución de las cuencas sedimentarias. In A. E. Artabe, E. M. Morel, & A. B. Zamuner (Eds.), *El Sistema Triásico en la Argentina* (pp. 81–101). La Plata: Fundación Museo de La Plata "Francisco Pascasio Moreno".
Spalletti, L. A., & Zavattieri, A. M. (2009). El sistema lacustre de la Formación Mollar en el depocentro triásico de Santa Clara (provincia de Mendoza, Argentina). *Andean Geology, 36*, 236–263.
Spalletti, L. A., Artabe, A. E., & Morel, E. M. (2003). Geological factors and evolution of south-western Gondwana Triassic plants. *Gondwana Research, 6*, 119–134.
Stappenbeck, R. (1910). La Precordillera de San Juan y Mendoza. Ministerio de Agricultura de la Nación. *Sección Geología, Mineralogía y Minería, Anales, 4*(3), 1–187. Buenos Aires.
Stipanicic, P. N. (1947). *Estudio geológico, estratigráfico y tectónico de la Precordillera, al este del río de Los Patos en Sorocayense* (San Juan). Tesis Doctoral. Universidad de Buenos Aires. Buenos Aires, 270 p. (Inédito).
Swart, P. K., Burns, S. J., & Leder, J. J. (1991). Fractionation of the stable isotopes of oxygen and carbon in carbon dioxide during the reaction of calcite with phosphoric acid as a function of temperature and technique. *Chemical Geology, 86*, 89–96.
Talbot, M. R. (1990). A review of the palaeohydrological interpretation of carbon and oxygen isotopic ratios in primary lacustrine carbonates. *Chemical Geology, 80*, 261–279.
Teixeira, B. M. N., Astini, R. A., Gomez, F. J., Morales, N., & Pimentel, M. M. (2018). Source-to-sink analysis of continental rift sedimentation: Triassic Cuyo basin, Precordillera Argentina. *Sedimentary Geology, 376*, 164–184.
Thompson, J. B., Schultze-Lam, S., Beveridge, T. J., & Des Marais, D. J. (1997). Whiting events: Biogenic origin due to the photosynthetic activity of cyanobacterial picoplankton. *Limnology and Oceanography, 42*(1), 133–141.
Torsvik, T. H., & Cocks, L. R. M. (2013). Gondwana from top to base in space and time. *Gondwana Research, 24*, 999–1030.
Torsvik, T. H., Van Der Voo, R., Preeden, U., Niocaill, C. M., Steinberger, B., Doubrovine, P. V., Van Hinsbergen, D. J. J., Domeier, M., Gaina, C., Tohver, E., Meert, J. G., Mccausland, P. J. A., & Cocks, L. R. M. (2012). Phanerozoic polar wander, palaeogeography and dynamics. *Earth-Science Reviews, 114*, 325–338.
Uliana, M. A., & Biddle, K. (1988). Mesozoic–Cenozoic paleogeographic and geodynamic evolution of southern South America. *Revista Brasileira de Geociencias, 18*, 172–190.
Van Hinsbergen, D. J. J., De Groot, L. V., Van Schaik, S. J., Spakman, W., Bijl, P. K., & Sluijs, A. (2015). A Paleolatitude Calculator for Paleoclimate Studies. *PLoS One, 10*(6), e0126946. https://doi.org/10.1371/journal.pone.0126946.
Veizer, J., & Prokoph, A. (2015). Temperatures and oxygen isotopic composition of Phanerozoic oceans. *Earth-Science Reviews, 146*, 92–104.
Wang, Y., Momohara, A., & Huang, Y.-J. (2019). Monsoon-induced errors in paleotemperature estimation based on leaf morphology analysis in central Japan. *Palaeogeography, Palaeoclimatology, Palaeoecology*, 533. https://doi.org/10.1016/j.palaeo.2019.109237.
Zavattieri, A. M. (1990). Palinología de la Formación Las Cabras (Triásico) en su localidad tipo, Cuenca Cuyana (Provincia de Mendoza, Argentina). Parte 1. Esporas triletes. *Ameghiniana, 27*(1–2), 107–129.
Zavattieri, A. M. (1991). Granos de polen de la Formación Las Cabras (Triásico) en su localidad tipo, Provincia de Mendoza, Argentina. Parte 2. *Ameghiniana, 28*(1–2), 3–29.
Ziegler, A. M., Scotese, C. R., & Barrett, S. F. (1983). Mesozoic and Cenozoic paleogeographic maps. In P. Broche & J. Sundermann (Eds.), *Tidal friction and the earth's rotation II*. Berlin: Springer.

Part VI
Asia

Modern Sedimentary Systems of Qinghai Lake

Jiang Zaixing and Liu Chao

Abstract Studies of modern lake sedimentary systems can guide the research and understanding of ancient lake basins. Qinghai Lake formed in the Miocene is the largest inland-faulted brackish lake in China with different types of terrigenous clastic sediments distributed around the lake and is an ideal location to study modern lake sedimentary systems. Observation of outcrops and exploratory trenches, bathymetry, remote sensing images analysis as well as meteorological data analysis forms the basis of this study. Five types of sedimentary systems were identified around the lake, including alluvial fan – fan delta system located south of the lake, north of Qinghai South Mountain and north of the lake, south of Datong Mountain; Delta system located northwest of the lake; barrier island – lagoon coastal system located northeast of the lake and nonbarrier coastal system in the south of the lake; eolian sedimentary system in the east of the lake; as well as deep lake system. The study revealed modern sedimentary systems are the product of the systematic combination of "wind (wind field) – source (provenance) – lake (lake basin)." The controlling factors include sediment source area lithology, topography, drainage, vegetation, power and direction of winds, coastal currents, and lake-level fluctuations. The characteristics, distribution, formation conditions, and controlling factors of the sedimentary systems provide new ideas and methods for studying the distribution of sand bodies in ancient lacustrine basins.

Keywords Qinghai Lake · Modern sedimentary systems · Controlling factors · Wind field

J. Zaixing (✉) · L. Chao
School of Energy Resources, China University of Geosciences (Beijing), Beijing, China

Key Laboratory of Marine Reservoir Evolution and Hydrocarbon Enrichment Mechanism, Ministry of Education, China University of Geosciences (Beijing), Beijing, China
e-mail: jiangzx@cugb.edu.cn

M. R. Rosen et al. (eds.), *Limnogeology: Progress, Challenges and Opportunities*, Syntheses in Limnogeology, https://doi.org/10.1007/978-3-030-66576-0_17

Introduction

Recent studies of Qinghai Lake mostly focused on single sedimentary facies (Shi et al. 1996, 2008; Song et al. 1999, 2000, 2001; Han et al. 2015). Tertiary continental rift basins in eastern China (e.g., Bohai Bay Basin) are similar to Qinghai Lake in turns of geological structure (rift lacustrine basins) and water chemistry (brackish-salt water). The study of modern sedimentary system in Qinghai Lake can provide an insight into the sediments of ancient continental-rift lacustrine basins, and help guide oil and gas exploration.

In recent years, many geologists have applied source-sink analysis to study the dynamic process of surface sedimentation in modern and ancient marine strata (Moore 1969; Moreno and Romero Segura 1997; Anthony and Julian 1999; Somme et al. 2009, 2013; Somme and Jackson 2013) and ancient continental-rift lacustrine basins (Xu 2013; Lin et al. 2015; Wu et al. 2015). Few scholars have carried out similar research on modern rift lakes. Qinghai Lake is surrounded by orogenic belts, with a number of source-sink depositional systems. The analysis of Qinghai Lake peripheral water system, topography, and sedimentary facies will help to elucidate modern sedimentation dynamics and serves as a modern example of a wind-source-basin system.

Geological Background

Qinghai Lake is the largest inland plateau brackish lake in China. It is located in the northeastern part of the Qinghai-Tibet Plateau between 99°36′E to 100°16′E and 36°32′N to 37°15′N. The lake is nearly oval with a long axis of 106 km and a short axis of 63 km. The area is 4473km^2 and the average depth is 21 m. Qinghai Lake is surrounded by high mountains and has nearly 40 inland-closed rivers in which Buha River is the largest (Chinese Academy of Sciences Lanzhou Geological Institute et al. 1979). Three sub-lakes (Gahai, New Gahai, and Erhai) are located in the east side of Qinghai Lake (Figs. 1 and 2).

Qinghai Lake has a typical arctic continental climate. Although the lake has a role in regulating the temperature, the Qinghai Lake area is characterized by large seasonal temperature differences (average − 12.7 °C in winter and 12.4 °C in summer), low precipitation (annually average 355.5 mm), hot rainy season (June–August), and well-defined wet and dry seasons (Chinese Academy of Sciences Lanzhou Geological Institute et al. 1979; Li et al. 2008). Over the past 60 years, lake level decreased (before 2004) and then increased (after 2004), which greatly influenced the sedimentary system of Qinghai Lake.

The main lithology of mountains surrounding Qinghai Lake is volcanic, metamorphic rocks, and a few sedimentary rocks, including low-grade metamorphic rocks, gneiss, granites, granodiorite, sandstone, shale, limestone, metamorphic

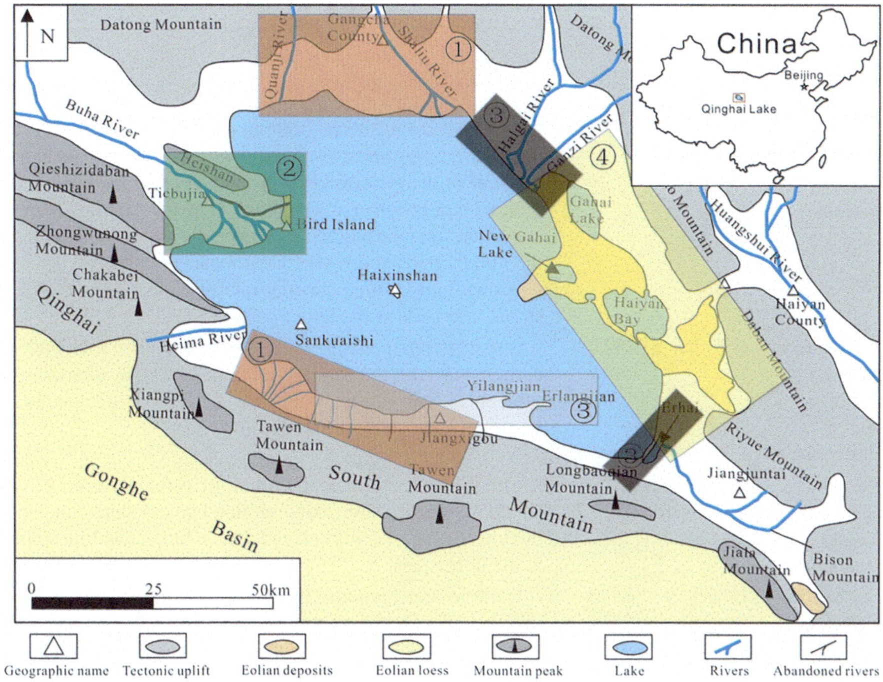

Fig. 1 Geographical overview of Qinghai Lake. ①: Alluvial fan—fan delta sedimentary system; ②: Delta sedimentary system; ③: Coastal sedimentary system; ④: Eolian sedimentary system

Fig. 2 Satellite image of Qinghai Lake

Table 1 Main lithology of Datong Mountain

Location	Strata	Age (ma)	Lithology
The western and middle part of Datong Mountain	Upper Triassic	201.3–237	A layer of stable off-white thick-layered conglomerate or pebbled sandstone at the bottom, gray thick-layered feldspar-quartz sandstone, quartz-feldspar sandstone, and siltstone at the lower part; gray thick-layered medium-grained quartz-feldspar sandstone, feldspar-quartz sandstone, and carbonaceous shale, siltstone, coal streak at the middle part; yellow-kelly carbonaceous shale, siltstone, fine-grained feldspar-quartz sandstone, and coal streak interbeds at the upper part.
The western and middle part of Datong Mountain	Permian	252.17–298.9	It is marine-terrigenous facies (layers of clastic rock and carbonate rock), the lithology is mainly arkose, quartz sandstone, fine-grained sandstone, siltstone, argillaceous siltstone, shale, and bioclastic limestone.
The eastern part of Datong Mountain	Middle and lower Triassic	237–252.17	It is marine-clastic rock, the lithology is mainly arkose, siltstone, sandy shale, coarse-grained quartz sandstone at the lower part; clastic rock and a little bioclastic limestone at the upper part.
The eastern part of Datong Mountain	Lower Silurian	433.4–443.8	It is marine-terrigenous clastic rock, the lithology is mainly slate, clay slate, and feldspar-quartz sandstone, schistose siltstone, arkose, and gray-purple conglomerate.
The eastern part of Datong Mountain	Intrusive rock	/	The adamellite formed in the late Caledonian invaded the lower Silurian strata.

Modified from Chinese Academy of Sciences Lanzhou Geological Institute et al. (1979)

sandstone, slate, and diorite (Table 1) (Chinese Academy of Sciences Lanzhou Geological Institute et al. 1979). Quaternary sediments are well-developed in the Qinghai Lake Basin, mostly lacustrine (Fig. 3). Middle-lower Pleistocene Hadatan Formation, Upper Pleistocene Erlangjian Formation, and Holocene Buhahe Formation are developed from the bottom up (Qinghai Geology and Mineral Resources Bureau 1991).

Methods and Materials

This study is mainly based on field survey, and remote sensing images. Nearly 5900 meteorological measurements of Qinghai Lake over the past 60 years (1956–2015) from several weather stations (e.g., Gangcha Weather Station) were analyzed for the study.

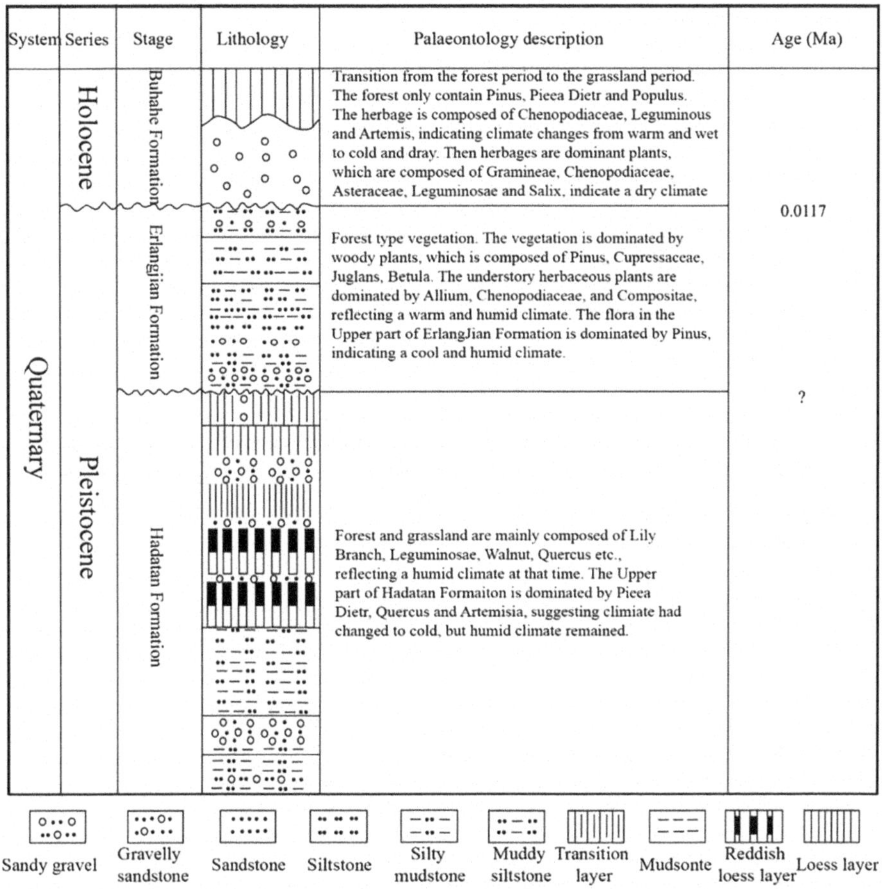

Fig. 3 Quaternary stratigraphy of Qinghai Lake. (Modified from Chinese Academy of Sciences Lanzhou Geological Institute et al. 1979)

Field Survey and Sample Collections

A total of 107 investigation points were selected for study, including 43 key outcrops, 32 trenches, and 32 observation points. These observation points are distributed in the main sedimentary systems around Qinghai Lake, such as Alluvial fan—fan delta in the north and south bank, Buha River Delta in the northwestern part of the lake, and coastal systems in the south and east coast of the lake. About 58 samples of bottom sediments in front of the Buha River Delta are collected by sampling boat and piston corer for particle size, 26 samples of lagoon muddy sediments located in the south coast of the lake are collected for dating analysis, as well as 45 sand samples of eolian dunes are collected for SEM observation.

Remote Sensing Imagery

To better understand the impact of lake-level changes on how the sedimentary system evolved, over 1000 remote sensing images over 30 years (1976–2015) were extracted from Google Earth. For example, the remote sensing imagery on April 23 in 2007 and April 13 in 2013 was chosen to show the influence of lake-level rise on the mouth of the Shaliu River. The remote sensing pictures in 2003 and 2010 revealed the migration of coastal eolian dunes on the east bank of the lake.

Particle Size

Particle-size analysis was conducted using a Malvern Mastersizer 3000 laser particle analyzer. The device was checked and adjusted to confirm that the error was not greater than 0.1%. The particle size of the samples was separated into segments based on particle size: (1) >63 μm, (2) 4–63 μm, and (3) <4 μm, which represent sand, silt, and clay. Particle size was analyzed at the Institute of Geology and Geophysics of the Chinese Academy of Sciences and the Key Laboratory of Mineral Processing of CUMTB (Beijing).

Results

Alluvial Fan—Fan Delta Sedimentary System

The alluvial fan deposits around Qinghai Lake are mainly developed in front of the Datong Mountain in the north bank and Qinghai Nanshan in the south bank (Fig. 1). Many alluvial fans of different sizes (radius range from 1 km to 16 km) are formed at the foot of the mountain. As the alluvial fans oscillated back and forth, combined and superimposed, they entered the lake, forming fan delta. The alluvial fans in front of the Datong Mountain on the north bank are discussed as an example below:

The alluvial fans of the Datong Mountain are a fan group with the mountain pass as the apex, in which Gangcha alluvial fan is the biggest. A seasonal river had developed at the mountain pass of each alluvial fan, but most of them dried up. Presently, only the Shaliu River has water all year round.

Gangcha Alluvial Fan

Gangcha alluvial fan has a slope of 21°and a radius of 16 km. Channel deposits and sheet-flood deposits are identified in the section.

Channel deposits are mainly composed of gravel and sand. The gravels are subangular and subcircular, poorly sorted with diameters between 5 cm and 30 cm.

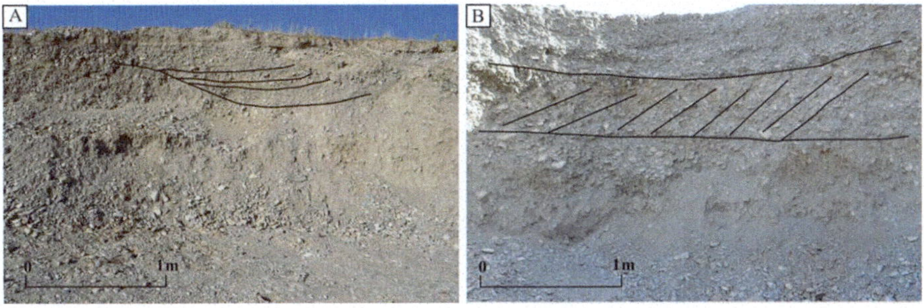

Fig. 4 Trough cross-bedding (**a**) and tabular cross-bedding (**b**) in channel deposits

Fig. 5 Nonconcentrated flow deposits (**a**) and its parallel bedding (**b**)

They dip to 5°–20°at angles of 30° to 40°. A single layer of channel deposits ranges 60 cm to 100 cm. Trough cross-beddings (Fig. 4a) and tabular cross-beddings (Fig. 4b) are observed in the section, indicating the strong erosion of flow and lateral migration of small gravel dams (Miall 1978; Bridge 1993).

Sheet-flood deposits mainly consist of thinly bedded (20–30 cm) coarse sand, occasionally fine gravel (Fig. 5a). Parallel bedding is well-developed inside (Fig. 5b). Sheet-flood deposits often coexist with channel deposits in the section.

Gangcha Fan Delta

While the alluvial fan entered into the lake, the content of gravel gradually decreased, and the content of sand and mud increased. Through the high-resolution remote sensing image, it was found that a large number of distributary-braided channels are visible at the end of the fan delta plain (Fig. 6a). With the abandonment of the channel, and the migration of the mouth bar, the fan delta plain has ceased to develop. At present, the abandoned-braided channel is covered by crops and vegetation, leaving only a small amount of mouth bars parallel to the coastline (Fig. 6b).

Because the fan delta is a transition from an alluvial fan, it belongs to the fan-type fan delta according to its position. The sedimentary characteristics of the fan delta are: (1) Coarser terrigenous clastic sediments (mostly gravel and coarse sand);

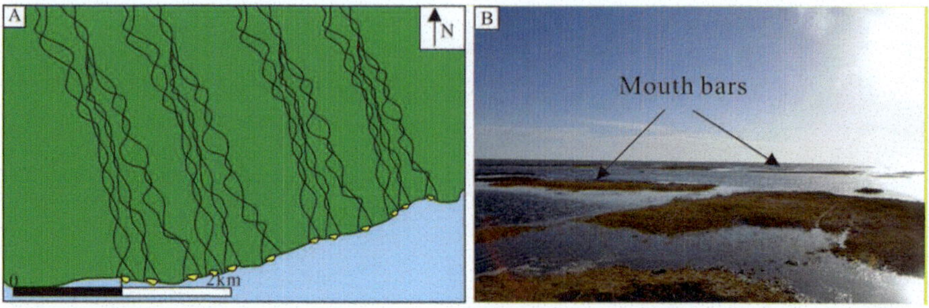

Fig. 6 The abandoned-braided channels on the Shaliuhe fan-delta plain (**a**); the mouth bar on Shaliuhe fan delta (**b**)

(2) Well-developed distributary-braided channels on fan delta plain which are easily abandoned.

Delta Sedimentary System

Delta deposits in Qinghai Lake are developed in the northwestern part of the lake, and are formed by the Buha River in the direction of the long axis of Qinghai Lake (Fig. 1), which has the largest annual runoff (about $1.1*10^9$ m^3annually, accounted for more than 67% of the total runoff into the lake) in the lake (Li et al. 2010). Based on the sedimentary environment and sedimentary characteristics, the Buha River Delta can be divided into delta plain, delta front, and prodelta.

Delta Plain

Distributary channels, abandoned-distributary channels, natural levee, and swamp are observed on the delta plain.

Distributary channel: It is the skeleton part of the delta plain, mainly composed of sandy sediments. The channel has a few branches, and is nearly straight, with few channel bars developed (Fig. 7a). Because of the flat terrain, the flow rate in distributary channels is much slower than that of the middle and upper reaches.

Natural levee: It is located on both sides of the distributary channel, but most of them are incompletely preserved and appear as hummocky bodies on topography (Fig. 7b). The sediments are mainly sand and mud, and the grain size is finer than the channel sand, coarser than the swamp deposits. Its surface is mostly covered with plant remains carried by the flood.

Abandoned-distributary channel: Consisting of sand and clay, are finer than sediments of the distributary channel. Some of the abandoned channels have been buried by sand dunes (Fig. 7c).

Fig. 7 The distributary channel (**a**), nature levee (**b**) on Buha River delta plain, the eolian dunes (**c**), and swamp (**d**) on north of Bird Island

Swamp: Developed on the delta plain near the lake, its surface is close to the average high-level of the lake. The swamp is periodically submerged by floods or changes in lake level. The swamp is rich in plants (Fig. 7d), contains stagnant water, and is mostly a reducing environment. Most of the sediments are clay, and the organic carbon content is ~0.8%.

Delta Front

Previous study shows the underwater distributary channel extends 12 km to the deep lake area (Chinese Academy of Sciences Lanzhou Geological Institute et al. 1979). Based on the water depth, particle size, and sedimentary microfacies, geomorphological characteristics of the submerged Buha River Delta front are identified and include distributary channels, underwater natural levee, mouth bar, sheet sand, and foreslope.

Underwater Distributary Channel and Natural Levee

Previous studies on the underwater distributary channel in the Poyang Lake Delta and the Daihai Delta (Jin et al. 2014; Shi et al. 2014) found no underwater distributary channel in the delta front. Field investigation of the Buha River Delta

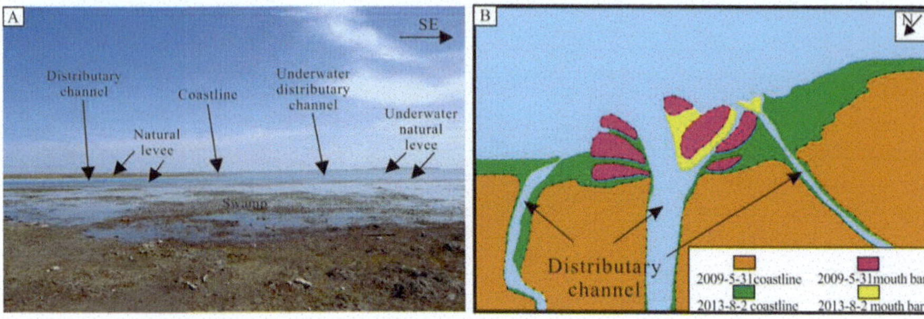

Fig. 8 (**a**) Field photo of Buha River mouth; (**b**) Comparison of the Buha River mouth in 2009.5.31 and 2013.8.2. Two triangle-mouth bars at the mouth of the Buha River in dry season were cut and transformed forming a divergent finger-shaped sand bar and underwater distributary channels. Previous delta plain close to the coastal is submerged to become swamp

front, document underwater distributary channels, and natural levees is shown (Fig. 8a, b).

By analyzing the remote sensing photos of the Buha River mouth from May 31, 2009 and August 2, 2013, we found the distributary channel and underwater distributary channel are interchangeable. An important factor in controlling this conversion is the periodic change of the lake level. The submerged distributary channel is actually formed by being submerged by the lake.

Mouth Bar

The mouth bar is formed by the accumulation of sand and mud material at the river mouth by a reduced flow rate (Edmonds and Slingerland 2007). The mouth bar can be transformed by channel and changes in lake level, resulting in changes in morphology and lateral migration. The reconstructed mouth bar may vertically overlap distributary channels or swamp deposits. In addition, bars are not developed in all river mouths, which are related to the amount of sediment supplied by the river (Wright and Coleman 1971, 1978).

Frontal Sheet Sand and Foreslope

A clear foreslope can be seen from the 3D topography map from 1.5–23.5 m (Fig. 9). Because the area is very seldomly affected by tectonic movement (Chen et al. 1964; Bian et al. 2000), the foreslope is well-formed by sedimentation. The slope extension to the southern depression is narrower, with a water depth variation of 2–23 m and a slope of 0.054%, while to the Tiebujia Bay is wider, with a water depth of 2–15 m and a slope of 0.025%. The bottom of the foreslope is flat.

The surface sediments are mainly clayey silt, silt, silty sand, sandy silt, and fine sand according to the classification principle of Shepard (Fig. 10).

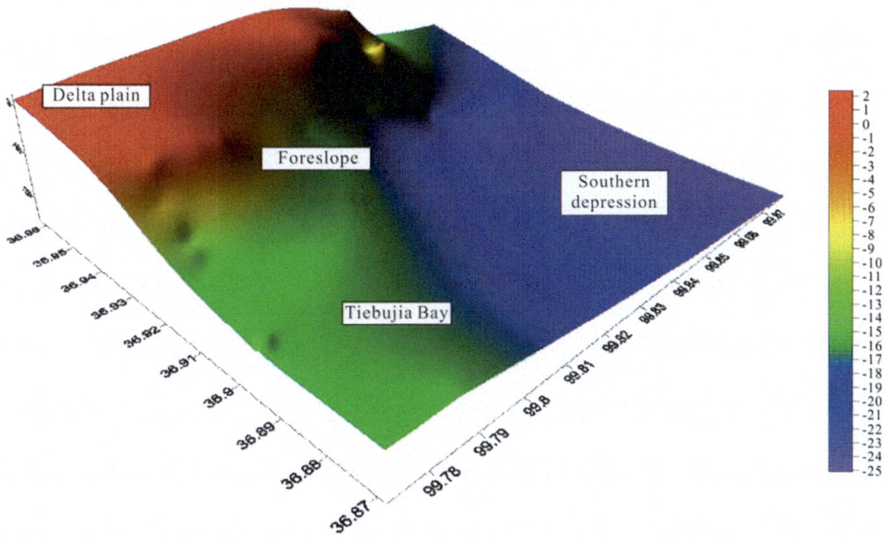

Fig. 9 Underwater geomorphological map of Buha River delta front. Development of the foreslope is uneven. The bottom of the foreslope is flat

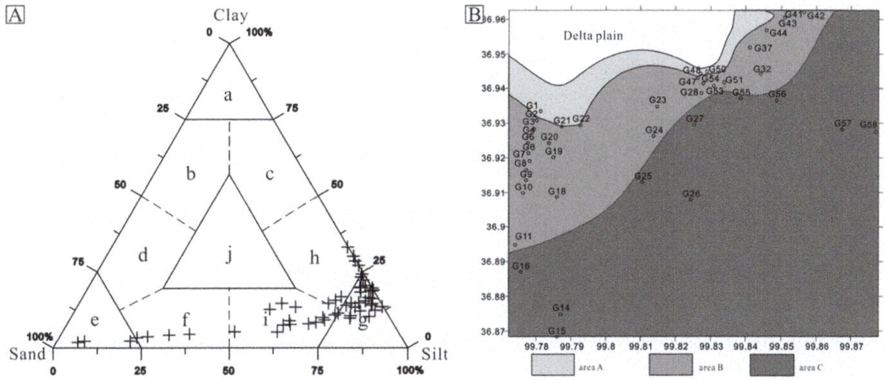

Fig. 10 (**a**) Triangle diagram of Shepard sediment nomenclature, the three units are clay minerals, sand, and silt. a:clay; b: sandy clay; c: silty clay; d: clayey sand; e: sand; f: silty sand; g: silt; h: clayey silt; i: sandy silt; j: sand-silt-clay. (**b**) Distribution of sampling point

The area can be roughly divided into three parts: A, B, and C (Fig. 10). The particle size of the sediments in each area is shown in Table 2.

In summary, the major sediments in study area are silt, clayey silt, and silty sand. The nearshore areas are mainly fine sand and silty sand.

The sheet sand is a thin and large sand layer formed by the mouth bar and modified by the waves and coastal currents (Jiang 2010). The sheet sand is widely distributed outside the delta plain, where the water depth is less than 4 m (A area,

Table 2 The statistical table of particle size of the surface sediments

Area	Mud			Silt			Fine sand			Md		
A	3.20	4.56	1.82	21.24	36.44	6.1	75.56	92.08	58.99	3.41	3.81	3.01
B	10.47	14.69	5.31	59.98	62.53	55.19	29.55	33.68	24.49	5.12	6	4.68
C	17.07	30.09	8.02	76.74	86.1	66.7	6.2	23.61	0.57	6.34	7.37	5.14

Area	Mz			σ1			SK1			KG		
A	3.5	3.91	3.03	0.96	1.2	0.73	0.35	0.39	0.28	1.86	2.26	1.54
B	5.18	5.37	4.65	1.76	2.19	1.55	0.11	0.57	-0.19	0.96	1.23	0.71
C	6.42	7.37	5.29	1.32	1.56	0.99	0.12	0.31	-0.05	1.06	1.26	0.94

P.S. In each form, the left is the average, the upper right is the maximum, and the lower right is the minimum

Fig. 11). The foreslope is developed at a water depth of 4–23.5 m and bounded by 14 m. Sediments shallower than 14 m are mainly silty-fine sand deposits with sand-mud interbeds (B area, Fig. 11); deeper than 12 m are silty-clay deposits rich in plant remains with no obvious stratification(C area, Fig. 11).

Prodelta

The deepest sampling point G58 has a water depth of 23.5 m, and the sediment is still mainly sand. Previous study revealed that the deep-lake deposits were preserved under a water depth of 25 m (Chinese Academy of Sciences Lanzhou Geological Institute et al. 1979), and thus we believe that the prodelta deposits are located under the water depth of 25 m.

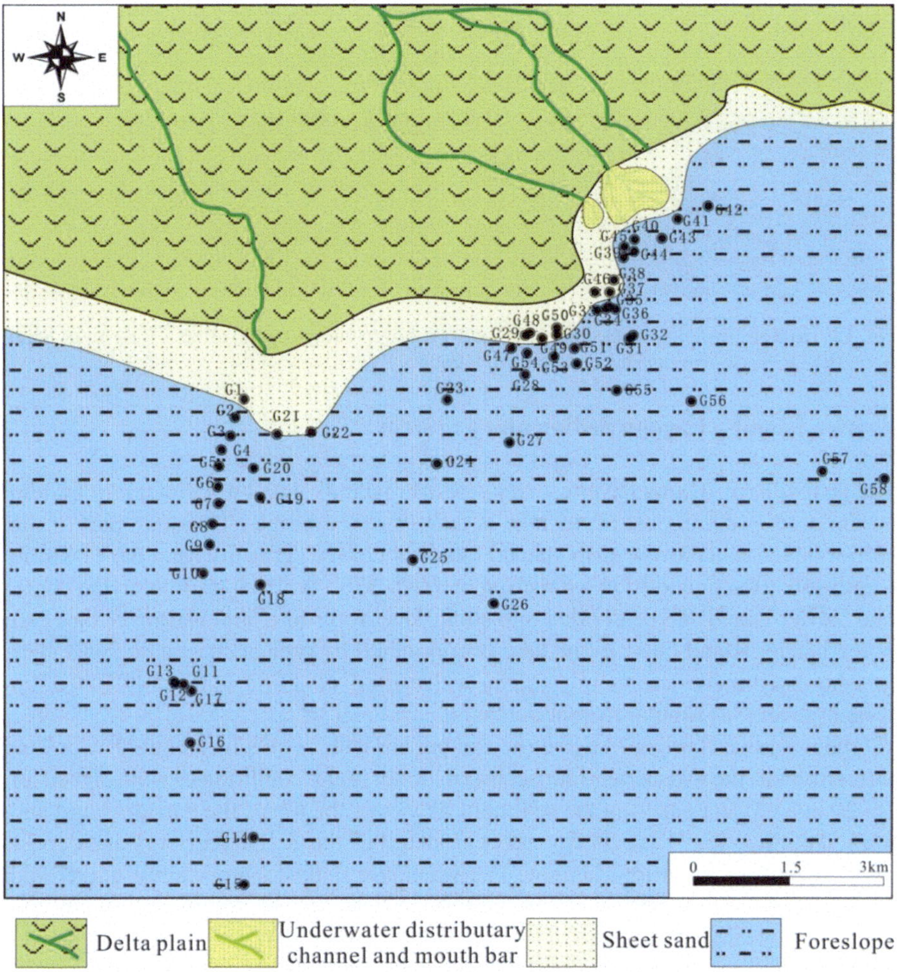

Fig. 11 Sedimentary facies map of Buha River delta

Coastal Sedimentary System

The coastal sedimentary system is widely developed around the lake where the provenance is weak. Based on the presence or absence of the barrier islands, the coastal deposits of Qinghai Lake can be divided into barrier island – lagoon and nonbarrier coastal sedimentary systems. The barrier island – lagoon system is situated on the northeastern and southeastern coast of the lake, while nonbarrier sedimentary system is mainly distributed on the southern coast of the lake where beach deposits are developed (Fig. 1).

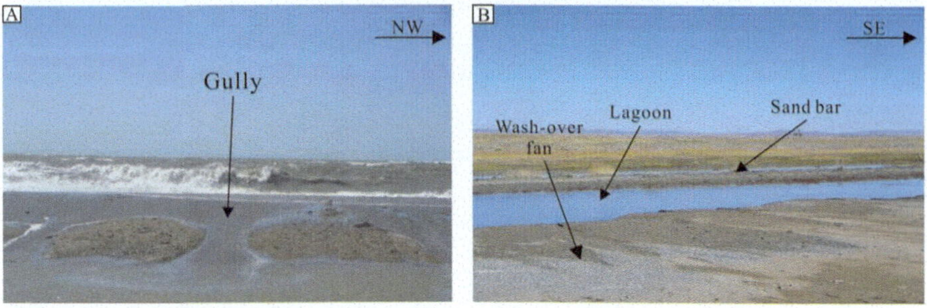

Fig. 12 (**a**) The forming process of gullies and washover fan under the effect of storm, at point HN7; (**b**) The washover fan and small bars in lagoon

Barrier Island – Lagoon Sedimentary System

The Halgai River/Ganzi River estuary – barrier island – lagoon sedimentary system on the northeastern coast of the lake is taken as an example. The main part of the barrier island is a narrow strip-shaped (slightly curved) gravelly sand bar (1 km in length and 40 m in width) in a direction of 140°–330°, which is parallel to the coast. The barrier island can be divided into shore face, foreshore, and washover fan. Part of the barrier island close to the lake is dominated by typical coastal deposits. The foreshore sediments are mainly (65%) well-sorted sand and some (30%) medium sorted gravel (subangular to subcircular).

The washover fan is more common in barrier islands (Hudock et al. 2014). The formation of the washover fan occurred under gale conditions (wind speed of 18–20 m/s). Water carrying the sediment flew over the top of the barrier island and formed gullies (Fig. 12a). Storm waves eroded the coastal and barrier island deposits, transporting them into lagoon. The washover deposits are mainly gravel and sand with a scoured bottom while the lagoon deposits are dominated by silt and clay. Small-scale bars composed of coarse debris were observed, formed by gravel and sand brought into the lagoon by large waves (Fig. 12b).

Barrier bars are characterized by reverse-graded bedding (Fig. 13). Two thinly gravel layers (1–2 cm) occurred at the front and top of the bar, and the gravels were arranged parallel to the coast. The gravel layer is interpreted as the stagnant layer deposited by the storm wave washing over the bar (Hudock et al. 2014).

The Ganzi River is located on the northeastern coast of Qinghai Lake, where the wind backwater formed by the northwest wind directly acts on the estuary, forming two-way flow (normal river flow to the SW and backwater by wind to the NE) here (Fig. 14a). A large number of overflow gullies were observed on the coastal bar, behind which a swampy lagoon was developed (Fig. 14b). There are several small waterways near the estuary, and here a barrier island – waterway – lagoon system was characterized. The planeview of the waterway is an asymmetric trumpet (Fig. 15a), with the lake side being wider than the land side. Waves from the northwest act directly at point G1 forming a distinct-scoured surface (Fig. 15b); waves of different heights leave significant marks at point G3 (Fig. 15d). The backwater

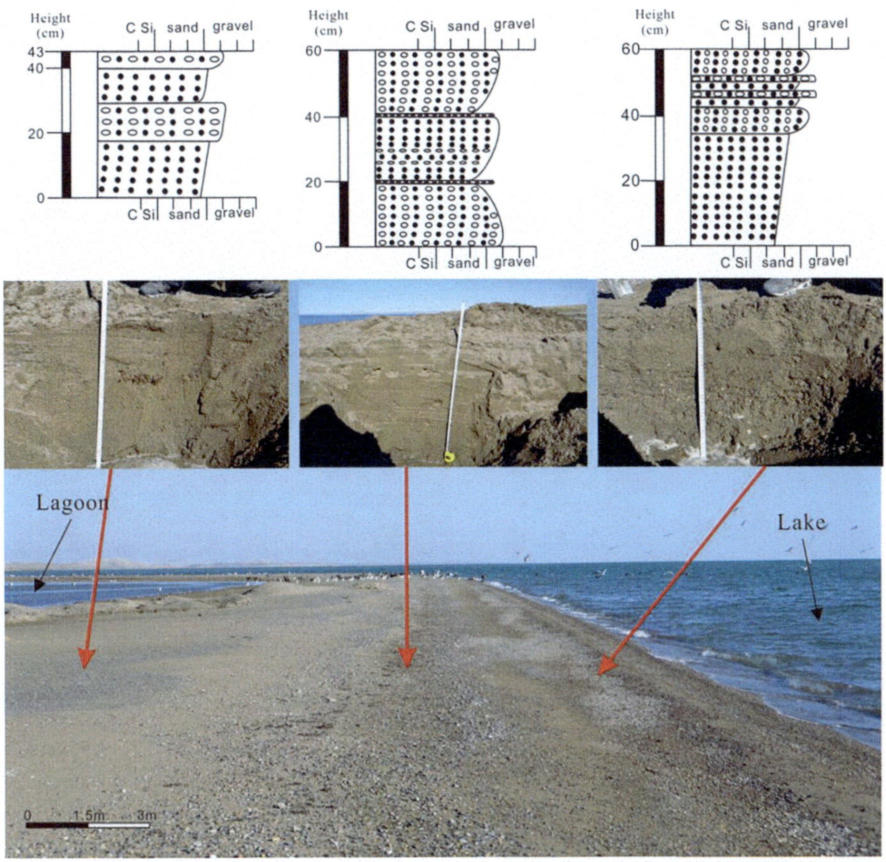

Fig. 13 The section picture of barrier bar

retains coarse gravel at point G2 (Fig. 15e), carrying and stacking the water at point G4 (Fig. 15d). The lake water flows along the waterway back into the lagoon behind the bar, leaving ripples at point G5 (the ripples indicating landward flow direction) (Fig. 15f). Several shallow beaches form in the lagoon by water exiting the waterway and spreading out (Fig. 15g).

Nonbarrier Coastal Sedimentary System

Modern Coastal Facies

Beach sediments parallel to the shoreline are usually sheet-shaped and have a large distribution area (Zhu and Xin 1994). Based on lithology, particle size, and micro-landforms, the beach can be divided into beach ridge, inter beach ridge, sheet sand, and lagoon.

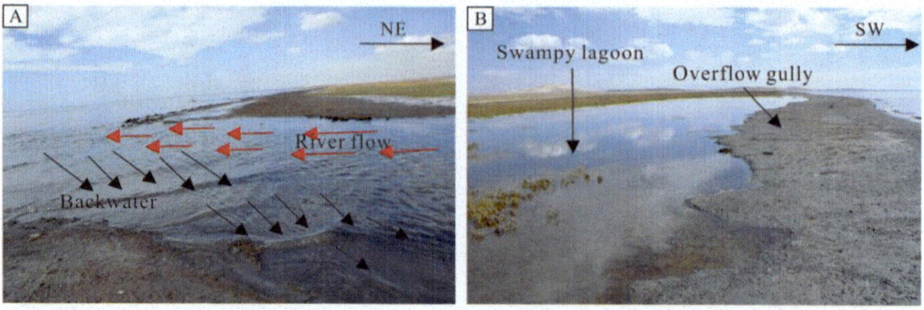

Fig. 14 (**a**) Field photo of normal river flow and backwater; (**b**) Coastal bar and the swampy lagoon behind the bar

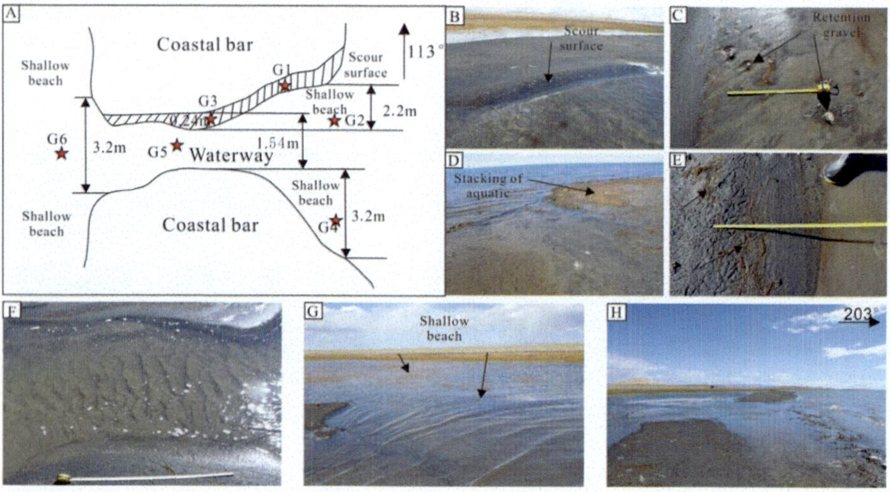

Fig. 15 (**a**) Drawing of barrier island – waterway – lagoon system; (**b**) Photo of scoured surface at G1; (**c**) Retention gravel sediment at G2; (**d**) Stacking of the aquatic at G4; (**e**) Lake-level mark at G5; (**f**) Flow ripples at G6; (**g**) The shallow beaches in the lagoon; (**h**) Field photo of barrier island – waterway – lagoon

A beach ridge is a wave-swept or wave-deposited ridge running parallel to the shoreline. It is often linear (Fig. 16a), fin or hook-shaped (Fig. 16b).

Beach ridge sediments are mainly gravel and sand. Some lake beaches have several beach ridges. Low-angle bedding (dip angle of ~6°) is commonly developed in the lake side of ridges, with fine gravel. Inter beach ridge is between ridges and is dominated by silt and fine sand. Obstacle marks due to the swash/backwash of lake waves were observed on the lake side of newly formed ridges. The finer end of obstacle mark points in the direction of flow (Fig. 17a). Sheet sands are located on the shore side of the ridge and are dominated by sand deposits. An erosional gully, widens landward, is commonly observed and developed in the sheet sand (Fig. 17b). The lagoon behind the beach is in the low area on the shore side of the sheet sand (Fig. 17b). Sediments in the lagoon are dominated by muddy sand, and water in the lagoon is from the lake during higher lake levels.

Fig. 16 (**a**) Linear beach ridge. (**b**) Hook-shaped beach ridge

Fig. 17 (**a**) Linear beach ridge. (**b**) Hook-shaped beach ridge

Fig. 18 (**a**) Plant-residue band on the Heima River gravel beach. (**b**) Section of the gravel beach

The beach can be divided into gravel and sand beaches based on composition. Gravel beaches are mainly distributed beside the fan delta south of Heima River or along the eroded shore. Plant remains are often found on gravel beaches, which were transported by swash (Fig. 18a). These beaches can be divided into upper and lower gravel beaches based on hydrodynamic energy and grain size

Fig. 19 (**a**) Section of a sand beach in the Erlangjian-Jiangxigou area. (**b**) Plant remains and gravel band on the beach

(Fig. 18b). The gravel content of upper gravel beach is about 60% (maximum gravel size is 15 cm), and poorly sorted sand content is about 35%. Lower gravel beach is composed of 28% gravel (maximum gravel size is 4 mm). Sand accounts for 69% and is well sorted. The flat surface of gravels dips towards the lake, whereas the long axis is parallel to the shoreline. In the vertical section, gravel beach displays reversed-graded bedding. The composition of the gravel is dominated by limestone, reflecting a provenance from the Triassic strata of the Zhongwunong Mountain.

Sand beach is mainly distributed in the Erliangjian-Jiangxigou area. The beach sediments are composed of gravelly-sand and sand. Sand beach can be divided into upper sand beach and lower sand beach (Fig. 19a). The upper sand beach contains about 35% gravel (maximum grain size 3 cm) and 62% sand (poor-sorted). The lower sand beach has about 12% gravel (maximum grain size 4 mm) and 90% sand (medium-sorted).

Both gravel and sand beaches have reversed-graded bedding, with coarse clasts deposited by swash on top and fine clasts deposited by backwash at bottom. Such reversed sequences appear rapidly in a vertical section and is shown with gravel or plant remains in linear belts parallel to shoreline. These features indicate the changes of lake level and the migration of shoreline (Fig. 19b).

Ancient Coastal Facies

The foreshore subfacies of the modern costal facies in this area are dominated by beach deposits. There is no obvious sedimentary structure in most deposits. A representative section was characterized in detail (N5) to understand the lithology, sedimentary structure, and evolution of these deposits. The N5 section is located at the east side of Erlangjian, about 150 m from the current shoreline. The section exhibits two sets of reversed sequences. Typical sedimentary structures are:

Fig. 20 Sedimentary structure on the Erlangjian sandpit section. (**a**) The lower part is characterized by massive bedding (mud sediments), whereas the upper part is characterized by reversed-gradient bedding (silt-sand-gravel sediments). (**b**) Parallel bedding (sand and gravel sediments). (**c**) The lower part is characterized by low-angle cross-bedding dipping towards the lake, whereas the upper part is characterized by oblique bedding dipping towards the land. (**d**) Swash mark, which is overlaid by low-angle cross-bedding dipping towards the land

(1) Massive bedding: Located at the bottom of the entire section. Sediments are mainly dark-gray mud, with no obvious structure, interpreted as rapid deposition of suspended sediment (Fig. 20a).
(2) Reversed-graded bedding: Located above the massive bedding. The layers are basically parallel (Fig. 20a).
(3) Parallel bedding: Often appears on sections parallel to the shoreline. The sediments are mainly coarse sand and gravel indicating a high-flow regime (Miall 1978; Bridge 1993) (Fig. 20b).
(4) Low-angle cross-bedding: Appears in the middle of the section. Sediments are dominated by coarse sand and gravel. The low-angle cross-bedding can be divided into two groups (Fig. 20c): low-angle cross-bedding dipping towards the lake (dipping angle 12°), and oblique bedding dipping landward (dipping angle 28°). The former is formed by swash/backwash when the lake level is constant or increasing. The latter indicates beach ridges migrating towards the lake when the water level decreases.
(5) Flow marks: Located between the two groups of low-angle cross-bedding. Sediments are mainly muddy silt. Flow marks are formed by differential erosion (Fig. 20d).

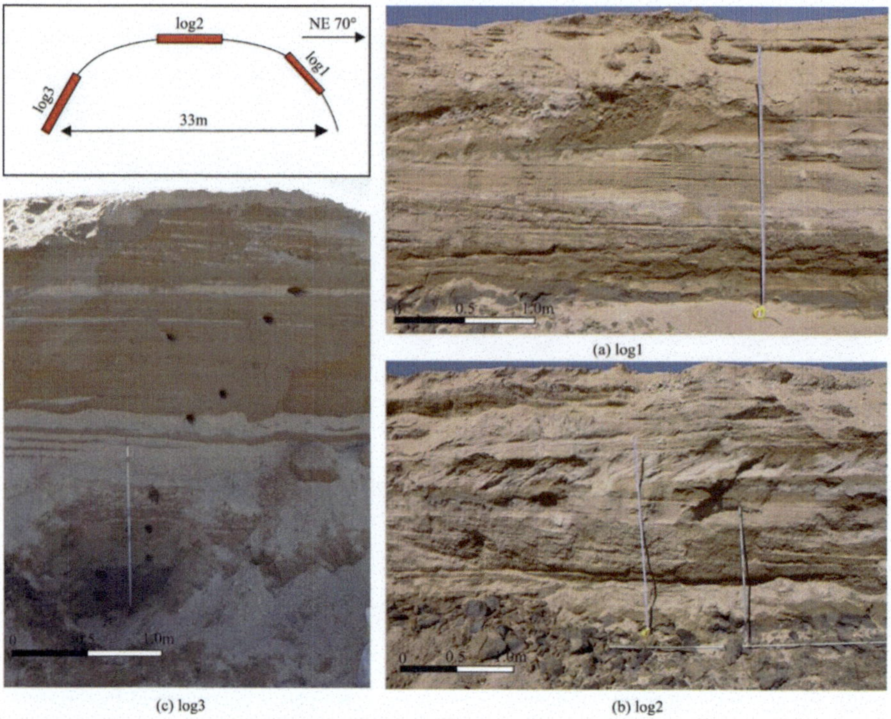

Fig. 21 The location and field pictures of Logs 1, 2, and 3 at observation point N5 (Erlangjian sandpit section)

As the section is arc-shaped, it is divided into three logs (Log1, Log2, and Log3), which are representative of the multistage beach deposition from lake to land (Fig. 21).

Log1: The black mud layer is the thinnest (8 cm) among all logs (Fig. 22) and is covered by about thinly layered (6 cm)brown muddy silt. This gradually transitions into sandy sediments with reversed-graded bedding, indicating an increasing flow energy. Up section is a layer of low-angle oblique bedding dipping towards the lake, indicating the repeated swash of the beach ridge as lake level increases (Jiang 2010; Wang and Jiang 2018). Above this, a thin layer of muddy sediments with flow marks occurs, showing the erosion of beach ridge during high water level period. Next is a set of oblique bedding dipping landward, suggesting the migration of the beach ridge towards the lake as lake level decreases (Wang and Jiang 2018). At the top is a set of sands (50 cm) with reverse grading formed during a rapid decrease of lake level.

Log2: The black mud layer at the base is 30 cm thick (Fig. 22), overlapped by a brown muddy slit of 15 cm, grading into sand and gravel. The sand-gravel has low-angle cross-bedding and the muddy sediment contains flow marks are significantly thicker than those in Log1, indicating that this is the main part of the multistage beach deposits. The flow conditions reflected in Log2 are similar to Log1.

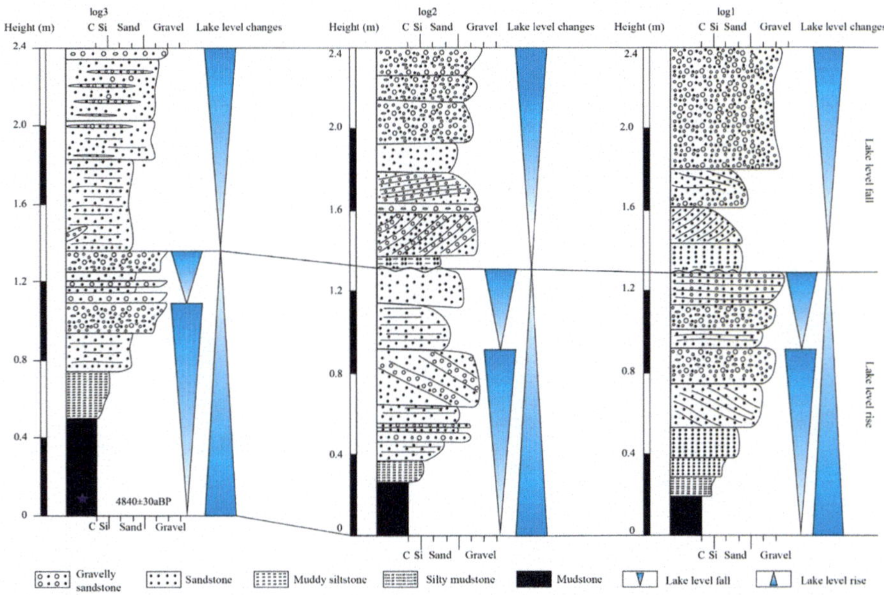

Fig. 22 Lithological log and lake-level variation of Logs 1, 2, and 3

Log3: The black mud layer at the bottom is the thickest among all logs (50 cm), reflecting a location close to the depocenter of the lagoon (Fig. 22). Upward is a 30 cm thick brown muddy silt, overlapped by parallel bedding interlayered with sandy sediments enriched in heavy minerals. The section shows lake level did not rise continually, but had some minor breaks.

A model of beach deposits in the study area is established (Fig. 23). The multi-stage deposition of beach is clearly controlled by the changing lake level. As the lake level rose, the beach ridge continued to migrate landward (Fig. 23c, d). When the lake reached the highstand, the beach ridge was scoured by waves during storms, forming a thin layer of muddy silt (Fig. 23e). Lake level then decreased, the beach was eroded, leaving dipping beds landward (Fig. 23f, g).

Eolian Sedimentary System

Large areas of eolian deposits are located on the eastern side of Qinghai Lake, and cover about one-fifth of the lake's perimeter (Fig. 1). The wind-dominated sedimentary system can be divided into dunes and interdunes. Most dunes share a common profile, or cross section, with a long shallow angle (stoss) facing into the wind, a peak (crest), and a steep lee side (McKee 1966). In the study of eolian deposits, the wind direction can be determined by the steep side resulting from the high-angle cross-bedding in the dunes (Fig. 24).

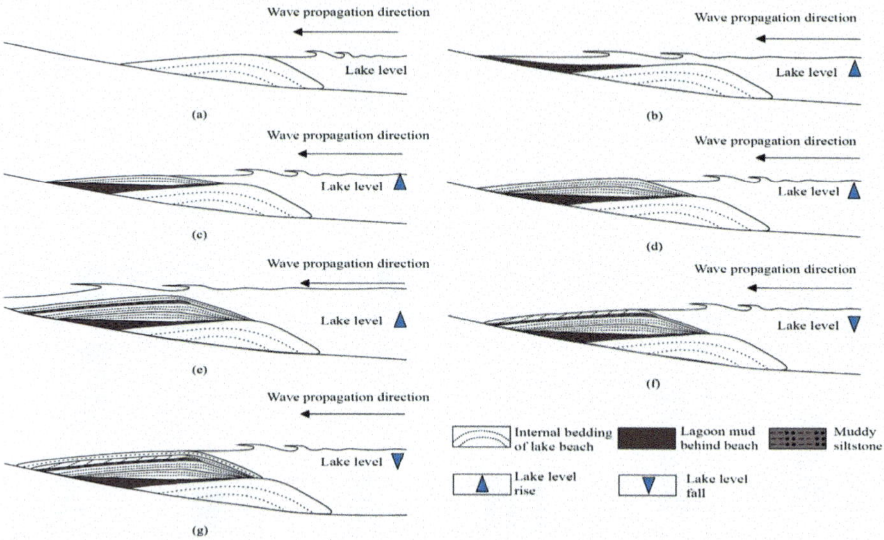

Fig. 23 Multistage lake beach depositional model

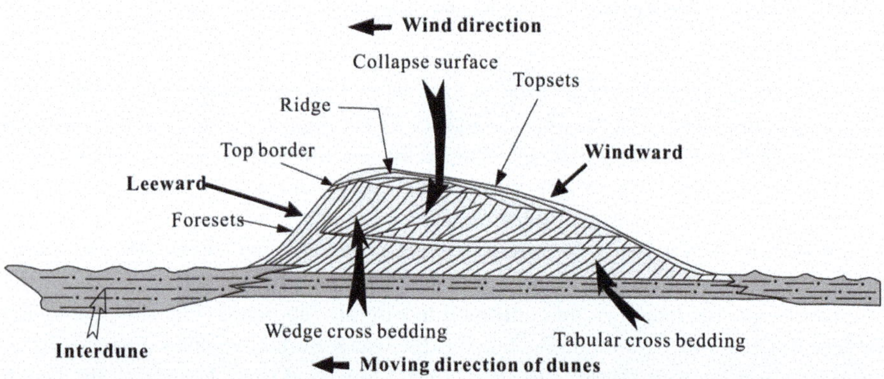

Fig. 24 Sedimentary model of eolian dunes. (Modified from Jiang 2003). The tendency of high-angle crossing-bedding or the tilt direction of foreset is exactly the same as the direction of prevailing wind

The wind acts on the sandy surface forming a microgeomorphology of eolian ripples, which are similar to water ripples. The shape of wind ripples corresponds to wind direction (similar to dunes, Fig. 25a). The sand tail structure is formed by changing the stacking direction after the sand encounters an obstacle (Fig. 25b).

The depressions between dunes are called interdune (Ahlbrandt and Fryberger 1981). In some area near the foothills, bedrocks or coarse-retained sediments may exposed between the dunes (Fig. 26a). In most cases, the interdune is totally covered by sand, where the water table is the base level and the lower limit of sand migration. Sandy sediments can be driven by wind above the water table, while those below it are wetted and bonded to each other which cannot move.

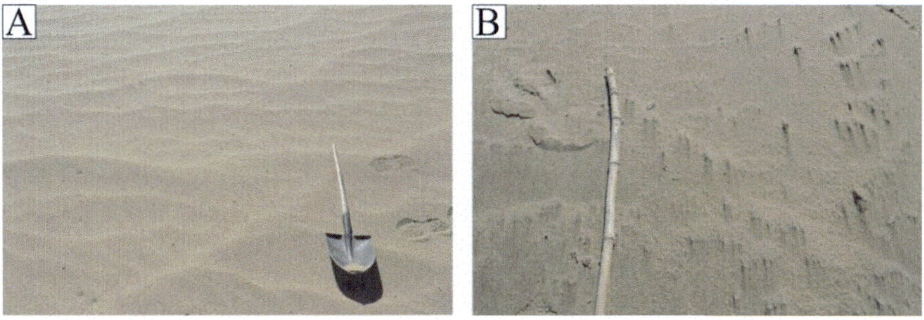

Fig. 25 Surface structure of eolian dunes (**a**: wind ripples, the windward side of the ripple is gentle while the leeward side is steep; **b**: sand tail, the tapered side indicates the leeward direction)

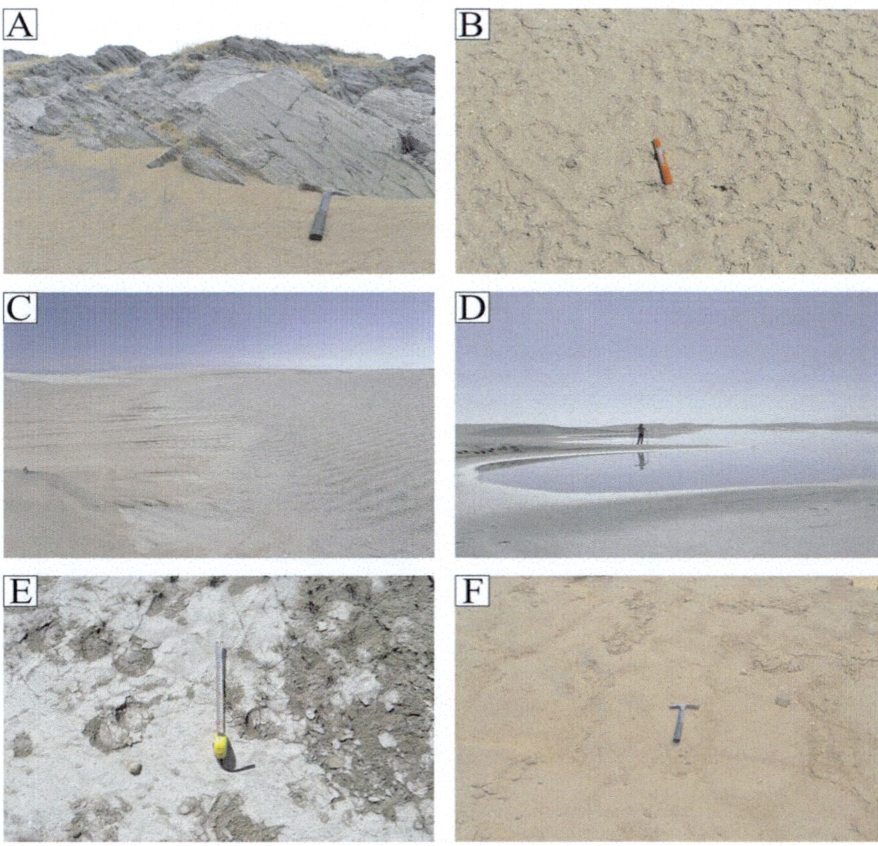

Fig. 26 Field photos of interdune deposits (**a**: at early stage of the dunes, the bedrock is exposed; **b**: the arid interdune is composed of fine gravel and coarse sand; **c**: adhesion ripples at the humid interdune; **d**: watery interdune; **e**: paste deposits at the surface of interdune; **f**: cracks)

Interdune can be divided into arid, humid, and watery types according to its surface environment. The arid interdune is in a dry environment where finer debris are blown away by wind, leaving a gravel desert (Fig. 26b).The humid interdune has a wet surface. The particles are often stuck to the surface, forming ripples (Fig. 26c). The watery interdune often exhibits a little lake between dunes (Fig. 26d). When the little lake has dried up, the mud layer exposed is covered with paste deposits (Fig. 26e) and cracks can be seen (Fig. 26f).

The Classification of Eolian Dunes

Eolian dunes of Qinghai Lake are divided into three categories: barchan dunes, barchan dune chains, and pyramidal dunes according to their shapes.

Barchan Dunes

Barchan dunes are the most widely distributed types in the eastern side of the lake (Fig. 27a). Barchan dunes often appear on the coasts of sandy lakes lacking vegetation. The direction of dunes is vertical to the main wind direction. The plane view is crescent-shaped, with two horns pointing to the leeward direction (Fig. 27b). The windward slope (toward the east) is convex and gentle (5°–20°), depending on the wind, amount of sand, particle size, and their proportion; the leeward side (westward) is a concave and steep slope (28°–34°) where grain size is finer than that on the windward side.

Fig. 27 Barchan dunes (**a**: barchan dunes above the water; **b**: high-angle oblique bedding of barchan dunes)

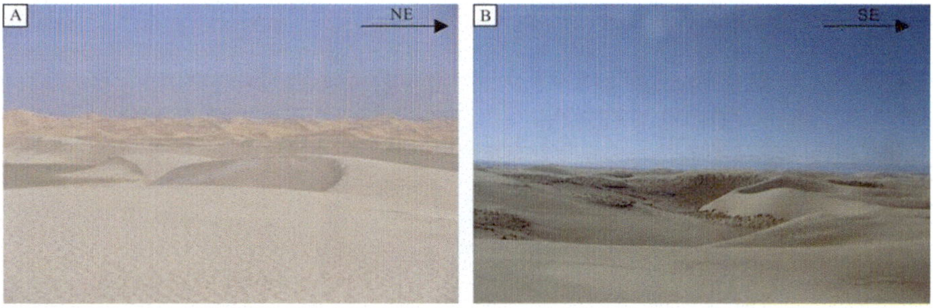

Fig. 28 Barchanoid ridges

Fig. 29 3D geomorphologic map (**a**) of pyramidal dune and the field photo (**b**)

Barchanoid Ridges

Barchanoid ridges are widely developed in the center of the eolian area (Fig. 28). Due to the relatively abundant sand source in central area, it is easy for prevailing wind to connect the densely arranged barchan dunes into barchanoid ridges.

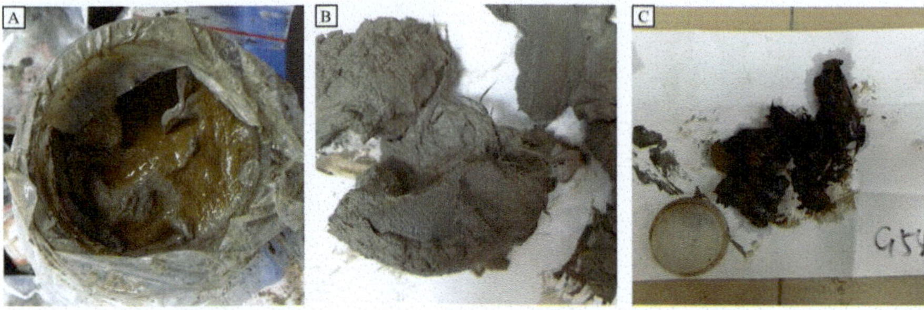

Fig. 30 (**a**) Calcareous silty mud, the surface is covered with ash-colored oxidized gel; (**b**) calcareous silty clay; (**c**) calcareous mud

Pyramidal Dunes

The pyramidal dunes in the study area are distributed near the foothills. Pyramidal dunes generally have multiple facets (slopes of 25° to 30°). The base of the dune is generally flat, while the top is sharp with distinct ridges. The highest pyramidal dunes in the area can reach 150 m (Fig. 29). As pyramidal dunes are mainly distributed near the foothills, their formation resulted from changing wind directions associated with the prevailing and valley winds and lake-land breezes.

Formation Mechanism of Eolian Dunes

Yao et al. (2015) proposed that the Qinghai Lake eolian dunes are actually the products of ancient sand dune reactivation. By comparing the water level in 1957 and 2002 (continuing decreasing), sand entering the lake is thought to increase the area of the sand. Through the study of the piedmont of Daban Mountain, it is proposed that the erosion of ancient dunes by flowing water and wind led to the reactivation of ancient sand dunes. Previous study showed that the climate in early-middle Pleistocene was warm and dry (China Academy of Sciences Lanzhou Division et al. 1994), compared to modern times, which indicated that the water level at that time might be much lower than now. Thus, it is speculated that the ancient sand dunes discussed above may form at early-middle Pleistocene.

Deep Lake Sedimentary System

The semideep lake subfacies are located between the normal wave base and the storm wave base; the deep lake subfacies are located at the bottom of the lake, below the storm wave base (Jiang 2003). The sediments of the semideep lake-deep lake subfacies of Qinghai Lake are mainly calcareous silty mud (Fig. 30a), calcareous

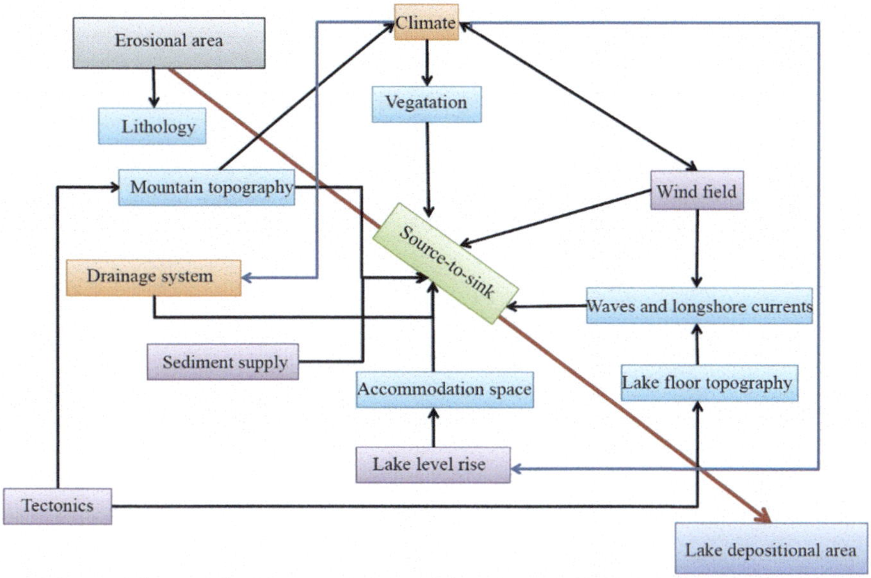

Fig. 31 The controlling factors of the modern sedimentary system of the Qinghai Lake

silty clay (Fig. 30b), and calcareous mud (Fig. 30c). The deep lake area of Qinghai Lake has a wide distribution and accounts for more than half of the lake.

Discussion

The factors controlling the modern sedimentary system of Qinghai Lake include topography, geomorphology, lithology of the source area, lake system, wind field, wave and lake currents, and water level (Fig. 31). The mountains affect the distribution of the lake system and regional wind field. The morphometry of the lake bottom affects the direction of currents. The type of sediments eroded and transported depends on the lithological characteristics of the provenance area. Vegetation types and coverage determine the stability of sediments in these areas. The runoff of the water system, the type of sediment supply, and seasonal climate change affect the sediment transportation process. The ancient coastal sediments are eroded, transported, and accumulated by wind to form eolian sand dunes. The characteristics of regional wind field are closely related to precipitation, which indirectly affects the vegetation type and coverage. Wave and the lake current control the plane distribution of sand bodies. Varying lake water level affects the accommodating space, and thus affects the depositional pattern.

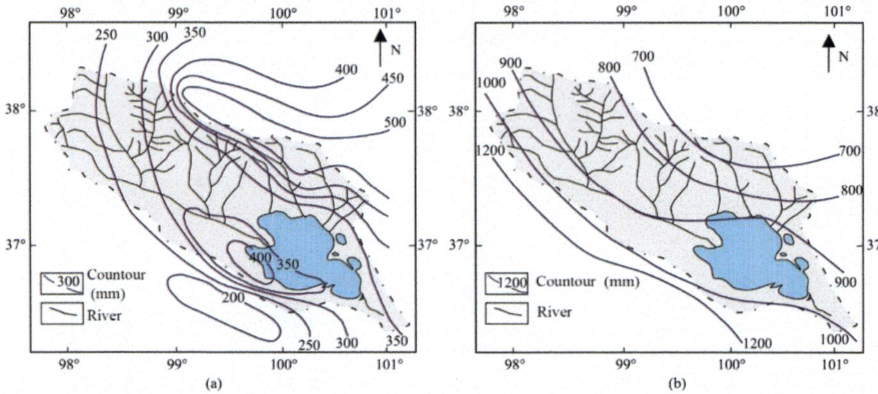

Fig. 32 The annual mean precipitation (**a**) and evaporation (**b**) contour map of the Qinghai lake basin in recent years. (Modified from China Academy of Sciences Lanzhou Division et al. 1994)

To briefly summarize, the modern sedimentary system of Qinghai Lake is contemporarily influenced by the three factors: "wind field–sediment source–lake basin". The following sections will discuss these three factors individually.

Wind Field

Climate refers to the state of the atmosphere of a region for several years. Climate factors include temperature, precipitation, evaporation, wind, etc. There are many previous studies of the temperature, precipitation, and evaporation of the Qinghai Lake area (Li et al. 2002, 2005, 2008). However, few studies had been done on the effects of wind field on the distribution of modern sediments at Qinghai Lake.

Although the Qinghai Lake and its drainage basin are located inland, precipitation in this area is higher than other inland areas due to the warm and wet air from southwest and monsoonal flow from the plateau, coupled with lake effect precipitation and the frequent transit of the westerly belt. The annual precipitation gradually decreases from the mountainous area north of the lake to south (Fig. 32a). Precipitation at Heima river is high (Fig. 32a) because the terrain is associated with ascending air. Evaporation is the main path of water loss in the Qinghai Lake drainage basin. The average annual evaporation is 1000–1800 mm (20 cm caliber evaporation value) (China Academy of Sciences Lanzhou Division et al. 1994). The trend in evaporation is opposite to that of precipitation, gradually increasing from north to south (Fig. 32b).

By comparison, the average annual evaporation of the Heima River area is similar to that of the eastern lake sand area. Average annual precipitation and humidity of the former are much higher than that of the latter. This is also reflected in the vegetation coverage and growth. Temperate steppe, which favors humid condition, is widely developed in the Heima River area. In contrast, the eastern lake sand area is dominated by shrub and sea buckthorn.

Fig. 33 (**a**) Outcrop of the western bank of Qinghai Lake; (**b**) Outcrop of the eastern bank of Qinghai Lake

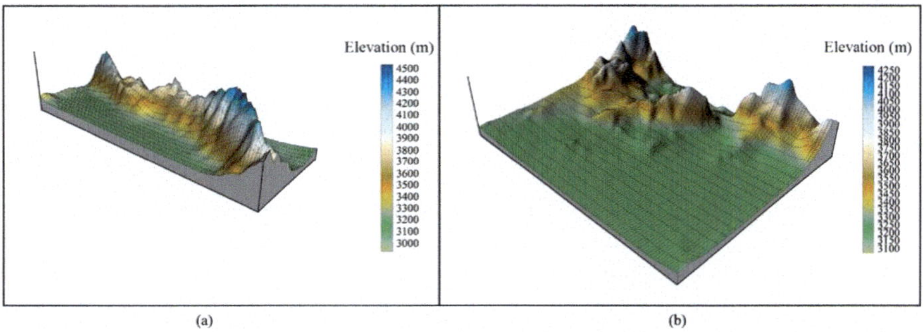

Fig. 34 (**a**) The three-dimensional geomorphology of western bank of Qinghai Lake. (**b**) The three-dimensional geomorphology of eastern bank of Qinghai Lake

Because of the differences in surface humidity and vegetation coverage, and the sediment produced by weathering is also different. The west coast has higher vegetation coverage and is more resistant to weathering resulting in weathering products that are mostly gravel. The east coast, with a lower vegetation coverage, has weathering products that are mostly sand. These differences in weathering products resulted in the deposition of gravel-sand lake beach sediments on the west coast (Fig. 33a), and the development of large area eolian sand dunes on the east coast (Fig. 33b). The humid climate in the west coast leads to the development of gray-black muddy sediments (e.g., alluvial fan in front of Qinghai Nanshan) in mudslide deposits and coastal deposits. The dry climate in the east coast leads to the dominance of reddish-brown mud deposits and the presence of calcareous nodules.

Changes in the wind field are controlled by the geomorphology of the landscape and lake effect winds, in addition to the prevailing wind and changes in the monsoon. There are obvious regional differences in the recent tectonism since the Late Tertiary. The uplift of Qinghai Nanshan is much greater than that of the TuanBao-Daban Mountain area. Qinghai Nanshan on the west coast of Qinghai Lake has relatively high landscape, with an overall shape like a wall (Fig. 34a). The Riyue-Tuanbao

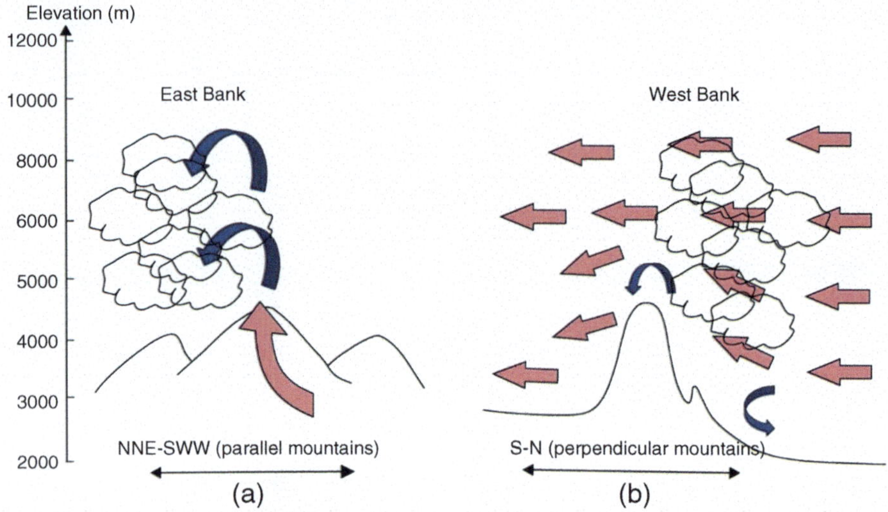

Fig. 35 Conceptual model showing the interaction between cloud and landscape in the Qinghai Lake

Mountain on the east coast of Qinghai Lake is a discontinuous landform with breaks resulting from regional uplift (Fig. 34b).

The difference in topography and geomorphology, coupled with the winds affects the regional precipitation (Fig. 35). On the south side of Qinghai Nanshan is the vast Gonghe basin. When the northwest wind dominates, southward airflow prevails on the ground of south side. The airflow carrying water vapor leads to much higher precipitation in the Jiangxigou-Heima River area. The Tuanbao-Riyue Mountain area is a discontinuous landform. Water vapor is driven westward by high-altitude northwest airflow before the formation of rainfall, resulting precipitation in the Huangshui River valley on the east side.

The effect of wind on the modern coastal sediments is reflected by long-term wind wave, wind setup, and longshore current generated by the prevailing northwest and north wind. Wind wave and coastal flow transport sediments from the Jiangxigou area to Yilangjian and Erlangjian and form sand spits.

Wind setup affects the estuary of the Ganzi River, and forms barrier island-channel-lagoon deposition. Long-term effects of wave and longshore current also lead to closure of the Gahai and Erhai. As the lake level decreases, new Gahai and Haiyan Bay also gradually close. Therefore, the northeast-east-south coast of Qinghai Lake is widely impacted by the wave and longshore currents generated by northwest and northern wind.

The wind also affects the coastal zone and piedmont area above the lake level. Large area of eolian sand dunes located at the piedmont and plain before the Tuanbao-Daban Mountain, is formed by the long-term effects of wind, combined with low precipitation, long sunshine duration, and high evaporation. The eolian sand dune on the west side of the Bird Island was formed after the original plain was abandoned due to the avulsion of the Buha River. The warm and dry climate lead to the development of Langma Shegang sand area on the southeastern margin of the Daotanghe River valley, which is dominated by crescent-shaped sand dunes.

The influence of wind direction and speed on the sedimentary system of Qinghai Lake are reflected by the local dry versus wet climate of the area, sediment supply, and the intensity of wave and coastal flow. The difference in local climate is seen in the vegetation coverage, resulting in different resistances to weathering and erosion. The areas that are less resistant to weathering provide abundant sediment supply to eolian sand accumulation. The wave and longshore currents generated by prevailing winds have strong ability to transport and transform sand bodies in the lake.

Source

There are significant differences in the source area of different sedimentary systems. Lithology, geomorphology, climate (wind, precipitation, evaporation, etc.), vegetation type and coverage rate, water system, etc. of source area are all significant factors.

(1) Characteristics and controlling factors of source areas in western and middle Qinghai Nanshan: carbonate rocks are exposed at western Qinghai Nanshan; conglomerate, sandstone, dacite, and granite are exposed at middle Qinghai Nanshan; Silurian shale and slate are exposed at northeastern Qinghai Nanshan. The precipitation and vegetation coverage in western Qinghai Nanshan are higher than those of middle Qinghai Nanshan. The water system in western Qinghai Nanshan is composed of relatively stable mountain rivers. In contrast, the water system in middle Qinghai Nanshan is ephemeral. These features lead to formation of fan deltas in front of western Qinghai Nanshan and alluvial fans in front of middle Qinghai Nanshan.

(2) Characteristics and controlling factors of source areas in the Buha River drainage basin: the catchment flows through granite strata; braided river and meandering river flow through limestone, sandstone, and shale strata. The NW high and SE low topography is the direct cause of the NW-SE flow direction of the Buha River. Ice and snow melt water, precipitation, and groundwater are the important water sources of Buha River. Soil along the river is mostly loose and easy to erode, which leads to high sediment load and provides sufficient sediment for the Buha River delta.

(3) Characteristics and controlling factors of the source areas in the western and eastern Datong Mountain: marine-clastic rocks (mostly sandstone) are exposed at the outlet of Shaliu River; terrigenous marine-clastic rocks (mostly slate and clay slate) are exposed near the Guoluo village. The sediments are derived from the water flow and are dominated by the Gangcha alluvial fan. Higher annual evaporation rates lead to lower runoff of the Shaliu River. The relatively low

vegetation coverage on both banks leads to relatively low resistance to weathering, making the river channel easy to divert.

(4) Characteristics and controlling factors of the source areas in the northeastern Datong Mountain: Haergai River and Ganzi River flow through the middle Jurassic coal-bearing and clastic strata, which provide sediments for the wetlands and swamps near the estuary. Low annual evaporation helps preserve water in soil, and provide favorable conditions for formation of wetlands and swamps.

(5) The characteristics and controlling factors of the source areas in Tuanbao Mountain / Daban Mountain / western Riyue Mountain: the mainly amphibolite facies that are susceptible to weathering. Low precipitation and high evaporation lead to relatively dry climate in this area. The relatively unitary vegetation type and low vegetation coverage lead to very week sand-fixing capacity. The lack of water results in development of large area of eolian deposits.

Lake Basin

The term "lake basin" contains two folds of meanings: lake basin topography and lake basin evolution. Lake basin topography can be divided into mountain landform and lake bottom landform. Mountain landform affects the distribution of water system and regional wind field. Lake bottom landform controls the propagation direction of lake current.

The direction of the main fault structure in the lake basin is NNW-NW-NWW, resulting in the mountainous region having strike direction of NW-SE. The upstream of rivers in the lake area are mostly NW-SE (except the Daotanghe River and Heima River). The terrain (northwest high and southeast low) determines the slope of the rivers. The slope from the upstream to the downstream of Buha River is large, which is conducive to river undercutting. The slopes of the Haergai River, Ganzi River, and Daotanghe River drainage basin are smaller, which lead to lower undercutting capacity and less obvious valley shape. The greater elevation difference and longer transportation distance from source to estuary result in more fine-grain sediments are at the floodplain and delta plain of the Buha-Shaliu drainage basin than that of the Ganzi-Daotanghe drainage basin.

In terms of the geomorphology of the lake base, three depressions in the lake basin impact the sediments transportation in the lake, which is characterized by three main clockwise transportation currents and other secondary counterclockwise transportation currents (e.g., lake bay, river). In addition, the slope of the lake bottom is steeper on the south side due to tectonism, which leads to coarser costal sediments on the south side than on the north side (Fig. 36).

The lake basin topography is the result of the combined effect of sedimentary background and tectonic activity. The role of tectonic activity in forming basin topography is as following: (1) determining the topographic height difference between sediment-carrying rivers and catchment basins, thus determining the rate

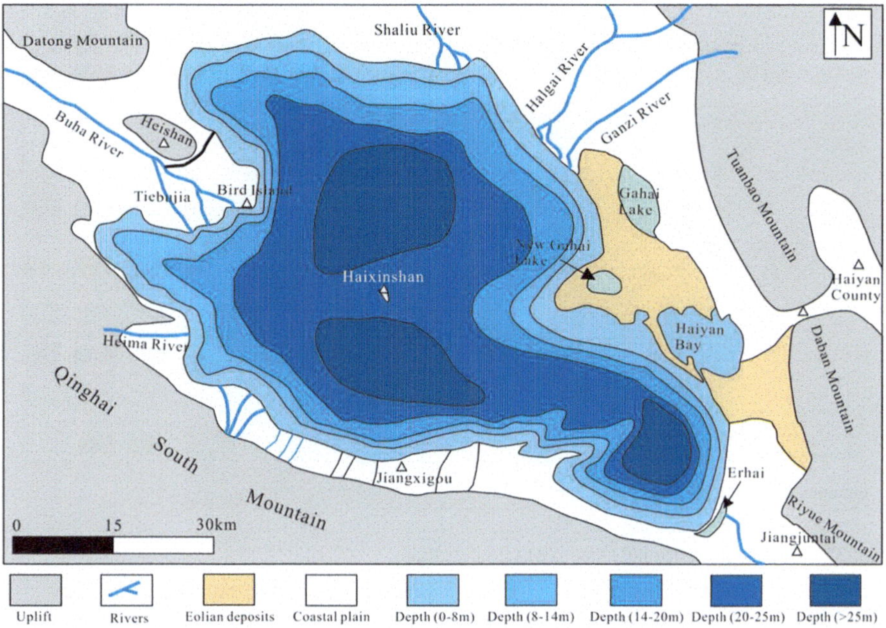

Fig. 36 Bathymetric map of Qinghai Lake

and velocity of flow; (2) determining the abundance of sediment source; (3) determining the plane distribution of lake sediments.

The evolution of lake basin affects the temporal variation of sedimentation. In terms of modern sedimentation, changing lake level can led to changing capacity, and thus affects depositional style. When the increase of accommodation space is higher than the rate of sediment supply, the Gangcha alluvial fan migrates towards the lake and forms Gangcha delta. When the rate of capacity growth is close to the rate of sediment supply, the growth of Gangcha delta suspends. When the rate of capacity growth is lower than the rate of sediment supply, the abandoned Gangcha delta is cut by the Shaliu River, and transforms into Shaliu River delta that migrates towards the lake bay.

Conclusion

This study found that when the strata of the provenance area are Silurian (marine-clastic rocks dominated by shale and sandstone); the strata are susceptible to erosion and transportation. Rivers passing those areas are enriched in sand and mud sediments, leading to formation of deltas by stable water flow (e.g., Buha River delta) and alluvial fans by ephemeral flow (e.g., Guoluo alluvial fan). When the strata of the provenance area are granite and Triassic sandstone and conglomerate;

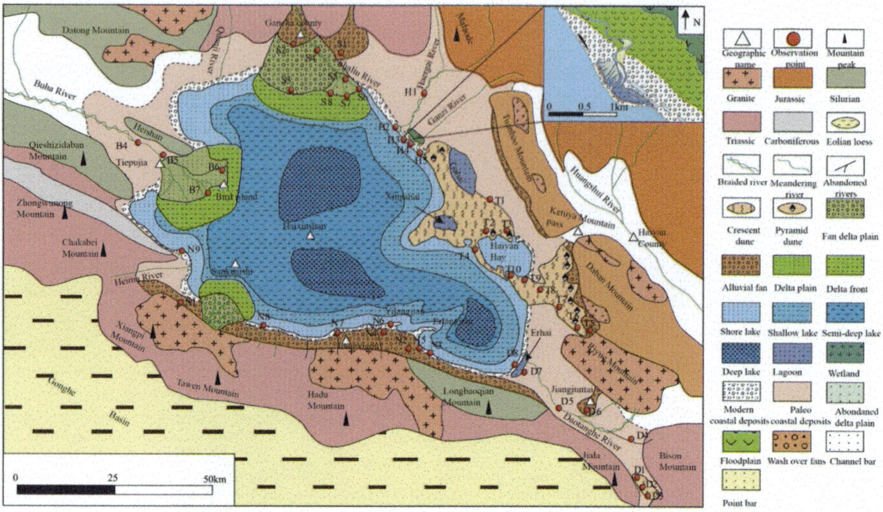

Fig. 37 Modern depositional system of Qinghai Lake

the rocks are resistant to weathering and erosion. Sediments transported by rivers are dominated by gravels, and thus gravel-rich fan delta and alluvial fan are formed. When the strata of the provenance area are Jurassic (coal-bearing strata and sand/gravel strata), the eroded coal-bearing strata are rich in carbon and clay minerals, providing source sediments for downstream wetlands and swamps. When the strata of the provenance area are Mesoproterozoic (metamorphic rocks of amphibolite facies), the rocks are rich in unstable amphibole and biotite, and are susceptible to weathering and erosion.

Previous studies also failed to take into consideration of the migration of lake shoreline. In this study, we divided the costal deposition in modern coastal facies and ancient coastal facies by the first lake terrace (Fig. 37). The modern coastal facies are further divided into foreshore and shoreface subfacies.

Based on field observation and water depth measurement, we revised some previous description of sedimentary facies: (1) The original Shaliu River fan delta is now divided into Gangcha alluvial fan, Gangche delta, and Shahe River delta; (2) the distribution of Buha River delta front is reoutlined; (3) the sedimentary facies at the lake inlet of the distributary north of Haergai River and their distribution and formation mechanism are refined.

From the 1960s to the early twentieth century, the warm and dry climate in the Qinghai Lake area led to a decline in the water level and shrinking of lake area. The type and distribution of sedimentary facies have changed as following. (1) Buha River, which originally flows towards the northern depression, diverted to the south of the island, towards the southern depression and Tiebujia Bay. (2) As the Buha River delta continue to migrate towards the lake, the bird foot delta evolved to a lobate delta, and connected with Bird Island and Egg Island. (3) The abandoned Buha delta plain (north of Bird Island) has formed aeolian sand dune. (4) Shaliu River delta continuously migrates towards the lake bay on the south side. (5) The originally isolated Sand Island has closed to form new Gahai, and the Haiyan Bay and Qinghai Lake is only connected by a long and narrow channel. (6) The sand spits at Erlangjian and Yilangjian migrate eastward.

References

Ahlbrandt, T. S., & Fryberger, S. G. (1981). Sedimentary features and significance of interdune deposits. In Frank G. Ethridge & Romeo M. Flores (Eds), *Recent and ancient nonmarine depositional environments: models for exploration* (SEPM Special Publication). 31. 293–314. https://doi.org/10.2110/pec.81.31.0293

Anthony, E. J., & Julian, M. (1999). Source-to-sink sediment transfers, environmental engineering and hazard mitigation in the steep Var River catchment, French Riviera, southeastern France. *Geomorphology, 31*(1), 337–354.

Bian, Q. T., Liu, J. Q., Luo, X. Q., et al. (2000). Geotectonic setting, formation and evolution of the Qinghai Lake[J]. *Seismology and Geology, 22*(1), 20–26.

Bridge, J. S. (1993). Description and interpretation of fluvial deposits: A critical perspective. *Sedimentology, 40*, 801–810.

Chen, K. Z., Huang, D. F., & Liang, D. G. (1964). Formation and development of Qinghai Lake[J]. *Acta Geographica Sinica, 30*(3), 214–233.

China Academy of Sciences Lanzhou Division, Research Center for Eco-Environmental Sciences, & Chinese Academy of Sciences. (1994). *The evolution and prediction of the modern environment of Qinghai Lake* (pp. 69–149). Beijing: Science Press.

Chinese Academy of Sciences Lanzhou Geological Institute, Chinese Academy of Sciences Shuisheng Biology Institute, Institute of Microbiology Chinese Academy of Sciences, et al. (1979). *Comprehensive investigation report of Qinghai Lake* (pp. 1–165). Beijing: Science Press.

Edmonds, D. A., & Slingerland, R. L. (2007). Mechanics of river mouth bar formation: Implications for the morphodynamics of delta distributary networks[J]. *Journal of Geophysical Research, 112*, 1–14.

Han, Y. H., Li, X. Y., Wang, Q., et al. (2015). Hydrodynamic control of sedimentary systems in Shore Zone of Qinghai Lake[J]. *Acta Sedimentologica Sinica, 33*(1), 97–104.

Hudock, J., Flaig, P., & Wood, L. (2014). Washover fans: A modern geomorphologic analysis and proposed classification scheme to improve reservoir models[J]. *Journal of Sedimentary Research, 84*(10), 854–865.

Jiang, Z. X. (2003). *Sedimentology [M]* (pp. 70–100). Beijing: Petroleum Industry Press.

Jiang, Z. X. (2010). *Sedimentology [M]* (pp. 1–80). Beijing: Petroleum Industry Press.

Jin, Z. K., Li, Y., Gao, B. S., et al. (2014). Depositional model of modern gentle-slope delta: A case study from Ganjiang Delta in Poyang Lake[J]. *Acta Sedimentologica Sinica, 32*(4), 710–723.

Li, L., Wang, Z. Y., Qin, N. S., et al. (2002). Climate change and its impact on desertization around Qinghai Lake[J]. *Plateau Meteorology, 21*(1), 59–65.

Li, L., Zhu, X. D., Wang, Z. Y., et al. (2005). Impacting factors and changing tendency of water level in Qinghai Lake in recent 42 years[J]. *Journal of Desert Research, 25*(5), 689–696.

Li, F. X., Fu, Y., Yang, Q., et al. (2008). Climate change and its environmental effects in the surrounding area of Qinghai Lake[J]. *Resources Science, 30*(3), 348–353.

Li, Y. T., Li, X. Y., Cui, B. L., et al. (2010). Trend of stream flow in Lake Qinghai Basin during the past 50 years (1956-2007)- take Buha River and Shaliu River for examples[J]. *Journal of Lake Sciences, 22*(5), 757–766.

Lin, C. S., Xia, Q. L., Shi, H. S., et al. (2015). Geomorphological evolution, source to sink system and basin analysis[J]. *Earth Science Frontiers, 22*(1), 9–20.

McKee, E. D. (1966). Structures of dunes at white sands national monument, New Mexico (and a comparison with structures of dunes from other selected areas)[J]. *Sedimentology, 7*, 3–69.

Miall, A. D. (1978). Lithofacies types and vertical profile models in braided river deposits: a summary. In A. D. Miall (Ed.), *Fluvial sedimentology, memoir 5* (pp. 597–604). Calgary: Canadian Society of Petroleum Geologists.
Moore, G. T. (1969). Interaction of rivers and oceans: Pleistocene petroleum potential. *AAPG Bulletin, 53*(12), 2421–2430.
Moreno, C., & Romero Segura, M. J. (1997). The development of small-scale sandy alluvial fans at the base of a modern coastal cliff: Process, observations and implications. *Geomorphology, 18*(2), 101–118.
Qinghai Geology and Mineral Resources Bureau. (1991). *Regional geology of Qinghai Province* (pp. 1–379). Beijing: Geological Publishing House.
Shi, Y. M., Wang, X. M., & Song, C. H. (1996). Aeolian deposition on the Qinghai Lake region [J]. *Acta Sedimentologica Sinica, 14*(S1), 234–238.
Shi, Y. M., Dong, P., Zhang, Y. G., et al. (2008). Revelation of modern deposits in Qinghai Lake to precise exploration of lithologic hydrocarbon reservoirs[J]. *Natural Gas Industry, 28*(1), 54–57.
Shi, L., Jin, Z. K., Li, G. Z., et al. (2014). Depositional characteristics and models of the modern-braided river delta in the Daihai Lake, Inner Mongolia[J]. *Natural Gas Industry, 34*(9), 33–39.
Somme, T. O., & Jackson, C. A.-L. (2013). Source-to-sink analysis of ancient sedimentary systems using a subsurface case study from the Mor-Trondelag area of southern Norway: Part 2-sediment dispersal and forcing mechanisms. *Basin Research, 25*(5), 512–531.
Somme, T. O., Helland-Hansen, W., Martunsen, O. J., et al. (2009). Relationships between morphological and sedimentological parameters in source-to-sink systems: A basis for predicting semi-quantitative characteristics in subsurface system. *Basin Research, 21*(4), 361–287.
Somme, T. O., Jackson, C. A.-L., & Vaksdal, M. (2013). Source-to-sink analysis of ancient sedimentary systems using a subsurface case study from the Mor-Trondelag area of southern Norway: Part 1-desponsitional setting and fan evolution. *Basin Research, 25*(5), 489–511.
Song, C. H., Wang, X. M., Shi, Y. M., et al. (1999). Sedimentary characteristics and microfacies of shore zone in Qinghai Lake[J]. *Acta Sedimentologica Sinica, 17*(1), 51–57.
Song, C. H., Fang, X. M., Shi, Y. M., et al. (2000). Characteristics and formation of Aeolian dunes on Western shore of the Qinghai Lake[J]. *Journal of Desert Research, 20*(4), 443–446.
Song, C. H., Fang, X. M., Shi, Y. M., et al. (2001). Sedimentary characteristics of modern lacustrine deltas in Qinghai Lake and their controlling factors[J]. *Journal of Lanzhou University (Natural Sciences), 37*(3), 112–120.
Wang, X. Y., & Jiang, Z. X. (2018). Sedimentary characteristics of beach-bar in the 3rd member of Paleogene Funing formation in Hai'an Sag, North Jiangsu Basin[J]. *Petroleum Geology and Recovery Efficiency, 25*(5), 57–64.
Wright, L. D., & Coleman, J. M. (1971). Effluent expansion and interfacial mixing in the presence of a salt wedge, Mississippi River delta[J]. *Journal of Geophysical Research, 76*, 8649–8661.
Wright, L. D., & Coleman, J. M. (1978). River deltamorphology: Wave climate and the role of subaqueous profile. In D. J. P. Swift & H. D. Palmer (Eds.), *Coastal sedimentation* (pp. 87–89). New York: Academic Press.
Wu, D., Zhu, X. M., Liu, C. N., et al. (2015). Discussion on depositional models of fan deltas in steep Slope Belt of the Rift Basin under the guidance of source-to-sink system theory: A case study from the Fula sub-basin, Muglad Basin, Sudan[J]. *Geological Journal of China Universities, 21*(4), 653–663.
Xu, C. G. (2013). Controlling sand principle of source-sink coupling in time and space in continental rift basins: Basin idea, conceptual systems and controlling sand models[J]. *China Offshore Oil and Gas, 25*(4), 1–21.
Yao, Z. Y., Li, X. Y., & Xiao, J. H. (2015). Driving mechanism of Sandy desertification around the Qinghai Lake[J]. *Journal of Desert Research, 35*(6), 1429–1437.
Zhu, X. M., & Xin, Q. L. (1994). Sedimentary characteristics and models of the beach-bar reservoirs in faulted down lacustrine basins[J]. *Acta Sedimentologica Sinica, 12*(2), 20–27.

Freshwater Microbialites in Early Jurassic Fluvial Strata of the Pranhita-Godavari Gondwana Basin, India

Suparna Goswami and Parthasarathi Ghosh

Abstract The role of microbes in the precipitation of carbonates in various modern freshwater environments like lakes, rivers, and springs is currently being actively researched. Extensive research over the last few decades has documented some of the distinctive features of the freshwater microbial carbonates. However, identification of the signatures of microbes in carbonate precipitation remains a challenging task in the case of geologically ancient deposits. It is especially difficult for deposits older than the Neogene, because of diagenetic imprints on the primary fabric. Due to this limitation, the habitats of microbes are poorly constrained for a significant part of the Phanerozoic. In this study, we report the field occurrences and micro-morphological features of thin carbonate bodies encased in a thick siliciclastic alluvial succession. The succession is Early Jurassic in age and formed in a continental rift basin, the Pranhita-Godavari Gondwana Basin of peninsular India.

It has been demonstrated that in spite of diagenetic modifications, a number of microscopic and sub-microscopic features of these carbonates can be equated to those found in modern freshwater microbialites. Seven facies have been identified from these freshwater carbonates. Among them, the first three are mainly autochthonous deposits, and the others show incorporation of detrital components (partly or completely allochthonous). The geometry of the carbonate bodies and the mode of association of the various textural features suggest three main types of primary carbonate accumulation and a fourth type representing coeval reworking of the primary deposits. The stratigraphic relationship between the carbonate and siliciclastic components of this succession reveals the mode of mutual interactions between the siliciclastic and carbonate depositing processes.

Keywords Early Jurassic · Freshwater carbonate · Gondwana · Microbialite · Rift basin

S. Goswami (✉) · P. Ghosh
Geological Studies Unit, Indian Statistical Institute, Kolkata, India
e-mail: pghosh@isical.ac.in

Michael R. Rosen et al. (eds.), *Limnogeology: Progress, Challenges and Opportunities*, Syntheses in Limnogeology, https://doi.org/10.1007/978-3-030-66576-0_18

Introduction

Microbes have played an important role in forming carbonate precipitates, called microbialites, since the early Precambrian to present (Burne and Moore 1987). Cyanobacteria are considered to be the first form of life on the Earth, and the microbial mats are considered to be the earliest form of ecosystem (Noffke et al. 2006; Dupraz et al. 2009). These microbes constructed the laminated variety of microbialites, i.e., stromatolites, that are found in abundance in Proterozoic marine deposits. It has been postulated that in the earliest part of the Phanerozoic, with the advent of multicellular organisms, aided by elevated level of atmospheric oxygen, these microbes abandoned their marine niche and colonized continental environments (Garrett 1970; Awramik 1971; Monty 1973; Riding 2000). However, the record of microbes colonizing freshwater environments is imperfect through much of the Phanerozoic, especially during Paleozoic and Mesozoic times. This can be attributed to the poor preservation potential of the continental strata and diagenetic effects.

Freshwater microbial carbonates in the form of lacustrine-palustrine deposits, spring deposits (i.e., tufa and travertine), and fluvial deposits have been extensively reported from the Late Cenozoic, particularly from Pleistocene to Holocene (Platt and Wright 1992; Pedley 2009; Arenas-Abad et al. (2010). However, there are relatively few reports of microbially precipitated freshwater carbonates from older strata (Bradley 1929; Zamarreño et al. 1997; Strasser 1986; Leslie et al. 1992; Arenas et al. 2000, 2007, 2015; Parrish et al. 2017; Dasgupta and Ghosh 2018). The microbial carbonates preserved in ancient sedimentary successions are rich repositories of information on physical, chemical, and biotic environments of the past (Freytet and Plaziat 1982; Retallack 1994, 2001; Ford and Pedley 1996; Kraus 1999; Riding 2000; Freytet and Verrecchia 2002; Alonso-Zarza 2003; Pedley et al. 2003; Jones and Renaut 2010; Gierlowski-Kordesch 2010). More studies are required to understand the mode of colonization or the environmental preferences of microbes in different terrestrial environments and how they interacted with alluvial sedimentary processes.

In this paper, we describe microbial constructions preserved in thin beds of impure limestone and grainstones in the Early Jurassic fluvial deposits of the Pranhita-Godavari rift basin of India (from now on, P-G Gondwana Basin). Based on the mode of association of the freshwater carbonates with siliciclastic deposits, we attempt to reconstruct different paleoenvironments where microbially induced carbonate precipitation occurred in this continental rift basin. This information will enrich the database of the freshwater microbialites through time.

Geological Background

The NNW-SSE-oriented P-G Gondwana Basin was formed due to the extensional forces related to the fragmentation of the Pangean landmass, in the beginning of Triassic (Biswas 2003; Chakraborty et al. 2003; Chaudhuri et al. 2012). Since that time and at least until the beginning of the Middle Jurassic, the P-G Basin was a half-graben with the basin bounding fault running along its eastern margin. Within this basin, 2–3-km-thick fluvial strata (Table 1) were accumulated between the early Triassic and the early Middle Jurassic (Fig. 1) (Robinson 1967; Sengupta 1970; Kutty and Sengupta 1989). The study area is situated in the northern part of the exposure belt of the syn-rift strata of the P-G Basin, south-central India (Fig. 2). The Early Jurassic part of the succession, which is the aim of the present study, is represented by the upper part of the Dharmaram Formation, the lower part of the Kota Formation conformably overlying it, and the succeeding Kota limestone (Fig. 3).

Based on the sharp distinction between the faunal assemblages of the lower and the upper parts, the lower part of the Dharmaram Formation has been assigned a late Late Triassic age, while Bandyopadhyay and Sengupta (2006) assigned a Hettangian age for the upper part of the Dharmaram Formation (Bandyopadhyay and Sengupta 2006). The lower part of the Kota Formation that conformably overlies the Dharmaram Formation yields fossils of sauropod dinosaurs, turtles, and other reptiles as well as fish and mammals (Kutty 1969; Yadagiri 1986; Jain 1973, 1983; Evans et al. 2002). Based on the faunal analysis, a Sinemurian to Pliensbachian age has been assigned to the lower part of the Kota Formation (Jain 1973; Jain 1983; Bandyopadhyay and Sengupta 2006). The fossil fish assemblage of the overlying Kota limestone indicates a Toarcian age (Bandyopadhyay and Roy Chowdhury 1996).

The Dharmaram Formation (Fig. 3) has been interpreted as braided fluvial deposits (Rudra 1982). The red mudstone and sandstone of the lower member of the Kota Formation has been interpreted as meandering river deposits (Rudra and Maulik 1994). On the basis of the fossil assemblage and presence of algal structures, the limestone beds (Kota Limestone) have been interpreted as freshwater lacustrine deposits (Robinson 1967; Tasch et al. 1975; Rudra 1982; Maulik and Rudra 1986). The siliciclastic succession of the upper member of the Kota Formation, overlying the limestone interval (Fig. 3), was interpreted as braided river deposits (Rudra and Maulik 1994).

Materials and Methods

The studied succession (Fig. 3) comprises around 450 m of Lower Jurassic siliciclastic strata underlying the Kota limestone (i.e., the upper part of the Dharmaram Formation and the lower part of the Kota Formation).

The occurrence of the carbonate beds in this siliciclastic succession was documented through field observations by measuring lithological logs (Fig. 3). Thin sections were made from more than 80 carbonate rock samples to study the microfabric. These thin sections were studied under the optical microscope, stained with Alizarin Red S, where necessary. These samples are of marls, impure limestones, sandy limestones, and calcirudite-calcarenite occurring at various stratigraphic levels. Carbonate samples were etched in HCl (1%, 1 min) and coated with a thin layer of platinum before being observed under the scanning electron microscope (EVO 18; CARL ZEISS, Germany) and subjecting them to the energy dispersive spectroscopic analysis or EDS (XFlash 6 I 30; BRUKER, Germany).

Table 1 The stratigraphy of the Gondwana succession of the Pranhita-Godavari rift basin, Central India

Formation	Lithologies	Important fossils	Age
Chikiala	Highly ferruginous sandstones and conglomerates	?	?
Gangapur	Coarse gritty sandstones, grayish white to pinkish mudstones with interbedded ferruginous sandstones and concretions	*Gleichenia, Pagiophyllhum, Ptilophyllum, Elatoclacdus*	Early Cretaceous
Kota	Sandstone, siltstone, mudstone, and limestone bands	"Holosteans," sauropods, sphenodontians, lepidosaurids, crocodylomorphs, cryptodire, "symmetrodonts"	Early to Middle Jurassic
Dharmaram	Coarse sandstone and mudstone	Sauropodomorpha, Neotheropoda, sphenosuchians, phytosaurids, aetosaurids	Late Late Triassic to early Early Jurassic
Maleri	Mudstone, fine to medium sandstone and calcirudite	Metoposaurid, chigutisaurids, phytosaurids, rhynchosaurid, prolecertid, aetosaurids, dicynodont, cynodont, basal dinosauroformes, sauropodomorpha	Early Late Triassic
Bhimaram	Coarse to fine sandstone, ferruginous or calcareous at places and mudstone		Late Middle Triassic
Yerrapalli	Red and violet mudstone and calcirudite	Capitosaurids, therapsids, cynodont, rhynchosaur, prolacertid, rauisuchid, erythrosuchid	Early Middle Triassic
Kamthi	Sandstone and siltstone	Brachyopid, therapsid, *Glossopteris flora*	Early Triassic
Kundaram	Ferruginous shale, mudstone and sandstone	*Endothiodon, Emydops,* "*Kingoria,*" *Oudenodon, Pristerodon*, cistecephalid, captohrinid, gorgonopsian	Late Late Permian
Barakar	Sandstone, carbonaceous shale and coal	*Glossopteris* flora	Early Late Permian
Talchir	Tillites, greenish shale and sandstones		Early Early Permian

The general lithological attributes, faunal records, and the depositional environments are taken from Bandyopadhyay and Rudra (1985), Kutty et al. (1987), Kutty and Sengupta (1989), Ray and Bandyopadhyay (2003), Bandyopadhyay and Sengupta (2006), and Bandyopadhyay et al. (2010)

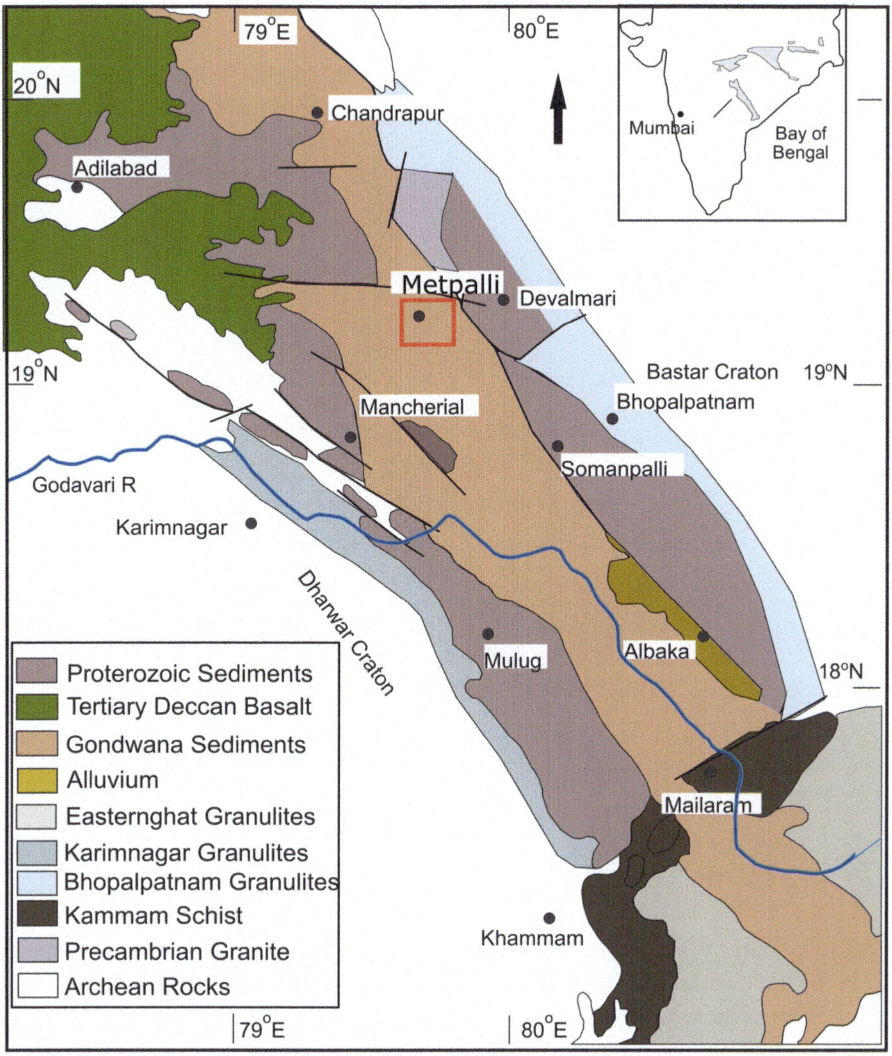

Fig. 1 Map showing the exposure belt of the Pranhita-Godavari Gondwana Supergroup, India, flanked on either side by Precambrian rocks. (Modified after Chaudhuri et al. 2012). The red square marks the study area. Inset shows the major Gondwanabasin of the Indian peninsula – the Pranhita-Godavari (P-G) basin

Stratigraphy of the Studied Succession

The 450 m succession may be divided in a lower part (0-285 m in Fig. 3) and an upper part (285 – 440 m in Fig. 3).

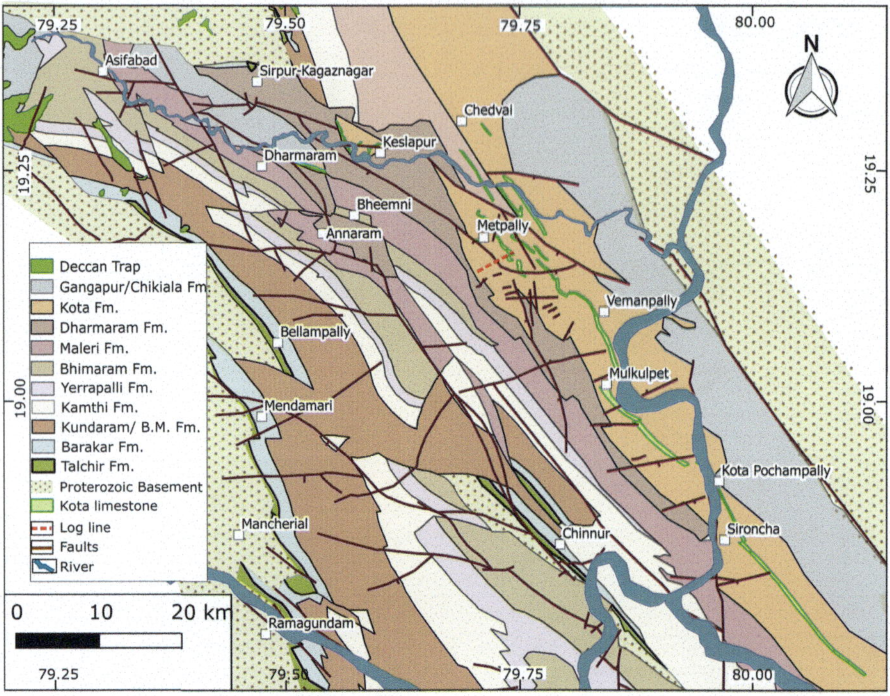

Fig. 2 Detailed geological map of the study area showing the Lower and Upper Gondwana formations. The red, dashed line represents the log line, along which the sedimentological log (Fig. 3) has been measured

The Lower Succession

The lower part of the studied succession (0-285 m in Fig. 3) is dominated by sandstone with intercalations of green and red mudstone with some thin carbonate beds (Fig. 4a–c). The sandstone bodies (12–20-m-thick tabular bodies) are characterized by an erosional basal bounding surface and by the presence of rip-up mud clasts and extrabasinal pebbles in their lower part. Grain size gradually decreases upward, and the sandstone bodies generally have a gradational contact with the overlying mudstone intervals (1–4-m-thick). Sets of planar cross-beds, organized in cosets, are the dominant primary structure. Cosets of trough cross-beds occur locally, in the basal part of the bodies. The thickness of cross-bed sets gradually decreases upward.

The cross-bed foreset azimuths are unidirectionally oriented and indicate that the direction of paleo-transport was from the western margin toward the eastern margin, across this NNW-SSE-oriented graben (Goswami 2019). This transverse drainage possibly existed on the surface of low-gradient alluvial fans originating from the hinterland of the western margin (Goswami 2019). Absence of features

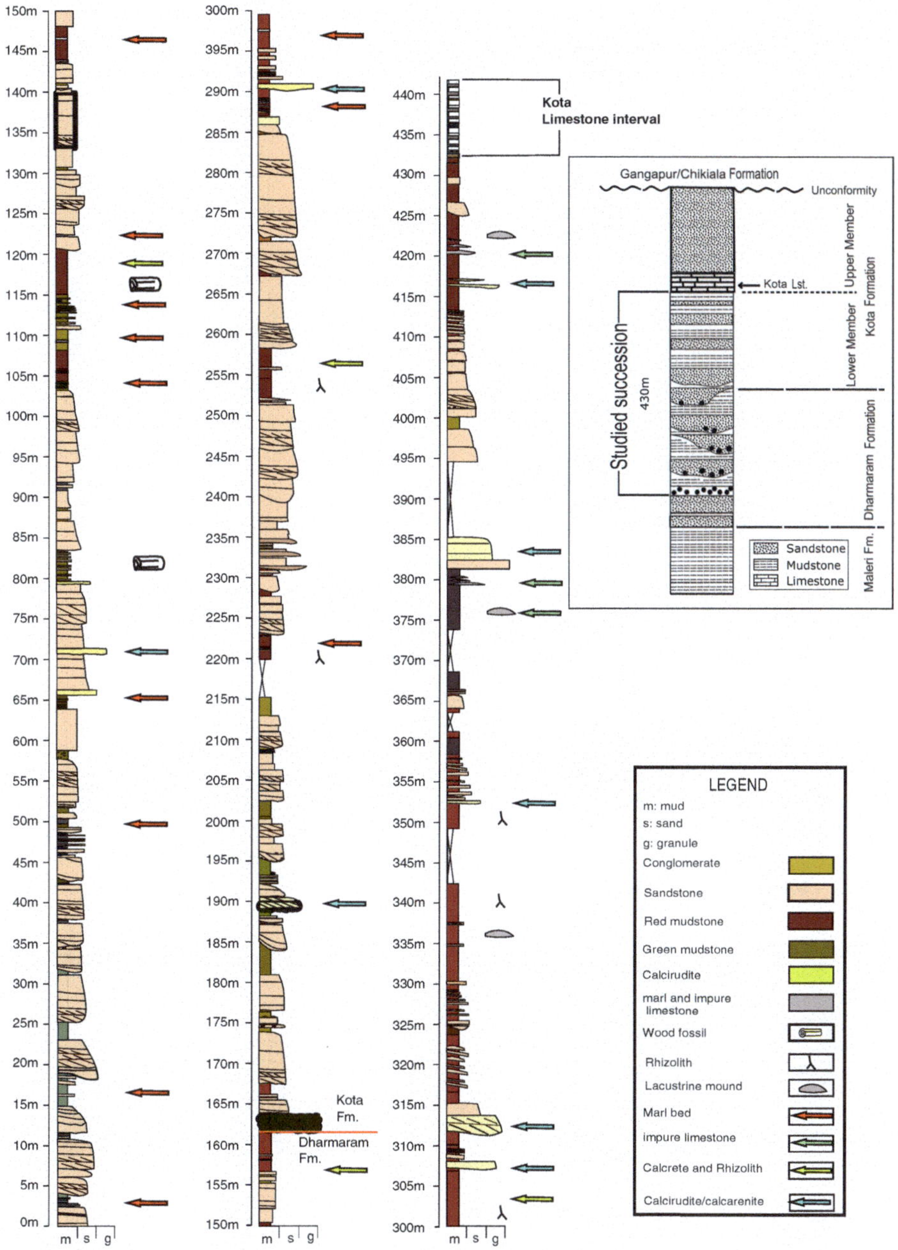

Fig. 3 Sedimentological log of the studied succession (including the upper part of the Dharmaram Formation and the lower part of the Kota Formation). The vertical scale has been graduated in 5-meter intervals. The horizontal scale represents the grainsize of the sediments. A schematic litholog of the studied succession is presented in the inset. The arrows indicate the stratigraphic positions of different types of carbonates within this siliciclastic succession

Fig. 4 (**a**) Field photograph of sheet-like marl bed (marked by arrow) in red mudstone. (**b**) Closer view of the marl bed of Figure (**a**); note the sharp lower bounding surface and up-arched upper surface. (**c**) Thin marl bed in a green mudstone unit. (**d**) Photomicrograph (using an optical microscope, under plane polarized light) of the marl shown in Figure (**b**); note the circular and elliptical spar-filled vugs marked by arrows. The three red squares correspond to zoomed areas of the sample, shown below, and are labelled as 1, 2, and 3. Number 1 shows massive micrites (Lpl) with irregular, diffused patches of darker clotted micrites separated by lighter patches of microspar. Number 2 presents brown, flaky organic matter associated with the dark, micritic laminae (marked by arrow). Number 3 shows laminated microfacies similar to Ls, comprised of alternate, crenulated laminae of micrite and sparite

comparable to the lateral accretion surface in the channel-fill bodies, together with the dominance of planar cross-bed sets, suggests the existence of low sinuosity sandy streams. The mudstones separating the channel-fill bodies can be compared to the overbank deposits (Goswami 2019). The presence of aquatic fossils like ostracods, fish, and turtles suggests the existence of swamp-like environments in the overbank area.

The Upper Succession

The upper part (285–440 m of Fig. 3) of the succession is dominated by 4–7-m-thick massive, red mudstones along with sheet-like bodies of plane-parallel stratified, fine-grained, sandstones (2–5 m thick) and lenticular bodies of inclined heterolithic strata (fine sand-silt). These heterolithic strata are vertically and laterally stacked, co-directionally inclined, sigmoidal in shape with an offlapping stacking pattern. They often found to be pinched out into the greenish laminated mudstones. Internally the heterolithic units show combined flow ripple cross-lamination and climbing ripple cross-laminations. Meter-thick, lenticular bodies of impure limestone occur encased in the red mudstones (section "Sedimentology of the Carbonate Bodies in the Succession", type II). On the other hand, dm-thick lenticular bodies of calcirudite-calcarenite occur in association with the bodies of heterolithic strata (section "Sedimentology of the Carbonate Bodies in the Succession", type IV).

The characteristics of the upper succession suggest that, with time, the channelized fluvial system was replaced by a non-channelized, sheet flood-dominated sediment dispersal system (Goswami 2019). The co-directionality in the inclination of the heterolithic units along with offlaping stacking pattern and pinching-out nature of them into greenish laminated mudstone intervals suggests progradation of macroform within a water body or a shallow lacustrine environment (Goswami 2019). The overall decrease in grain size and increased proportion of mudstones in the succession, as well as the presence of combined flow ripples, indicate a lacustrine coastal environment (Goswami 2019; Goswami and Ghosh, 2020, under review).

Further toward the top, the siliciclastic succession gradationally passes to the palustrine carbonate beds of the Kota limestone interval, early Middle Jurassic in age. The sedimentological characteristics of this limestone interval suggest the existence of a carbonate wetland system (cf. Goswami et al. 2018).

Sedimentology of the Carbonate Bodies in the Succession

The freshwater carbonates that occur in association with this siliciclastic syn-rift succession (0–435 m, Fig. 3) show four modes of occurrence (Table 2). Each of these four types internally shows various types of microfabrics. Based on the

Table 2 Mode of occurrences of freshwater carbonates

Carbonate bodies	Dimensions of the bodies	Sedimentary structures	Lithofacies associations	Interpretation
Type I: sheet-like bodies of marl	th: 5–25 cm le: few tens of meters	Massive, in places very crude, discontinuously laminated	Lpl, ls, Cs, Lc	Precipitation of carbonate mud in shallow, ephemeral ponds
Type II: lenticular bodies of impure limestone	th: 2–5 m le: 7–20 m	Generally massive. Crude, lamination present in places	La, Lg	Isolated carbonate depositing spring mounds
Type III: sheet-like bodies of sandy limestone	th: 20–50 cm le: tens of meters	Massive	Lpd	Pedogenic calcrete
Type IV: lenticular bodies of calcirudite-calcarenite	th: 70–120 cm le: 10–25 m	Cross-bedded	Lpl, La, Lpd	Shallow, short-lived, distributary channels

detailed textural study, seven types of facies have been identified from these freshwater carbonates, and they have been summarized in Table 3. To describe the different facies identified in each type of carbonate bodies, the nomenclature followed by authors like Arenas-Abad et al. (2010), García-García et al. (2014), and Pla-Pueyo et al. (2009, 2017) has been used. The common factor in all these publications is the use of a capital letter to describe the lithology ("L" for limestone, "M" for marl, "C" for calcilutite) and a small letter to specify further the nature of the facies. It has been confirmed from the staining that all these carbonates are low-Mg calcite.

Freshwater Carbonate Sedimentary Bodies

Based on the field observations, four major types of freshwater carbonate deposits (Types I to IV) have been identified within the siliciclastic succession (Table 2), i.e., (I) sheet-like bodies of marl (Fig. 4a–c), (II) lenticular bodies of impure limestone (Fig. 5), (III) sheet-like bodies of sandy limestone (Fig. 6), and (IV) tabular and lenticular bodies of calcirudite-calcarenite (Fig. 7). Detailed photographs of a selection of lithofacies present within these bodies are shown in Figs. 8a–h and 9a–f, and for some of them, SEM photographs are provided as well in Fig. 10a–d.

Table 3 Carbonate facies description

Facies with representative diagrams	Textural description	Interpretation
Lpl: massive, dense-clotted micrite 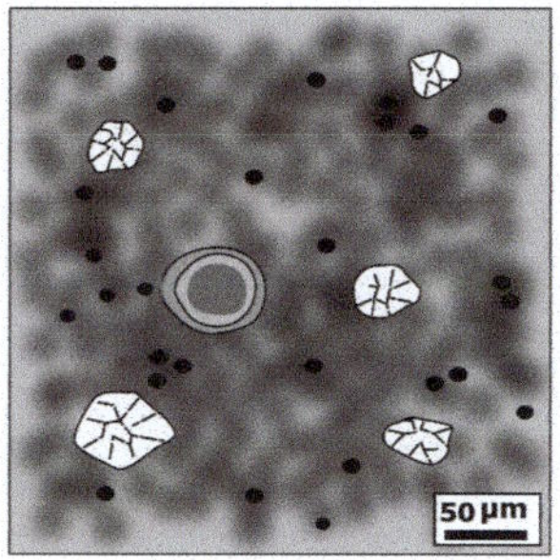 	Under optical microscope: Massive, almost devoid of any siliciclastics. Dark micrite, clotted in most places (Fig. 8a). Few spar-filled vugs (20–70 μm in diameter). In places, clotted peloidal fabric is made up of clusters of dark, small (3–5 μm in diameter) pellets (Fig. 8b). A few in situ oncoid grains (Fig. 8c) Under scanning electron microscope or SEM: Rhombohedron-shaped, subhedral crystals (1–3 micron across) comprise the dark micrite. Some of the rhombs show smooth edge and gently curved faces. Crystals are similar in size. Individual crystals appear to be perforated by tubular pores. Pores are circular to polygonal in cross section (250 nm in diameter; Fig. 10a)	Similar clotted and peloidal micritic fabrics have been reported from microbially precipitated carbonates (see Chafetz and Buczynski 1992; Reitner et al. 1995; Riding 1991; Riding 2000). Riding (2000) suggested that dense peloidal micrite are possibly calcified bacterial aggregates. The micron-scale smooth rhombohedral crystals resemble those formed during whiting events ("algal" blooms) (cf. Riding 2000, Dupraz et al. 2004). Freytet and Verrecchia (1998) and Bosak et al. (2004) demonstrated that rhombic crystals form around filamentous microbes, and leave a tubular space in the crystals after decay of the filament. The spar-filled vugs possibly represent original pore spaces in the microbial mat
Ls: Alternating lamina of micrite and filament-like structures 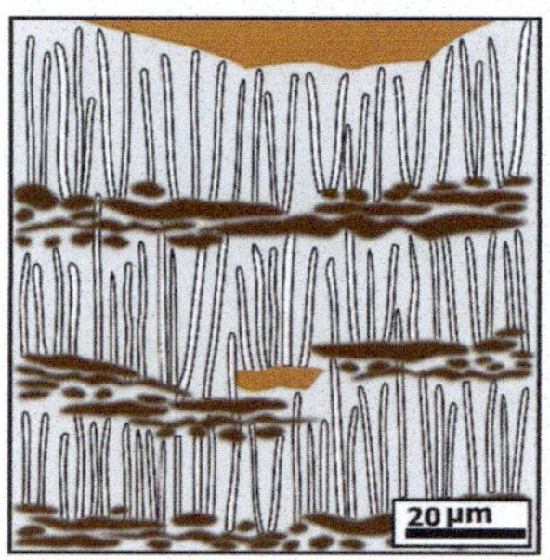 	Under optical microscope: Alternating lamina of darker micrite and lighter sparite. Siliciclastic grains are typically absent (Fig. 8d). Thickness of micritic lamina varies between 1 and 20 μm. They are laterally impersistent and wavy. The upper and lower boundaries are diffused. They internally show clotted micritic fabric Lighter laminae show the presence of numerous upright micrite-lined filament-like structures, originating from underlying micritic lamina (outward, in case of the oncoids; Fig. 8d). The filaments are 20–30 μm long and half μm across. They are slightly curved, uniform in thickness, and unbranched. The filament-like structures are entombed in poikilotopic calcite spars	The wavy laminae made up of dark clotted micrite possibly represent carbonate precipitation in microbial mats, whereas the transversely oriented micritic filament-like structures resemble calcified filamentous microbes growing on mat layers (see Freytet and Verrecchia 1998; Arp 1995). Therefore, the laminated character could be a reflection of repeated change in microbial community. The spar-rich nature of the lighter layers may suggest that the porous parts of the mat were readily cemented during early diagenesis

(continued)

Table 3 (continued)

Facies with representative diagrams	Textural description	Interpretation
La: Network of micritic flakes and fibrous calcite crystals 	Under optical microscope: Discontinuous, sinuous and curled, dark micritic flakes (20–50 µm thick), coated on either side by light colored, orthogonally disposed, fibrous calcite (Fig. 8e, f), define crude lamination. Coatings are 50–150 µm thick. Coated, curled, dark, micritic laminae, in places, constitute a complex mosaic. The intervening spaces are filled with either an equigranular mosaic of calcite spars or yellowish siliciclastic mud. The dark, massive micrite is traversed by spindle shaped spar-filled cracks (Fig. 8e). The facies is mostly free from sand-size or larger siliciclastic grains and bioclasts. Only, in places, a few fine sand-sized quartz clasts are present	Thin, curved micritic laminae of this microfacies are comparable to the planar stromatolite described from the Holocene Lagoa Salgadan Microbialites (Rezende et al. 2003). Laminar fabric consisting couplets of dark micritic laminae and light microsparitic/sparitic laminae is considered to be the basic morphology of planar stromatolite (Zamarreño et al. 1997). The coatings made up of fibrous crystals resemble abiotic precipitates formed due to rapid crystal growth (cf. Jones et al. 2005; Gandin and Capezzuoli 2014) during the phases of higher evaporative enrichment. The cracks in the dark micrite could be desiccation features formed during those phases when the growth of microbial colonies was supressed (cf. Gandin and Capezzuoli 2008; Rainey and Jones 2009; Pola et al. 2014). Absence of bioclasts may also suggest that condition was not suitable for larger fauna The occurrence of fibrous crusts and the presence of polygonal vugs filled either with spar or siliciclastics suggest the presence of high amount of porosity the mats, similar to those described by Gandin and Capezzuoli (2014), from travertine deposits. The paucity of coarse siliciclastics suggest that these stromatolites formed in areas away from the conduits of siliciclastic input, i.e., in the central part of the shallow water bodies
Lc: Alternating lamina of sandy detritus and filament-like structures 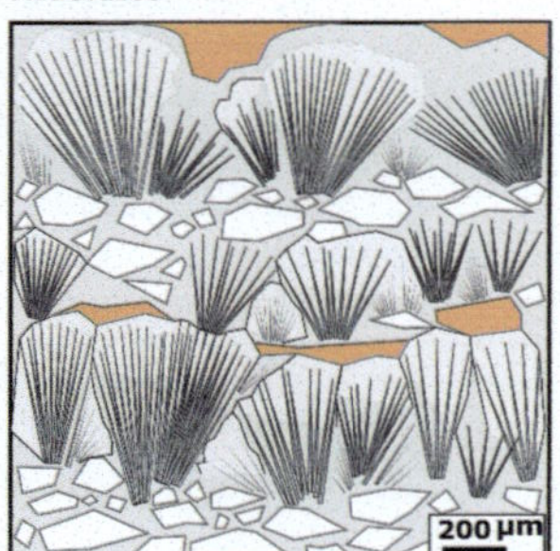	Under optical microscope: Characterized by siliciclastic-rich laminae (50–250 µm thick) alternating with laminae containing crystal fans (200–300 µm thick) (Fig. 8g). The siliciclastic-rich laminae are made up of fine sand-sized, angular quartz grains in a muddy matrix. The crystal fan calcite crystals contain embedded dark, micrite-lined, filamentous, cyanobacteria-like structures. The filament-like structures are non-branching. They are oriented transverse to the lamination and fan-out towards the top. The vugs are filled with yellowish siliciclastic mud	Siliciclastic-rich laminae represent influx of clastics in a water body. The radiating filament-like structures suggest formation of microbial mats over clastic-rich substrate. Carbonate precipitation was induced by filamentous microbes during phases of low clastic supply. Similar filament-like structures have been represented from freshwater microbialites by Freytet and Verrecchia (1998), Arp (1995), Zamarreño et al. (1997), Della Porta (2015). The pores present in the carbonate deposits were filled by siliciclastic mud during the early phase of diagenesis

Lg: Matrix-rich grainstones 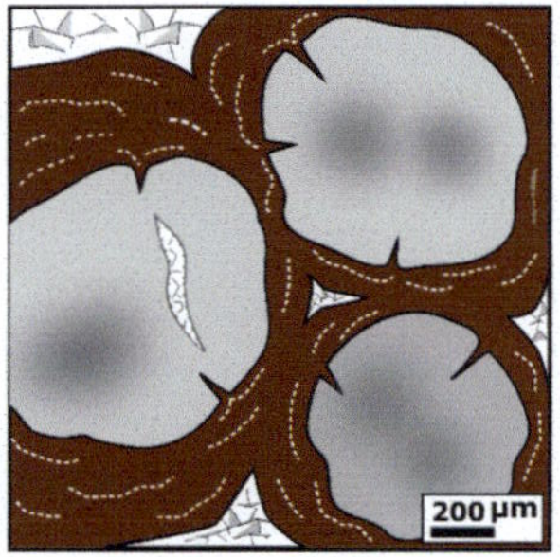	Under optical microscope: Sand to granule-size, sub-spherical carbonate grains coated by crudely and discontinuously laminated dark micrite (Fig. 9a). The intergranular space is filled with calcite spars. The grains internally show either massive, micrite, or clotted micrite fabric, with mosaic of microspar in patches. Internal fabric is similar to facies Lpl, though many of them contain a higher proportion of neo-formed micro-spars. In some grains, spindle-shaped cracks, filled with drusy mosaic of calcite spars, are present. Grains are free from siliciclastics and bioclasts. The grain outlines are smooth and show small, tapering inside desiccation cracks (Fig. 9b). The cracks are filled with fine silt-sized siliciclastic grains floating in micrite Grain coatings comprise a few non-isopachous and discontinuous micritic lamina, each 50–100 μm thick (Fig. 9c). Laminae internally show dark, peloidal micritic fabric (Fig. 9c). A few spar-filled vugs, perfectly circular in outline occur within these micritic cortical laminae (Fig. 9c) Under SEM: Dark micrite contain rod-like (~150 nm long and 10 nm in diameter) and sheet-like, nanometer-scale globular calcareous structures (20–30 nm in diameter; Fig. 10b–d). Tips of the rod-like structures are rounded, and margin of the sheet-like structures is jagged. These two types coalesce to form bean-bag like aggregates, 700 nm to a few microns in diameter (Fig. 10b). All the rod-like structures in an aggregate are similar in size	This microfacies is indicative of multiple phases of desiccation, dissolution, and re-sedimentation in a palustrine condition. The desiccation along the boundary of micritic grains indicates subaerial pedogenic modification of earlier carbonate deposit. The smooth, irregular outline suggests dissolution. The grains were possibly formed due to “grainification” that results in in situ formation of diagenetic grainstones (cf. Freytet and Plaziat 1982; Mazzullo and Birdwell 1989; Alonso-Zarza et al. 1992). The micritic coatings suggest development of microbial-mats on the intraclasts and subsequent carbonate precipitation by the microbes in the mat. The circular spar-filled vugs in the clotted micritic lamina possibly represent trapped gas bubbles formed in the microbial mat. Some of the interstitial spaces that were not filled during the sedimentation were filled later by calcite spar cement Presence of nanometer-scale globular masses observed under SEM also suggests microbial influence in carbonate precipitation (cf. Dupraz et al. 2009; Perri and Tucker 2007). In modern lithified microbial mats, similar autochthonous peloids have been found and interpreted to have formed due to the coalescence of sub-micron spherical structures that replace organic substrates, i.e., extracellular polymeric substances (EPS) (Dupraz et al. 2004). The rod-like features observed under SEM are similar to the nanobacterial cells reported by Folk (1999),. whereas the flaky structures could be fragments of calcified EPS (see Dupraz et al. 2004; Dupraz et al. 2009). The bean-shaped aggregates of rod-like and sheet-like structures possibly represent bacterial colonies
Cs: Crudely laminated detritus-rich micrite 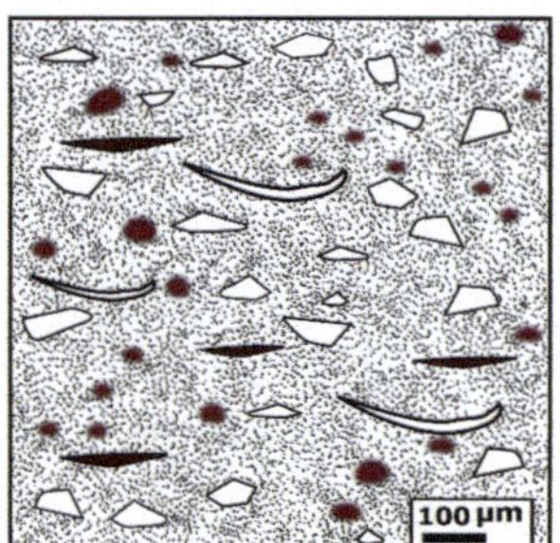	Under optical microscope: Detrital, siliciclastic grains and bioclasts floating in a matrix of massive or very crudely laminated, micrite and neoformed microsparite (Fig. 9d). Irregular patches of brown iron-rich mottles occur in micrite. Sub-circular, fine sand-size, patches of iron-poor micrite occur locally. The detrital components comprise silt-sized sub-angular quartz grains, rounded ferruginous concretions, flakes of organic matter, and broken ostracod shells	This facies indicates influx of clastic detritus in a lime-mud precipitating environment. This is similar to the “Marsh-Pool facies” described by Guo and Riding (1998) and “silty lime-mudstone” by Gandin and Cappezuolli (2014) and also comparable with the calcilutite lithofacies Cs described by Pla Pueyo et al. 2009. The ferruginous concretions indicate derivation from pedogenically modified source. Variable concentration of iron in the micrite suggests local hydromorphic changes indicating fluctuating water level in a shallow waterbody (cf. Kraus and Hasiotis 2006). The poor development of lamination possibly suggests bioturbation

(continued)

Table 3 (continued)

Facies with representative diagrams	Textural description	Interpretation
Lpd: Floating grains in massive microspar 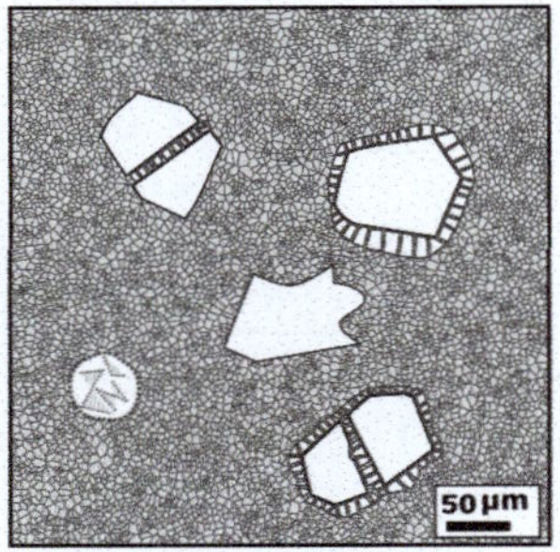	Under optical microscope: Sand-sized, angular to sub-angular detrital grains (>40%) of quartz and feldspar floating in a matrix of microspar (Fig. 9e). In places, spar-filled vugs with circular outline are present. The voids are lined with crudely-laminated, dark micrite (Fig. 9f). Individual clastic grains are surrounded by a corona of equidimensional calcite spar. Some of the clastic grains show embayed boundary. Many of the grains are broken. Parts of such grains have matching boundary and remain separated by spar (Fig. 9e). Shells of aquatic invertebrates are typically absent	This facies resemble the microfabric of pedogenic calcrete (cf. Alonso Zarza 2003). The floating grain fabric suggests early diagenetic enrichment of calcium carbonate in a siliciclastic deposit, which is supported by the presence of displacive calcite crystal (cf. Ghosh 1997). The spar-filled vugs with circular outline resemble root-cavities. They resemble vugs left behind by decayed roots later filled by sparry calcite. The laminated micritic linings around vugs could have formed during decay of the root. The micrite *was possibly precipitated by the microbes involved in the* process of decay

Index

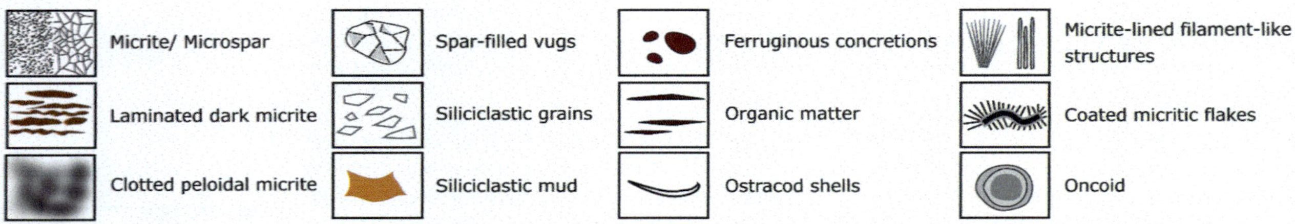

Fig. 5 (**a**) Field photograph of lenticular units of impure limestones within red mudstone interval. (**b**) Closer view of the impure limestones. Note the beds of grainstones in the upper and the middle part (marked by "g")

Type I: Sheet-Like Bodies of Marl

These marl beds are white to yellowish-gray in color. The sheets are 5- to 25-cm-thick and alternate with red/green mudstones (Fig. 4a–c). Individual sheets can be traced laterally for a few tens of meters, with sharp and flat basal surfaces. The upper surface, in places, arches up in the form of low relief domes (Fig. 4b). The marls are internally massive, though in places very crudely, contorted, and discontinuous lamination can be observed.

From the detailed study of thin sections, several lithofacies (Lpl, Ls, Cs, Lc) are distinguished (Figs. 4d, 8a–h, 9a–f, and 10a–d). Massive micrites (Lpl) are the dominant facies (Fig. 8a–c) characterized by irregular, diffused patches of darker clotted micrites, separated by lighter patches of microspars. In places brown, flaky organic matter is found to be associated with the dark, clotted micritic laminae. The upper parts of the marl beds are mostly made up with laminated facies Ls (Figs. 8d and 10a, d).

Fig. 6 (**a**) Field photograph of sheet-like beds of calcrete (marked by arrow) on the top of the channel-fill sandstone beds. (**b**) Field photograph of calcareous rhizoliths (marked by arrows) in massive red mudstone interval. (**c**) Closer view of one of the rhizoliths

Some sheet-like marl beds are entirely made up of massive to very crudely laminated Cs lithofacies (Fig. 9c, d). In places, laminated facies Lc (Figs. 8g, h and 10c) are present in the lateral parts of these sheet-like bodies. Disarticulated and articulated ostracod shells are dispersed within the matrix. A few cracks filled with drusy mosaic of spar traverse the samples.

Interpretation

The thin sheet-like geometry of the beds, their overall massive character, and the presence of the shells of aquatic invertebrates suggest precipitation of carbonate mud in shallow, ephemeral ponds. The presence of spar-filled cracks in the micritic laminae suggests subaerial desiccation of the mats. The thinness of the marl beds and their limited lateral extent along with the presence of desiccation cracks suggest that the water bodies were shallow and ephemeral.

Presence of coarser siliciclastic detritus in the lateral part of the marl beds suggests input of coarse detritus from the lateral margins of the shallow water bodies, similar to "marsh-pool" deposits (see Guo and Riding 1998). These marl beds suggest palustrine conditions in shallow ephemeral ponds close to the margins of a large lake (cf. Alonso-Zarza and Wright, 2010). The lack of sedimentary structures is interpreted as the result of bioturbation in these nearshore facies.

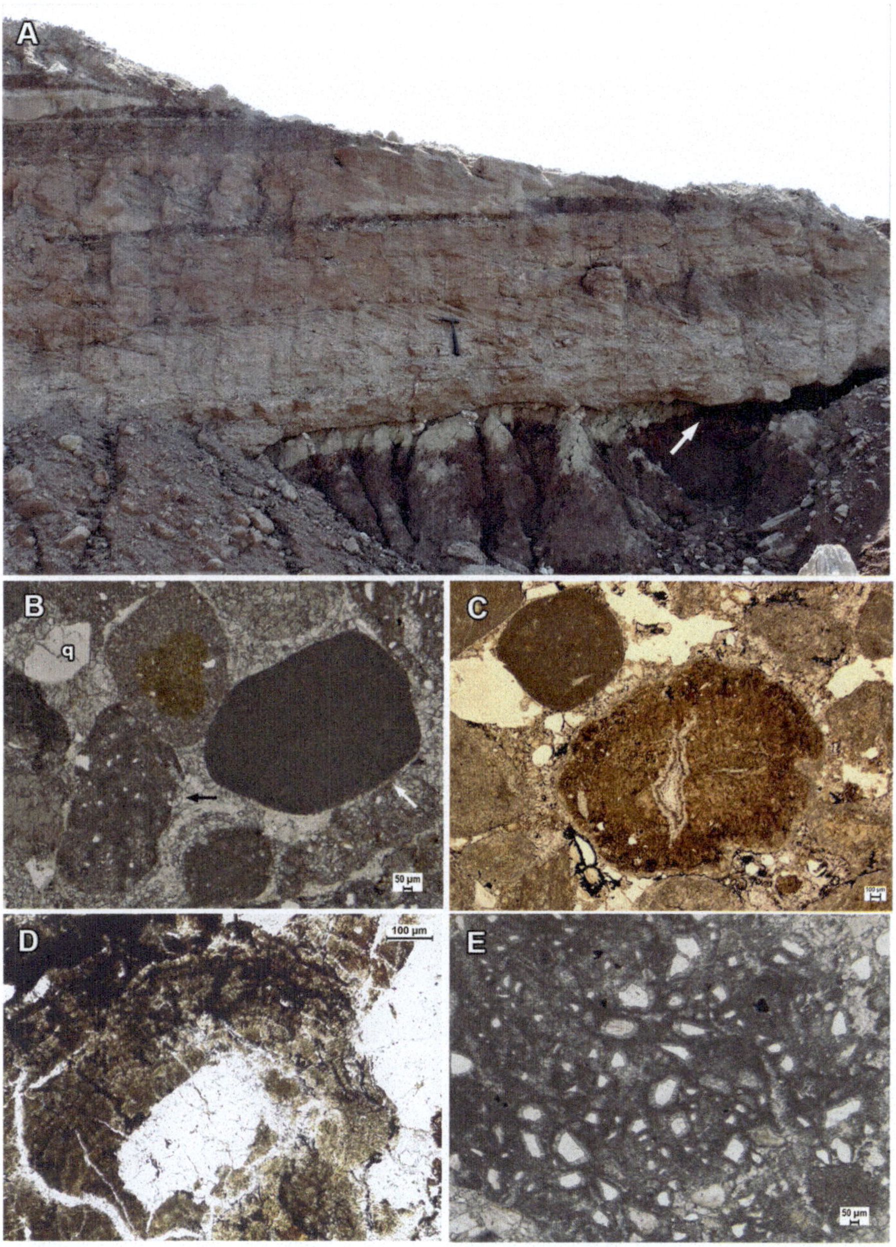

Fig. 7 (**a**) Field photograph of cross-bedded calcirudite/calcarenite from the upper part of the succession. Note the sharp, irregular basal bounding surface (marked by arrow) (**b**) Photomicrograph (in optical microscope, under PPL) of calcirudite/calcarenite with massive micritic peloid grains (white arrow), intraclast of marl (black arrow) and quartz grains (marked as "q"). (**c**) Coated grain from calcirudite/calcarenite; cortical layers are made up of thinner dark micrite, alternate with thicker layers of radiating fibrous calcite. Note the up-arched, non-isopachous nature of the laminae and the wedge-shaped, spar-filled crack in the central part of the grain. (**d**) Coated grain from calcirudite/calcarenite; outer layers are made up of thin, wavy/deformed laminac. Iron oxide-stained rims within dark massive micrite are also present. Nucleus of the grain is recrystallized into spar. (**e**) Closer view of an intraclast of sandy limestone. Note, the corona structures made up with fibrous calcite crystals around the siliciclastic grains

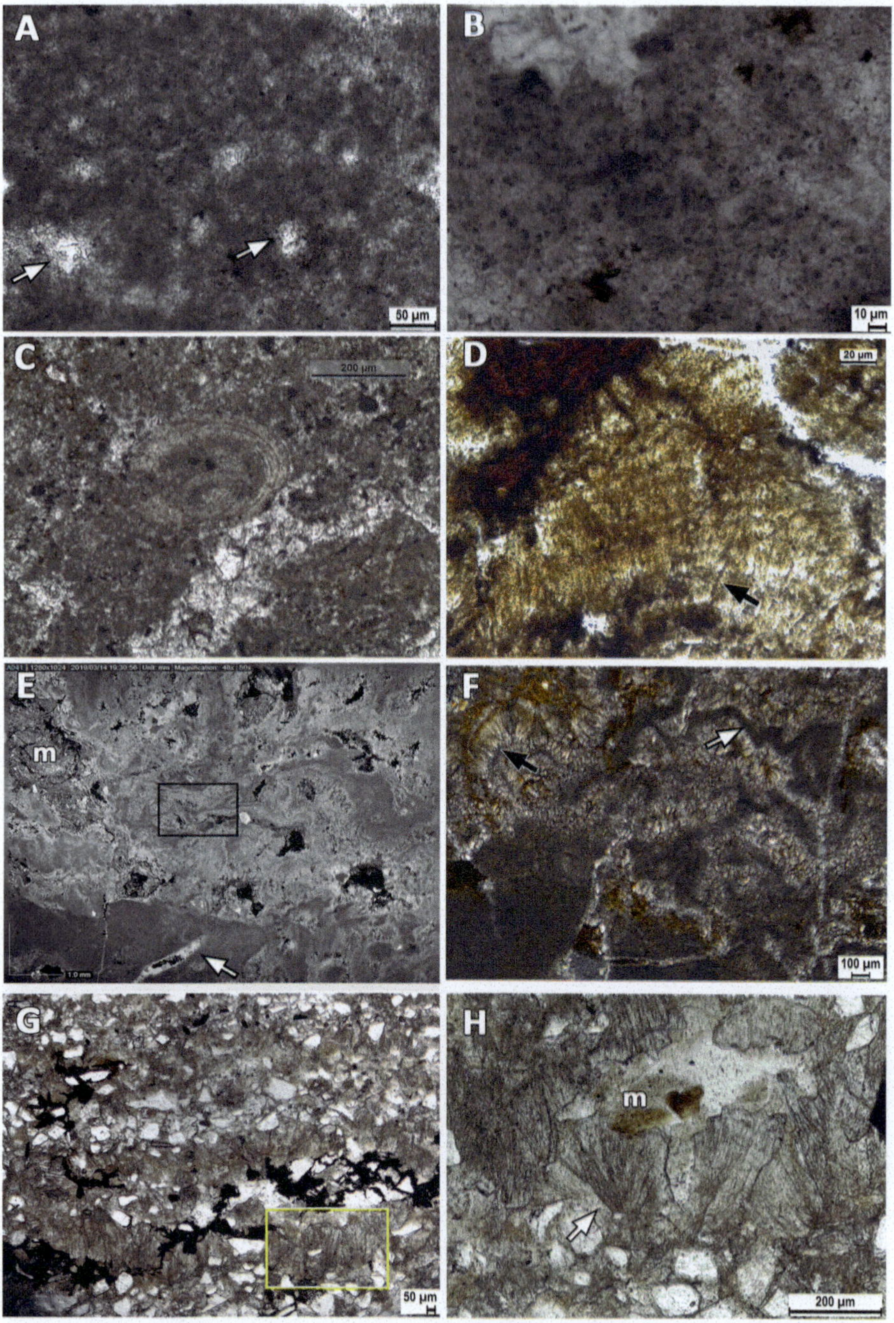

Fig. 8 Photomicrographs of facies of freshwater carbonates in the study area. Pictures have been taken using an optical microscope, under plane polarized light. (**a**) Lpl: dark massive micrite, clotted in places. Note the patches of spar-filled vugs (marked by white arrows). (**b**) Lpl: Peloids

Type II: Lenticular Bodies of Impure Limestone

The lenticular (bi-convex) variety of marl (Table 2) occurs as 2–5-m-thick units encased in very thick package of massive, red mudstone (Fig. 5a, b). This variety of carbonate beds is not very common in the study area. The concave-up, basal contact with the siliciclastic mudstone is transitional. The contact with the overlying mudstone is sharp and convex-up. Crude lamination is present in places, in these impure limestone bodies. The basal part of the bodies is dominated by the laminated lithofacies La (Fig. 8e, f). Discontinuous beds (10–30 cm) of in situ grainstones (Lg) (Figs. 9a, b and 10b) occur in the upper part of the bodies (Fig. 5b). The grainstone beds are devoid of any internal stratification.

Interpretation
Restricted lateral extent of these lenticular bodies of impure limestone along with the typical lenticular geometry and their occurrence within the mudstone suggests that they are comparable with the modern lacustrine microbial bioherms and lacustrine spring mounds described by Arp (1995), Perri et al. (2012), and Della Porta (2015), among others. Absence of pedogenic features in the enclosing mudstone succession also supports this view. The overall massive appearance of both impure limestones and grainstones may be an indicative of syngenetic formation and no physical transport. Individual grains possibly resulted from "grainification" that leads to the formation of this in situ formation of diagenetic grainstones (Freytet and Plaziat 1982; Mazzullo and Birdwell 1989; Alonso-Zarza et al. 1992).

Type III: Sheet-Like Bodies of Sandy Limestone

Sandy limestones are yellowish to brownish in color, sheet-like beds (Table 2). They are found on top of the channel-fill sandstone bodies (Fig. 6a) and are overlain by massive red mudstone containing upright rhizoliths. The sandy limestone beds are made up with facies Lpd (Fig. 9e, f). Shells of aquatic invertebrates are typically absent in these samples. The enclosing mudstone shows colored mottles as well as wedge-shaped peds with pedogenic slickensides on the ped surfaces. In places,

Fig. 8 (continued) in massive micrite. (**c**) Lpl: Partly dissolved oncoid (in situ) grain in the massive clotted micrite. (**d**) Ls: Alternating darker and lighter laminae. Darker laminae are made up with clotted micrite and lighter laminae are characterized by dark, micritic filament-like structures (marked by arrow) embedded in poikilotopic calcite spars. (**e**) La: Marl showing sinuous and deformed, dark micritic flakes lined on both sides with fan-shaped, aggregates of radially-arranged, lighter-colored, fibrous calcite crystals. Spindle-shaped spar-filled crack in massive clotted micritic flake (marked by white arrow). Picture taken in plane polarized light. "m" denotes the patches of siliciclastic mud. (**f**) La: Enlarged image of the rectangle of Figure (**c**). Sinuous and deformed, dark micritic flake is marked by white arrow and the fibrous calcite crystals are marked by black arrow. Picture taken in PPL. (**g**) Lc: Crudely laminated marl, lamination is defined by alternation of fine sand-sized siliciclastic-rich laminae and carbonate laminae made up of crystal fan calcite crystals. (**h**) Lc: Enlarged image of the rectangle of Figure (**g**). Note the dark, filamentous, cyanobacteria-like structures marked by white arrow. "m" denotes the patches of siliciclastic mud

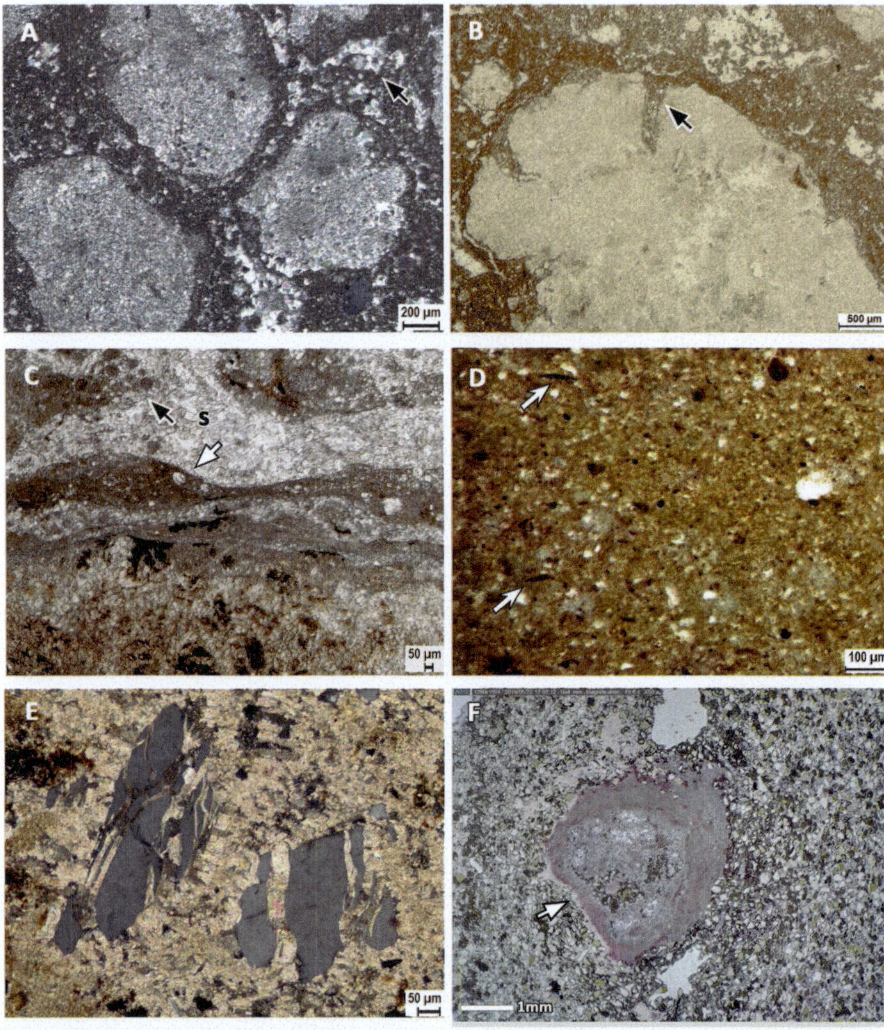

Fig. 9 Photomicrographs of facies of freshwater carbonates in the study area. Pictures taken using an optical microscope, under plane polarized light. (**a**) Lg: Sub-spherical micritic/microsparitic grains floating in dark micritic matrix. Note the spar-filled fenestral vugs marked by black arrows. Picture taken in crossed polars. (**b**) Lg: Desiccation cracks (marked by black arrows) along the grain boundary of the micritic/microsparitic grains. Picture taken in PPL. (**c**) Cs: Enlarged picture of micritic coatings of the micritic/microsparitic grains. Note the dark, peloidal micritic fabric in these coatings (marked by black arrow) and the circular spar-filled vugs within these cortical laminae (marked by white arrow). (**d**) Cs: Massive marl with few siliciclastic grains and parallel-arranged flakes of organic materials (marked by arrows). Picture taken in PPL. (**e**) Lpd: Angular to sub-angular detrital grains of quartz are floating in a sparitic matrix. Note the growth of displacive calcite crystals and matching boundary of the broken clastic grains. (**f**) Lpd: Transverse section of a rhizolith showing large, spar-filled alveoli lined by micrite (marked by arrow)

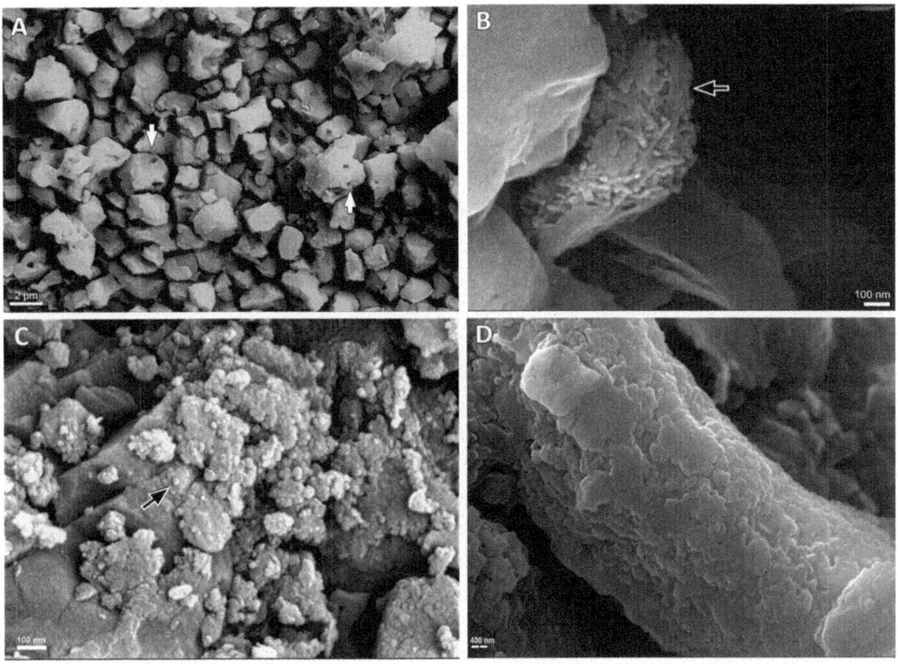

Fig. 10 Scanning electron microscopic (SEM) images of different microfacies. (**a**) Impure limestone with micro-pores within micron-sized smooth calcite crystals of Ls. Note the pore arrangements (marked by arrows) within single crystals. (**b**) Nanometer-scale rod-like aggregates of calcite associated with mineralized EPS (marked by arrow) in the micritic matrix of Lg. (**c**) Nanometer-scale spheroid (arrow) of calcium carbonates in peloidal micritic layers of Lc. (**d**) Enlarged view of calcitic filaments from the micrite-lined fibrous structures of Ls. Note the nanometer-scale globular masses that comprise the wall of the filament

large, upright, cylindrical calcareous rhizoliths (sensu Klappa 1980) (Fig. 6b) are found in these red massive mudstones overlying the sandy limestones. The rhizoliths are 4–30 cm long and 0.5–5 cm in diameter. The muddy host sediment displays circular (diameter 1–4 cm) patches of drab haloes in shades of white and yellow. Under the microscope they show the presence of centrally located, spar-filled voids (Fig. 9f). They are surrounded by crudely laminated, dark, micrite with a diffused outer boundary.

Interpretation

The sheet-like beds of sandy limestone, overlying the channel-fill sand bodies, could be interpreted as pedogenically modified sandy beds formed in crevasse splay. Presence of the calcareous rhizoliths in red mudstone and their microfabrics suggests precipitation of pedogenic carbonate under semi-arid to arid conditions (Allen 1986) within the floodplain muds. The carbonates were precipitated through dissolution-precipitation mechanism (Wright and Tucker 1991), with some likely contribution from respiration of living roots in the case of the rhizoliths. However, absence of properly developed soil profiles along with sporadic occurrence of rhizoliths indicates that carbonate precipitation takes place only in discontinuous areas in close association with roots and related microorganisms (similar to stage III of calcrete formation as described by Alonso-Zarza 2003).

Type IV: Lenticular Bodies of Calcirudite-Calcarenite

The calcirudite-calcarenites are usually cross-bedded. They are abundant in the upper part of the succession. They occur as 70–120-cm-thick, lenticular and tabular bodies (Fig. 7a) associated with the complexes made up of inclined heterolithic strata. In places, these calcirudite-calcarenite units occurring at the base of the IHS (inclined heterolithic strata) bodies overlie massive red mudstones. The lower bounding surface of the unit is gently concave-up, with small-scale erosional scour present, in places. Pebble-sized, rip-up mud clasts occur in the basal part of these units. The size of the clastic grains decreases upward. The upper bounding surface of the bodies is planar and sharp. Planar cross-beds dominate over trough cross-beds, with set thicknesses (10–25 cm) decreasing upward. The mean paleocurrent direction is similar to that of the associated inclined heterolithic beds, i.e., toward the NE.

The calcirudite-calcarenites show grain-supported fabric, and their framework is made up of medium sand to granule-sized, well-rounded carbonate grains admixed with medium to coarse sand-sized siliciclastic grains of mainly quartz along with a few grains of orthoclase feldspar (Fig. 7b). The intergranular space is filled with poikilotopic spar displaying a drusy spar mosaic. In parts, the intergranular space is filled with gypsum cement. The carbonate grains in the framework show a variety of internal fabrics. There are four main types, i.e., massive micritic peloids, coated grains, intraclasts of marls, and intraclasts of sandy limestones.

Grains of massive micritic peloids are more numerous than other grain types (Fig. 7b). Internally they are made up of dense and dark micrite, clotted in places, comparable with Lpl. Most of the peloids are free from siliciclastic grains and bioclasts.

Another common grain type is coated grains. Internally these coated grains show alternate, concentric laminae of La facies. They are very coarse sand to granule-sized, moderate to well-rounded grains. A crudely laminated cortex is defined by an alternation of dense, brownish thinner micrite and thicker lighter layers of radiating fibrous calcite crystals (10–20 μm and 50–80 μm, respectively) (Fig. 7c, d). Micrite-lined fibrous structures are found embedded in these fibrous crystals. The cortex encircles a large nucleus. The nucleus consists of dark peloids and the cortex usually contains three to six layers. The micritic and/or calcitic layers of the cortex are non-isopachous and, in places, wavy or deformed. Wedge-shaped, spar-filled cracks (ca. 0.8 mm long) traverse the coated grains (Fig. 7c).

Intraclasts of marl also occur as grains in these calcirudite-calcarenite beds. They have irregular to subhedral boundaries and internally made up of micrite with a few silt to very fine sand-sized quartz grains. Neomorphic calcite spar and microspars are present in irregular patches (Fig. 7b).

Intraclasts of sandy limestones are of granule size and sub-angular (Fig. 7e). The internal fabric of these clasts is characterized by silt to very fine sand-sized quartz grains floating in iron-stained microsparite or in an equigranular mosaic of microspar (Lpd).

Interpretation

The erosional basal bounding surface with rip-up mud clasts in the lower part and the fining upward of grain-size trend of the cross-bedded calcarenite-calcirudites denotes that these bodies were deposited by small, shallow channels. Thinning upward trend of the cross-stratal sets indicate that the channel flow was decelerating. Abrupt vertical change of facies, from granule-sized calcarenite-calcirudites to mudstone, denotes that these channels were abandoned suddenly. These channel-fills possibly represent shallow, short-lived, distributary channels.

The intraclasts of marl and calcretes have been derived from the marl beds and calcrete beds. The micritic peloids and the coated grains with well-defined nucleus and a coarsely laminated cortex are similar to the modern and ancient freshwater oncoids formed in fluvial channels, in floodplain ponds, and in shallow lacustrine conditions (see Strasser 1986; Zamarreño et al. 1997; Arenas et al. 2000; Arenas-Abad et al. 2010; Arenas et al. 2015; Della Porta 2015). The wavy, non-isopachous nature of the cortical laminae and radially disposed fibrous structures distinguish them from ooids and pisoids (cf. Tucker 2001). The wavy and up-arched lamination in the micritic layers suggests repeated formation of microbial mats on the grain surface similar to those reported by Chafetz and Guidry (1999), Sanders et al. (2006), Shiraishi et al. (2008), and Dupraz et al. (2009). Radially arranged, curved, non-branching, micrite-lined fibrous structures are similar to those found embedded in fibrous crystals and clotted micritic dendrites of Pyramid Lake microbialites and are comparable to filamentous cyanobacteria (see Della Porta 2015).

Depositional Setting

The occurrence of the sheet-like bodies of marl (Type I deposits) in the laminated mudstones that alternate with the channel-fill sandstone bodies suggest that the condition for carbonate precipitation prevailed in the shallow water bodies of the overbank areas (Fig. 11). It may be suggested that the depositional sites were partially sheltered from the input of siliciclastic detritus escaping the confines of the channel. However, the local development of Lc and Cs facies suggest periodic influx of intraformational and extraformational detritus during floods. A flourishing microbial community dwelling in these water bodies perhaps helped in sustaining the carbonate precipitation through biological mediation. The features observed in facies Lpl, Ls, La, and Lc (Table 3) provide evidence in support. Furthermore, the up-arched upper contact of the marl bodies that internally contain features indicative of biomat

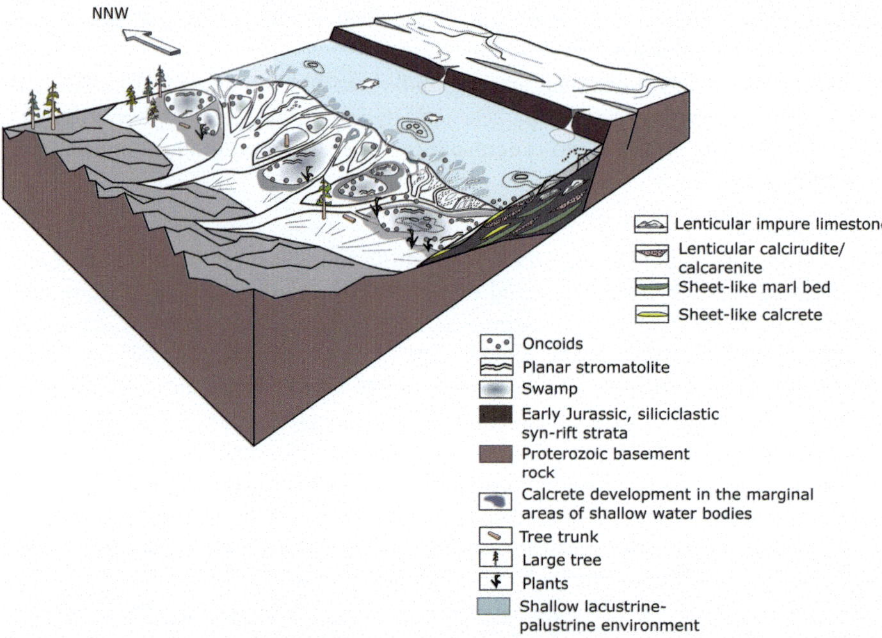

Fig. 11 Schematic block diagram showing depositional environments associated with a braided fluvial system and carbonate precipitating swamps in the interfluve area, in the rift basin during Early Jurassic deposition. The deposits of the large water body at the basin trough (shallow lacustrine and palustrine environment) are not represented in the studied siliciclastic-dominated succession. However, they have been included to show all the environments of the half-graben

formation (Lpl and Ls) provides further support to the existence of low-relief stromatolitic mounds in these small water bodies.

The meter-thick lenticular bodies of carbonates (Type II) that show prolific development of La possibly formed in more sheltered locations, i.e., in deeper part or in centrally located areas. The features of La provide evidences for the existence of porous biomats in clear water condition. The minor amount of siliciclastic detritus found in this facies might have been incorporated later during the fluvial flooding events. These lensoid bodies can be compared to the modern lacustrine microbial bioherms and lacustrine spring mounds described by Arp (1995), Perri et al. (2012), and Della Porta (2015), among others. The part of the water bodies encircled by the mounds was likely to be sheltered from siliciclastic inputs. These provided suitable sites for development of planar stromatolites (Ls) and oncoids (Lpl). On the other hand, the karstic features preserved in the upper part of these bodies (Lg) indicate repeated lowering of water level and subaerial exposure of the deposits.

The occurrence of the sheet-like calcrete bodies or Type III deposits (Lpd) and associated rhizoliths on top of the channel-fill sandstone bodies suggests early diagenetic carbonate enrichment in the fabric of siliciclastic deposits by soil-forming process at the sites of abandoned channels and floodplains.

The small channel-fill bodies of calcirudite-calcarenite (Type IV) possibly deposited at the marginal part of the basinal environment (possibly a wetland system) of the half-graben (Fig. 11). The erosion of the various freshwater carbonate deposits by fluvial flows possibly generated the sand-sized intraformational grains found in the framework of these clastic channel-fill deposits. The pedogenic modification of the deposits possibly aided production of carbonate grains. The mixing of these intraformational detritus with the extraformational detritus took place during the fluvial transport process. However, the good preservation of cortical lamina of the oncoid grains found in the framework of the calcirudite-calcarenites goes against energetic transportation of these grains over a long distance. One possibility is that the cortex might have developed during the transportation of these grains in the fluvial channels.

Apart from the presence of microbes, the ionic concentration in the water and rate of evaporation are some of the critical factors for carbonate precipitation. This continental rift basin possibly received calcium-rich water from the hinterlands. The dissolution of the Proterozoic limestones that constituted the rift shoulders (cf. Chaudhuri et al. 2012, 2015) can be a major supplier of calcium in the groundwater. The fluvial system is interpreted as a part of low gradient hanging wall alluvial fans (Goswami 2019), and the shallow water bodies or the main cites of carbonate precipitation was present in the floodplain areas. The points of intersection of the groundwater table and the alluvial fan surface could be the preferred sites for the formation of the spring carbonate mounds in this case.

Conclusions

The following conclusions have been derived from the study presented and discussed in this article:

1. This study of the Jurassic syn-rift sediments from the P-G Gondwana Basin, India, recognizes different freshwater carbonate-precipitating environments that were preserved in the siliciclastic depositing fluvial system.
2. The Proterozoic limestones at the rift shoulders served as the provenance that saturated the water with calcium and allowed the carbonate to precipitate.
3. This fluvial system belongs to the braided transverse drainage of the hanging wall block of the P-G half-graben. In the floodplain areas of the channels, there is evidence of presence of shallow ponds. Presence of aquatic invertebrates in the carbonates also supports this view. These shallow water bodies in between the fluvial channels were the main sites of carbonate deposition, where microbial activity was profuse.
4. Apart from depositing in the shallow ponds, the freshwater carbonates were precipitated as spring mounds, possibly along the marginal areas of the ponds.
5. Secondary carbonate enrichment as pedogenic calcrete formation also took place in this succession.

6. Besides these three types of freshwater carbonates, there were shallow distributary channels in the distal part of the basin. These small channels mainly carried the intraformational detritus of the three other types of carbonates and formed lenticular deposits of calcirudite-calcarenites.
7. Carbonates with microbial structures are well-preserved in the four different types of carbonate beds that record the episodes when microorganisms played an important role in the sedimentation of carbonates during the Early Jurassic period. The microfabric of the carbonates shows that a complex interplay between the biotic and abiotic modes concurs in freshwater carbonate precipitation. Microbially induced carbonate precipitation was punctuated by phases of siliciclastic input within these water bodies.

Acknowledgments The authors thank the Indian Statistical Institute, Kolkata, for financial and infrastructural support for this research. The fond memories of Prof. Elizabeth Gierlowski-Kordesch, who visited the field sites in India along with us in the January of 2016, will always remain with us. We are extremely grateful to her for introducing us to the sedimentology of freshwater carbonates. We would like to acknowledge Dr. Maurice Tucker, Dr. Ildefonso Armenteros, and an anonymous reviewer for their thorough revision of the original manuscript. We are extremely grateful to the editor, Dr. Sila Pla Pueyo, for her valuable suggestions in the final revision.

References

Allen, R. L. (1986). Calcretes in the old red sandstone facies (Late Silurian-Eady of the Anglo-Welsh area, Southern Britain). In V. P. Wright (Ed.), *Paleosols: Their and interpretation* (pp. 58–86). Oxford: Blackwell Scientific Publications.

Alonso-Zarza, A. M. (2003). Palaeoenvironmental significance of palustrine carbonates and calcretes in the geological record. *Earth-Science Reviews, 60*, 261–298.

Alonso-Zarza, A. M., & Wright, V. P. (2010). Palustrine carbonates. In A. M. Alonso-Zarza & L. H. Tanner (Eds.), *Carbonates in continental settings: Facies, environments and processes: Developments in sedimentology 61* (pp. 133–175). Amsterdam: Elsevier.

Alonso-Zarza, A. M., Calvo, J. P., & Garcla del Cura, M. A. (1992). Palustrine sedimentation and associated features grainification and pseudo-microkarst in the middle Miocene (intermediate unit) of the Madrid Basin, Spain. *Sedimentary Geology, 76*, 43–61.

Arenas, C., Gutierrez, F., Osacar, C., & Sancho, C. (2000). Sedimentology and geochemistry of fluvio-lacustrine tufa deposits controlled by evaporite solution subsidence in the Central Ebro Depression, NE Spain. *Sedimentology, 47*, 883–909.

Arenas, C., Cabrera, L., & Ramos, E. (2007). Sedimentology of tufa facies and continental microbialites from the Palaeogene of Mallorca Island (Spain). *Sedimentary Geology, 197*, 1–27.

Arenas, C., Pinuela, L., & Garcia-Ramos, J. (2015). Climatic and tectonic controls on carbonate deposition in syn-rift siliciclastic fluvial systems: A case of microbialites and associated facies in the Late Jurassic. *Sedimentology, 62*, 1149–1183.

Arenas-Abad, C., Vazquez-Urbez, M., Pardo-Tirapu, G., & Sancho-Marcen, C. (2010). Fluvial and associated carbonate deposits. In A. M. Alonso-Zarza & L. H. Tanner (Eds.), *Carbonates in continental settings: Facies, Environments and processes* (Developments in sedimentology 61) (pp. 133–175). Amsterdam: Elsevier.

Arp, G. (1995). Lacustrine bioherms, spring mounds, and marginal carbonates of the Ries-Impact-Crater (Miocene, southern Germany). *Facies, 33*, 35–90.

Awramik, S. M. (1971). Precambrian columnar stromatolite diversity: Reflection of Metazoan appearance. *Science, 174*, 825–827.

Bandyopadhyay, S., & Roy Chowdhury, T. (1996). Beginning of the continental Jurassic in India – A palaeontological approach: Museum of Northern Arizona. *Bulletin, 60*, 371–378.

Bandyopadhyay, S., & Rudra, D. (1985). Upper Gondwana stratigraphy, North of the Pranhita-Godavari confluence, Southern India. *Journal of the Geological Society of India, 26*, 261–266.

Bandyopadhyay, S., & Sengupta, D. P. (2006). Vertebrate faunal turnover during the Triassic-Jurassic transition: An Indian scenario. In J. D. Harris, S. G. Lucas, J. D. Kirkland, & A. R. C. Milner (Eds.), *Terrestrial Triassic-Jurassic transition* (New Mexico museum of natural history and science bulletin 37) (pp. 77–85).

Bandyopadhyay, S., Gillette, D. D., Ray, S., Sengupta, D.P. (2010). Osteology of Barapasaurus tagorei (Dinosauria: Sauropoda) from the Early Jurassic of India. *Palaeontology 53*(3), 533–569.

Biswas, S. K. (2003). Regional tectonic framework of the Pranhita-Godavari basin, India. *Journal of Asian Earth Sciences, 21*, 543–551.

Bosak, T., Souza-Egipsy, A., Corsetti, F. A., & Newman, D. (2004). Micrometre-scale porosity as a biosignature in carbonate crusts. *Geology, 32*, 781–784.

Bradley, W. H. (1929). Algal reefs and oolites of the Green River formation. *Geological Survey Professional Paper, 154*, 203–223.

Burne, R. V., & Moore, L. S. (1987). Microbialites: Organosedimentary deposits of benthic microbial communities. *PALAIOS, 2*, 241–254.

Capezzuoli, E., Gandin, A., & Pedley, M. (2014). Decoding tufa and travertine (fresh water carbonates) in the sedimentary record: The state of the art. *Sedimentology, 61*, 1–21.

Chafetz, H. S., & Buczynski, C. (1992). Bacterially induces lithification of microbial mats. *PALAIOS, 7*, 277–293.

Chafetz, H. S., & Guidry, S. A. (1999). Bacterial shrubs, crystal shrubs and ray-crystal shrubs: Bacterial vs. abiotic precipitation. *Sedimentary Geology, 126*, 57–74.

Chakraborty, C., Mondal, N., & Ghosh, K. (2003). Kinematics of the Gondwana basins of peninsular India. *Tectonophysics, 377*, 299–324.

Chaudhuri, A. K., Deb, G. K., Patranabis-Deb, S., & Sarkar, S. (2012). Paleogeographic and tectonic evolution of the Pranhita-Godavari valley, Central India: A stratigraphic perspective. *American Journal of Science, 312*, 766–815.

Chaudhuri, A. K., Deb, G. K., & Patranabis-Deb, S. (2015). Conflicts in stratigraphic classification of the Puranas of the Pranhita–Godavari Valley: Review, recommendations and status of the "Penganga" sequence. In R. Mazumder & P. G. Eriksson (Eds.), *Precambrian Basins of India: Stratigraphic and Tectonic Context* (pp. 165–183). Geological Society of London, Memoir 43.

Dasgupta, S., & Ghosh, P. (2018). Freshwater carbonates within a Late Triassic siliciclastic fluvial system in a Gondwana rift basin: The Maleri formation, India. *Sedimentary Geology, 373*, 254–271.

Della Porta, G. (2015). Carbonate build-ups in lacustrine, hydrothermal and fluvial settings: Comparing depositional geometry, fabric types and geochemical signature. *Geological Society, London, Special Publications, 418*, 17–68.

Dupraz, C., Visscher, P. T., Baumgartner, L. K., & Reid, R. P. (2004). Microbe–mineral interactions: Early carbonate precipitation in a hypersaline lake (Eleuthera Island, Bahamas). *Sedimentology, 51*, 745–765.

Dupraz, C., Reid, R. P., Braissant, O., Decho, A. W., Norman, R. S., & Vissche, P. T. (2009). Processes of carbonate precipitation in modern microbial mats. *Earth-Science Reviews, 96*, 141–162.

Evans, S. E., Prasad, G. V. R., & Manhas, B. K. (2002). Fossil lizards from the Jurassic Kota formation of India. *Journal of Vertebrate Paleontology, 22*, 299–312.

Folk, R. L., Chafetz, H. S., & Tiezzi, P. A. (1985). Bizzarre forms of depositional and diagenetic calcite in hot-spring travertines, Central Italy, the Society of Economic Palaeontologists rind mineralogists. *Special Publication, 36*, 349–356.

Ford, T. D., & Pedley, H. M. (1996). A review of tufa and travertine deposits of the world. *Earth-Science Reviews, 41*, 117–175.

Freytet, P., & Plaziat, J. C. (1982). Continental carbonate sedimentation and pedogenesis – late cretaceous and early tertiary of Southern France. *Contributions to Sedimentology, 12*, 213.

Freytet, P., & Verrecchia, E. P. (1998). Freshwater organisms that build stromatolites: A synopsis of biocrystallization by prokaryotic and eukaryotic algae. *Sedimentology, 45*, 535–563.

Freytet, P., & Verrecchia, E. P. (2002). Lacustrine and palustrine carbonate petrography: An overview. *Journal of Paleolimnology, 27*, 221–237.

Gandin, A., & Capezzuoli, E. (2008). Travertine versus calcareous tufa: Distinctive petrologic features and stable isotopes signatures. *Quaternario – Italian Journal of Quaternary Sciences, 21*, 125–136.

Gandin, A., & Capezzuoli, E. (2014). Travertine: Distinctive depositional fabrics of carbonates from thermal spring systems. *Sedimentology, 61*, 264–290.

García-García, F., Pla-Pueyo, S., Nieto, L. M., & Viseras, C. (2014). Sedimentology of geomorphologically controlled Quaternary tufas in a valley in southern Spain. *Facies, 60*, 53–72.

Garrett, P. (1970). Phanerozoic stromatolites: Non-competitive ecologic restriction by grazing and burrowing animals. *Science, 169*, 171–173.

Ghosh, P. (1997). Geomorphology and Palaeoclimatology of some upper cretaceous palaeosols in Central India. *Sedimentary Geology, 110*, 25–49.

Gierlowski-Kordesch, E. H. (2010). Lacustrine carbonates. In A. M. Alonso-Zarza & L. H. Tanner (Eds.), *Carbonates in continental settings: Facies, environments and processes* (Developments in sedimentology 61) (pp. 1–102). Amsterdam: Elsevier.

Goswami, S. (2019). *Sedimentology of the limestone interval and its underlying siliciclastic succession of the Kota Formation, Pranhita-Godavari Gondwana Basin, India*, Ph.D thesis submitted to University of Calcuttta.

Goswami, S., Gierlowski-Kordesch, E. H., & Ghosh, P. (2018). Sedimentology of the early Jurassic limestone beds of the Kota formation: Record of carbonate wetlands in a continental rift basin of India. *Juornal of Palaeolimnology, 59*(1), 21–38.

Goswami, S. & Ghosh, P. (2020). Evolution of sedimentation pattern in a continental rift basin of India, between the Late Triassic and the early Middle Jurassic: Tectonic and climatic controls. *Sedimentary Geology, 405*, https://doi.org/10.1016/j.sedgeo.2020.105679.

Guo, L., & Riding, R. (1998). Hot-spring travertine facies and sequences, Late Pleistocene, Rapolano Terme, Italy. *Sedimentology, 45*, 163–180.

Jain, S. L. (1973). New specimens of lower Jurassic holostean fishes from India. *Palaeontology, 16*, 149–177.

Jain, S. L. (1983). A review of the genus Lepidotes (Actinopteryhii: Semionotiformes) with special reference to the species from Kota formation (lower Jurassic), India. *Journal of the Palaeontological Society of India, 28*, 7–42.

Jones, B., & Renaut, R. W. (2010). Calcareous spring deposits in continental settings. In A. M. Alonso-Zarza & L. H. Tanner (Eds.), *Carbonates in continental settings: Facies, environments and processes* (Developments in sedimentology 61) (pp. 177–224). Amsterdam: Elsevier.

Jones, B., Renaut, R. W., Owen, R. B., & Torfason, H. (2005). Growth patterns and implications of complex dendrites in calcite travertines from Lysuholl, Snaefellsnes, Iceland. *Sedimentology, 52*, 1277–1301.

Klappa, C. F. (1980). Rhizoliths in terrestrial carbonates: Classification, recognition, genesis and significance. *Sedimentology, 27*, 613–629.

Kraus, M. J. (1999). Paleosols in clastic sedimentary rocks: Their geologic applications. *Earth-Science Reviews, 47*, 41–70.

Kraus, M. J., & Hasiotis, S. T. (2006). Significance of different modes of rhizolith preservation to interpreting paleoenvironmental and paleohydrologic settings: Examples from Paleogene paleosols, Bighorn Basin, Wyoming, U.S.A. *Journal of Sedimentary Research, 76*, 633–646.

Kutty, T. S. (1969). Some contributions to the stratigraphy of the upper Gondwana formations of the Pranhita-Godavari Valley, Central India. *Journal of The Geological Society of India, 10*(1).

Kutty, T. S., & Sengupta, D. P. (1989). The Late Triassic formations of the Pranhita-Godavari valley and their vertebrate faunal succession – a reappraisal. *Indian Journal of Earth Sciences, 16*, 189–206.

Kutty, T. S., Jain, S. L., & Roy Chowdhury, T. (1987). Gondwana sequence of the northern Pranhita-Godavari Valley: Its stratigraphy and vertebrate faunas. *Palaeobotanist, 36*, 214–229.

Leslie, A. B., Tucker, M. E., & Spiro, B. (1992). A sedimentological and stable isotopic study of travertines and associated sediments within upper Triassic lacustrine limestone, South Wales, UK. *Sedimentology, 39*, 613–629.

Maulik, P. K., & Rudra, D. K. (1986). Trace fossils from the freshwater Kota Limestone of the Pranhita-Godavari valley, South Central India. *Bulletin Geological, Mining & Metallurgical Society of India, 54*, 114–123.

Mazzullo, S.J., & Birdwell. B.A. (1989). Syngenetic Formation of Grainstones and Pisolites from Fenestral Carbonates in Peritidal Settings, *Journal of Sedimentary Petrology, 59*(4), p. 605–611.

Monty, C. L. V. (1973). Remarques sur la nature, la morphological distributions patiale des Stromatolites. *Sciences de la Terre, 18*, 189–212.

Noffke, N., Eriksson, K. A., Hazen, R. M., & Simpson, E. L. (2006). A new window into early Archean life: Microbial mats in Earth's oldest siliciclastic tidal deposits (3.2 Ga Moodies Group, South Africa). *Geology, 34*(4), 253–256.

Parrish, J. T., Hasiotis, S. T., & Chan, M. A. (2017). Carbonate deposits in the lower Jurassic Navajo sandstone, southern Utah and Northern Arizona, U.S.A. *Journal of Sedimentary Research, 87*, 740–762.

Pedley, H. M. (2009). Tufas and travertines of the Mediterranean region: A test in ground for freshwater carbonate concepts and developments. *Sedimentology, 56*, 221–246.

Pedley, H. M., Martin, J. A. G., Delgado, S. O., & Del Cura, A. G. (2003). Sedimentology of quaternary perched springline and paludal tufas: Criteria for recognition, with examples from Guadalajara Province, Spain. *Sedimentology, 50*, 23–44.

Perri, E., & Tucker, M. (2007). Bacterial fossils and microbial dolomite in Triassic stromatolites. *Geology, 35*, 207–210.

Perri, E., Tucker, M. E., & Spadafora, A. (2012). Carbonate organo-mineral micro- and ultrastructures in sub-fossil stromatolites: Marion lake, South Australia. *Geobiology, 10*, 105–117.

Pla-Pueyo, S., Gierlowski-Kordesch, E. H., Viseras, C., & Soria, J. M. (2009). Major controls on carbonate deposition during the evolution of a continental basin: Pliocene-Pleistocene of the Guadix basin (Betic cordillera, southern Spain). *Sedimentary Geology, 219*, 97–114.

Pla-Pueyo, S., Viseras, C., Henares, S., Yeste, L. M., & Candy, I. (2017). Facies architecture, geochemistry and palaeoenvironmental reconstruction of a barrage tufa reservoir analogue (Betic cordillera, S. Spain). *Quaternary International, 437*, 15–36.

Platt, N. H., & Wright, V. P. (1992). Palustrine carbonates and the Florida Everglades: Towards an exposure index for the fresh-water environment? *Journal of Sedimentary Petrology, 62*, 1058–1071.

Pola, M., Gandin, A., Tuccimei, P., Soligo, M., Deiana, R., Fabbri, P., & Zampieri, D. (2014). A multidisciplinary approach to understanding carbonate deposition under tectonically controlled hydrothermal circulation: A case study from a recent travertine mound in the Euganean hydrothermal system, northern Italy. *Sedimentology, 61*, 172–199.

Rainey, D. K., & Jones, B. (2009). Abiotic v. biotic controls on the development of the Fairmont Hot Springs carbonate deposit, British Columbia, Canada. *Sedimentology, 56*, 1832–1857.

Ray, S., & Bandyopadhyay, S., (2003). Late Permian vertebrate community of the Pranhita–Godavari valley, India. *Journal of Asian Earth Sciences 21*(6), 643–654.

Reitner, J., Gautret, P., Marin, F., & Neuweiler, F. (1995). Automicrites in modern marine microbialite. Formation model via organic matrices (Lizard Island, Great Barrier Reef,Australia). *Bulletin de l'Institut Océanographique (Monaco) Numéro Spécial, 14*, 237–264.

Retallack, G. J. (1994). The environmental factor approach to the interpretation of palaeosols. In R. Amundson, J. Harden, & M. Singer (Eds.), *Factors of soil formation: A fiftieth anniversary retrospective* (Soil science Society of America, special publication 33) (pp. 31–64).
Retallack, G. J. (2001). Soils of the past: *An introduction to Paleopedology* (2nd ed., p. 600). Blackwell, Oxford, UK.
Rezende, M. F., Tonietto, S. N., & Pope, M. C. (2003). Three-dimensional pore connectivity evaluation in a Holocene and Jurassic microbialite build up. *AAPG Bulletin, 97*(11), 2085–2101.
Riding, R. (1991). Classification of microbial carbonates. In R. Riding (Ed.), *Calcareous algae and stromatolites* (pp. 21–52). Berlin: Springer.
Riding, R. (2000). Microbial carbonates: The geological record of calcified bacterial-algalmats and biofilms. *Sedimentology, 47*, 179–214.
Robinson, P. L. (1967). The Indian Gondwana Formations – A review: 1st International Symposium on Gondwana stratigraphy. *International Union of Geological Sciences Argentina*, pp. 201–268.
Rudra, D. (1982). Upper Gondwana stratigraphy and sedimentation in the Pranhita-Godavari Valley, India. *Quarterly Journal Geological, Mining and Metallurgical Society of India, 54*, 56–79.
Rudra, D. K., & Maulik, P. K. (1994). Lower Jurassic Kota limestone of India. *Global Geological Record of Lake Basins, 1*, 185–191.
Sanders, D., Unterwurzacher, M., & Ruf, B. (2006). Microbially induced calcium carbonate in tufas of the western eastern Alps: A first overview. *Alpine Geology (Geo Alp), 3*, 167–189.
Sengupta, S. (1970). Gondwana sedimentation around Bheemaram (Bhimaram), Pranhita-Godavari Valley, India. *Journal of Sedimentary Petrology, 40*, 140–170.
Shiraishi, F., Reimer, A., Bissett, A., Beer, D., & Arp, G. (2008). Microbial effects on biofilm calcification, ambient water chemistry and stable isotope records in a highly supersaturated setting (Westerhöfer Bach, Germany). *Palaeogeography, Palaeoclimatology, Palaeoecology, 262*, 91–106.
Strasser, A. (1986). Ooids in Purbeck limestones (lowermost cretaceous) of the Swiss and French Jura. *Sedimentology, 33*, 711–727.
Tasch, P., Sastry, M. V. A., Shah, S. C., Rao, B. R. J., Rao, C. N., & Ghosh, S. C. (1975). Estherids of the Indian Gondwanas: Significance for continental fit. In *Advances in stratigraphy and palaeontology* (pp. 443–452).
Tucker, M. E. (2001). *Sedimentary petrology: An introduction to the origin of sedimentary rocks* (3rd ed., p. 272). Blackwell Scientific Publication.
Wright, V. P., & Tucker, M. E. (1991). Calcretes: An introduction. In *Calcrete* (International Association; Sedimentologists Reprint Series 2) (pp. 1–22).
Yadagiri, P. (1986). Lower Jurassic lower vertebrates from Kota formation, Pranhita-Godavari valley, India. *Journal of the Palaeontological Society of India, 31*, 89–96.
Zamarreño, I., Anadon, P., & Utrilla, R. (1997). Sedimentology and isotopic composition of upper Palaeocene to Eocene non-marine stromatolites, eastern Ebro Basin, NE Spain. *Sedimentology, 44*, 159–176.

Index

M. R. Rosen et al. (eds.), *Limnogeology: Progress, Challenges and Opportunities*, Syntheses in Limnogeology, https://doi.org/10.1007/978-3-030-66576-0

M

S

The manufacturer's authorised representative in the EU is Springer Nature Customer Service Centre GmbH, Europaplatz 3, 69115 Heidelberg, Germany. If you have any concerns regarding our products, please contact ProductSafety@springernature.com

Printed and bound by CPI Group (UK) Ltd, Croydon, CR0 4YY
23/09/2026
02221471-0011